ELSEV
DICTIONA
AUTOMA
TECHN

T0226892

ELSEVIER'S DICTIONARY OF AUTOMATION TECHNICS

in
English, German, French and Russian

compiled by

B. ZHELYAZOVA
Sofia, Bulgaria

2005

ELSEVIER

Amsterdam – Boston – Heidelberg – London – New York – Oxford
Paris – San Diego – San Francisco – Singapore – Sydney – Tokyo

ELSEVIER B.V.
Radarweg 29
P.O. Box 211
1000 AE Amsterdam
The Netherlands

ELSEVIER Inc.
525 B Street, Suite 1900
San Diego, CA 92101-4495
USA

ELSEVIER Ltd
The Boulevard, Langford Lane
Kidlington, Oxford OX5 1GB
UK

ELSEVIER Ltd
84 Theobalds Road
London WC1X 8RR
UK

First edition 2005

Library of Congress Cataloging in Publication Data
A catalog record is available from the Library of Congress.

British Library Cataloguing in Publication Data
A catalogue record is available from the British Library.

ISBN: 0-444-51533-X

⊚ The paper used in this publication meets the requirements of ANSI/NISO Z39.48-1992 (Permanence of Paper).

Printed and bound in the United Kingdom
Transferred to Digital Print 2010

PREFACE

The dictionary contains 13,000 terms with more than 4,000 cross-references used in the following fields: automation, technology of management and regulation, computing machine and data processing, computer control, automation of industry, laser technology, theory of information and theory of signals, theory of algorithms and programming, cybernetics and mathematical methods.

Automation pertains to the theory, art, or technique of making a machine, a process, or a device more fully automatic. Computers and information processing equipment play a large role in the automation of a process because of the inherent ability of a computer to develop decision that will, in effect, control or govern the process from the information received by the computer concerning the status of the process. Thus automation pertains to both the theory, and techniques of using automatic systems in industrial applications and the processes of investigation, design, and conversion to automatic methods. Automatic control, automatic materials handling, automatic testing, automatic packaging, for continuous as well as batch processing, are all considered parts of the overall or completely automatic process.

The Dictionary consists of two parts – *Basic Table* and *Indexes*. In the first part the English terms are listed alphabetically, numbered consecutively and followed by its German, French and Russian equivalents. English synonyms appear as cross-references to the main entries in their proper alphabetical order. The second part of the Dictionary, the *Indexes*, contains separate alphabetical indexes of the German, French and Russian terms. The reference number(s) with each term stands for the number of the English term(s) in the basic table.

Elsevier's Dictionary of Automation Technics will be a valuable tool for specialists, scientists, students and everyone who takes interest in the problems of investigation devoted to the design, development, and application of methods and techniques for rendering a process or group of machines self-actuating, self-moving, or self controlling.

Dr. Boyanka Zhelyazova

CONTENTS

EXPLANATION OF SPECIAL SIGNS

1. The italic *d, f, s,* and *r* in the basic table stand respectively for the German, French and Russian equivalents of the English terms.

2. The gender of nouns is indicated as follows:

f	feminine	*fpl*	feminine plural
m	masculine	*mpl*	masculine plural
n	neuter	*npl*	neuter plural
pl	plural	*m/f*	masculine or feminine

3. The symbol v designates a verb.

4. The symbol *adj* designates an adjective.

5. Synonyms and abbreviations are separated by semicolons.

6. The abbreviation (US) means American usage.

7. Two kinds of brackets are used:

   [ ] the information can be either included or left out;

   ( ) the information does not form an integral part of expression, but helps to clarify it.

Basic Table

A

1 Abbe's sine theorem
d Abbescher Sinussatz *m*; Abbesche
    Sinusbedingung *f*
f condition *f* des sinus d'Abbe
r синусовый закон *m* Аббея

2 abbreviated addressing
d abgekürzte Adressierung *f*
f adressage *m* abrégé
r сокращенная адресация *f*

3 Abel integral equation
d Abelsche Integralgleichung *f*
f équation *f* intégrale d'Abel
r интегральное уравнение *n* Абеля

4 aberration constant
d Aberrationskonstante *f*
f constante *f* d'aberration
r постоянная *f* аберрации

5 ability to respond; response capacity
d Ansprechvermögen *n*
f pouvoir *m* de réponse
r способность *f* срабатывания;
    чувствительность *f*

6 ability to withstand handling
d Griffestigkeit *f*
f résistance *f* au toucher
r сопротивление *n* захватыванию

7 abnormal end of task
d anormales Ende *n* der Aufgabe
f fin *f* abnormale de la tâche
r аварийное прекращение *n* задачи

8 abnormal system end; system crash
d Systemabsturz *m*; Systemzusammenbruch *m*
f effondrement *m* de système
r авария *f* в системе

* **abort → 6948**

* **abort** *v* → 6933

9 above-threshold operation
d Betrieb *m* oberhalb der Schwelle
f fonctionnement *m* au-dessus du seuil
r работа *f* выше порога

10 abrupt change
d plötzliche Änderung *f*; sprungartige
    Änderung
f changement *m* abrupt; modification *f* subite
r внезапное изменение *n*

11 abscissa of absolute convergence
d Abszisse *f* der absoluten Konvergenz
f abscisse *f* de convergence absolue
r абсцисса *f* абсолютной сходимости

12 abscisse axis; X-axis
d Abszissenachse *f*; X-Achse *f*
f axe *m* des abscisses; axe des X
r ось *f* абсцисс; ось X

* **absence of failures → 5675**

**13 absence of feedback; absence of
    interaction**
d Rückwirkungsfreiheit *f*
f absence *f* de réaction
r однонаправленость *f*

* **absence of interaction → 13**

14 absolute activity
d absolute Aktivität *f*
f activité *f* absolue
r абсолютная активность *f*

15 absolute address
d absolute Adress *f*
f adresse *f* absolue
r истинный адрес *m*; абсолютный адрес

16 absolute altimeter
d Absoluthöhenmesser *m*
f altimètre *m* absolu
r абсолютный высотомер *m*

17 absolute assembler
d Absolutassembler *m*
f assembleur *m* absolu
r абсолютный ассемблер *m*

18 absolute bolometric magnitude
d absolute bolometrische Größe *f*
f magnitude *f* bolométrique absolue
r абсолютная болометрическая величина *f*

19 absolute calibration
d Absoluteichung *f*
f étalonnage *m* absolu
r абсолютная градуировка *f*

20 absolute code
d absoluter Kode *m*

    f code *m* absolu
    r абсолютный код *m*

21 absolute coding
    d absolute Kodierung *f*
    f codage *m* absolu
    r абсолютное кодирование *n*

22 absolute coordinate system
    d absolutes Koordinatensystem *n*
    f système *m* absolu de coordonnées
    r абсолютная система *f* координат

23 absolute counter
    d Absolutzähler *m*
    f compteur *m* absolu
    r счётчик *m* для абсолютных измерений

24 absolute cross section
    d absoluter Wirkungsquerschnitt *m*
    f section *f* efficace absolue
    r истинное поперечное сечение *n*; абсолютное поперечное сечение

25 absolute damping
    d absolute Dämpfung *f*
    f amortissement *m* absolu
    r полное демпфирование *n*; полное гашение *n* колебаний

26 absolute delay
    d absolute Verzögerung *f*
    f retard *m* absolu
    r абсолютная задержка *f*

27 absolute digital control
    d absolute Digitalsteuerung *f*
    f commande *f* numérique absolue; commande digitale absolue
    r абсолютное цифровое управление *n*

28 absolute disintegration rate
    d absolute Zerfallsrate *f*
    f vitesse *f* absolue de désintégration
    r абсолютная скорость *f* распада

29 absolute effector orientation
    d absolute Effektororientierung *f*
    f orientation *f* d'effecteur absolue
    r абсолютная ориентировка *f* эффектора

30 absolute electrometer
    d absolutes Elektrometer *n*
    f électromètre *m* absolu
    r абсолютный электрометр *m*

31 absolute energy scale
    d absolute Energieskale *f*

    f échelle *f* absolue d'énergie
    r абсолютная шкала *f* энергии

32 absolute error
    d absoluter Fehler *m*
    f erreur *f* absolue
    r абсолютная погрешность *f*; абсолютная ошибка *f*

33 absolute extremum optimizer
    d Globaloptimisator *m*
    f optimaliseur *m* à extremum absolu
    r глобальный оптимизатор *m*

34 absolute format
    d absolutes Format *n*
    f format *m* absolu
    r абсолютный формат *m*

35 absolute frequency
    d absolute Häufigkeit *f*
    f fréquence *f* absolue
    r абсолютная частота *f*

36 absolute heating effect
    d absolute Wärmetönung *f*; absoluter Wärmeeffekt *m*
    f effet *m* calorifique absolu; chaleur *f* de réaction absolue
    r абсолютный тепловой эффект *m*

37 absolute humidity
    d absolute Feuchte *f*
    f humidité *f* absolue
    r абсолютная влажность *f*

38 absolute index
    d absoluter Index *m*
    f index *m* absolu
    r абсолютный индекс *m*

39 absolute loader
    d absoluter Lader *n*; Absolutlader *m*
    f chargeur *m* absolu
    r абсолютный загрузчик *m*

40 absolute loading
    d absolutes Laden *n*; Laden mit absoluten Adressen
    f chargement *m* absolu
    r абсолютная загрузка *f*; загрузка с абсолютным адресом

41 absolutely convergent
    d absolut konvergent
    f convergent absolument
    r абсолютно сходимый

42 absolute maximum
d absolutes Maximum *n*
f maximum *m* absolu
r абсолютный максимум *m*

43 absolute measuring method
d absolute Messmethode *f*;
Absolutmessverfahren *n*
f méthode *f* absolue de mesure; procédé *m*
absolu de mesure
r абсолютный метод *m* измерения

44 absolute minimum
d absolutes Minimum *n*
f minimum *m* absolu
r абсолютный минимум *m*

45 absolute moisture content
d absoluter Feuchtegehalt *m*
f teneur *m* en humidité absolu; taux *m*
d'humidité absolu
r абсолютное влагосодержание *n*

46 absolute motion
d absolute Bewegung *f*
f mouvement *m* absolu
r абсолютное движение *n*

47 absolute position of manipulator
d Manipulatorabsolutposition *f*;
Absolutposition *f* eines Manipulators
f position *f* absolue d'un manipulateur
r абсолютная позиция *f* манипулятора

48 absolute pressure
d absoluter Druck *m*
f pression *f* absolue
r абсолютное давление *n*

49 absolute probability
d absolute Wahrscheinlichkeit *f*
f probabilité *f* absolue
r абсолютная вероятность *f*

50 absolute programming
d absolutes Programmieren *n*
f programmation *f* absolue
r абсолютное программирование *n*

51 absolute sensitivity
d absolute Empfindlichkeit *f*
f sensibilité *f* absolue
r абсолютная чувствительность *f*

52 absolute stability
d absolute Stabilität *f*
f stabilité *f* absolue
r абсолютная устойчивость *f*

53 absolute system
d absolutes System *n*
f système *m* absolu
r абсолютная система *f*

54 absolute temperature
d absolute Temperatur *f*; Kelvin-Temperatur *f*
f température *f* absolue
r абсолютная температура *f*

55 absolute temperature scale
d absolute Temperaturskale *f*
f échelle *f* absolue de température
r абсолютная температурная шкала *f*; шкала
Келвина

56 absolute value
d absoluter Wert *m*
f valeur *f* absolue
r абсолютная величина *f*

57 absolute value representation
d Absolutwertdarstellung *f*
f représentation *f* de valeur absolue
r представление *n* абсолютной величины;
воспроизведение *n* абсолютного значения

58 absolute vector
d absoluter Vektor *m*
f vecteur *m* absolu
r абсолютный вектор *m*

59 absolute zero
d absoluter Nullpunkt *m*
f zéro *m* absolu
r абсолютный нуль *m*

60 absorb *v*
d absorbieren; aufsaugen; aufnehmen;
anziehen
f absorber
r поглощать; абсорбировать; всасывать

61 absorbable
d absorbierbar; aufnahmefähig
f absorbable
r поглощаемый

62 absorbed horsepower
d aufgenommene Leistung *f*
f puissance *f* absorbée
r потребляемая мощность *f*

63 absorbent filter; absorption filter
d Absorptionsfilter *n*
f filtre *m* d'absorption
r поглащающий фильтр *m*

64 **absorber; absorbing apparatus;
absorption apparatus**
d Absorber *m*; Absorptionsapparat *m*
f absorbeur *m*; installation *f* d'absorption
r абсорбер *m*; поглотитель *m*

* **absorbing** → 69

* **absorbing apparatus** → 64

65 **absorbing medium**
d Absorptionsmedium *n*
f milieu *m* absorbant
r поглащающая среда *f*; абсорбирующая
  среда

66 **absorbing temperature**
d Absorptionstemperatur *f*
f température *f* d'absorption
r температура *f* абсорбции

67 **absorb power** *v*
d Leistung verbrauchen; Leistung aufnehmen
f absorber la puissance; consommer la
  puissance
r поглащать энергию

68 **absorptiometer**
d Absorptionsmesser *m*
f appareil *m* de mesure à absorption
r абсорбциометр *m*; измеритель *m*
  поглащения

69 **absorption; absorbing**
d Absorption *f*; Absorbieren *n*; Einsaugen *n*
f absorption *f*; aspiration *f*
r абсорбция *f*; поглощение *n*

70 **absorption analysis**
d Absorptionsanalyse *f*
f analyse *f* par absorption
r абсорбционный анализ *m*

* **absorption apparatus** → 64

71 **absorption band**
d Absorptionsband *n*
f bande *f* absorptive
r полоса *f* поглощения

72 **absorption capacity**
d Absorptionsfähigkeit *f*;
  Absorptionsvermögen *n*
f pouvoir *m* d'absorption
r абсорбционная способность *f*;
  поглощающая способность; всасывающая
  способность

73 **absorption chromatography**
d Absorptionschromatografie *f*
f chromatographie *f* à absorption
r абсорбционная хроматография *f*

74 **absorption circuit**
d Absorptionskreis *m*
f circuit *m* d'absorption
r абсорбционный контур *m*; поглащающий
  контур

75 **absorption coefficient**
d Absorptionskoeffizient *m*
f coefficient *m* d'absorption
r коэффициент *m* поглощения

76 **absorption column**
d Absorptionssäule *f*
f colonne *f* d'absorption
r абсорбционная колонна *f*

77 **absorption control**
d Regelung *f* durch Absorption;
  Absorptionsregelung *f*
f réglage *m* par absorption
r управление *n* методом поглощения;
  регулирование *n* методом поглощения

78 **absorption curve**
d Absorptionskurve *f*
f courbe *f* d'absorption
r кривая *f* поглощения

79 **absorption dehumidifier**
d Absorptionstrockner *m*
f sécheur *m* à absorption
r абсорбционный влагопоглотитель *m*

80 **absorption discontinuity**
d Absorptionssprung *m*
f discontinuité *f* d'absorption
r скачок *m* поглощения

81 **absorption dynamometer**
d Bremsdynamometer *n*
f dynamomètre *m* de frein; amortisseur *m* à
  moulinet
r абсорбционный динамометр *m*

82 **absorption edge**
d Absorptionskante *f*
f seuil *m* d'absorption; bord *m* d'absorption
r край *m* [полосы] поглощения

83 **absorption equilibrium**
d Absorptionsgleichgewicht *n*
f équilibre *m* d'absorption
r абсорбционное равновесие *n*

84 absorption equivalent
 d Absorptionsäquivalent *n*
 f équivalent *m* d'absorption
 r эквивалент *m* поглощения

85 absorption experiment
 d Absorptionsversuch *m*
 f essai *m* d'absorption
 r испытание *n* на поглощение

* **absorption filter** → 63

86 absorption frequency meter
 d Absorptionsfrequenzmesser *m*
 f fréquencemètre *m* à absorption
 r частотомер *m* поглощающего типа

87 absorption index
 d Absorptionsgrad *m*
 f facteur *m* d'absorption; coefficient *m*
     d'absorption; absorptance *f*
 r показатель *m* поглощения;
     коэффициент *m* поглощения

88 absorption line
 d Absorptionslinie *f*
 f raie *f* d'absorption
 r линия *f* поглощения

89 absorption loss
 d Aufsaugverlust *m*
 f perte *f* par imbibage initial
 r абсорбционные потери *fpl*

90 absorption measuring method
 d absorptiometrische Methode *f*
 f méthode *f* absorptiométrique
 r метод *m* измерения абсорбции

91 absorption modulation
 d Absorptionsmodulation *f*
 f modulation *f* à absorption
 r модуляция *f* поглощением

92 absorption of infrared radiation
 d Infrarotstrahlungsabsorption *f*
 f absorption *f* de rayonnement infrarouge
 r поглощение *n* инфракрасного излучения

93 absorption peak
 d Absorptionsspitze *f*
 f pic *m* d'absorption
 r максимум *m* поглощения

94 absorption photometer
 d Absorptionsfotometer *n*
 f photomètre *m* à absorption
 r абсорбционный фотометр *m*

95 absorption plane
 d Absorptionsfläche *f*
 f surface *f* absorbante
 r поверхность *f* поглощения

96 absorption probability
 d Absorptionswahrscheinlichkeit *f*
 f probabilité *f* d'absorption
 r вероятность *f* поглощения

97 absorption process
 d Absorptionsprozess *m*;
     Absorptionsverfahren *n*
 f procédé *m* d'absorption
 r абсорбционный процесс *m*

98 absorption saturation
 d Absorptionssättigung *f*
 f saturation *f* de l'absorption
 r насыщение *n* поглощения

99 absorption signal
 d Absorptionssignal *n*
 f signal *m* d'absorption
 r сигнал *m* поглощения

100 absorption spectrophotometer
 d Absorptionsspektralfotometer *n*
 f spectrophotomètre *m* à absorption
 r абсорбционный спектрофотометр *m*

101 absorption spectrum
 d Absorptionsspektrum *n*
 f spectre *m* d'absorption
 r спектр *m* поглощения

102 absorption spectrum of X-rays
 d Absorptionsröntgenspektrum *n*
 f spectre *m* d'absorption de rayons X
 r спектр *m* поглощения рентгеновских
     лучей

103 absorption unit
 d Absorptionseinheit *f*
 f unité *f* d'absorption
 r абсорбционная установка *f*

104 absorption wavemeter
 d Absorptionswellenmesser *m*
 f ondemètre *m* à absorption
 r абсорбционный волномер *m*

105 absorption wave trap
 d Wellenabsorptionssaugkreis *m*
 f circuit *m* aspirateur d'ondes d'absorption
 r абсорбционный фильтр *m* волн

106 abstract automaton
 d abstrakter Automat *m*

 f automate *m* abstrait
 r абстрактный автомат *m*

107 abstract code; pseudo-code
 d Pseudokode *m*; Pseudobefehl *m*; abstrakter
   Kode *m*
 f pseudocode *m*
 r абстрактный код *m*; псевдокод *m*

108 abstract computer
 d Pseudorechner *m*; abstrakte
   Rechenmaschine *f*
 f pseudo-calculateur *m*
 r абстрактный компьютер *m*

109 abstract connection
 d abstrakte Verbindung *f*
 f connexion *f* abstraite
 r абстрактная связь *f*

110 abstract design
 d abstrakter Entwurf *m*
 f plan *m* abstrait
 r абстрактное проектирование *n*

111 abstraction
 d Abstraktion *f*
 f abstraction *f*
 r абстракция *f*

112 abstract model
 d abstraktes Modell *n*
 f modèle *m* abstrait
 r абстрактная модель *f*

113 abstract number
 d unbenannte Zahl *f*; abstrakte Zahl
 f nombre *m* non défini
 r абстрактное число *n*

114 abstract theory of automata
 d abstrakte Automatentheorie *f*; abstrakte
   Theorie *f* der Automaten
 f théorie *f* abstraite des automates
 r абстрактная теория *f* автоматов

115 abstract type concept
 d abstraktes Typekonzept *n*
 f concept *m* de type abstrait
 r абстрактная типовая концепция *f*

116 abstract value
 d abstrakter Wert *m*
 f valeur *f* abstraite
 r абстрактное значение *n*

117 abundance ratio
 d Häufigkeitsverhältnis *n*

 f rapport *m* de fréquences
 r относительное содержание *n*

118 abuse; operation error
 d Fehlbehandlung *f*; Betriebsfehler *m*;
   betriebsbedingter Fehler *m*
 f erreur *f* d'exploitation
 r неправильная эксплоатация *f*; ошибка *f* в
   эксплуатации

 * **AC → 701**

119 accelerate *v*
 d beschleunigen
 f accélérer
 r ускорять

120 accelerated life test
 d zeitraffende Lebensdauerprüfung *f*
 f test *m* accéléré de longévité
 r ускоренное испытание *n* на долговечность

 * **accelerated memory adapter → 121**

**121 accelerated storage adapter; accelerated
   memory adapter**
 d beschleunigter Speicheradapter *m*
 f adapteur *m* de mémoire accéléré
 r ускоренный накопительный адаптер *m*

122 accelerated test technique
 d beschleunigte Prüftechnik *f*
 f technique *f* de test accéléré
 r способ *m* ускоренного испытания

123 accelerating electrode
 d Beschleunigungselektrode *f*
 f électrode *f* d'accélération
 r ускоряющий электрод *m*

124 accelerating relay
 d Beschleunigungsrelais *n*
 f relais *m* accélérateur
 r реле *n* ускорения

125 accelerating voltage
 d Beschleunigungsspannung *f*
 f tension *f* accélératrice
 r ускоряющее напряжение *n*

126 acceleration
 d Beschleunigung *f*
 f accélération *f*
 r ускорение *n*

127 acceleration constant
 d Anlaufkonstante *f*;
   Beschleunigungskonstante *f*

   f constante *f* d'accélération; constante de
      vitesse; gain *m* statique
   r постоянная *f* ускорения

128 acceleration controller
   d Beschleunigungsregler *m*; Anlaufregler *m*
   f organe *m* de commande d'accélération
   r регулятор *m* ускорения

**129 acceleration gauge; acceleration pickup;
      acceleration sensitive element**
   d Beschleunigungsfühler *m*;
      Beschleunigungsaufnehmer *m*
   f tâteur *m* d'accélération; transmetteur *m*
      d'accélération; organe *m* sensible
      d'accélération
   r датчик *m* ускорения; чувствительный
      элемент *m* ускорения

130 acceleration indicator
   d Beschleunigungsanzeiger *m*
   f indicateur *m* d'accélération
   r индикатор *m* ускорения

131 acceleration lag
   d Beschleunigungsverzögerung *f*;
      Beschleunigungsträgheit *f*
   f retard *m* d'accélération
   r запаздывание *n* по ускорению

132 acceleration measurement
   d Beschleunigungsmessung *f*
   f mesure *f* d'accélération
   r измерение *n* ускорения

133 acceleration misalignment
   d Beschleunigungsstörung *f*;
      Beschleunigungsabweichung *f*
   f désalignement *m* d'accélération
   r рассогласование *n* ускорения

   * **acceleration pickup** → **129**

   * **acceleration sensitive element** → **129**

134 acceleration space
   d Beschleunigungsraum *m*
   f espace *m* d'accélération
   r пространство *n* ускорения; область *f*
      ускорения

135 acceleration transducer
   d Beschleunigungswandler *m*
   f transducteur *m* d'accélération
   r преобразователь *m* ускорения

136 acceleration value
   d Beschleunigungswert *m*;

      Beschleunigungsbetrag *m*
   f valeur *f* d'accélération
   r величина *f* ускорения

137 accelerator
   d Beschleuniger *m*
   f accélérateur *m*
   r ускоритель *m*

138 accelerometer
   d Beschleunigungsmesser *m*
   f accéléromètre *m*
   r акселерометр *m*

139 accentuate *v*; emphasize *v*
   d betonen; hervorheben
   f accentuer
   r выделять

140 accept *v*
   d annehmen; akzeptieren
   f accepter; prendre
   r принимать; воспринимать

141 acceptability criterion
   d Akzeptierbarkeitskriterium *n*
   f critère *m* d'acceptabilité
   r критерий *m* приемлемости

142 acceptable
   d annehmbar; akzeptabel
   f acceptable
   r приемлемый

143 acceptable design
   d akzeptabler Entwurf *m*
   f dessin *m* acceptable
   r приемлемый проект *m*

144 acceptable deviation
   d zulässige Abweichung *f*
   f écart *m* toléré
   r допустимое отклонение *n*

145 acceptable deviation of controlled variable
   d zulässige Regelabweichung *f*
   f écart *m* de réglage toléré
   r допустимое отклонение *n* регулируемой
      величины

146 acceptable limit
   d akzeptierbarer Grenzwert *m*; zulässiger
      Grenzwert
   f limite *f* acceptable; limite admissible
   r приемлимый предел *m*

147 acceptable program
   d annehmbares Programm *n*

f programme *m* acceptable
r приемлемая программа *f*

148 acceptable quality level
d annehmbare Qualitätsstufe *f*
f degré *m* de qualité acceptable
r доступный приемлемый уровень *m* качества

149 acceptable reliability level; acceptance reliability level
d Zuverlässigkeitsniveau *n*; annehmbare Zuverlässigkeitsstufe *f*
f niveau *m* de fiabilité; degré *m* de fiabilité d'acceptation
r [доступный приемлемый] уровень *m* надёжности

150 acceptance
d Annahme *f*
f acceptation *f*; admission *f*
r принятие *n*

151 acceptance angle
d Akzeptanzwinkel *m*; Öffnungswinkel *m*
f angle *m* d'admission; angle d'acceptance
r угол *m* приемлемости

152 acceptance checkout equipment
d Annahmeprüfeinrichtung *f*; Abnahmeprüfeinrichtung *f*
f équipement *m* de contrôle d'acceptation
r устройство *n* приёмно-сдаточного испытания

153 acceptance control
d Übernahmekontrolle *f*
f contrôle *m* à la réception; contrôle d'acceptation
r приёмный контроль *m*

154 acceptance of equipment
d Ausrüstungsabnahme *f*
f réception *f* du matériel
r приёмка *f* оборудования

155 acceptance pattern
d Akzeptanzdiagramm *n*
f diagramme *m* d'acceptation
r диаграмма *f* приемлемости

* acceptance reliability level → 149

156 acceptance test
d Abnahmeprüfung *f*; Abnahmeversuch *m*
f essai *m* de réception
r приёмное испытание *n*

157 accept data state
d Datenübernahmezustand *m*
f état *m* d'acceptation des données
r состояние *n* приёма данных

158 accepted
d angenommen; übernommen
f accepté
r принятый

159 accepting station
d annehmbare Station *f*
f station *f* acceptée
r принимающая станция *f*

160 accept of request
d Annahme *f* der Anforderung
f acceptation *f* de la requête
r приём *m* запроса

161 accept of response
d Antwortannahme *f*
f acceptation *f* de réponse
r приём *m* ответа

162 acceptor
d Akzeptor *m*
f accepteur *m*
r акцептор *m*

163 acceptor control
d Annahmesteuerung *f*
f commande *f* d'accepteur
r приёмочный контроль *m*

164 acceptor density
d Akzeptorendichte *f*
f densité *f* d'accepteur
r плотность *f* акцептора

165 acceptor level
d Akzeptorniveau *n*
f niveau *m* d'accepteur
r акцепторный уровень *m*

166 acceptor of data
d Datenempfänger *m*; Datenabnehmer *m*
f récepteur *m* des données
r потребитель *m* данных

167 access
d Zugriff *m*
f accès *m*
r доступ *m*; выборка *f*; обращение *n*

168 access *v*
d zugreifen

f accéder
r обращаться

169 access address
d Zugriffsadresse *f*
f adresse *f* d'accès
r адрес *m* доступа

170 access control
d Zugriffssteuerung *f*
f commande *f* d'accès
r управление *n* доступом

171 access control facility
d Zugriffssteuereinrichtung *f*
f dispositif *m* de commande d'accès
r устройство *n* управления доступом

172 access cover
(of a disk store)
d Zugriffsabdeckung *f*
f accès *m* de recouvrement
r перекрытие *n* доступа

173 access cycle
d Zugriffszyklus *m*
f cycle *m* d'accès
r цикл *m* доступа; цикл обращения

174 access file attribute
d Zugriffsdateiattribut *n*
f attribut *m* de fichier d'accès
r описание *n* обращения к файлу

175 accessibility; availability
d Zugänglichkeit *f*; Erreichbarkeit *f*;
Verfügbarkeit *f*
f accessibilité *f*; disponibilité *f*
r доступность *f*

176 accessible
d zugänglich; erreichbar
f accessible
r доступный

177 access instructions
d Kommunikationsbefehl *m*
f instruction *f* d'accès; commande *f* d'accès
r команда *f* доступа

178 accession number
d Zugriffszahl *f*
f nombre *m* d'accession
r число *n* [возможных] считываний

179 access mechanism
(of a disk store)
d Zugriffsmechanismus *m*

f mécanisme *m* d'accès
r механизм *m* выборки

180 access method
d Zugriffsmethode *f*
f méthode *f* d'accès
r метод *m* доступа

181 access mode
d Zugriffsart *f*
f mode *m* d'accès
r режим *m* доступа

* access of air → 587

182 accessorial service
d zusätzliche Dienstleistung *f*
f service *m* accessoire
r дополнительное обслуживание *n*

183 accessories
d Zubehör *n*; Zubehörteile *npl*
f accessoires *mpl*
r принадлежности *fpl*; запасные части *fpl*

184 access point
d Zugriffspunkt *m*
f point *m* d'accès
r точка *f* доступа

185 access switching circuit
d Zugriffsschaltkreis *m*
f circuit *m* de commutation d'accès
r переключательная схема *f* обращения

186 access technique
d Zugriffstechnik *f*
f technique *f* d'accès
r техника *f* доступа

187 access time; read-out time
d Zugriffszeit *f*
f temps *m* d'accès
r время *n* доступа; время выборки

188 access violation
d Zugriffsverletzung *f*;
Zugriffsverbotverletzung *f*
f violation *f* d'accès
r нарушение *n* обращения

189 accidental error; random error
d zufälliger Fehler *m*; Zufallsfehler *m*
f erreur *f* accidentelle; erreur fortuite
r случайная ошибка *f*; случайная
погрешность *f*

190 accident prevention
d Unfallschutz *m*

f sécurité f industrielle; protection f contre les accidents
r аварийная защита f

191 **accordance; concordance; coincidence**
d Übereinstimmung f
f concordance f; coïncidence f
r соответствие n; совпадение n

192 **accumulate** v
d akkumulieren
f accumuler
r аккумулировать; накапливать

193 **accumulated error; stored error**
d akkumulierter Fehler m
f erreur f cumulée; erreur d'accumulation
r накопленная ошибка f; суммарная ошибка

194 **accumulating speed**
d Sammelganggeschwindigkeit f
f vitesse f d'accumulation
r скорость f накопления

195 **accumulation**
d Akkumulation f; Aufhänfung f
f accumulation f
r аккумуляция f; накопление n

196 **accumulation coefficient**
d Akkumulationskoeffizient m
f coefficient m d'accumulation
r коэффициент m накопления

197 **accumulator register**
d Akkumulatorregiste n
f registre m accumulateur
r накопительный регистр m

198 **accumulator stage**
d Speicherstufe f
f étage m accumulateur
r накопительный каскад m

199 **accuracy; exactness; exactitude; precision**
d Exaktheit f; Genauigkeit f; Präzision f
f exactitude f; précision f
r точность f; верность f

200 **accuracy check**
d Genauigkeitsprüfung f
f vérification f de précision
r контроль m точности

201 **accuracy class; class of accuracy**
d Genauigkeitsklasse f
f classe f de précision
r класс m точности

202 **accuracy control system**
d Genauigkeitsüberwachungssystem n
f système m de surveillance de précision
r система f контроля точности

203 **accuracy grade; degree of accuracy**
d Genauigkeitsgrad m
f degré m de précision
r степень f точности

204 **accuracy limit**
d Genauigkeitsgrenze f
f limite f de précision
r предел m точности

205 **accuracy of control; accuracy of regulation; regulation precision**
d Regelgenauigkeit f; Genauigkeit f der Regelung
f précision f de réglage; exactitude f de régulation
r точность f регулирования

206 **accuracy of manipulation system**
d Genauigkeit f des Manipulationssystems
f exactitude f de système de manipulation
r точность f манипуляционной системы

207 **accuracy of manufacture**
d Herstellungsgenauigkeit f
f précision f de manufacture
r точность f изготовления

* **accuracy of measurement → 7944**

208 **accuracy of reading**
d Ablesegenauigkeit f
f précision f de lecture
r точность f отсчёта

* **accuracy of regulation → 205**

209 **accuracy of tacho-generator**
d Tachogeneratorgenauigkeit f
f exactitude f de la génératrice tachymétrique
r точность f тахогенератора

210 **accuracy study**
d Genauigkeitsuntersuchung f
f étude f d'exactitude
r исследование n точности; анализ m точности

211 **accurate current range of a meter**
d Bereich m höchster Strommessergenauigkeit
f domaine m de précision des courants d'un compteur
r диапазон m точных значений тока измерительного прибора

212 **accurate positioning**
  d Feinpositionierung *f*
  f positionnement *m* de précision
  r прецизионное позиционирование *n*

213 **accurate scanning**
  d Feinabtastung *f*; Feinortung *f*
  f balayage *m* précis; exploration *f* précise
  r точное сканирование *n*

214 **achievable reliability**
  d erreichbare Zuverlässigkeit *f*
  f fiabilité *f* accessible
  r реальная [действительная] надёжность *f*

  * **ACIA → 1075**

215 **acidimeter**
  d Säuremesser *m*
  f acidimètre *m*
  r ацидометр *m*

216 **acknowledged run flag; display for non-validation**
  d Anzeige *f* für die Nichtbestätigung
  f affichage *m* pour la non-validation
  r индикатор *m* неподтверждения

217 **acknowledge enable**
  d Rückmeldebefähigung *f*
  f aptitude *f* d'annonce en retour
  r готовность *f* обратной сигнализации

218 **acknowledgement priority**
  d Rückmeldepriorität *f*
  f priorité *f* de quittance
  r приоритет *m* подтверждения

219 **acknowledge[ment] signal**
  d Betätigungszeichen *n*; Quittungszeichen *n*
  f signal *m* de confirmation
  r сигнал *m* подтверждения

  * **acknowledge signal → 219**

220 **acknowledging relay**
  d Empfangsbestätigungsrelais *n*; Steuerquittungsschalter *m*
  f relais *m* de réception
  r приёмное реле *n*

221 **acoustical channel**
  d akustischer Kanal *m*
  f canal *m* acoustique
  r акустический канал *m*

222 **acoustical interferometer**
  d akustisches Interferometer *n*
  f interféromètre *m* acoustique; réfractomètre *m* interférentiel
  r акустический интерферометр *m*

223 **acoustic altimeter**
  d akustischer Höhenmesser *m*; akustisches Echolot *n*
  f altimètre *m* acoustique
  r акустический высотомер *m*

224 **acoustic bridge**
  d akustische Brücke *f*
  f pont *m* acoustique
  r акустический мост *m*

225 **acoustic calibrator**
  d akustische Eichvorrichtung *f*
  f appareil *m* étalon acoustique
  r акустический калибратор *m*

226 **acoustic communications**
  d akustische Kommunikation *f*
  f communication *f* acoustique
  r звуковая связь *f*

227 **acoustic computer signal**
  d akustisches Rechnersignal *n*
  f signal *m* acoustique de l'ordinateur
  r акустический сигнал *m* компьютера

228 **acoustic deflection circuit**
  d akustischer Ablenkkreis *m*
  f circuit *m* acoustique de déviation
  r акустическая схема *f* отклонения

229 **acoustic delay line; sonic delay line**
  d akustische Verzögerungsstrecke *f*; akustische Verzögerungsleitung *f*
  f ligne *f* à retard acoustique
  r акустическая линия *f* задержки

230 **acoustic delay line memory**
  d akustischer Laufzeitspeicher *m*; Speicher *m* mit akustischem Laufzeitglied
  f mémoire *f* à ligne à retard acoustique
  r память *f* на акустической линии задержки

231 **acoustic dispersion**
  d Schalldispersion *f*
  f dispersion *f* acoustique
  r акустическая дисперсия *f*

232 **acoustic excitation**
  d Schallanregung *f*
  f excitation *f* acoustique
  r акустическое возбуждение *n*

233 **acoustic feedback**
  d akustische Rückkopplung *f*

f réaction *f* acoustique
r акустическая обратная связь *f*

234 acoustic identification system
  d akustisches Erkennungssystem *n*
  f système *m* d'identification acoustique
  r акустическая система *f* распознавания

235 acoustic image
  d akustisches Bild *n*
  f image *f* acoustique
  r акустическое изображение *n*

236 acoustic impedance measurement
  d Schallimpedanzmessung *f*;
     Schallwellenwiderstandsmessung *f*
  f mesure *f* d'impédance acoustique
  r измерение *n* акустического импеданса

**237 acoustic memory; acoustic storage;
     acoustic store**
  d akustischer Speicher *m*
  f mémoire *f* acoustique
  r акустический накопитель *m*; акустическая
     память *f*

238 acoustic overload signal
  d akustisches Überlastsignal *n*
  f signal *m* de surcharge acoustique
  r акустический сигнал *m* перегрузки

239 acoustic quantity
  d akustische Größe *f*
  f grandeur *f* acoustique
  r акустическая величина *f*

240 acoustic radiometer
  d akustisches Radiometer *n*
  f sonomètre *m* acoustique
  r акустический радиометр *m*

241 acoustic refraction
  d Schallbrechung *f*
  f réfraction *f* acoustique
  r акустическая рефракция *f*

242 acoustic relay
  d akustisches Relais *n*
  f relais *m* acoustique
  r акустическое реле *n*

243 acoustic sensor
  d akustischer Aufnehmer *m*; akustischer
     Sensor *m*
  f capteur *m* acoustique; senseur *m* acoustique
  r акустический датчик *m*

244 acoustic signal

  d akustisches Signal *n*
  f signal *m* acoustique
  r акустический сигнал *m*

* **acoustic storage → 237**

* **acoustic store → 237**

245 acoustic transmission
  d akustische Übertragung *f*
  f transmission *f* acoustique
  r акустическая передача *f*

* **acoustooptic → 246**

246 acoustooptic[al]
  d akustooptisch
  f acousto-optique
  r акустооптический

247 acoustooptical deflection device
  d akustooptisches Ablenkungsgerät *n*
  f dispositif *m* acousto-optique de déviation
  r оптикоакустическое устройство *n*
     отклонения

**248 acoustooptical modulation device;
     acoustooptical modulator**
  d akustooptischer Modulator *m*
  f modulateur *m* acousto-optique
  r оптикоакустический модулятор *m*

* **acoustooptical modulator → 248**

249 acoustooptic effect
  d akustooptischer Effekt *m*
  f effet *m* acousto-optique
  r акустооптический эффект *m*

250 acoustooptic interaction
  d akustooptische Wechselwirkung *f*
  f interaction *f* acousto-optique
  r акустооптическое взаимодействие *n*

251 acoustooptics
  d Akustooptik *f*
  f acoustooptique *f*
  r акустооптика *f*

252 acoustooptic signal processing
  d akustooptische Signalverarbeitung *f*
  f traitement *m* acousto-optique du signal
  r акустооптическая обработка *f* сигнала

253 acoustooptic system
  d akustooptisches System *n*
  f système *m* acousto-optique
  r акустооптическая система *f*

254 **acquisition of handling process**
  d Erfassung *f* eines Handhabungsvorgangs
  f acquisition *f* d'un procédé de manutention
  r сбор *m* данных о процессе
       манипулирования

255 **actinograph**
  d Aktinograf *m*
  f actinographe *m*
  r актинограф *m*

256 **actinometer**
  d Aktinometer *m*; Strahlenmesser *m*;
       Strahlungsmesser *m*
  f actinomètre *m*
  r актинометр *m*

257 **action**
  d Aktion *f*; Wirkung *f*; Bedienungsmaßnahme *f*
  f action *f*
  r действие *n*; воздействие *n*

258 **action acceptor**
  d Aktionsakzeptor *m*
  f accepteur *m* d'action
  r акцептор *m* действия

259 **action centre**
  d Aktionszentrum *n*
  f centre *m* d'action
  r центр *m* действия

260 **action correction**
  d Funktionskorrektur *f*; Einflußkorrektur *f*
  f correction *f* de l'action
  r коррекция *f* воздействия

261 **action field**
  d Aktionsfeld *n*
  f champ *m* d'action
  r поле *n* действия

262 **action in error condition**
  d Maßnahme *f* bei Fehlerbedingung
  f action *f* par condition d'erreur
  r действие *n* при наличии ошибок

263 **action limited by absolute value**
  d modulbeschränkte Einwirkung *f*
  f action *f* limitée par module
  r действие *n* ограниченное по абсолютной
       величине

264 **action line**
  d Wirkungslinie *f*
  f ligne *f* d'action
  r активная линия *f*

265 **action macro**
  d Arbeitsmakro *n*
  f macro *m* de travail
  r макродействие *n*

266 **action period; action phase**
  d Wirkungsperiode *f*; Wirkungsphase *f*;
       Wirkungsdauer *f*
  f temps *m* d'action
  r активный период *m*; время *n* действия

*   **action phase → 266**

267 **action potential**
  d Aktionspotential *n*
  f potentiel *m* d'action
  r потенциал *m* действия

268 **action principle**
  d Wirkungsprinzip *n*
  f principe *m* de fonctionnement
  r принцип *m* действия

269 **action quantity**
  d Wirkungsgröße *f*
  f quantité *f* active; quantité d'action
  r активная величина *f*; действующая
       величина

270 **action simulation of a switch**
  d Wirkungssimulation *f* eines Schalters
  f simulation *f* d'action d'un commutateur
  r моделирование *n* действия переключателя

271 **action spot**
  d Wirkungspunkt *m*; Abtastpunkt *m*
  f point *m* d'exploration
  r развёртывающее пятно *n*

272 **action system**
  d Aktionssystem *n*
  f système *m* d'action
  r система *f* воздействия

273 **activate** *v*
  d aktivieren; anregen
  f activer; agir
  r действовать; воздействовать;
       активи[зи]ровать

*   **activated → 284**

274 **activation**
  d Aktivierung *f*; Anreizung *f*; Erregung *f*;
       Anregung *f*
  f activation *f*; excitation *f*; attaque *f*
  r актив[из]ация *f*

275 **activation analysis**
  d Aktivierungsanalyse *f*
  f analyse *f* par activation
  r активационный анализ *m*

276 **activation detector**
  d Aktivierungsdetektor *m*
  f détecteur *m* par activation
  r активационный детектор *m*

277 **activation energy**
  d Aktivierungsenergie *f*
  f énergie *f* d'activation
  r энергия *f* активации

278 **activation heat**
  d Aktivierungswärme *f*
  f chaleur *f* d'activation
  r теплота *f* активирования

279 **activation integral**
  d Aktivierungsintegral *n*
  f intégrale *f* d'activation
  r интеграл *m* активации

280 **activation machanism**
  d Aktivierungsmechanismus *m*
  f mécanisme *m* d'activation
  r механизм *m* активации

281 **activation speed**
  d Aktivierungsgeschwindigkeit *f*
  f vitesse *f* d'activation
  r скорость *f* активации

282 **activation yield**
  d Aktivierungsausbeute *f*
  f rendement *m* d'activation
  r эффективность *f* активации; результат *m* активации

283 **activator**
  d Aktivator *m*
  f activateur *m*
  r активатор *m*; возбудитель *m*

284 **active; activated**
  d in Betrieb
  f en service; en fonctionnement
  r активный

285 **active abonent**
  d aktiver Teilnehmer *m*
  f abonné *m* actif
  r активный абонент *m*

286 **active area**
  d Wirkfläche *f*
  f surface *f* active
  r действующая поверхность *f*

287 **active check**
  d aktive Kontrolle *f*
  f contrôle *m* actif
  r активный контроль *m*

288 **active circuit**
  d aktiver Stromkreis *m*
  f circuit *m* actif
  r активная цепь *f*

289 **active current**
  d Wirkstrom *m*
  f courant *m* actif
  r активный ток *m*

290 **active data record**
  d aktiver Datensatz *m*
  f enregistrement *m* actif des données
  r активная запись *f* данных

291 **active electrode**
  d Wirkelektrode *f*
  f électrode *f* active
  r активный электрод *m*

292 **active element**
  d aktives Element *n*; aktives Glied *n*
  f élément *m* actif; organe *m* actif
  r активный элемент *m*

293 **active filter**
  d aktives Filter *n*
  f filtre *m* actif
  r активный фильтр *m*

294 **active guidance; active homing**
  d aktive Führung *f*; aktive Ziellenkung *f*
  f guidage *m* actif
  r активное наведение *n*

* **active homing → 294**

295 **active infrared detection system**
  d aktives Ultrarotstrahlenerfassungssystem *n*
  f système *m* détecteur actif aux rayons infrarouges
  r активная инфракрасная система *f* обнаружения

296 **active input device**
  d aktives Eingabegerät *n*
  f dispositif *m* d'entrée actif
  r активное устройство *m* ввода

297 **active interference filter laser amplifier**
   d Laserverstärker *m* mit aktivem
      Interferenzfilter
   f laser *m* amplificateur à filtre actif à
      interférence
   r лазерный усилитель *m* с активным
      интерференционным фильтром

298 **active joint mechanism**
   d aktiver Fügemechanismus *m*
   f mécanisme *m* de jointage actif
   r активный механизм *m* сопряжения

299 **[active] laser medium**
   d aktives Lasermedium *n*
   f millieu *m* [actif] laser
   r лазерная активная среда *f*

300 **active laser tracking system**
   d aktiver Laserkursverfolger *m*; aktives
      Laserkursfolgesystem *n*
   f traceur *m* actif à laser; système *m* actif de
      poursuite à laser
   r активная лазерная система *f*
      сопровождения

301 **actively coupled**
   d mit aktiver Kopplung; aktiv gekoppelt
   f activement couplé
   r с активной связью; активно связанный

302 **active medium gain**
   d Gewinn *m* im aktiven Lasermedium
   f gain *m* dans le milieu actif du laser
   r коэффициент *m* усиления активного
      вещества лазера

303 **active optical component**
   d aktives optisches Element *n*
   f composant *m* optique actif
   r активный оптический элемент *m*

304 **active organ**
   d Wirkorgan *n*
   f organe *m* actif
   r активный [исполнительный] орган *n*

305 **active output device**
   d aktives Ausgabegerät *n*
   f dispositif *m* de sortie actif
   r активное устройство *n* вывода

306 **active power; real power**
   d Wirkleistung *f*; Nutzleistung *f*
   f puissance *f* réelle; puissance active
   r активная мощность *f*

307 **active-power meter**

   d Wirkverbrauchsmesser *m*
   f compteur *m* d'énergie active
   r счётчик *m* активной энергии

308 **active-power relay**
   d Wirkleistungsrelais *n*
   f relais *m* de puissance active
   r реле *n* активной мощности

309 **active pressure feedback**
   d aktive Druckrückführung *f*
   f réaction *f* de pression active
   r активная обратная связь *f* по давлению

310 **active satellite**
   d aktiver Satellit *m*
   f satellite *m* actif
   r активный спутник *m*

311 **active sensor**
   d arbeitender Sensor *m*; aktiver Sensor
   f senseur *m* actif
   r активный сенсор *m*; действующий сенсор

312 **active store**
   d Aktivspeicher *m*
   f mémoire *f* active
   r активный накопитель *m*

313 **active surface**
   d wirksame Oberfläche
   f surface *f* active
   r активная поверхность *f*

314 **active system**
   d aktives System *n*
   f système *m* actif
   r активная система *f*

315 **active task list**
   d Liste *f* aktiver Aufgaben
   f liste *f* des tasks actifs
   r список *m* активных задач

316 **active transducer**
   d Aktivwandler *m*; Aktivsender *m*; aktiver
      Wandler
   f transmetteur *m* actif; détecteur *m* actif;
      transducteur *m* actif
   r активный преобразователь *m*;
      действующий передатчик *m*

317 **active waveguide**
   d aktiver Wellenleiter *m*
   f guide *m* d'onde actif
   r активный волновод *m*

318 **activity analysis**
   d Aktivitätsanalyse *f*

f analyse *f* d'activité
r анализ *m* деятельности

319 activity coefficient
d Aktivitätskoeffizient *m*
f coefficient *m* d'activité
r коэффициент *m* активности

320 activity curve
d Aktivitätskurve *f*
f courbe *f* d'activité
r кривая *f* активности

321 activity decay
d Aktivitätsabfall *m*
f chute *f* d'activité
r спад *m* активности

322 activity distribution
d Aktivitätsverteilung *f*
f distribution *f* d'activité
r распределение *n* активности

323 activity level; level of activity; level of process
d Aktivitätsniveau *n*; Aktivitätspegel *m*; Prozessniveau *n*
f niveau *m* d'activité
r уровень *m* активности; уровень процесса

324 activity measurement
d Aktivitätsmessung *f*
f mesure *f* de l'activité
r измерение *n* активности

325 activity unit
d Aktivitätseinheit *f*
f unité *f* d'activité
r единица *f* активности

326 actual address; effective address
d wirkliche Adresse *f*
f adresse *f* effective
r действительный адрес *m*

327 actual address attribution
d Zuerkennung *f* realer Adressen; Zuordnung *f* realer Adressen
f attribution *f* d'adresses réelles
r присвоение *n* истинных адресов

328 actual capacity
d tatsächliche Leistung *f*; effektive Leistung
f puissance *f* effective
r действительная производительность *f*

329 actual cycle
d realer Prozess *m*

f cycle *m* actuel; processus *m* réel
r реальный цикл *m*

330 actual design of object identification
d Istmuster *n* einer Objekterkennung
f dessin *m* réel d'une identification d'objet
r действительный образ *m* в процессе распознавания объекта

331 actual displacement
d tatsächlicher Förderstrom *m*
f débit *m* effectif; volume *m* de transport réel
r действительный объёмный поток *m*

332 actual efficiency
d tatsächlicher Wirkungsgrad *m*
f rendement *m* effectif
r действительный коэффициент *m* полезного действия

333 actual gas; real gas
d reales Gas *n*
f gaz *m* réel
r реальный газ *m*

334 actual hardware system simulation
d Analogtechnik *f*; analoge Gerätetechnik *f*
f technique *f* analogique
r аналоговая техника *f*; моделирующая техника

335 actual hours
d tatsächliche Abeitsstunden *fpl*; effektive Arbeitszeit *f*
f heures *fpl* de travail effectives
r эффективное рабочее время *n*

336 actual manipulator state
d aktueller Manipulatorzustand *m*
f état *m* de manipulateur actuel
r текущее состояние *n* манипулятора

337 actual movement path
d Istbewegungsbahn *f*
f chemin *m* de mouvement réel
r текущая траектория *f* движения

338 actual output
d Ist-Leistung *f*
f puissance *f* réelle; production *f* effective
r эффективная производительность *f*; фактическая мощность *f*

339 actual parameter
d Aktualparameter *m*; aktueller Parameter *m*
f paramètre *m* effectif
r действительный параметр *m*

340 actual parameter area
  d aktueller Parameterbereich *m*
  f domaine *m* de paramètre actuel
  r область *f* фактических параметров

341 actual position
  d Istposition *f*
  f position *f* réelle
  r текущее положение *n*

342 actual process temperature
  d tatsächliche Prozesstemperatur *f*
  f température *f* réelle de procédé
  r температура *f* действительного процесса

343 actual quantity
  d Istmenge *f*
  f quantité *f* actuelle
  r действительная величина *f*

344 actual range
  d wirksamer Bereich *m*; effektive Spanne *f*
  f gamme *f* d'utilisation
  r действительный диапазон *m*

345 actual system state
  d aktueller Systemzustand *m*
  f état *m* de système actuel
  r текущее состояние *n* системы

346 actual value; real value; desired value
  d Realwert *m*; Istwert *m*
  f valeur *f* réelle; valeur effective; valeur de
    consigne
  r фактическая величина *f*; действительное
    значение *n*

347 actual value comparison
  d Istwertvergleich *m*
  f comparaison *f* de valeur réelle
  r сравнение *n* действительного значения

348 actual value of controlled variable
  d Istwert *m* der Regelgröße
  f valeur *f* réelle de la grandeur réglée; valeur
    instantanée de la grandeur réglée
  r действительное значение *n* регулируемой
    величины

349 actual value transmitter
  d Istwertgeber *m*
  f indicateur *m* de valeur nominale
  r датчик *m* действительного значения;
    датчик истинного значения

350 actuate *v*
  d betätigen
  f actionner

  r запускать

**351 actuating appliance; actuating unit;
    regulating element; effector; regulating
    unit**
  d Betätigungsorgan *n*; Effektor *m*;
    Stelleinheit *f*; Stellorgan *n*
  f organe *m* de réglage; élément *m* de réglage;
    dispositif *m* de commande; organe
    correcteur; organe de positionnement
  r устройство *n* управления; управляющий
    элемент *m*; орган *m* управления

352 actuating device
  d Betätigungsvorrichtung *f*; Stelleinrichtung *f*
  f dispositif *m* de manœuvre
  r исполнительное устройство *n*

353 actuating path; control path
  d Wirkungsweg *m* in Steuerungs- und
    Regelungssystemen
  f trajectoire *f* de réponse
  r направление *n* воздействия

**354 actuating pulse; control pulse; driving
    pulse**
  d Betätigungsimpuls *m*; Antriebsimpuls *m*;
    Verstellimpuls *m*; Steuerimpuls *m*
  f impulsion *f* de commande; impulsion de
    déclenchement
  r управляющий импульс *m*; пусковой
    импульс

355 actuating quantity
  d Betätigungsgröße *f*; erregende Größe *f*;
    Stellgröße *f*
  f grandeur *f* d'influence
  r величина *f* воздействия

356 actuating signal
  d Stellsignal *n*; Betätigungssignal *n*
  f signal *m* d'influence; signal d'activation
  r управляющий сигнал *m*

357 actuating system
  d Betätigungssystem *n*
  f système *m* d'entraînement
  r система *f* [силовых] приводов

358 actuating transfer function
  d Stellübertragungsfunktion *f*;
    Übertragungsfunktion *f* der Stelleinrichtung
  f fonction *f* de transfert de commande
  r передаточная функция *f* по
    регулирующему воздействию

  * **actuating unit → 351**

359 **actuating variable; influencing variable**
 d Einflußgröße *f*; Steuersignal *n*
 f grandeur *f* d'influence; grandeur de commande; signal *m* d'action
 r действующая переменная *f*; регулирующая переменная

360 **actuator**
 d Stellantrieb *m*; Stellmotor *m*; Wirkungsglied *n*
 f organe *m* moteur; moteur *m* de commande; actionneur *m*
 r исполнительный механизм *m*; привод *m*

361 **actuator mechanism**
 d Antriebsmechanismus *m*
 f mécanisme *m* de commande; système *m* opérateur
 r механизм *m* привода

362 **acyclic process**
 d azyklischer Vorgang *m*
 f processus *m* acyclique
 r непериодический процесс *m*

363 **adaptability**
 d Anpassungsfähigkeit *f*
 f facultéf d'adaptation
 r адаптивность *f*

364 **adaptable**
 d verwendbar
 f utilisable
 r адаптивный; применимый

365 **adaptable assembly condition**
 d anpassbare Montagebedingung *f*
 f condition *f* d'assemblage adaptable
 r адаптивное условие *n* сборки

366 **adaptable programmable assembly system**
 d anpassungsfähiges programmierbares Montagesystem *n*
 f système *m* d'assemblage programmable adaptable; système de montage comparable et programmable
 r адаптивная программируемая сборочная система *f*

367 **adaptable sensor**
 d anpassungsfähiger Sensor *m*
 f senseur *m* adaptable; capteur *m* adaptable
 r [само]приспосабливающийся датчик *m*

368 **adaptable system; adaptive system**
 d anpassungsfähiges System *n*; adaptives System
 f système *m* adaptable; système adaptatif

 r адаптивная система *f*

369 **adaptable system of equipment**
 d anpassungsfähiges Anlagensystem *n*
 f système *m* adaptable de l'équipement
 r адаптивное оборудование *n*

370 **adaptation; adapting; matching**
 d Anpassung *f*
 f adaptation *f*
 r адаптация *f*; согласование *n*

371 **adaptation mechanism**
 d Anpassungsmechanismus *m*
 f mécanisme *m* d'adaptation
 r механизм *m* приспособления

372 **adaptation of circuits**
 d Stromkreisanpassung *f*
 f adaptation *f* des circuits
 r согласование *n* контуров

373 **adaptation parameter**
 d Anpassungsparameter *m*
 f paramètre *m* d'adaptation
 r параметр *m* настройки

374 **adapter**
 d Adapter *m*; Anpassungseinheit *f*
 f adapteur *m*
 r переходное устройство *n*; адаптер *m*

* **adapting → 370**

375 **adapting machine**
 d adaptive Maschine *f*
 f machine *f* adaptable
 r адаптивная машина *f*

376 **adaptive architecture**
 d ampassungsfähige Rechnerstruktur *f*; selbstanpassende Struktur *f*
 f structure *f* adaptative
 r адаптивная архитектура *f*

377 **adaptive behaviour of manipulation systems**
 d adaptives Verhalten *n* von Handhabungssystemen
 f comportement *m* adaptif de systèmes de manipulation
 r адаптивное поведение *n* манипуляционных систем

378 **adaptive character reader**
 d adaptiver Zeichenleser *m*
 f lecteur *m* adaptif de caractères
 r адаптивное устройство *n* считывания знаков

379 **adaptive control optimization**
  d adaptive Steuerungsoptimierung *f*;
     Optimierung *f* adaptiver Steuerung
  f réglage *m* d'optimisation
  r оптимизация *f* адаптивного управления

380 **adaptive control process**
  d adaptiver Steuerprozess *m*
  f processus *m* de commande adaptif
  r процес *m* адаптивного управления

381 **adaptive control system; self-adjusting
    system**
  d selbsteinstellendes System *n*; adaptives
     Regelsystem *n*
  f système *m* d'auto-régulation; système d'auto-
     adaptation
  r самонастраивающаяся система *f*
     управления

382 **adaptive converter of learning system**
  d adaptiver Umformer *m* des lernenden
     Systems
  f convertisseur *m* adaptif du système apprenant
  r адаптивный преобразователь *m*
     самообучающейся системы

383 **adaptive differential pulse code
    modulation; ADPCM**
  d adaptive differentielle Pulskodemodulation *f*
  f modulation *f* par impulsions et codage
     différentielle adaptive; MICDA
  r адаптивная дифференциальная
     импульсно-кодовая модуляция

384 **adaptive digital element**
  d adaptives digitales Element *n*
  f élément *m* adaptif digital
  r адаптивный цифровой элемент *m*

385 **adaptive digital network**
  d adaptive digitale Schaltung *f*; adaptives
     digitales Netzwerk *n*
  f circuit *m* digital adaptif; réseau *m* digital
     adaptif
  r самонастраивающаяся цифровая схема *f*

386 **adaptive element**
  d adaptives Glied *n*
  f élément *m* adaptif
  r самонастраивающийся елемент *m*

387 **adaptive equalization**
  d adaptiver Ausgleich *m*
  f équilibrage *m* adaptif
  r адаптивное выравнивание *n*

388 **adaptive learning controller**

  d adaptiver lernender Regler *m*
  f régulateur *m* apprenant adaptif
  r самонастраивающийся регулятор *m*

389 **adaptive learning system**
  d adaptives Lernsystem *n*
  f système *m* d'auto-apprentissage adaptif
  r самонастраивающаяся самообучающаяся
     система *f*

390 **adaptive logic**
  d anpassungsfähige Logik *f*
  f logique *f* adaptive
  r адаптивная логика *f*

391 **adaptive manipulation**
  d adaptive Handhabung *f*
  f manipulation *f* adaptive
  r адаптивное манипулирование *n*

392 **adaptive model**
  d adaptives Modell *n*
  f modèle *m* adaptif
  r адаптивная модель *f*

393 **adaptive optimum filtration**
  d adaptive Optimalfilterung *f*
  f filtration *f* optimale adaptative
  r адаптивная оптимальная фильтрация *f*

394 **adaptive predistortion**
  d adaptive Vorverzerrung *f*
  f prédistorsion *f* adaptive
  r адаптивное предискажение *n*

395 **adaptive process**
  d adaptiver Prozess *m*; selbstanpassender
     Prozess
  f procédé *m* adaptif; processus *m* à auto-
     adaptation
  r адаптивный процесс *m*

396 **adaptive robot control**
  d adaptive Robotersteuerung *f*
  f commande *f* de robot adaptive
  r адаптивное управление *n* роботом

397 **adaptive sensor-guided system**
  d adaptive sensorgeführtes System *n*
  f système *m* adaptif commandé par capteur
  r адаптивная сенсорная система *f*

398 **adaptive speed regulation**
  d selbstanpassende Drehzahlregelung *f*
  f réglage *m* de vitesse auto-adaptif
  r самонастраивающееся регулирование *n*
     скорости

* **adaptive system** → 368

399 adaptive threshold elements
  d adaptive Schwellwertelemente *npl*
  f éléments *mpl* adaptifs de seuil
  r адаптивные пороговые элементы *mpl*

400 add carry
  d Additionsübertrag *m*
  f retenue *f* d'addition
  r добавочный перенос *m*

401 added facility
  d Zusatzeinrichtung *f*; Zusatzausrüstung *f*;
     Zusatzmöglichkeit *f*
  f dispositif *m* supplémentaire; équipement *m*
     supplémentaire
  r дополнительное оборудование *n*

402 added filter
  d Zusatzfilter *n*
  f filtre *m* additionnel
  r дополнительный фильтр *m*

403 added instruction kit
  d Zusatzbefehlsbausatz *m*
  f jeu *m* d'instructions supplémentaires
  r дополнительный набор *m* команд

404 adder; adding element; summator
  d Additionsstelle *f*; Add[ier]er *m*;
     Addierwerk *n*; Summationsstelle *f*
  f addeur *m*; addit[ionn]eur *m*; élément *m*
     sommateur; sommateur *m*; totalisateur *m*
  r сумматор *m*; суммирующее устройство *n*

405 adder circuit; summation circuit
  d Summationskette *f*
  f circuit *m* additionneur; circuit sommateur;
     chaîne *f* de sommation
  r суммирующая цепь *f*

406 adder stage; counting stage
  d Additionsstelle *f*
  f étage *m* additionneur; étage de sommateur
  r ячейка *f* счёта; счётчик *m*

407 add gate
  d Additionsgatter *n*; Additionstor *n*
  f élément *m* d'addition
  r элемент *m* сложения

408 add impulse
  d Additionsimpuls *m*
  f impulsion *f* d'addition
  r импульс *m* сложения

* **adding element** → 404

409 adding relay
  d Summierungsrelais *n*
  f relais *m* intégrateur
  r суммирующее реле *n*

410 add instruction
  d Additionsbefehl *m*
  f instruction *f* d'addition
  r добавочная инструкция *f*

411 additional circuit
  d Zusatzschaltung *f*
  f circuit *m* additionnel
  r дополнительная схема *f*

412 additional code
  d Zuschlagskode *m*; Komplementärkode *m*
  f code *m* complémentaire
  r дополнительный код *m*

413 additional coupling
  d zusätzliche Kopplung *f*
  f couplage *m* additionnel
  r дополнительная связь *f*

414 additional error
  d zusätzlicher Fehler *m*
  f erreur *f* complémentaire
  r дополнительная погрешность *f*

415 additional filtration
  d Zusatzfilterung *f*; Filterung *f* durch
     Zusatzfilter
  f filtration *f* additionnelle
  r дополнительная фильтрация *f*

416 additional information
  d zusätzliche Information *f*
  f information *f* additionnelle
  r дополнительная информация *f*

417 additional peripheral unit
  d zusätzliche Randeinheit *f*; zusätzliche
     periphere Einheit *f*
  f unité *f* périphérique complémentaire
  r дополнительное оконечное устройство *n*

418 additional pulse
  d Zusatzimpuls *m*
  f impulsion *f* additionnelle
  r дополнительный импульс *m*

419 additional resistance
  d Zusatzwiderstand *m*
  f résistance *f* additionnelle
  r добавочное сопротивление *n*

420 additional switching unit
  d zusätzliches Schaltglied *n*

f élément *m* supplémentaire de commutation
r дополнительный элемент *m* включения

421 additional translation movement
  d zusätzliche translatorische Bewegung *f*;
     zusätzliche Translationsbewegung *f*
  f mouvement *m* de translation additionnel
  r дополнительное поступательное
     движение *n*

**422 additional treatment; subsequent
     treatment**
  d Nachbehandlung *f*; Nachbearbeitung *f*
  f retraitement *m*; traitement *m* ultérieur
  r окончательная обработка *f*

423 addition cascade
  d Addierstufe *f*
  f étage *m* additionneur
  r суммирующий каскад *m*

424 addition electrical conductivity
  d Gesamtelektroleitfähigkeit *f*
  f conductibilité *f* électrique d'additions
  r добавочная электропроводность *f*

  * **addition of energy** → 5053

425 addition theorem
  d Additionstheorem *n*
  f théorème *m* d'addition
  r теорема *f* сложения

426 additive action
  d additive Wirkung *f*
  f action *f* additive; effet *m* additif
  r аддитивное действие *n*; аддитивный
     эффект *m*

427 additive circuit
  d Additivkreis *m*; additive Mischstufe *f*
  f circuit *m* additif
  r суммирующая схема *f*

428 additive group
  d additive Gruppe *f*
  f groupe *m* additif
  r аддитивная группа *f*

429 additive inverse
  d additive Inverse *f*; inverses Element *n* der
     Addition
  f inverse *m* additif; réciproque *f* additive
  r инверсный элемент *m* сложения

430 additive process
  d Additivverfahren *n*
  f technologie *f* additive

r аддитивная технология *f*

431 additive quantity
  d additive Größe *f*
  f quantité *f* additive
  r аддитивная величина *f*

432 additivity property
  d Additivitätseigenschaft *f*
  f propriété *f* d'additivité
  r свойство *n* аддитивности

433 address
  d Adresse *f*
  f adresse *f*
  r адрес *m*

434 addressable control module
  d adressierbarer Steuerungsbaustein *m*
  f module *m* de commande adressable
  r адресуемый модуль *m* управления

435 addressable latch
  d adressierbarer Zwischenspeicher *m*;
     adressierbares Auffangregister *n*
  f mémoire *f* intermédiaire adressable
  r адресуемый регистр-фиксатор *m*

436 addressable memory
  d adressierbarer Speicher *m*
  f mémoire *f* adressable
  r адресуемая память *f*

437 addressable store
  d adressierbarer Speicher *m*
  f mémoire *f* à adresser
  r адресующий накопитель *m*

438 address arithmetic[al] element
  d Adressenrechenwerk *n*
  f organe *m* de calcul d'adresse
  r адреснос арифметико-логическое
     устройство *n*

  * **address arithmetic element** → 438

439 address array
  d Adressenfeld *n*
  f zone *f* d'adresse
  r адресная зона *f*; адресная матрица *f*

440 address calculation; address computation
  d Adressenrechnung *f*; Adressenberechnung *f*;
     Adressenbildung *f*
  f calcul *m* d'adresse
  r вычисление *n* адреса

441 address capability
  d Adressierungsvermögen *n*

f capacité *f* d'adresses
r возможности *fpl* задания адреса

442 address code; address part
d Adressenkode *m*; Adressenteil *m* im
Befehlswort
f code *m* d'adresse; section *f* d'adresse; partie *f*
d'adresse
r адресный код *m*; адресная часть *f*

* address computation → 440

443 address data strobe; address reading pulse
d Adressenabtastimpuls *m*
f impulsion *f* de lecture d'adresses
r адресный импульс *m* считывания

444 address decoder
d Adressenlöser *m*
f traducteur *m* d'adresse codifiée; décodeur *m*
d'adresse
r дешифратор *m* адреса; адресное
декодирующее устройство *n*

445 address delay time
d Adressverzögerungszeit *f*;
Adresssignalverzögerungszeit *f*
f temps *m* de retardement d'adresse
r время *n* запаздывания [сигналов] адреса

446 addressed device
d adressierte Einrichtung *f*
f dispositif *m* à désignation fixée des adresses;
dispositif adressé
r адресованное устройство *n*

447 address expression
d Adressenausdruck *m*
f expression *f* d'adresse
r адресное выражение *n*

448 addressing concept
d Adressierungskonzeption *f*
f concept *m* d'adressage
r концепция *f* адресации

449 addressing machine
d Adressiermaschine *f*
f machine *f* à adresser
r устройство *n* адресации

450 addressing system
d Adressiersystem *n*
f système *m* d'adressage
r система *f* адресации

451 address language
d Adressensprache *f*

f langage *m* d'adresses
r адресный язык *m*

452 address line
d Adress[ier]leitung *f*
f ligne *f* d'adresse
r адресная линия *f*

453 address main line
d Adressenhauptlinie *f*
f ligne *f* principale d'aresses
r адресная магистраль *f*

454 address modification
d Adressenänderung *f*
f changement *m* d'adresse
r модификация *f* адреса

455 address number
d Adressenzahl *f*
f nombre *m* d'adresses
r число *n* адресов

* address part → 442

456 address range
d Adressbereich *m*
f champ *m* d'adresse
r диапазон *m* адресов

* address reading pulse → 443

457 address register
d Adressenregister *n*
f registre *m* d'adresses
r адресный регистр *m*

458 address selection switch
d Adressenwahlschalter *m*
f interrupteur *m* sélecteur d'adresses
r адресный селектор *m*

459 address substitution
d Adressensubstitution *f*
f substitution *f* d'adresse
r переадресация *f*

460 add time
d Additionszeit *f*
f temps *m* d'addition
r время *n* суммирования; такт *m* сложения

461 adequate approximation
d genügende Näherung *f*
f approximation *f* suffisante
r достаточное приближение *n*

462 adherence
d Haftfestigkeit *f*

f adhérence f
r адгезионная прочность f

463 adherence test
d Haftfestigkeitsprüfung f
f test m d'adhérence
r испытание n на адгезионную прочность

464 adhesion coefficient
d Adhäsionsbeiwert m
f coefficient m d'adhésion
r коэффициент m адгезии

465 adhesive principle
d Haftprinzip n
f principe m de collage
r принцип m адгезии

466 adiabatic equivalent temperature
d adiabatische Äquivalenztemperatur f
f température f adiabatique équivalente
r эквивалентная адиабатическая
   температура f

467 adiabatic process
d adiabatisches Verfahren n; adiabatischer
   Prozess m
f procédé m adiabatique
r адиабатный процесс m

468 adiabatic saturation process
d adiabatischer Sättigungsprozess m
f processus m de saturation adiabatique
r процесс m адиабатического насыщения

469 adjacent channel
d Nachbarkanal m
f canal m adjacent
r смежный канал m; соседний канал

470 adjacent-channel attenuation
d Trennschärfe f gegen Nachbarkanal
f affaiblissement m du canal adjacent
r ослабление n сигнала в соседнем канале

471 adjacent element
d Nachbarelement n
f élément m adjacent
r соседний элемент m

472 adjacent-mode beat frequency
d Nachbarmodenüberlagerungsfrequenz f;
   Nachbarkanalschwebungsfrequenz f
f battement m entre modes adjacents
r частота f биений между соседними
   типами колебаний

473 adjacent states

d Nachbarzustände mpl
f états mpl voisins
r соседние состояния npl

474 adjoint
d adjungiert
f adjoint
r сопряжённый

475 adjoint computing technique
d adjungierte Rechentechnik f
f technique f de calcul adjointe
r метод m параллельных вычислений

476 adjoint function
d konjugierte Funktion f
f fonction f conjuguée
r функция f сопряжения

477 adjoint matrix
d adjungierte Matrix f
f matrice f adjointe
r присоединенная матрица f

478 adjoint system
d adjungiertes System n
f système m adjoint
r присоединённая система f

479 adjust v; align v; tune v; trim v
d einstellen; justieren; abstimmen
f ajuster; mettre au point
r настраивать

480 adjustable attenuator; variolosser
d variables Dämpfungsglied n; regelbares
   Dämpfungsglied
f atténuateur m variable
r регулируемое демпфирующее звено n;
   переменное демпфирующее звено

481 adjustable collimator
d verstellbarer Kollimator m
f collimateur m réglable
r регулируемый коллиматор m

482 adjustable contact
d stellbarer Kontakt m; einstellbarer Kontakt
f contact m ajustable; contact variable
r регулируемый контакт m

483 adjustable controller
d einstellbarer Regler m
f régulateur m variable
r регулятор m с настройкой

484 adjustable counter-balance
d einstellbares Gegengewicht n

f contrepoids *m* ajustable
r регулируемый противовес *m*

485 adjustable current setting
d veränderliche Stromeinstellung *f*
f mise *f* au point du courant
r регулируемая настройка *f* тока

486 adjustable drive
d einstellbarer Antrieb *m*; einstellbare
   Aussteuerung *f*
f entraînement *m* ajustable; traction *f* ajustable;
   excitation *f* ajustable
r привод *m* с регулировкой

487 adjustable electric contact thermometer
d Thermometer *n* mit verstellbarem Kontakt
f thermomètre *m* à contact électrique réglable
r термометр *m* с регулируемым
   электрическим контактом

488 adjustable exciter forces
d einstellbare Erregerkräfte *fpl*
f forces *fpl* d'excitation ajustables
r регулируемые возбуждающие силы *fpl*

489 adjustable gain
d regelbare Verstärkung *f*
f gain *m* réglable
r регулируемый коэффициент *m* усиления

**490 adjustable impulse counter with automatic
     rerun**
d einstellbarer Impulszähler *m* mit
   automatischer Wiederholung
f compteur *m* d'impulsions réglable à itération
   automatique
r регулируемый счётчик *m* импульсов с
   автоматическим повторным действием

491 adjustable inductor; variometer
d Drehdrossel *f*; Variometer *n*
f variomètre *m*
r вариометр *m*

492 adjustable length
d einstellbare Länge *f*
f longueur *f* ajustable
r регулируемая длительность *f*

493 adjustable limit moment
d einstellbares Grenzmoment *n*
f moment *m* limite ajustable
r регулируемый предельный момент *m*

494 adjustable parameter
d verstellbarer Parameter *m*; variierbarer
   Parameter

f paramètre *m* ajustable
r регулируемый параметр *m*

495 adjustable reference mismatch
d Reflexionsnormaleinstellung *f*
f déréglage *m* ajustable de référence
r регулируемое исходное рассогласование *n*

496 adjustable resistance
d Einstellwiderstand *m*; veränderlicher
   Widerstand *m*
f résistance *f* ajustable; résistance variable
r регулируемое сопротивление *n*

497 adjustable restriction valve
d einstellbares Drosselventil *n*
f soupape *f* à étranglement ajustable
r регулируемый дроссельный клапан *m*

498 adjustable scale
d regulierbare Skaleneinteilung *f*
f échelle *f* réglable
r регулируемая шкала *f*

499 adjustable sensor
d einstellbarer Sensor *m*
f senseur *m* ajustable
r регулируемый сенсор *m*; регулируемый
   датчик *m*

500 adjustable striking
d einstellbarer Anschlag *m*
f butée *f* réglable
r регулируемое усилие *n*

501 adjustable threshold
d einstellbarer Schwellwert *m*
f valeur *f* de seuil réglable
r регулируемой порог *m*

502 adjustable time constant
d einstellbare Zeitkonstante *f*
f constante *f* de temps ajustable
r регулируемая постоянная *f* времени

503 adjustable voltage control
d verstellbare Spannungsregelung *f*
f régulation *f* de tension ajustable
r регулируемый контроль *m* напряжения

504 adjustable voltage divider
d verstellbarer Spannungsteiler *m*
f potentiomètre *m* réglable
r регулируемый делитель *m* напряжения

505 adjustable voltage rectifier
d verstellbarer Spannungsgleichrichter *m*

f redresseur *m* à tension réglable
r регулируемый выпрямитель *m*
   напряжения

506 adjustable voltage stabilizer
d einstellbarer Spannungsstabilisator *m*
f stabiliseur *m* ajustable de tension
r стабилизатор *m* регулируемого
   напряжения

507 adjuster
d Justiereinrichtung *f*; Einsteller *m*
f ajusteur *m*
r устройство *n* настройки

* **adjusting → 519**

508 adjusting capacitor
d Regelkondensator *m*
f condensateur *m* de réglage
r конденсатор *m* переменной ёмкости

509 adjusting characteristic
d Einstellungskennwert *m*
f caractéristique *f* de réglage
r характеристика *f* настройки

510 adjusting device
d Abstimmeinheit *f*; Einstellvorrichtung *f*
f bloc *m* de mise au point
r блок *m* настройки

511 adjusting element; setting device
d Einstellelement *n*; Einstellgerät *n*
f organe *m* d'ajustement; dispositif *m*
   d'ajustage
r задающее устройство *n*

512 adjusting key
d Einstellungsschlüssel *m*
f clef *f* de réglage
r установочный ключ *m*

513 adjusting method
d Einstellverfahren *n*
f méthode *f* d'ajustage
r способ *m* регулировки; способ
   настройки

514 adjusting moment
d Stellmoment *m*
f moment *m* d'ajustage
r исполнительный момент *m*

515 adjusting of mode of operation
d Betriebsarteneinstellung *f*
f réglage *m* de mode d'operation
r установка *f* режима работы

516 adjusting of the cycle
d Zyklusregelung *f*
f réglage *m* du cycle
r регулировка *f* цикла

517 adjusting operation
d Einstelloperation *f*; Justageoperation *f*
f opération *f* d'ajustement
r операция *f* настройки

518 adjusting speed; regulation speed; control speed
d Stellgeschwindigkeit *f*;
   Regel[ungs]geschwindigkeit *f*
f vitesse *f* d'ajustage; vitesse de régulation
r скорость *f* регулирования

519 adjustment; adjusting
d Justierung *f*; Adjustierung *f*
f ajustement *m*; ajustage *m*; réglage *m*
r настройка *f*; регулирование *n*; юстировка *f*

520 adjustment curve
d Einstellkurve *f*
f courbe *f* de mise au point
r кривая *f* настройки

521 adjustment instruction
d Regulierbefehl *m*; Einstellbefehl *m*;
   Berichtigungsbefehl *m*
f instruction *f* d'ajustement
r команда *f* настройки

522 adjustment of measuring-channels
d Einstellen *n* der Messkanäle
f ajustage *m* des canaux de mesure
r настройка *f* каналов измерения

523 adjustment range; tuning range
d Einstellbereich *m*
f plage *m* de réglage
r диапазон *m* настройки

524 adjustment value by manipulation of components
d Einstellwert *m* bei Teilehandhabung
f valeur *f* d'ajustage en manipulation de pièces
r установка *f* при манипулировании
   изделиями

525 admissible attenuation
d zulässige Dämpfung *f*
f affaiblissement *m* admissible
r допустимое затухание *n*

* **admissible basis → 5417**

526 admissible deviation domain
d zulässiger Abweichungsbereich *m*

f domaine *m* d'écarts admissible
r допустимая область *f* отклонений

527 admissible error
d zulässiger Fehler *m*
f erreur *f* admissible
r допустимая погрешность *f*

528 admissible gripper displacement
d zulässige Greiferverlagerung *f*
f déplacement *m* de grappin admissible
r допустимое смещение *n* захватного
устройства

529 admissible perturbation
d zulässige Störung *f*
f perturbation *f* admissible
r допустимое возмущение *n*

530 admissible position deviation
d zulässige Lageabweichung *f*
f déviation *f* de position admissible
r допустимое отклонение *n* позиции

531 admissible repeater spacing
d zulässige Regeneratorfeldlänge *f*
f distance *f* autorisée entre répéteurs
r допустимая длина *f* регенераторного
участка

**532 admissible solution; feasible solution;
permissible solution**
d zulässige Lösung *f*
f solution *f* admissible
r допустимое решение *n*

533 admissible value
d zulässiger Wert *m*
f valeur *f* admissible
r допустимое значение *n*; допустимая
величина *f*

534 admission controller
d Zuflußregler *m*
f régulateur *m* d'afflux
r регулятор *m* подачи

535 admittance
d Admittanz *f*
f admittance *f*
r полная проводимость *f*

536 admittance function
d Leitwertfunktion *f*
f fonction *f* d'admittance; fonction de
conductibilité
r функция *f* проводимости

* **ADPCM** → 383

**537 adsorption dehumidification
system**
d Adsorptionstrocknungsanlage *f*
f installation *f* de séchage par adsorption
r адсорбционная осушительная
установка *f*

538 adsorption efficiency
d Adsorptionswirkungsgrad *m*
f rendement *m* d'adsorption
r коэффициент *m* полезного действия
адсорбции

539 adsorption measurement
d Adsorptionsmessung *f*
f mesure *f* d'adsorption
r адсорбционный метод *m*
измерения

540 adsorption phenomenon
d Adsorptionserscheinung *f*;
Adsorptionsvorgang *m*
f phénomène *m* d'adsorption
r адсорбционное явление *n*

541 adsorption process
d Adsorptionsprozeß *m*;
Adsorptionsverfahren *n*
f procédé *m* d'adsorption
r адсорбционный процесс *m*

542 advace angle
d Voreilungswinkel *m*
f angle *m* d'avance
r угол *m* опережения

543 advance control
d Vorschubsteuerung *f*
f commande *f* d'avancement
r управление *n* подачей

544 advanced circuit technique
d fortschrittliche Schaltkreistechnik *f*
f technique *f* de circuits avancée
r современная схемотехника *f*

545 advanced communication technology
d moderne Kommunikationstechnik *f*
f technique *f* de communication moderne
r современная техника *f* связи

546 advanced concept
d fortschrittliche Konzeption *f*
f conception *f* avancée
r прогрессивная концепция *f*

547 **advanced data-communications control procedure**
d   modernes Datenübertragungssteuerverfahren *n*
f   procédure *f* moderne de commande de télécommunication
r   процедура *f* с расширенными возможностями управления передачей данных

548 **advanced data management**
d   erweiterte Datenverwaltung *f*
f   gestion *f* avancée des données
r   расширенная система *f* для обработки цифровых данных

549 **advanced data organization**
d   weiterentwickelte Datenorganisation *f*; moderne Datenorganisation *f*
f   organisation *f* avancée des données
r   современная организация *f* данных

550 **advanced linear programming system**
d   modernes Programmsystem *n* zur linearen Optimierung; modernes lineares Programmsystem
f   système *m* de programme moderne pour optimisation linéaire
r   современная система *f* линейного программирования

551 **advanced logic processing system**
d   weiterentwickeltes Logik-Verarbeitungssystem *n*
f   système *m* de traitement logique moderne
r   усовершенствованная логическая система *f* обработки

552 **advanced micro devices**
d   moderne Mikrobausteine *mpl*
f   micro-dispositifs *mpl* modernes
r   современные микроэлементы *mpl*

553 **advanced operating system**
d   modernes Betriebssystem *n*
f   système *m* d'exploitation avancé
r   современная операционная система *f*

554 **advanced priority scheduling**
d   Vorrangsteuerung *f*
f   réglage *m* par priorité
r   обработка *f* по приоритету

555 **advanced testing technique**
d   fortschrittliche Testungstechnik *f*; fortschrittliche Prüftechnik *f*
f   technique *f* progressive d'essai
r   прогресивная техника *f* испытаний

556 **advance pulse**
d   Vorgabeimpuls *m*; Voreilimpuls *m*
f   impulsion *f* d'avance
r   опережающий импульс *m*

557 **aerodynamic properties**
d   aerodynamische Eigenschaften *fpl*
f   propriétés *fpl* aérodynamiques
r   аэродинамические характеристики *fpl*; аэродинамические свойства *npl*

*   **AFC → 1279**

558 **after point alignment**
d   stellengerechte Anordnung *f*
f   disposition *f* correcte
r   поразрядная установка *f*

559 **after-threshold laser behaviour**
d   Überschwellen-Laserzustand *m*
f   régime *m* du laser au-dessus du seuil
r   послепороговый режим *m* работы лазера

560 **aggressivity of object**
d   Objektaggressivität *f*
f   aggressivité *f* d'objet
r   агресивность *f* объекта

561 **aided tracking**
d   halbautomatische Bahnverfolgung *f*
f   poursuite *f* semi-automatique
r   полуавтоматическое слежение *n*

562 **aids of positioning**
d   Positionierhilfen *fpl*
f   aides *fpl* de positionnement
r   вспомагательные средства *npl* для позицирования

563 **aids-oriented data**
d   hilfsmittelorientierte Daten *pl*
f   données *fpl* orientées vers le moyen
r   данные *pl*, ориентированные на вспомагательные средства

564 **Aiken code**
d   Aiken-Kode *m*
f   code *m* d'Aiken
r   код *m* Айкена

565 **air accumulator**
d   Druckluftspeicher *m*
f   accumulateur *m* aérohydraulique
r   пневматический аккумулятор *m*

566 **air actuator**
d   Druckluftantrieb *m*; pneumatisches Glied *n*; pneumatische Stelleinrichtung *f*

f vérin *m* pneumatique
r пневматический исполнительный
механизм *m*

567 air-brake dynamometer
d Luftbremsdynamometer *n*
f dynamomètre *m* à frein aérodynamique;
dynamomètre à frein pneumatique
r динамометр *m* с воздушным тормозом

568 air-break contactor
d Luftschütz *n*
f contacteur *m* pneumatique
r воздушный контактор *m*

569 air chamber
d Luftbehälter *m*; Luftkammer *f*
f chambre *f* d'air
r воздушная камера *f*

570 air circulation; air cycle
d Luftkreislauf *m*; Luftzirkulation *f*
f circulation *f* d'air
r воздушная циркуляция *f*

571 air cleaning plant
d Luftreinigungsanlage *f*
f installation *f* d'épuration de l'air
r воздухоочистительное устройство *n*

572 air-conditioning
d Klimatisierung *f*
f climatisation *f*; conditionnement *m* d'air
r кондиционирование *n* воздуха

573 air-conditioning plant
d Klimaanlage *f*
f conditionneur *m* d'air; climatiseur *m*
r установка *f* для кондиционирования
воздуха

574 air-conditioning process
d Klimatisierungsprozess *m*
f processus *m* de conditionnement d'air
r процесс *m* кондиционирования воздуха

575 air-conditioning system
d Klimaanlage *f*
f système *m* à air conditionné
r система *f* кондиционирования воздуха

576 air conditions
d Luftparameter *mpl*
f paramètres *mpl* d'air; valeurs *fpl*
caractéristiques de l'air
r параметры *mpl* воздуха

577 air contamination indicator

d Luftkontaminationsanzeiger *m*
f signaleur *m* de contamination atmosphérique
r индикатор *m* загрязнённости воздуха

578 air contamination meter
d Luftkontaminationsmessgerät *n*;
Luftkontaminationsmesser *m*
f contaminamètre *m* atmosphérique
r радиометр *m* загрязнённости воздуха

579 air conveyor
d pneumatischer Förderer *m*
f installation *f* de transport pneumatique
r пневматический транспортер *m*

580 air-cooled reactor
d luftgekühlter Reaktor *m*
f réacteur *m* à refroidissement à l'air
r реактор *m* с воздушным охлаждением

581 air cooling
d Luftkühlung *f*
f refroidissement *m* de l'air
r воздушное охлаждение *n*

582 air cooling channel
d Kühlluftschacht *m*
f canal *m* à refroidissement par air
r канал *m* воздушного охлаждения

583 air cooling system
d Luftkühlsystem *n*
f système *m* de refroidissement à air
r воздушная система *f* охлаждения

* **air cycle → 570**

584 air damping
d Luftdämpfung *f*; pneumatische Dämpfung *f*
f amortissement *m* à air; amortissement par
l'air
r воздушное демпфирование *n*

585 air drying plant
d Lufttrocknungsanlage *f*
f installation *f* pour le séchage de l'air
r воздушная сушилка *f*

586 air duct circuit
d Luftführungssystem *n*
f circuit *m* de gaines d'air; système *m* de
conduites aériennes
r схема *f* воздушных каналов

587 air entry; access of air; air inlet
d Lufteintritt *m*; Luftzuführung *f*;
Luftzuleitung *f*; Luftzutritt *m*

f admission *f* de l'air; entrée *f* d'air
r поступление *n* воздуха; доступ *m* воздуха; подача *f* воздуха

588 air-flow measurement
d Messen *n* des Luftdurchflusses
f mesure *f* du débit de l'air
r измерение *n* расхода воздуха

589 air-gap method
d Luftspaltmethode *f*
f méthode *f* d'entrefer
r метод *m* воздушного зазора

590 air handler
d Luftaufbereiter *m*
f épurateur *m* pneumatique
r камера *f* обработки воздуха

591 air handling plant
d Luftaufbereitungsanlage *f*
f installation *f* de traitement de l'air; installation de conditionnement d'air
r установка *f* для подготовки воздуха

* **air inlet** → 587

592 air lift
d pneumatische Förderung *f*
f transport *m* pneumatique
r пневматическая подача *f*

593 air monitor
d Luftüberwachungsgerät *n*
f moniteur *m* d'air
r прибор *m* для контроля воздуха

594 air-operated
d pneumatisch; druckluftgesteuert
f pneumatique
r пневматический

595 air-operated amplifier; pneumatic amplifier
d pneumatischer Verstärker *m*
f amplificateur *m* pneumatique
r пневматический усилитель *m*

596 air-operated control; pneumatic control
d pneumatische Regelung *f*
f réglage *m* pneumatique
r пневматическое регулирование *n*

597 air-operated controller; pneumatically operated regulator; pneumatic controller
d pneumatischer Regler *m*
f régulateur *m* pneumatique
r пневматический регулятор *m*

598 air-operated control system; pneumatic control system
d pneumatisches Regelsystem *n*
f système *m* de réglage pneumatique
r пневматическая система *f* регулирования

599 air-operated digital computer
d pneumatischer Digitalrechner *m*
f calculateur *m* digital pneumatique
r пневматическая цифровая вычислителная машина *f*

600 air-operated drive; pneumatic actuator
d pneumatischer Antrieb *m*; pneumatischer Effektor *m*
f commande *f* pneumatique
r пневматический привод *m*

601 air-operated logical element
d pneumatisches Verknüpfungselement *n*; pneumatisches logisches Glied *n*
f membre *m* logique pneumatique
r пневматический логический элемент *m*

602 air-operated power cylinder
d pneumatischer Steuerzylinder *m*
f servomoteur *m* à cylindre pneumatique
r пневматический сервомотор *m*

603 air-operated remote transmission
d pneumatische Fernübertragung *f*
f transmission *f* pneumatique à distance
r пневматическая система *f* передачи на расстояние

604 air-operated telemetering system
d pneumatisches Fernmesssystem *n*
f système *m* de télémesure pneumatique
r пневматическая телеметрическая система *m*

605 air pilot valve
d pneumatisches Steuerventil *n*
f soupape *f* de manœuvre pneumatique
r [автоматический регулирующий] клапан *m* с пневматическим управлением

606 air pneumatic motor
d Druckluftmotor *m*
f moteur *m* à air comprimé
r пневмодвигатель *m*

607 air-powered
d luftangetrieben
f alimenté par air
r с пневматическим приводом

608 air-pressure regulator
d Luftdruckregler *m*

f régulateur *m* de la pression d'air
r регулятор *m* давления воздуха

609 air-pressure test
d Luftdruckprüfung *f*
f épreuve *f* à air comprimé; essai *m* de pression
à air
r испытание *n* сжатым воздухом

610 air-processing system
d Luftaufbereitungssystem *n*
f système *m* de traitement de l'air
r система *f* обработки воздуха

611 air regulating valve
d Luftregulierungsventil *n*
f soupape *f* de réglage d'air
r воздухорегулирующий клапан *m*

612 air regulator
d Luftregler *m*
f régulateur *m* d'air
r регулятор *m* подачи воздуха

613 air sampler
d Luftprobensammler *m*; Luftprobenehmer *m*
f échantillonneur *m* d'air
r прибор *m* для отбора проб воздуха

* **air separation plant → 614**

614 air separator; air separation plant
d Luftzerlegungsanlage *f*;
Lufttrennungsanlage *f*
f installation *f* pour la séparation de l'air;
unité *f* de fractionnement de l'air
r воздухоразделительная установка *f*

**615 air throttle; governor valve; regulating
valve**
d Regelventil *n*
f vanne *f* de réglage
r регулирующий клапан *m*

616 air throttling damper
d Luftdrosselklappe *f*
f clapet *m* d'étranglement d'air; papillon *m*
d'air
r воздушный дроссельный клапан *m*

617 air turbulence
d Luftturbulenz *f*
f turbulence *f* de l'air
r турбулентность *f* потока воздуха

618 air ventilation system; fan system
d Lüftungsanlage *f*; Belüftungsanlage *f*
f installation *f* de ventilation; installation

d'aération; dispositif *m* d'aération
r вентиляционная установка *f*

619 Aitken's interpolation method
d Aitkensche Interpolationsmethode *f*
f méthode *f* d'interpolation d'Aïtken
r метод *m* интерполяции Эйткена

620 alarm annunciator
d Warnanlage *f*; Alarmeinrichtung *f*
f avertisseur *m*
r аварийный сигнализатор *m*

621 alarm circuit; alarm set
d Alarmanlage *f*
f circuit *m* d'alarme
r сигнальная цепь *f*; схема *f* сигнализации

622 alarm contact
d Grenzkontakt *m*; Alarmkontakt *m*
f contact *m* terminal
r контакт *m* сигнала тревоги

623 alarm device
d Alarmgerät *n*
f dispositif *m* d'alarme
r аварийный прибор *m*

624 alarm function
d Störmeldung *f*
f avertissement *m* de perturbation
r сигнал *m* повреждения

625 alarm message
d Fehlermeldung *f*; Gefahrenmeldung *f*
f message *m* d'alarme
r аварийное сообщение *n*

626 alarm monitor
d Warngerät *n*
f moniteur *m* avertisseur
r сигнальный монитор *m*

627 alarm program
d Alarmprogramm *n*
f programme *m* d'alarme
r аварийная программа *f*

628 alarm relay
d Alarmrelais *n*
f relais *m* avertisseur
r сигнализационное реле *n*

* **alarm set → 621**

629 alarm setting
d Einstellpunkt *m* der Warnung;
Warnungseinstellung *f*

f réglage *m* de niveau d'alarme
r регулирование *n* уровня
    предупредительной сигнализации

630 alarm signal system
d Alarmsignalsystem *n*
f système *m* [de signaux] d'alarme
r система *f* аварийной сигнализации

631 alert condition
d Alarmzustand *m*
f condition *f* d'alerte
r аварийное состояние *n*

632 algebraic adder
d algebraischer Addierer *m*; algebraische
    Addiereinrichtung *f*
f addeur *m* algébrique; totalisateur *m*
    algébrique
r алгебраический сумматор *m*

633 algebraic automaton theory
d algebraische Automatentheorie *f*
f théorie *f* algébrique d'automates
r алгебраическая теория *f* автоматов

634 algebraic complement
d algebraisches Komplement *n*
f complément *m* algébrique
r алгебраическое дополнение *n*

635 algebraic equation of higer degree
d algebraische Gleichung *f* höheren Grades
f équation *f* algébrique de haut degré
r алгебраическое уравнение *n* высшего
    порядка

636 algebraic error-correcting code
d algebraischer fehlerkorrigierender Kode *m*
f code *m* algébrique autocorrectif
r код *m* с исправлением алгебраических
    ошибок; алгебраический
    корректирующий код

637 algebraic function
d algebraische Funktion *f*
f fonction *f* algébrique
r алгебраическая функция *f*

638 algebraic selection method
d algrbraische Auswahlmethode *f*
f méthode *f* de sélection algébrique
r алгебраический метод *m* выборки

639 algebraic sign control circuit
d Vorzeichenkontrollkreis *m*
f circuit *m* de contrôle du signe
r схема f, контролирующая знаки

    слагаемых

640 algebraic stability criterion
d algebraisches Stabilitätskriterium *n*
f critère *m* algébrique de stabilité
r алгебраический критерий *m* устойчивости

641 algebraic structure
d algebraische Struktur *f*
f structure *f* algébrique
r алгебраическая структура *f*

642 algebraic sum of impulses
d algebraische Summe *f* der Impulse
f somme *f* algébrique des impulsions
r алгебраическая сумма *f* импульсов

643 algebra of logic; Boolean algebra
d Algebra *f* der Logik; Boolesche Algebra
f algèbre *f* logique; algèbre de Boole
r алгебра *f* логики; булевая алгебра

644 algorithm
d Algorithmus *m*
f algorithme *m*
r алгоритм *m*

645 algorithm equivalence
d Äquivalenz *f* von Algorithmen
f équivalence *f* d'algorithmes
r равносильность *f* алгоритмов

646 algorithm for arithmetic operation
d Algorithmus *m* für arithmetische Operation
f algorithme *m* pour opération arithmétique
r алгоритм *m* арифметической операции

647 algorithm for division; division algorithm
d Algorithmus *m* für Division;
    Divisionsalgorithmus *m*
f algorithme *m* de division
r алгоритм *m* деления

648 algorithm for pattern recognition
d Mustererkennungsalgorithmus *m*
f algorithme *m* de reconnaissance de figures
r алгоритм *m* распознавания образов

649 algorithm for quadratic interpolation
d Algorithmus *m* für quadratische Interpolation
f algorithme *m* pour interpolation quadratique
r алгоритм *m* квадратичной интерполяции

650 algorithm for the inversion
d Algorithmus *m* für die Inversion
f algorithme *m* de l'inversion
r алгоритм *m* инверсии

651 algorithmic
 d algorithmisch
 f algorithmique
 r алгоритмический

652 algorithmic elaboration
 d algorithmische Ausarbeitung *f*
 f élaboration *f* algorithmique
 r разработка *f* алгоритма

653 algorithmic language
 d algorithmische Sprache *f*
 f langue *f* algorithmique
 r алгоритмический язык *m*

654 algorithmic manipulation
 d algorithmische Betätigung *f*
 f manipulation *f* algorithmique
 r алгоритмическая манипуляция *f*

655 algorithmic minimizing
 d algorithmische Minimisierung *f*
 f minimisation *f* algorithmique
 r минимизация *f* алгоритма

656 algorithmic system
 d algorithmisches System *n*
 f système *m* algorithmique
 r алгоритмическая система *f*

657 algorithm insolubility
 d algorithmische Unlösbarkeit *f*
 f insolubilité *f* algorithmique
 r алгоритмическая неразрешимость *f*

658 algorithm of grip force
 d Greifkraftalgorithmus *m*
 f algorithme *m* d'une force de grippion
 r алгоритм *m* усилия захвата

659 algorithm theory
 d Algorithmentheorie *f*
 f théorie *f* des algorithmes
 r теория *f* алгоритмов

 * **align** *v* → **479**

660 alignment
 d Ausrichtung *f*; Abgleich *m*
 f alignement *m*; équilibrage *m*
 r наладка *f*

661 alignment mechanism
 d Ausrichtungmechanismus *m*
 f mécanisme *m* d'alignement
 r механизм *m* ориентации

662 alignment requirements

 d Justierungsanforderungen *fpl*
 f prescriptions *fpl* d'exactitude d'alignement
 r требования *npl* к точности совмещения

663 alimentation of parts; feeding of parts
 d Teilezuführung *f*; Bauteilezuführung *f*
 f alimentation *f* de pièces
 r подача *f* [обрабатываемых] деталей

664 all-channel decoder
 d Mehrkanaldekodierer *m*
 f décodeur *m* à plusieurs canaux
 r многоканальный дешифратор *m*

665 all-inertial guidance
 d Vollinertiallenkung *f*; reine Trägheitsführung *f*
 f guidage *m* totalement inertiel
 r полностью инерциальное наведение *n*

666 all-magnetic system
 d ganzmagnetisches System *n*
 f système *m* entièrement magnétique
 r [полностью] магнитная система *f*

667 allocate *v*
 d zuordnen; zuweisen
 f attribuer; assigner
 r занимать; назначать; размещать

668 allocating
 d Speicherverteilung *f*
 f répartition *f* de la mémoire; assignation *f* de la mémoire
 r распеределение *n* [памяти]

669 allocation; assignment
 d Zuordnung *f*; Zuweisung *f*
 f attribution *f*
 r размещение *n*

670 allocation problem
 d Zuteilungsproblem *n*
 f problème *m* d'allocation
 r задача *f* размещения

671 all-optical computer
 d volloptischer Rechner *m*
 f ordinateur *m* tout optique
 r полностью оптический компьютер *m*

672 allowed increment
 d zulässiges Inkrement *n*
 f incrément *m* permis
 r допустимое приращение *n*

 * **all-pass element** → **673**

673 **all-pass filter; all-pass element**
  d Allpassfilter *n*; Breitbandfilter *n*;
       Allpassglied *n*
  f filtre *m* universel; filtre à large bande
  r фазовый фильтр *m*; фазосдвигающий
       фильтр

674 **all-purpose controller; all-regime**
     controller
  d Allzweckregler *m*; Allbetriebregler *m*
  f régulateur *m* universel; régulateur à régimes
       multiples
  r универсальный регулятор *m*

675 **all-purpose meter; multimeter**
  d Allzweckmesser *m*; Mehrzweckmesser *m*
  f appareil *m* de mesure universel;
       multimètre *m*
  r универсальный измерительный прибор *m*

  * **all-regime controller** → 674

676 **all-type sensor control**
  d Alltypsensorsteuerung *f*
  f commande *f* pour tous types de capteurs
  r универсальное сенсорное управление *n*

677 **almost linear system**
  d fastlineares System *n*
  f système *m* le plus linéaire
  r почти линейная система *f*

678 **almost periodic behaviour; quasi-periodic**
     behaviour
  d quasiperiodisches Verhalten *n*
  f conduite *f* quasi-périodique
  r квазипериодическое поведение *n*

679 **almost periodic solution**
  d fastperiodische Lösung *f*
  f solution *f* quasi-périodique
  r почти периодическое решение *n*

680 **alphabet**
  d Alphabet *n*; Menge *f* von Zeichen
  f alphabet *m*
  r алфавит *m*

681 **alphabetic character**
  d alphabetisches Zeichen *n*; Buchstabe *m*
  f signe *m* alphabétique; caractère *m*
  r алфавитный знак *m*; буква *f*

682 **alphabetic code**
  d alphabetischer Kode *m*
  f code *m* alphabétique
  r алфавитный код *m*

683 **alphabetic customer's information system**
  d alphabetisches Kundeninformationssystem *n*
  f service *m* d'informations alphabétique de la
       clientèle
  r алфавитная информационная система *f*

684 **alphabetic information**
  d alphabetische Information *f*
  f information *f* alphabétique
  r буквенная информация *f*

  * **alphameric coding** → 688

685 **alphanumeric**
  d alphanumerisch
  f alphanumérique
  r алфавитно-цифровой

686 **alphanumeric character**
  d alphanumerisches Zeichen *n*
  f caractère *m* alphanumérique
  r буквено-цифровой знак *m*

687 **alphanumeric-coded geometric**
     information
  d alphanumerisch kodierte geometrische
       Information *f*
  f information *f* géométrique codée
       alphanumériquement
  r геометрическая информация *f* в
       алфавитно-цифровом коде

688 **alpha[nu]meric coding**
  d alphanumerisches Kodieren *n*;
       alphanumerisches Verschlüsseln *n*
  f codification *f* alphanumérique
  r алфавитно-цифровое кодирование *n*

689 **alphanumeric data**
  d alphanumerische Daten *pl*
  f données *fpl* alphanumériques
  r алфавитно-цифровые данные *pl*

690 **alphanumeric display**
  d alphanumerisches Sichtanzeigegerät *n*
  f appareil *m* de visualisation alphanumérique
  r алфавитно-цифровой дисплей *m*

691 **alphanumeric indicating unit**
  d alphanumerische Anzeigeeinheit *f*
  f unité *f* d'indication alphanumérique
  r алфавитно-цифровой блок *m* индикации

692 **alphanumeric keyboard**
  d alphanumerische Tastatur *f*
  f clavier *m* alphanumérique
  r алфавитно-цифровая клавиатура *f*

693 alphanumeric output
 d alphanumerische Ausgabe *f*
 f sortie *f* alphanumérique
 r алфавитно-цифровой выход *m*

694 alphanumeric reader
 d alphanumerischer Leser *m*
 f lecteur *m* alphanumérique
 r буквено-цифровое устройство *n* ввода

695 alphanumeric representation
 d alphanumerische Darstellung *f*
 f représentation *f* alphanumérique
 r буквено-цифровое представление *n*

696 alphanumeric store
 d alphanumerischer Speicher *m*;
   Textspeicher *m*
 f mémoire *f* alphanumérique
 r алфавитно-цифровая память *f*

697 alternate code
 d Austauschkode *m*
 f code *m* alternatif
 r переменны код *m*

698 alternate instruction
 d Sprungbefehl *m*; Verzweigungsbefehl *n*
 f instruction *f* de saut; commande *f* de saut;
   instruction *m* de branchement
 r команда *f* перехода; альтернативная
   инструкция *f*

699 alternate route
 d alternativer Übertragungsweg *m*;
   Ersatzleitweg *m*
 f voie *f* alternative
 r альтернативный маршрут *m*

700 alternating behaviour
 d alternierendes Verhalten *n*
 f comportement *m* alternatif
 r альтернативное поведение *n*

* alternating contact → 2251

701 alternating current; AC
 d Wechselstrom *m*
 f courant *m* alternatif
 r переменный ток *m*

702 alternating-current amplifier
 d Wechselstromverstärker *m*
 f amplificateur *m* [à courant] alternatif
 r усилитель *m* переменного тока

703 alternating-current balancer
 d Wechselstromausgleichsvorrichtung *f*;
   Wechselstromsymmetrierschaltung *f*
 f compensateur *m* de courant alternatif
 r уравнитель *m* переменного тока

704 alternating-current dump
 d Wechselspannungsausfall *m*; Ausfall *m* der
   Wechselspannungsversorgung
 f défaillance *f* de tension alternative
 r внезапное отключение *n* напряжения
   переменного тока

705 alternating-current interruption
 d Wechselstromunterbrechung *f*
 f interruption *f* du circuit de courant alternatif
 r размыкание *n* цепи переменного тока

706 alternating-current line voltage
 d Netzwechselspannung *f*
 f tension *f* de ligne de courant alternatif
 r переменное напряжение *n* сети

707 alternating-current long-distance dialling
 d Wechselstromfernwahl *f*
 f sélection *f* interurbaine à courant alternatif
 r дистанционное искание *n* переменным
   током

708 alternating-current [power] supply
 d Wechselspannungsstromversorgung *f*
 f alimentation *f* en courant alternatif
 r источник *m* питания переменного тока

709 alternating-current relay
 d Wechselstromrelais *n*
 f relais *m* pour courant alternatif
 r реле *n* переменного тока

* alternating-current supply → 708

710 alternating direction method
 d Methode *f* mit alternierender Richtung
 f méthode *f* à direction alternée
 r метод *m* альтернативного выбора
   направления

711 alternating manipulator sequence
 d wechselnde Manipulatorsequenz *f*
 f suite *f* de manipulateur alternative
 r переменная последовательность *f*
   манипулятора

712 alternating operation
 d Wechselbetrieb *m*
 f travail *m* en bascule
 r переключаемый режим *m*

713 alternating quantity
 d Wechselgröße *f*

f grandeur *f* alternative
r переменная величина *f*

714 alternating series
d alternierende Reihen *fpl*
f séries *fpl* alternées
r знакопеременные ряды *mpl*

715 alternative
d Alternative *f*
f alternative *f*
r альтернатива *f*

716 alternative *adj*
d alternativ; wechselweise
f alternatif
r альтернативный

717 alternative branching
d Alternativverzweigung *f*
f branchement *m* alternatif
r альтернативное ветвление *n*

718 alternative channel
d Alternativkanal *m*
f canal *m* alternatif
r альтернативный канал *m*

719 alternative machine arrangement
d alternative Maschinenanordnung *f*
f arrangement *m* de machine alternatif;
groupement *m* de machine alternatif
r альтернативное размещение *n* машин

720 alternative machine order
d alternative Maschinenordnung *f*
f ordre *m* de machine alternatif
r альтернативная последовательность *f*
машин

* **ALU → 988**

721 ambient conditions
d Umgebungsbedingungen *fpl*
f conditions *fpl* de l'ambiance
r условия *npl* окружающей среды

722 ambient disturbances
d Umgebungsstörungen *fpl*
f perturbations *fpl* d'ambiance
r внешние возмущения *npl*

723 ambient temperature information
d Umgebungstemperaturinformationen *fpl*
f informations *fpl* de température d'ambiance
r информация *f* о температуре окружающей
среды

724 ambiguity diagram
d Unbestimmtheitsdiagramm *n*;
Mehrdeutigkeitsdiagramm *n*
f diagramme *m* d'ambiguïté
r диаграмма *f* неопределённости

725 ambiguity of relay characteristic
d Mehrdeutigkeit *f* der Relaiskennlinie
f ambiguïté *f* de caractéristique de relais
r неоднозначность *f* релейной
характеристики

726 ambiguous coordinate transformation
d mehrdeutige Koordinatentransformation *f*
f transformation *f* ambiguë de coordonnées
r неоднозначное преобразование *n*
координат

727 ambiguous function
d zweideutige Funktion *f*; mehrdeutige
Funktion
f fonction *f* ambiguë
r неоднозначная функция *f*

**728 American standard code of information
interchange; ASCII**
d amerikanischer Standardkode *m* für
Informationsaustausch; ASCII
f code *m* standard américain pour échange
d'informations
r американский стандартный код *m* обмена

729 amount of energy
d Energiebetrag *m*; Energiemenge *f*
f quantité *f* d'énergie; montant *m* d'énergie
r количество *n* энергии

730 amplidyne
d Amplidyne *f*
f amplidyne *f*
r амплидин *m*

731 amplidyne servosystem
d Amplidynservosystem *n*
f système *m* asservi à amplidyne
r амплидинная сервосистема *f*

732 amplification class
d Verstärkerbetriebsart *f*; Verstärkerklasse *f*
f classe *f* d'amplification
r класс *m* усиления

**733 amplification coefficient; amplification
constant; amplification factor; gain
coefficient**
d Verstärkungsfaktor *m*;
Verstärkungskoeffizient *m*

f coefficient *m* d'amplification; facteur *m*
   d'amplification; gain *m*
r коэффициент *m* усиления

* **amplification constant** → 733

* **amplification factor** → 733

734 **amplification relay**
   d Verstärkungsrelais *n*
   f relais *m* amplificateur
   r усилительное реле *n*

735 **amplification stage**
   d Verstärkerstufe *f*
   f étage *m* d'amplification
   r каскад *m* усиления

736 **amplified signal**
   d verstärktes Signal *n*
   f signal *m* amplifié
   r усиленный сигнал *m*

737 **amplifier; amplifying element**
   d Verstärker *m*
   f amplificateur *m*
   r усилитель *m*

738 **amplifier bandwidth**
   d Verstärkerbandbreite *f*
   f largeur *f* de bande de l'amplificateur
   r полоса *f* пропускания усиления

739 **amplifier chain**
   d Verstärkerkette *f*
   f chaîne *f* d'amplification
   r цепь *f* усилителя

740 **amplifier circuit**
   d Verstärkerschaltung *f*
   f circuit *m* amplificateur
   r усилительная схема *f*

741 **amplifier for micrologic circuit**
   d Verstärker *m* für Mikrologikschaltung
   f amplificateur *m* pour circuit micrologique
   r усилитель *m* для логических микросхем

742 **amplifier response**
   d Verstärkerfrequenzgang *m*
   f réponse *f* harmonique d'amplificateur
   r характеристика *f* усилителя

743 **amplifier stage**
   d Verstärkerstufe *f*
   f étage *m* d'amplificateur
   r усилительный каскад *m*

744 **amplifier valve**
   d Verstärkerventil *n*
   f soupape *f* d'amplificateur
   r усилительный клапан *m*

745 **amplifier winding**
   d Verstärkerwicklung *f*
   f enroulement *m* amplificateur
   r обмотка *f* усиления

746 **amplify** *v*
   d verstärken
   f amplifier
   r усиливать

747 **amplifying circuit**
   d Verstärkerschaltung *f*
   f circuit *m* amplificateur
   r усилительное звено *n*

748 **amplifying electron tube; amplifying valve**
   d Verstärker[elektonen]röhre *f*
   f tube *m* amplificateur
   r усилительная электронная лампа *f*

* **amplifying element** → 737

* **amplifying valve** → 748

749 **amplistat**
   d Amplistatverstärker *m*
   f amplistat *m*; amplificateur *m*
      électromagnétique
   r амплистат *m*; электромагнитный
      усилитель *m* с обратной связью

750 **amplitron; platinotron**
   d Platinotron *n*
   f platinotron *m*
   r платинотрон *m*

751 **amplitude**
   d Amplitude *f*
   f amplitude *f*
   r амплитуда *f*

752 **amplitude adjustment**
   d Amplitudeneinstellung *f*
   f ajustement *m* d'amplitude
   r настройка *f* амплитуды

753 **amplitude analysis; pulse amplitude
      analyzing**
   d Amplitudenanalyse *f*; Impulshöhenanalyse *f*;
      Impulsamplitudenanalyse *f*
   f analyse *f* en amplitude; analyse d'amplitudes
      des impulsions
   r амплитудный анализ *m* [импульсов]

754 amplitude analyzer
 d Amplitudenanalysator *m*
 f analyseur *m* d'amplitude
 r амплитудный анализатор *m*

**755 amplitude analyzing assembly; pulse
     height analizing assembly**
 d Amplitudenanalysator-Baugruppe *f*
 f ensemble *m* analyseur d'amplitude
 r установка *f* для амплитудного анализа

**756 amplitude characteristic; amplitude
     response**
 d Amplitudencharakteristik *f*;
     Amplitudenkennlinie *f*
 f réponse *f* en amplitude
 r амплитудная характеристика *f*

757 amplitude delay
 d Amplitudenverzögerung *f*
 f délai *m* d'amplitude; retard *m* en amplitude
 r запаздывание *n* по амплитуде;
     амплитудная задержка *f*

758 amplitude-density spectrum
 d Amplitudendichte-Spektrum *n*
 f spectre *m* de la densité de modulation;
     spectre de la densité d'amplitude
 r спектр *m* плотности амплитуды

759 amplitude discriminator
 d Amplitudendiskriminator *m*
 f discriminateur *m* d'amplitudes
 r амплитудный дискриминатор *m*

**760 amplitude distortion; harmonic distortion;
     waveform distortion**
 d Amplitudenverzerrung *f*
 f distorsion *f* harmonique; distorsion
     d'amplitude
 r искажение *n* формы сигнала; амплитудное
     искажение

761 amplitude distribution
 d Amplitudenverteilung *f*
 f distribution *f* des amplitudes; répartition *f* des
     amplitudes
 r распределение *n* амплитуд

762 amplitude envelope
 d Signalhüllkurve *f*
 f enveloppe *f* de signal
 r огибающая *f* сигнала

763 amplitude error
 d Amplitudenfehler *m*
 f erreur *f* d'amplitude
 r амплитудная ошибка *f*

764 amplitude factor
 d Überschwingfaktor *m*; Amplitudenfaktor *m*
 f facteur *m* de crête; facteur de pointe
 r коэффициент *m* амплитуды

765 amplitude-frequency correction
 d Amplitudenfrequenzentzerrung *f*
 f correction *f* de courbe amplitudes-fréquence
 r амплитудно-частотная коррекция *f*

766 amplitude-frequency spectrum
 d Amplitudenfrequenzspektrum *n*
 f spectre *m* de fréquence d'amplitude
 r амплитудный спектр *m* частоты

767 amplitude level
 d Amplitudenniveau *n*; Höhenpegel *m*
 f niveau *m* d'amplitude
 r амплитудный уровень *m*

768 amplitude limitation
 d Amplitudenbegrenzung *f*
 f limitation *f* d'amplitude
 r ограничение *n* по амплитуде

769 amplitude locus
 d Amplitudenkennlinie *f*
 f lieu *m* d'amplitudes
 r амплитудный годограф *m*

770 amplitude margin
 d Amplitudenrand *m*; Amplitudenreserve *f*
 f marge *f* d'amplitude
 r запас *m* по амплитуде

771 amplitude-modulated carrier
 d amplitudenmodulierter Träger *m*
 f porteuse *f* modulée en amplitude
 r амплитудно-модулированная несущая *f*

772 amplitude-modulated oscillations
 d amplitudenmodulierte Schwingungen *fpl*
 f oscillations *fpl* modulées en amplitude
 r амплитудно-модулированные
     колебания *npl*

773 amplitude-modulated pulse
 d amplitudenmodulierter Impuls *m*
 f impulsion *f* à modulation en amplitude
 r импульс m, модулированный по
     амплитуде

774 amplitude-modulated remote transmission
 d Fernübertragung *f* mit
     Amplitudenmodulation
 f transmission *f* à distance à modulation
     d'amplitude
 r дистационная передача *f* с амплитудной
     модуляцией

775 **amplitude-modulated signal tracer**
  d amplitudenmodulierter Signalverfolger *m*
  f traceur *m* du signal à modulation d'amplitude
  r следящее устройство *n* с амплитудной
   модуляцией

776 **amplitude-modulated transmitter**
  d amplitudenmodulierter Sender *m*;
   amplitudenmodulierter Geber *m*
  f émetteur *m* à modulation en amplitude
  r передатчик *m* сигналов амплитудной
   модуляции

777 **amplitude modulation**
  d Amplitudenmodulation *f*
  f modulation *f* d'amplitude
  r амплитудная модуляция *f*

778 **amplitude noise limiter**
  d Amplitudengeräuschbegrenzer *m*
  f limiteur *m* d'amplitude du bruit de fond
  r ограничитель *m* уровня помех

779 **amplitude of an alternating quantity**
  d Scheitelwert *m* einer Wechselgröße
  f amplitude *f* d'une grandeur alternative
  r амплитуда *f* переменной величины

780 **amplitude quantization**
  d Niveauquantisierung *f*
  f découpage *m* en niveau
  r квантование *n* по уровню

781 **amplitude resonance**
  d Amplitudenresonanz *f*
  f résonance *f* d'amplitude
  r амплитудный резонанс *m*

 * **amplitude response** → 756

782 **amplitude scale factor**
  d Amplitudennormierungsfaktor *m*;
   Amplitudenskalierungsfaktor *m*
  f facteur *m* d'échelle des amplitudes; facteur de
   proportionnalité des amplitudes
  r множитель *m* нормирования амплитуды

783 **amplitude selector**
  d Amplitudenwähler *m*
  f sélecteur *m* d'amplitude
  r амплитудный селектор *m*

784 **amplitude spectrum**
  d Amplitudenspektrum *n*
  f spectre *m* des amplitudes
  r амплитудный спектр *m*

785 **amplitude stability margin**
  d Stabilitätsresrve *f* der Amplitude;
   Stabilitätsrand *m* der Amplitude
  f marge *f* de stabilité en amplitude
  r запас *m* устойчивости по амплитуде

786 **amplitude-stabilized laser**
  d amplitudenstabilisierter Laser *m*
  f laser *m* à amplitude stabilisée
  r стабилизированный по амплидуде лазер *m*

787 **amplitude telemetering system**
  d Amplitudenfernmesssystem *n*
  f système *m* télémétrique à amplitude
  r амплитудная телеметрическая система *f*

788 **amplitude vector**
  d Amplitudenvektor *m*
  f vecteur *m* d'amplitude
  r амплитудный вектор *m*

 * **analog** → 834

 * **analog** → 835

789 **analog-active sensor**
  d analog arbeitender Sensor *m*
  f senseur *m* actif analogique
  r аналоговый сенсор *m*

790 **analog amplifier**
  d Analogverstärker *m*
  f amplificateur *m* analogique
  r аналоговый усилитель *m*

791 **analog channel**
  d Analogkanal *m*; Kanal *m* für analoge Signale
  f canal *m* analogique
  r канал *m* аналоговых сигналов

792 **analog circuit**
  d analoge Schaltung *f*
  f réseau *m* analogique
  r аналоговая цепь *f*

793 **analog circuitry**
  d Analogschaltung *f*
  f montage *m* analogique
  r аналоговая схемотехника *f*

794 **analog code**
  d analoger Kode *m*
  f code *m* analogique
  r аналоговый код *m*

795 **analog comparator**
  d Analogkomparator *m*; Analogvergleicher *m*
  f comparateur *m* analogique
  r аналоговый компаратор *m*

* **analog computer** → 4857

796 analog computer logical element
d logisches Element *n* der
   Analogrechenmaschine
f élément *m* logique de calculateur analogique
r логический элемент *m* аналоговой
   вычислительной машины

797 analog computer simulation
d Analogrechner-Simulation *f*;
   Echtzeitmodellierung *f* auf dem
   Analogrechner
f simulation *f* à calculateur analogique
r моделирование *n* на аналоговой
   вычислительной машине

798 analog computing technique
d analoge Rechentechnik *f*;
   Analogrechentechnik *f*
f technique *f* de calcul analogique
r аналоговая вычислительная техника *f*

799 analog control
d Analogregelung *f*; Analogsteuerung *f*
f réglage *m* analogique
r управление *n* с помощью моделирующего
   устройства

800 analog controller; analog regulator
d analoger Regler *m*
f régulateur *m* analogique; contrôleur *m*
   analogique
r аналоговый регулятор *m*

801 analog converter
d Analogumwandler *m*
f convertisseur *m* analogique
r аналоговый преобразователь *m*

802 analog correction
d Analogkorrektur *f*
f correction *f* analogique
r непрерывная коррекция *f*

803 analog data
d Analogdaten *pl*
f données *fpl* analogiques
r аналоговые данные *pl*

804 analog data recorder
d Registrator *m* stetiger Daten
f enregistreur *m* de données analogiques
r регистратор *m* непрерывных данных

805 analog detector
d analoger Detektor *m*
f détecteur *m* analogique

r аналоговый детектор *m*

806 analog-digital conversion
d Analog-Digital-Umwandlung *f*
f conversion *f* analogique-digitale
r преобразование *n* аналоговых данных в
   цифровые

807 analog-digital converter
d Analog-Digital-Konverter *m*; Analog-
   Digital-Umsetzer *m*; Analog-Digital-
   Wandler *m*
f convertisseur *m* analogique-numérique;
   convertisseur analogique-digital
r аналого-цифровой преобразователь *m*

808 analog-digital loop
d Analog-Digital-Schleife *f*
f boucle *f* analogique digitale
r аналого-цифровой контур *m*

**809 analog-digital simulation; analog-
     numerical simulation**
d analog-digitale Simulation *f*; analog-
   numerische Nachbildung *f*
f simulation *f* analogique-digitale; simulation
   analogique-numérique
r аналого-цифровое моделирование *n*

810 analog-digital technique
d Analog-Digital-Technik *f*
f technique *f* analogique-numérique
r аналого-цифровая техника *f*

811 analog display
d analoge Darstellung *f*; Analogdarstellung *f*
f représentation *f* analogique
r аналоговое представление *n*

812 analog element
d Analogelement *n*
f élément *m* analogique
r аналоговый элемент *m*

813 analog extremal system
d analoges Extremalsystem *n*
f système *m* extrémal analogue
r аналоговая экстремальная система *f*

814 analog input
d analoge Eingabe *f*; Eingabe analoger
   Informationen
f entrée *f* analogique
r аналоговый вход *m*; ввод *m* аналогового
   сигнала

815 analog input scanner
d analoger Eingabe-Abtaster *m*; analoges
   Eingabeabtastglied *n*

f explorateur *m* d'entrée analogique
r сканирующее устройство *n* аналоговых сигналов

816 analog interface
d Analogschnittstelle *f*
f interface *f* analogique
r аналоговый интерфейс *m*

817 analog interpolation
d Analogieinterpolation *f*
f interpolation *f* analogique
r аналоговая интерполяция *f*

818 analog mathematical simulation
d analoge mathematische Simulation *f*
f simulation *f* de calcul analogique
r аналоговое математическое моделирование *n*

819 analog measurement
d analoge Messung *f*
f mesure *f* analogique
r аналоговое измерение *n*

820 analog model
d Analogmodell *n*; analoges Modell *n*
f modèle *m* analogique
r аналоговая модель *f*

821 analog modulation
d Analogmodulation *f*
f modulation *f* analogique
r аналоговая модуляция *f*

* **analog-numerical simulation → 809**

822 analog optical transmission
d optische Analog[signal]übertragung *f*
f transmission *f* optique analogique
r оптическая передача *f* аналоговых сигналов

823 analog output
d analoge Ausgabe *f*; Ausgabe analoger Informationen
f sortie *f* analogique
r вывод *m* аналогового сигнала

824 analog point-to-point control
d analoge Punkt-Punkt-Steuerung *f*
f commande *f* analogique point à point
r аналоговое позиционное управление *n*

825 analog process computer
d analoger Prozessrechner *m*; analogarbeitender Prozessrechner
f calculateur *m* de processus analogique;

calculatrice *f* de processus analogique
r аналоговая вычислительная машина *f* для управления промышленными процессами

826 analog processing unit
d analoge Verarbeitungseinheit *f*
f unité *f* de traitement analogique
r блок *m* аналоговой обработки

827 analog processor
d analoger Prozessor *m*
f processeur *m* analogique
r аналоговый процессор *m*

828 analog quantity
d Analoggröße *f*; analogische Größe *f*
f quantité *f* analogique
r непрерывная величина *f*

* **analog regulator → 800**

829 analog representation
d analoge Darstellung *f*
f représentation *f* analogique
r аналоговое представление *n*

830 analog signal
d analoges Signal *n*
f signal *m* analogique
r аналоговый сигнал *m*

831 analog study
d Analogiestudie *f*
f étude *f* analogique
r исследование *n* с помощью моделирования

832 analog switch
d Analogschalter *m*; analoger Schalter *m*
f commutateur *m* analogique
r аналоговый выключатель *m*

833 analog transmission
d Analogübertragung *f*
f transmission *f* analogique
r передача *f* аналоговых сигналов

834 analog[ue]
d Analog *n*
f analogue *m*
r аналог *m*

835 analog[ue]
d analog
f analog[iq]ue
r аналоговый

836 analog unit
d Analogeinheit *f*

f unité *f* analogique
r аналоговый блок *m*; аналоговое
устройство *n*

837 analogy
d Analogie *f*
f analogie *f*
r аналогия *f*; сходство *n*

838 analyse *v*; analyze *v*
d analysieren
f analyser
r анализировать

839 analyser; analyzer
d Analysengerät *n*
f appareil *m* d'analyse
r анализатор *m*

840 analysis
d Analyse *f*
f analyse *f*
r анализ *m*

841 analysis of behaviour
d Verhaltensanalyse *f*
f analyse *f* de comportement
r поведенческий анализ *m*

842 analysis of control system
d Analyse *f* von Regelungs- und Steuerungs-
Systemen
f analyse *f* des systèmes de réglage et de
commande
r анализ *m* систем управления

843 analysis of geometric objects
d Analyse *f* geometrischer Objekte
f analyse *f* des objets géométriques
r анализ *m* геометрических объектов

844 analysis of oscillation
d Schwingungsanalyse *f*
f analyse *f* d'oscillation
r анализ *m* колебаний

845 analysis of stability; stability analysis
d Stabilitätsanalyse *f*; Stabilitätsuntersuchung *f*
f analyse *f* de stabilité; étude *f* de la stabilité
r анализ *m* устойчивости; исследование *n*
стабильности

846 analysis of the manufacturing process
d Fertigungsprozessanalyse *f*
f analyse *f* de processus de production
r анализ *m* производственного процесса

847 analysis procedures

d Methoden *fpl* der Analyse
f procédés *mpl* d'analyse
r методы *mpl* анализа

* **analytic → 848**

848 analytic[al]
d analytisch
f analytique
r аналитический

849 analytical control
d analytische Steuerung *f*; analytische
Regelung *f*
f commande *f* analytique
r аналитический контроль *m*

850 analytical data reflection
d analytische Datenreflexion *f*
f réflexion *f* des données analytique
r аналитическое отображение *n* данных

851 analytical design
d analytischer Entwurf *m*
f conception *f* analytique
r аналитическая разработка *f*

852 analytical function
d analytische Funktion *f*
f fonction *f* analytique
r аналитическая функция *f*

853 analytical research method
d analytische Untersuchungsmethode *f*
f méthode *f* analytique de recherche
r аналитический метод *m* исследования

854 analytical structure
d analytische Struktur *f*
f structure *f* analytique
r аналитическая структура *f*

855 analytic expression
d analytischer Ausdruck *m*
f expression *f* analytique
r аналитическое выражение *n*

856 analytic method
d analytisches Verfahren *n*; analytische
Methode *f*
f méthode *f* analytique
r аналитический метод *m*

857 analytic relationship
d analytische Beziehung *f*
f relation *f* analytique
r аналитическое соотношение *n*

858 **analytic simulation**
   d analytische Simulation *f*
   f simulation *f* analytique
   r аналитическое моделирование *n*

   * **analyze** *v* → 838

   * **analyzer** → 839

859 **AND-circuit; AND-gate**
   d UND-Schaltung *f*; UND-Tor *n*
   f circuit *m* ET
   r схема *f* И

860 **AND-element**
   d UND-Glied *n*
   f élément *m* ET
   r элемент *m* И

   * **AND-gate** → 859

861 **AND-operation**
   d UND-Operation *f*
   f opération *f* ET
   r операция *f* И

862 **anemometer**
   d Windmesser *m*; Windstärkemesser *m*
   f anémomètre *m*
   r анемометр *m*

863 **anemostat**
   d Luftregler *m*
   f anémostat *m*
   r анемостат *m*

864 **angle coder**
   d Winkelkodierer *m*
   f encodeur *m* angulaire
   r кодирующее устройство *n* для углов

865 **angle comparator**
   d Winkelkomparator *m*
   f comparateur *m* d'angles
   r угловой компаратор *m*

866 **angle feedback**
   d Winkelrückkopplung *f*
   f réaction *f* angulaire
   r обратная связь *f* по углу

867 **angle indicator**
   d Winkelindikator *m*
   f indicateur *m* d'angle
   r индикатор *m* угла

   * **angle-integrated intensity** → 1510

868 **angle modulation**
   d Winkelmodulation *f*
   f modulation *f* angulaire
   r угловая модуляция *f*

869 **angle-to-digit converter**
   d Winkel-Zahl-Umsetzer *m*
   f codeur *m* angle-arithmétique
   r преобразователь *m* углового положения в цифровую форму

870 **angular acceleration**
   d Winkelbeschleunigung *f*
   f accélération *f* angulaire
   r угловое ускорение *n*

871 **angular coefficient**
   d Winkelkoeffizient *m*
   f coefficient *m* angulaire
   r угловой коэффициент *m*

872 **angular coordinates**
   d Winkelkoordinaten *fpl*
   f coordonnées *fpl* angulaires
   r угловые координаты *fpl*

873 **angular displacement**
   d Winkelversetzung *f*; Winkelverschiebung *f*
   f déplacement *m* angulaire
   r угловое перемещение *n*

874 **angular distance**
   d Winkelabstand *m*
   f écart *m* angulaire
   r угловое расстояние *n*

875 **angular divergence**
   d Divergenzwinkel *m*
   f angle *m* de divergence
   r угловое расхождение *n*

   * **angular division multiplex** → 876

876 **angular division multiplex[ing]**
   d Winkelmultiplex *m*
   f multiplexage *m* par répartition angulaire
   r угловое уплотнение *n*

877 **angular frequency**
   d Kreisfrequenz *f*; Winkelgeschwindigkeit *f*
   f vitesse *f* angulaire
   r круговая частота *f*

878 **angular measuring system**
   d Winkelmesssystem *n*
   f système *m* de mesure angulaire
   r система *f* измерения углов

879 **angular misalignment**
  d Winkelversatz *m*; Winkelfehler *m*
  f défaut *m* d'alignement angulaire
  r угловое отклонение *n*

880 **angular motion**
  d Winkelbewegung *f*
  f mouvement *m* angulaire
  r угловое движение *n*

881 **angular position**
  d Winkelstellung *f*
  f position *f* angulaire
  r угловое положение *n*

882 **angular resolution**
  d Winkelauflösung *f*
  f pouvoir *m* séparateur angulaire
  r угловое разрешение *n*

883 **angular resonant frequency**
  d Resonanzkreisfrequenz *f*
  f fréquence *f* angulaire de résonance
  r резонансная круговая частота *f*

884 **angular velocity**
  d Winkelgeschwindigkeit *f*
  f vitesse *f* angulaire
  r угловая скорость *f*

885 **angular velocity indicator**
  d Winkelgeschwindigkeitsmesser *m*
  f mesureur *m* de vitesse angulaire
  r измеритель *m* угловой скорости

886 **anharmonic relation**
  d anharmonisches Verhältnis *n*
  f rapport *m* anharmonique
  r ангармоническое соотношение *n*;
     нелинейное соотношение

  * **anisochronous connection** → 1077

887 **annular flow**
  d Ringströmung *f*; Ringraumströmung *f*
  f courant *m* annulaire; écoulement *m* circulaire
  r кольцевой [пространственный] поток *m*

888 **annunciator board**
  d Signaltafel *f*
  f tableau *m* indicateur
  r индикаторный щит *m*

  * **annunciator relay** → 6463

889 **anode bend detector**
  d Anodendetektor *m*
  f détecteur *m* par la plaque
  r анодный детектор *m*

890 **anode correction**
  d Anodenkorrektion *f*
  f correction *f* anodique
  r анодная коррекция *f*

891 **anode load**
  d Anodenbelastung *f*
  f charge *f* anodique
  r анодная нагрузка *f*

892 **anomalous**
  d anomal
  f anormal
  r аномальный

893 **anomalous propagation**
  d anomale Ausbreitung *f*
  f propagation *f* anormale
  r аномальное распространение *n*

894 **answering equipment**
  d Abfrageeinrichtung *f*
  f poste *m* d'opérateur
  r устройство *n* опроса

895 **antiblocking device**
  d Sicherung *f* gegen Störung
  f dispositif *m* d'antibourrage
  r противопомеховое устройство *n*

896 **anticipated value**
  d Erwartungswert *m*
  f valeur *f* prévue
  r предполагаемое значение *n*

897 **anticipatory control**
  d Vorsteuerung *f*
  f précommande *f*
  r предупредительный контроль *m*

898 **anticipatory signal**
  d Antizipationssignal *n*
  f signal *m* d'anticipation
  r упреждающий сигнал *m*

899 **anticoincidence circuit**
  d Antikoinzidenzkreis *m*;
     Antikoinzidenzschaltung *f*
  f circuit *m* d'anticoïncidence
  r схема *f* антисовпадений

900 **anticoincidence element**
  d Antivalenzglied *n*
  f élément *m* anticoïncidence
  r звено *n* неравнозначности

901 anticoincidence method
d Antikoinzidenzmethode f
f méthode f d'anticoïncidence
r метод m антисовпадений

902 antihunting control
d pendelfreie Regelung f
f commande f antipompage
r компенсация f регулятора

903 antilogarithm
d Antilogarithmus m
f antilogarithme m
r антилогарифм m

904 antinode
d Schwingungsbauch m
f ventre m d'oscillation
r пучность f колебаний

905 antiparallel
d antiparallel
f antiparallèle
r антипараллельный

906 antiparallel connexion; inverse-parallel connexion
d Antiparallelschaltung f
f connexion f inverse-parallèle
r встречно-параллельное включение n

907 antiresonance; parallel phase resonance
d Antiresonanz f; Parallelresonanz f
f antirésonance f
r антирезонанс m

* **antivibrator → 3663**

908 aperiodic; dead-beat
d aperiodisch
f apériodique; amorti
r успокоенный; апериодичный

909 aperiodic amplifier
d aperiodischer Verstärker m
f amplificateur m apériodique
r апериодический усилитель m

* **aperiodic attenuation → 911**

910 aperiodic circuit
d aperiodischer Kreis m
f circuit m apériodique
r апериодический контур m

911 aperiodic damping; aperiodic attenuation
d aperiodische Dämpfung f; unabgestimmte Dämpfung
f amortissement m apériodique; atténuation f apériodique
r апериодическое демпфирование n; апериодическое затухание n

912 aperiodic element
d aperiodisch gedämpftes Glied n
f élément m apériodique
r апериодический элемент m

913 aperiodic exponential signal
d aperiodisches Exponentialsignal n
f signal m exponentiel apériodique
r апериодический экспоненциальный сигнал m

914 aperiodic frequency divider
d aperiodischer Frequenzteiler m
f diviseur m apériodique de fréquence
r апериодический делитель m частоты

915 aperiodic instrument
d aperiodisch gedämpftes Instrument n
f appareil m apériodique
r прибор m с апериодическим демпфированием

916 aperiodic link
d aperiodisches Glied n
f membre m apériodique
r апериодическое звено n

917 aperiodic motion
d aperiodische Bewegung f
f mouvement m apériodique
r апериодическое движение n

918 aperiodic phenomenon
d aperiodescher Vorgang m
f phénomène m apériodique
r апериодическое явление n

919 aperiodic regime
d aperiodischer Betriebszustand m; aperiodische Arbeitsweise f
f régime m apériodique; état m de fonctionnement apériodique
r апериодический режим m

920 aperiodic stability
d aperiodische Stabilität f
f stabilité f apériodique
r апериодическая устойчивость f

921 apparent contact surface
d Scheinberührungsfläche f
f surface f apparente de contact
r истинная контактная поверхность f

922 apparent value
 d Scheinwert *m*
 f valeur *f* apparente
 r кажущееся значение *n*

 * **appliance** → 7024

923 application; use; employment
 d Anwendung *f*; Einsatz *m*; Verwendung *f*
 f application *f*
 r использование *n*

924 application area of sensor
 d Sensoreinsatzbereich *m*
 f domaine *m* d'application d'un senseur
 r область *f* применения сенсоров

925 application control
 d Anwendungssteuerung *f*
 f commande *f* d'application
 r управление *n* прикладными процесами

926 application-dedicated terminal
 d Einzweckterminal *n*
 f terminal *m* attribué
 r специализированное оконечное
     устройство *n*

927 application field; area of application
 d Anwendungsgebiet *n*
 f domaine *m* d'application; zone *f* d'emploi
 r область *f* применения

928 application flexibility
 d Einsatzflexibilität *f*
 f flexibilité *f* d'application
 r гибкость *f* эксплуатации

929 application of contact measuring system
 d Kontaktmesssystemanwendung *f*
 f utilisation *f* de système de mesure par contact
 r применение *n* контактной системы
     измерения

**930 application of measuring system; use of
     measuring system**
 d Messsystemnutzung *f*;
     Messsystemanwendung *f*
 f utilisation *f* de système de mesure
 r применение *n* измерительной техники

 * **application of sensorics** → 931

**931 application of sensor technique;
     application of sensorics**
 d Sensortechnik-Applikation *f*;
     Sensorikanwendung *f*; Applikation *f* der
     Sensortechnik

 f application *f* de la technique de senseur;
     application de la sensorique
 r применение *n* сенсорной техники

932 application of special machines
 d Sondermaschineneinsatz *m*
 f insertion *f* de machine spéciale
 r использование *n* специальных машин

933 application-oriented
 d anwendungsorientiert
 f orienté application
 r ориентированный на прикладное
     программирование

934 application problem
 d Anwendungsproblem *n*; Einsatzproblem *n*
 f problème *m* d'application
 r прикладная задача *f*

935 application program
 d Anwendungsprogramm *n*
 f programme *m* d'application
 r прикладная программа *f*

**936 applications of robots; [industrial] robot
     operation; use of robots**
 d Robotereinsatz *m*; Einsatz *m* der
     Industrieroboter; Robotereinbeziehung *f*
 f fonctionnement *m* de robot [industriel]
 r применение *n* [промышленных] роботов

937 application study
 d Anwendungsstudie *f*
 f étude *f* d'application
 r исследование *n* [области] применения

938 applied circuit synthesis
 d angewandte Schaltungssynthese *f*
 f synthèse *f* de circuit appliquée
 r прикладные методы *mpl* синтеза схем

939 applied signal
 d angelegtes Signal *n*
 f signal *m* appliqué
 r действующий сигнал *m*; приложенный
     сигнал

940 applied voltage
 d angelegte Spannung *f*
 f tension *f* appliquée
 r приложенное напряжение *n*

941 applying of syntactic algorithms
 d Anwendung *f* syntaktischer Algorithmen
 f application *f* des algorithmes syntactiques
 r применение *n* синтактических алгоритмов

942 **appreciation; evaluation**
 d Einschätzung *f*; Abschätzung *f*; Bewertung *f*
 f évaluation *f*
 r оценка *f*

943 **approach**
 d Herangehen *n*; Annäherung *f*;
    Betrachtungsweise *f*
 f approche *f*
 r подход *m*

944 **approach mode**
 d Einfahrverhalten *n*
 f mode *m* d'approche
 r асимптотический процесс *m*

945 **approach speed**
 d Annäherungsgeschwindigkeit *f*
 f vitesse *f* d'approche
 r скорость *f* приближения

946 **approximate calculation; approximate
    computation**
 d approximierte Berechnung *f*;
    Näherungsrechnung *f*
 f calcul *m* approximatif; calcul
    d'approximation
 r приближённое вычисление *n*

* **approximate computation** → 946

947 **approximate design**
 d Näherungsentwurf *m*
 f projet *m* approximatif
 r приближённый расчёт *m*

948 **approximate equation**
 d Näherungsgleichung *f*
 f équation *f* approximative
 r приближённое уравнение *n*

949 **approximate formula**
 d Näherungsformel *f*
 f formule *f* d'approximation
 r формула *f* приближения

950 **approximate integration**
 d Näherungsintegration *f*
 f intégration *f* approximative
 r приближённое интегрирование *n*

951 **approximate quantity; approximate value**
 d Näherungswert *m*; Annäherungswert *m*;
    Näherungsgröße *f*
 f grandeur *f* approchée; valeur *f* approchée
 r приближённое значение *n*; приближённая
    величина *f*

952 **approximate sensitive element**
 d Annäherungsfühler *m*
 f palpeur *m* d'approximation
 r аппроксимативный чувствительный
    элемент *m*

953 **approximate solution**
 d Annäherungslösung *f*; angenäherte Lösung *f*;
    Näherungslösung *f*
 f solution *f* approchée; solution approximative
 r приближённое решение *n*

* **approximate value** → 951

954 **approximation**
 d Annäherung *f*; Approximation *f*; Näherung *f*
 f approximation *f*
 r аппроксимация *f*

955 **approximation in the frequency domain**
 d Approximation *f* im Frequenzbereich
 f approximation *f* dans le domaine de
    fréquence
 r аппроксимация *f* в области частот

956 **approximation in the time domain**
 d Approximation *f* im Zeitbereich
 f approximation *f* dans le domaine de temps
 r аппроксимация *f* во временной области

957 **approximation method**
 d Approximationsmethode *f*;
    Näherungsverfahren *n*
 f méthode *f* approchée; méthode
    d'approximation
 r метод *m* аппроксимации

958 **approximation of exponential functions**
 d Approximation *f* von Exponentialfunktionen
 f approximation *f* de fonctions exponentieles
 r аппроксимация *f* экспоненциальных
    функций

959 **approximation of time functions**
 d Approximation *f* von Zeitfunktionen
 f approximation *f* de fonctions temporelles
 r аппроксимация *f* функций времени

960 **approximation sensor system**
 d Näherungssensorsystem *n*
 f système *m* de senseur d'approximation;
    système de capteur d'approximation
 r система *f* сенсоров приближения

961 **approximation simulation**
 d Näherungssimulation *f*
 f simulation *f* d'approximation
 r приближённое моделирование *n*

962 **approximation theory**
  d Approximationstheorie *f*
  f théorie *f* d'approximation
  r теория *f* аппроксимации

963 **approximative determination of the overshooting**
  d angenäherte Bestimmung *f* der Überregelung
  f détermination *f* approximative du surréglage
  r приближённое определение *n* перерегулирования

964 **approximative representation**
  (of a physical phenomenon)
  d approximative Darstellung *f*
  f représentation *f* approximative
  r приближённое представление *n*

965 **arbitrary constant**
  d beliebige Konstante *f*; willkürliche Konstante
  f constante *f* arbitraire; constante choisie arbitrairement
  r произвольная постоянная *f*

966 **arbitrary function**
  d willkürliche Funktion *f*; eigenmächtige Funktion
  f fonction *f* arbitraire
  r произвольная функция *f*

967 **arbitrary function generator**
  d Funktionstestgenerator *m*
  f générateur *m* de fonctions aléatoires
  r генератор *m* произвольных функции

968 **arbitrary number**
  d willkürliche Zahl *f*
  f nombre *m* arbitraire
  r произвольное число *n*

969 **arbitrary parameter**
  d freier Parameter *m*
  f paramètre *m* arbitraire
  r свободный параметр *m*

970 **arbitrary sequence**
  d willkürliche Folge *f*; beliebige Folge
  f séquence *f* arbitraire
  r произвольная последовательность *f*

971 **arbitrate** *v*
  d entscheiden; schiedsrichtern
  f arbitrer
  r принимать решение

972 **arborescent data structure**
  d baumartige Datenstruktur *f*
  f structure *f* arborescente des données
  r древовидная структура *f* данных

973 **architectural compatibility**
  d strukturelle Kompatibilität *f*; strukturelle Verträglichkeit *f*
  f compatibilité *f* structurelle
  r структурная совместимость *f*

974 **area monitor**
  d Raumüberwachungsgerät *n*
  f moniteur *m* spatial
  r монитор *m* для контроля пространства

*   **area of application** → 927

975 **area of permissible errors**
  d zugelassener Fehlerbereich *m*
  f plage *f* admissible d'erreurs
  r область *f* допустимых ошибок

976 **argon laser**
  d Argonlaser *m*
  f laser *m* à l'argon
  r аргоновый лазер *m*

977 **argument**
  d Argument *n*
  f argument *m*
  r аргумент *m*

978 **argument of a function**
  d Argument *n* einer Funktion
  f argument *m* d'une fonction
  r аргумент *m* функции

979 **argument principle**
  d Prinzip *n* des Argumentes
  f principe *m* d'argument
  r принцип *m* аргумента

*   **arithmetical check** → 980

980 **arithmetical check[ing]; mathematical check[ing]**
  d Rechenprüfung *f*; arithmetische Kontrolle *f*; arithmetische Prüfung *f*
  f vérification *f* arithmétique
  r арифметическая проверка *f*

981 **arithmetic[al] element**
  d arithmetisches Element *n*; rechnerischer Grundteil *m*
  f élément *m* arithmétique; organe *m* de calcul
  r арифметический элемент *m*

982 **arithmetical function**
  d arithmetische Funktion *f*

    f fonction *f* arithmétique
    r арифметическая функция *f*

983 arithmetical operation
    d arithmetische Operation *f*; Rechenoperation *f*
    f opération *f* arithmétique
    r арифметическая операция *f*

984 arithmetical shift
    d arithmetische Verschiebung *f*;
       Stellenverschiebung *f*;
       Stellenwertverschiebung *f*
    f décalage *m* arithmétique; décalage
       numérique
    r арифметический сдвиг *m*

985 arithmetic assignment statement
    d arithmetische Ergibtanweisung *f*
    f instruction *f* d'affectation arithmétique
    r арифметическая команда *f* присвоения

986 arithmetic data
    d arithmetische Daten *pl*
    f données *fpl* arithmétiques
    r арифметические данные *pl*

    * **arithmetic element** → 981

987 arithmetic-logic processor
    d Arithmetik-Logik-Prozessor *m*
    f processeur *m* arithmétique-logique
    r арифметическое и логическое
       процессорное устройство *n*

988 arithmetic-logic unit; ALU
    d Arithmetik-Logik-Einheit *f*; Rechenwerk *n*;
       ALU
    f unité *f* arithmétique et logique
    r арифметико-логическое устройство *n*;
       АЛУ

989 arithmetic mean
    d arithmetisches Mittel *n*
    f moyenne *f* arithmétique
    r арифметическое среднее *n*

990 arithmetic notation
    d arithmetische Schreibweise *f*
    f notation *f* arithmétique
    r арифметическая запись *f*

991 arithmetic processor
    d Arithmetikprozessor *m*
    f processeur *m* arithmétique
    r арифметический процессор *m*

992 arithmetic register; calculating register
    d Rechenregister *n*

    f registre *m* arithmétique; registre de calcul
    r регистр *m* арифметического устройства;
       регистр вычислительного устройства

993 arithmetic unit
    d Arithmetikeinheit *f*
    f unité *f* arithmétique
    r арифметическое звено *n*

994 arrange *v*
    d einordnen; einreihen
    f arranger
    r устраивать; размещать; расставлять

**995 arrangement of the memory system; store
       system arrangement**
    d Anordnung *f* des Speichersystems
    f agencement *m* du système de mémoire
    r компоновка *f* системы памяти

996 arrangement planning
    d Anordnungsplanung *f*; Aufstellungsprojekt *n*
    f planification *f* de disposition; projet *m*
       d'arrangement
    r планировка *f* размещения; планировка
       компоновки

997 arrangement principle
    d Anordnungsprinzip *n*
    f principe *m* d'arrangement; principe de
       disposition
    r принцип *m* размещения; принцип
       компоновки

998 arrangement variant
    d Anordnungsvariante *f*
    f variante *f* d'arrangement
    r вариант *m* размещения

999 arresting device
    d Arretiervorrichtung *f*
    f dispositif *m* de blocage
    r арретирующее устройство *n*

1000 artificial carry
    d künstlicherÜbertrag *m*
    f report *m* artificiel
    r искусственный перенос *m*

1001 artificial dielectric
    d künstliches Dielektrikum *n*
    f diélectrique *m* artificiel
    r искусственный диэлектрик *m*

1002 artificial intelligence
    d künstliche Intelligenz *f*
    f intelligence *f* artificielle
    r искусственный интеллект *m*

1003 **artificial language**
   d künstliche Sprache *f*
   f langage *m* artificiel
   r искусственный язык *m*

   * **ASCII → 728**

1004 **assemblage device; assembly fixture**
   d Montageeinrichtung *f*;
      Zusammensetzungsgerät *n*; Montagehilfe *f*
   f dispositif *m* d'assemblage
   r сборочное устройство *n*;
      приспособление *n* для сборки

   * **assemblage technique → 8363**

1005 **assembling technology**
   d Montagetechnologie *f*
   f technologie *f* de montage
   r технология *f* монтажа; технология сборки

1006 **assembling unit**
   d Montageeinheit *f*
   f unité *f* d'assemblage
   r монтажная единица *f*

1007 **assembly automaton**
   d Montageautomat *m*
   f automate *m* d'assemblage
   r сборочный автомат *m*

1008 **assembly centre**
   d Montagezentrum *n*
   f centre *m* d'assemblage
   r сборочный участок *m*

   * **assembly check → 1009**

1009 **assembly check[ing]**
   d Baugruppenprüfung *f*
   f essai *m* d'ensemble
   r контроль *m* узлов

1010 **assembly control systems**
   d Montagesteuersystem *n*;
      Fertigungssteuersystem *n*
   f système *m* de commande d'assemblage
   r система *f* управления сборочными
      процессами

1011 **assembly data**
   d Montagedaten *pl*
   f données *fpl* d'assemblage
   r данные *pl* сборки

   * **assembly difficulty → 8362**

1012 **assembly draft investigation**

   d Montagekonzeptuntersuchung *f*;
      Untersuchung *f* von Montagekonzepten
   f exploration *f* de conception d'assemblage
   r исследование *n* концепции сборочного
      процесса

1013 **assembly equipment**
   d Montageausrüstung *f*
   f équipement *m* d'assemblage
   r сборочное оборудование *n*

1014 **assembly feed device**
   d Montagezubringervorrichtung *f*;
      Zubringervorrichtung *f* für Montage
   f dispositif *m* d'apport d'assemblage
   r транспортное устройство *n* для сборки

   * **assembly fixture → 1004**

1015 **assembly handling**
   d Baugruppenhandhabung *f*
   f manutention *f* d'assemblage
   r манипулирование *n* узлами

1016 **assembly instruction**
   d Montagehinweis *m*
   f instruction *f* d'assemblage; instruction de
      montage
   r монтажная инструкция *f*

1017 **assembly instrument**
   d Montagemittel *n*
   f outillage *m* de montage
   r монтажный инструмент *m*

1018 **assembly line**
   d Montageband *n*
   f chaîne *f* de montage
   r линия *f* сборки

1019 **assembly-line control**
   d Steuerung *f* des Montagebandes
   f contrôle *m* de la chaîne d'assemblage;
      contrôle de la chaîne de montage
   r управление *n* линией сборки

1020 **assembly machine configuration**
   d Montagemaschinenkonfiguration *f*
   f configuration *f* de machine d'assemblage
   r конфигурация *f* сборочных машин

1021 **assembly method**
   d Montagemethode *f*
   f méthode *f* de montage
   r метод *m* монтажа

1022 **assembly model**
   d Montagemodell *n*

    f modèle *m* de montage
    r монтажная модель *f*

1023 assembly object
    d Montageobjekt *n*
    f objet *m* d'assemblage
    r сборочный объект *m*

1024 assembly of manipulator control
    d Aufbau *m* einer Manipulatorsteuerung
    f assemblage *m* d'une commande de manipulateur
    r компоновка *f* системы управления манипулятором

1025 assembly operation
    d Montageoperation *f*
    f opération *f* de montage
    r операция *f* монтажа; операция сборки

1026 assembly organization
    d Montageorganisation *f*
    f organisation *f* de montage
    r организация *f* монтажа; организация сборки

1027 assembly positioning error
    d Montagepositionierfehler *m*
    f erreur *f* de positionnement d'assemblage
    r погрешность *f* позиционирования при сборке

1028 assembly process
    d Montagevorgang *m*
    f procédé *m* d'assemblage
    r сборочный процесс *m*

1029 assembly process periphery
    d Montageprozessperipherie *f*
    f périphérie *f* pour procédé d'assemblage
    r внешнее оборудование *n* для сборочного процесса

1030 assembly program
    d Montageprogramm *n*
    f programme *m* d'assemblage
    r собирающая программа *f*

1031 assembly simulation process
    d Montagesimulationsverfahren *n*; Simulationsverfahren *n* für Montage
    f procédé *m* de simulation pour le montage; procédé de simulation pour l'assemblage
    r метод *m* моделирования сборочных процессов

1032 assembly system; mounting system
    d Montagesystem *n*

    f système *m* d'assemblage; système de montage
    r сборочная система *f*

1033 assembly system layout
    d Layout *n* eines Montagesystems
    f installation *f* d'un système d'assemblage
    r схема *f* сборочной системы

1034 assembly system parameter
    d Montagesystemparameter *m*; Parameter *m* eines Montagesystems
    f paramètre *m* d'un système d'assemblage
    r параметр *m* сборочной системы

1035 assembly technique
    d Montagetechnik *f*
    f technique *f* de montage
    r техника *f* монтажа; техника сборки

* assembly technology → 8364

1036 assembly test
    d Montageversuch *m*
    f test *m* d'assemblage
    r сборочное испытание *n*

1037 assembly unit
    d Montageeinheit *f*
    f unité *f* d'assemblage
    r сборочная единица *f*

1038 assembly unit solution
    d Baukastenlösung *f*
    f solution *f* de boîte de construction
    r решение *n* агрегатного принципа

1039 assembly valuation
    d Montagebewertung *f*
    f évaluation *f* d'assemblage
    r оценка *f* процесса сборки

* assignment → 669

1040 associate *v*; join *v*
    d anschließen; vereinigen
    f associer; joindre
    r ассоциировать; [при]соединять

1041 associated [channel] signaling
    d assoziierte Zeichengabe *f*
    f signalisation *f* associée
    r связанная передача *f* сигналов

* associated signaling → 1041

1042 associated structure
    d assoziierte Struktur *f*

f   structure f associée
r   ассоциативная структура f

1043 association
d   Assoziation f
f   association f
r   ассоциация f; связывание n

1044 associative
d   assoziativ
f   associatif
r   ассоциативный

1045 associative index method
d   assoziative Indexmethode f
f   méthode f d'index associative
r   ассоциативный индексный метод m

1046 associative law
d   assoziatives Gesetz n
f   loi f associative
r   ассоциативный закон m

1047 associative link
d   assoziativer Link m
f   lien m associatif
r   ассоциативная связь f

1048 associative principe
     (of a memory)
d   Assoziativprinzip n
f   principe m associatif
r   ассоциативный принцип m

1049 associative programming
d   assoziative Programmierung f
f   programmation f par association
r   ассоциативное программирование n

*   **assume** v → 12049

1050 astable multivibrator
d   astabiler Multivibrator m
f   multivibrateur m astable
r   неустойчивый мультивибратор m

1051 astatic controlled system
d   astatische Regelstrecke f; Regelstrecke f ohne
     Ausgleich
f   système m réglé astatique
r   астатическая система f управления

**1052 astatic controller; integral controller;
     floating-action controller; reset controller**
d   Integralregler m; astatischer Regler m
f   régulateur m astatique
r   астатический регулятор m

1053 astatic device
d   astatisches Gerät n
f   appareil m astatique
r   астатический прибор m

1054 astatic system
d   astatisches System n
f   système m astatique; système sans erreur de
     position
r   астатическая система f

1055 asymmetrical conductivity
d   richtungsabhängige Leitfähigkeit f
f   conductibilité f unidirectionnelle
r   несимметрическая проводимость f

1056 asymmetric[al] function
d   asymmetrische Funktion f
f   fonction f asymétrique
r   несимметричная функция f

1057 asymmetrical heterostatic circuit
d   asymmetrisch-heterostatische Schaltung f;
     ungleichförmiger heterostatischer
     Stromkreis m
f   montage m hétérostatique dissymétrique
r   несимметричная гетеростатическая цепь f

1058 asymmetrical non-linearity
d   unsymmetrische Nichtlinearität f
f   non-linéarité f asymétrique
r   асимметричная нелинейность f

1059 asymmetric biprocessor system
d   asymmetrisches Biprozessorsystem n
f   système m biprocesseur asymétrique
r   несимметричная бипроцессорная
     система f

*   **asymmetric function** → 1056

1060 asymmetric mass force
d   unsymmetrische Massenkraft f
f   force f de masse dissymétrique
r   несимметричная сила f инерции

1061 asymmetric modulation
d   asymmetrische Modulation f;
     unsymmetrische Modulation
f   modulation f asymétrique
r   асимметричная модуляция f

1062 asymmetric network
d   asymmetrisches Netzwerk n
f   réseau m asymétrique
r   несиммтричная схема f

1063 asymmetric non-linear unit
d   asymmetrisches nichtlineares Element n

f organe m non linéaire asymétrique
r несимметричный нелинейный элемент m

1064 asymmetric work piece
d asymmetrisches Werkstück n
f pièce f à usiner asymétrique
r асимметричное изделие n

1065 asymptote
d Asymptote f
f asymptote f
r асимптота f

1066 asymptotic
d asymptotisch
f asymptotique
r асимптотический

1067 asymptotical optimum
d asymptotisches Optimum n
f optimum m asymptotique
r асимптотический оптимум m

1068 asymptotic expansion
d asymptotische Entwicklung f
f développement m asymptotique
r асимптотическое разложение n

1069 asymptotic flux
d asymptotischer Fluss m
f flux m asymptotique
r асимптотический поток m

1070 asymptotic method
d asymptotische Methode f
f méthode f asymptotique
r асимптотический метод m

1071 asymptotic relation
d asymptotische Beziehung f
f relation f asymptotique
r асимптотическое отношение n

1072 asymptotic stability
d asymptotische Stabilität f
f stabilité f asymptotique
r асимптотическая устойчивость f

1073 asymptotic value
d asymptotischer Wert m; asymptotische
   Grenze f
f valeur f asymptotique
r асимптотическое значение n

1074 asynchronous communication adapter
d asynchroner Kommunikationsadapter m
f adaptateur m de communication asynchrone
r асинхронный адаптер m коммутации

1075 asynchronous communications interface adapter; ACIA
d Adapter m für asynchrone Datenübertragung;
   Interfaceadapter m für asynchrone
   Kommunikation
f adaptateur m pour communication
   asynchrone des données
r адаптер m асинхронной передачи данных

1076 asynchronous computer
d Asynchronrechner m; asynchrone
   Rechenanlage f
f calculatrice f asynchrone
r асинхронная вычислительная машина f

1077 asynchronous connection; anisochronous connection
d asynchrone Verbindung f; anisochrone
   Verbindung
f connexion f asynchrone; connexion
   anisochrone
r асинхронная связь f

1078 asynchronous control
d asynchrone Steuerung f
f commande f asynchrone
r асинхронное управление n

1079 asynchronous disconnected mode
d asynchrone unterbrochene Betriebsart f
f mode m de service asynchrone séparé
r асинхронный разъединительный режим m

1080 asynchronous excitation
d asynchrone Erregung f
f excitation f asynchrone
r асинхронное возбуждение n

1081 asynchronous logic
d asynchrone Logikschaltung f
f montage m logique asynchrone
r асинхронная логическая схема f

1082 asynchronous quenching
d asynchrone Unterdrückung f;
   Asynchrondämpfung f
f amortissement m asynchrone; élimination f
   non synchrone
r асинхронное гашение n

1083 asynchronous relay system
d asynchrones Relaissystem n
f système m de relais asynchrone
r асинхронная релейная система f

1084 asynchronous response mode
d asynchrone Antwortbetriebsart f

f mode *m* de réponse asynchrone
r режим *m* асинхронного ответа

1085 asynchronous sequential circuit
d asynchrone Folgeschaltung *f*
f circuit *m* séquentiel asynchrone
r асинхронная следящая схема *f*

1086 asynchronous servomotor
d asynchroner Stellmotor *m*
f servomoteur *m* asynchrone
r асинхронный сервомотор *m*

1087 asynchronous transmission system
d asynchrones Übertragungssystem *n*
f système *m* de transmission non synchrone
r асинхронная система *f* передачи

1088 asynchronous working
d asynchrone Betriebsweise *f*
f manière *f* d'opérer asynchrone
r асинхронный режим *m* работы

1089 atmospheric braking
d atmosphärisches Bremsen *n*
f freinage *m* atmosphérique
r атмосферное торможение *n*

1090 atmospheric optics
d atmosphärische Optik *f*
f optique *f* atmosphérique
r атмосферная оптика *f*

1091 atmospheric perturbation
d atmosphärische Störung *f*
f perturbation *f* atmosphérique
r атмосферные помехи *fpl*

1092 atmospheric propagation
d Ausbreitung *f* in der Atmosphäre
f propagation *f* dans l'atmosphère
r распространение *n* в атмосфере

* atomic energy → 8817

1093 atomic heat
d Atomwärme *f*
f chaleur *f* atomique
r атомная теплоёмкость *f*

1094 atomic pile; nuclear reactor
d Kernreaktor *m*
f réacteur *m* nucléaire; pile *f* nucléaire
r атомный реактор *m*

1095 attenuating medium
d Dämpfungsmedium *n*
f milieu *m* affaiblissant

r ослабляющая среда *f*

1096 attenuation characteristic
d Dämpfungscharakteristik *f*
f caractéristique *f* d'affaiblissement
r характеристика *f* затухания

1097 attenuation compensator
d Dämpfungsentzerrer *m*;
   Dämpfungsausgleicher *m*
f compensateur *m* d'affaiblissement
r компенсатор *m* затухания

1098 attenuation degree; degree of attenuation
d Dämpfungsgrad *m*; Schwächungsgrad *m*
f degré *m* d'affaiblissement; degré
   d'atténuation
r степень *f* затухания; степень отслабления

1099 attenuation factor; damping coefficient;
       damping factor
d Dämpfungsfaktor *m*; Abklingkonstante *f*
f facteur *m* d'affaiblissement; coefficient *m*
   d'amortissement
r коэффициент *m* затухания

1100 attenuation-limited operation
d dämpfungsbegrenzter Betrieb *m*
f fonctionnement *m* limité par l'affaiblissement
r работа f, ограниченная затуханием

1101 attenuation measurement
d Dämpfungsmessung *f*
f mesure *f* de l'affaiblissement
r измерение *n* затухания

1102 attenuation peak frequency
d dämpfende Spitzenfrequenz *f*
f fréquence *f* de crête d'amortissement
r ослабление *n* пика частоты

1103 attenuation regime
d Dämpfungsbetrieb *m*
f régime *m* d'affaiblissement
r режим *m* затухания

1104 attenuation region
d Dämpfungsbereich *m*
f domaine *m* d'amortissement; zone *f*
   d'amortissement
r область *f* затухания

1105 attenuation value
d Dämpfungswert *m*
f valeur *f* d'affaiblissement
r величина *f* затухания

1106 attenuator
d Dämpfungsnetzwerk *n*

   f affaiblisseur *m*
   r аттенюатор *m*

1107 audio channel
   d Niederfrequenzkanal *m*
   f voie *f* audiofréquence
   r низкочастотный канал *m*

1108 audio-frequency amplifier
   d Niederfrequenzverstärker *m*;
      Tonfrequenzverstärker *m*
   f amplificateur *m* basse fréquence;
      amplificateur d'audiofréquences
   r усилитель *m* низкой частоты

1109 audio-frequency circuit
   d Niederfrequenzkreis *m*
   f circuit *m* basse fréquence
   r цепь *f* звуковой частоты

1110 audio-frequency generator
   d Tonfrequenzoszillator *m*; Tongenerator *m*
   f générateur *m* à basse fréquence
   r генератор *m* звуковой частоты

1111 audio-frequency multiplex system
   d Tonfrequenzmultiplexsystem *n*
   f système *m* multiplex à audiofréquence
   r мултиплексная система *f* в диапазоне
      звуковых частот

1112 audio signal; sound signal
   d Hörsignal *n*; Tonsignal *n*; Schallsignal *n*
   f signal *m* audible; signal phonique; signal *m*
      sonore
   r звуковой сигнал *m*

   *   **audiovideo** → 1113

1113 audiovisual; audiovideo
   d audiovisuell
   f audiovisuel; audiovidéo
   r аудиовизуальный

1114 audiovisual presentation
   d audiovisuelle Präsentation *f*
   f présentation *f* audiovisuelle
   r аудиовизуальное представление *n*

1115 aural warning device
   d akustische Warnvorrichtung *f*; akustische
      Warnanlage *f*
   f dispositif *m* d'alarme sonore
   r устройство *n* звуковой сигнализации

1116 auto-adaptive model
   d Selbstanpassungsmodell *n*
   f modèle *m* auto-adaptif

   r модель *f* самонастройки

1117 auto-bias circuit
   d Schaltung *f* für automatische Vorspannung
   f circuit *m* de polarisation automatique
   r цепь *f* автоматического смещения

   *   **auto-checking** → 11285

1118 autochemogram; autochemograph
   d Autochemogramm *n*
   f autochémogramme *m*
   r автохемограмма *f*

   *   **autochemograph** → 1118

1119 autocode
   d Autokode *m*
   f autocode *m*
   r автокод *m*; автоматический код *m*

1120 autocorrection
   d automatische Berichtigung *f*
   f autocorrection *f*
   r автоматическое исправление *n*;
      автоматическая коррекция *f*

1121 autocorrelation
   d Autokorrelation *f*
   f autocorrélation *f*
   r автокорреляция *f*

1122 autocorrelation coefficient
   d Autokorrelationskoeffizient *m*
   f coefficient *m* d'autocorrélation
   r коэффициент *m* автокорреляции

1123 autocorrelation function; self-correlation
      function
   d Autokorrelationsfunktion *f*; selbsttätige
      Korrelationsfunktion *f*
   f fonction *f* d'autocorrélation
   r автокорреляционная функция *f*

1124 autocorrelator
   d Autokorrelator *m*
   f autocorrélateur *m*
   r автокоррелятор *m*

1125 autocorrelogram
   d Autokorrelogramm *n*
   f autocorrélogramme *m*
   r автокоррелограмма *f*

1126 autocovariance
   d Autokovarianz *f*
   f autocovariace *f*
   r автоковариация *f*

1127 autodiagnostic system
  d Autodiagnostiksystem *n*
  f système *m* d'autodiagnostic
  r система *f* автодиагностики

1128 autoexcitation; self-excitation
  d Selbsterregung *f*; Selbstanregung *f*
  f autoexcitation *f*
  r самовозбуждение *n*

1129 autofluoroscope
  d Autofluoroskop *n*
  f autofluoroscope *m*
  r автофлюороскоп *m*

1130 autographic recording apparatus
  d Registriergerät *n*
  f enregistreur *m*
  r автографический прибор *m* записи

1131 autoinductive coupling
  d autoinduktive Kopplung *f*
  f couplage *m* auto-inductif
  r автотрансформаторная связь *f*

1132 auto-ionization; preionization
  d Autoionisation *f*; Präionisation *f*
  f auto-ionisation *f*; préionisation *f*
  r автоионизация *f*; предионизация *f*

1133 automated assembly
  d automatisierte Montage *f*
  f assemblage *m* automatisé
  r автоматизированная сборка *f*

1134 automated assembly mounting
  d automatisierte Baugruppenmontage *f*
  f montage *m* d'ensemble automatisé; montage d'assemblage automatisé
  r автоматизированная сборка *f* узлов

1135 automated construction engineering
  d automatisierte Konstruktionstechnik *f*
  f technique *f* de construction automatisée
  r техника *f* автоматизированного проектирования

1136 automated control; automated direction
  d automatisierte Steuerung *f*; automatisierte Leitung *f*
  f commande *f* automatisée; gestion *f* automatisée
  r автоматизированное управление *n*

1137 automated data entry system
  d automatisiertes Dateneingabesystem *n*
  f système *m* d'entrée automatisé des données
  r автоматизированная система *f* ввода данных

1138 automated data processing
  d automatisierte Datenverarbeitung *f*
  f traitement *m* automatisé des données
  r автоматизированная обработка *f* данных

1139 automated data system
  d automatisiertes Datensystem *n*
  f système *m* automatisé des données
  r автоматизированная система *f* обработки данных

1140 automated design engineering principle
  d automatisiertes Entwurfstechnikprinzip *n*
  f principe *m* de la technique de dessin automatisée
  r принцип *m* автоматизированного проектирования

*   **automated direction → 1136**

1141 automated element
  d automatisiertes Element *n*
  f élément *m* automatisé
  r автоматизированный элемент *m*

1142 automated engineering design
  d automatisierter ingenieurtechnischer Entwurf *m*
  f dessin *m* automatisé génietechnique
  r автоматизированное техническое проектирование *n*

1143 automated handling
  d automatisierte Handhabung *f*
  f manutention *f* automatisée
  r автоматизированное манипулирование *n*

1144 automated information flow
  d automatisierter Informationsfluss *m*
  f flux *m* automatisé des informations
  r информационный поток *m* в автоматизированной системе

1145 automated information processing
  d automatisierte Informationsverarbeitung *f*
  f traitement *m* de l'information automatisé
  r автоматизированная переработка *f* информации

1146 automated information system
  d automatisiertes Informationssystem *n*
  f système *m* d'information automatisé
  r автоматизированная информационно-поисковая система *f*

1147 automated learning process
  d automatisierter Lernprozess *m*
  f procédé *m* à apprentissage automatisé
  r автоматический процесс *m* обучения

1148 automated logic design
 d automatisierter Logikentwurf *f*
 f dessin *m* logique automatisé
 r автоматизированный логический синтез *m*

1149 automated logic diagram
 d automatisiertes logisches Diagramm *n*
 f diagramme *m* logique automatisé
 r автоматическая логическая диаграмма *f*

1150 automated machine assembly line
 d Taktmontagestrecke *f*
 f phase *f* de la chaîne de montage
 r поточная сборочная линия *f*

1151 automated management information system
 d automatisiertes Leitungsinformationssystem *n*
 f système *m* d'information de direction automatisé
 r автоматическая система *f* обработки данных и управления [производством]

1152 automated manipulator
 d automatisierter Manipulator *m*
 f manipulateur *m* automatisé
 r автоматизированный манипулятор *m*

1153 automated manufacturing planning
 d automatisierte Fertigungsplanung *f*
 f planning *m* automatisé de production
 r автоматическое планирование *n* производства

1154 automated monitoring of manufacturing
 d automatisierte Fertigungsüberwachung *f*
 f surveillance *f* de fabrication automatisée
 r автоматизированный производственный контроль *m*

1155 automated precision device engineering
 d automatisierte Präzisionsgerätetechnik *f*
 f technique *f* automatisée des appareils de précision
 r автоматизированная прецизионная приборная техника *f*

1156 automated processing method
 d automatisierte Verarbeitungsmethode *f*
 f méthode *f* de traitement automatisée
 r автоматизированный метод *m* обработки

1157 automated production
 d automatisierte Produktion *f*
 f production *f* automatisée
 r автоматизированное производство *n*

1158 automated production development
 d automatisierte Produktionsentwicklung *f*
 f développement *m* de la production automatisé
 r автоматизированное развитие *n* производства

1159 automated route management
 d automatisierte Leitwegverwaltung *f*
 f direction *f* d'acheminement automatisée
 r автоматическое обслуживание *n* маршрутов

1160 automated selective precision mounting
 d automatisierte selektive Präzisionsmontage *f*
 f montage *m* de précision sélectif et automatisé
 r автоматизированная селективная прецизионная сборка *f*

1161 automated single machine
 d automatisierte Einzelmaschine *f*
 f machine *f* unique automatisée
 r автоматизированная машина *f*

1162 automated transport and storage system
 d automatisiertes Transport- und Lagersystem *n*
 f système *m* de transport et de magasin automatisé
 r автоматизированная система *f* транспортировки и складирования

1163 automated welding process
 d automatisiertes Schweißverfahren *n*
 f procédé *m* de soudage automatisé
 r автоматизированный способ *m* сварки

1164 automated workshop
 d automatisierter Produktionsbereich *m*
 f zone *f* de production automatisée
 r автоматизированный производственный участок *m*

1165 automatic; self-acting
 d automatisch; selbsttätig
 f automatique
 r автоматический; самодействующий

1166 automatic acceptance of the data
 d automatische Datenübernahme *f*
 f acceptation *f* automatique des données
 r автоматический приём *m* данных

1167 automatic accumulation
 d automatische Akkumulation *f*
 f accumulation *f* automatique
 r автоматическое накопление *n*

1168 automatic accuracy check
   d automatische Richtigkeitsprüfung *f*
   f contrôle *m* d'exatitude automatique
   r автоматическая проверка *f* правильности

1169 automatic-actuated protective equipment
   d automatisch betätigte Schutzeinrichtung *f*
   f protecteur *m* actionné automatiquement
   r автоматическое устройство *n* защиты

1170 automatic addition
   d automatische Addition *f*
   f addition *f* automatique
   r автоматическое сложение *n*

1171 automatic address substitution
   d selbsttätige Umspeicherung *f*
   f substitution *f* automatique d'adresse
   r автоматическая замена *f* адресов

1172 automatic adjustment of exposure
   d automatische Belichtungseinstellung *f*
   f ajustement *m* automatique d'exposition
   r автоматическая установка *f* экспозиции

1173 automatic air control system
   d automatisches Luftraumkontrollsystem *n*
   f système *m* de contrôle de l'air automatique
   r автоматическая система *f* контроля
       воздушного пространства

1174 automatic alarm
   d automatische Störsignalisation *f*;
       automatische Meldeanlage *f*; automatische
       Warnanlage *f*
   f avertissement *m* automatique; avertisseur *m*
       automatique
   r автоматическая аварийная сигнализация *f*

1175 automatically balancing digital instrument
   d automatisch abgleichendes
       Digitalmessgerät *n*
   f appareil *m* de mesure nimérique à
       équilibrage automatique
   r цифровой [измерительный] прибор *m* с
       автоматическим уравновешиванием

1176 automatically controlled drawing equipment
   d automatisch gesteuerte Zeicheneinrichtung *f*
   f appareil *m* à dessiner à commande
       automatique
   r автоматически управляемое чертёжное
       устройство *n*

1177 automatically operating
   d selbsttätig arbeitend
   f à fonctionnement automatique
   r действующий автоматически

1178 automatically operating gas analyzer
   d automatischer Gasanalysator *m*
   f analyseur *m* de gaz à fonctionnement
       automatique
   r автоматический газоанализатор *m*

1179 automatically programmed tools
   d automatisch programmierte Werkzeuge *npl*
   f outils *mpl* programmés automatiquement
   r автоматически запрограммированные
       инструменты *mpl*

1180 automatically registering electronic photocamera
   d automatisch registrierende elektronische
       Fotokamera *f*
   f appareil *m* photographique électronique avec
       enregistrement automatique
   r электронная фотокамера *f* с
       автоматической регистрацией

1181 automatically taught system; self-teaching system of automatic optimization
   d selbstoptimierendes System *n*; optimierendes
       System
   f système *m* d'optimisation automatique
   r самооптимизирующая система *f*

1182 automatic alternating-current compensator
   d selbsttätiger Wechselstromkompensator *m*
   f compensateur *m* automatique à courant
       alternatif
   r автоматический компенсатор *m*
       переменного тока

1183 automatic alternation of grip units
   d automatischer Wechsel *m* der Greifeinheiten
   f changement *m* automatique des unités de
       grippion
   r автоматическая смена *f* захватных
       устройств

1184 automatic amplitude control
   d automatische Amplitudensteuerung *f*
   f commande *f* automatique d'amplitude
   r автоматическое регулирование *n*
       амплитуды

1185 automatic answering
   d automatische Beantwortung *f*
   f réponse *f* automatique
   r автоматический ответ *m*

1186 automatic arc welding
   d Lichtbogenautomatenschweißen *n*;
       automatisches Lichtbogenschweißen *n*

f soudage *m* automatique à l'arc [électrique]
r автоматическая дуговая сварка *f*

**1187 automatic arc welding equipment;
       automatic arc welding machine**
d automatische Lichtbogenschweißanlage *f*;
   Lichtbogenschweißautomat *m*; selbsttätige
   Lichtbogenschweißmaschine *f*
f installation *f* de soudage automatique à l'arc;
   équipement *m* de soudage automatique à l'arc
r автоматическая установка *f* для дуговой
   сварки

*   **automatic arc welding machine** → 1187

1188 automatic assembly
d automatische Montage *f*
f assemblage *m* automatique
r автоматическая сборка *f*

1189 automatic assembly process
d automatischer Montagevorgang *m*
f procédé *m* d'assemblage automatique
r процесс *m* автоматической сборки

1190 automatic backup and recovery
d automatischer Wiederanlauf *m* und
   Wiederherstellung *f*
f redémarrage *m* automatique et restauration *f*
r автоматический повторный пуск *m* и
   восстановление *n* начального состояния

**1191 automatic balancing; automatic
       compensation**
d Selbstabgleich *m*; automatische Korrektur *f*;
   automatischer Abgleich *m*
f balancement *m* automatique; compensation *f*
   automatique
r автоматическое согласование *n*

1192 automatic balancing machine
d selbsttätige Auswuchtmaschine *f*
f dispositif *m* de balance automatique
r устройство *n* автоматической
   балансировки

1193 automatic binary data link
d automatische binäre Datenverbindung *f*
f liason *f* de données automatique binaire
r канал *m* передачи двоичной информации

**1194 automatic blocking; automatic
       interlocking; automatic lockout**
d automatische Blockierung *f*; automatische
   Sperrung *f*; automatische Verriegelung *f*;
   selbsttätige Sperrung
f blocage *m* automatique
r автоматическая блокировка *f*

1195 automatic branch control
d automatische Verzweigungssteuerung *f*;
   automatische Zweigstellensteuerung *f*
f commande *f* de branchement automatique
r автоматическое управление *n* ветвлением

1196 automatic breaker
d selbsttätiger Unterbrecher *m*
f disjoncteur *m* automatique; interrupteur *m*
   automatique
r автоматический выключатель *m*

1197 automatic calibration
d automatische Eichung *f*
f étalonnage *m* automatique
r автоматическая калибровка *f*

1198 automatic call
d automatischer Ruf *m*
f appel *m* automatique
r автоматический вызов *m*

1199 automatic central proccessing
d automatische zentrale Bearbeitung *f*
f traitement *m* central automatique
r автоматическая централизованная
   обработка *f*

1200 automatic change-over control
d Umschaltregelautomatik *f*
f dispositif *m* de commande automatique par
   commutation
r коммутационная регулирующая
   автоматика *f*

*   **automatic check** → 1201

**1201 automatic check[ing]; automatic test[ing];
       automatic inspection**
d automatische Kontrolle *f*; Selbstkontrolle *f*;
   automatische Prüfung *f*
f contrôle *m* automatique; essai *m*
   automatique; essai automatisé; examen *m*
   automatique
r автоматический контроль *m*;
   автоматическое испытание *n*

1202 automatic checking actuating unit
d Selbstüberwachungsstellglied *n*; Stellglied *n*
   für automatische Kontrolle
f dispositif *m* commandé de contrôle
   automatique
r исполнительное устройство *n*
   автоматического контроля

1203 automatic checkout system
d automatisches Ausprüfungssystem *n*

f   système *m* automatique de tests
r   автоматическая система *f* испытаний

1204   automatic check system
d   Automatik-Kontrollsystem *n*; automatisches
    Kontrollsystem *n*
f   système *m* de contrôle automatique
r   система *f* автоматического контроля

1205   automatic circuit exchange
d   automatische Wählervermittlung *f*
f   échange *m* automatique de circuit
r   автоматическая коммутация *f*

1206   automatic clamping technique
d   automatische Spanntechnik *f*
f   technique *f* de serrage automatique
r   техника *f* автоматического крепления

1207   automatic clearing
d   automatische Löschung *f*
f   effaçage *m* automatique
r   автоматическое стирание *n*;
    автоматический сброс *m*

**1208   automatic closed-loop control system;
         automatic monitored control system**
d   selbsttätiges Regelsystem *n*
f   système *m* de réglage automatique; système
    asservi de commande
r   замкнутая система *f* автоматического
    управления; система автоматического
    регулирования с обратной связью

1209   automatic closed-loop servosystem
d   automaische Servoanlage *f* mit geschlossener
    Schleife
f   système *m* de réglage automatique à boucle
    fermée
r   атоматическая замкнутая сервосистема *f*

1210   automatic coding
d   selbsttätige Kodierung *f*; automatische
    Verschlüsselung *f*
f   codage *m* automatique
r   автоматическое кодирование *n*

1211   automatic coding system
d   automatisches Kodierungssystem *n*
f   système *m* de codage automatique
r   система *f* автоматического кодирования

1212   automatic collating
d   automatisches Mischen *n*
f   interclassement *m* automatique
r   автоматический подбор *m* и
    автоматическая раскладка *f*

1213   automatic colour matching

d   automatische Farbeinstellung *f*
f   mise *f* au point de la teinte automatique
r   автоматическое регулирование *n*
    цветового тона

*   **automatic compensation → 1191**

1214   automatic compensator
d   automatischer Kompensator *m*
f   compensateur *m* automatique
r   автоматический компенсатор *m*

1215   automatic component handler
d   Manipulator *m* zur automatischen
    Handhabung von Bauelementen
f   manipulateur *m* automatique pour
    composants
r   автоматический манипулятор *m* для
    компонентов

1216   automatic computing
d   automatisches Rechnen *n*
f   calcul *f* automatique
r   автоматическое вычисление *n*

1217   automatic connection
d   Selbstanschluss *m*
f   raccord *m* automatique; jonction *f*
    automatique
r   автоматическое включение *n*

1218   automatic constant
d   automatische Konstante *f*
f   constante *f* automatique
r   автоматическая константа *f*

1219   automatic contour digitizer
d   automatisches
    Konturendigitalisierungsgerät *n*
f   dispositif *m* automatique pour la
    digitalisation de contours
r   автоматическое устройство *n* квантования
    кривых

1220   automatic contrast control
d   automatische Kontrastregelung *f*
f   réglage *m* automatique du contraste
r   автоматическая регулировка *f*
    контрастности

**1221   automatic control; automatic regulation;
         self-regulation**
d   Regelung *f*; automatische Regelung;
    selbsttätige Regelung
f   réglage *m* automatique; régulation *f*
    automatique; commande *f* automatique;
    autoréglage *m*
r   автоматическое регулирование *n*;
    автоматическое управление *n*

1222 automatic control device
  d automatische Regeleinrichtung *f*; selbsttätige
     Steuerungseinrichtung *f*
  f équipement *m* de commande automatique;
     dispositif *m* de contrôle automatique
  r автоматическое устройство *n* управления

1223 automatic control engineering
  d automatische Regelungstechnik *f*
  f technique *f* de régulation automatique
  r техника *f* автоматического регулирования

1224 automatic control gear
  d automatische Regelungsvorrichtung *f*
  f équipement *m* de réglage automatique
  r механизм *m* для автоматической
     регулировки

1225 automatic-controlled machine system
  d automatisch gesteuertes Maschinensystem *n*
  f système *m* de machine commandé
     automatiquement
  r машинная система *f* с автоматическим
     управлением

1226 automatic controller
  d Regler *m*; Einheit *f* für die Regelung
  f régulateur *m* [automatique]; unité *f* de
     réglage automatique
  r устройство *n* автоматического
     регулирования

1227 automatic control loop
  d automatischer Regelkreis *m*
  f boucle *f* à réglage automatique
  r контур *m* автоматического управления

1228 automatic control of exposure time
  d automatische Belichtungszeitregelung *f*
  f régulation *f* automatique du temps de pose
  r автоматическое регулирование *n* времени
     экспозиции

1229 automatic control system
  d selbsttätiges Regelungssystem *n*
  f système *m* de réglage automatique
  r система *f* автоматического регулирования

1230 automatic control system for electric drive
  d selbsttätige Elektroantriebssteuerung *f*
  f commande *f* automatique de moteur
     électrique
  r автоматическое управление *n*
     электроприводом

1231 automatic control system stability
  d Stabilität *f* automatischer Regelkreise
  f stabilité *f* des circuits de commande

  r устойчивость *f* систем автоматического
     регулирования

1232 automatic control theory
  d Regelungstheorie *f*
  f théorie *f* de régulation automatique
  r теория *f* автоматического регулирования

1233 automatic control valve
  d selbsttätiges Steuerventil *n*; automatisches
     Regelventil *n*
  f vanne *f* de réglage automatique
  r клапан *m* с автоматическим управлением

1234 automatic conveying
  d selbsttätige Förderung *f*
  f transfert *m* automatique
  r автоматическая транспортировка *f*

1235 automatic curve scanning
  d automatische Kurvenabtastung *f*
  f exploration *f* automatique de courbes
  r автоматическое сканирование *n* кривых

1236 automatic cut-out
  d Selbstunterbrecher *m*; Sicherungsautomat *m*
  f interrupteur *m* automatique
  r автоматический предохранитель *m*

1237 automatic cutting equipment
  d automatische Schneideanlage *f*
  f installation *f* automatique pour le taillage
  r устройство *n* для автоматического резания

1238 automatic cycle
  d automatischer Kreislauf *m*; automatischer
     Zyklus *m*
  f cycle *m* automatique
  r автоматический цикл *m*

1239 automatic data acquisition system
  d automatisches Datenerfassungssystem *n*
  f système *m* d'acquisition des données
     automatique
  r автоматическая система *f* регистрации
     данных

1240 automatic data collecting system
  d automatisches Datensammlungssystem *n*
  f système *m* de collection des données
     automatique
  r автоматическая система *f* сбора данных

1241 automatic data exchange
  d selbsttätige Datenvermittlung *f*
  f échange *m* automatique des données
  r автоматический обмен *m* данными

1242 automatic data input
d automatische Dateneingabe f
f alimentation f automatique en données;
   introduction f automatique de données
r автоматический вход m данных

1243 automatic data interchange system
d automatisiertes Datenaustauschsystem n
f système m d'échange automatisé des données
r автоматизированная система f обмена
   данными

1244 automatic data processing
d automatische Datenverarbeitung f
f traitement m automatique de données
r автоматическая обработка f данных

1245 automatic data recording
d automatische Datenaufzeichnung f
f enregistrement m automatique des données
r автоматическая запись f данных

1246 automatic data retrieval
d automatisches Datenwiederauffinden n
f rétablissement m automatique des données;
   rappel m automatique des données
r автоматический поиск m данных

1247 automatic data surveillance
d automatische Datenüberwachung f
f surveillance f automatique des données
r автоматический контроль m данных

1248 automatic data transmission system
d automatisches Datenübertragungssystem n
f système m automatique de transmission des
   données
r автоматическая система f передачи
   данных

1249 automatic defect localization
d automatische Defektlokalisierung f
f localisation f de défaut automatique
r автоматическое обнаружение n дефектов

1250 automatic demagnetization device
d Entmagnetisierautomat m
f dispositif m de désaimantation automatique
r автоматическое размагничивающее
   устройство n

1251 automatic diaphragm setting
d automatische Blendeneinstellung f
f mise f au point automatique du diaphragme
r автоматическая установка f диафрагмы

1252 automatic digital input-output system
d automatisiertes digitales Ein- und
   Ausgabesystem n
f système m entrée-sortie digital automatisé
r автоматизированная система f ввода-
   вывода цифровых данных

1253 automatic digitizing system
d automatisches Digitalisiersystem n
f système m de digitalisation automatique
r автоматическая система f преобразования
   в цифровую форму

1254 automatic digit recognizer
d automatischer Ziffernerkenner m; Gerät n zur
   automatischen Ziffernerkennung
f dispositif m automatique à reconnaissance
   des chiffres
r устройство n автоматического
   распознавания цифр

1255 automatic discharge
d automatische Entladung f
f vidange f automatique; déchargement m
   automatique; défournement m automatique
r автоматический выпуск m;
   автоматическая разгрузка f

1256 automatic disconnection
d automatische Abschaltung f
f déconnexion f automatique
r автоматическое отключение n

1257 automatic discriminator switching
d automatische Frequenzdetektorschaltung f
f commutation f automatique de discriminateur
r автоматическая коммутация f частотного
   детектора

1258 automatic distortion correction
d automatische Verzerrungskorrektur f
f correction f de distorsion automatique
r автоматическая коррекция f искажений

1259 automatic distribution subsystem
d automatisches Verteilungsuntersystem n
f sous-système m de distribution automatique
r автоматическая распределительная
   подсистема f

1260 automatic division
d automatische Division f
f division f automatique
r автоматическое деление n

*   **automatic dosage → 1353**

1261 automatic drive
d automatischer Antrieb m
f commande f automatique
r автоматический привод m

1262 **automatic duplication**
  d automatisches Doppeln *n*; automatisches
     Duplizieren *n*
  f duplication *f* automatique
  r автоматическое дублирование *n*

1263 **automatic ejection**
  d automatische Ausrückung *f*; automatisches
     Auswerfen *n*
  f éjection *f* automatique
  r автоматический выброс *m*

1264 **automatic electronic data processing**
  d automatische elektronische
     Datenverarbeitung *f*
  f traitement *m* des données électronique
     automatique
  r автоматическая электронная обработка *f*
     данных

1265 **automatic electroslag welder; automatic
     electroslag welding machine**
  d Unterschlackeschweißautomat *m*;
     Automat *m* für das Elektro-Schlacke-
     Schweißen; ES-Schweißautomat *m*; ES-
     Automat *m*
  f machine *f* à souder automatique sous laitier
  r автомат *m* для электрошлаковой сварки

  * **automatic electroslag welding
     machine** → **1265**

1266 **automatic embossing machine**
  d automatische Prägemaschine *f*
  f machine *f* à gaufrer automatique
  r автоматическая штамповочная машина *f*

1267 **automatic emergency shutdown**
  d automatische Schnellabschaltung *f*
  f arrêt *m* d'urgence automatique
  r автоматический аварийный останов *m*;
     автоматическое аварийное отключение *n*

1268 **automatic energy flux**
  d selbsttätiger Energiefluss *m*
  f flux *m* énergétique automatique
  r автоматический поток *m* энергии

1269 **automatic equalizer**
  d automatischer Entzerrer *m*
  f compensateur *m* de distorsion automatique
  r автоматический компенсатор *m*
     искажений

1270 **automatic error detection**
  d automatische Fehlererkennung *f*
  f reconnaissance *f* automatique d'erreur
  r автоматическое распознавание *n* ошибок

1271 **automatic error recovery**
  d automatische Fehlerbeseitigung *f*
  f élimination *f* automatique du défaut
  r автоматическое устранение *n* ошибок

1272 **automatic fault signalling**
  d selbsttätige Signalisierung *f* von Störungen
  f signalisation *f* automatique des pannes
  r автоматическая сигнализация *f*
     неисправностей

1273 **automatic feed**
  d automatische Speisung *f*; automatischer
     Vorschub *m*
  f alimentation *f* automatique; avance *f*
     automatique
  r автоматическое питание *n*

1274 **automatic feeding of work pieces**
  d automatisches Zubringen *n* von Werkstücken
  f alimentation *f* automatique de pièces à
     travailler
  r автоматическая подача *f* деталей

1275 **automatic flame-cutting machine;
     automatic gas-cutting machine**
  d Brennschneidautomat *m*; automatische
     Brennschneidmaschine *f*
  f machine *f* d'oxycoupage automatique
  r автомат *m* для кислородной резки;
     машина *f* для автоматической
     кислородной резки

1276 **automatic flow-charting package**
  d automatisches Flussdiagrammpaket *n*
  f paquet *m* automatique d'organigramme
  r пакет *m* программ для автоматического
     составления блок-схем

1277 **automatic flow of informations**
  d selbsttätiger Informationsfluss *m*
  f flux *m* automatique des informations
  r автоматический поток *m* информации

1278 **automatic focussing action**
  d automatische Bündelung *f*; automatische
     Konzentration *f*
  f focalisation *f* automatique; concentration *f*
     automatique
  r автоматическая фокусировка *f*

1279 **automatic frequency control; AFC**
  d automatische Frequenzregelung *f*
  f commande *f* automatique de fréquence
  r автоматическое регулирование *n* частоты

1280 **automatic function**
  d automatische Funktion *f*

f fonction *f* automatique
r автоматическая функция *f*

1281 automatic fusion welding
  d automatisches Schmelzschweißen *n*
  f soudage *m* par fusion automatique
  r автоматическая сварка *f* плавлением

1282 automatic gain control
  d automatische Verstärkungsregelung *f*
  f commande *f* automatique de gain
  r автоматическая регулировка *f* усиления

1283 automatic gain controller; automatic volume controller
  d automatischer Verstärkungsregler *m*; automatischer Lautstärkeregler *m*; automatischer Volumregler *m*
  f régulateur *m* automatique de gain
  r автоматический регулятор *m* усиления

*   automatic gas-cutting machine → 1275

1284 automatic gas-shielded arc welding equipment; automatic gas-shielded welding unit; automatic inert-gas arc welder; automatic inert-gas welding unit
  d Schutzgasschweißautomat *m*; automatische Schutzgasschweißanlage *f*
  f automate *m* de soudage par points en atmosphère inerte
  r автомат *m* для дуговой сварки в среде защитного газа; автомат для газоэлектрической сварки

*   automatic gas-shielded welding unit → 1284

1285 automatic gas welding machine
  d Gasschmelzschweißautomat *m*
  f soudeuse *f* automatique pour le soudage au gaz; automate *m* de soudage au gaz
  r аппарат *m* для [автоматической] газовой сварки

1286 automatic generation
  d automatische Generierung *f*
  f génération *f* automatique
  r автоматическое генерирование *n*

1287 automatic generator
  d Generator *m* mit selbsttätiger Steuerung
  f générateur *m* à commande automatique
  r генератор *m* с автоматическим управлением

1288 automatic grading; automatic sorting method

  d automatische Sortierung *f*; automatische Sortiermethode *f*
  f triage *m* automatique; méthode *f* automatique de triage
  r автоматическая сортировка *f*

1289 automatic guidance system
  d Nachführautomatik *f*
  f guidage *m* automatique
  r система *f* автоматического слежения

1290 automatic guided vehicle
  d automatisch geführtes Fahrzeug *n*
  f véhicule *m* guidé automatiquement
  r транспортное средство *n* с автоматическим управлением

1291 automatic handling device
  d automatische Handhabeeinrichtung *f*
  f dispositif *m* de manutention automatique
  r автоматическое устройство *n* манипулирования

1292 automatic idling control
  d automatische Stillstandsüberwachung *f*
  f surveillance *f* automatique de la suspension
  r автоматический контроль *m* холостого хода

1293 automatic indexing
  (of informations)
  d automatisches Indizieren *n*
  f indexation *f* automatique
  r автоматическая индексация *f*

1294 automatic indication
  d automatische Anzeige *f*
  f indication *f* automatique
  r автоматическая индикация *f*

1295 automatic indicator
  d automatischer Anzeiger *m*
  f indicateur *m* automatique
  r автоматический индикатор *m*

*   automatic inert-gas arc welder → 1284

*   automatic inert-gas welding unit → 1284

1296 automatic information acquisition
  d automatische Informationserfassung *f*
  f rassemblement *m* automatique des informations
  r автоматический сбор *m* информации

1297 automatic information processing
  d automatische Informationsverarbeitung *f*

f   traitement *m* automatique de l'information
r   автоматическая обработка *f* информации

1298   automatic information reduction
d   automatische Informationsreduzierung *f*
f   réduction *f* automatique d'information
r   автоматическое сжатие *n* информации

1299   automatic information retrival
d   automatisches Wiederauffinden *n* von
    Informationen
f   rappel *m* automatique des informations
r   автоматический поиск *m* информаций

*   **automatic inspection → 1201**

*   **automatic interlocking → 1194**

1300   automatic internal diagnosis
d   automatische interne Diagnose *f*
f   diagnose *f* interne automatique
r   автоматическое обнаруживание *n*
    неисправностей

1301   automatic ionization chamber
d   automatische Ionisationskammer *f*
f   chambre *f* d'ionisation automatique
r   автоматическая ионизационная камера *f*

1302   automatic joining process
d   automatischer Fügevorgang *m*
f   procédé *m* d'assemblage automatique
r   автоматическое стыкование *n*

1303   automatic layout technique
d   automatisches Layout-Design *n*
f   technique *f* automatique de topologie
r   автоматический метод *m* разработки
    топологии

1304   automatic level control
d   automatische Niveauregelung *f*
f   réglage *m* de niveau automatique
r   автоматическое регулирование *n* уровня

1305   automatic load limitation
d   automatische Lastbegrenzung *f*
f   limitation *f* automatique de charge
r   автоматическая ограничение *n* нагрузки

1306   automatic locking circuit
d   automatische Synchronisierschaltung *f*
f   circuit *m* synchronisant automatique
r   схема *f* автоматической синхронизации

*   **automatic lockout → 1194**

1307   automatic logging

d   automatische Messwerterfassung *f*
f   enregistrement *m* automatique des résultats
r   автоматическая регистрация *f* результатов

1308   automatic logic design
d   rechnergestütztes Logik-Design *n*
f   conception *f* logique automatisée
r   автоматизированное логическое
    проектирование *n*

1309   automatic machine control
d   automatische Maschinensteuerung *f*
f   pilotage *m* automatique des machines
r   автоматическое управление *n* машиной

1310   automatic machine operation
d   automatische Maschinenoperation *f*
f   opération *f* de la machine automatique
r   автоматическая машинная операция *f*

1311   automatic machine processing
d   automatische Maschinenverarbeitung *f*
f   traitement *m* automatique par machine
r   автоматическая машинная обработка *f*

1312   automatic [machine] welding
d   Automatenschweißen *n*; automatisches
    Schweißen *n*; selbstablaufendes Schweißen
f   soudage *m* automatique
r   автоматическая сварка *f*; сварка
    автоматом

1313   automatic machining
    (of control disks)
d   automatisches Bearbeiten *n*
f   usinage *m* automatique
r   автоматическая обработка *f*

1314   automatic manipulation system
d   automatisches Manipulationssystem *n*
f   système *m* de manipulation automatique
r   автоматическая манипуляционная
    система *f*

1315   automatic mean value determination
d   automatische Mittelwertbildung *f*
f   formation *f* automatique de valeur moyenne
r   автоматическое определение *n* средней
    величины

1316   automatic measuring and checking device
d   automatische Mess- und
    Kontrolleinrichtung *f*
f   équipement *m* de mesure et de contrôle
    automatique
r   автоматическое контрольно-
    измерительное устройство *n*

1317 **automatic measuring station**
  d automatische Messstation *f*
  f station *f* de mesure automatique
  r автоматическая измерительная станция *f*

1318 **automatic mechanism**
  d automatischer Mechanismus *m*
  f mécanisme *m* automatique
  r автоматический механизм *m*

1319 **automatic mechanism with
     electromagnetic control**
  d automatischer Mechanismus *m* mit
     elektromagnetischer Steuerung
  f mécanisme *m* automatique à commande
     électromagnétique
  r автоматический механизм *m* с
     электромагнитным управлением

1320 **automatic message registering**
  d automatische Informationsregistrierung *f*
  f enregistrement *m* d'information automatique
  r автоматическая регистрация *f* сообщений

1321 **automatic model recognition**
  d automatische Modellerkennung *f*
  f reconnaissance *f* de modèle automatique
  r автоматическая идентификация *f* модели

1322 **automatic mode of operation; automatic
     operation mode**
  d automatische Betriebsart *f*
  f mode *m* opératoire automatique
  r автоматический режим *m* работы

  * **automatic monitored control
     system → 1208**

1323 **automatic monitoring**
  d automatische Überwachungsanlage *f*
  f surveillance *f* automatique
  r автоматический надзор *m*

1324 **automatic multiple-electrode machine**
  d Mehrfachelektrodenautomat *m*
  f automate *m* à électrodes multiples; machine *f*
     automatique à électrodes multiples
  r многоэлектродный автомат *m*

1325 **automatic multiplication**
  d automatische Multiplikation *f*
  f multiplication *f* automatique
  r автоматическое умножение *n*

1326 **automatic network analyzer**
  d selbsttätiger Netzwerkanalysator *m*
  f analyseur *m* automatique des réseaux
  r автоматическое устройство *n*

     моделирования цепей

1327 **automatic network voltage regulator**
  d automatischer Netzspannungsregler *m*
  f régulateur *m* automatique de tension de
     réseau
  r автоматический регулятор *m* напряжения
     сети

1328 **automatic noise gate**
  d automatische Störsperre *f*
  f élimineur *m* automatique du bruit
  r автоматический шумозаграждающий
     фильтр *m*

1329 **automatic numerical control data backup**
  d automatische NC-Datensicherung *f*
  f sûreté *f* automatique CN de données
  r автоматическое резервирование *n* данных
     числового управления

1330 **automatic operation**
  d automatischer Arbeitsablauf *m*
  f fonctionnement *m* automatique
  r автоматическое функционирование *n*

  * **automatic operation mode → 1322**

1331 **automatic operation of testing**
  d automatischer Prüfvorgang *m*
  f opération *f* automatique d'essai
  r процесс *m* автоматического контроля

1332 **automatic operations at manipulation
     processes**
  d automatische Abläufe *mpl* bei
     Handhabungsprozessen
  f suite *f* automatique en processus de
     manipulation
  r автоматическое протекание *n*
     манипуляционных процессов

1333 **automatic optimization system**
  d System *n* automatischer Optimierung
  f système *m* automatique d'optimalisation
  r система *f* автоматической оптимизации

1334 **automatic output**
     (of data)
  d automatische Ausgabe *f*
  f extraction *f* automatique
  r автоматическая выдача *f*

  * **automatic parts flow → 1410**

1335 **automatic pattern programming**
  d automatische Musterprogrammierung *f*

f programmation f d'échantillon automatique
r автоматическое программирование n
образов

1336 automatic phase adjustment
d automatische Phaseneinstellung f
f réglage m automatique de phase
r автоматическая настройка f фазы

1337 automatic phase comparison circuit
d automatischer Phasenvergleichskreis m
f circuit m à comparaison automatique de
phase
r схема f для автоматического сравнения
фаз

1338 automatic phase control
d automatische Phasenregelung f
f mise f en phase automatique
r автоматическое регулирование n фазы

1339 automatic plant
d automatische Fabrik f
f usine-automate f; usine f automatique
r автоматический завод m

1340 automatic pneumatic devices; pneumatic automation installations
d automatische pneumatische Geräte npl;
Geräte der Pneumatik
f appareils mpl automatiques pneumatiques;
dispositifs mpl automatiques pneumatiques
r устройства npl пневмоавтоматики

1341 automatic positioning system
d automatisches Positioniersystem n
f système m de positionnement automatique
r система f автоматической установки
координаты

1342 automatic position telemetering
d automatische Fernmessung f der Position
f télémesure f automatique de la position
r автоматическое телеметрирование n
позиции

1343 automatic power plant control
d automatische Kraftwerksteuerung f
f commande f automatique d'une centrale
électrique
r автоматическое управление n
электростанции

1344 automatic precision spectrophotometer
d automatisches Präzisionsspektrofotometer n
f spectrophotomètre m de précision
automatique
r автоматический точный
спектрофотометр m

1345 automatic precision welding equipment
d Präzisionsschweißautomat m
f automate m du soudage de précision
r автомат m для прецизионной сварки;
автомат для сварки мелких деталей

1346 automatic pressure suppression system
d automatisches Druckentlastungssystem n
f système m de suppression de pression
automatique
r автоматическая система f разгрузки
давления

1347 automatic priority group
d automatische Prioritätengruppe f
f groupe m de priorité automatique
r автоматическая приоритетная группа f

1348 automatic process control
d automatische Prozesssteuerung f
f commande f automatique d'un processus
r автоматическое управление n процессом

1349 automatic process correlator
d selbsttätiger Betriebskorrelator m
f corrélateur m automatique d'exploitation
r автоматический производственный
коррелятор m

1350 automatic process cycle controller
d automatische Verfahrenskreislauf-
Regelvorrichtung f
f dispositif m de commande automatique du
cycle de travail
r автоматический регулятор m рабочего
цикла

1351 automatic program control
d automatische Programmregelung f
f réglage m automatique suivant un
programme
r автоматическое программное
управление n

1352 automatic programming; self-programming
d automatische Programmierung f;
Selbstprogrammierung f
f programmation f automatique
r автоматическое программирование n

1353 automatic proportioning; automatic dosage
d automatische Dosierung f
f dosage m automatique
r автоматическое дозирование n

1354 automatic protection
d automatische Schutzeinrichtung *f*
f protection *f* automatique
r автоматическая защита *f*

1355 automatic quality control
d automatische Qualitätskontrolle *f*
f contrôle *m* de qualité automatique
r автоматический контроль *m* качества

1356 automatic range selection
d automatische Reihenauswahl *f*
f sélection *f* séquentielle automatique
r автоматическая последовательная
   выборка *f*

1357 automatic reading
d selbsttätiges Lesen *n*
f lecture *f* automatique
r автоматическое чтение *n*

1358 automatic receiving bunker
d automatischer Aufnahmebunker *m*
f trémie *f* de réception automatique; soute *f* de
   réception automatique
r автоматический приёмный бункер *m*

1359 automatic reclosing
d selbsttätige Wiedereinschaltung *f*
f réenclenchement *m* automatique; refermeture
   f automatique
r автоматическое повторное включение *n*

1360 automatic reclosing circuit-breaker
d Schalter *m* mit selbsttätiger
   Wiedereinschaltung
f disjoncteur *m* à refermeture automatique
r выключатель *m* с автоматическим
   повторным включением

1361 automatic recorder; recording meter
d Registrierzählinstrument *n*;
   Registriermessgerät *n*; selbstschreibendes
   Instrument *n*
f compteur *m* enregistreur; instrument *m* de
   mesure enregistreur
r самопишущий прибор *m*; рекордер *m*

*   **automatic regulation** → 1221

1362 automatic remote control
d automatische Fernsteuerung *f*
f télécommande *f* automatique
r автоматическое дистационное
   управление *n*

1363 automatic repeat request
d automatische Wiederholanforderung *f*

f requête *f* de répétition automatique
r автоматический запрос *m* повторной
   передачи

**1364 automatic request; automatic return
question**
d automatische Rückfrage *f*
f requête *f* automatique; question *f* de retour
   automatique
r автоматическая справка *f*; автоматический
   запрос *m*

1365 automatic reserve equipment switching
d automatische Einschaltung *f* der
   Reserveeinrichtung
f enclenchement *m* automatique d'installation
   de réserve
r автоматическое включение *n* резервного
   оборудования

1366 automatic reset
d automatischer Rückgang *m*; automatische
   Rückführung *f*
f retour *m* automatique
r автоматический возврат *m*

1367 automatic resistance welder
d automatische Widerstandsschweißanlage *f*;
   Widerstandsschweißautomat *m*
f soudeuse *f* automatique par résistance
r установка *f* для автоматической
   контактной сварки; автомат *m* для
   контактной сварки

1368 automatic retransmission request
d automatische
   Rückübertragungsanforderung *f*
f requête *f* de la retransmission automatique
r автоматическое требование *n* обратной
   передачи

*   **automatic return question** → 1364

1369 automatic rotary line
d automatische Rotationslinie *f*
f chaîne *f* automatique de rotation
r автоматическая вращающаяся линия *f*

1370 automatics
d Regelungs- und Steuerungstechnik *f*
f technique *f* de commande et régulation
   automatique; contrôle *m* automatique
r техника *f* контроля; техника
   регулирования

1371 automatic sample changer
d automatischer Probenwechsler *m*

f passeur *m* automatique [d'échantillons]; changeur *m* automatique [d'échantillons]
r автоматическое устройство *n* для смены проб

1372 automatic scaler; autoscaler
d automatischer Zähler *m*; automatisches Zählgerät *n*
f échelle *f* automatique de comptage
r автоматичное пересчётное устройство *n*

1373 automatic search
d selbsttätiges Suchen *n*
f recherche *f* automatique
r автоматический поиск *m*

1374 automatic search circuit
d automatischer Suchkreis *m*
f circuit *m* de recherche automatique
r схема *f* автоматического поиска

* **automatic selection control** → 1376

1375 automatic selection unit
d automatische Auswahleinheit *f*
f unité *f* de sélection automatique
r автоматический искатель *m*

* **automatic selective control** → 1376

1376 automatic selectivity control; automatic selection control; automatic selective control
d automatische Auswahlsteuerung *f*; Anwahlautomatik *f*
f commande *f* de sélection automatique; commande sélective automatique
r автоматическое регулирование *n* избирательности; избирательная автоматика *f*

1377 automatic sensitivity control
d automatische Empfindichkeitsregelung *f*
f réglage *m* automatique de sensibilité
r автоматическая регулировка *f* чувствительности

1378 automatic sequence control
d selbsttätige Folgesteuerung *f*; bedingte Steuerung *f*
f automatisme *m* de séquence; commande *f* séquentielle automatique
r автоматическое следящее управление *n*

1379 automatic sequence manufacture
d automatisierte Serienfertigung *f*
f fabrication *f* en série automatisée

r автоматизированное серийное производство *n*

1380 automatic sequencing procedure
d automatisches Reihenfolgeverfahren *n*
f procédure *f* de séquence automatique
r процедура *f* автоматического упорядочивания

1381 automatic shifting
d automatische Umstellbarkeit *f*
f déplacement *m* automatique
r возможность *f* автоматической перестройки

* **automatic shim rod follow-up** → 1382

1382 automatic shim rod follow-up [control]
d Trimmnachfolgesteuerung *f*
f compensation *f* automatique
r следящее регулирование *n* компенсирующих стержней

* **automatic sorting method** → 1288

1383 automatic stabilization
d automatische Stabilisierung *f*
f stabilisation *f* automatique
r автоматическая стабилизация *f*

1384 automatic stabilization system
d System *n* mit selbsttätiger Stabilisierung
f système *m* de stabilisation automatique
r система *f* автоматической стабилизации

* **automatic standby start** → 1385

1385 automatic standby start[-up]
d automatische Reserveeinschaltung *f*
f mise *f* en marche automatique de réserve
r автоматическое включение *n* резерва

* **automatic start** → 1386

1386 automatic start[-up]
d automatische Anlassen *n*; Selbstanlassung *f*
f démarrage *m* automatique; automarche *f*
r автоматический запуск *m*; автопуск *m*

1387 automatic stopping
d automatische Stillsetzung *f*
f arrêt *m* automatique
r автоматическая остановка *f*

1388 automatic supervision room
d automatischer Überwachungsraum *m*
f poste *m* de surveillance automatique
r пункт *m* автоматического контроля

1389 automatic synchronizer
 d automatische Synchronisiereinrichtung *f*
 f mécanisme *m* de synchronisation
      automatique
 r устройство *n* автоматической
      синхронизации

1390 automatic synthesis
 d automatische Synthese *f*
 f synthèse *f* automatique
 r автоматический синтез *m*

1391 automatic system
 d Automatiksystem *n*
 f système *m* d'automaticité
 r автоматическая система *f*

1392 automatic target recognition
 d automatische Zielerkennung *f*
 f discrimination *f* automatique du but
 r автоматическое распознавание *n*

1393 automatic test equipment
 d automatische Prüfeinrichtung *f*
 f dispositif *m* automatique de test
 r автоматическое испытательное
      оборудование *n*

1394 automatic tester
 d Prüfautomat *m*
 f automate *m* de verification
 r устройство *n* автоматического контроля

1395 automatic tester network
 d Prüfautomatennetz *n*
 f réseau *m* de l'automate d'essai
 r сеть *m* контролирующих автоматов

 *   **automatic test → 1201**

 *   **automatic testing → 1201**

1396 automatic testing system
 d automatisches Prüfsystem *n*
 f système *m* de vérification automatique
 r автоматическая система *f* для испытаний

1397 automatic time sharing
 d automatisches Zeitteilungsverfahren *n*;
      automatische Zeitschachtelung *f*
 f simultanéité *f* par partage automatique du
      temps d'opération; répartition *f* temporelle
      automatique
 r автоматическая система *f* разделения
      времени

1398 automatic tool change; automatic tool
 exchange

 d automatischer Werkzeugwechsel *m*
 f échange *m* d'outils automatique; changement
      m d'instruments automatique
 r автоматическая смена *f* инструмента

 *   **automatic tool exchange → 1398**

1399 automatic tool flow
 d selbsttätiger Werkzeugfluss *m*
 f flux *m* automatique des outils
 r автоматический поток *m* инструментов

1400 automatic tracking
 d selbsttätige Nachführung *f*
 f poursuite *f* automatique; repérage *m*
      automatique; guidage *m* automatique
 r автоматическое сопровождение *n*

1401 automatic translation
 d automatische Übersetzung *f*
 f traduction *f* automatique
 r автоматическая трансляция *f*

1402 automatic translation device
 d automatische Übersetzungseinrichtung *f*
 f dispositif *m* automatique de traduction
 r устройство *n* автоматического перевода

1403 automatic treatment process
 d automatischer Bearbeitungsvorgang *m*
 f procédé *m* de traitement automatique
 r автоматический технологический
      процесс *m*

1404 automatic tuning control
 d automatische Scharfabstimmung *f*
 f accord *m* automatique
 r автоматическая точная настройка *f*

1405 automatic valve
 d selbsttätiges Ventil *n*
 f soupape *f* automatique
 r автоматический клапан *m*

1406 automatic video noise limiting
 d automatische Video-Rauschbegrenzung *f*
 f limitation *f* automatique de bruit vidéo
 r автоматическое ограничение *n* видеошума

1407 automatic viscosity controller
 d automatischer Viskositätsregler *m*
 f régulateur *m* automatique de viscosité
 r автоматический регулятор *m* вязкости

1408 automatic voltage regulation
 d automatische Spannungsregelung *f*
 f réglage *m* automatique de tension
 r автоматическое регулирование *n*
      напряжения

  * **automatic volume controller** → 1283

  * **automatic welding** → 1312

1409 automatic wiring design
  d automatischer Entwurf *m* der Verdrahtung
  f croquis *m* automatique de câblage
  r автоматическое составление *n* монтажной схемы

1410 automatic work flow; automatic parts flow
  d selbsttätiger Werktückfluss *m*
  f flux *m* automatique de pièces
  r автоматический поток *m* деталей

1411 automatic working matching
  d automatische Arbeitsanpassung *f*; automatische Anpassung *f* an Arbeitsfunktionen
  f ajustage *m* de travail automatique
  r автоматическая адаптация *f* рабочих функции

1412 automatic working system
  d automatisches Arbeitssystem *n*
  f système *m* de travail automatique
  r автоматическая рабочая система *f*

1413 automatic work piece handling
  d automatische Werkstückhandhabung *f*
  f manutention *f* automatique d'une pièce à usiner
  r автоматическое манипулирование *n* обрабатываемых деталей

1414 automatic X-ray spectrograph
  d automatischer Röntgenspektrograf *m*
  f spectrographe *m* à rayons X automatique
  r автоматический рентгеновский спектрограф *m*

  * **automatic zero adjustment** → 1415

1415 automatic zero balancing; automatic zero adjustment
  d selbsttätige Nulleinstellung *f*
  f mise *f* à zéro automatique
  r автоматическая установка *f* нуля

1416 automatic zero check
  d automatische Nullkontrolle *f*
  f contrôle *m* automatique de zéro
  r автоматическая проверка *f* нуля

1417 automatic zero step
  d automatischer Nullschritt *m*
  f avance *f* décimale automatique
  r автоматический нулевой шаг *m*

1418 automating reverse switching
  d automatische Umschaltung *f*
  f commutation *f* automatique
  r автоматическое переключение *n*

  * **automation** → 1431

  * **automation degree** → 3884

1419 automation elements for production lines
  d Automatisierungselemente *npl* für Fertigungsstraßen
  f éléments *mpl* d'automatisation pour les lignes de production
  r элементы *mpl* автоматизации для производственных линий

1420 automation equipment
  d Automatisierungsanlage *f*
  f installation *f* d'automatisation
  r система *f* автоматики

  * **automation-friendly design** → 3985

1421 automation means in measuring circuits
  d Automatisierungsmittel *npl* in Messkreisen
  f appareillage *m* d'automatisme dans les circuits de mesure
  r средства *npl* автоматики в измерительных контурах

1422 automation medium
  d Automatisierungsmittel *n*
  f moyen *m* d'automatisation
  r средство *n* автоматизации

1423 automation of discontinuous processes
  d Automatisierung *f* diskontinuierlicher Prozesse
  f automatisation *f* des procédés discontinus
  r автоматизация *f* прерывистых процессов

1424 automation of drawing work; drawing work automation
  d Automatisierung *f* der Zeichenarbeiten
  f automation *f* des traveaux de dessin
  r автоматизация *f* чертежных работ

  * **automation of project** → 3979

1425 automation of supervising process
  d Automatisierung *f* des Überwachungsprozesses
  f automatisation *f* du processus d'inspection
  r автоматизация *f* супервизорного процесса

1426 automation of technology
  d Automatisierung *f* der Technologie

f automatisation *f* de la technologie
r автоматизация *f* технологии

1427 automation plan
d Automationsplan *m*
f plan *m* d'automation
r план *m* автоматизации

1428 automation task
d Automatisierungsaufgabe *f*
f tâche *f* d'automatisation
r задача *f* автоматизации

1429 automation term
d Automatisierungsterminus *m*
f terme *m* d'automatisation
r термин *m* автоматики

1430 automatism
d Automatismus *m*
f automatisme *m*
r автоматизм *m*

1431 automati[zati]on
d Automatisierung *f*; Automatisation *f*
f automation *f*; automatisation *f*
r автоматизация *f*

1432 automatization degree of assembly
d Automatisierungsgrad *m* der Montage
f degré *m* d'automatisation de l'assemblage
r степень *f* автоматизации монтажа

1433 automatization of auxiliary operations
d Automatisierung *f* von Hilfsoperationen
f automatisation *f* d'opérations auxiliaires
r автоматизация *f* вспомагательных
    операций

1434 automatize *v*
d automatisieren
f automatiser
r автоматизировать

1435 automatized working place
d automatisierter Arbeitsplatz *m*
f place *f* de travail automatisée
r автоматизированное рабочее место *n*

1436 automaton algebra of events
d Automatenalgebra *f* der Ereignisse
f algèbre *f* d'automate des événements
r алгебра *f* событий для автоматов

**1437 automaton for decoding; decoding
      automaton**
d Dekodierungsautomat *m*; Automat *m* für
    Dekodierung; Entschlüsselungsautomat *m*

f automate *m* de décodage
r декодирующий автомат *m*

1438 automaton network
d Automatennetz *n*
f réseau *m* d'automate
r сеть *f* автоматов

1439 automaton with final memory
d Automat *m* mit Endspeicher
f automate *m* à mémoire finale
r автомат *m* с конечной памятью

1440 automaton without outputs
d Automat *m* ohne Ausgaben
f automate *m* sans sorties
r автомат *m* без выходов

1441 automaton with reduced inputs
d Automat *m* mit reduzierten Eingaben
f automate *m* à entrées réduites
r автомат *m* с обобщенными входами

1442 automode
d automatische Betriebsweise *f*
f régime *m* automatique; mode *m* automatique
r автоматический режим *m*

1443 automonitor
d automatisches Kontrollgerät *n*; automatischer
    Monitor *m*; Automonitor *m*
f moniteur *m* automatique
r автоматическое контрольное устройство *n*

1444 automorphism
d Automorphismus *m*; automorphe
    Abbildung *f*
f automorphisme *m*
r автоморфизм *m*

*   **autonomous** → 8892

1445 autonomous linear automaton
d autonomer linearer Automat *m*
f automate *m* autonome linéaire
r автономный линейный автомат *m*

**1446 autonomous portable gamma-ray
      spectrometer; independent portable
      gamma-ray spectrometer**
d autonomes tragbares Gammaspektrometer *n*;
    unabhängiges tragbares Gammaspektrometer
f spectromètre *m* gamma portatif autonome
r независимый переносный гамма-
    спектрометр *m*

1447 autonomous system
d autonomes System *n*

f système *m* autonome
r автономная система *f*

1448 autonomy
  d Autonomie *f*
  f autonomie *f*
  r автономия *f*

1449 autooscillation; self-oscillation; self-vibration; free oscillation; natural oscillation
  d Eigenschwingung *f*; oszillierende Eigenbewegung *f*; freie Schwingung *f*; ungezwungene Schwingung
  f oscillation *f* propre; oscillation naturelle; vibration *f* libre; résonance *f* propre
  r собсвенное колебательное движение *n*; свободное колебание *n*

1450 autooscillation link
  d Selbstschwingungsglied *n*
  f réseau *m* à auto-oscillations
  r автоколебательное звено *n*

1451 autooscillations in servosystems
  d Eigenschwingungen *fpl* in Servosystemen
  f auto-oscillations *fpl* des servosystèmes
  r автоколебания *npl* в следящих системах

1452 autooscillation system; self-sustained oscillation system
  d Selbstschwingungssystem *n*
  f système *m* des auto-oscillations
  r автоколебательная система *f*

1453 autopath
  d automatische Programmkorrektur *f*
  f correction *f* de programme automatique
  r автоматическая отладка *f* программы

1454 autopolarization
  d Autopolarisation *f*
  f autopolarisation *f*
  r автополяризация *f*

* **autopoll** → 1455

1455 autopoll[ing]
  d automatische Abfrage *f*
  f interrogation *f* automatique
  r автоматический опрос *m*

1456 autopower spectral density
  d Autospektraldichte *f*; autospektrale Leistungsdichte *f*; Autoleistungsdichte *f*
  f densité *f* autospectrale [de puissance]
  r автоспектральная плотность *f* [мощности]

1457 autopressuregraph; autopressuregrapm
  d Autoradiografie *f* als Resultat einer Druckreaktion
  f autoradiogramme *m* comme résultat d'une réaction de pression
  r авторадиограмма *f* в результате реакции давления

* **autopressuregrapm** → 1457

1458 autoradiation
  d Autostrahlung *f*
  f autoradiation *f*
  r собственное излучение *n*

1459 autoradiogram; autoradiograph; radioautograph; radioautogram
  d Autoradiogramm *n*; Autoradiograf *m*; Radioautogramm *n*; Radioautografie *f*
  f autoradiogramme *m*; autoradiographe *m*; radio-autogramme *m*; radio-autographe *m*
  r авторадиограмма *f*; авторадиограф *m*; радиавтограф *m*; радиоавтограмма *f*

* **autoradiograph** → 1459

1460 autoradiographic method
  d autoradiografisches Verfahren *n*; autoradiografische Regelung *f*
  f commande *f* autoradiographique; méthode *f* autoradiographique
  r авторадиографический метод *m*

1461 autoradiography
  d Autoradiografie *f*
  f autoradiographie *f*
  r авторадиография *f*

1462 autoradiography in electron microscopy
  d Autoradiografie *f* in der Elektronenmikroskopie
  f autoradiographie *f* en microscopie électronique
  r авторадиография *f* в электронной микроскопии

1463 autoregressive series
  d autoregressive Folge *f*; autoregressive Reihe *f*
  f séries *fpl* autorégressives
  r авторегресивная последовательность *f*

* **autoscaler** → 1372

1464 autospectrum
  d Autospektrum *n*; Autoleistungsspektrum *n*
  f autospectre *m*
  r автоспектр *m*

1465 autostabilizer
d automatischer Stabilisator *m*; automatischer Gleichschalter *m*
f stabilisateur *m* automatique
r автостабилизатор *m*

* **auto-testing** *adj* → **11285**

1466 autotest-system
d Autotest-System *n*
f système *m* Autotest
r автотест-система *f*

1467 auxiliary air regulator
d Hilfsluftregler *m*
f dispositif *m* auxiliaire de réglage d'air
r вспомагательный регулятор *m* воздуха

1468 auxiliary algorithm
d Hilfsalgorithmus *m*
f algorithme *m* auxiliaire
r вспомагательный алгоритм *m*

1469 auxiliary bridge
d Hilfsbrücke *f*
f pont *m* auxiliaire
r вспомагательный мостик *m*

1470 auxiliary circuit
d Hilfskreislauf *m*
f circuit *m* auxiliaire
r вспомагательная схема *f*; вспомагательный цикл *m*

1471 auxiliary console
d Zusatzkonsole *f*
f console *f* auxiliaire
r вспомагательный пульт *m*

1472 auxiliary contact for photoelectric sensor
d Hilfskontakt *m* für fotoelektrischen Sensor
f contact *m* auxiliaire pour capteur photoélectrique
r вспомогательный контакт *m* фотоэлектрического датчика

1473 auxiliary control function
d Hilfssteuerfunktion *f*
f fonction *f* de commande auxiliaire
r вспомагательная управляющая функция *f*

1474 auxiliary control system
d Hilfssteuersystem *n*
f système *m* de commande auxiliaire
r вспомагательная управляющая система *f*

1475 auxiliary corrector
d Hilfsregler *m*; Hilfskorrektor *m*
f correcteur *m* auxiliaire; correcteur secondaire
r вспомагательный корректор *m*; вспомагательное корректирующее устройство *n*

1476 auxiliary device
d Hilfseinrichtung *f*
f dispositif *m* auxiliaire
r вспомагательное устройство *n*

1477 auxiliary [electrical] system
d [elektrische] Eigenbedarfsanlage *f*
f système *m* [électrique] des besoins auxiliaires; installation *f* [électrique] des besoins intérieurs
r [электрическая] установка *f* собственных нужд

1478 auxiliary equipment
d Hilfsausrüstung *f*; Zusatzausrüstung *f*
f équipement *m* auxiliaire; équipement additionnel; appareillage *m* complémentaire
r вспомагательное оборудование *n*

1479 auxiliary equipment and supplies
d Zusatzeinrichtungen *fpl* und Zubehör *n*
f équipement *m* auxiliaire et accessoires *mpl*
r комплект *m* дополнительного оборудования и принадлежностей

1480 auxiliary equipment compartment
d Hilfsanlagenraum *m*
f compartiment *m* de l'équipement auxiliaire
r помещение *n* вспомагательного оборудования

1481 auxiliary fabrication process
d Hilfsprozess *m*; Hilfsfertigungsprozess *m*
f procédé *m* de fabrication auxiliaire
r вспомагательный производственный процесс *m*

1482 auxiliary feature
d Abfragepuffer *m*
f interrogation *f* intermédiaire
r буфер *m* запроса

1483 auxiliary fluid system
d Hilfsarbeitsmittelsystem *n*
f système *m* des substances actives auxiliaire
r система *f* вспомагательных рабочих сред

1484 auxiliary generator
d Eigenbedarfsgenerator *m*; Hilfsgenerator *m*
f génératrice *f* auxiliaire
r генератор *m* собственного расхода

1485 auxiliary machine
d Ergänzungsmachine *f*

f machine f auxiliaire
r вспомагательная машина f

1486 auxiliary operation
d Hilfsoperation f
f opération f auxiliaire
r вспомагательная операция f

1487 auxiliary process
d Hilfsprozess m
f processus m auxiliaire
r вспомагательный процесс m

1488 auxiliary processing system
d Hilfsverabeitungssystem n
f système m de traitement auxiliaire
r вспомагательная обрабатывающая
система f

1489 auxiliary pump
d Hilfspumpe f
f pompe f auxiliaire
r вспомагательный насос m

1490 auxiliary quantity
d Hilfsgröße f
f quantité f auxiliaire
r вспомогательная величина f

1491 auxiliary selector
d Hilfsselektor m
f sélecteur m auxiliaire
r вспомагательный селектор m

1492 auxiliary steam supply system
d Hilfsdampf[versorgungs]system n
f système m d'alimentation en vapeur
auxiliaire
r система f вспомогательного
пароснабжения

1493 auxiliary store
d Hilfsspeicher m
f mémoire f auxiliare
r вспомагательный накопитель m

* auxiliary system→ 1477

1494 auxiliary unit
d Zusatzeinheit f; Zusatzblock m
f unité f auxiliaire
r вспомагательный блок m

* availability → 175

1495 availability control unit
d Zugangs-Steuereinheit f
f unité f de commande d'accessibilité

r блок m управления доступом

1496 availability factor
d Arbeitsverfügbarkeit f;
Verfügbarkeitsfaktor m
f facteur m de disponibilité
r коэффициент m эксплуатационной
готовности

1497 available power
d verfügbare Leistung f; abgebbare Leistung
f puissance f disponible
r мощность f на согласованной нагрузке

1498 available sensor system
d verfügbares Sensorsystem n
f système m de senseur disponible; système de
capteur disponible
r имеющаяся сенсорная система f

1499 available unit
d verfügbare Einheit f
f unité f disponible
r предоставленное устройство n

1500 average; average value
d Mittelwert m
f valeur f moyenne
r среднее значение n

1501 average access time
d mittlere Zugriffszeit f
f temps m moyen de découpage
r среднее время n выборки

1502 average calculating speed
d mittlere Rechengeschwindigkeit f
f vitesse f de calcul moyenne
r средняя скорость f вычисления

1503 average deviation
d mittlere Abweichung f; Mittelabweichung f
f écart m moyen
r среднее отклонение n

1504 average error
d mittlerer Fehler m
f erreur f moyenne
r средняя ошибка f

1505 average excitation energy
d mittlere Anregungsenergie f
f énergie f d'excitation moyenne
r средняя энергия f возбуждения

1506 average execution time
d mittlere Ausführungszeit f
f temps m moyen d'exécution
r среднее время n выполнения

**1507 average fuel rating; mean fuel rating;
average specific power**
  d mittlere spezifische Leistung *f*;
       durchschnittliche spezifische Leistung
  f puissance *f* massique moyenne
  r средняя удельная мощность *f*

1508 average gain coefficient
  d mittlerer Verstärkungskoeffizient *m*
  f gain *m* moyen
  r коэффициент *m* среднего усиления

1509 average impulse power
  d mittlere Impulsleistung *f*
  f puissance *f* moyenne d'impulsion
  r средняя импульсная мощность *f*

**1510 average intensity; angle-integrated
intensity**
  d mittlere Intensität *f*; winkelintegrierte
       Intensität
  f intensité *f* moyenne; intensité intégrée sur
       l'angle
  r средняя интенсивность *f*;
       интегрированная по углу интенсивность

1511 average logarithmic energy decrement
  d [mittleres] logarithmisches
       Energiedekrement *n*; mittlerer
       logarithmischer Energieverlust *m*
  f paramètre *m* de ralentissement; décrément *m*
       logarithmique moyen de l'énergie
  r [средний] логарифмический декремент *m*
       энергии; средняя логарифмическая
       потеря *f* энергии

1512 average noise
  d Mittelwert *m* des Rauschens;
       Mittelrauschen *n*
  f valeur *f* moyenne de bruit
  r среднее значение *n* шума

1513 average operation
  d mittlere Operation *f*
  f opération *f* moyenne; prise *f* de la moyenne
  r операция *f* усреднения

1514 average operation time
  d durchschnittliche Operationszeit *f*
  f temps *m* moyen d'opération
  r среднее время *n* операции

1515 average optical power
  d mittlere optische Leistung *f*
  f puissance *f* optique moyenne
  r средняя оптическая мощность *f*

  * **average power → 7921**

  * **average specific power → 1507**

1516 average supply current
  d mittlere Stromaufnahme *f*
  f consommation *f* moyenne de courant
  r среднее значение *n* потребляемого тока

1517 average transfer rate
  d mittlere Übertragungsgeschwindigkeit *f*
  f vitesse *f* moyenne de transmission
  r средняя скорость *f* передачи [данных]

  * **average value → 1500**

**1518 average value pulse indicator; mean-pulse
indicator**
  d Mittelwertanzeiger *m*
  f indicateur *m* d'impulsions moyennes
  r указатель *m* средных значений импульсов

1519 average voltmeter
  d Mittelwertspannungsmesser *m*;
       Mittelwertvoltmeter *n*
  f voltmètre *m* de valeur moyenne
  r усредняющий вольтметр *m*

1520 average waiting time
  d mittlere Wartezeit *f*
  f temps *m* d'attente moyen
  r среднее время *n* ожидания

1521 averaging relay; integrating relay
  d Summierrelais *n*; Integralrelais *n*
  f relais *m* intégrateur; relais d'addition
  r усредняющее реле *n*; реле-индикатор *m*

1522 awake process
  d aktiver Prozess *m*
  f processus *m* actif
  r активный процесс *m*

1523 axial adjustment
  d Achseneinstellung *f*; Achsenregelung *f*
  f reglage *m* axial
  r осевая регулировка *f*

1524 axial diffusion coefficient
  d axialer Diffusionskoeffizient *m*
  f coefficient *m* de diffusion axial
  r осевой коэффициент *m* диффузии

1525 axial displacement
  d axiale Verschiebung *f*
  f décalage *m* axial
  r осевое смещение *n*

1526 axial distribution
  d axiale Verteilung *f*

    f distribution *f* axiale; répartition *f* axiale
    r аксиальное распределение *n*

1527 axial expansion coefficient
    d axialer Ausdehnungskoeffizient *m*
    f coefficient *m* d'expansion axiale
    r коэффициент *m* расширения вдоль оси

1528 axial flow
    d axiale Stömung *f*
    f écoulement *m* axial
    r осевой поток *m*

1529 axial flow fan
    d Axialgebläse *n*; Axialventilator *m*
    f soufflante *f* axiale; ventilateur *m* hélicoïdal
    r осевой вентилятор *m*

    * **axial flow transonic compressor → 1530**

1530 axial flow tran[s]sonic compressor
    d Axial-Transsonikkompressor *m*; Axial-Transsonikverdichter *m*
    f compresseur *m* transsonique axial
    r осевой околозвуковой компрессор *m*

1531 axial flow turbine
    d Axialstrahlturbine *f*
    f turbine *f* axiale
    r реактивный двигатель *m* с осевым компрессором

1532 axiom; postulate
    d Axiom *n*; Postulat *n*
    f axiome *m*; théorème *m*; postulat *m*
    r аксиома *f*; постулат *m*

1533 axiomatics of theory of probabilities
    d Axiomatik *m* der Theorie der Wahrscheinlichkeiten
    f axiomatique *f* de la théorie des probabilités
    r аксиоматика *f* теории вероятностей

1534 axiomatic system
    d axiomatisches System *n*
    f système *m* axiomatique
    r аксиоматическая система *f*

1535 axis of the ordinates; Y-axis
    d Ordinatenachse *f*; Y-Achse *f*
    f axe *m* des ordonnées; axe des Y
    r ось *f* ординат; ось Y; ордината *f*

1536 axis regulation
    d Achsregelung *f*
    f régulation *f* d'axe
    r регулирование *n* осей

1537 axis regulation structure
    d Achsregelungsstruktur *f*
    f structure *f* de réglage d'axe
    r структура *f* регулирования оси

1538 axisymmetric
    d achsensymmetrisch
    f symétrique par rapport à l'axe
    r осесимметрично

B

r процессор *m* обслуживания базы данных

1539 back conductance
d Sperrleitwert *m*
f conductance *f* inverse
r проводимость *f* в обратном направлении

1540 back contact; normally open contact
d Ruhekontakt *m*; Arbeitskontakt *m*
f contact *m* de repos; contact à ouverture
r нормально разомкнутый контакт *m*

1541 back-coupling; feedback; inverse coupling
d Rückkopplung *f*; Rückführung *f*
f réaction *f*; couplage *m* de réaction; contre-réaction *f*
r обратная связь *f*

1542 back-coupling condition
d Rückkopplungsbedingung *f*
f condition *f* de réaction
r условие *n* создания обратной связи

1543 back-coupling connection
d Rückkopplungsleitung *f*
f connexion *f* de réaction
r схема *f* включения обратной связи

1544 back-coupling effect
d Rückkoppungseffekt *m*
f effet *m* de réaction
r влияние *n* обратной связи

1545 back-current bracking
d Gegenstrombremsung *f*
f freinage *m* en contre-courant
r торможение *n* противотоком

1546 back-current principle; countercurrent principle
d Gegenstromprinzip *n*; Rückstromprinzip *n*
f principe *m* de contre-courant
r принцип *m* противотока

1547 back drive
d Rücktrieb *m*
f marche *f* arrière
r обратное вращение *n*

1548 back-end processor
d Nachschaltprozessor *m*; nachgeschalteter Prozessor *m*
f processeur *m* back-end

1549 back flow
d Rückstrom *m*
f courant *m* de retour
r обратный поток *m*

1550 background computer
d Hintergrundprozessor *m*
f calculateur *m* de fond
r фоновая вычислительная машина *f*

1551 background data
d Hintergrunddaten *pl*
f données *fpl* de fond
r данные *pl* для фоновой обработки

1552 background measurement radiometer
d Hintergrundstrahlungsmesser *m*
f radiomètre *m* de fond
r радиометр *m* для измерения излучения фона

1553 background noise
d Störgeräusch *n*; Grundrauschen *n*
f perturbations *fpl*; bruit *m* perturbateur; bruit *m* de fond
r помехи *fpl*; фоновый шум *m*

1554 background-noise level
d Hintergrundrauschpegel *m*
f niveau *m* du bruit de fond
r уровень *m* шумов фона

1555 background processing
d Hintergrundverarbeitung *f*
f traitement *m* en arrière-plan
r фоновая обработка *f*

1556 background pulse rate
d Nulleffektimpulsrate *f*
f taux *m* d'impulsions parasites
r скаваженость *f* фона импульса

1557 background return
d Grundecho *n*; Hintergrundecho *n*
f écho *m* de fond
r местные помехи *fpl* радиолокатора

1558 background signal
d Hintergrundsignal *n*
f signal *m* de fond
r сигнал *m* фона

1559 back indication
d Rückanzeige *f*
f indication *f* en retour
r обратное показание *n*

1560 **backing store**
d Ergänzungsspeicher *m*
f mémoire *f* additionnelle
r дополнительный накопитель *m*

1561 **backlash**
d Spiel *n*; toter Gang *m*; Totgang *m*
f jeu *m*; intervalle *m* mort
r люфт *m*; мёртвый ход *m*

1562 **backlash of hradware signal**
d Hardwaresignalspiel *n*
f jeu *m* d'un signal de hardware
r мёртвый ход *m* сигнала аппаратных
средств

1563 **backmix reactor**
d Rückvermischungsreaktor *m*
f réacteur *m* à remélangeage
r реактор *m* с обратным перемешиванием

1564 **back pulse front**
d hintere Impulsflanke *f*
f front *m* arrière d'impulsion
r задный фронт *m* импульса

* **backscatter** → 1566

1565 **backscattered power**
d rückgestreute Leistung *f*
f puissance *f* rétrodiffusée
r обратно рассеянная мощность *f*

1566 **backscatter[ing]**
d Rückstreuung *f*
f rétrodiffusion *f*
r обратное рассеяние *n*

1567 **backscattering analysis**
d Rückstreuanalyse *f*
f analyse *f* par rétrodiffusion
r анализ *m* обратного рассеяния

1568 **backscattering measurement**
d Rückstreumessung *f*
f mesure *f* par rétrodiffusion
r измерение *n* методом обратного рассеяния

1569 **backscattering process**
d Rückstreuvorgang *m*
f processus *m* de rétrodiffusion
r процесс *m* обратного рассеяния

1570 **backscattering response**
d Rückstreuungsverlauf *m*;
Rückstreuungscharakteristik *f*
f réponse *f* de rétrodiffusion

r характеристика *f* обратного рассеяния

1571 **backscattering signal**
d Rückstreusignal *n*; rückgestreutes Signal *n*
f signal *m* rétrodiffusé
r сигнал *m* обратного рассеяния

1572 **backspace of hardware system; reset of hardware system**
d Hardwaresystemrücksetzen *n*
f recul *m* d'un système d'hardware
r возврат *m* системы аппаратных средств

1573 **backspace of software system; reset of software system**
d Softwaresystemrücksetzen *n*
f recul *m* d'un système de software
r возврат *m* системы программного
обеспечения

1574 **backspace statement**
d Rücksetzanweisung *f*
f instruction *f* de rappel
r команда *f* возврата

1575 **back stroke; reversed motion; return trace; reverse run**
d Rücklauf *m*; Rückgang *m*
f course *f* de retour; retour *m*
r обратный ход *m*

1576 **back-to-back method**
d Gegenverfahren *n*
f méthode *f* d'opposition
r метод *m* взаимной нагрузки

* **back-up connection** → 11815

1577 **backup equipment**
d Reserveausstattung *f*
f équipement *m* de réserve
r резервное оборудование *n*

1578 **back-up protection; reserve protection**
d Reserveschutz *m*; überlagerter Schutz *m*
f protection *f* de réserve
r резервная защита *f*

1579 **backup system; standby system**
d Reservesystem *n*; Aushilfssystem *n*;
Bereitschaftssystem *n*
f système *m* de réserve
r резервная система *f*

1580 **back voltage**
d Sperrspannung *f*
f tension *f* inverse
r обратное напряжение *n*

1581 **backward differences**
  d rückwärtige Differenzen *fpl*;
      rückwärtsgenommene Differenzen
  f différences *fpl* en arrière
  r нисходящая разность *f*

1582 **backward transfer characteristic**
  d Rückkopplungscharakteristik *f*
  f caractéristique *f* de la réaction
  r характеристика *f* обратной связи

1583 **backward-wave parametric amplifier**
  d parametrischer Rückwärtswellenverstärker *m*
  f amplificateur *m* paramétrique à onde
      pétrograde
  r параметрический усилитель *m* обратной
      волны

1584 **bagging machine**
  d Absackvorrichtung *f*
  f dispositif *m* d'ensachage
  r автоматический мешконаполнитель *m*

1586 **balance** *v*
  d abgleichen; ausgleichen
  f équilibrer; égaliser; niveler
  r выравнивать

1585 **balance; equilibrium**
  d Gleichgewicht *n*
  f balance *f*; balancement *m*; équilibre *m*
  r равновесие *n*

1587 **balance attenuation**
  d Fehlerdämpfung *f*
  f affaiblissement *m* d'équilibrage
  r балансное затухание *n*

1588 **balance check; null balance; zero balance**
  d Nullpunktabgleich *m*; Nullabgleich *m*
  f balance *f* de zéro; contrôle *m* d'ajustage
  r настройка *f* нуля; балансировка *f* нуля

  * **balance conditions** → 5118

1589 **balanced bipolar circuit**
  d abgeglichene bipolare Schaltung *f*
  f circuit *m* bipolaire équilibré
  r уравновешенная биполярная схема *f*

1590 **balanced bridge**
  d abgeglichene Brücke *f*
  f pont *m* équilibré
  r уравновешенный мост *m*

1591 **balanced circuit; symmetrical circuit**
  d symmetrische Schaltung *f*
  f circuit *m* symétrique

  r симметричная схема *f*; симметричная
      цепь *f*

1592 **balanced control**
  d ausgeglichene Regelung *f*; kompensierte
      Regelung
  f réglage *m* compensé; réglage équilibré
  r уравновешенное управление *n*

1593 **balanced current**
  d Ausgleichsstrom *m*
  f courant *m* équilibré
  r уравновешенный ток *m*

1594 **balanced-current telemetry system**
  d Fernmesssystem *n* mit Ausgleichsstrom
  f système *m* de télémesure à compensation de
      courant
  r телеизмерительная токоуравновешенная
      система *f*

1595 **balance detector**
  d Abgleichdetektor *m*
  f détecteur *m* d'équilibrage
  r балансный детектор *m*

1596 **balanced input**
  d symmetrischer Eingang *m*
  f entrée *f* symétrique
  r симметричный вход *m*

1597 **balanced line**
  d abgeglichene Leitung *f*; symmetrische
      Leitung
  f ligne *f* équilibrée; ligne symétrique
  r симметричная линия *f*

1598 **balanced load**
  d symmetrische Belastung *f*
  f charge *f* équilibrée
  r уравновешанная нагрузка *f*

1599 **balanced method**
  d Ausgleichsmethode *f*;
      Ausgleichsverfahren *n*; Nullmethode *f*
  f méthode *f* de compensation
  r балансный метод *m*

1600 **balanced mixer**
  d symmetrische Mischstufe *f*
  f mélangeur *m* équilibré
  r балансный смеситель *m*

1601 **balanced modulator**
  d Gergentaktmodulator *m*
  f modulateur *m* équilibré
  r балансный модулятор *m*

1602 **balanced output**
  d symmetrischer Ausgang *m*; symmetrische
       Ausgangsspannung *f*
  f sortie *f* symétrique; tension *f* symétrique de
       sortie
  r симметричный выход *m*

1603 **balanced phase discriminator**
  d symmetrischer Phasendiskriminator *m*
  f discriminateur *m* de phase équilibré
  r балансный фазовый дискриминатор *m*

1604 **balanced pulses**
  d symmetrische Impulse *mpl*
  f impulsions *fpl* symétriques
  r симметричные импульсы *mpl*

1605 **balanced push-pull amplifier**
  d symmetrischer Gegentaktverstärker *m*
  f amplificateur *m* symétrique push-pull
  r симметричный двухтактный усилитель *m*

  * **balanced relay** → 4141

1606 **balanced signal**
  d symmetrisches Signal *n*
  f signal *m* symétrique
  r симметричный сигнал *m*

1607 **balanced voltage**
  d erdsymmetrische Spannung *f*
  f tension *f* symétrique [par rapport à la terre]
  r напряжение n, симметричное
       относительно земли

1608 **balance equation**
  d Bilanzgleichung *f*
  f équation *f* de bilan
  r уравнение *n* баланса

1609 **balance point**
  d Abgleichpunkt *m*
  f point *m* d'équilibrage
  r точка *f* согласования

1610 **balancing**
  d Ausgleichen *n*; Abgleicheinstellung *f*;
       Auswuchtung *f*; Kompensation *f*
  f équilibrage *m*; compensation *f*; réglage *m*
  r балансировка *f*

1611 **balancing bridge circuit**
  d selbstabgleichende Brückenschaltung *f*
  f circuit *m* en pont à auto-équilibrage
  r мостовая самоуравновешивающая схема *f*

1612 **balancing circuit**
  d Abgleichschaltung *f*

  f circuit *m* d'équilibrage
  r схема *f* компенсации

1613 **balancing dynamometer**
  d Ausgleichsdynamometer *n*
  f dynamomètre *m* à compensation
  r уравновещивающий динамометр *m*

1614 **balancing element**
  d Ausgleichselement *n*
  f élément *m* de balance
  r компенсатор *m*

1615 **balancing indicator**
  d Ausgleichsindikator *m*
  f indicateur *m* de position
  r индикатор *m* баланса

1616 **balancing movement of gripper**
  d Ausgleichsbewegung *f* eines Greifers
  f mouvement *m* de balancement d'un grappin
  r выравнивающее движение *n* захвата

1617 **balancing network**
  d Ausgleichsleitung *f*
  f réseau *m* d'équilibrage
  r балансная схема *f*

1618 **balancing of positioning inaccuracy**
  d Ausgleich *m* von Positionierungenauigkeiten
  f balancement *m* d'inexactitude de
       positionnement
  r компенсация *f* погрешности
       позиционирования

1619 **balancing potentiometer**
  d Abgleichpotentiometer *n*
  f potentiomètre *m* ajustable
  r уравновесивающий потенциометр *m*;
       компенсирующий потенциометр

1620 **balancing resistance**
  d Abgleichwiderstand *m*
  f résistance *f* d'ajustement; résistance de
       compensation
  r юстировочное сопротивление *n*

1621 **balancing speed**
  d Ausgleichsgeschwindigkeit *f*
  f vitesse *f* compensatrice
  r установившаяся скорость *f*

1622 **balancing transformer**
  d Ausgleichsumspanner *m*
  f transformateur *m* d'équilibrage
  r уравновешивающий трансформатор *m*

1623 **balancing unit**
  d Ausgleichseinheit *f*

   f unité *f* d'équilibrage
   r устройство *n* выравнивания

1624 balancing unit of gripper
   d Greiferausgleichseinheit *f*; Ausgleichseinheit
     f eines Greifers
   f unité *f* d'équilibrage d'un grappin; unité de
     balance d'un grappin
   r блок *m* выравнивания захватного
     устройства

1625 balancing unit with elastic elements
   d Ausgleichseinheit *f* mit eleastischen Gliedern
   f unité *f* d'équilibrage avec éléments élastiques
   r блок *m* выравнивания с эластичными
     элементами

1626 balancing variable
   d Bilanzierungsvariable *f*
   f variable *f* de bilan
   r переменная *f* баланса

1627 balayage; scanning; exploration
   d Abtasten *n*; Absuchen *n*; Bildfeldzerlegung *f*
   f balayage *m*; sondage *m*; exploration *f*
   r квантование *n*; сканирование *n*

1628 ball spindle drive
   d Kugelspindelantrieb *m*
   f entraînement *m* de tige filetée à billes
   r привод *m* шарикового винта

1629 band adjustment; range adjustment
   d Bereichsinstellung *f*; Umfangsregelung *f*
   f réglage *m* de la bande
   r регулировка *f* диапазона; настройка *f*
     диапазона

1630 band elimination filter; band rejection
     filter
   d Sperrfilter *n*; Sperrkreis *m*; Bandsperre *f*
   f filtre *m* coupe-bande; filtre de blocage; filtre
     à élimination de bande
   r полосно-заграждающий фильтр *m*

1631 bandlimited; frequency-band limited
   d bandbegrenzt
   f limité à la bande de fréquence
   r ограниченный по полосе

1632 band of entrainment
   d Mitnehmerband *n*
   f bande *f* d'entraînement
   r полоса *f* захвата

1633 band-pass amplifier
   d Bandpassverstärker *m*
   f amplificateur *m* passe-bande

   r полосовой усилитель *m*

1634 band-pass filter
   d Bandfilter *n*; Bandpass *m*
   f filtre *m* passe-bande; filtre de bande
   r полосовой фильтр *m*

*   **band rejection filter → 1630**

1635 band-selective filter
   d selektives Bandfilter *n*
   f filtre *m* passe-bande sélectif
   r полосовой избирательный фильтр *m*

1636 band switch
   d Bandumschalter *m*
   f commutateur *m* de bande
   r переключатель *m* диапазона

1637 band theory
   d Zonentheorie *f*
   f théorie *f* de zones
   r зонная теория *f*

1638 bandwidth
   d Bandbreite *f*
   f largeur *f* de bande; largeur passante; bande *f*
     passante
   r ширина *f* полосы

1639 bandwidth control
   d Bandbreitensteuerung *f*;
     Bandbreitenregelung *f*
   f réglage *m* de la bande passante
   r регулировка *f* полосы пропускания

1640 bandwidth-limited operation
   d bandbreitebegrenzter Betrieb *m*
   f fonctionnement *m* limité par largeur de
     bande
   r работа f, ограничиваемая шириной
     полосы

1641 bang-bang control; two-stage control;
     two-step control
   d Zweipunktregelung *f*;
     Bang-Bang-Regelung *f*; Relaisregelung *f*
   f réglage *m* par tout ou rien; réglage à deux
     position; réglage à relais
   r двухпозиционное регулирование *n*;
     релейное регулирование

1642 bang-bang relay
   d Relais *n* mit zwei festen Lagen
   f relais *m* à deux positions
   r реле *n* с двумя устойчивыми
     положениями

1643 **bang-bang servo**
   d Zweipunktservo *n*
   f système *m* asservi à deaux paliers
   r двухпозиционный сервопривод *m*;
       релейный сервомеханизм *m*

   * **bar** *v* → **2465**

1644 **barrier capacity; barrier-layer**
     capacitance
   d Barrierekapazität *f*; Sperrschichtkapazität *f*
   f capacité *f* de barrière
   r барьерная ёмкость *f*; ёмкость запорного
       слоя

   * **barrier-layer capacitance** → **1644**

1645 **barrier-layer rectifier**
   d Sperrschichtgleichrichter *m*
   f redresseur *m* à couche d'arrêt
   r выпрямитель *m* с запирающим слоем

1646 **base; basis**
   d Basis *f*
   f base *f*
   r базис *m*; основа *f*

1647 **base address; basic address**
   d Basisadresse *f*
   f adresse *f* de base
   r базовый адрес *m*

1648 **base algorithm**
   d Basisalgorithmus *m*
   f algorithme *m* de base
   r базовый алгоритм *m*

1649 **baseband**
   d Basisband *n*
   f bande *f* de base
   r базисная полоса *f*

1650 **baseband signalling**
   d Grundbandsignalgabe *f*
   f signalisation *f* de bande de base
   r прямая передача *f*

1651 **baseband transfer function**
   d Basisband-Übertragungsfunktion *f*
   f fonction *f* de transfert en bande de base
   r передаточная функция *f* базисной полосы

1652 **baseband video signal**
   d Basisband-Videosignal *n*
   f signal *m* vidéo en bande de base
   r видеосигнал *m* базисной полосы

1653 **base coordinate**

   d Basiskoordinate *f*
   f coordonnée *f* de base
   r базовая координата *f*

1654 **based variable**
   d basisbezogene Variable *f*
   f variable *f* basée
   r базовая переменная *f*

1655 **base equation; fundamental equation**
   d Grundgleichung *f*; Grundbeziehung *f*
   f équation *f* fondamentale
   r основное уравнение *n*; базовое
       соотношение *n*

1656 **base frequency**
   d Grundfrequenz *f*
   f fréquence *f* de base
   r основная частота *f*

1657 **base movement**
   d Grundbewegung *f*
   f mouvement *m* de base
   r элементарное движение *n*

1658 **base of code**
   d Kodebasis *f*
   f base *f* de code
   r базис *m* кода

1659 **base of realization**
       (theory of automata)
   d Realisierungsbasis *f*
   f base *f* de réalisation
   r базис *m* реализации

1660 **base register; index accumulator**
   d Adressenregister *n*; Indexregister *n*
   f registre *m* de base; registre d'index
   r базисный регистр *m*; регистр индекса

   * **basic address** → **1647**

1661 **basic approach**
   d Basisverfahren *n*
   f méthode *f* de base
   r базовый метод *m*

1662 **basic assembling procedure**
   d Montagegrundverfahren *n*
   f procédé *m* fondamental de montage
   r основная монтажная процедура *f*

1663 **basic circuit**
   d Grundkreis *m*
   f circuit *m* fondamental
   r основная схема *f*

1664 basic clock pulse frequency
 d Grundtaktfrequenz *f*
 f fréquence *f* d'impulsions de base
 r основная последовательность *f* тактовых
    импульсов

1665 basic code
 d Grundkode *m*
 f code *m* de base
 r основной код *m*

1666 basic computing element
 d Grundrechenelement *n*
 f élément *m* fondamental de calcul
 r основной решающий элемент *m*

1667 basic configuration of manipulator
 d Grundkonfiguration *f* eines Manipulators
 f configuration *f* de base d'un manipulateur
 r базовая конфигурация *f* манипулятора

1668 basic constant
 d Grundkonstante *f*
 f constante *f* fondamentale
 r основная постоянная *f*

1669 basic data representation
 d Grunddatendarstellung *f*;
    Basisdatendarstellung *f*
 f représentation *f* des données de base
 r представление *n* исходных данных

1670 basic element
 d Grundbauelement *n*
 f élément *m* de base
 r основной узел *m*

1671 basic equipment
 d Grundausrüstung *f*
 f équipement *m* de base
 r основное оборудование *n*

1672 basic execution
 d Grundausführung *f*
 f exécution *f* de base
 r основное исполнение *n*

1673 basic format
 d Grundformat *n*
 f format *m* de base
 r базовый формат *m*

1674 basic function
 d Grundfunktion *f*
 f fonction *f* de base
 r базовая функция *f*

1675 basic hardware

 d Grundausrüstung *f*; Grundausstattung *f*;
    Basishardware *f*
 f hardware *m* de base
 r основной комплект *m* оборудования

1676 basic instruction
 d Grundbefehl *m*
 f instruction *f* de base
 r основная инструкция *f*

1677 basic logic element
 d logisches Grundelement *n*
 f élément *m* logique fondamental
 r элемент *m* базиса

1678 basic logic function
 d logische Grundfunktion *f*
 f fonction *f* logique fondamentale
 r основная функция *f* алгебры логики

1679 basic machine
 d Grundmaschine *f*
 f machine *f* de base
 r основная машина *f*; базовая машина

1680 basic mapping support
 d Basis-Abbildungsunterstützung *f*
 f support *m* de représentation de base
 r базовая поддержка *f* отображения

1681 basic model
 d Grundtyp *m*; Grundmodell *n*
 f modèle *m* de base
 r базовая модель *f*

1682 basic noise
 d Eigenrauschen *n*
 f bruit *m* propre
 r собственный шум *m*

1683 basic operating system
 d Basisbetriebssystem *n*
 f système *m* de service de base
 r основная операционная система *f*

1684 basic operation
 d Grundoperation *f*
 f opération *f* de base
 r базовая операция *f*

1685 basic oscillation
 d Grundschwingung *f*
 f oscillation *f* fondamentale
 r основное колебание *n*

1686 basic process; elementary process
 d Elementarvorgang *m*

f processus *m* élémentaire
r элементарный процесс *m*

* **basic rate system → 5603**

1687 basic sequential access method
d sequentielle Basiszugriffsmethode *f*
f méthode *f* d'accès séquentielle fondamentale
r базисный последовательный метод *m* доступа

1688 basic speed
d Grundgeschwindigkeit *f*
f vitesse *f* de base
r расчётное быстродействие *n*

1689 basic symbol
d Grundsymbol *n*
f symbole *m* fondamental; symbole de base
r основной символ *m*

1690 basic synchronous tools
d Basissynchronwerkzeuge *npl*
f outils *m* de synchronisation basés
r базисные синхронные инструменты *mpl*

1691 basic system
d Basissystem *n*
f système *m* de base
r базовая система *f*

1692 basic teleprocessing access method
d Datenfernverarbeitungs-Zugriffsmethode *f*
f méthode *f* d'accès de base au traitement à distance
r базисный телекоммуникационный метод *m* доступа

1693 basic time
d Grundzeit *f*
f temps *m* de base
r основное время *n*

1694 basic transmission unit
d Basisübertragungseinheit *f*
f unité *f* de transmission de base
r базисный блок *m* передачи

1695 basic unit
d Grundbaugruppe *f*
f unité *f* de base
r основная часть *f*

* **basis → 1646**

1696 basis for n-dimensional state space
d System *n* der Basisvektoren eines n-dimensionalen Zustandsraumes; Basis *f*
eines n-dimensionalen Zustandsraumes
f base *f* des espaces d'état de dimension n; système *m* des espaces d'état de dimension n
r базис *m* n-мерного пространства состояний

1697 basis function
d Basisfunktion *f*; Fundamentalfunktion *f*
f fonction *f* de base
r базисная функция *f*; фундаментальная функция

1698 basis solution
d Basislösung *f*
f solution *f* de base
r базисное решение *n*

1699 basis vector
d Basisvektor *m*
f vecteur *m* de base
r базисный вектор *m*

1700 batching
d Chargierung *f*; Dosierung *f*; Zumessung *f*
f dosage *m*; chargement *m*
r дозирование *n*

1701 batch meter; dosimeter
d Dosierer *m*; Dosiermessgerät *n*
f doseur *m*; dosimètre *m*; appareil *m* doseur
r дозатор *m*; дозирующее устройство *n*; дозиметр *m*

1702 batch operation
d Reihenbetrieb *m*; Satzbetrieb *m*
f opération *f* discontinue
r периодическая операция *f*

1703 batch process
d Satzverfahren *n*; Reihenverfahren *n*
f procédé *m* discontinu
r периодический процесс *m*

1704 batch processing
d Stapelverarbeitung *f*; schubweise Verarbeitung *f*; blockweise Verarbeitung
f traitement *m* en groupes; traitement par étapes
r обработка *f* партии

1705 batch pulse
d Vorwahlimpuls *m*
f impulsion *f* de présélection
r импульс *m* предварительной установки

1706 batch reactor
d diskontinuierlich arbeitender Reaktor *m*

f réacteur *m* discontinu
r реактор *m* периодического действия

1707 batch system
d diskontinuierliches System *n*
f système *m* à fonctionnement discontinu
r периодическая система *f*

1708 batch testing
d Reihenprüfung *f*
f test *m* de série
r групповые испытания *npl*

1709 batchwise
d chargenweise
f par charges périodiques
r периодический

1710 beam control
d Strahleinstellung *f*
f commande *f* du faisceau
r управление *n* лучом

1711 beam coupling
d Strahlenkopplung *f*
f couplage *m* à faisceaux
r лучевая связь *f*

1712 beam deflection
d Strahlablenkung *f*
f déviation *f* du faisceau
r отклонение *n* луча

1713 beam forming
d Strahlenformierung *f*
f formation *f* du faisceau
r лучеобразование *n*

1714 beam intensity
d Strahlenintensität *f*
f intensité *f* du faisceau
r интензивность *f* луча

* **beam pen** → 7256

1715 beam-scanning method
d Strahlabtastmethode *f*
f méthode *f* d'exploration par faisceau;
méthode de balayage par faisceau
r растровый метод *m*

* **beat** → 1717

1716 beat frequency
d Schwebungsfrequenz *f*
f fréquence *f* de battement
r частота *f* биений

1717 beat[ing]; pulsing; pulsation
d Takt *m*; Schlagen *n*; Schwebung *f*;
Impulsgabe *f*; Pulsen *n*; Pulsation *f*
f battement *m*; pulsation *f*; flottement *m*
r биение *n*; пульсация *f*

**1718 before-threshold behaviour;
below-threshold laser state;
below-threshold laser mode**
d Unterschwellenlaserzustand *m*;
Unterschwellenlasermode *m*
f régime *m* subliminal de laser; mode *m* de
laser sous-seuil
r допороговый режим *m* работы лазера

1719 begin *v*
d einsetzen; einleiten
f amorcer; provoquer
r вводить в действие; пускать

1720 behaviour; way of behaviour
d Verhalten *n*; Verhaltensweise *f*
f conduite *f*
r поведение *n*

1721 behaviour analysis of automaton
d Verhaltensanalyse *f* eines Automaten
f analyse *f* de comportement d'un automate
r поведенческий анализ *m* автомата

* **behaviour during breakdown** → 5365

1722 behaviour of force in the joint
d Kraftverlauf *m* im Gelenk
f allure *f* de force dans le joint
r характеристика *f* усилий в шарнире

1723 behaviour of motion
d Bewegungsverhalten *n*
f comportement *m* de mouvement
r характеристика *f* движений

1724 behaviour strategy
d Verhaltensstrategie *f*
f béhaviourisme *m*
r стратегия *f* поведения

1725 bell character
d Warnzeichen *n*
f caractère *m* de sonnerie
r символ *m* сигнализации

1726 bell-shaped curve
d Glockenkurve *f*; Gausskurve *f*
f courbe *f* de Gauss; courbe en forme de
cloche
r кривая *f* распределения Гауса

* **below-threshold laser mode** → 1718

* **below-threshold laser state** → 1718

1727 **belt feed**
  d Förderbandzuführung *f*
  f alimentation *f* par convoyeur à bande
  r подача *f* на ленточном транспортере

1728 **benchboard; operation console**
  d Steuerpult *n*; Operationssteuerpult *n*
  f pupitre *m* de commande; console *f*
    d'opération
  r пульт *m* управления

1729 **bench-scale equipment**
  d Labormaßstabsausrüstung *f*; kleintechnische
    Ausrüstung *f*
  f installation *f* pilote; installation de laboratoire
  r оборудование *n* лабораторного масштаба

1730 **Bernoulli equation**
  d Bernoullische Gleichung *f*
  f équation *f* de Bernoulli
  r уравнение *n* Бернули

1731 **Bessel function**
  d Bessel-Funktion *f*
  f fonction *f* de Bessel
  r функция *f* Бесселя

1732 **bias control**
  d Vorspannungsregelung *f*
  f réglage *m* de polarisation
  r регулировка *f* смещения

1733 **bias distortion**
  d unsymmetrische Verzerrung *f*; einseitige
    Verzerrung
  f distorsion *f* dissymétrique; distorsion biaise
  r несимметричное искажение *n*

1734 **biased flip-flop**
  d vorgespannte Kippschaltung *f*;
    Kippschaltung *f* mit Vorspannung
  f basculeur *m* polarisé
  r ждущий мультивибратор *m*; запертый
    мультивибратор

1735 **bias error; systematic error**
  d systematischer Fehler *m*
  f erreur *f* systématique
  r систематическая ошибка *f*

1736 **bias storage elements for program
    registers**
  d Biaxspeicherelemente *npl* für
    Programmspeicher
  f éléments *mpl* d'enregistrement biax pour
    programmeurs
  r биакс-накопительные элементы *mpl* для
    программных регистров

1737 **bias theorem**
  d Verschiebungssatz *m*
  f théorème *m* de la dérive
  r теорема *f* смещения

1738 **bias winding**
  d Vorspannungswicklung *f*
  f bobinage *m* de polarisation
  r обмотка *f* смещения

1739 **biconical taper**
  d bikonischer Taper *m*; Bitaper *m*
  f dispositif *m* à raccord progressif biconique
  r биконическое согласующее устройство *n*

1740 **bidecimal code**
  d directer Kode *m*; dezimal-binärer Kode
  f code *m* direct; code binaire-décimal
  r двоично-десятичный код *m*

1741 **bidimensional optical switch**
  d zweidimensionaler optischer Schalter *m*
  f commutateur *m* optique bidimensionnel
  r двухмерный оптический переключатель *m*

1742 **bidirectional coupler**
  d bidirektionaler Koppler *m*
  f coupleur *m* bidirectionnel
  r двунаправленный элемент *m* связи

1743 **bidirectional pulses**
  d Impulsfolge *f* mit positiven und negativen
    Impulsen
  f impulsions *fpl* bidirectionnelles
  r двунаправленные импульсы *mpl*

1744 **bidirectional transducer**
  d Zweirichtungswandler *m*; zweiseitiger
    Geber *m*
  f transducteur *m* bidirectionnel; capteur *m*
    bidirectionnel
  r двунаправленный датчик *m*

1745 **bilateral Laplace transformation; two-
    sided Laplace transformation**
  d zweiseitige Laplace-Transformation *f*
  f transformation *f* de Laplace bilatérale;
    transformation de Laplace à deux cotés
  r двухстороннее преобразование *n* Лапласа;
    прямое и обратное преобразование
    Лапласа

1746 **bilateral servosystem**
  d bilaterales Servosystem *n*

f servosystème *m* bilatéral
r симметрическая сервосистема *f*

1747 bilateral switching element
d bilaterales Schaltelement *n*
f élément *m* de commutation bilatéral
r переключательный элемент *m* на два
     направления

1748 bilateral transducer
d bilateraler Wandler *m*
f trasducteur *m* bilatéral
r двусторонный преобразователь *m*

1749 bilinear form
d Bilinearform *f*
f forme *f* bilinéaire
r билинейная форма *f*

1750 binaries
d Binärwerte *mpl*
f valeurs *fpl* binaires
r двоичные значения *npl*

1751 binary arithmetic computer
d binärer arithmetischer Rechner *m*
f calculateur *m* arithmétique binaire
r двойчное арифметическое
     вычислительное устройство *n*

1752 binary automaton
d binärer Automat *m*
f automate *m* binaire
r двойчный автомат *m*

1753 binary chain
d Binärkette *f*
f chaîne *f* binaire
r двоичная цепь *f*

1754 binary code
d Binärkode *m*
f code *m* binaire
r двоичный код *m*

  * **binary-coded number → 4524**

1755 binary command; binary signal
d binäres Signal *n*; binärer Befehl *m*
f signal *m* binaire; commande *f* binaire
r двоичный сигнал *m*; двоичная команда *f*

1756 binary data transmission technique
d Übertragungstechnik *f* binärer Daten
f technique *f* de transmission des données
     binaires
r техника *f* передачи двоичных данных

  * **binary-decimal conversion → 1785**

1757 binary digit; bit
d Binärziffer *f*; Bit *n*; binäre Ziffer *f*
f chiffre *m* binaire; bit *m*; digit *m* binaire
r двоичная цифра *f*; бит *m*

1758 binary digital computer
d binärer Digitalrechner *m*
f calculatrice *f* digitale binaire
r двоичная вычислительная машина *f*

1759 binary element
d Binärelement *n*
f élément *m* binaire
r двоичный элемент *m*

1760 binary feed
d Zweistoffeinspritzung *f*;
     Zweistoffeinspeisung *f*
f alimentation *f* binaire
r двухкомпонентное питание *n*

1761 binary function; logic function
d Schaltfunktion *f*; logische Funktion *f*
f fonction *f* binaire; fonction logique
r двоичная функция *f*; логическая функция

1762 binary image
d Binärbild *n*
f image *f* binaire
r двоичное изображение *n*

1763 binary information
d binäre Information *f*
f information *f* binaire
r двоичная информация *f*

1764 binary information and control signal
d binäres Informations- und Steuersignal *n*
f signal *m* d'information binaire et signal de
     commande
r двоичный информационный и
     управляющий сигнал *m*

1765 binary look-up
d binäres Suchen *n*
f recherche *f* binaire
r двоичный поиск *m*

1766 binary notation
d binäre Zahlendarstellung *f*; binäre
     Schreibweise *f*
f notation *f* binaire
r представление *n* чисел в двоичной
     системе

1767 binary number system
d Binärsystem *n*; Dualsystem *n*

f système *m* binaire; système dual
r двоичная система *f* счисления

1768 binary operation
d binäre Operation *f*
f opération *f* binaire
r двоичная операция *f*

1769 binary output
d binärer Ausgang *m*
f sortie *f* binaire
r двоичный выход *m*

1770 binary point
d Binärkomma *n*
f virgule *f* binaire
r двоичная запятая *f*

1771 binary pulse-code modulation
d binäre PCM-Modulation *f*
f modulation *f* PCM binaire
r двоичная импульсно-кодовая модуляция *f*

1772 binary recording
d binäre Aufzeichnung *f*
f energistrement *m* binaire
r запись *f* в двоичном коде

1773 binary reflex-code
d Binär-Reflex-Kode *m*
f code *m* binaire réfléchi
r двоичный рефлексный код *m*

1774 binary representation
d Binärdarstellung *f*
f représentation *f* binaire
r двоичное представление *n*

1775 binary scale
d Binärskale *f*
f échelle *f* binaire
r двоичная шкала *f*

1776 binary scaling circuit
d binäre Skalenschaltung *f*
f circuit *m* d'échelle binaire
r двоичная пересчётная схема *f*

1777 binary search method
d binäres Suchverfahren *n*
f méthode *f* binaire de recherche
r двоичный метод *m* поиска

1778 binary sensor
d Binärgeber *m*; Binärsensor *m*
f capteur *m* binaire; senseur *m* binaire
r двоичный датчик *m*

* **binary signal** → 1755

1779 binary storage element
d binäres Speicherglied *n*
f organe *m* binaire de mémoire
r двоичный запоминающий элемент *m*

1780 binary structure
d binäre Struktur *f*
f structure *f* binaire
r двоичная структура *f*

1781 binary symbol
d binäres Symbol *n*
f symbole *m* binaire
r двоичный символ *m*

1782 binary symmatric channel
d symmetrischer Binärkanal *m*; binärer symmetrischer Kanal *m*
f canal *m* symétrique binaire
r двоичный симметричный канал *m*

1783 binary synchronous communication
d binäre synchrone Kommunikation *f*
f communication *f* binaire synchrone
r двоичная синхронная связь *f*

1784 binary synchronous control
d binäre synchrone Steuerung *f*
f commande *f* binaire synchrone
r двоичное синхронное управление *n*

1785 binary-[to-]decimal conversion
d Binär-Dezimalkonvertierung *f*; Binär-Dezimal-Umwandlung *f*
f conversion *f* binaire-décimale
r двоично-десятичное преобразование *n*

1786 binary touch sensor
d binärer Tastsensor *m*
f capteur *m* de toucher binaire; senseur *m* de toucher binaire
r двоичный тактильный сенсор *m*

1787 binary translation
d binäre Übertragung *f*
f transfert *m* binaire
r двоичное преобразование *n*

1788 binary weight
d Binärgewicht *n*
f poids *m* d'une position binaire
r двоичный вес *m*

1789 binding energy; linking energy
d Bindungsenergie *f*

f énergie f de liaison
r энергия f связи

1790 binding property
d Bindefähigkeit f
f pouvoir m agglutinant
r способность f к сцеплению

1791 Bingham model
d Binghamsches Modell n
f modèle m de Bingham
r модель f Бингама

1792 binomial coefficient
d Binomialkoeffizient m; Koeffizient m der
Binomialreihe; binomischer Koeffizient
f coefficient m binomial
r биномный коэффициент m

1793 binomial theorem
d Binomialsatz m
f loi f de distribution binomiale
r биноминальная теорема f

1794 biochemical sensor
d biochemischer Sensor m
f senseur m biochimique
r биохимический сенсор m

*   **biocomputer → 1797**

1795 biocybernetics; bionics
d Biokybernetik f; Bionik f
f bionique f
r биокибернетика f; бионика f

1796 biofilm reactor
d Biofilmreaktor m
f bioréacteur m à film
r биофильмреактор m

1797 biological computer; biocomputer
d Biocomputer m
f ordinateur m biologique
r биокомпьютер m

1798 biological information signal
d biologisches Informationssignal n
f signal m d'information biologique
r биологический информационный сигнал n

1799 biological sensor
d biologischer Sensor m
f senseur m biologique
r биологический датчик m

1800 biologic simulation
d biologische Simulation f

f simulation f biologique
r биологическое моделирование n

1801 biologic system
d biologisches System n
f système m biologique
r биологическая система f

1802 biometric identification systems
d biometrische Identifizierungssysteme npl
f systèmes mpl d'identification biométriques
r биометрические идентификационные
системы fpl

*   **bionics → 1795**

1803 bionic simulation
d bionische Simulation f
f simulation f bionique
r бионическое моделирование n

1804 bionic system
d bionisches System n; Bioniksystem n
f système m bionique
r бионическая система f

1805 biperiodical regime
d doppeltperiodischer Vorgang m
f régime m bipériodique
r двухпериодный режим m

1806 bipolar device
d Bipolarbaustein m; Bauelement n in bipolarer
Technik
f module m bipolaire
r биполярное устройство n

1807 bipolar integrated circuit
d Bipolarschaltkreis m
f circuit m intégré bipolaire
r биполярная интегральная схема f

**1808 bipolar microprogrammed
microcomputer**
d bipolarer mikroprogrammierter
Mikrorechner m
f microcalculateur m microprogrammé
bipolaire
r биполярный микропрограммированный
микрокомпьютер m

1809 bipolar technique
d Bipolartechnik f
f technique f bipolaire
r биполярная технология f

1810 biprocessor system
d Biprozessorsystem n

f système *m* biprocesseur
r бипроцессорная система *f*

1811 bistable circuit
   d bistabile Schaltung *f*
   f montage *m* bistable
   r схема *f* с двумя устойчивыми
      состояниями

1812 bistable device
   d bistabile Einrichtung *f*
   f dispositif *m* bistable
   r устройство *n* с двумя устойчивыми
      состояниями

1813 bistable element
   d bistabiles Element *n*
   f élément *m* bistable
   r бистабильный элемент с двумя
      устойчивыми состояниями

1814 bistable multivibrator
   d bistabiler Multivibrator *m*
   f multivibrateur *m* bistable; multivibrateur
      symétrique
   r мультивибратор *m* с двумя устойчивыми
      состояниями

1815 bistable optical element
   d bistabiles optisches Element *n*
   f élément *m* optique bistable
   r бистабильный оптический элемент *m*

1816 bistable pulse relay
   d bistabiles Impulsrelais *n*
   f relais *m* bistable à impulsions
   r импульсное реле *n* с двумя устойчивыми
      состояниями

   * bit → 1757

1817 bit check
   d Bitkontrolle *f*
   f essai *m* de bit
   r контроль *m* по двоичным знакам

1818 bit density
   d Bitdichte *f*
   f densité *f* de bits
   r плотность *f* битов

1819 bit level
   d Bitniveau *n*; Bitstand *m*
   f niveau *m* de bit
   r уровень *m* двоичного сигнала

1820 bit rate
   d Bit-Übertragungsgeschwindigkeit *f*

f vitesse *f* de tranfert de bit
r скорость *f* передачи бита

1821 bit traffic
   d Bitverkehr *m*
   f trafic *m* de bits
   r поток *m* двоичной информации

1822 bivariate normal distribution
   d Normalverteilung *f* von zwei Größen
   f répartition *f* normale de deux grandeurs
   r двумерное нормальное распределение *n*

1823 black box
   d Blackbox *f*; schwarzer Kasten *m*
   f black-box *m*; boîte *f* noire
   r черный ящик *m*

1824 black-box method
   d Blackbox-Methode *f*; Blackbox-Analyse *f*
   f méthode *f* de boîte noire; analyse *f* à
      black-box
   r метод *m* черного ящика

1825 blackout pulse
   d Austastimpuls *m*
   f impulsion *f* d'extinction
   r затемняющий импульс *m*

1826 blank column detection device
   d Leerspaltensucher *m*
   f détecteur *m* de colonnes vierges
   r устройство *n* для детектирования
      свободных колонок

1827 blank cycle; idle cycle; null cycle
   d Leerlauf *m*; Leergang *m*
   f cycle *m* blanc
   r пустой цикл *m*; свободный цикл *m*

1828 blanking circuit; quenching circuit
   d Ausschaltkreis *m*; Löschkreis *m*
   f circuit *m* d'effacement; circuit de
      suppression; circuit d'extinction
   r схема *f* гашения

1829 blank instruction
   d Leerbefehl *m*; Verweisungsauftrag *m*
   f instruction *f* à vide; instruction de référence
   r команда *f* пропуска

1830 blast process; converter process
   d Windfrischverfahren *n*; Konverterverfahren *n*
   f affinage *m* au vent; affinage par soufflage
   r конвертерный процесс *m*

1831 blend *v*; mix *v*; mingle *v*
   d mischen; [ver]mengen; versetzen;
      durchmischen

    f mélanger; mêler; melaxer
    r смешивать

1832 blip
    d Echozeichen *n*; Echoanzeiger *m*;
       Radarecho *n*
    f blip *m*; pip *m*
    r отметка *f*; выброс *m*

1833 block
    d Block *m*; Glied *n*
    f bloc *m*; ordinogramme *m*
    r блок *m*; звено *n*

1834 block adaptation
    d Blockanpassung *f*
    f adaptation *f* de bloc
    r приспособление *n* блока

1835 block address
    d Blockadresse *f*
    f adresse *f* en bloc
    r адрес *m* блока

1836 block changing
    d Blockaustausch *m*
    f changement *m* de bloc
    r блочная передача *f*

1837 block check sequence
    d Blockprüffolge *f*
    f séquence *f* de vérification de bloc
    r последовательность *f* проверки блока

1838 block diagram; schematic diagram; functional block diagram
    d Blockschema *n*; Blockbild *n*;
       Blockschaltbild *n*; Blockzeichnung *f*;
       Funktionsdiagramm *n*
    f diagramme *m* fonctionnel; schéma *m*
       fonctionnel; organigramme *m*
    r блок-схема *f*; функциональная
       диаграмма *f*; функциональная блок-схема

1839 block diagram compiler
    d Blockbild-Compiler *m*
    f compilateur *m* en schéma bloc
    r компилятор *m* структурной схемы

1840 blocking; interlock[ing]
    d [gegenseitige] Verriegelung *f*; Sperrung *f*;
       Blockierung *f*
    f blocage *m*; couplage *m* interlock
    r взаимное запирание *n*; взаимная
       блокировка *f*

1841 blocking capacitor
    d Sperrkondensator *m*; Blockierkondensator *m*

    f condensateur *m* d'arrêt
    r блокировочный конденсатор *m*

1842 blocking characteristic
    d Sperrkennlinie *f*
    f caractéristique *f* d'arrêt
    r характеристика *f* запирающего слоя

1843 blocking circuit; interlock circuit
    d Blockstromkreis *m*; Sperrstromkreis *m*
    f circuit *m* de blocage; circuit de verrouillage
    r блокирующая цепь *f*

1844 blocking contact
    d Blockierkontakt *m*
    f contact *m* de verrouillage
    r блок-контакт *m*; запирающий контакт *m*

1845 blocking contactor
    d Sperrschütz *n*
    f contacteur *m* de verrouillage
    r запирающий контактор *m*

1846 blocking direction
    d Sperrichtung *f*
    f sens *m* du blocage; sens d'arrêt
    r запирающее направление *n*

1847 blocking generator; blocking oscillator
    d Block[ier]generator *m*; Sperrschwinger *m*;
       Sperrkippsender *m*
    f oscillateur *m* surcouplé; oscillateur de
       blocage
    r блокинг-генератор *m*; запирающий
       генератор *m*

1848 blocking impulse; disabling pulse
    d Blockierimpuls *m*
    f impulsion *f* de blocage
    r запирающий импульс *m*; блокирующий
       импульс *m*

1849 blocking magnet
    d Sperrmagnet *m*
    f aimant *m* de blocage
    r запирающий электромагнит *m*

1850 blocking order
    d Blockierbefehl *m*
    f instruction *f* de blocage; ordre *m* de blocage
    r блокирующая команда *f*

*   **blocking oscillator → 1847**

1851 blocking period
    d Sperrzyklus *m*
    f temps *m* de blocage
    r период *m* блокировки

1852 **blocking relay; guard relay; interlocking relay**
  d Sperrelais *n*; Verriegelungsrelais *n*; Halterelais *n*
  f relais *m* de verrouillage; relais de blocage
  r блокирующее реле *n*

1853 **blocking resistance**
  d Sperrwiderstand *m*
  f résistance *f* de blocage
  r запирающее сопротивление *n*

1854 **blocking signal**
  d Blocksignal *n*
  f signal *m* de blocage
  r сигнал *m* блокировки

1855 **blocking state**
  d Sperrzustand *m*
  f état *m* bloqué
  r запрещенное состояние *n*

1856 **block name**
  d Blockname *m*
  f nom *m* de bloc
  r название *n* блока

  * **block of instruction** → 6730

1857 **block of microelectronics**
  d Mikroelektronikblock *m*
  f bloc *m* de micro-électronique
  r микроэлектронный блок *m*

1858 **block-oriented random access**
  d blockorientierter wahlfreier Zugriff *m*
  f accès *m* libre orienté sur le bloc
  r блочно-ориентированный свободный доступ *m*

1859 **block-oriented simulation**
  d blockorientierte Simulation *f*
  f simulation *f* orientée sur les blocs
  r блочное моделирование *n*

1860 **block-oriented structure**
  d blockorientierte Struktur *f*
  f structure *f* orientée sur les blocs
  r блочная структура *f*

1861 **block size**
  d Blocklänge *f*
  f longueur *f* de bloc
  r длина *f* блока

1862 **blow-down system**
  d Entspannungssystem *n*; Abblasesystem *n*
  f système *m* de détente; système de
  décompression; système d'évacuation
  r система *f* сбрасывания давления

1863 **blower efficiency**
  d Gebläsewirkungsgrad *m*
  f rendement *m* de la soufflerie; rendement du ventilateur
  r коэффициент *m* полезного действия вентилятора

1864 **Bode diagram**
  d Bode-Diagramm *n*; Amplitudengang *m*
  f diagramme *m* de Bode
  r диаграмма *f* Боде

1865 **Bode method**
  d Bode-Methode *f*; Stabilitätsuntersuchung *f* nach Bode
  f méthode *f* de Bode; analyse *f* de stabilité de Bode
  r метод *m* Боде

1866 **body-centered**
  d raumzentriert
  f centré dans l'espace
  r объёмноцентрированный

1867 **body force**
  d Volumenkraft *f*; Raumkraft *f*
  f force *f* volumétrique
  r объёмная сила *f*

1868 **boiler efficiency**
  d Kesselwirkungsgrad *m*
  f rendement *m* de chaudière
  r коэффициент *m* использования котлоагрегата

1869 **boiler testing**
  d Kesselkontrolle *f*; Kesseluntersuchung *f*
  f visite *f* de chaudière
  r котлонадзор *m*

1870 **bolometric instrument**
  d bolometrisches Instrument *n*; Bolometer *n*
  f bolomètre *m*
  r болометрический прибор *m*; болометр *m*

1871 **Boltzmann constant**
  d Boltzmann-Konstante *f*
  f constante *f* de Boltzmann
  r постоянная *f* Больцмана

1872 **Boltzmann's superposition principle**
  d Boltzmannsches Superpositionsprinzip *n*
  f principe *m* de superposition de Boltzmann
  r принцип *m* суперпозиции Больцмана

1873 **bond failure**
 d Lösen *n* einer Bondverbindung
 f détérioration *f* de connexion
 r нарушение *n* соединения

1874 **bonding speed**
 d Bondgeschwindigkeit *f*
 f productivité *f* de machine pour soudage à
    thermocompression
 r производительность *f* установки
    термокомпрессионной сварки

 * **Boolean algebra** → 643

1875 **Boolean calculation**
 d Boolesche Rechnungsart *f*
 f calcul *m* de Boole
 r вычисление *n* методом булевой алгебры

1876 **Boolean function**
 d Boolesche Funktion *f*
 f fonction *f* booléenne; fonction de Boole
 r булевая функция *f*

1877 **Boolean processor**
 d Boolescher Prozessor *m*
 f processeur *m* booléen
 r процессор *m* обработки булевых функций

1878 **Boolean representation**
 d Boolesche Darstellung *f*
 f représentation *f* booléenne
 r описание *n* на языке булевых функций

1879 **Boolean value**
 d Boolescher Wert *m*
 f valeur *f* booléenne
 r булевая величина *f*

1880 **Boolean variable**
 d Boolesche Variable *f*
 f variable *f* booléenne
 r булевая переменная *f*

1881 **boost** *v*
 d verstärken; aufladen
 f amplifier; renforcer
 r усиливать; усилить

1882 **booster**
 d Spannungsverstärker *m*
 f amplificateur *m* de tension
 r усилитель *m* напряжения

1883 **booster mechanism**
 d Verstärkermechanismus *m*;
    Verstärkeranlage *f*
 f mécanisme *m* d'asservissement

 r бустерный механизм *m*

1884 **boost-pressure controller**
 d Ladedruckregler *m*
 f régulateur *m* de suralimentation
 r регулятор *m* давления питания

1885 **boost-pressure gauge**
 d Landedruckmesser *m*
 f manomètre *m* de suralimentation
 r манометр *m* давления

1886 **bootstrap** *v*
 d aufladen; anheben
 f charger; hausser
 r загружать[ся] автоматически

1887 **bootstrap integrator**
 d Integrator *m* mit parametrischer
    Fehlerkompensation
 f intégrateur *m* à compensation paramétrique
    d'erreur
 r интегратор *m* с параметрической
    компенсацией погрешности

1888 **bootstrapping**
    (of program)
 d Ladeprogrammaufbautechnik *f*
 f technique *f* d'établissement de programme
    chargeur
 r автоматическая загрузка *f* программ

1889 **bootstrapping circuitry**
 d Aufladeschaltung *f*
 f montage *m* de chargement
 r средства *npl* управления начальной
    загрузкой

1890 **bottom voltage**
 d untere Spannung *f*
 f tension *f* inférieure
 r нижний уровень *m* напряжения

1891 **boundary; limit**
 d Grenzwert *m*
 f limite *f*
 r предел *m*

1892 **boundary conditions**
 d Randbedingungen *fpl*; Grenzbedingungen *fpl*
 f conditions *fpl* limites
 r предельные условия *npl*

1893 **boundary element**
 d Randelement *n*
 f élément *m* marginal
 r краевой элемент *m*

1894 **boundary equation**
  d Randwertproblem *n*
  f problème *m* de valeur limite
  r краевая задача *f*

1895 **boundary maximum**
  d Randmaximum *n*; Maximum *n* auf dem
     Rande eines Intervalls
  f maximum *m* limité
  r максимум *m* на границе области

1896 **boundary minimum**
  d Randminimum *n*; Minimum *n* auf dem
     Rande eines Intervalls
  f minimum *m* limité
  r минимум *m* на границе области

1897 **boundary parameter value**
  d Parameterrandwert *m*
  f valeur *f* limite de paramètre
  r граничное значение *n* параметра

1898 **boundary perturbation**
  d Grenzstörung *f*
  f perturbation *f* aux limites
  r предельная помеха *f*

1899 **boundary utility theory; marginal utility
     theory**
  d Grenznutzentheorie *f*
  f théorie *f* de limite d'utilité
  r теория *f* предельной полезности

1900 **boundary value problem**
  d Randwertproblem *n*
  f problème *m* de valeur limite
  r краевая задача *f*

1901 **boundary value theorem; threshold
     theorem**
  d Grenzwertsatz *m*
  f théorème *m* de valeur limite
  r теорема *f* о предельном значении

1902 **bounded**
  d beschränkt
  f limité
  r ограниченный

1903 **bounded input**
  d begrenzter Eingang *m*; amplitudenbegrenzter
     Eingang; beschränkte Eingangsgröße *f*;
     begrenztes Eingangssignal *n*
  f entrée *f* limitée d'amplitude; signal *m* limité
     d'entrée d'amplitude
  r ограниченный вход *m*

1904 **bounded output**

  d begrenzter Ausgang *m*; amplitudenbegrenzter
     Ausgang; beschränkte Ausgangsgröße *f*;
     begrenztes
     Ausgangssignal *n*
  f sortie *f* limitée d'amplitude; signal *m* limité
     de sortie d'amplitude; sortie à amplitude
     limitée
  r ограниченный выход *m*

  * **boundless** → 12626

1905 **bound pair**
  d Grenzenpaar *n*; Indexgrenzenpaar *n*
  f paire *f* de bornes; couple *m* de bornes
  r граничная пара *f*

1906 **Box's complex method**
  d Komplexverfahren *n* nach Box
  f méthode *f* de Box
  r комплексный метод *m* Бокса

1907 **brake dynamometer**
  d Bremsdynamometer *n*
  f dynamomètre *m* à frein
  r тормозной динамометр *m*

1908 **braking element**
  d Bremselement *n*
  f élément *m* de freinage
  r тормозной элемент *m*

1909 **branch address**
  d Verzweigungsadresse *f*
  f adresse *f* d'aiguillage
  r адрес *m* разветвления

1910 **branch condition**
  d Verzweigungsbedingung *f*
  f condition *f* de branchement
  r условие *n* ветвления

1911 **branch current**
  d Teilstrom *m*
  f courant *m* de branchement
  r поток *m* разветвления

1912 **branch decision**
  d Verzweigungsentscheidung *f*
  f décision *f* de branchement
  r анализ *m* условия ветвления

1913 **branching**
  d Verzweigung *f*; Abzweigen *n*
  f branchement *m*; embranchement *m*;
     aiguillage *m*
  r разветвление *n*

1914 **branching control**
  d Verzweigungssteuerung *f*

f commande *f* d'aiguillage; commande de dérivation
r управление *n* при разветвлении

1915 branching index
d Verzweigungsindex *m*
f indice *m* de ramification
r индекс *m* разветвления

1916 branching instruction
d Verzweigungsbefehl *m*
f instruction *f* de branchement
r команда *f* разветвления

1917 branching method
d Verzweigungsmethode *f*
f méthode *f* de branchement
r метод *m* разветвления

1918 branching process
d Verzweigungsprozess *m*
f procédé *m* de branchement
r ветвящийся процесс *m*

1919 branching program
d Programm *n* des bedingten Übergangs
f programme *m* de transfert conditionnel
r программа *f* условного перехода

1920 branch on zero
d Verzweigen *n* bei Null
f aiguillage *m* par zéro
r разветвление *n* по нулю

1921 branch point
d Verzweigungspunkt *m*
f embranchement *m*
r точка *f* разветвления; узловая точка

1922 Braun electrometer
d Braunsches Elektrometer *n*
f électromètre *m* de Braun
r электрометр *m* Брауна

1923 breadbord circuit
d Brettschaltung *f*
f montage *m* expérimental d'un circuit
r опытный монтаж *m* схемы

* **break** → 6948

* **break** *v* → 6933

1924 break contact; resting contact
d Öffnungskontakt *m*; Ruhekontakt *m*
f contact *m* à ouverture; contact rupteur
r размыкающийся контакт *m*

* **breakdown** → 5364

1925 breakdown puncture; slugging
d Durchschlag *m*
f poinçon *m*; perçoir *m*; broche *f*
r пробой *m*

* **breakdown voltage** → 4401

1926 break impulse
d Öffnungsimpuls *m*
f impulsion *f* d'ouverture
r импульс *m* прерывания

* **breaking** → 6948

1927 breaking capacity
d Abschaltleistung *f*
f puissance *f* de rupture
r разрывная мощность *f*

1928 breaking capacity of a circuit-breaker
d Schaltleistung *f* eines Ausschalters
f pouvoir *m* de coupure d'un disjoncteur
r разрывная мощность *f* выключателя

1929 breaking capacity of relay contacts
d Kommutationsfähigkeit *f* von Relaiskontakten
f pouvoir *m* de commutation des contacts de relais
r коммутационная способность *f* контактов

1930 breaking down temperature
d Abbautemperatur *f*; Zersetzungstemperatur *f*
f température *f* de décomposition
r температура *f* распада

1931 breaking time; turn-off time
d Abschaltzeit *f*; Ausschaltzeit *f*
f durée *f* de coupure; temps *m* de mise au repos
r время *n* выключения

1932 breaking unit
d Ausschalteinheit *f*
f élément *m* de coupure
r выключающий блок *m*

1933 break length
d Ausschaltstrecke *f*
f longueur *f* de coupure
r длина *f* разрыва

1934 break-make ratio; duty ratio
d Tastverhältnis *n*
f rapport *m* de duitage
r скважность *f*

1935 **break-point instruction**
  d Zwischenstoppbefehl *m*
  f instruction *f* d'arrêt de contrôle
  r команда *f* контрольного останова

  *   **break signal** → 6946

1936 **bridge circuit; bridge connection**
  d Brückenschaltung *f*
  f circuit *m* en pont; montage *m* en pont
  r мостовая схема *f*

  *   **bridge connection** → 1936

1937 **bridge detector**
  d Brückendetektor *m*; Stromanzeiger *m* in Brückenschaltung
  f détecteur *m* en pont
  r мостовой указатель *m*

1938 **bridge equilibrium**
  d Brückengleichgewicht *n*
  f équilibre *m* du pont
  r равновесие *n* моста

1939 **bridge feedback**
  d Messbrückenrückkopplung *f*
  f réaction *f* du pont de mesure
  r мостовая обратная связь *f*

1940 **bridge measurements**
  d Brückenmessungen *fpl*
  f mesures *fpl* par pont
  r измерения *npl* по мостовой схеме

1941 **bridge method**
  d Brückenmethode *f*
  f méthode *f* de pont
  r метод *m* моста

1942 **bridge structure**
  d Brückenstruktur *f*
  f structure *f* de pont
  r мостовая структура *f*

1943 **bridge transition**
  d Reihenparallelschaltung *f* mit Brückenschaltung
  f transition *f* série-parallèle à l'aide d'un pont
  r мостовой переход *m*

1944 **bridging contacts**
  d Überbrückungskontakte *mpl*
  f contacts *mpl* de court-circuit
  r шунтирующие контакты *mpl*

1945 **brightness control; luminance control**
  d Leuchtdichteeinstellung *f*;

Helligkeitseinstellung *f*
  f commande *f* de la luminosité; commande de la luminance
  r регулировка *f* диапазона яркости

1946 **brightness controller**
  d Helligkeitsregler *m*
  f dispositif *m* de réglage de la luminosité
  r регулятор *m* яркости

1947 **broadband amplifier**
  d Breitbandverstärker *m*
  f amplificateur *m* à large bande
  r широкополосный усилитель *m*

1948 **broadband modulation**
  d Breitbandmodulation *f*
  f modulation *f* à large bande
  r широкополосная модуляция *f*

1949 **broadband optical transmission system**
  d optisches Breitband-Übertragungssystem *n*
  f système *m* de transmission optique à large bande
  r оптическая широкополосная система *f* передачи

1950 **broadband stationary noise**
  d stationäres Breitbandrauschen *n*
  f bruit *m* stationnaire à large bande
  r широкополосный стационарный шум *m*

1951 **broken-line approximation**
  d stückweise lineare Approximation *f*
  f approximation *f* par degrés linéaires
  r частично линейная аппроксимация *f*

1952 **buckling load**
  d Knickbelastung *f*
  f chargement *m* de flambage
  r нагрузка *f* при продольном изгибе

1953 **buckling point**
  d Knickpunkt *m*
  f point *m* de flexion
  r точка *f* изгиба

1954 **buckling test**
  d Knickversuch *m*
  f essai *m* de flambage
  r испытание *n* на продольный изгиб

1955 **Bueche model**
  d Buechesches Modell *n*
  f modèle *m* de Bueche
  r модель *f* Бики

1956 **buffer**
  d Pufferspeicher *m*

f mémoire *f* temporaire; mémoire
intermédiaire
r буфер *m*

1957 buffer amplifier
  d Pufferverstärker *m*; Trennverstärker *m*
  f amplificateur *m* tampon
  r буферный усилитель *m*

1958 buffer cascade; buffer stage
  d Pufferstufe *f*; Trennstufe *f*
  f étage *m* intermédiaire; cascade *f* à tampon
  r буферный каскад *m*

1959 buffer circuit
  d Pufferkreis *m*
  f circuit *m* intermédiaire
  r буферная цепь *f*

1960 buffered communication adapter
  d Adapter *m* für gepufferte Übertragung
  f adaptateur *m* pour communication
     tamponnée
  r адаптер *m* передачи данных с
     буферизацией

1961 buffer function
  d Pufferfunktion *f*
  f fonction *f* intermédiaire
  r согласующая функция *f*

**1962 buffer memory; buffer storage; buffer
     store**
  d Puffer *m*; Pufferspeicher *m*;
     Pufferspeicher *m*; Übergangsspeicher *m*
  f mémoire *f* temporaire; mémoire
     intermédiaire
  r промеждуточное запоминающее
     устройство *n*; буфернос запоминающее
     устройство *n*

1963 buffer module
  d Pufferbaustein *m*; Puffermodul *m*
  f module *m* intermédiaire; organe *m* tampon
  r буферный модуль *m*

  *   **buffer stage → 1958**

  *   **buffer storage → 1962**

  *   **buffer store → 1962**

  *   **building-block concept → 1965**

1964 building-block gripper
  d Baukastengreifer *m*
  f grappin *m* de construction par blocs
  r схват *m* модульного типа

**1965 building-block principle; building-block
     concept**
  d Baukastenprinzip *n*
  f principe *m* de block-éléments
  r принцип *m* составных элементов;
     блочный принцип констуирования

**1966 building-block system; construction unit
     system**
  d Baukastensystem *n*; Baueinheitensystem *n*
  f système *m* de construction par blocs; système
     d'unité de construction
  r агрегатная система *f*; система готовых
     блоков; модульная система

**1967 build[ing]-up time; transient period; rise
     time**
  d Anlaufzeit *f*; Anstiegszeit *f*; Einschwingzeit *f*
  f période *f* transitoire; durée *f* d'établissement;
     temps *m* de croissance
  r время *n* нарастания; характеристическое
     время

  *   **build-up time → 1967**

1968 built-in check
  d eingebaute Prüfung *f*; eingebaute Kontrolle *f*
  f contrôle *m* incorporé; essai *m* incorporé
  r схемный контроль *m*

1969 built-in gauge line
  d eingebaute Eichleitung *f*
  f ligne *f* d'étalonnage incorporée
  r встроенная эталонная линия *f*

1970 built-in manipulator
  d eingebauter Manipulator *m*
  f manipulateur *m* incorporé
  r встроенный манипулятор *m*

1971 built-in microprocessor
  d eingebauter Mikroprozessor *m*
  f microprocesseur *m* incorporé
  r встроенный микропроцессор *m*

1972 built-in minicomputer
  d eingebauter Minirechner *m*
  f ordinateur *m* petit incorporé
  r встроенный миникомпьютер *m*

1973 built-in power fail protection
  d eingebauter Spannungsausfallschutz *m*
  f protection *f* contre manque de tension
     incorporée
  r встроенная защита *f* от отказов питания

1974 built-in protection mechanism
  d eingebaute Schutzvorrichtung *f*

f dispositif *m* de protection incorporé
r внутренный механизм *m* защиты

1975 built-in repetition automatism
   d eingebaute Weiderholautomatik *f*
   f système *m* incorporé de répétition automatique
   r встроенная автоматика *f* повторения

1976 bulk data processing
   d Massendatenverarbeitung *f*
   f traitement *m* de grande capacité des données
   r массовая обработка *f* данных

1977 bulk factor
   d Schüttfaktor *m*; Füllfaktor *m*
   f coefficient *m* de déversement; facteur *m* de remplissage
   r коэффициент *m* уплотнения

1978 bulk information
   d große Informationsmenge *f*
   f information *f* de grande capacité
   r большое количество *n* информации

   * **bulk memory** → 1981

1979 bulk processing
   d Massenverarbeitung *f*
   f traitement *m* de masses
   r обработка *f* массивов информации

1980 bulk shielding
   d Kompaktabschirmung *f*; kompakte Abschirmung *f*
   f blindage *m* compact
   r компактный экран *m*

1981 bulk store; bulk memory
   d Großraumspeicher *m*
   f mémoire *f* à grande capacité
   r накопитель *m* большой ёмкости

1982 bulk transmission of data
   d Übertragung *f* großer Datenmengen
   f transfert *m* d'un grand nombre de données; transmission *f* d'un grand nombre de données
   r передача *f* большого объёма данных

1983 Burgers-Frenkel model
   d Burgers-Frenkelsches Modell *n*
   f modèle *m* de Burgers-Frenkel
   r модель *f* Бургерса-Френкеля

1984 buried contact
   d vergrabener Kontakt *m*
   f contact *m* caché
   r скрытый контакт *m*

1985 burst gating circuit
   d Chrominanz-Austastkreis *m*
   f porte *f* déclenchant le signal de synchronisation de la sous-porteuse de chrominance
   r схема *f* стробирования сигнала вспышки

1986 burst mode
   d Einpunktbetrieb *m*
   f régime *m* à coups
   r монопольный режим *m*

1987 bus clock
   d Bustakt *m*; Bussynchronisationstakt *m*
   f rythme *m* de bus
   r синхронизация *f* шины

1988 bus connection
   d Busverbindung *f*
   f connexion *f* de bus
   r соединение *n* шин

1989 bus control circuit
   d Bussteuerschaltung *f*; Bussteuerschaltkreis *m*
   f circuit *m* de commande d'un bus
   r контроллер *m* шины

1990 bus exchange signal
   d Busvermittlungssignal *n*
   f signal *m* d'échange d'un bus
   r сигнал *m* управления коммутацией шин

1991 bushing transformer
   d Durchführungsumformer *m*; Durchgangstransformator *m*
   f transformateur *m* de traversée
   r встроенный трансформатор *m*

1992 bus logic
   d Buslogik *f*
   f logique *f* de bus
   r шинная логика *f*

1993 bus operation
   d Busoperation *f*
   f opération *f* de bus
   r шинная операция *f*

1994 bus-organized structure
   d busorganisierte Struktur *f*
   f structure *f* organisée bus
   r шинная структура *f*

1995 bus-out
   d Ausgangssammelleitung *f*; Datenausgangsleitung *f*
   f barre *f* de sortie; sortie *f* commune
   r выходная шина *f* канала данных

1996 **bus priority control**
  d Busprioritätssteuerung *f*
  f commande *f* de priorité d'un bus
  r приоритетное управление *n* шиной

1997 **bus-structured system**
  d busstrukturiertes System *n*
  f système *m* structuré bus
  r система *f* с магистральной структурой

1998 **bus system**
  d Bussystem *n*
  f système *m* de bus
  r система *f* шины

1999 **bus vectored interrupt logic**
  d Buslogik *f* zur Behandlung vektorisierter
       Unterbrechungen
  f logique *f* de bus à vecteur d'interruption
  r логика *f* обработки векторного
       прерывания

2000 **busy relay**
  d Besetztrelais *n*; Belegungsrelais *n*
  f relais *m* d'occupation; relais de ligne occupée
  r реле *n* занятия

2001 **busy signal**
  d Besetztsignal *n*
  f signal *m* d'occupation
  r сигнал *m* занятости

2002 **butt contact; pressure contact**
  d Druckkontakt *m*
  f contact *m* à pression
  r нажимный контакт *m*

2003 **bypass** *v*; **shunt** *v*
  d umgehen; übergehen; überbrücken
  f contourner; dépasser; shunter
  r обходить; шунтировать

2004 **bypass circuit**
  d Umgehungsschaltung *f*
  f circuit *m* de dépassement
  r шунтирующая схема *f*

2005 **bypass system**
  d Umgehungssystem *n*
  f système *m* de dérivation; système by-pass
  r обходная система *f*

2006 **byte**
  d Byte *n*
  f byte *m*
  r байт *m*

2007 **byte device**

  d Gerät *n* mit byteweiser Übertragung
  f dispositif-byte *m*
  r устройство *n* с байтовый структурой

2008 **byte mode**
  d Bytebetrieb *m*; byteweiser Betrieb *m*
  f régime *m* à octets
  r побайтовый режим *m* [обмена]

2009 **byte of data; data byte**
  d Datenbyte *n*
  f byte *m* de données
  r байт *m* данных

* **byte of zero** → 12959

2010 **byte-oriented computer**
  d byteorientierter Rechner *m*; Byterechner *m*
  f calculateur *m* orienté octets
  r компьютер *m* с побайтовой обработкой
       данных

2011 **byte rate**
  d Bytegeschwindigkeit *f*
  f vitesse *f* de bytes
  r скорость *f* передачи по байтам

2012 **byte serial data**
  d Byte-Seriendaten *pl*
  f données *fpl* à représentation séquentielle des
       bytes
  r данные *pl* с последовательным
       представлением по байтам

2013 **byte size control**
  d Bytegrößensteuerung *f*
  f commande *f* de grandeur d'un byte
  r изменение *n* длины байта

C

2014 **cable distribution system**
  d Kabelverteilsystem *n*
  f réseau *m* de distribution par câble
  r кабельная распределительная система *f*

2015 **cache control register**
  d Cachespeicher-Steuerregister *n*
  f registre *m* de commande de cache
  r регистр *m* управления кэш-памяти

  * **CAD** → 2847

  * **CAD/CAM** → 2870

2016 **cage relay**
  d Käfigrelais *n*
  f relais *m* à cage
  r клеточное реле *n*

2017 **calculate** *v*; **compute** *v*
  d berechnen; ausrechnen
  f calculer
  r вычислять; расчитывать

2018 **calculated gas velocity**
  d berechnete Gasgeschwindigkeit *f*
  f vitesse *f* calculée des gaz
  r расчётная скорость *f* истечения газа

2019 **calculated power**
  d berechnete Leistung *f*
  f puissance *f* calculée
  r расчётная мощность *f*

2020 **calculating frequency**
  d Rechenfrequenz *f*
  f fréquence *f* de calcul
  r тактовая частота *f* вычислений

2021 **calculating function**
  d Rechenfunktion *f*
  f fonction *f* de calcul
  r вычислительная функция *f*

  * **calculating machine** → 2845

2022 **calculating operation**
  d Rechenoperation *f*
  f opération *f* de calcul
  r вычислительная операция *f*

  * **calculating register** → 992

2023 **calculation; computation**
  d Rechnung *m*; Berechnung *m*
  f calcul *m*; calculation *f*
  r вычисление *n*; подсчёт *m*; расчёт *m*

2024 **calculation algorithm**
  d Berechnungsalgorithmus *m*
  f algorithme *m* de calcul
  r алгоритм *m* вычисления

2025 **calculation initialization; computation initiation**
  d Einleitung *f* der Berechnung; Berechnungsinitialisierung *f*
  f initialisation *f* de calcul
  r инициирование *n* вычислений

2026 **calculation method**
  d Berechnungsmethode *f*
  f méthode *f* de calcul
  r метод *m* вычисления

2027 **calculation of grip force**
  d Greifkraftberechnung *f*
  f calcul *m* de force de grippion
  r расчёт *m* усилия захвата

2028 **calculation of joint reaction**
  d Berechnung *f* der Gelenkreaktion
  f calcul *m* de la réaction de joint
  r расчёт *m* силы реакции шарнира

2029 **calculation of modes**
  d Modenberechnung *f*
  f calcul *m* des modes
  r модовый расчёт *m*

2030 **calculation principle; computing principle**
  d Berechnungsprinzip *n*
  f pricipe *m* de calcul
  r принцип *m* вычисления; принцип расчёта

  * **calculator** → 2845

2031 **calculus mathematics**
  d Rechenmathematik *f*; Berechnungsmathematik *f*
  f mathématique *f* des calculs
  r вычислительная математика *f*

2032 **calculus of differences**
  d Differenzenrechnung *f*; Rechnung *f* mit Differenzen
  f calcul *m* de différences
  r дифференциальное исчисление *n*

2033 **calculus of residues**
  d Residuentheorie *f*

f théorie f des résidus
r теория f остатков

2034 calculus of variation
d Variationsrechnung f
f calcul m des variations
r вариационное исчисление n

*   **calibrate** v → **5872**

2035 calibrated accuracy
d bezogene Genauigkeit f
f exactitude f relative
r относительная точность f

2036 calibrated dial
d geeichte Skale f; Skalenscheibe f
f cadran m étalonné
r калиброванная шкала f; градуированный
   диск m

2037 calibrated voltage-level pulses
d geeichte Spannungsimpulse mpl
f impulsions fpl étalonnées de niveau de
   tension
r калиброванные импульсы mpl [уровня]
   напряжения

2038 calibrating potentiometer
d Eichpotentiometer n; Eichspannungsteiler m
f potentiomètre m d'étalonnage
r калиброванный потенциометр m

2039 calibrating signal
d Eichsignal n
f signal m étalon
r калиброванный сигнал m

2040 calibration
d Eichung f; Eichen n
f étalonnage m; jaugeage m
r калибровка f; юстировка f;
   эталонирование n

2041 calibration accuracy
d Eichungsgenauigkeit f
f précision f d'étalonnage
r точность f калибровки; точность
   градуировки

2042 calibration accuracy of control system
d Eichgenauigkeit f eines Steuerungssystems
f précision f d'étalonnage de système de
   commande
r точность f калибровки системы
   управления

2043 calibration circuit

d Eichkreis m
f circuit m d'étalonnage
r калиброванная схема f; эталонная схема

2044 calibration current; calibration flow
d Eichstrom m
f courant m étalon
r эталонный ток m; эталонный поток m

2045 calibration curve
d Eichkurve f
f courbe f d'étalonnage
r градуировочная кривая f

*   **calibration flow** → **2044**

2046 calibration frequency
d Eichfrequenz f
f fréquence f d'étalonage
r калиброванная частота f

2047 calibration instrument
d Eichgerät n
f appareil m d'étalonnage; instrument m
   d'étalonnage; étalon m
r эталонный прибор m

2048 calibration pulse
d Eichimpuls m
f impulsion f d'étalonnage
r калибровочный импульс m

2049 calibration resistance
d Eichwiderstand m
f résistance f d'étalonnage
r калиброванное сопротивление n

2050 calibration scale
d Eichmessteilung f; Eichskale f
f échelle f étalonnée
r градуированная шкала f

2051 calibration temperature
d Eichtemperatur f
f température f d'étalonnage
r температура f калибровки

2052 call address
d Rufadresse f
f adresse f d'appel
r адрес-вызов m

2053 call distributor
d Rufverteiler m
f distributeur m d'appel
r распределитель m вызова

2054 call form
d Aufrufform f

f forme *f* de requête
r форма *f* обращения

2055 calling instruction; call statement
  d Abrufbefehl *m*; Aufrufbefehl *m*
  f commande *f* d'appel; instruction *f* d'appel
  r команда *f* вызова

2056 calling relay; ringing relay
  d Anrufrelais *n*; Linienrelais *n*
  f relais *m* d'appel; relais de ligne
  r реле *n* вызова; линейное реле

2057 call register
  d Abrufregister *n*
  f registre *m* d'appel
  r регистр *m* вызова

2058 call-reply system
  d Ruf-Antwort-System *n*
  f système *m* appel-réponse
  r система *f* типа запрос-ответ

2059 call signal
  d Rufsignal *n*
  f signal *m* d'appel
  r сигнал *m* вызова

  * **call statement** → 2055

2060 caloric conductibility measuring apparatus
  d Wärmeleitfähigkeitsmessgerät *n*
  f appareil *m* pour mesurer la conductibilité calorique
  r прибор *m* для измерения теплопроводности

2061 calorific value; heating value
  d Heizwert *m*
  f puissance *f* calorifique
  r теплотворная способность *f*

2062 calorimetric gas-traces analyzer
  d kalorimetrischer Gasspurenanalysator *m*
  f analyseur *m* calorimétrique de traces de gaz
  r калориметрический анализатор *m* следов газа

2063 cam control; cam-set control
  d Nockensteuerung *f*
  f commande *f* à came
  r управление *n* кулачком

2064 camera image
  d Kamerabild *n*
  f image *f* de caméra
  r изображение *n* от камеры

2065 camera information
  d Kamerainformation *f*
  f information *f* de caméra
  r информация *f* от камеры

2066 camera installation
  d Kamerainstallation *f*
  f installation *f* de caméra
  r установка *f* камеры

2067 camless automatic machine
  d nockenloser Automat *m*
  f machine *f* automatique sans cames
  r автоматическое оборудование *n* кулачкового управления

2068 cam positioning
  d Nockenpositionierung *f*
  f positionnement *m* par cames
  r позиционирование *n* кулачков

  * **cam-set control** → 2063

2069 cancel *v*
  d ausstreichen; ungültig machen
  f annuler
  r аннулировать

2070 cancellation ratio
  d Unterdrückungskoeffizient *m*
  f coefficient *m* de suppression
  r коэффициент *m* подавления

2071 canonical distribution
  d kanonische Verteilung *f*
  f distribution *f* canonique
  r каноническое распределение *n*

2072 canonical equation
  d kanonische Gleichung *f*
  f équation *f* canonique
  r каноническое уравнение *n*

2073 canonical form
  d kanonische Form *f*
  f forme *f* canonique
  r каноническая форма *f*

2074 canonical transformation
  d kanonische Transformation *f*
  f transformation *f* canonique
  r каноническое преобразование *n*

2075 capacitance load
  d kapazitive Last *f*; kapazitive Belastung *f*
  f charge *f* capacitive
  r ёмкостная нагрузка *f*

2076 **capacitance meter**
  d Kapazitätsmesser *m*
  f capacimètre *m*
  r измеритель *m* ёмкости

2077 **capacitance-operated**
  d kapazitätsgesteuert
  f commandé par capacité
  r с ёмкостным действием

2078 **capacitance-resistance oscillator**
  d RC-Generator *m*
  f oscillateur *m* résistance-capacité;
     générateur *m* à résistance-capacité
  r ёмкостно-резистивный генератор *m*

2079 **capacitance strain gauge**
  d kapazitive Dehnungsmessstreifen *m*
  f jauge *f* de contrainte à capacité
  r ёмкостной тензометр *m*

2080 **capacitive coupling; capacity coupling**
  d kapazitive Kopplung *f*; Kapazitätskopplung *f*
  f couplage *m* capacitif
  r ёмкостная связь *f*

2081 **capacitive level gauge**
  d kapazitiver Niveaumesser *m*
  f indicateur *m* capacitif de niveau
  r ёмкостный уровнемер *m*

2082 **capacitive measuring cell**
  d kapazitive Messzelle *f*
  f cellule *f* capacitive de mesure
  r [фото]элемент *m* для измерения ёмкости

2083 **capacitive transducer; capacity pick-up**
  d Kapazitätsübertrager *m*;
     Kapazitätsumformer *m*
  f capteur *m* capacitif; palpeur *m* capacitif
  r ёмкостный преобразователь *m*; ёмкостный
     датчик *m*

2084 **capacitor electrometer**
  d Kondensatorelektrometer *n*
  f électromètre *m* [à] condensateur
  r конденсаторный электрометр *m*

2085 **capacitor store**
  d Kondensatorspeicher *m*
  f mémoire *f* à condensateur
  r конденсаторный накопитель *m*

2086 **capacitor-type measuring cell**
  d Kondensatormesszelle *f*
  f cellule *f* de mesure à condensateur
  r измерительный элемент *m*
     конденсаторного типа

2087 **capacity**
  d Kapazität *f*; Leistungsvermögen *n*;
     Leistungsfähigkeit *f*; Fassungsvermögen *n*;
     Tragfähigkeit *f*; Tragvermögen *n*
  f capacité *f*; productivité *f*
  r производительность *f*; мощность *f*;
     ёмкость *f*; способность *f*

2088 **capacity altimeter**
  d kapazitiver Höhenmesser *m*; kapazitives
     Altimeter *n*
  f altimètre *m* capacitif
  r ёмкостной высотомер *m*; альтиметр *m*

2089 **capacity bridge**
  d Kapazitätsmessbrücke *f*
  f pont *m* capacitif
  r ёмкостный мост *m*

2090 **capacity change**
  d Kapazitätsänderung *f*
  f changement *m* de capacité
  r изменение *n* ёмкости

2091 **capacity controller**
  d Leistungsregler *m*
  f régleur *m* de rendement
  r регулятор *m* мощности

*   **capacity coupling** → 2080

2092 **capacity demand**
  d Leistungsbedarf *m*
  f puissance *f* nécessaire
  r потребление *n* мощности

2093 **capacity exceeding number**
  d Überlaufzahl *f*
  f nombre *m* dépassant la capacité
  r число n, превышающее ёмкость

2094 **capacity factor**
  d Kapazitätsfaktor *m*
  f facteur *m* de capacité
  r фактор *m* производственной мощности

2095 **capacity limit**
  d Kapazitätsgrenze *f*
  f limite *f* de capacité
  r предел *m* ёмкости

2096 **capacity-loading fine planning**
  d Kapazitätsbelastungsfeinplanung *f*
  f planification *f* fine pour la capacité de charge
  r точный прогноз *m* загрузки оборудования

2097 **capacity manometer**
  d Kapazitätsmanometer *n*

    f manomètre *m* à capacité
    r ёмкостный манометр *m*

2098 capacity of an oscillating circuit
    d Kapazität *f* eines Schwingkreises
    f capacité *f* d'un circuit oscillant
    r ёмкость *f* колебательного контура

2099 capacity of connection
    d Verknüpfungskapazität *f*
    f capacité *f* d'enchaînement
    r ёмкость *f* соединения

2100 capacity of element of automatic control
    system
    d Speicherfähigkeit *f* der Regelkreisglieder
    f capacité *f* des éléments du système asservi
    r ёмкость *f* звеньев системы
    автоматического регулирования

2101 capacity of remote-control systems
    d Kapazität *f* der Fernwirksysteme
    f capacité *f* des systèmes télémécaniques
    r ёмкость *f* систем телеуправления

    * capacity pick-up → 2083

2102 capacity relay
    d Kapazitätsrelais *n*
    f relais *m* capacitif
    r ёмкостное реле *n*

2103 capacity-type analog-to-digital converter
    d kapazitiver Analog-Digital-Wandler *m*
    f convertisseur *m* capavitif analogique-digital
    r ёмкостный аналого-цифровой
    преобразователь *m*

2104 capacity-type micromanometer
    d Kapazitätsmikromanometer *n*
    f micromanomètre *m* capacitif
    r конденсаторный микроманометр *m*

2105 capacity-type sensing element
    d Kapazitätsfühler *m*
    f organe *m* sensible capacitif
    r ёмкостный чувствительный элемент *m*

2106 capacity unbalance
    d Kapazitätsasymmetrie *f*
    f déséquilibre *m* de capacité
    r ёмкостная асимметрия *f*

2107 capacity value
    d Kapazitätswert *m*
    f valeur *f* de capacité
    r величина *f* ёмкости

2108 capillary electrometer

    d Kapillarelektrometer *n*
    f électromètre *m* capillaire
    r капиллярный электрометр *m*

2109 capillary flowmeter
    d Kapillardurchflussmesser *m*
    f débitmètre *m* capillaire
    r капиллярный расходомер *m*

2110 capillary pressure
    d Kapillardruck *m*
    f pression *f* capillaire
    r капиллярное давление *n*

2111 capillary system
    d Kapillarsystem *n*
    f système *m* capillaire
    r капиллярная система *f*

2112 capillary viscometer
    d Kapillarviskosimeter *n*
    f viscosimètre *m* capillaire
    r капиллярный вискозиметр *m*

2113 capture conditions
    d Mitnahmebedingungen *fpl*
    f conditions *fpl* d'etraînement
    r условия *npl* захвата

2114 capture interface sequence
    d Signalfolge *f* zum Belegen der Koppeleinheit
    f séquence *f* de capture d'échange
    r последовательность *f* сигналов для
    занятия интерфейса

2115 Carnot efficiency
    d Carnot-Wirkungsgrad *m*
    f rendement *m* de Carnot
    r коэффициент *m* полезного действия цикла
    Карно

2116 carrier-actuated relay
    d Trägerfrequenzrelais *n*
    f relais *m* à fréquence porteuse
    r реле n, работающее на несущей частоте

2117 carrier amplitude
    d Trägeramplitude *f*
    f amplitude *f* porteuse
    r амплитуда *f* несущей

2118 carrier current
    d Trägerstrom *m*
    f courant *m* porteur
    r несущий ток *m*; несущий поток *m*

2119 carrier-current protection
    d Trägerfrequenzstreckenschutz *m*

    f protection *f* par courant porteur
    r высокочастотная защита *f*

2120 carrier current signal
    d Träger[strom]signal *n*
    f signal *m* à courant porteur
    r носитель *m* сигнала

2121 carrier-current telemetering equipment
    d Trägerstromfernmessgerät *n*
    f dispositif *m* télémétrique à courants porteurs
    r устройство *n* для измерения на расстоянии токами несущей частоты

2122 carrier frequency
    d Trägerfrequenz *f*
    f fréquence *f* porteuse
    r несущая частота *f*

2123 carrier-frequency amplifier
    d Trägerfrequenzverstärker *m*
    f amplificateur *m* de fréquence porteuse
    r усилитель *m* несущей частоты

2124 carrier-frequency signal transmission
    d Trägerfrequenzsignalübertragung *f*
    f transmission *f* des signaux par courant porteur
    r передача *f* сигналов токами несущей частоты

2125 carrier frequency transmission
    d trägerfrequente Übertragung *f*
    f transmission *f* par courants porteurs
    r передача *f* на несущей частоте

2126 carrier of reaction
    d Reaktionsträger *m*
    f porteur *m* de réaction
    r носитель *m* реакции

2127 carrier telemetring
    d Trägerfrequenzfernmessung *f*
    f télémesure *f* à courant porteur
    r телеизмерение *n* несущими токами

2128 carrier-to-noise ratio
    d Träger-Rausch-Verhältnis *n*; Träger-Rausch-Abstand *m*
    f rapport *m* porteuse-bruit
    r отношение *n* несущей частоты к шуму

2129 carrier wave
    d Trägerwelle *f*
    f onde *f* porteuse
    r несущая волна *f*

2130 carrier wave amplification

    d Trägerwellenverstärkung *f*
    f amplification *f* de l'onde porteuse
    r усиление *n* на несущей волне

2131 carrier wave component
    d Trägerfrequenzkomponente *f*
    f composante *f* de l'onde porteuse
    r составляющая *f* несущей волны

2132 carrier wave supply equipment
    d Trägerfrequenzspeiseanlage *f*
    f appareillage *m* d'alimentation en courants porteurs
    r блок *m* питания током несущей частоты

2133 carry; carry over; transfer
    d Übertragen *n*; Übertrag *m*
    f transfert *m*; report *m*; retenue *f*
    r перенос *m*; передача *f*

2134 carry digit
    d Übertragsziffer *f*; Übertragsstelle *f*
    f chiffre *m* de transfert
    r цифра *f* переноса

2135 carry flip-flop
    d Trigger *m* der Übertragung
    f basculeur *m* de transfert
    r мультивибратор *m* переноса

2136 carrying air
    d Tragluft *f*
    f air *m* de transport
    r транспортирующий воздух *m*

2137 carrying capacity
    d Belastbarkeit *f*; Tragkraft *f*
    f capacité *f* de charge; portée *f*
    r нагружаемость *f*

2138 carrying channel
    d Trägerkanal *m*
    f canal *m* porteur
    r несущий канал *m*

2139 carry initiating signal
    d Signal *n* zum Übertrag
    f signal *m* de transfert
    r передаточный сигнал *m*; сигнал для начала передачи

2140 carry latch
    d Übertragsselbsthalteschaltung *f*
    f bascule *f* de report
    r схема *f* переноса с самоблокировкой

  * carry over → 2133

2141 **carry-over function**
 d Übertragungsfunktion *f*
 f fonction *f* de report
 r функция *f* переноса

2142 **carry register**
 d Übertragsregister *n*
 f registre *m* de transfert
 r регистр *m* переноса

2143 **carry storage**
 d Übertragsspeicherung *f*
 f accumulation *f* du report; emmagasinage *m*
    du report
 r хранение *n* переноса

 * **Cartesian components → 10754**

 * **Cartesian coordinates → 10754**

2144 **Cartesian working room**
 d kartesischer Arbeitsraum *m*
 f espace *m* de travail cartésien
 r рабочая зона *f* в декартовых координатах

2145 **cascade action control**
 d Folgeregelung *f*
 f réglage *m* en cascade
 r каскадное регулирование *n*

2146 **cascade [action] controller**
 d Kaskadenregler *m*
 f régulateur *m* en cascade
 r задающий регулятор *m*

2147 **cascade amplifier**
 d Kaskadenverstärker *m*; Stufenverstärker *m*;
    mehrstufiger Verstärker *m*
 f amplificateur *m* en cascade
 r каскадный усилитель *m*

2148 **cascade connection; series connection**
 d Serienschaltung *f*; Reihenschaltung *f*;
    Hintereinanderschaltung *f*
 f connexion *f* en série; connexion en cascade;
    montage *m* en série
 r последовательное соединение *n*

2149 **cascade control; concatenated control**
 d Kaskadensteuerung *f*; überlagerte
    Regelung *f*; kaskadengeschaltete
    Steuerung *f*; verkettete Steuerung
 f commande *f* en cascade; régulation *f* en
    cascade; réglage *m* en cascade
 r каскадное управление *n*

 * **cascade controller → 2146**

2150 **cascade control system**

 d Kaskadenregelsystem *n*
 f système *m* de réglage en cascade
 r каскадная система *f* управления

2151 **cascade coupling**
 d Kaskadenschaltung *f*
 f couplage *m* en cascade
 r каскадное соединение *n*

2152 **cascaded carry**
 d Kaskadenübertrag *m*
 f report *m* accélére
 r последовательный перенос *m*

2153 **cascaded stages**
 d kaskadenartig geschaltete Stufen *fpl*
 f étages *mpl* en cascade
 r последовательно соединенные
    каскады *mpl*

2154 **cascade electrooptic modulator**
 d elektrooptischer Kaskadenmodulator *m*
 f modulateur *m* électro-optique en cascade
 r каскадный электрооптический
    модулятор *m*

2155 **cascade exciter**
 d Erregerkaskade *f*
 f excitatrice *f* en cascade
 r каскадный возбудитель *m*

2156 **cascade operation**
 d Kaskadenbetrieb *m*
 f opération *f* en cascade
 r каскадная операция *f*

2157 **cascade process**
 d Kaskadenprozess *m*
 f processus *m* en cascade
 r каскадный процесс *m*

2158 **cascade refrigeration system**
 d Kältekaskade *f*
 f système *m* de refrigération en cascade
 r каскадная система *f* охлаждения

2159 **cascade relay**
 d Kaskadenrelais *n*
 f relais *m* à cascade
 r каскадное реле *n*

2160 **cascade scrubber**
 d Kaskadenwäscher *m*
 f scrubber *m* en cascade; laveur *m* à cascade
 r каскадная промывная установка *f*

2161 **cascade system**
 d Kaskadensystem *n*; stufenartiges System *n*

f système *m* en chaîne; système en cascade
r каскадная система *f*

2162 case of application
  d Anwendungsfall *m*
  f éventualité *f* d'application
  r случай *m* [практического] применения

2163 case study
  d Fallstudie *f*
  f étude *f* de cas
  r исследование *n* конкретных условий
     [применения]

2164 catalytic activity
  d katalytische Aktivität *f*
  f activité *f* catalytique
  r каталитическая активность *f*

2165 catalytic process
  d katalytischer Prozess *m*
  f procédé *m* catalytique
  r каталитический процесс *m*

2166 catalytic reactor
  d Kontaktapparat *m*
  f appareil *m* de contact; réacteur *m* catalytique
  r контактный аппарат *m*

2167 cathode detector
  d Katodengleichrichter *m*
  f détecteur *m* cathodique
  r катодный детектор *m*

2168 cathode dispersion
  d Katodenzerstäubung *f*
  f dispersion *f* cathodique
  r катодное распыление *n*

2169 cathode feedback
  d Katodenrückkopplung *f*
  f réaction *f* cathodique; réaction à travers de la
     cathode
  r катодная обратная связь *f*

2170 cathode feedback circuit
  d Katodenrückkopplungskreis *m*
  f circuit *m* à réaction cathodique
  r цепь *f* катодной обратной связи

2171 cathode follower
  d Katodenfolger *m*
  f amplificateur *m* à charge cathodique
  r катодный повторитель *m*

2172 cathode ray
  d Katodenstrahl *m*
  f rayon *m* cathodique

  r катодный луч *m*

2173 cathode-ray coder
  d Katodenstrahlkodierer *m*
  f codeur *m* à rayon cathodique
  r кодирующая электроннолучевая трубка *f*

2174 cathode-ray function generator
  d Katodenstrahlfunktionsgenerator *m*
  f générateur *m* de fonctions à rayons
     cathodiques
  r электронный генератор *m* функции

2175 cathode-ray oscillograph
  d Katodenstrahloszillograf *m*
  f oscillographe *m* cathodique; oscillographe à
     rayons cathodiques
  r электронный осциллограф *m*; катодный
     осциллограф

2176 cathode-ray oscilloscope
  d Katodenstrahloszilloskop *n*
  f oscilloscope *m* à rayons cathodiques
  r электронный осциллоскоп *m*; катодный
     осциллоскоп

2177 cathode-ray switch
  d Katodenstrahlschalter *m*
  f commutator *m* à rayons cathodiques
  r электронный коммутатор *m*; электронный
     переключатель *m*

2178 cathode-ray tube
  d Katodenstrahlröhre *f*; Elektronenstrahlröhre *f*
  f tube *m* cathodique
  r электроннолучевая трубка *f*

2179 cathode screen
  d Katodenschirm *m*
  f écran *m* cathodique
  r экран *m* катода

2180 cathodic inhibitor
  d Katodeninhibitor *m*; Katodenverzögerer *m*
  f inhibiteur *m* cathodique
  r катодный замедлитель *m*

2181 cathodic polarization
  d katodische Polarisation *f*
  f polarisation *f* cathodique
  r катодная поляризация *f*

2182 cathodic protection
  d katodischer Schutz *m*
  f protection *f* cathodique
  r катодная защита *f*

2183 cathodic reaction
  d Katodenreaktion *f*

f réaction *f* cathodique
r катодная реакция *f*

2184 Cauchy-Lipschitz theorem
  d Existenz- und Eindeutigkeitssatz *m* von
     Cauchy-Lipschitz
  f théorème *m* de Cauchy-Lipschitz
  r теорема *f* Коши-Липшица

2185 Cauchy-Schwarz inequality
  d Cauchy-Schwarzsche Ungleichung *f*
  f inégalité *f* de Cauchy-Schwarz
  r неравенство *n* Коши-Шварца

2186 Cauchy sequence
  d Cauchy-Folge *f*; Fundamentalfolge *f*
  f séquence *f* de Cauchy
  r последовательность *f* Коши;
     фундаментальная последовательность

2187 Cauchy's linear equation
  d Cauchysche Lineargleichung *f*
  f équation *f* linéaire de Cauchy
  r линейное уравнение *n* Коши

2188 causality
  d Kausalität *f*
  f causalité *f*
  r причинность *f*

2189 cause of defects
  d Fehlerursache *f*
  f origine *f* de défauts
  r причина *f* ошибок

2190 cause of disturbance
  d Störungsursache *f*
  f cause *f* de dérangement; source *f* de
     dérangement
  r причина *f* отказа; причина аварии

2191 cavitational erosion
  d Kavitationserosion *f*
  f érosion *f* de cavitation
  r кавитационная эрозия *f*

2192 cavity magnetron
  d Hohlraummagnetron *n*
  f magnétron *m* à cavités
  r резонаторный магнетрон *m*

2193 cavity resonator
  d Hohlraumresonator *m*
  f cavité *f* résonnante
  r объёмный резонатор *m*

2194 cavity wavemeter
  d Hohlraumwellenmesser *m*

f ondemètre *m* à cavité résonnante
r резонаторный волномер *m*

2195 Cayley-Hamilton theorem
  d Cayley-Hamiltonsches Theorem *n*
  f théorème *m* de Cayley-Hamilton
  r теорема *f* Кэлей-Гамильтона

  *   **CCD-sensor** → 2311

2196 cell
  d Zelle *f*
  f cellule *f*
  r ячейка *f*; клетка *f*

2197 cell construction
  d Zellenaufbau *m*
  f structure *f* de cellule
  r клеточная структура *f*

2198 cell model
  d Zellenmodell *n*
  f modèle *m* cellulaire
  r клеточная модель *f*

2199 centering device
  d Zentriereinrichtung *f*
  f dispositif *m* de centrage
  r центрирующее приспособление *n*

2200 central acquisition
  d zentrale Überwachung *f*
  f surveillance *f* centrale
  r централизованный сбор *m* данных

2201 central contactless control
  d kontaktlose Zentralsteuerung *f*
  f commande *f* centrale sans contact
  r бесконтактное центральное управление *n*

**2202 central control board; central control
     desk; master control panel; main control
     board**
  d Zentralsteuerungspult *n*; Hauptleitstand *m*;
     Hauptschalttafel *f*
  f pupitre *m* de commande central; poste *m* de
     commande central; poste de commande
     principal
  r центральный пульт *m* управления

2203 central control computer
  d zentraler Steuerrechner *m*
  f calculateur *m* de commande central
  r центральный управляющий компьютер *m*

  *   **central control desk** → 2202

2204 central control device
  d zentrale Regelanlage *f*

f réglage *m* central
r центральное регулирующее устройство *n*

2205 central control room; central measuring station
d Zentralmesswarte *f*; zentrale Messwarte *f*
f salle *f* de contrôle centrale; poste *f* de contrôle centrale
r центральный контрольно-измерительный пункт *m*

2206 central control station
d Steuerungszentrale *f*; Kommandoraum *m*
f centrale *f* de commande
r центральная станция *f* управления

2207 central control unit
d zentrale Steuereinheit *f*
f unité *f* de commande centrale
r центральный блок *m* управления

2208 central data processing system
d zentrales Datenverarbeitungssystem *n*
f système *m* central du traitement des données
r централизованная система *f* обработки данных

2209 central data registration equipment
d zentrale Datenerfassungsanlage *f*
f installation *f* centrale d'introduction de données
r машина *f* [для] централизованного сбора данных

2210 central element
d Zentralelement *n*
f organe *m* central
r центральное звено *n*

2211 central guide column
d zentrale Führungssäule *f*
f colonne *f* de guidage centrale
r центальная направляющая колонна *f*

2212 centralized check
d Zentralkontrolle *f*
f contrôle *m* centralisé
r централизованный контроль *m*

2213 centralized control system
d zentralisiertes Steuersystem *n*
f système *m* de commande centralisée
r система *f* централизованного управления

2214 centralized direction
d zentralisierte Leitung *f*; zentralisierte Führung *f*
f gestion *f* centralisée

r централизованное управление *n*

2215 centralized gripper; centred gripper
d zentrierender Greifer *m*
f grappin *m* centré
r центрирующий схват *m*

* **central measuring station → 2205**

2216 central processor
d Zentralrechenanlage *f*; Zentralprozessor *m*
f unité *f* centrale de traitement
r центральный процессор *m*; центральное устройство *n* для обработки информации

2217 central register
d Zentralregister *n*
f registre *m* central
r центральный регистр *m*

2218 central remote control
d Zentralfernsteuerung *f*
f télécommande *f* centrale
r централизованное телеуправление *n*

2219 central support
d Zentralstütze *f*
f support *m* central
r центральная опора *f*

2220 central temperature
d Mittelpunktstemperatur *f*
f température *f* centrale; température de centre
r температура *f* в центре

* **centre → 8615**

* **centred gripper → 2215**

2221 centre of gravity of gripper
d Greiferschwerpunkt *m*
f centre *m* de gravité d'un grappin
r центр *m* тяжести захвата

2222 centre stable relay
d Relais *n* mit stabiler Mittelstellung; polarisiertes Relais mit Neutralstellung
f relais *m* polarisé à repos central
r [поляризованное] реле *n* с нейтральным состоянием

2223 centre zero measurement
d Messung *f* mit Mittennullpunkt
f mesure *f* au point neutre
r измерение *n* относительно нуля

2224 centre zero relay
d Wechselrelais *n*

f relais *m* commutateur
r трехпозиционное реле *n*

2225 centrifugal acceleration
d Zentrifugalbeschleunigung *f*
f accélération *f* centrifuge
r центробёжное ускорение *n*

2226 centrifugal efficiency; centrifugal force
d Schleuderwirkung *f*; Zentrifugalkraft *f*
f effet *m* centrifuge; force *f* centrifuge
r центробёжное действие *n*; центробёжная сила *f*

* **centrifugal force** → 2226

2227 centrifugal governor; gravity regulator
d Fliehkraftregler *m*
f régulateur *m* centrifuge
r центробёжный регулятор *m*

2228 centrifugal relay
d Fliehkraftrelais *n*; Zentrifugalrelais *n*
f relais *m* centrifuge
r центробёжное реле *n*

2229 centring adjustment
d Zentrieren *n*
f alignement *m* du centre; centrage *m*; cadrage *m*
r регулировка *f* центрирования; центровка *f*

2230 centring controller
d Zentrierregler *m*
f régulateur *m* de centrage
r регулятор *m* центрирования

2231 centripetal acceleration
d Zentripetalbeschleunigung *f*
f accélération *f* entripète
r центростремительное ускорение *n*

* **ceramic pressing handler** → 2232

2232 ceramic pressing manipulator; ceramic pressing handler
d Manipulator *m* für Keramikpressen
f manipulateur *m* pour pressage de céramique
r манипулятор *m* для прессования керамических материалов

2233 ceramic sensor
d keramischer Sensor *m*
f senseur *m* céramique
r керамический сенсор *m*

2234 certain data processing
d zuverlässige Datenverarbeitung *f*

f traitement *m* certain de données
r надёжная обработка *f* данных

2235 certain information
d zuverlässige Information *f*
f information *f* certaine
r надёжная информация *f*

2236 chain
   (of action)
d Reihenfolge *f*; Kette *f*
f chaîne *f*
r цепь *f*

2237 chain command
d Kettenbefehl *m*
f commande *f* en chaîne
r команда f, включённая в цепочку

2238 chained drive
d Kettenantrieb *m*
f entraînement *m* par chaîne
r цепной привод *m*

* **chained machines** → 2240

2239 chained manipulator
d verketteter Manipulator *m*
f manipulateur *m* enchaîné
r манипулятор m, встроенный в автоматическую линию

2240 chained [work] machines
d verkettete Arbeitsmaschinen *fpl*
f machines *fpl* opératrices enchaînées
r технологические машины fpl, встроенные в автоматическую линию

2241 chain elevator monitoring
d Kettenelevatorüberwachung *f*
f surveillance *f* d'élévateurs à chaînes
r контроль *m* работы цепного подъемника

2242 chance decision
d Zufallsentscheidung *f*
f décision *f* aléatoire
r вероятностное решение *n*

2243 change *v*
d ändern
f changer
r изменять

2244 changeable workpiece collet
d auswechselbare Werkstückaufnahme *f*
f montage *m* de pièce échangeable
r сменное базирующее устройство *n*

2245 change gripper
  d Wechselgreifer *m*
  f grappin *m* de changement
  r сменный схват *m*

2246 change in load
  d Belastungsschwankung *f*
  f fluctuation *f* de charge
  r флуктуация *f* нагрузки

2247 change of manipulation kind
  d Veränderung *f* der Manipulationsart
  f variation *f* de type de manipulation
  r изменение *n* вида манипулирования

2248 change of mass
  d Masseänderung *f*
  f changement *m* de masse
  r изменение *n* массы

2249 change of structure
  d Strukturveränderung *f*
  f changement *m* de structure; variation *f*
       structurale
  r структурное изменение *n*

2250 change of variables
  d Variablensubstitution *f*;
       Variablenvertauschung *f*
  f changement *m* des variables; substitution *f*
       des variables
  r замена *f* переменных

**2251 change-over contact; transfer contact;
       alternating contact**
  d Umschaltkontakt *m*; Wechselkontakt *m*
  f contact *m* de basculement; contact *m*
       alternant
  r переключающий контакт *m*

2252 change-over flexibility
       (of gripper)
  d Umrüstflexibilität *f*
  f flexibilité *f* de rééquipement
  r способность *f* к переналадке

2253 change-over gate; commutation switch
  d Umschalttor *n*; Serienschalter *m*
  f commutateur *m*; aiguillage *m*; commutateur
       séquentiel
  r переключатель *m*

2254 change-speed motor
  d Motor *m* mit regelbarer Drehzahl;
       Umschaltmotor *m*
  f moteur *m* à vitesses multiples
  r двигатель *m* с регулируемыми оборотами

2255 change system
  d Wechselsystem *n*
  f système *m* de changement
  r сменная система *f*

2256 change time
  d Wechselzeit *f*
  f temps *m* de changement
  r время *n* замены

2257 channel
  d Kanal *m*
  f piste *f*; canal *m*; voie *f*
  r канал *m*

2258 channel activity
  d Kanalaktivität *f*
  f activité *f* de canal
  r активность *f* канала

2259 channel adapter
  d Kanaladapter *m*
  f adapteur *m* de canal
  r адаптер *m* канал

2260 channel byte
  d Kanalbyte *n*
  f byte *m* de canal
  r канальный байт *m*

  * **channel capacity**→ 2275

**2261 channel check handler; channel error
       routine**
  d Kanalfehlerroutine *f*
  f routine *f* d'erreur de canal
  r программа *f* обработки ошибок канала

2262 channel coding
  d Kanalkodierung *f*
  f codage *m* de voie
  r кодирование *n* канала

2263 channel control
  d Kanalsteuerung *f*
  f commande *f* de canal
  r управление *n* каналом

2264 channel control unit
  d Kanalsteuereinheit *f*
  f unité *f* de commande de canal
  r контроллер *m* канала

2265 channel-coupled multiprocessor
  d Multiprozessor *m* mit Verbindungskanälen
  f multiprocesseur *m* aux canaux unifiés
  r мультипроцессор *m* со связанными
       каналами

* **channel data check** → 2266

2266 **channel data check[ing]**
    d Kanaldatenprüfung *f*
    f contrôle *m* de données de canal
    r контроль *m* данных канала

2267 **channel design**
    d Aufbau *m* eines Kanals; Kanalkonstruktion *f*
    f construction *f* d'un canal
    r структура *f* канала

* **channel error routine** → 2261

2268 **channel filter**
    d Kanalfilter *m*
    f filtre *m* de canal
    r фильтр *m* канала

2269 **channel-guide structure**
    d kanalgeführte Struktur *f*
    f structure *f* de guide sillonné de canaux
    r каналоведущая структура *f*

2270 **channel input-output device**
    d Kanal-Eingabe-Ausgabegerät *n*
    f dispositif *m* d'entrée-sortie d'un canal
    r канальное устройство *n* ввода-вывода

2271 **channelling**
    d Kanalbildung *f*; Bachbildung *f*
    f formation *f* de canaux; formation de
       ruissellements
    r каналообразование *n*

2272 **channel reliability**
    d Kanalzuverlässigkeit *f*
    f sécurité *f* du canal
    r надёжность *f* канала

2273 **channel separation**
    d Kanalabstand *m*
    f séparation *f* de canaux
    r интервал *m* между каналами

2274 **channel spacing**
    d Kanalabstand *m*
    f écart *m* de canal; écart de piste; écart de voie
    r полоса *f* пропускания канала

2275 **channel [transmission] capacity**
    d Kanalkapazität *f*;
       Kanalübertragungskapazität *f*
    f capacité *f* d'information d'une voie; capacité *f*
       de transmission d'un canal
    r пропускная способность *f* канала

2276 **channel width**

    d Kanalbreite *f*
    f largeur *f* de voie
    r ширина *f* канала

2277 **Chapman-Kolmogoroff equation**
    d Chapman-Kolmogoroffsche Gleichung *f*
    f équation *f* de Chapman-Kolmogoroff
    r уравнение *n* Чепмена-Колмогорова

2278 **character; sign**
    d Zeichen *n*
    f caractère *m*
    r знак *m*

2279 **character code**
    d Zeichenkode *m*
    f code *m* de caractère
    r знак *m* кода; символ *m* кода

2280 **character coding**
    d Zeichenverschlüsselung *f*
    f codage *m* de caractères
    r кодирование *n* знаков

2281 **characteristic; characteristic curve**
    d Charakteristik *f*; Kennlinie *f*
    f caractéristique *f*; courbe *f* caractéristique
    r характеристика *f*; характеристическая
       кривая *f*

2282 **characteristic correction**
    d Korrektur *f* der Kennlinie
    f correction *f* de la caractéristique
    r коррекция *f* характеристики

* **characteristic curve** → 2281

2283 **characteristic data of computers**
    d charakteristische Angaben *fpl* des Rechners
    f données *fpl* caractéristique du calculateur
    r характеристические данные *pl*
       счётнорешающих устройств

2284 **characteristic energy**
    d Eigenenergie *f*
    f énergie *f* propre
    r собственная энергия *f*

2285 **characteristic equation system**
    d charakteristisches Gleichungssystem *n*
    f système *m* d'équations caractéristique
    r характеристическая система *f* уравнений

2286 **characteristic function**
    d charakteristische Funktion *f*
    f fonction *f* caractéristique
    r характеристическая функция *f*

2287 **characteristic impedance; wave impedance**
  d Wellenwiderstand *m*
  f impédance *f* caractéristique
  r характеристический импеданс *m*

2288 **characteristic infrared radiation detection**
  d Erfassung *f* der charakteristischen Infrarotstrahlung
  f détection *f* du rayonnement caractéristique infrarouge
  r обнаружение *n* характерного инфракрасного излучения

2289 **characteristic line of grip power**
  d Greifkraftkennlinie *f*
  f ligne *f* caractéristique d'une force de grippion
  r характеристика *f* захватывающего усилия

2290 **characteristic line of tactile sensor**
  d Kennlinie *f* eines taktilen Sensors
  f ligne *f* caractéristique d'un senseur tactile
  r характеристика *f* тактильного сенсора

2291 **characteristic method**
  d Charakteristikenmethode *f*; Kennlinienverfahren *n*
  f méthode *f* des caractéristiques
  r метод *m* характеристик

2292 **characteristic of drive**
  d Antriebscharakteristik *f*
  f caractéristique *f* d'entraînement
  r характеристика *f* привода

2293 **characteristic of limit moment**
  d Grenzmomentcharakteristik *f*
  f caractéristique *f* d'un moment limite
  r характеристика *f* предельного момента

2294 **characteristic polynom**
  d charakteristisches Polynom *n*
  f polynôme *m* caractéristique
  r характеристический многочлен *m*

2295 **characteristics**
  d Kenndaten *pl*
  f données *fpl* caractéristiques
  r характеристические данные *pl*

2296 **characteristic spacing**
  d charakteristischer Abstand *m*
  f espacement *m* caractéristique
  r характеристический интервал *m*

2297 **characteristic state**
  d Eigenzustand *m*
  f état *m* propre
  r характеристическое состояние *n*

2298 **characteristic time**
  d charakteristische Zeit *f*
  f temps *m* caractéristique
  r характеристическое время *n*

2299 **characteristic under load**
  d Belastungskennlinie *f*
  f caractéristique *f* sous charge
  r характеристика *f* недогрузки

2300 **characteristic value**
  d Kennwert *m*
  f valeur *f* caractéristique
  r характеристическое значение *n*

2301 **characteristic value problem**
  d Eigenwertproblem *n*
  f problème *m* de valeur propre
  r задача *f* нахождения собственного значения

2302 **characterization of logic circuits**
  d Charakterisierung *f* von logischen Schaltungen
  f caractérisation *f* des circuits logiques
  r характеристика *f* логических схем

2303 **character of the surface**
  d Oberflächenbeschaffenheit *f*
  f constitution *f* de la superficie; état *m* de surface
  r состояние *n* [обработанной] поверхности

2304 **character reading**
  d Zeichenlesen *n*
  f lecture *f* de caractères
  r считывание *n* знаков

2305 **character recognition**
  d Schriftzeichenerkennung *f*; Schriftzeichenunterscheidung *f*
  f reconnaissance *f* de caractères
  r распознавание *n* символов

2306 **character recognition circuits**
  d Schaltkreise *mpl* für die Zeichenerkennung
  f circuits *mpl* pour la reconnaissance des caractères
  r схема *f* распознавания образов

2307 **character recognition logic**
  d Zeichenerkennungslogik *f*
  f logique *f* de la reconnaissance des caractères; logique de l'identification des caractères
  r логика *f* распознавания символов

2308 **chargeable**
  d belastbar
  f capable d'être chargé
  r допускающий нагрузку

2309 **chargeable-time indicator**
  d Gesprächszeitmesser *m*; Zeitzähler *m*
  f chronotaximètre *m*
  r счётчик *m* времени

2310 **charge coefficient**
  d Beschickungskoeffizient *m*; Füllungsgrad *m*
  f coefficient *m* de charge
  r степень *f* заполнения

2311 **charge-coupled device sensor; CCD-sensor**
  d CCD-Sensor *m*
  f capteur *m* CCD
  r сенсор *m* с зарядовой связью

2312 **charge emission detector**
  d Ladungsemissionsdetektor *m*
  f détecteur *m* à émission de charge
  r эмиссионный детектор *m*

2313 **charger reader device**
  d Lade- und Ablesegerät *n*
  f chargeur-lecteur *m*
  r зарядно-контрольное устройство *n*

2314 **charging cycle**
  d Beschickungszyklus *m*
  f cycle *m* de chagement
  r цикл *m* загрузки

2315 **charging device**
  d Fülleinrichtung *f*
  f dispositif *m* de chargement; chargeur *m*
  r устройство *n* для загрузки

2316 **charging manipulator; loading handler**
  d Beschickungsmanipulator *m*;
       Zuführungsmanipulator *m*;
       Entnahmemanipulator *m*
  f manipulateur *m* de chargement
  r манипулятор *m* для загрузки и разгрузки

2317 **charging program**
  d Beschickungsprogramm *n*
  f programme *m* de chargement
  r программа *f* загрузки

2318 **chart recorder**
  d Registriergerät *n*; Registrierinstrument *n*
  f appareil *m* enregistreur
  r записывающий прибор *m*

2319 **Chebyshev approximation**

  d Tschebyscheffsche Approximation *f*;
       Approximation mit der Tschebyscheffschen Metrik
  f approximation *f* de Tchebychev
  r аппроксимация *f* Чебышева

2320 **Chebyshev inequality**
  d Tschebyscheffsche Ungleichung *f*
  f inégalité *f* de Tchebychev
  r неравенство *n* Чебышева

2321 **check addition**
  d Kontrollsummierung *f*
  f addition *f* de contrôle
  r контрольное суммирование *n*

*   **check algorithm → 2331**

2322 **check automaton**
  d Kontrollautomat *m*
  f automate *m* de contrôle
  r контрольный автомат *m*

2323 **check bit; test bit**
  d Prüfbit *n*; Kontrollbit *n*; Testbit *n*
  f bit *m* de vérification; bit de test
  r контрольный бит *m*

2324 **check circuit**
  d Kontrollschaltung *f*
  f circuit *m* de contrôle
  r контрольная схема *f*

*   **check code → 5174**

2325 **check computations**
  d Kontrollrechnungen *fpl*
  f calculs *mpl* de contrôle
  r контрольные расчёты *mpl*

2326 **check digit**
  d Kontrollbit *m*; Kontrollziffer *f*;
       Kontrollzeichen *n*
  f chiffre *m* de contrôle; chiffre d'essai
  r контрольный символ *m*; контрольная цифра *f*

2327 **check digit system**
  d Prüfziffernsystem *n*
  f système *m* de chiffres de contrôle
  r цифровая система *f* контроля

2328 **checked component**
  d kontrolliertes Bauteil *n*
  f composant *m* contrôlé
  r контролированное изделие *n*

2329 **checked quantity**
  d geprüfte Große *f*

f grandeur f vérifiée
r контролируемая величина f

2330 check indicator
d Prüfanzeiger m; Anzeigelampe f;
   Signallampe f
f indicateur m de contrôle; avertisseur m
   lumineux
r контрольный указатель m

* **check information → 2336**

2331 check[ing] algorithm
d Prüfalgorithmus m
f algorithme m de vérification
r алгоритм m контроля

2332 checking bus
d Kontrollbus m
f bus m de contrôle
r контрольная шина f

2333 checking data
d Prüfdaten pl
f données fpl d'essai
r данные pl испытания

2334 checking equipment
d Prüfeinrichtung f
f dispositif m d'essai
r испытательное устройство n

2335 checking function of grip force
d Kontrollfunktion f einer Greifkraft
f fonction f de contrôle d'une force de
   grippion
r контрольная функция f усилия захвата

2336 check[ing] information
d Prüfinformation f
f information f d'essai
r контрольная информация f

2337 checking of assembly process
d Kontrolle f eines Montagevorgangs
f contrôle m d'un procédé d'assemblage
r контроль m процесса сборки

2338 checking of process
d Vorgangskontrolle f
f contrôle m de procédé
r контроль m процессов

**2339 checking procedure; test procedure;
      testing technique**
d Prüfverfahren n; Testverfahren n
f procédé m de vérification; procédé de test

r тестовая процедура f

2340 checking program
d Prüfprogramm n
f programme m d'essai
r программа f испытаний

2341 check[ing] screen
d Kontrollschirm m
f écran m de contrôle
r контрольный экран m

2342 checking sequence; test sequence
d Prüffolge f
f séquence f de vérification; séquence de test
r контрольная последовательность f

2343 checking station
d Prüfstation f
f station f d'essai
r контрольная станция f

2344 checking subroutine
d kontrollierendes Unterprogramm n
f sous-programme m de contrôle
r контролирующая подпрограмма f

2345 checking system
d Prüfsystem n
f système m de vérification
r система f контроля

2346 checking time of indications
d Kontrollzeit f der Anzeigen
f durée f de contrôle des indications
r время n проверки показаний

2347 checking unit
d Prüfeinheit f; Kontrolleinheit f
f unité f de vérification
r устройство n контроля

* **check instrument → 12300**

2348 check message
d Kontrollmeldung f
f message m de vérification
r контрольное сообщение n

2349 check modulation
d Prüfmodulation f; Kontrollmodulation f
f modulation f de contrôle
r контрольная модуляция f

2350 check problem
d Kontrollproblem n; Prüfaufgabe f
f problème m de contrôle
r контрольная задача f

2351 **check program beginning; test program beginning**
    d Prüfprogrammbeginn *m*; Beginn *m* des Testprogramms
    f commencement *m* du programme de contrôle
    r начало *n* тестовой программы

2352 **check program end; test program end**
    d Prüfprogrammende *n*
    f fin *f* du programme de contrôle
    r конец *m* тестовой программы

*   **check program error → 12319**

2353 **check program time; test program time**
    d Prüfprogrammzeit *f*
    f temps *m* de programme de test
    r время *n* выполнения тестовой программы

2354 **check run**
    d Prüflauf *m*
    f cours *m* de test
    r контрольный пуск *m*

*   **check screen → 2341**

2355 **check solution**
    d Kontrollösung *f*
    f résolution *f* de contrôle
    r контрольное решение *n*

2356 **check stop**
    d Prüfstopp *m*
    f arrêt *m* de test
    r контрольный останов *m*

2357 **check test**
    d Gegenprobe *f*
    f essai *m* de vérification
    r контрольное испытание *n*

*   **check value → 12748**

2358 **chemical process engineering**
    d chemische Verfahrenstechnik *f*
    f génie *m* chimique; technique *f* des procédés chimiques
    r процессы *mpl* и аппараты *mpl* химической технологии

*   **chemical process system → 2359**

2359 **chemical-technological system; chemical process system**
    d chemisch-technologisches System *n*; verfahrenstechnisches System
    f système *m* chimico-technologique
    r химико-технологическая система *f*

2360 **chemical treatment process**
    d Verfahren *n* zur chemischen Behandlung
    f procédé *m* de traitement chimique
    r процесс *m* химической обработки

2361 **chip select signal**
    d Chipauswahlsignal *n*
    f signal *m* de sélection de chip
    r сигнал *m* выбора микросхемы

2362 **choice criterion of gripper**
    d Greiferauswahlkriterium *n*
    f critère *m* de sélection de grappin
    r критерий *m* выбора захвата

2363 **choke-coupled amplifier**
    d Drosselverstärker *m*
    f amplificateur *m* à bobines de self
    r дроссельный усилитель *m*

2364 **choking factor; throttling index**
    d Drosselungskennwert *m*
    f degré *m* d'atténuation
    r показатель *m* степени дросселирования

2365 **chopped beam**
    d zerhackter Strahl *m*
    f rayon *m* intermittent; faisceau *m* interrompu
    r прерывистый луч *m*

2366 **chopped impulse voltage**
    d zerhackte Stoßspannung *f*
    f tension *f* de choc découpée
    r прерывистое импульсное напряжение *n*

2367 **chopper**
    d Zerhacker *m*
    f interrupteur *m*; hacheur *m*
    r прерыватель *m*

2368 **chopper amplifier; contact-modulated amplifier**
    d Zerhackerverstärker *m*
    f amplificateur *m* avec interrupteur
    r усилитель *m* с прерывателем

2369 **chopper-bar recording**
    d Fallbügelregistrierung *f*
    f enregistrement *m* à pointe par étrier mobile
    r регистрация *f* точечным методом

2370 **chopper circuit**
    d Zerhackerschaltung *f*
    f circuit *m* de découpage
    r цепь *f* прерывателя

2371 **chopper modulation**
    d Zerhackermodulation *f*

    f modulation *f* par interrupteur
    r прерывистая модуляция *f*

2372 chopper phase
    d Zerhackungsphase *f*; Zerhackerphase *f*
    f phase *f* d'interruption; phase de l'interrupteur
    r фаза *f* прерывания

2373 chopping frequency
    d Zerhackerfrequenz *f*
    f fréquence *f* d'interruption
    r частота *f* прерывания

2374 chopping relay
    d Unterbrecherrelais *n*; Zerhackerrelais *n*
    f relais *m* interrupteur
    r реле *n* с прерывателем

2375 chromatographic analyzer
    d chromatografisches Analysiergerät *n*
    f analyseur *m* chromatographique
    r хроматографический анализатор *m*

2376 circle diagram
    d Kreisdiagramm *n*
    f diagramme *m* circulaire
    r круговая диаграмма *f*

2377 circuit; connection diagram circuit
    d Schaltung *f*; Schaltkreis *m*
    f circuit *m*; montage *m*; connection *f*
    r схема *f*; цепь *f*; контур *m*

2378 circuit adjustment; circuit setting
    d Schwingkreiseinstellung *f*
    f ajustement *m* de circuit; syntonisation *f* de circuit
    r настройка *f* контура

2379 circuit analog
    d Analogstromkreis *m*; Stromkreisäquivalent *n*
    f circuit *m* analogique
    r моделирующая цепь *f*

2380 circuit analysis
    d Netzwerkanalyse *f*
    f analyse *f* de circuit
    r анализ *m* схемы

2381 circuit analysis program
    d Schaltungsanalyseprogramm *n*; Programm *n* zur Analyse von Schaltungen
    f programme *m* pour analyse des connexions
    r программа *f* для анализа схем

2382 circuit analyzer
    d Schaltkreisanalysator *m*
    f analyseur *m* de circuit
    r прибор *m* для анализа цепей

2383 circuit and program technique
    d Schaltungs- und Programmtechnik *f*
    f technique *f* des circuits et de programme
    r схемотехника *f* и техника *f* программного обеспечения

2384 circuit-breaker
    d Stromkreisunterbrecher *m*
    f interrupteur *m*
    r выключатель *m*; прерыватель *m*; разединитель *m*

2385 circuit capacity
    d Schaltungskapazität *f*
    f capacité *f* de montage
    r монтажная ёмкость *f*

2386 circuit characteristic
    d Schaltungskenngröße *f*
    f caractéristique *f* de circuit
    r характеристика *f* схемы

2387 circuit-closing connection
    d Arbeitsstromschaltung *f*
    f alimentation *f* par circuit normalement ouvert
    r схема *f* на срабатывание рабочим током

2388 circuit complexity
    d Schaltungskomplexität *f*
    f complexité *f* de circuit
    r сложность *f* схемы

2389 circuit design technique
    d Schaltungsentwurfstechnik *f*
    f technique *f* de projet de circuit
    r техника *f* проектирования схем

2390 circuit development
    d Schaltkreisentwicklung *f*; Schaltungsentwicklung *f*
    f développement *m* de circuit
    r разработка *f* схемы

2391 circuit efficiency
    d Kreiswirkungsgrad *m*
    f rendement *m* de circuit
    r эффективность *f* схемы

2392 circuit element; switching element; network element
    d Schaltelement *n*; Baustein *m*; Netzwerkelement *n*
    f élément *m* de circuit; élément de commutation; élément du réseau
    r элемент *m* цепи; элемент схемы

2393 circuit logic
  d Schaltungslogik *f*
  f logique *f* de circuit
  r схемная логика *f*

2394 circuit mean
  d Schaltmittel *n*
  f moyen *m* de commutation
  r средства *npl* коммутации

2395 circuit modification
  d Schaltungsabwandlung *f*;
     Schaltungsmodifizierung *f*
  f modification *f* de circuit
  r схемная модификация *f*

2396 circuit module
  d Schaltungsmodul *m*
  f circuit *m* modulaire
  r цепь-модуль *m*

2397 circuit node
  d Schaltungsknotenpunkt *m*
  f nœud *m* de circuit
  r узел *m* схемы

2398 circuit noise level
  d Rauschpegel *m* eines Stromkreises
  f niveau *m* de bruit d'un circuit
  r уровень *m* шума в схеме

2399 circuit noise meter
  d Stromkreisrauschmesser *m*;
     Schaltkreisrauschmesser *m*
  f psophomètre *m* de bruit de ligne
  r измеритель *m* шумов контура

2400 circuit-opening connection
  d Ruhestromschaltung *f*
  f connexion *f* à interruption
  r схема *f* на срабатывание током покоя;
     включение *n* на возврат

2401 circuit optimization
  d Schaltungsoptimierung *f*
  f optimisation *f* de circuit
  r оптимизация *f* схемы

2402 circuit parameter
  d Schaltungsparameter *m*
  f paramètre *m* de circuit
  r параметр *m* схемы

2403 circuit point
  d Schaltstelle *f*
  f point *m* de mise en circuit
  r точка *f* схемы

2404 circuit reliability
  d Schaltungszuverlässigkeit *f*
  f sécurité *f* de circuit
  r надёжность *f* схемы

2405 circuit requirements
  d Schaltungsanforderungen *fpl*
  f exigence *f* aux circuits
  r схемотехнические требования *npl*

2406 circuitry
  d Schaltungstechnik *f*
  f technique *f* de montage
  r техника *f* схем

2407 circuit selector
  d Stromkreiwähler *m*
  f sélecteur *m* de circuit
  r селектор *m* цепи

  * **circuit setting → 2378**

2408 circuit simulation
  d Schaltungssimulation *f*
  f simulation *f* de circuit
  r моделирование *n* схемы

**2409 circuit speed; operation speed; switching
     speed**
  d Schaltgeschwindigkeit *f*; Geschwindigkeit *f*
     der Schaltung; Arbeitsgeschwindigkeit *f*
  f rapidité *f* de fonctionnement; vitesse *f* de
     circuit; vitesse d'opération
  r скорость *f* коммутаций

2410 circuit switching system
  d Leitungsvermittlungssystem *n*
  f système *m* de communication de lignes
  r система *f* коммутации каналов

2411 circuit tester
  d Schaltkreistester *m*
  f vérificateur *m* de circuit
  r контрольно-измерительный прибор *m* для
     проверки схем

2412 circular buffer memory
  d Umlaufpufferspeicher *m*
  f mémoire *f* de tampon circulaire
  r буферное динамическое запоминающее
     устройство *n*

2413 circular interpolation
  d Kreisinterpolation *f*
  f interpolation *f* circulaire
  r круговая интерполяция *f*

2414 circular scanning
  d Kreisabtastung *f*

f balayage *m* circulaire
r круговой обзор *m*

2415 circulating storage
d Umlaufspeicher *m*
f mémoire *f* à circulation
r запоминающее устройство *n*
  динамического типа

* circulation → 3627

2416 circulation counter
d Umdrehungszähler *m*
f compteur *m* de circulation
r счётчик *m* [числа] оборотов

2417 circulation direction
d Drehrichtung *f*; Umlaufrichtung *f*
f sens *m* de rotation; direction *f* de circulation
r направление *n* вращения

2418 circulation integral
d Umlaufintegral *n*; Randintegral *n*
f intégrale *f* de circulation
r интеграл *m* по замкнотому контуру

2419 circulation potentiometer
d Umlaufpotentiometer *n*
f potentiomètre *m* de circulation
r поворотный потенциометр *m*

2420 circulation system
d Kreislaufschaltung *f*; Umlaufsystem *n*;
  Zirkulationssystem *n*
f système *m* de circulation; circuit *m* fermé
r система *f* циркуляции

2421 circulation transmission
d Übertragung *f* der Drehbewegung
f transmission *f* de rotation
r передача *f* вращения

2422 clamp *v*
d spannen
f serrer
r зажать

2423 clamping circuit
d Klemmschaltung *f*; Blockierschaltung *f*;
  Randwertschaltung *f*
f circuit *m* de verrouillage
r фиксирующая цепь *f*

2424 clamping device
d Verriegelungsvorrichtung *f*;
  Festspannvorrichtung *f*; Klemmvorrichtung *f*
f dispositif *m* de verrouillage

r фиксирующее устройство *n*

2425 clamping elements
d Spannelemente *npl*
f éléments *mpl* de serrage
r зажимные приспособления *npl*

2426 clamping principle
d Klemmprinzip *n*
f principe *m* de serrage
r принцип *m* зажатия

2427 clamping technique
d Spanntechnik *f*
f technique *f* de serrage
r техника *f* зажима; техника крепления

2428 classical electromagnetic theory
d klassische Theorie *f* des Elektromagnetismus
f théorie *f* classique de l'électromagnétisme
r класическая электромагнитная теория *f*

2429 classical information processing system
d klassisches
  Informationsverarbeitungssystem *n*
f système *m* classique de traitement de
  l'information
r класическая система *f* обработки
  информации

2430 classical switching element
d klassisches Schaltelement *n*
f élément *m* de commutation classique
r классический переключательный
  элемент *m*

2431 classification keyboard
d Fachwählertastenfeld *n*; Fachwählertastatur *f*
f clavier-trieur *m*
r классификационная коммутационная
  панель *f*

2432 classification of assembly operations
d Klassifizierung *f* von Montageoperationen
f classification *f* des opérations d'assemblage
r классификация *f* сборочных операций

2433 classification of manipulator controls
d Klassifizierung *f* der Manipulatorsteuerungen
f classification *f* de commandes de
  manipulateur
r классификация *f* систем управления
  манипуляторами

2434 classification system
d Klassifikationssystem *n*
f système *m* de classification
r система *f* классификации

2435 **classifier of time-variable patterns**
   d Klassierer *m* von zeitlich veränderlichen Bildern
   f classificateur *m* des images variables en temps
   r кассификатор *m* изменяемых изображений по времени

2436 **classify** *v*
   d klassifizieren; klassieren; sortieren
   f classer; cribler; trier
   r класифицировать

   * **class of accuracy** → 201

2437 **Clausius-Rankine process**
   d Clausius-Rankine-Prozess *m*
   f procédé *m* Clausius-Rankine
   r Клаузиус-Ранкин процесс *m*

2438 **clearing control before commencing**
   d Klarkontrolle *f* vor Arbeitsbeginn
   f vérification *f* avant de commencer le travail
   r контроль *m* сброса перед началом работы

2439 **clearing device**
   d Störungsbeseitigungseinrichtung *f*
   f dispositif *m* de dépannge
   r устройство *n* для стирания

2440 **clearing relay**
   d Schlußzeichenrelais *n*
   f relais *m* terminal; relais de fin de conversation
   r отбойное реле *n*

2441 **climatic conditions**
   d klimatische Bedingungen *fpl*
   f conditions *fpl* climatiques
   r климатические условия *npl*

2442 **clinometer**
   d Klinometer *n*; Neigungswinkelmesser *m*
   f clinomètre *m*; indicateur *m* de pente
   r клинометр *m*; уклономер *m*

2443 **clipped noise**
   d abgeschnittenes Rauschen *n*; abgekapptes Rauschen
   f bruit *m* écrêté
   r ограниченный шум *m*

2444 **clipper; limiter; restrictor; delimiter**
   d Begrenzer *m*; Begrenzerstufe *f*; Clipper *m*; Begrenzungssymbol *n*
   f limiteur *m*; délimiteur *m*
   r ограничитель *m*

2445 **clipper amplifier**
   d Abkappungsverstärker *m*
   f amplificateur *m* écrêteur
   r ограничивающий усилитель *m*

2446 **clipper circuit**
   d Abkappkreis *m*; Amplitudenbegrenzer *m*
   f circuit *m* d'écrêtage
   r ограничивающая схема *f*

   * **clock-actuated** → 2447

2447 **clock-controlled; clock-actuated**
   d taktgesteuert
   f commandé par rythmeur
   r управляемый тактовыми импульсами

2448 **clock cycle**
   d Taktzyklus *m*
   f cycle *m* d'horloge
   r цикл *m* синхронизации

2449 **clocked system**
   d getaktetes System *n*
   f système *m* rythmé
   r синхронизированная система *f*

2450 **clock error**
   d Zeitgeberfehler *m*
   f erreur *f* de minuterie
   r ошибка *f* датчика синхроимпульсов

2451 **clock frequency**
   d Taktgeberfrequenz *f*
   f fréquence *f* d'horloge
   r тактовая частота *f*; частота синхронизации

   * **clock generator** → 2458

   * **clocking** → 12459

   * **clocking error** → 12463

2452 **clock input**
   d Takteingang *m*
   f entrée *f* de ruthme
   r синхронизирующий вход *m*

2453 **clock operation**
   d Zeitgeberoperation *f*
   f opération *f* de minuterie
   r функционирование *n* датчика тактовых импульсов

2454 **clock output**
   d Taktausgang *m*
   f sortie *f* de rythme
   r тактовый выход *m*

2455 clock period
d Taktperiode f
f période f d'horloge
r период m тактовых сигналов

2456 clock pulse
d Taktimpuls m
f impulsion f d'horloge
r синхронизирующий тактовый импульс m

2457 clock pulse distributor
d Taktimpulsverteiler m
f distributeur m d'impulsion d'horloge
r распределитель m синхронизирующих
   импульсов

2458 clock[-pulse] generator
d Taktimpulsgenerator m; Taktgenerator m
f générateur m d'impulsions de rythme
r генератор m тактовых импульсов

2459 clock read-out control
d Zeitgebersteuerung f für Auslesen
f commande f minuterie pour lecture
r управление n считывания при помощи
   тактовых импульсов

2460 clock relay
d Uhrrelais n; Zeitmesserrelais n; Schaltuhr f
f interrupteur m horaire
r реле n с часовым механизмом

2461 clock selector
d Zeitgeberselektor m
f sélecteur m de minuterie
r переключатель m тактового генератора

2462 clock signal
d Taktsignal n
f signal m d'horloge
r тактовый сигнал m

2463 clock stage
d Taktgeberstufe f
f étage m de chronoréglage
r каскад m тактового генератора

2464 clock transition
d Taktimpulsübergang m
f transition f d'impulsion de rythme
r переключение n синхронизации

2465 close v; bar v
d schließen; unterbrechen; absperren
f fermer; arrêter; stopper; couper
r замыкать; закрывать; запирать

2466 closed circuit

d geschlossener Stromkreis m
f circuit m électrique fermé; circuit électrique
   bouclé
r замкнутая электрическая цепь f

2467 closed curve; closed graph
d geschlossene Kurve f
f courbe f fermée
r замкнутая кривая f

2468 closed cycle
d geschlossener Kreislauf m
f cycle m fermé
r замкнутый цикл m

2469 closed-cycle control; closed-loop control
d Regelung f mit geschlossenem Zyklus;
   Regelung mit geschlossenem Regelkreis
f réglage m à cycle fermé; réglage à circuit
   fermé; réglage à boucle fermée; réglage à
   réaction
r регулирование n по замкнутому циклу

2470 closed-cycle cooling
d Kreislaufkühlung f
f refroidissement m à cycle fermé
r охлаждение n по замкнутому циклу

* **closed graph → 2467**

2471 closed interval
d abgeschlossenes Intervall n
f intervalle m fermé
r замкнутый интервал m

2472 closed loop; closed path
d geschlossener Zyklus m; geschlossene
   Schleife f
f circuit m fermé; boucle f fermée
r замкнутый контур m

2473 closed-loop circuit
d Schaltung f mit geschlossener Schleife;
   Kreisschaltung f; Rückführungsschaltung f
f circuit m à boucle fermée
r контур m с обратной связью

* **closed-loop control → 2469**

2474 closed-loop control system
d Regelsystem n
f système m asservi; système de commande en
   boucle fermée
r замкнутая система f управления

2475 closed-loop gain
d Verstärkung f in geschlossenen Regelkreis

f gain m en boucle fermée
r коэффициент m усиления замкнутого
контура

2476 closed-loop phase angle
d Phasenwinkel m in geschlossenem Kreis
f déphasage m en boucle fermée
r фазовый сдвиг m замкнутого контура

2477 closed loop position control
d geschlossener Lageregelkreis m
f régulation f en boucle fermée de position
r контур m позиционного регулирования с
обратной связью

2478 closed-loop pulse system
d geschlossenes Impulssystem n
f système m impulsionnel bouclé
r замкнутая импульсная система f

2479 closed-loop stability
d Stabilität f der geschlossenen Schleife
f stabilité f de la boucle fermée
r устойчивость f замкнутой системы

2480 closed-loop telemetry
d Fermessverfahren n mit geschlossenem
Messkreis
f télémesure f à boucle fermée
r телеметрическая система f работающая по
замкнутой схеме

**2481 closed-loop transfer function; output
transfer function** (US)
d Übertragungsfunktion f des geschlossenen
Regelkreises
f fonction f de transfert en boucle fermée
r переходная функция f замкнутой системы

* closed path → 2472

2482 closed subroutine
d geschlossenes Unterprogramm n
f sous-routine f fermée; sous-programme m
fermé
r замкнутая подпрограмма f

2483 closed system
d geschlossenes System n
f système m fermé
r замкнутая система f

2484 close sensors
d Nahsensoren mpl
f capteurs mpl de proximité
r сенсоры mpl контакта

2485 closing delay

d Einschaltverzug m
f retard m de la fermeture
r замедление n при замыкании

2486 closing equation
d Schlußgleichung f
f équation f de bouclage
r уравнение n замыкания

2487 closing force
d Schließkraft f
f force f de fermeture
r запорное усилие n

2488 closing order
d Einschaltbefehl m
f ordre m d'enclenchement
r команда f замыкания

2489 closing relay
d Schließrelais n; Sperrelais n
f conjoncteur m
r включающее реле n

2490 closing time
d Schließzeit f
f temps m de fermeture
r время n перекрытия

2491 closing voltage; pull-in voltage
d Einschaltspannung f
f tension f d'enclenchement
r включающее напряжение n

2492 clustered errors
d Fehlerhäufung f
f accumulation f des erreurs
r накопление n ошибок

2493 clutch control
d Kupplungssteuerung f
f commande f d'embrayage
r управление n сцеплением

2494 Coanda effect
d Coanda-Effekt m
f effet m Coanda
r эффект m Коанда

2495 coarse adjustment
d Grobeinstellung f
f réglage m grossier
r грубая наладка f

2496 coarse adjustment of gripper
d Greifergrobeinstellung f
f mise f au point grossière d'un grappin
r грубая настройка f захвата

2497 **coarse control**
  d grobe Regelung *f*
  f réglage *m* approximatif
  r грубое регулирование *n*

2498 **coarse positioning**
  d Grobpositionierung *f*
  f positionnement *m* grossier
  r грубое позиционирование *n*

2499 **coarse position of manipulator**
  d Manipulatorgrobposition *f*; Grobposition *f*
       eines Manipulators
  f position *f* grossière d'un manipulateur
  r глобальная позиция *f* манипулятора

2500 **coarse presetting of displacements**
  d Grobwegvorgabe *f*
  f avantage *m* de parcours gros
  r задание *n* регионального перемещения

2501 **coarse structure; macrostructure**
  d Grobstruktur *f*
  f structure *f* grossière; macrostructure *f*
  r макроструктура *f*

2502 **coarse tuning**
  d Grobabstimmung *f*
  f accord *m* approximatif
  r грубая настройка *f*

2503 **coating thickness measurement**
  d Belagsdickenmessung *f*
  f mesure *f* d'épaisseur de revêtement
  r измерение *n* толщины покрытия

2504 **coaxial line**
  d Koaxialleitung *f*
  f ligne *f* coaxiale
  r коаксиальная линия *f*

2505 **coaxial relay**
  d Koaxialrelais *n*
  f relais *m* coaxial
  r коаксиальное реле *n*

2506 **coaxial resonator**
  d Koaxialresonator *m*
  f résonateur *m* coaxial
  r коаксиальный резонатор *m*

2507 **coaxial transmission line**
  d konzentrische Übertragungsleitung *f*
  f ligne *f* de transmission coaxiale
  r коаксиальная линия *f* связи

2508 **code check**
  d Kodeprüfung *f*

  f vérification *f* de code
  r проверка *f* кода

2509 **code checking time**
  d Kodeprüfzeit *f*
  f temps *m* de vérification de code
  r период *m* проверки кода

2510 **code combination**
  d Kodekombination *f*
  f combinaison *f* de code
  r кодовая комбинация *f*

2511 **code control system**
  d Kodesteuerungssystem *n*
  f système *m* de commande à codage
  r кодовая система *f* управления

2512 **code conversion; code translation**
  d Kodeumsetzung *f*
  f convertissement *m* du code; traduction *f* du
       code
  r преобразование *n* кода

2513 **code converter; code translator;
       transcoder**
  d Kodewandler *m*; Umkodierer *m*;
       Kodeumsetzer *m*
  f convertisseur *m* de code; traducteur *m* de
       code
  r преобразователь *m* кода

2514 **coded**
  d verschlüsselt
  f codifié
  r кодированный

2515 **coded communication system**
  d kodiertes Nachrichtenübertragungssystem *n*
  f système *m* de communication codé
  r система *f* передачи кодированных
       сообщений

2516 **coded decimal digit**
  d kodierte Dezimalziffer *f*
  f chiffre *m* décimal codé
  r кодированная десятичная цифра *f*

2517 **coded decimal notation**
  d kodierte dezimale Schreibweise *f*
  f notation *f* décimale codifiée
  r кодираванная запись *f* в десятичной
       системе

2518 **coded identification**
  d Kodebezeichnung *f*; kodierte Bezeichnung *f*
  f désignation *f* codée
  r кодированное обозначение *n*

2519 **code digit**
  d Kodeziffer *f*
  f chiffre *m* de code
  r кодовый разряд *m*; цифра *f* кода

2520 **coded instruction**
  d kodierter Befehl *m*
  f instruction *f* codée; instruction codifiée
  r кодированная инструкция *f*

2521 **coded instruction range**
  d kodierte Befehlsreihe *f*
  f série *f* d'instructions codées
  r закодированная последовательность *f* команд

2522 **code discriminator**
  d Kodediskriminator *m*
  f discriminateur *m* de codes
  r кодовый дискриминатор *m*

2523 **coded program**
  d kodiertes Programm *n*
  f programme *m* codifié
  r кодированная программа *f*

2524 **coded signal**
  d kodiertes Signal *n*
  f signal *m* codé
  r кодированный сигнал *m*

2525 **code element**
  d Kodeelement *n*
  f élément *m* de code
  r кодовый элемент *m*

  * **code equipment → 2530**

2526 **code generating unit; code investigation unit**
  d Schlüsselermittlereinheit *f*
  f unité *f* de recherche de code; unité d'investigation de code
  r блок *m* генерации кодов

  * **code investigation unit → 2526**

2527 **code point**
  d Kodepunkt *m*
  f point *m* de code
  r кодовая точка *f*

2528 **code protection**
  d Kodesicherung *f*
  f protection *f* de code
  r защита *f* кода

2529 **code pulse train**

  d Kodeimpulsfolge *f*; kodierte Impulsfolge *f*
  f train *m* d'impulsions codées
  r кодированная последовательность *f* импульсов

2530 **coder; encoder; code equipment**
  d Schlüssler *m*; Verschlüssler *m*; Kodierer *m*; Kodiereinrichtung *f*
  f codeur *m*; traducteur *m*; dispositif *m* de codification
  r кодирующее устройство *n*

2531 **code regeneration**
  d Koderegeneration *f*
  f régénération *f* de code
  r перезапись *f* кода

2532 **code structure**
  d Kodestruktur *f*; Kodeaufbau *m*
  f structure *f* de code
  r структура *f* кода

  * **code translation → 2512**

  * **code translator → 2513**

2533 **code word**
  d Kodewort *n*
  f mot *m* du code
  r кодовое слово *n*

2534 **code-word distance**
  d Kodewortabstand *m*
  f distance *f* de mot du code
  r расстояние *n* кодового слова

2535 **coding**
  d Kodierung *f*; Verschlüsselung *f*
  f codage *m*; codification *f*
  r кодирование *n*

2536 **coding circuit**
  d Kodierungskreis *m*
  f circuit *m* codeur; circuit de codage
  r кодирующая цепь *f*

2537 **coding hypothesis**
  d Kodierungshypothese *f*
  f hypothèse *f* de codage
  r гипотеза *f* кодирования

2538 **coding line**
  d kodierte Befehlszeile *f*
  f ligne *f* d'instruction codée
  r одиночная инструкция *f*

  * **coding of numbers → 8842**

*   coding of states → 11839

2539 coding relay
d Kodierungsrelais *n*
f relais *m* de codage
r кодирующее реле *n*

2540 coding section
d Kodierungsbschnitt *m*; Befehlsreihe *f* eines
   Teilprogrammausschnitts
f section *f* de codage; série *f* d'instructions
   codifiées
r код *m* команд части программы

2541 coding theory
d Theorie *f* der Kodierung;
   Kodierungstheorie *f*
f théorie *f* de codification
r теория *f* информаций

2542 coefficent of inherent regulation
d Selbstregelungsfaktor *m*; Ausgleichsgrad *m*
f coefficient *m* d'autoréglage
r коэффициент *m* саморегулирования

2543 coefficient of difficulty
d Schwierigkeitsgrad *m*
f coefficient *m* de difficulté
r коэффициент *m* сложности

2544 coefficient of expansion
d Ausdehnungszahl *f*; Expansionsgrad *m*
f coefficient *m* d'expansion; degré *m* de
   détente
r коэффициент *m* расширения

2545 coefficient of heat radiation
d Wärmestrahlungskoeffizient *m*
f coefficient *m* de radiation thermique
r коэффициент *m* теплоизлучения

2546 coefficient of linear thermal expansion
d linearer Wärmeausdehnungskoeffizient *m*
f coefficient *m* linéaire de dilatation thermique
r коэффициент *m* линейного теплового
   расширения

2547 coefficient of proportionality
d Proportionalitätsfaktor *m*
f coefficient *m* de proportionnalité
r коэффициент *m* пропорциональности

2548 coefficient of trust
d Vertrauenskoeffizient *m*
f coefficient *m* de confiance
r доверительный коэффициент *m*

2549 coefficient-setting potentiometer
d Koeffizienten-Einstellpotentiometer *n*

f potentiomètre *m* de coefficient
r потенциометр *m* настройки коэффициента

2550 coherence time
d Kohärenzzeit *f*
f temps *m* de cohérence
r время *n* когерентности

2551 coherent carrier
d kohärenter Träger *m*; kohärente
   Trägerwelle *f*
f onde *f* porteuse cohérente
r когерентная несущая *f*

2552 coherent detection
d kohärente Demodulation *f*
f démodulation *f* cohérente
r когерентное детектирование *n*

2553 coherent electromagnetic oscillations
d kohärente elektromagnetische
   Schwingungen *fpl*
f oscillations *fpl* électromagnétiques
   cohérentes
r когерентные электромагнитные
   колебания *npl*

2554 coherent infrared radar
d Kohärent-Infrarotstrahlenradar *n*
f radar *m* à rayonnement infrarouge cohérent
r когерентный инфракрасный
   [радио]локатор *m*

2555 coherent laser signal
d kohärentes Lasersignal *n*
f signal *m* cohérent du laser
r когерентный сигнал *m* лазера

2556 coherent optical radar
d optisches Kohärentstrahlenradar *n*
f radar *m* optique cohérent
r когерентный оптический
   [радио]локатор *m*

2557 coherent radiation
d kohärente Strahlung *f*
f rayonnement *m* cohérent
r когерентное излучение *n*

2558 coherent source
d kohärente Quelle *f*
f source *f* cohérente
r когерентный источник *m*

2559 coherent transmission
d kohärente Übertragung *f*
f transmission *f* cohérente
r когерентная передача *f*

2560 **cohesive energy density**
d Bindungsenergiedichte *f*
f densité *f* de l'énergie de liaison
r энергетическая плотность *f* связи

2561 **coil pitch**
d Spulensteigungsmaß *n*
f pas *m* d'enroulement
r шаг *m* намотки

* **coincidence** → 191

2562 **coincidence amplifier**
d Koinzidenzverstärker *m*
f amplificateur *m* du circuit à coïncidence
r усилитель *m* схемы совпадений

2563 **coincidence and anticoincidence pulse counting assembly**
d Koinzidenz-Antikoinzidenz-Zählanordnung *f*
f ensemble *m* de mesure à coïncidences et anticoïncidences
r установка *f* для счёта совпадений и антисовпадений

2564 **coincidence circuit; coincidence gate; gate [circuit]; gating circuit**
d Koinzidenzschaltung *f*; Koinzidenzgatter *n*; Koinzidenzregister *n*; Torschaltung *f*
f circuit *m* à coïncidence; porte *f* à coïncidence; gatter *m*
r схема *f* совпадений

2565 **coincidence correction**
d Koinzidenzberichtigung *f*
f correction *f* due à la coïncidence
r поправка *f* на совпадение

2566 **coincidence counter**
d Koinzidenzzähler *m*; Gleichzeitigkeitszähler *m*
f compteur *m* à coïncidence
r счётчик *m* совпадений

2567 **coincidence curve**
d Koinzidenzkurve *f*
f courbe *f* des coïncidences
r кривая *f* совпадений

2568 **coincidence element**
d Koinzidenzelement *n*
f élément *m* de coïncidence
r элемент m, работающий по принципу совпадения

2569 **coincidence error**
d Koinzidenzfehler *m*
f erreur *f* de coïncidence

r ошибка *f* совпадений

* **coincidence gate** → 2564

2570 **coincidence gate**
d Koinzidenzgatter *n*; Koinzidenzregister *n*
f porte *f* à coïncidence
r схема *f* совпадений

2571 **coincidence loss**
d Koinzidenzverlust *m*
f perte *f* par coïncidences
r просчёты *mpl* совпадений

2572 **coincidence method**
d Koinzidenzmethode *f*; Koinzidenzverfahren *n*
f méthode *f* de coïncidence
r метод *m* совпадений

2573 **coincidence mode**
d Koinzidenzbetrieb *m*
f mode *m* de coïncidence
r режим *m* совпадений

2574 **coincidence range finder**
d Koinzidenzentfernungsmesser *m*
f télémètre *m* à coïncidence
r дальномер *m* с совпадением изображений

2575 **coincidence resolving power**
d Koinzidenzauflösungsvermögen *n*
f pouvoir *m* de résolution du montage à coïncidences
r разрешающая способность *f* схемы совпадений

2576 **coincidence selector**
d Koinzidenzselektor *m*; Gleichzeitigkeitswähler *m*
f sélecteur *m* à coïncidence
r селектор *m* совпадений

2577 **coincidence signal**
d Koinzidenzsignal *n*
f signal *m* de coïncidence
r сигнал *m* совпадения

2578 **coincidence spectroscopy**
d Koinzidenzspektroskopie *f*
f spectroscopie *f* à coïncidence
r спектроскопия *f* совпадений

2579 **coincidence synchro**
d Synchrowinkelvergleicher *m*
f synchro-comparateur *m* d'angle
r сельсин *m* совпадений

2580 **coincident current selection**
   d Auswahl *f* mittels Koinzidenzströmen
   f sélection *f* par coïncidence de courants
   r выборка *f* методом совпадения токов

2581 **coincident current selection system**
   d Koinzidenzstrom-Auswahlsystem *n*
   f système *m* de sélectio par courants
        coïncidents
   r система *f* выборки по методу совпадения
        токов

2582 **coincident current store**
   d Koinzidenzstromspeicher *m*
   f mémoire *f* à courants coïncidents
   r накопитель *m* с использованием принципа
        совпадения токов

2583 **coincident flux device**
   d Koinzidenzflusselement *n*;
        Speicherelement *n* mit Vormagnetisierung
   f élément *m* à flux coïncidence
   r элемент m, работающий по принципу
        совпадения магнитных потоков

2584 **collecting circuit**
   d Sammelschaltung *f*
   f montage *m* collecteur
   r собирательная схема *f*

2585 **collecting electrode; collector electrode**
   d Kollektorelektrode *f*; Sammelelektrode *f*
   f électrode *f* collectrice
   r коллекторный электрод *m*; собирающий
        электрод

2586 **collector circuit**
   d Kollektorschaltung *f*; Kollektorstromkreis *m*
   f circuit *m* collecteur
   r цепь *f* коллектора

2587 **collector current**
   d Kollektorstrom *m*
   f courant *m* collecteur
   r ток *m* коллектора

2588 **collector dissipation**
   d Kollektorverlustleistung *f*
   f dissipation *f* de collecteur
   r рассеяние *n* на коллекторе

   * **collector electrode → 2585**

2589 **collector junction**
   d Kollektorübergang *m*
   f jonction *f* de collecteur
   r коллекторный переход *m*

2590 **collector terminal**
   d Kollektorklemme *f*; Kollektorpol *m*
   f borne *f* collectrice
   r вывод *m* коллектора

2591 **collector voltage**
   d Kollektorspannung *f*
   f tension *f* de collector
   r напряжение *n* коллектора

2592 **collision phase**
   d Kollisionsphase *f*
   f phase *f* de la collision
   r фаза *f* столкновения

2593 **colorimeter**
   d Kolorimeter *n*; Farbmesser *m*
   f colorimètre *m*
   r колориметр *m*

2594 **colorimetric dosimetry**
   d Farbdosimetrie *f*
   f dosimétrie *f* colorimétrique
   r колориметрическая дозиметрия *f*

2595 **colour-coded communication**
   d farbenkodierte Verbindung *f*
   f communication *f* à code couleurs;
        communication encodée en couleurs
   r связь *f* с использованием цветового
        кодирования

2596 **colour control**
   d Farbregelung *f*
   f réglage *m* chromatique; commande *f* couleur
   r регулирование *n* цвета

2597 **colour correction filter**
   d Korrektionsfarbfilter *n*
   f filtre *m* correcteur de couleurs
   r цветной корректирующий светофильтр *m*

2598 **colour identification**
   d Farberkennung *f*
   f identification *f* de couleur
   r распознавание *n* цвета

2599 **colour laser display**
   d farbige Abbildung *f* mittels Laser
   f représentation *f* en couleurs au moyen de
        laser
   r цветной лазерный индикатор *m*

2600 **colour pyrometer**
   d Farbpyrometer *n*
   f pyromètre *m* de couleur
   r цветовой пирометр *m*

colour sensor   130

2601 colour sensor
  d Farbsensor *m*
  f senseur *m* de couleur
  r датчик *m* цвета

2602 column indicating device
  d Spaltenanzeiger *m*
  f index *m* de colonne
  r блок *m* индикаций столбца

2603 column shift unit
  d Kolonnenschalter *m*
  f unité *f* de saut de colonne
  r блок *m* сдвига столбца

2604 column vector
  d Vektorspalte *f*
  f vecteur-colonne *m*
  r вктор *m* столбца

2605 comb filter
  d Kammfilter *n*; Filter *n* mit kammförmigem
     Frequenzspektrum
  f filtre *m* en peige; filtre à spectre de fréquence
     en forme de peigne
  r гребенчатый фильтр *m*

2606 combination
  d Kombination *f*
  f combinaison *f*
  r комбинация *f*; сочетание *n*

2607 combination actuator
  d kombinierter Effektor *m*; kombiniertes
     Antriebsglied *n*
  f élément *m* combiné de commande
  r комбинированный привод *m*

2608 combinational circuit
  d Verknüpfungsglied *n*
  f circuit *m* combinatoire
  r комбинированная цепь *f*

2609 combinational logic element
  d logisches Kombinationselement *n*
  f élément *m* logique combinatoire
  r комбинированный логический элемент *m*

2610 combination automatic controller
  d verkoppelte Regelkreise *mpl*
  f circuits *mpl* couplés de réglage
  r многоконтурная система *f*
     автоматического регулирования

2611 combination control
  d kombinierte Steuerung *f*
  f commande *f* combinée

  r комбинированное регулирование *n*

*   combination graph → 2622

2612 combination logic function
  d logische Kombinationsfunktion *f*
  f fonction *f* logique combinatoire
  r комбинированная логическая функция *f*

2613 combination of control loop elements
  d Kombination *f* von Regelelementen
  f combinaison *f* des éléments du système
     asservi
  r соединение *n* звеньев в замкнутой системе
     регулирования

2614 combination of the states; joining of the
      states
  d Vereinigung *f* der Zustände
  f union *f* des états; jonction *f* des états
  r комбинация *f* состояний

2615 combination table
  d Zustandstabelle *f*
  f tableau *m* des états
  r таблица *f* состояний

2616 combined computing system
  d kombiniertes Rechensystem *n*
  f système *m* de calcul combiné
  r комбинированная вычислительная
     система *f*

2617 combined controller
  d kombinierter Regler *m*
  f régulateur *m* à action composée
  r комбинированный регулятор *m*

2618 combined cycle
  d kombinierter Arbeitsablauf *m*
  f cycle *m* combiné
  r комбинированный [рабочий] цикл *m*

2619 combined data and address bus
  d kombinierter Daten-Adress-Bus *m*
  f bus *m* combiné données-adresses
  r комбинированная шина *f* адрес-данные

2620 combined dosimetric system
  d kombiniertes Dosimetersystem *n*
  f système *m* dosimétrique combiné
  r комбинированная дозиметрическая
     система *f*

2621 combined electrochemical and ultrasonic
      machining
  d kombinierte elektrochemische und
     Ultraschallbearbeitung *f*

f usinage *m* par ultrasons et électrochimique combiné
r комбинированная электрохимическая и ультразвуковая обработка *f*

2622 combined graph; combination graph
d Vereinigungsgraf *m*
f graphe *m* de combinaison
r комбинированный граф *m*

2623 combined joint mechanism
d kombinierter Fügemechanismus *m*
f mécanisme *m* de jointage combiné
r комбинированный механизм *m* сопряжения

2624 combined non-linearity
d kombinierte Nichtlinearität *f*
f non-linéarité *f* combinée
r комбинированная нелинейность *f*

2625 combined systems of automatons
d kombinierte Automatensysteme *npl*
f systèmes *mpl* combinant des automates
r комбинированная система *f* автоматов

2626 combined tasks
d kombinierte Aufgaben *fpl*
f problèmes *mpl* combinés
r комбинированные задачи *fpl*

2627 combiner
d Kombinator *m*; Übersetzer *m*
f combinateur *m*
r комбинирующее устройство *n*

2628 combustion control equipment
d Verbrennungsregelungsanlage *f*
f dispositif *m* de réglage de la combustion
r оборудование *n* для управления процессом горения

2629 combustion controller
d Verbrennungsregler *m*
f régulateur *m* de combustion
r регулятор *m* процесса горения

2630 combustion system
d Verbrennungssystem *n*
f système *m* de combustion
r топливная система *f*

2631 coming into step
d Intrittfallen *n*
f accrochage *m*
r вступление *n* в синхронизм

* **command → 6726**

2632 command control
d Befehlssteuerung *f*
f commande *f* par instruction
r командное управление *n*

* **command decoder → 6734**

2633 command device
d Befehlsanlage *f*
f dispositif *m* de commande
r командное устройство *n*

2634 command guidance
d Leitstrahlsteuerung *f*
f guidage *m* télécommandé
r командное наведение *n*

* **command message → 3300**

* **command modification → 6741**

2635 command module
d Führungseinheit *f*; Befehlseinheit *f*
f cellule *f* de commande
r командный отсек *m*

2636 command processor
d Befehlsprozessor *m*
f processeur *m* à instructions
r процессор *m* для обработки команд

2637 command realization time
d Befehlsausführungszeit *f*
f temps *m* de réalisation d'instruction
r время *n* выполнения команды

2638 command resolution
d Steuerungsunterscheidungsvermögen *n*; Steuerungsempfindlichkeit *f*
f sensibilité *f* de commande
r чувствительность *f* по управляющему воздействию

2639 command signal; control signal
d Führungssignal *n*; Regelsignal *n*
f signal *m* de commande
r сигнал *m* управления

2640 command signals with audio frequency
d Kommandosignale *npl* mit Tonfrequenz
f signaux *mpl* de commande à audiofréquence
r сигналы *mpl* управления звуковой частоты

2641 command statement
d Kommandoanweisung *f*
f instruction *f* d'ordre
r командная инструкция *f*

* command structure → 6745

* command system → 6746

2642 **command variable; control variable**
   d Führungsgröße *f*; Regelvariable *f*
   f grandeur *f* de référence; grandeur de
       commande
   r управляющая переменная *f*; управляющая
       переменная величина *f*

2643 **commensurability; comparability**
   d Vergleichbarkeit *f*
   f commensurabilité *f*; comparabilité *f*
   r сравнимость *f*; сопоставимость *f*

* commensurable → 2687

2644 **commitment**
   d Einsatz *m*
   f emploi *m*
   r применение *n*; эксплуатация *f*

2645 **common base circuit**
   d Basisschaltung *f*
   f circuit *m* à base commune
   r принципиальная базисная цепь *f*

2646 **common communication interface**
   d gemeinsame Kommunikationsschnittstelle *f*
   f interface *f* de communication commune
   r общий интерфейс *m* связи

2647 **common control system**
   d gemeinsames Regelungssystem *n*
   f système *m* commun de réglage
   r система *f* с общим управлением

2648 **common data line**
   d gemeinsame Datenleitung *f*
   f ligne *f* commune des données
   r общая информационная линия *f*

2649 **common information carrier**
   d einheitlicher Datenträger *m*
   f porteur *m* commun d'information
   r универсальный носитель *m* информации

2650 **common language**
   d gemeinsame Sprache *f*
   f langage *m* commun
   r универсальный язык *m*

2651 **common machine language**
   d einheitliche Maschinensprache *f*
   f langage *m* de machine commun
   r универсальный машинный язык *m*

2652 **common mode**

   d Gleichtaktbetrieb *m*
   f régime *m* push-push
   r синфазный режим *m*

2653 **common mode rejection**
   d Gleichtaktunterdrückung *f*
   f suppression *f* de synchronisme
   r подавление *n* синфазного сигнала

2654 **common peripherals**
   d gemeinsame Peripherie *f*
   f périphérie *f* commune
   r общие периферийные устройства *npl*

2655 **communication**
   d Kommunikation *f*
   f communication *f*
   r коммуникация *f*

2656 **communication and control in the animal
     and in the machine**
     (of cybernetics)
   d Nachrichtenübertragung *f* und Steuerung *f* im
       Tier und in der Maschine
   f communication *f* et commande *f* en animal et
       en machine
   r связь *f* и управление *n* в живом организме
       и в машине

2657 **communication application**
   d Nachrichtenverkehrsanwendung *f*;
       Übertragungstechnik-Anwendung *f*
   f application *f* de communication
   r применение *n* в связи

2658 **communication channel**
   d Nachrichtenkanal *m*
   f canal *m* de télécommunications
   r канал *m* связи

2659 **communication control**
   d Kommunikationssteuerung *f*
   f commande *f* de communication
   r управление *n* передачей

2660 **communication control program**
   d Übertragungssteuerprogramm *n*
   f programme *m* de commande de
       communication
   r программа *f* управления передачей

2661 **communication engineering; engineering
     cybernetics**
   d technische
       Kommunikationswissenschaften *fpl*;
       technische Kybernetik *f*
   f cybernétique *f* technique
   r теория *f* техники связи; техническая
       кибернетика *f*

2662 **communication line adapter**
 d Fernmeldeleitungsanschluss *m*
 f adapteur *m* pour ligne de commutation
 r адаптер *m* линии связи

2663 **communication medium**
 d Übertragungsmedium *n*; Nachrichtenmittel *n*
 f milieu *m* de communication; moyen *m* de communication
 r средства *npl* связи

2664 **communication mode**
 d Kommunikationsmodus *m*; Betriebsart *f* des Informationsaustausches
 f mode *m* de communication
 r режим *m* передачи

2665 **communication module**
 d Kommunikationsmodul *m*
 f module *m* de communication
 r набор *m* устройств передачи данных

2666 **communication network**
 d Kommunikationsnetz *n*; Nachrichtenübertragungsnetz *n*
 f réseau *m* de communications
 r сеть *f* связи

2667 **communication network processor**
 d Kommunikationsnetzwerkprozessor *m*
 f processeur *m* de réseau de communication
 r процессор *m* сети связи

2668 **communication procedure**
 d Übertragungsprozedur *f*; Übertragungsverfahren *n*
 f procédé *m* de communication
 r процедура *f* передачи данных

2669 **communication process**
 d Nachrichtenübertragungsprozess *m*
 f procédé *m* de transmission des communications
 r процесс *m* коммуникации

2670 **communication signal**
 d Kommunikationssignal *n*
 f signal *m* de communication
 r сигнал *m* связи

2671 **communication system**
 d Übertragungssystem *n*
 f système *m* de communication
 r система *f* связи

2672 **communication system mode**
 d Betriebsart *f* eines Nachrichtensystems
 f mode *m* d'exploitation d'un système de communication
 r режим *m* работы системы связи

2673 **communication theory**
 d Nachrichtentheorie *f*; Kommunikationstheorie *f*
 f théorie *f* de l'information; théorie de communication
 r теория *f* связи

2674 **communication unit**
 d Kommunikationseinheit *f*
 f unité *f* de communication
 r блок *m* связи

2675 **commutated error signal**
 d umgeschaltetes Fehlersignal *n*; kommutiertes Fehlersignal
 f signal *m* commuté d'erreur
 r коммутированный сигнал *m* рассогласования

2676 **commutated network; switching network**
 d Umschaltkreis *m*; Umschaltungskette *f*
 f circuit *m* de commutation; commutateur *m*
 r переключающая сеть *f*

 *   **commutation switch** → 2253

2677 **commutative algebra**
 d kommutative Algebra *f*; wechselseitige Algebra
 f algèbre *f* commutative
 r коммутативная алгебра *f*

2678 **commutative algebra of events**
 d wechselseitige Algebra *f* der Ereignisse
 f algèbre *f* commutative des événements
 r коммутативная алгебра *f* событий

2679 **commutative automaton**
 d kommutativer Automat *m*
 f automate *m* commutatif
 r коммутативный автомат *m*

2680 **commutative law**
 d kommutatives Gesetz *n*
 f loi *f* commutative
 r коммутативный закон *m*

2681 **compact computing instrument**
 d kompaktes Recheninstrument *n*
 f instrument *m* de calcul compact
 r компактное вычислительное устройство *n*

2682 **compact construction; compact design**
 d Kompaktbauweise *f*; raumsparende Ausführung *f*

   f construction *f* compacte
   r компактная конструкция *f*; компактное
    исполнение *n*

2683 compact construction unit
   d kompakte Baueinheit *f*
   f unité *f* de montage compacte
   r компактный конструктивный элемент *m*

* **compact design** → **2682**

2684 compactness
   d Kompaktheit *f*
   f compacité *f*
   r компактность *f*

2685 compact oscilloscope
   d Kompaktoszilloskop *n*
   f oscilloscope *m* compact
   r компактный осциллоскоп п

2686 compact system
   d Kompaktsystem *n*
   f système *m* compact
   r компактная система *f*

* **comparability** → **2643**

2687 comparable; commensurable
   d vergleichbar
   f commensurable; comparable
   r сравнимый; сопоставимый

* **comparator** → **2688**

2688 comparator device; comparator
   d Vergleichsglied *n*; Vergleichsorgan *n*
   f dispositif *m* comparateur; organe *m*
    comparateur; comparateur *m*
   r орган *m* сравнения; устройство *n* для
    сравнения; компаратор *m*; сравниватель *m*

2689 compare *v*
   d vergleichen
   f comparer
   r сравнивать

2690 compare instruction
   d Vergleichsinstruktion *f*
   f instruction *f* de comparaison
   r команда *f* сравнения

2691 compare logic
   d Vergleichslogik *f*; Vergleicherlogik *f*
   f logique *f* de comparaison
   r аппаратные средства *npl* связи

2692 comparison
   d Komparation *f*; Vergleich *m*; Vergleichung *f*

   f comparaison *f*
   r сравнение *n*

2693 comparison circuit
   d Vergleichsschaltung *f*
   f circuit *m* de comparaison
   r схема *f* сравнения; цепь *f* сравнения

2694 comparison element
   d Vergleichselement *n*
   f élément *m* comparateur
   r элемент *m* сравнения

2695 comparison method; method of comparison
   d Vergleichsmethode *f*
   f méthode *f* de comparaison
   r метод *m* сравнения

2696 comparison operation
   d Vergleichsoperation *f*
   f opération *f* de comparaison
   r операция *f* сравнения

2697 comparison system
   d Vergleichssystem *n*
   f système *m* de comparaison
   r система *f* сравнения

2698 comparison theorem
   d Vergleichssatz *m*; Vergleichstheorem *n*
   f théorème *m* de comparaison
   r теорема *f* сравнения

2699 compatibility
   d Kompatibilität *f*; Verträglichkeit *f*;
    Anpassungsfähigkeit *f*
   f compatibilité *f*; faculté *f* d'adaptation
   r совместимость *f*

2700 compatibility attachment
   d Verträglichkeitsanschluss *m*
   f attachement *m* de compatibilité
   r приставка *f* для обеспечения
    совместимости

2701 compatibility condition
   d Verträglichkeitsbedingung *f*
   f condition *f* de compatibilité
   r условие *n* совместимости

2702 compatible
   d kompatibel
   f compatible
   r совместимый

2703 compatible distribution function; congruent distribution function
   d komportable Verteilungsfunktion *f*

f   fonction f de répartition de probabilité
r   совместимая функция f распределения

2704 compatible state
(theory of automata)
d   vereinbarer Zustand m
f   état m compatible
r   совместимое состояние n

2705 compatible subassembly
d   kompatible Baugruppe f
f   sous-ensemble m compatible
r   совместимые модули mpl

2706 compatible system
d   kompatibles System n
f   système m compatible
r   совместимая система f

2707 compensated character recognition logic
d   ausgleichende Zeichenerkennungslogik f
f   logique f compensatrice de la reconnaissance
    des caractères
r   компенсационная логика f распознавания
    знаков

2708 compensated instrument transformer
d   kompensierter Messwandler m
f   transformateur m de mesure compensé
r   компенсированный измерительный
    трансформатор m

2709 compensated thermal overload relay
d   thermische Relais n mit
    Temperaturkompensation
f   relais m thermique compensé
r   компенсированное термореле n
    перегрузки

2710 compensating action; compensation signal
d   Abgleichsignal n; Kompensationssignal n
f   signal m de compensation
r   сигнал m компенсации; компенсирующее
    действие n

2711 compensating circuit
d   Kompensationsschaltung f
f   circuit m de compensation
r   компенсирующая схема f

2712 compensating controller
d   Kompensationsregler m
f   régulateur m compensateur
r   компенсирующий регулятор m

2713 compensating curve
d   Ausgleichslinie f
f   courbe f égalisatrice

r   линия f регулирования; кривая f
    выравнивания

2714 compensating errors
d   sich aufhebende Fehler mpl
f   erreur fpl compensatrices
r   [взаимно] компенсирующиеся ошибки fpl

2715 compensating feedback
d   Kompensationsrückkopplung f;
    Ausgleichsrückkopplung f
f   réaction f de compensation
r   компенсирующая обратная связь f

2716 compensating feedforward
d   Kompensationsvorkopplung f
f   réaction f positive de compensation
r   компенсирующий сдвиг m вперед

2717 compensating line; compensation lead
d   Ausgleichsleitung f; Kompensationsleitung f
f   ligne f de compensation
r   компенсационные провода mpl

2718 compensating network
d   Kompensationsschaltung f;
    Stabilisierschaltung f
f   réseau m correcteur; élément m
    compensateur
r   компенсирующий контур m;
    стабилизирующее устройство n

2719 compensating polarimeter
d   Kompensationspolarimeter n
f   polarimètre m à compenation
r   корректирующий поляриметр m

2720 compensating-pressure transducer
d   Kompensationsdruckgeber m
f   capteur m de pression à compensation
r   компенсационный датчик m давления

2721 compensating process
d   Ausgleichsvorgang f;
    Ausgleichserscheinung f
f   phénomène m d'évolution
r   процесс m выравнивания

2722 compensating resistor
d   Kompensationswiderstand m
f   résistance f de compensation
r   компенсирующий резистор m

2723 compensating selfrecording instrument
d   Kompensationsschreibgerät n
f   enregistreur m à compensation
r   компенсационный самопишущий
    прибор m

2724 **compensating tank**
d Ausgleichsbehälter *m*
f réservoir *m* de compensation; récipient *m* de détente
r регулирующий резервуар *m*; выравнивающий резервуар

* **compensation** → 5102

2725 **compensation adjustment**
d Kompensationseinstellung *f*
f mise *f* au point de compensation; réglage *m* compensateur
r установка *f* степени компенсации

2726 **compensation element**
d Kompensationselement *n*
f élément *m* compensateur
r компенсационный элемент *m*

2727 **compensation factor**
d Kompensationsgrad *m*
f facteur *m* de compensation
r коэффициент *m* коменсации

* **compensation lead** → 2717

2728 **compensation measuring method; null point method of measurement**
d Kompensationsmessmethode *f*; Nullmessmethode *f*
f méthode *f* de compensation de mesure
r компенсационный метод *m* измерений

* **compensation of distortion** → 4417

2729 **compensation of errors**
d Fehlerausgleich *m*
f compensation *f* des erreurs
r компенсация *f* ошибок

2730 **compensation ratio**
d Kompensationsverhältnis *n*
f rapport *m* de compensation
r степень *f* компенсации

* **compensation signal** → 2710

2731 **compensation transmitter**
d Kompensationsferngeber *m*
f transmetteur *m* à compensation
r компенсированный передатчик *m*

2732 **compensation voltage**
d Kompensationsspannung *f*
f tension *f* de compensation
r компенсирующее напряжение *n*

2733 **compensation winding**

d Kompensationswicklung *f*
f enroulement *m* de compensation
r компенсационная обмотка *f*

2734 **compensograph**
d Kompensograf *m*
f compensographe *m*
r компенсограф *m*

2735 **compile** *v*
d zusammentragen
f compiler
r компилировать

2736 **compiled program; compiler**
d Compiler *m*; Übersetzer *m* in maschinenorientierte Programme; kompilierendes Programm *n*
f compilateur *m*; autoprogrammeur *m*; programme *m* compilateur
r компилятор *m*; компилирующая программа *f*

* **compiler** → 2736

2737 **compiling method**
d Kompilationsmethode *f*
f méthode *f* de compilateurs
r метод *m* компиляций

2738 **compiling simulation**
d Compiler-Simulation *f*; kompilierende Simulation *f*
f simulation *f* de compilateur
r компилирующее моделирование *n*

2739 **complementary circuit**
d Ergänzungsschaltung *f*
f circuit *m* complémentaire
r схема *f* на дополняющих структурах

2740 **complementary code**
d Komplementärkode *m*
f code *m* complémentaire
r дополнительный код *m*

2741 **complementary event**
d komplementäres Ereignis n
f événement m complémentaire
r противоположное событие n

2742 **complementary function**
d Komplementärfunktion *f*
f fonction *f* complémentaire
r дополнительная функция *f*

2743 **complementary non-linearity**
d komplementäre Nichtlinearität *f*

f non-linéarité f complémentaire
r дополнительная нелинейность f

2744 complement pulse
d Komplementimpuls m
f impulsion f complémentaire
r импульс m дополнения

2745 complement representation
d Komplementdarstellung f
f représentation f de complément
r дополнительное представление n

2746 complete v; finish v
d vervollkommnen; vervollständigen;
   vollenden
f perfectionner; achever; compléter
r дополнять; завершать

2747 complete assembly station
d vollständige Montagestation f
f station f d'assemblage complète
r комплексный участок m сборки

2748 complete attenuation
d Volldämpfung f
f affaiblissement m total
r полное затухание n

2749 complete carry
d vollständiger Übertrag m
f transfert m complet
r полный перенос m

2750 complete circuit
d geschlossener Stromkreis m; vollständiger
   Stromkreis
f circuit m fermé; circuit complet
r замкнутая цепь f

2751 complete connected automaton
d vollständig angeschlossener Automat m
f automate m connecté complet
r полный присоединенный автомат m

2752 complete derivative
d vollständige Ableitung f
f dérivation f complète
r полная производная f

2753 complete Fourier series
d vollständige Fourier-Reihe f
f série f pleine de Fourier
r полный ряд m Фурье

2754 complete integration of information
      processing
d vollständige Integration f der

Informationsverarbeitung
f intégration f complète de traitement
   d'informations
r полная интеграция f обработки
   информации

2755 complete mechanization; full
      mechanization; complex mechanizing
d Vollmechanisierung f;
   Komplexmechanisierung f
f mécanisation f complète; mécanisation
   complexe
r комплексная механизация f

2756 completeness
d Vollständigkeit f
f complet m; intégralité f
r полнота f; комплектность f

2757 complete operation
d Volloperation f
f opération f complète
r полный цикл m работы

* complete plant → 9294

2758 complete regulating algorithm
d vollständiger Regelalgorithmus m
f algorithme m de régulation complet
r полный алгоритм m регулирования

2759 complete reversibility
d vollständige Umkehrbarkeit f
f réversibilité f complète
r полная обратимость f

2760 complete subquasi-automaton
d vollständiger Unterquasiautomat m
f sous-quasi-automate m complet
r полный квазиподавтомат m

2761 complete system
d Komplettsystem n
f système m complet
r законченная система f

2762 complete table of states
d vollständige Zustandsliste f
f table f d'état complète
r полная таблица f состояний

2763 completion
d Vollendung f; Vervollständigung f
f complètement m; achèvement m
r завершение n; окончание n

2764 complex
d mehrteilig; komplex

f multiple; à plusieurs parties; complexe
r комплексный; составной

2765 complex admittance
d komplexer Scheinleitwert *m*
f admittance *f* complexe
r комплексная полная проводимость *f*

2766 complex amplitude
d Komplexamplitude *f*
f amplitude *f* complexe
r комплексная амплитуда *f*

2767 complex assembly sequence
d komplexer Montageablauf *m*
f séquence *f* d'assemblage complexe
r комплексная последовательность *f* сборки

2768 complex assembly task
d komplexe Montageaufgabe *f*
f tâche *f* d'assemblage complexe
r комплексная задача *f* сборки

2769 complex automatic control system
d selbsttätiges komplexes Regelungssystem *n*
f système *m* automatique de réglage complexe
r комплексная система *f* автоматического
регулирования

*   **complex automation** → 6781

*   **complex automatization** → 6781

2770 complex circuit
d komplexer Schaltkreis *m*
f circuit *m* complexe
r сложная схема *f*

2771 complex component parts proportioning
d komplexe Bauelementedimensionierung *f*
f dimensionnement *m* complexe des
composants
r комплексный расчёт *m* размеров
конструктивных элементов

2772 complex connection instruction
d komplexe Verknüpfungsvorschrift *f*
f instruction *f* d'enchaînement complexe
r комплексная команда *f* соединения

2773 complex controller
d komplexer Regler *m*
f régulateur *m* complexe
r комплексный регулятор *m*

2774 complex control system
d komplexes Regelungssystem *n*
f système *m* de réglage complexe

r комплексная система *f* регулирования

2775 complex data
d komplexe Daten *pl*
f données *fpl* complexes
r комплексные данные *pl*

2776 complex data processing
d komplexe Datenverarbeitung *f*
f traitement *m* complexe de données
r комплексная обработка *f* данных

2777 complex data transposition
d komplexe Datenübertragung *f*
f transposition *f* complexe des données
r комплексная передача *f* данных

2778 complex electronic equipment
d komplexes elektronisches Gerät *n*
f appareil *m* électronique complexe
r комплексный электронный прибор *m*

2779 complex electronic subassembly
d komplexe elektronische Baugruppe *f*
f sous-ensemble *m* électronique complexe
r комплексный электронный блок *m*

2780 complex environmental model
d komplexes Umweltmodell *n*
f modèle *m* d'environnement complexe
r комплексная модель *f* окружающей среды

2781 complex factor
d komplexer Faktor *m*
f facteur *m* complexe
r комплексный фактор *m*

2782 complex force sensor
d komplexer Kraftsensor *m*
f capteur *m* de force complexe
r комплексный датчик *m* усилия

2783 complex group
d Komplexgruppe *f*
f groupe *m* complexe
r комплексная группа *f*

2784 complex handling process
d komplexer Handhabungsvorgang *m*
f procédé *m* de manutention complexe
r комплексный процесс *m*
манипулирования

2785 complex impedance
d komplexer Scheinwiderstand *m*
f impédance *f* complexe
r комплексное полное сопротивление *n*

2786 **complex information**
d komplexe Information f
f information f complexe
r комплексная информация f

2787 **complex information processing**
d komplexe Informationsverarbeitung f
f traitement m complexe des informations
r комплексная обработка f информации

2788 **complexity of the control process**
d Kompliziertheit f des Regelvorganges
f complexité f du processus de commande
r сложность f процесса регулирования

2789 **complex logical operation**
d komplexe logische Operation f
f opération f logique complexe
r сложная логическая операция f

* **complex mechanizing** → 2755

2790 **complex permeability**
d komplexe Permeabilität f
f perméabilité f complexe
r комплексная [магнитная] проницаемость f

2791 **complex plane**
d komplexe Ebene f
f plan m complexe
r комплексная плоскость f

2792 **complex power**
d komplexe Leistung f
f puisance f complexe
r комплексная мощность f

2793 **complex production plant**
d komplexe Produktionsanlage f
f installation f de production complexe
r сложная производственная установка f

2794 **complex root**
d komplexe Wurzel f
f racine f complexe
r комплексный корень m

2795 **complex structure**
d komplexe Struktur f
f structure f complexe
r комплексная структура f

2796 **complex system**
d komplexe System n
f système m complexe
r комплексная система f

2797 **complex transfer function**

d komplexe Übertragungsfunktion f
f fonction f de transfert complexe
r компелексная передаточная функция f

2798 **complex variable**
d komplexe Variable f
f variable f complexe
r комплексная переменная f

2799 **compliance device**
d Ausgleichseinrichtung f
f dispositif m de compensation
r уравнивающее устройство n

2800 **compliance level**
d Gewährungsebene f; Bewilligungsebene f
f niveau m de consentement
r степень f соответствия

2801 **complicated assembly object**
d kompliziertes Montageobjekt n
f objet m d'assemblage compliqué
r сложный объект m сборки

2802 **complicated feeding task**
d komplizierte Beschickungsaufgabe f
f tâche f de chargement compliqué
r сложная задача f загрузки

2803 **complicated sensor structure**
d komplizierter Sensoraufbau m
f structure f de senseur compliqué
r сложная структура f сенсора

2804 **component**
d Komponente f
f composante f
r компонента f; составляющая f

2805 **component characteristics**
d Bauelementekennwerte mpl;
   Bauelementeeigenschaften fpl
f caractéristiques fpl de composants
r характеристики fpl компонентов

2806 **component data bank**
d Bauelementedatenbank f
f centrale f de données pour composants
r массив m данных о конструктивных
   элементах

2807 **component density**
d Bautieldichte f
f densité f des composants
r плотнасть f размещения деталей

2808 **component derating**
d Abnahme f von Nennwerten der
   Bauelementeparameter

f détérioration f des performances de
  composant
r ухудшение n характеристик компонента

2809 component fabrication
  d Teilefertigung f
  f fabrication f des composants
  r изготовление n деталей

2810 component file
  d Komponentendatei f
  f fichier m des composants
  r файл m составляющих

2811 component integration
  d Bauelementeintegration f; Integration f von
    Bauelementen
  f intégration f des composants
  r интеграция f компонентов

2812 component life
  d Bauelementelebensdauer f
  f longévité f de composants
  r долговечность f элементов

2813 component of assembly
  d Montagebauteil n
  f élément m de montage
  r монтажный элемент m

2814 component parameter variation
  d Änderung f des Bauteilparameters
  f variation f de paramètre d'un élément de
    construction
  r изменение n параметра блока

2815 component part; construction unit;
  structural member
  d Bauelement n; Baustein m
  f composant m; élément m de construction;
    bloc m fonctionnel; unité f [de construction]
  r деталь f; модуль m

2816 component reliability
  d Bauelementezuverlässigkeit f
  f fiabilité f de composants
  r надёжность f элементов

2817 component technology
  d Bauelementetechnologie f
  f technologie f de composants
  r технология f изготовления элементов

2818 component tester
  d Bauelementeprüfeinrichtung f;
    Bauelementetester m
  f dispositif m de test de composants
  r тестер m для проверки компонентов

2819 composed controller action
  d zusammengesetzte Reglerwirkung f
  f action f composée du régulateur
  r сложное действие n регулятора

2820 composite data stream
  d zusammengesetzter Datenstrom m
  f flot m de données composé
  r комбинированный поток m данных

2821 composite function
  d zusammengesetzte Funktion f
  f fonction f composée
  r сложная функция f

2822 composite junction
  d zusammengesetzter Übergang m
  f jonction f composée
  r составной переход m

2823 composite material
  d Verbundstoff m
  f matériau m composite; composite f
  r композиционный материал m; композит m

2824 composite signal
  d Signalgemisch n
  f signal m multiple
  r комбинированный сигнал m

2825 composition
  d Komposition f; Zusammensetzung f
  f composition f
  r композиция f; состав m

2826 composition potentiometer
  d Kompositionspotentiometer n
  f potentiomètre m à couche
  r композиционный потенциометр m

2827 compound action
  d gekoppeltes Verhalten n
  f action f composée
  r комбинированное воздействие n

2828 compound condition
  d Verbundbedingung f; zusammengesetzte
    Bedingung f; Mehrfachbedingung f
  f condition f compound
  r объединенное условие n

2829 compound connection
  d Verbundschaltung f
  f montage m d'interconnexion
  r смешанная схема f

2830 compound control action
  d Summierungsregelung f

f   réglage *m* composé
r   многокомпонентная система *f*
    регулирования

2831   compound controller
d   Kompoundregler *m*
f   régulateur *m* à action composée
r   комбинированный регулятор *m*

2832   compound excitation
d   Verbunderregung *f*
f   excitation *f* composée
r   смешанное возбуждение *n*

2833   compound expression
d   Verbundausdruck *m*; zusammengesetzter
    Ausdruck *m*
f   expression *f* compound
r   составное выражение *n*

2834   compound feedback
d   zusammengesetzte Rückkopplung *f*
f   réaction *f* composite
r   многокомпонентная обратная связь *f*

2835   compound feedforward
d   zusammengesetzte Vorwärtswirkung *f*
f   action *f* directe composite
r   многокомпонентная прямая связь *f*

2836   compound synthesis
d   Verbundsynthese *f*
f   synthèse *f* composée
r   связанный синтез *m*

2837   compound system; interconnected system; overall system
d   Verbundsystem *n*; Gesamtsystem *n*
f   système *m* de liaison; système compound;
    système total
r   объединенная система *f*; общая система

2838   compressed air
d   Druckluft *f*
f   air *m* comprimé
r   сжатый воздух *m*

2839   compressed coding
d   gedrängte Kodierung *f*
f   codage *m* comprimé
r   кодирование *n* со сжатием

2840   compulsory checking
d   zwangsläufig wirkende Überwachung *f*
f   contrôle *m* impératif
r   вынужденный контроль *m*

*   **computation → 2023**

2841   computation aids; computations means
d   Rechenhilfsmittel *npl*
f   moyens *mpl* de calcul
r   вычислительные средства *npl*

2842   computational algorithm
d   [numerischer] Algorithmus *m*
f   algorithme *m* [numérique]
r   вычислительный алгоритм *m*

2843   computational process; computing process
d   Rechenprozess *m*; Rechenvorgang *m*
f   processus *m* de calcul; procédé *m* de calcul
r   процесс *m* вычисления

*   **computation initiation → 2025**

2844   computation of error
d   Fehlerrechnung *f*
f   calcul *m* d'erreurs
r   вычисление *n* ошибок

*   **computations means → 2841**

*   **compute *v* → 2017**

2845   computer; computing machine; calculating machine; calculator
d   Computer *m*; elektronischer Rechner *m*;
    Rechenanlage *f*; Rechenmaschine *f*;
    elektronische Datenverarbeitungsanlage *f*
f   ordinateur *m*; calculatrice *f*; machine *f* à
    calculer; calculateur *m*; machine calculatrice;
    computer *m*
r   компьютер *m*; вычислительная машина *f*

*   **computer-aided conception → 2847**

2846   computer-aided control data generation
d   computergestützte Steuerdatenerzeugung *f*
f   génération *f* de données de commande à
    l'aide des ordinateurs
r   автоматизированная выработка *f*
    управляющих данных

2847   computer-aided design; CAD; computer-aided conception
d   computerunterstützter Entwurf *m*; CAD
f   conception *f* assistée par ordinateur; CAO
r   проектирование *n* с помощью
    компьютера; автоматизированное
    проектирование

*   **computer-aided design and manufacturing → 2870**

2848   computer-aided design facilities
d   rechnergestützte Enwurfsmittel *npl*

f moyens *mpl* de conception assistée par
   ordinateur
r средства *npl* автоматизированного
   проектирования

2849 computer-aided design library
d CAD-Bibliothek *f*
f bibliothèque *f* des programmes pour CAO
r библиотека *f* программ для
   автоматизированного проектирования

2850 computer-aided digital system analysis
d Analyse *f* eines digitalen Systems mit Hilfe
   eines Rechners
f analyse *f* d'un système digital par calculateur
r анализ *m* цифровой системы при помощи
   компьютера

2851 computer-aided digital system design
d Entwurf *m* eines digitalen Systems mittels
   Rechners
f projet *m* d'un système digital par calculatrice
r проектирование *n* цифровой системы при
   помощи компьютера

2852 computer-aided drawing
d rechnergestütztes Zeichnen *n*
f dessin *m* assisté par ordinateur
r изготовление *n* чертежей с помощью
   компьютера

2853 computer-aided engineering
d computerunterstütztes Ingenieurwesen *n*
f génie *m* civil à l'aide des ordinateurs
r методы *mpl* проектирования с
   использованием компьютера

2854 computer-aided investigation
d computergestützte Untersuchung *f*
f étude *f* assistée d'ordinateur
r исследование *n* с использованием
   компьютера

2855 computer-aided manufacturing
d computergestützte Fertigung *f*
f fabrication *f* à l'aide de calculateur;
   production *f* à l'aide des ordinateurs
r автоматизированное производство *n* с
   применением компьютера

2856 computer-aided process identification
d computerunterstützte Verfahrenserkennung *f*
f identification *f* d'un processus assistée par
   ordinateur
r распознавание *n* процессов с помощью
   компьютера

2857 computer-aided production

d computergestützte Herstellung *f*
f production *f* à l'aide de la calculatrice
r производство *n* с использованием
   компьютера

**2858 computer-aided programming; computer-
      assisted programming**
d rechnergestützte Programmierung *f*
f programmation *f* assistée par calculateur
r автоматизированное программирование *n*

2859 computer-aided structural detailing
d rechnergestütztes strukturelles Detaillieren *n*
f détaillage *m* structural à l'aide d'une
   calculatrice
r детализация *f* структуры с помощью
   компьютера

2860 computer-aided system design
d rechnergestützter Systementwurf *m*
f projet *m* de système à l'aide d'un ordinateur
r проектирование *n* системы с помощью
   компьютера

2861 computer-aided text processing
d rechnergestützte Textverarbeitung *f*
f traitement *m* des textes à l'aide d'un
   ordinateur
r анализ *m* текста с помощью компьютера

2862 computer analog input
d analoge Rechnereingabe *f*
f entrée *f* d'ordinateur analogique
r аналоговый вход *m* компьютера

2863 computer analysis
d Rechneranalyse *f*
f analyse *f* de calculateur
r компьютерный анализ *m*

2864 computer application
d Rechneranwendung *f*
f application *f* de calculateur
r применение *n* компьютера

2865 computer architecture
d Rechnerarchitektur *f*
f architecture *f* de calculateur
r компьютерная архитектура *f*

2866 computer-assisted diagnostics
d computergestützte Diagnostik *f*
f diagnostic *m* assisté par ordinateur
r автоматизированная диагностика *f*;
   компьютерная диагностика

2867 computer-assisted learning
d rechnergestütztes Lernen *n*

   f apprentissage *m* assisté par ordinateur
   r компьютерное обучение *n*

2868 computer-assisted manufacturing
   d rechnerunterstützte Fertigung *f*
   f fabrication *f* assistée par ordinateur
   r производство *n* с управлением от
      компьютера

   * **computer-assisted programming** → 2858

2869 computer-assisted signal evaluation
   d rechnerunterstützte Signalauswertung *f*
   f évaluation *f* de signal assistée par ordinateur
   r расшифровка *f* сигнала с использованием
      компьютера

**2870 computer augmented design and
manufacturing; CAD/CAM; computer-
aided design and manufacturing**
   d rechnerunterstütztes Entwerfen *n* und
      Fertigen *n*
   f dessin *m* et fabrication *f* à l'aide des
      ordinateurs
   r машинное проектирование *n* и
      автоматизированное производство *n* с
      применением компьютера

2871 computer backing store
   d Rechnerergänzungsspeicher *m*;
      Zusatzspeicher *m* eines Rechners
   f mémoire *f* additionnelle de calculateur
   r дополнительная память *f* компьютера

2872 computer-based education scenario
   d Unterrichtsgestaltung *f* auf Rechnerbasis
   f éducation *f* sur base d'ordinateur
   r методика *f* обучения с использованием
      компьютера

2873 computer block diagram
   d Rechnerblockdiagramm *n*;
      Rechnerblockschaltbild *n*;
      Rechnerblockschema *n*
   f diagramme *f* synoptique de calculatrice;
      diagramme générale de calculatrice
   r блок-схема *f* компьютера

2874 computer capacity
   d Rechnerkapazität *f*;
      Rechnerleistungsfähigkeit *f*
   f capacité *f* de calculateur
   r производительность *f* компьютера

2875 computer check
   d Rechnertest *m*
   f test *m* de calculateur
   r компьютерный контроль *m*

2876 computer-checked supplying
   d rechnerkontrollierte Zulieferung *f*;
      rechnerkontrollierte Lieferung *f*
   f livraison *f* contrôlée par ordinateur
   r доставка *f* с управлением от компьютера

2877 computer checking state
   d Rechnerkontrollzustand *m*
   f état *m* de contrôle de calculateur
   r контрольное состояние *n* компьютера

2878 computer control
   d Rechnersteuerung *f*
   f commande *f* de l'ordinateur
   r компьютерное управление *n*; управление
      с помощью вычислительной машины

2879 computer-controlled assembly
   d rechnergesteuerte Montage *f*
   f montage *m* commandé par ordinateur
   r процесс *m* сборки, управляемый
      компьютером

2880 computer-controlled automatic tester
   d rechnergesteuerter Prüfautomat *m*
   f automate *m* d'essai commandé par ordinateur
   r автоматическое устройство *n* контроля,
      управляемое компьютером

2881 computer-controlled display
   d rechnergesteuerte Anzeige *f*
   f display *m* commandé par ordinateur
   r дисплей m, управляемый с помощью
      компьютера

2882 computer-controlled distribution centre
   d rechnergesteuertes Verteilungszentrum *n*
   f centre *m* de distribution commandé par
      calculateur
   r распределительный узел m, управляемый
      компьютером

2883 computer-controlled drive
   d rechnergesteuerter Antrieb *m*
   f entraînement *m* commandé par ordinateur
   r привод m, управляемый с помощью
      компьютера

2884 computer-controlled error seek
   d rechnergesteuertes Fehlersuchen *n*
   f recherche *f* de défaut commandée par
      ordinateur
   r компьютерный поиск *m* неисправностей

2885 computer-controlled manipulator
   d rechnergesteuerter Manipulator *m*
   f manipulateur *m* commandé par ordinateur
   r манипулятор *m* с управлением от
      компьютера

2886 computer-controlled measuring device
d rechnergesteuertes Messgerät n
f appareil m de mesure commandé par
ordinateur
r измерительное устройство n, управляемое
с использованием компьютера

2887 computer-controlled production region
d rechnergesteuerter Fertigungsbereich m
f région f de fabrication commandée par
ordinateur
r производственный участок m,
управляемый с помощью компьютера

**2888 computer-controlled system for visual
processes**
d rechnergesteuertes System n für
Sehvorgänge
f système m commandé par calculateur pour
procédés visuels
r система f зрительных процессов с
управлением от компьютера

**2889 computer-controlled transformation of
coordinates**
d rechnergesteuerte
Koordinatentransformation f
f transformation f de coordonnées commandée
par ordinateur
r преобразование n координат, управляемое
компьютером

2890 computer-controlled unmanned factory
d rechnergesteuerte unbemannte Fabrik f
f manufacture f non habitée commandée par
ordinateur
r автоматическая фабрика f с управлением
от компьютера

2891 computer-controlled visual checking
d rechnergesteuerte Sichtkontrolle f
f contrôle m commandé par calculateur
r визуальный контроль m с управлением от
компьютера

**2892 computer control of manipulator
workplace**
d Rechnersteuerung f eines
Manipulatorarbeitsplatzes
f commande f de calculateur d'un
emplacement de travail de manipulateur
r компьютерное управление n рабочим
местом манипулятора

2893 computer designer
d Rechnerentwerfer m; Entwickler m für
Rechner
f constructeur m pour calculatrices
r разработчик m компьютерных систем

2894 computer efficiency
d Nutzungsgrad m eines Rechners
f coefficient m d'efficacité d'un calculateur
r коэффициент m использования
компьютера

2895 computer engineer
d Rechenautomatentechniker m
f technicien m des machines à calculer
r компьютерный инженер m

* **computer equation** → 7629

2896 computer feature; feature of the computer
d Rechnermerkmal n
f propriété f de calculateur; caractéristique f
d'une calculatrice
r характеристика f компьютера

2897 computer field
d Rechnergebiet n
f domaine m de calculatrice
r область f применения компьютерных
систем

2898 computer for diagnosis
d Rechner m für Diagnose
f calculateur m pour diagnose
r диагностический компьютер m

2899 computer for machine tool control
d Rechner m für Werkzeugmaschinensteuerung
f calculateur m pour commande des
machinesoutils
r компьютер m для управления станками

2900 computer for quality control
d Computer m für Qualitätskontrolle
f ordinateur m pour contrôle de qualité
r компьютер m применяемый для контроля
качества

2901 computer for reactor engineering
d Computer m für Reaktortechnik
f computer m pour technique des réacteurs
r компьютер m применяемый в
реакторостроении

2902 computer for testing
d Rechner m für Testung; Kontrollrechner m
f calculatrice f pour essai
r компьютер m для испытаний

2903 computer-generated artwork
d rechnererstellte Originalschablone f

f original *m* construit par ordinateur
r оргинал m, спроектированный с помощью
   компьютера

2904 computer-generated information
d rechnergenerierte Information *f*
f information *f* engendrée par l'ordinateur
r информация f, сгенерированная
   компьютером

2905 computer-graphical operation model
d computergrafisches Arbeitsmodell *n*
f modèle *m* de travail computergraphique
r операционная модель *f* с применением
   компьютерной графики

2906 computer graphics
d Rechnergrafik *f*; Computergrafik *f*
f graphique *m* de calculateur
r компьютерная графика *f*

2907 computer graphics application
d Computergrafikanwendung *f*
f application *f* du graphique d'ordinateur
r применение *n* компьютерной графики

2908 computer group
d Rechnergruppe *f*
f groupe *m* des calculatrices
r компьютерная группа *f*

2909 computer-guided image identification
d rechnergeführte Bilderkennung *f*
f identification *f* d'image guidée par ordinateur
r распознавание *n* изображений с помощью
   компьютера

2910 computer-guided numerical control
d rechnergeführte numerische Steuerung *f*
f commande *f* numérique guidée par ordinateur
r числовое управление *n* с использованием
   компьютера

2911 computer hardware
d Computerbauelemente *npl*
f pièces *fpl* constructives d'ordinateur
r аппаратные средства *npl* компьютера

2912 computer in biomedical research
d Rechner *m* in biomedizinischer Forschung
f calculateur *m* en recherche biomédicale
r компьютерная система *f* в
   биомедицинских исследованиях

2913 computer index register
d Rechnerindexregister *n*
f registre *m* d'index de calculateur
r компьютерный индексный регистр *m*

2914 computer indicator
d Rechnerindikator *m*; Anzeiger *m* eines
   Rechners
f indicateur *m* de calculateur
r компьютерный индикатор *m*

2915 computer industry
d Computerindustrie *f*
f industrie *f* de calculateur
r компьютерная индустрия *f*

2916 computer installation
d Rechnerinstallation *f*
f installation *f* de calculateur
r установка *f* компьютера

2917 computer instruction
d Rechnerbefehl *m*
f instruction *f* de calculatrice
r машинная команда *f*

2918 computer interface
d Rechnerinterface *n*
f interface *f* de calculateur
r интерфейс *m* компьютера

2919 computer interface adapter
d Rechnerschnittstellenadapter *m*
f adapteur *m* d'interface de calculateur
r адаптер *m* интерфейса компьютера

2920 computerized
d rechnerbestückt; rechnergeführt
f à calculateur
r оборудованный средствами
   компьютерной техники

2921 computerized automatic tester
d rechnerbestückter Prüfautomat *m*
f automate *m* de test à calculateur
r испытательный автомат m,
   оборудованный средствами
   компьютерной техники

2922 computerized design technique
d CAD-Technik *f*
f technique *f* de conception assistée par
   ordinateur
r метод *m* автоматизированного
   проектирования

2923 computerized measuring assembly
d rechnergeführter Messplatz *m*
f ensemble *m* de mesure guidé par ordinateur
r измерительное устройство n, управляемое
   компьютером

**2924 computerized measuring device;
computerized test rack**
d rechnerbestückter Messplatz m
f installation f de test à calculateur; place f de
mesure à calculateur
r установка f для измерения, оборудованная
средствами компьютерной техники

* **computerized test rack → 2924**

2925 computer literature
d Rechnerliteratur f
f littérature f de calculateur
r компьютерная литература f

2926 computer logic
d Rechnerlogik f
f logique f de calculatrice
r логика f компьютера

2927 computer memory drum
d Rechnerspeichertrommel f
f mémoire f à tambour magnétique de
calculatrice
r накопительный барабан m
вычислительного устройства

2928 computer method
d Rechnerverfahren n; Rechnermethode f
f méthode f de calculateur; procédé m
d'ordinateur
r метод m с применением компьютера

2929 computer micrographics
d rechnergestützte Mikrografik f
f micrographique m d'ordinateur
r автоматизированная микрография f

2930 computer model
d Rechnermodell n
f modèle m d'ordinateur
r модель f компьютера

2931 computer modules
d Rechnermodule mpl
f modules mpl de calculateur
r модули mpl компьютера

2932 computer operation; program flow
d Programmablauf m
f passage m de programme
r ход m программы

2933 computer operation model
d Computerarbeitsmodell n
f modèle m de travail de computer
r операционная модель f компьютера

2934 computer-oriented algorithm
d rechnerorientierter Algorithmus m
f algorithme m orienté sur calculateur
r машинный алгоритм m

2935 computer performance
d Rechnerleistung f; Leistung f eines Rechners
f performance f de calculateur
r производительность f компьютерной
системы

2936 computer peripherals
d Rechnerperipherie f
f périphérie f de calculateur
r периферийное оборудование n
компьютера

2937 computer printer
d Rechnerdrucker m
f imprimeuse f de calculateur
r печатающее устройство n компьютерной
системы

2938 computer program for structural analysis
d Rechenprogramm n für Satzbau-Analyse
f programme m de calcul pour analyse
structurale; programme d'ordinateur pour
analyse structurale
r компьютерная программа f для
структурного анализа

2939 computer programming system
d Rechnerprogrammierungssystem n
f système m de programmation de calculateur
r система f для компьютерного
программирования

2940 computer result
d Rechnerergebnis n;
Rechenmaschinenergebnis n
f résultat m de calculateur
r результат m компьютерных вычислений

2941 computer routine
d Maschinenprogramm n
f programme m machine
r компьютерная программа f

2942 computer simulation
d Rechnersimulierung f; Simulation f
f simulation f par calculateur
r имитационное моделирование n

2943 computer simulation of digital systems
d Rechnersimulation f digitaler Systeme;
Nachbildung f von digitalen Systemen auf
einem Rechner

    f simulation *f* des systèmes digitaux par
      ordinateur
    r имитационное моделирование *n*
      цифровых систем

2944 computer symbology
    d Symbolik *f* der Rechenmaschinen
    f symbolique *f* de machines calculatrices
    r компьютерная символика *f*

    * **computer system structure** → 11994

2945 computer technology
    d Rechnertechnologie *f*
    f technologie *f* de calculateur
    r технология *f* производства компьютера

2946 computer test routine
    d Rechnerprüfprogramm *n*
    f routine *f* d'essai de calculateur
    r тестовая компьютерная программа *f*

2947 computer time; machine time
    d Rechenzeit *f*; Maschinenzeit *f*
    f temps *m* de machine; temps de calcul
    r машинное время *n*

2948 computer transfer
    d Rechner-Übertragung *f*
    f transfert *m* par ordinateur
    r компьютерная передача *f*

2949 computer word
    d Maschinenwort *n*
    f mot *m* d'information; mot de machine; mot
      de calculateur
    r машинное слово *n*

2950 computing accuracy
    d Rechengenauigkeit *f*
    f précision *f* de calcul
    r точность *f* вычислений

2951 computing automaton technique
    d Rechenautomatentechnik *f*
    f technique *f* d'automate de calcul
    r техника *f* вычислительных автоматов

    * **computing centre** → 3733

2952 computing element
    d Rechenelement *n*
    f élément *m* de calcul
    r решающий элемент *m*

2953 computing interval
    d Rechentakt *m*; Rechenintervall *n*
    f intervalle *m* de calcul

    r период *m* вычисления

2954 computing law
    d Rechengesetz *n*
    f loi *f* de calcul
    r арифметическое правило *n*

    * **computing machine** → 2845

    * **computing principle** → 2030

2955 computing problem
    d Rechenaufgabe *f*
    f problème *m* de calcul
    r арифметическая задача *f*

    * **computing process** → 2843

2956 computing speed; computing velocity
    d Rechengeschwindigkeit *f*
    f vitesse *f* de calcul
    r скорость *f* вычисления

2957 computing system
    d Berechnungssystem *n*
    f système *m* de calcul
    r система *f* расчёта

2958 computing time
    d Rechenzeit *f*
    f temps *m* de calcul
    r время *n* вычисления

2959 computing-time minimization
    d Rechenzeitminimierung *f*
    f minimisation *f* de temps de calcul
    r оптимизация *f* машинного времени

    * **computing velocity** → 2956

    * **concatenated control** → 2149

2960 concatenation expression
    d Verkettungsausdruck *m*
    f expression *f* d'enchaînement
    r выражение *n* связи

2961 concatenation speed control
    d reihengeschaltete Drehzahlsteuerung *f*
    f réglage *m* en cascade de la vitesse
    r каскадное регулирование *n* скорости

2962 concentration controller
    d Konzentrationsregler *m*
    f régulateur *m* de concentration
    r регулятор *m* концентрации

2963 concentration overvoltage
    d Konzentrationsüberspannung *f*

f surtension *f* de concentration
r сосредоточение *n* перенапряжения

2964 concentrator system
   d Konzentratorsystem *n*
   f système *m* concentrateur
   r система *f* сжатия информации

2965 concept-coordination
   d Konzeptkoordinierung *f*; Koordinierung *f* nach Konzept
   f coordination *f* d'après concept
   r согласование *n* концепций

2966 concept formation
   d Konzeptbildung *f*; Begriffsbildung *f*; Invariantenbildung *f*
   f formation *f* de concept
   r формирование *n* понятия

2967 concept of computer
   d Rechnerkonzept *n*; Konzept *n* eines Rechners
   f concept *m* d'un calculateur; concept d'une calculatrice
   r концепция *f* компьютера

   * **concept of simultaneous operations** → 10090

2968 conclusion
   d Folgerung *f*
   f déduction *f*; conclusion *f*
   r заключение *n*; вывод *m*

   * **concordance** → 191

2969 concurrent
   d gleichzeitig existent; nebenlaufend
   f concurrent
   r совмещенный; совпадающий

2970 concurrently operating computer systems
   d Systeme *npl* zusammenarbeitender Rechner
   f ensembles *mpl* des calculatrices à fonctionnement simultané
   r системы *fpl* совместно работающих компьютеров

2971 concurrent testing
   d mitlaufende Testung *f*; simultan zur Aufgabenbearbeitung durchgeführte Prüfung *f*
   f test *m* concurrent
   r контроль m, выполняемый параллельно вычислениям

2972 conditional breakpoint instruction
   d bedingter Stoppbefehl *m*

f commande *f* d'arrêt conditionnel
r команда *f* условного останова

2973 conditional code
   d Pseudokode *m*
   f pseudocode *m*
   r условный код *m*

2974 conditional complete system
   d bedingt vollständiges System *n*
   f système *m* complet conditionnel
   r условно полная система *f*

2975 conditional disjunction
   d bedingte Disjunktion *f*
   f disjonction *f* conditionnelle
   r условная дизъюнкция *f*

2976 conditional distribution function
   d bedingte Vereilungsfunktion *f*
   f fonction *f* conditionnelle de répartition
   r условная функция *f* распределения

2977 conditional instruction; conditional order
   d bedingter Befehl *m*
   f instruction *f* conditionnelle
   r условная команда *f*

2978 conditional jump; conditional transfer
   d bedingter Sprung *m*; bedingter Übergang *m*
   f transfert *m* conditionnel; rupture *f* de séquence conditionnelle
   r условный переход *m*; условная передача *f*

2979 conditional jump instruction; conditional transfer instruction
   d bedingter Überleitungsbefehl *m*; bedingter Sprungbefehl *m*
   f instruction *f* de saut conditionnel
   r команда *f* условного перехода

2980 conditionally convergent
   d bedingt konvergent
   f convergent conditionnellement
   r условно сходимый

2981 conditional mathematical expectation
   d bedingte mathematische Erwartung *f*
   f probabilité *f* mathématique conditionnelle
   r условное математическое ожидание *n*

2982 conditional operation
   d bedingte Operation *f*
   f opération *f* conditionnelle
   r условная операция *f*

   * **conditional order** → 2977

2983 conditional probability
d bedingte Wahrscheinlichkeit f
f probabilité f conditionnelle
r условная вероятность f

2984 conditional stability
d bedingte Stabilität f
f stabilité f conditionnelle
r условная устойчивость f

2985 conditional state
d bedingter Zustand m
f état m conditionnel
r условное состояние n

2986 conditional statement
d bedingte Anweisung f
f instruction f conditionnelle
r условный оператор m

* **conditional transfer** → 2978

* **conditional transfer instruction** → 2979

2987 conditional transfer of control
d bedingter Übergang m der Steuerung
f transfer m conditionnel de commande
r условная передача f управления

2988 conditioning process
d konditionierender Prozess m
f procédé m de conditionnement
r подготвительный процесс m

2989 conditions of measurement
d Messbedingungen fpl
f conditions fpl de mesure
r условия npl проведения измерений

2990 conductance
d Leitwert m
f conductance f
r активная проводимость f

2991 conducting path
d Leiterzug m
f train m de conducteurs
r проводящая линия f

2992 conduction band
d Leitungsband n
f bande f de conduction
r полоса f проводимости

2993 conduction current
d Leitungsstrom m
f courant m de conduction
r ток m проводимости

2994 conductive coupling
d konduktive Kopplung f
f accouplement m conducteur; couplage m direct
r непосредственная связь f

2995 conductive medium of analog device
d Analogeinrichtung f mit leitendem Medium
f unité f analogique avec milieu conducteur
r аналоговое устройство n с проводящей средой

2996 conductivity
d Leitfähigkeit f
f conductivité f; conductibilité f
r проводимость f

2997 conductivity controller
d Leitfähigkeitsregler m
f régulateur m de conductibilité
r регулятор m проводимости

2998 conductivity measuring bridge
d Leitfähigkeitsmessbrücke f
f pont m de mesure de conductibilité
r мост m для измерения активных проводимостей

2999 conductivity transmitter
d Leitfähigkeitsgeber m
f capteur m de conductivité
r преобразователь m проводимости

3000 conductometric method of analysis
d konduktometrische Analysenmethode f
f méthode f conductométrique d'analyse
r кондуктометрический метод m анализа

3001 configurable
d konfigurierbar
f configurable
r конфигурируемый

3002 configuration of the program
d Bauweise f des Programmes
f configuration f de programme
r конфигурация f программы

3003 configuration unit
d Konfigurations[kontroll]einheit f
f unité f de configuration
r блок m управления конфигурацией

3004 conformable value
d übereinstimmender Wert m
f valeur f convenable
r согласующаяся величина f

3005 **conformity; similitude; similarity**
 d Ähnlichkeit *f*
 f similitude *f*; similarité *f*; conformité *f*
 r подобность *f*

3006 **congruence relation**
 d Kongruenzverhältnis *n*
 f rapport *m* de congruence
 r отношение *n* конгруэнтности

 * **congruent distribution function** → 2703

3007 **conical scanning**
 d konische Abtastung *f*; Quirlen *n*
 f exploration *f* conique; balayage *m* en cône
 r коническое сканирование *n*

3008 **conjuctive normal form**
 d konjunktive Normalform *f*
 f forme *f* normale conjonctive
 r конъюнктивная нормальная форма *f*

3009 **conjugate complex number**
 d konjugierte komplexe Zahl *f*
 f nombre *m* conjugué complexe
 r сопряжённое комплексное число *n*

3010 **conjugate complex value**
 d konjugierter komplexer Wert *m*
 f valeur *f* compkexe conjuguée
 r комплексно-сопряжённая величина *f*

3011 **conjugate root**
 d konjugierte Wurzel *f*
 f racine *f* conjuguée
 r сопряжённый корень *m*

3012 **conjugation frequency**
 d Verknüpfungsfrequenz *f*
 f fréquence *f* conjuguée
 r сопряжённая частота *f*

3013 **conjunction operation**
 d Konjunktionsoperation *f*; Bildung *f* der Konjunktion
 f opération *f* d'intersection
 r операция *f* конъюнкции

3014 **conjunctive search**
 d konjunktives Aufsuchen *n*
 f recherche *f* conjonctive
 r конъюнктивный поиск *m*

3015 **connect** *v*
 d verschalten; einschalten
 f connecter; raccorder
 r соединять; подключать; присоединять

3016 **connected**
 d verbunden; zusammenhängend
 f connexe; connecté
 r связной; связанный; соединенный

3017 **connected automaton**
 d angeschlossener Automat *m*
 f automate *m* connecté
 r связанный автомат *m*

3018 **connected graph**
 d zusammenhängender Graf *m*
 f graphe *m* connecté
 r связанный граф *m*

3019 **connected recording equipment**
 d angeschlossene Registrieranlage *f*
 f enregistreuse *f* couplée
 r подключенное регистрирующее устройство *n*

3020 **connecting compatibility**
 d Verbindungsverträglichkeit *f*; Zusammenschaltverträglichkeit *f*
 f compatibilité *f* de connexion
 r совместимость *f* по связям

3021 **connecting function**
 d Verbindungsfunktion *f*
 f fonction *f* de connexion
 r функция *f* объединения

3022 **connection buildup**
 d Verbindungsaufbau *m*
 f établissement *m* de connexion
 r установление *n* связи

3023 **connection diagram**
 d Schaltungsdiagramm *n*; Schaltplan *m*; Leitungsschema *n*; Verbindungsschema *n*
 f schéma *m* de câblage; plan *m* de couplage; schéma de connexion
 r схема *f* соединения

 * **connection diagram circuit** → 2377

 * **connection in oposition** → 3486

3024 **connection matrix**
 d Verbindungsmatrix *f*
 f matrice *f* de connexion
 r матрица *f* связей

3025 **connection point; point of connection**
 d Anschlusspunkt *m*
 f poste *m* de connexion; point *m* de raccordement
 r точка *f* подключения; точка включения

3026 connection-programmed control
 d verbindungsprogrammierte Steuerung f
 f commande f programmée par connexion
 r управление n по объединяющей
    программе

3027 connective
 d Satzband n; Verbindungsglied n; Koppler m
 f copule f; pièce-raccord m
 r связка f; соединительный элемент m

3028 connectives of two variables
 d Verknüpfungen fpl zwischen zwei Variablen
 f relations fpl entre deux variables
 r связи fpl двух переменных

3029 connectivity
 d Konnektivität f
 f connectivité f
 r связываемость f

3030 conscious error
 d realisierter Fehler m
 f erreur f réalisée
 r сознательно допускаемая ошибка f

3031 consecutive access
 d Zugriff m zu aufeinanderfolgenden Zellen
 f accès m consécutif
 r последовательный доступ m

3032 conservative system
 d konservatives System n
 f système m conservatif
 r консервативная система f

3033 conservative value
 d Beharrungswert m; ststionärer Wert m;
    Stationärwert m; Gleichgewichtswert m
 f valeur f permanente; valeur prescrite
 r установившееся значение n

3034 conserved information
 d konservierte Information f
 f information f conservée
 r законсервированная информация f

3035 considerable information
 d bedeutende Information f
 f information f considérable
 r значительная по объёму информация f

3036 consistence control
 d Konsistenzsteuerung f
 f réglage m de la consistance
 r регулирование n консистенции

3037 console

 d Pult n; Steuerpult n
 f pupitre m [de commande]
 r пульт m

3038 console auxiliary adapter
 d Konzolen-Zusatzadapter m
 f adapteur m auxiliaire de la console
 r вспомогательный адаптер m пульта

3039 console handler
 d Konsol-Handler m
 f handler-console m; manipulateur m console
 r пультовой манипулятор m

3040 console inquiry unit
 d Konsolenabfrageeinheit f
 f unité f d'interrogation de console
 r консольное опрашивающее устройство n

* console process unit → 7062

3041 constant area
 d konstanter Bereich m
 f domaine m constant; zone f constante
 r постоянная область f

* constant coefficient → 3047

**3042 constant-coefficient system; [linear]
       system of constant coefficients**
 d [lineares] System n mit konstanten
    Koeffizienten
 f système m aux coefficients constants;
    système linéaire aux coefficients invariables
 r [линейная] система f с постоянными
    коэффициентами

3043 constant component
 d konstante Komponente f
 f composante f constante
 r постоянная составляющая f

3044 constant-current regulator
 d Konstantstromregler m
 f régulateur m à courant constant
 r регулятор m постоянного тока

3045 constant-current source
 d Konstantstromquelle f
 f source f à intensité constante
 r источник m стабилизированного тока

3046 constant-data rate system
 d Datenübertragung n mit konstanter
    Übertragungsgeschwindigkeit
 f système m de transmission de données à
    vitesse constante
 r система f передачи данных с постоянной
    скоростью

3047 **constant factor; constant coefficient**
   d konstanter Faktor *m*
   f coefficient *m* constant
   r постоянный коэффициент *m*

3048 **constant grip force**
   d konstante Greifkraft *f*
   f force *f* de grippion constante
   r постоянное захватывающее усилие *n*

3049 **constant-length data unit**
   d Dateneinheit *f* konstanter Länge
   f unitè *f* des données à longueur constante
   r элемент *m* данных постоянной длины

* **constant of integration** → 6811

3050 **constant-pressure cycle**
   d Gleichdruckprozess *m*
   f processus *m* à pression constante
   r процесс *m* при постоянном давлении

3051 **constant-pressure drop**
   d konstanter Druckabfall *m*
   f chute *f* constante de pression
   r постоянный перепад *m* давления

3052 **constant radiation**
   d konstante Strahlung *f*
   f rayonnement *m* constant
   r постоянное облучение *n*

3053 **constant repetition interval**
   d stetiges Reihenfolgeintervall *n*
   f période *f* constante de répétition
   r постоянный интервал *m* чередования

3054 **constant resistance**
   d Konstantwiderstand *m*
   f résistance *f* constante
   r постоянное сопротивление *n*

3055 **constant source; constant supply**
   d Konstantspannungsquelle *f*
   f source *f* de tension constante
   r постоянный источник *m* питания

3056 **constant-speed floating regulator**
   d ausschlagunabhängiger astatischer Regler *m*
   f régulateur *m* astatique à vitesse constante
   r астатический регулятор *m* с постоянной скоростью

3057 **constant-speed playback**
   d Wiedergabe *f* mit gleichbleibender Geschwindigkeit
   f reproduction *f* à vitesse constante
   r воспроизведение *n* с постоянной скоростью

3058 **constant-speed scanning**
   d Abtastung *f* mit konstanter Geschwindigkeit
   f exploration *f* à vitesse constante; balayage *m* à vitesse constante
   r обзор *m* с постоянной скоростью

3059 **constant-stress life test**
   d Lebensdauerprüfung *f* bei konstanter Belastung
   f essai *m* de temps de vie par contrainte constante
   r испытание *n* на долговечность при постоянной нагрузке

* **constant supply** → 3055

3060 **constant time lag**
   d konstante Verzögerung *f*
   f retard *m* constant
   r постоянная выдержка *f* времени

3061 **constant value control; control with fixed set point**
   d Festwertregelung *f*
   f réglage *m* à valeur constante [de consigne]
   r стабилизация *f* регулируемого параметра

3062 **constant-velocity servomotor**
   d Stellmotor *m* mit konstanter Geschwindigkeit
   f servomoteur *m* à vitesse constante
   r сервомотор *m* с постоянной скоростью

3063 **constant-velocity system**
   d System *n* mit konstanter Stellgeschwindigkeit
   f système *m* à vitesse constante
   r система *f* с постоянной скоростью

3064 **constant-voltage regulator; voltage stabilizer**
   d Konstantspannungsregler *m*; Spannungsstabilisator *m*
   f stabilisateur *m* de tension
   r стабилизатор *m* напряжения; регулятор *m* постоянного напряжения

3065 **constituting components of numerical control**
   d Bausteine *mpl* für Digitaltechnik
   f éléments *mpl* constitutifs pour commande numérique
   r слагающие *npl* цифрового управления

* **construct** *v* → 3976

3066 **constructed assembly unit**
   d konstruierte Montageeinheit *f*

    f unité *f* d'assemblage construite
    r конструированная сборочная единица *f*

3067 construction
    d Konstruction *f*
    f konstruction *f*
    r конструкция *f*

3068 construction of computer system
    d Rechnersystemaufbau *m*
    f construction *f* du système de calculateur
    r конструкция *f* компьютерной системы

3069 construction of subassembly
    d Baugruppenkonstruktion *f*
    f construction *f* de sous-ensemble
    r конструкция *f* узла

3070 construction of the control system
    d Steuerungssystemaufbau *m*
    f montage *m* de système de commande
    r конструктивная структура *f* системы
       управления

3071 construction parts
    d Konstruktionsteile *npl*
    f pièces *fpl* de construction
    r элементы *mpl* конструкции

    * **construction unit → 2815**

3072 construction unit of manipulator
    d Manipulatorbaueinheit *f*
    f unité *f* de construction de manipulateur
    r узел *m* манипулятора

    * **construction unit system → 1966**

3073 constructive change of assembly unit
    d konstruktive Veränderung *f* der
       Montageeinheit
    f changement *m* constructif d'une unité
       d'assemblage
    r конструктивное изменение *n* сборочной
       единицы

3074 constructive drive design
    d konstruktive Antriebsgestaltung *f*
    f dessin *m* d'entraînement constructif
    r конструктивное исполнение *n* [системы]
       привода

3075 constructive interface condition
    d konstruktive Interfacebedingung *f*
    f condition *f* d'interface constructive
    r конструктивное условие *n* интерфейса

3076 constructive manipulator shaping
    d konstruktive Manipulatorgestaltung *f*

    f façonnage *m* de manipulateur constructif
    r конструктивное исполнение *n*
       манипулятора

3077 consumption
    d Verbrauch *m*
    f consommation *f*
    r потребление *n*

3078 contact
    d Kontakt *m*
    f contact *m*
    r контакт *m*

3079 contact arrangement
    d Kontaktanordnung *f*
    f montage *m* des contacts
    r расположение *n* контактов

3080 contact breaker
    d Kontaktunterbrecher *m*
    f interrupteur *m* de contact
    r прерыватель *m* контакта

3081 contact current
    d Kontaktstrom *m*
    f courant *m* de contact
    r контактный ток *m*; ток прикосновения

    * **contact element → 3098**

3082 contact feeler; contouring tracer
    d Fühler *m*; Taststift *m*
    f galet *m*; palpeur *m*
    r контактный щуп *m*

3083 contact force
    d Kontaktkraft *f*
    f force *f* de contact
    r контактное усилие *n*

3084 contact gap
    d Kontaktabstand *m*; Kontaktlücke *f*
    f intervalle *m* de contact
    r контактный зазор *m*

3085 contactless circuit
    d kontaktlose Schaltung *f*
    f circuit *m* sans contact
    r бесконтактная схема *f*

3086 contactless control
    d kontaktlose Steuerung *f*
    f commande *f* sans contacts
    r бесконтактное управление *n*

3087 contactless device
    d kontaktlose Einrichtung *f*

f dispositif *m* sans contacts
r бесконтактное устройство *n*

3088 contactless distributor in remote control
d kontaktloser fernwirktechnischer
   Vorwähler *m*
f distributeur *m* de télémécanique sans
   contacts
r бесконтактный распределитель *m* в
   системе телеуправления

3089 contactless impulse generator
d kontaktloser Impulsgeber *m*
f générateur *m* d'impulsions sans contacts
r бесконтактный генератор *m* импульсов

3090 contactless limit switch
d kontaktloser Endschalter *m*
f interrupteur *m* limite sans contacts
r бесконтактный концевой выключатель *m*

3091 contactless magnetic delay member
d kontaktloses magnetisches
   Verzögerungsglied *n*
f membre *m* magnétique de retard sans
   contacts
r бесконтактный элемент *m* магнитной
   линии задержки

3092 contactless measurement
d kontaktlose Messung *f*
f mesure *f* sans contacts
r бесконтактное измерение *n*

**3093 contactless pick-up; non-contact feeler
        device; non-contacting sensor**
d kontaktloser Geber *m*; kontaktloser
   Aufnehmer *m*; berührungslose
   Abtasteinrichtung *f*; berührungsloser
   Sensor *m*
f palpeur *m* sans contacts; capteur *m* sans
   contacts
r бесконтактный датчик *m*

3094 contactless scanning of pointers
d berührungslose Zeigerabtastung *f*
f balayage *m* sans contact des aiguilles
r бесконтактный поиск *m* положения
   стрелок [прибора]

3095 contactless selsyn
d kontaktloser Drehmelder *m*
f selsyn *m* sans contacts
r бесконтактный сельсин *m*

3096 contactless switch
d kontaktloser Schalter *m*
f commutateur *m* sans contacts

r бесконтактный выключатель *m*

3097 contactless system
d kontaktloses System *n*
f système *m* sans contact
r бесконтактная система *f*

**3098 contact member; contact element; contact
        piece**
d Kontaktglied *n*; Berührungselement *n*;
   Kontaktelement *n*; Kontaktstück *n*
f élément *m* de contact
r контактный элемент *m*

* **contact-modulated amplifier → 2368**

3099 contact noise
d Kontaktgeräusch *n*
f bruit *m* de contact
r шум *m* контакта

3100 contactor
d Kontaktgeber *m*; Steuerschütz *n*; Schütz *n*
f contacteur *m*; conjoncteur *m*
r контактор *m*; замыкатель *m*

3101 contactor servomechanism
d Relaisschalterservomechanismus *m*
f servomécanisme *m* à conjonctions
r релейный сервомеханизм *m*

3102 contact pick-off
d Kontaktabnahme *f*
f prise *f* à contact
r снимание *n* контактом; контактная
   съемка *f*

* **contact piece → 3098**

3103 contact-potential difference
d Kontaktspannung *f*
f différence *f* de potentiel au contact
r контактная разность *f* потенциалов

3104 contact pressure
d Kontaktdruck *m*
f pression *f* au contact
r контактное давление *m*

3105 contact profile
d Kontaktprofil *n*
f profil *m* de contact
r контактный профиль *m*

* **contact pyramid → 10898**

3106 contact sense point
d Kontaktabfragepunkt *m*

    f point *m* d'interrogation de contact
    r точка *f* контактного опроса

3107 contact sensor
    d Kontaktsensor *m*; Berührungssensor *m*;
       Kontaktfühler *m*
    f capteur *m* de contact; senseur *m* de contact
    r контактный сенсор *m*; датчик *m* наличия
       контакта

3108 contact sensor system
    d Berührungssensorsystem *n*
    f système *m* de contact d'un senseur
    r система *f* сенсоров касания

3109 contact set
    d Kontaktsatz *m*
    f groupe *m* des contacts
    r контактная группа *f*

3110 contact thermometer
    d Kontaktthermometer *n*
    f thermomètre *m* de contact
    r контактный термометр *m*

3111 contact travel
    d Schaltweg *m*; Kontaktstrecke *f*
    f longueur *f* de course de contact
    r расстояние *n* между контактами

3112 contact voltmeter
    d Kontaktvoltmeter *n*;
       Berührungsspannungsmesser *m*
    f voltmètre *m* à contacts
    r контактный вольтметр *m*

3113 contamination meter
    d Verseuchungsmessgerät *n*
    f mesureur *m* de contamination
    r аппарат *m* для измерения контаминации

3114 contamination monitor
    d Kontaminationsmonitor *m*
    f moniteur *m* de contamination
    r монитор *m* для контаминации

3115 continuity
    d Stetigkeit *f*; Kontinuität *f*
    f continuité *f*
    r непрерывность *f*

3116 continuity conditions
    d Stetigkeitsbedingungen *fpl*;
       Kontinuitätsbedingungen *fpl*
    f conditions *fpl* de continuité
    r условия *npl* непрерывности

3117 continuity equations

    d Kontinuitätsbeziehungen *fpl*
    f relations *fpl* de continuité
    r отношения *npl* непрерывности

3118 continuity tester
    d Durchgangsprüfer *m*
    f contrôleur *m* de continuité
    r прибор *m* для испытания на обрыв

3119 continuous
    d stetig
    f continu
    r непрерывный

3120 continuous action; continuous operation
    d kontinuierliche Wirkung *f*; Dauerbetrieb *m*;
       Fließbetrieb *m*
    f action *f* permanente; action continue;
       marche *f* continue; service *m* continu
    r непрерывное [воз]действие *n*

3121 continuous-action controller
    d kontinuierlicher Regler *m*
    f régulateur *m* à action continue
    r регулятор *m* непрерывного действия

3122 continuous-action optimization system
    d kontinuierlich wirkendes
       Optimisierungssystem *n*
    f système *m* d'optimalisation à action continue
    r система *f* оптимизации непрерывного
       действия

3123 continuous-action servomechanism
    d Dauerbetriebsservogerät *n*
    f servomécanisme *m* à action continue
    r сервомеханизм *m* непрерывного действия

3124 continuous action system
    d Daueraktionssystem *n*; System *n* mit stetiger
       Wirkung
    f système *m* à action continue
    r система *f* непрерывного действия

3125 continuous air monitor
    d kontinuierlich arbeitender Luftmonitor *m*;
       kontinuierlich arbeitendes
       Aerosolüberwachungsgerät *n*
    f moniteur *m* d'aérosols radioactifs en continu
    r прибор *m* для непрерывного контроля
       радиоактивности воздуха

3126 continuous analyzer
    d stetiger Analysator *m*
    f analyseur *m* continu
    r непрерывный анализатор *m*

3127 continuous approximation
    d stetige Annäherung *f*

    f approximation *f* continue
    r непрерывное приближение *n*

3128 continuous automatic measurement
    d stetig verlaufende selbsttätige Messung *f*
    f mesure *f* automatique permanente
    r автоматическое непрерывное измерение *n*

    * **continuous automatic viscometer** → 3129

3129 continuous automatic visco[si]meter
    d stetiges selbsttätiges Visko[si]meter *n*
    f viscosimètre *m* automatique à action continue
    r автоматический вискозиметр *m* непрерывного действия

3130 continuous control; infinitely fine control
    d stetige Regelung *f*; stetige Steuerung *f*
    f réglage *m* continu
    r непрерывное регулирование *n*; плавное регулирование

3131 continuous controller action
    d kontinuierliche Reglerwirkung *f*
    f action *f* permanente du régulateur
    r непрерывное регулирующее воздействие *n*

3132 continuous control system
    d stetig geregeltes System *n*
    f système *m* à commande continue
    r система *f* непрерывного регулирования

3133 continuous correction
    d kontinuierliche Korrektion *f*
    f correction *f* continue
    r непрерывная коррекция *f*

3134 continuous curve
    d stetige Kurve *f*
    f courbe *f* continue
    r непрерывная кривая *f*

3135 continuous curve distance-time protection
    d Distanzschutz *m* mit stetiger Auslösekennlinie; kontinuierliche Fernabschirmung *f*
    f dispositif *m* de protection de distance à caractéristique continue
    r дистанционная защита *f* с плавно-зависимой характеристикой выдержки времени

3136 continuous dependence
    d stetige Abhängigkeit *f*; kontinuierliche Abhängigkeit
    f dépendance *f* continue

    r непрерывная зависимость *f*

3137 continuous displacement operation
    d kontinuierlicher Verschiebungsarbeitsgang *m*
    f opération *f* de déplacement continu
    r операция *f* непрерывного перемещения

3138 continuous display
    d kontinuierliche Anzeige *f*
    f affichage *m* continu
    r дисплей m, работающий в непрерывном режиме

3139 continuous distribution
    d kontinuierliche Verteilung *f*
    f distribution *f* continue
    r непрерывное распределение *n*

3140 continuous exciting source
    d Dauerstricherreger *m*; Dauerstricherregungsquelle *f*
    f source *f* permanente d'excitation; source à excitation entretenue; source excitative à onde entretenue
    r источник *m* возбуждения, работающий в непрерывном режиме

3141 continuous fabrication; continuous manufacture
    d kontinuierliche Herstellung *f*
    f fabrication *f* continue
    r непрерывное изготовление *n*

3142 continuous frequency spectrum
    d kontinuierliches Frequenzspektrum *n*
    f spectre *m* continu de fréquences
    r непрерывный спектр *m* частот

3143 continuous function
    d stetige Funktion *f*
    f fonction *f* continue
    r непрерывная функция *f*

3144 continuous gas quantity measurement
    d kontinuierliche Gasmengenmessung *f*
    f mesure *f* continue de la quantité de gas
    r непрерывное измерение *n* расхода газа

3145 continuous humidity measurement
    d ununterbrochene Feuchtigkeitsmessung *f*
    f mesure *f* continue de l'humidité
    r непрерывное измерение *n* влажности

3146 continuous laser
    d Dauerstrichlaser *m*
    f laser *m* à onde entretenue
    r лазер *m* непрерывного излучения

3147 continuous liquid level measurement
d stetige Flüssigkeitsstandmessung *f*
f mesure *f* continue du niveau de liquide
r плавное измерение *n* уровня жидкости

3148 continuously adjustable inductivity
d regelbare Induktivität *f*
f inductance *f* variable
r переменная индуктивность *f*

* **continuous manufacture** → 3141

* **continuous mode** → 3150

3149 continuous monitoring; continuous supervision
d fortlaufende Kontrolle *f*; fortlaufende Überwachung *f*; Daueraufsicht *f*; Dauerüberwachung *f*
f contrôle *m* continu; surveillance *f* permanente
r непрерывный контроль *m*

* **continuous operation** → 3120

3150 continuous [operation] mode
d kontinuierlicher Modus *m*; kontinuierliche Betriebseise *f*
f mode *m* continu
r непрерывный режим *m*

3151 continuous oscillations
d kontinuierliche Schwingungen *fpl*; Dauerschwingungen *fpl*
f oscillations *fpl* continues
r незатухающие колебания *npl*

3152 continuous path control
d Stetigbahnsteuerung *f*
f positionnement *m* continu
r непрерывный контроль *m* траектории

3153 continuous phase
d kontinuierliche Phase *f*
f phase *f* continue
r непрерывная фаза *f*

3154 continuous pneumatic pressure regulation
d gleichmäßige Luftdruckregulation *f*
f régulation *f* continue de pression de l'air
r плавное регулирование *n* давления воздуха

3155 continuous process
d kontinuierlicher Prozess *m*
f procédé *m* continu
r непрерывный процесс *m*

3156 continuous process control
d Regelung *f* kontinuierlicher Prozesse
f régulation *f* continue de processus d'opération
r управление *n* поточным процессом

3157 continuous random variable
d stetige Zufallsgröße *f*
f variable *f* aléatoire continue
r непрерывная случайная величина *f*

3158 continuous rated current
d Dauernennstrom *m*
f courant *m* nominal continu
r номинальный продолжительный ток *m*

3159 continuous reading
d kontinuierliche Abtastung *f*
f lecture *f* continue; balayage *m* continu
r непрерывное снятие *n* сигнала

3160 continuous record
d Linienaufzeichnung *f*
f enregistrement *m* continu
r непрерывная запись *f*

3161 continuous request
d Daueranforderung *f*
f interrogation *f* continue
r непрерывный запрос *m*

3162 continuous signal
d kontinuierliches Signal *n*
f signal *m* permanent
r непрерывный сигнал *m*

3163 continuous stabilization
d kontinuierliche Stabilisierung *f*
f stabilisation *f* continue
r непрерывная стабилизация *f*

3164 continuous structure
d kontinuierliche Struktur *f*
f structure *f* continue
r непрерывная структура *f*

* **continuous supervision** → 3149

3165 continuous telemetring
d stetiges Fernmessverfahren *n*
f télémétrie *f* continue
r непрерывная телеметрия *f*; непрерывное дистанционное измерение *n*

3166 continuous temperature regulation
d gleichmäßige Temperaturregelung *f*
f régulation *f* continue de la température
r плавное регулирование *n* температуры

3167 continuous test signal controller
- *d* Regler *m* mit stetigem Prüfsignal
- *f* régulateur *m* à signal continu d'essai
- *r* регулятор *m* с непрерывным контрольным сигналом

3168 continuous variable
- *d* stetige Größe *f*; stetige Variable *f*; kontinuierliche Variable
- *f* grandeur *f* variable en continu
- *r* непрерывная переменная *f*

3169 continuous zooming
- *d* kontinuierliches Zoomen *n*
- *f* zoom *m* continu
- *r* непрерывное масштабирование *n*

3170 contour; outline
- *d* Kontur *m*
- *f* contour *m*
- *r* контур *m*; очертание *n*

3171 contour analysis
- *d* Konturenauswertung *f*
- *f* analyse *f* de contours
- *r* анализ *m* контура

3172 contour control system
- *d* Konturenregelungssystem *n*
- *f* système *m* de réglage des contours
- *r* контурная система *f* регулирования

3173 contour follower
- *d* Umrissfolgeregler *m*
- *f* profilomètre *m*
- *r* контурное копировальное следящее устройство *n*

3174 contouring control
- *d* Umrisssteuerung *f*
- *f* commande *f* du profilomètre
- *r* управление *n* копированием

* contouring tracer → 3082

3175 contour of grip surface
- *d* Greifflächenkontur *f*
- *f* contour *m* de la surface de grippion
- *r* контур *m* [внутренней] поверхности захвата

3176 control
- *d* Kontrolle *f*; Steuerung *f*
- *f* contrôle *m*; commande *f*
- *r* управление *n*

3177 control *v*
- *d* regeln; steuern; kontrollieren
- *f* commander; régler; contrôler
- *r* регулировать; контролировать

3178 control action
- *d* Regelvorgang *m*; Regelwirkung *f*; Steuereinwirkung *f*
- *f* action *f* de réglage; action de contrôle; action de commande
- *r* управляющее воздействие *n*; регулирующее воздействие

3179 control action coefficient
- *d* Regelfaktor *m*
- *f* coefficient *m* d'action réglante
- *r* коэффициент *m* регулирующего воздействия

3180 control adapter
- *d* Steuerungsanpassungsbaustein *m*
- *f* adapteur *m* de commande
- *r* адаптер *m* [схемы] управления

3181 control agent
- *d* Regelmedium *n*
- *f* agent *m* de réglage
- *r* регулирующее рабочее вещество *n*; регулирующий агент *m*

3182 control air
- *d* Steuerluft *f*
- *f* air *m* de commande
- *r* воздух *m* управления

3183 control algorithm
- *d* Steuerungsalgorithmus *m*
- *f* algorothme *m* de commande
- *r* алгоритм *m* управления

3184 control algorithm simulation
- *d* Steueralgorithmensimulation *f*
- *f* simulation *f* d'algorithme de commande
- *r* моделирование *n* алгоритма управления

3185 control and display panel
- *d* Steuer- und Anzeigetafel *f*
- *f* panneau *m* de contrôle et d'affichage
- *r* панель *f* управления и индикации

3186 control appliance
- *d* Steuergerät *n*; Befehlsgerät *n*
- *f* appareil *m* de manœuvre
- *r* управляющий прибор *m*

3187 control area; control domain
- *d* Regelfläche *f*; Regelbereich *m*
- *f* surface *f* de réglage; domaine *m* de réglage
- *r* область *f* управления; область регулирования; регулируемая зона *f*

3188 **control assembly**
    d Regelgruppe *f*; Regeleinrichtung *f*
    f montage *m* de réglage
    r узел *m* управления

3189 **control automaton**
    d Steuerautomat *m*
    f automate *m* de commande
    r управляющий автомат *m*

3190 **control band; regulation band**
    d Regelbereich *m*; Regelband *n*; Steuerband *n*
    f plage *f* de réglage; bande *f* de réglage;
       étendue *f* de réglage
    r полоса *f* регулирования

3191 **control block**
    d Kontrollblock *m*
    f bloc *m* de contrôle
    r контрольный блок *m*

3192 **control board; control panel**
    d Steuerpaneel *n*; Schalttafel *f*;
       Bedienungspult *n*
    f panneau *m* de commande; tableau *m* de
       contrôle
    r панель *f* управления; щит *m* управления

3193 **control break**
    d Gruppenunterbrechung *f*
    f coupure *f* de contrôle
    r регулирующий прерыватель *m*

3194 **control bus**
    d Regelbus *m*; Steuerbus *m*
    f bus *m* de commande
    r управляющая шина *f*

3195 **control by means of infinitely variable
      gears**
    d Regelung *f* mit stufenlos einstellbaren
       Getrieben
    f réglage *m* par engrenages à ajustage continu
    r регулирование *n* при помощи
       бесступенчатой передачи

3196 **control capacitance**
    d Steuerkapazität *f*
    f capacité *f* de commande
    r управляющая ёмкость *f*

3197 **control card**
    d Steuerkarte *f*
    f carte *f* de commande
    r управляющая карта *f*

3198 **control centre**
    d Regelwarte *f*; Kommandoraum *m*;

       Befehlsanlage *f*
    f poste *m* de commande; poste central de
       contrôle
    r пункт *m* управления

3199 **control chamber**
    d Steuerkammer *f*
    f chambre *f* de commande
    r управляющая камера *f*

3200 **control character**
    d Steuerzeichen *n*
    f caractère *m* de contrôle
    r управляющий символ *m*

3201 **control characteristic**
    d Steuercharakteristik *f*; Steuerkennlinie *f*
    f caractéristique *f* de réglage
    r характеристика *f* регулирования;
       характеристика управления

3202 **control circuit; steering circuit; driving
      circuit**
    d Steuerschaltung *f*; Antriebsstromkreis *m*
    f circuit *m* de commande; circuit de réglage;
       circuit de contrôle
    r управляющая схема *f*; цепь *f*
       регулирования

3203 **control circuit with prescribed overshoot**
    d Regelkreis *m* mit vorgeschiebener
       Überschwingweite
    f circuit *m* de réglage à dépassement prescrit
    r система *f* регулирования с предписанным
       пререгулированием

3204 **control circuit with transfer lag**
    d Regelkreis *m* mit Übertragungsverzögerung
    f circuit *m* de réglage à retard de transfert
    r контур *m* регулирования с отставанием в
       переходном процессе

3205 **control circuit with variable amplification**
    d Regelkreis *m* mit veränderlicher Verstärkung
    f circuit *m* de réglage à amplification variable
    r контур *m* регулирования с переменным
       усилением

*   **control clock → 3206**

3206 **control clock [pulse]**
    d Steuertakt *m*
    f impulsion *f* d'horloge de commande;
       horloge *f* de contrôle
    r управляющий тактовый импульс *m*

3207 **control code**
    d Steuerkode *m*

f code *m* de commande
r управляющий код *m*

3208 control coefficient
d Regelungskoeffizient *m*
f coefficient *m* de réglage
r коэффициент *m* регулирования

3209 control command
d Steuerbefehl *m*; Steuerinstruktion *f*
f instruction *f* de commande
r команда *f* управления

3210 control component
d Bauteil *n* der Regelung
f composante *f* de réglage
r регулирующий блок *m*

3211 control computer
d Steuerrechner *m*
f calculatrice *f* de commande
r управляющий компьютер *m*

3212 control constant
d Rückstellkonstante *f*
f constante *f* de rappel
r восстанавливающая постоянная *f*

3213 control contact
d Steuerkontakt *m*
f contact *m* de commande
r управляющий контакт *m*

3214 control convention
d Steuerungsregel *f*; Steuerungsvorschrift *f*
f convention *f* de commande
r инструкция *f* по управлению

3215 control counter; instruction counter
d Befehlszähler *m*
f compteur *m* d'instructions
r счётчик *m* команд

3216 control cut-off switch
d Steuerungsausschalter *m*;
   Steuerungsunterbrecher *m*
f interrupteur *m* de commande
r контрольный выключатель *m*

3217 control cycle
d Regelzyklus *m*
f cycle *m* de commande
r цикл *m* управления

* **control data → 3243**

3218 control data generation
d Steuerdatengeneration *f*

f génération *f* de données de commande
r выработка *f* управляющих данных

3219 control date
d Kontrolltermin *m*
f date *m* de contrôle
r контрольный срок *m*

3220 control decentralization
d Steuerungsdezentralisierung *f*
f décentralisation *f* de commande
r децентрализация *f* управления

3221 control design
d Regelungssystementwurf *m*
f projet *m* du système de réglage
r синтез *m* системы автоматического
   регулирования

3222 control deviation
d Regelabweichung *f*
f écart *m* de réglage
r отклонение *n* регулируемого
   параметра

* **control device → 3284**

3223 control dial
d Steuerscheibe *f*
f cadran *m* de commande
r диск *m* управления

* **control domain → 3187**

3224 control drive
d Regelantrieb *m*
f commande *f* de réglage
r регулирующий привод *m*

3225 control effectiveness
d Regelgüte *f*
f efficience *f* de réglage
r качество *n* регулирования

3226 control electrode
d Steuerelektrode *f*
f électrode *f* de commande
r регулирующий электрод *m*

3227 control electronics
d Steuerelektronik *f*
f électronique *f* de commande
r электронные устройства *npl*
   управления

3228 control element
d Regelelement *n*

    f élément *m* de réglage
    r регулирующий элемент *m*

3229 control engineering
    d Regelungstechnik *f*
    f technique *f* de réglage
    r техника *f* регулирования

3230 control equipment; controlling system
    d Regeleinrichtung *f*; Steuereinrichtung *f*
    f équipement *m* de commande
    r управляющее устройство *n*

3231 control error
    d Regelabweichung *f*
    f erreur *f* de réglage
    r ошибка *f* регулирования

3232 control error signal
    d Regelabweichungssignal *n*
    f signal *m* à la déviation
    r сигнал *m* рассогласования

3233 control feature of manipulator
    d Steuerungsmerkmal *n* eines Manipulators
    f repère *m* de commande de manipulateur
    r признак *m* системы управления манипулятором

3234 control field
    d Steuerzone *f*
    f zone *f* de commande
    r зона *f* управления

3235 control-flow computer
    d Steuerflussrechner *m*
    f ordinateur *m* à flux d'instructions
    r компьютер m, управляемый потоком команд

3236 control flux
    d Steuerfluss *m*
    f flux *m* de commande
    r управляющий поток *m*

3237 control for manipulation systems
    d Steuerung *f* für Manipulationssysteme
    f commande *f* pour systèmes de manipulation
    r управление *n* манипуляционными системами

3238 control function
    d Regelfunktion *f*
    f fonction *f* de commande
    r управляющая функция *f*

3239 control function operation
    d Steuerfunktionsoperation *f*
    f opération *f* de fonction de commande
    r операция *f* функции управления

3240 control grid; signal grid
    d Steuergitter *n*
    f grille *f* de contrôle; grille de commande
    r управляющая сетка *f*

3241 control hardware
    d Steuerungshardware *f*
    f hardware *m* de commande
    r аппаратура *f* управления

3242 control hierarchy principe
    d Prinzip *n* der Gruppenrangordnung
    f principe *m* d'hiérarchie de commande
    r принцип *m* иерархического управления

3243 control information; control data
    d Steuerdaten *pl*; Steuerinformation *f*
    f données *fpl* de commande; information *f* de commande
    r управляющая информация *f*; управляющие данные *pl*

3244 control input signal
    d Steuereingang *m*; Steuereingangsignal *n*
    f signal *m* d'entrée de commande; entrée *f* de commande
    r управляющий входной сигнал *m*

3245 control installation
    d Regelanlage *f*
    f installation *f* de réglage
    r установка *f* регулирования

3246 control installation with narrow dead zone
    d Regelanlage *f* mit schmaler Unempfindlichkeitszone
    f installation *f* de réglage à zone étroite d'insensibilité
    r регулирующая установка *f* с узким диапазоном нечувствительности

3247 control instruction
    d Regelbefehl *m*
    f commande *f* de réglage
    r команда *f* регулирования

3248 control interaction factor
    d Abhängigkeitsfaktor *m*
    f facteur *m* d'interaction
    r коэффициент *m* взаимодействия

3249 control-internal function
    d steuerungsinterne Funktion *f*

f fonction *f* interne de commande
r внутренняя функция *f* управления

3250 control interruption
d Steuerungsunterbrechung *f*
f interruption *f* de commande
r разрыв *m* в цепи управления

3251 control interval
d Regelungsintervall *n*; Regelabstand *m*
f intervall *m* de réglage
r интервал *m* регулирования

3252 control key
d Bedienungstaste *f*; Regelknopf *m*
f touche *f* de service
r ключ *m* управления

3253 controllability; readjustability
d Regelbarkeit *f*
f réglabilité *f*
r регулируемость *f*; управляемость *f*;
контролируемость *f*

3254 controllable
d verstellbar; lenkbar; regelbar; steuerbar
f contrôlable; réglable
r регулируемый; управляемый

3255 controllable approximation sensor
d steuerbarer Näherungssensor *m*
f senseur *m* d'approximation maniable
r управляемый датчик *m* ближней локации

3256 controllable coordinates
d steuerbare Koordinaten *fpl*
f coordonnées *fpl* maniables
r управляемые координаты *fpl*

3257 control law
d Regelungsgesetz *n*
f loi *f* de commande
r закон *m* управления

3258 controlled access system
d gesteuertes Zugriffssystem *n*
f système *m* d'accès commandé
r система *f* с контрольным доступом

3259 controlled addition of impurities
d gesteuerte Beimischung *f* von Unreinheiten
f addition *f* contrôlée d'impuretés
r контролируемое добавление *n* примесей

3260 controlled-atmosphere welding
d Schweißen *n* in kontrollierter Atmosphäre;
Schweißen in Kammern mit kontrollierter
Atmosphäre
f soudage *m* dans une atmosphère contrôlée;

soudage dans des chambres à atmosphère
contrôlée
r сварка *f* в камерах с контролируемой
атмосферой

3261 controlled by transistor
d transistorgesteuert
f commandé par transistor
r управляемый транзистором

3262 controlled damping
d gesteuerte Dämpfung *f*
f amortissement *m* commandé
r управляемое демпфирование *n*

3263 controlled delay pulse generator
d Impulsgenerator *m* mit gesteuerter
Impulsverzögerung
f générateur *m* d'impulsion à retard réglable
r генератор *m* импульсов с регулируемой
выдержкой времени

3264 controlled device; controlled member
d Regelobjekt *n*; Steuerobjekt *n*
f organe *m* commandé
r управляемое устройство *n*

3265 controlled effector state
d geregelter Effektorzustand *m*
f état *m* d'effecteur réglé
r регулированное эффекторное состояние *n*

3266 controlled function generator
d gesteuerter Funktionsgenerator *m*
f générateur *m* commandé de fonctions
r управляемый функциональный
преобразователь *m*

3267 controlled joint mechanism
d gesteuerter Fügemechanismus *m*
f mécanisme *m* de jointage commandé
r управляемый механизм *m* сопряжения

3268 controlled medium
d geregeltes Medium *n*
f médium *m* reglé
r регулируемая среда *f*

* **controlled member → 3264**

**3269 controlled member with interacted
conditions; controlled plant with
interacted parameters** (US)
d vermaschte Anlage *f*; vermaschte
Regelstrecke *f*
f élément *m* de régulation à plusieurs
paramètres interconnectés
r элемент *m* регулирования с несколькими
взаимосвязанными параметрами

3270 **controlled network**
  d geregeltes Netzwerk *n*; gesteuertes Netzwerk
  f réseau *m* réglé
  r управляемая схема *f*

3271 **controlled parameter**
  d Regelgröße *f*
  f quantité *f* réglée
  r регулируемый параметр *m*

3272 **controlled plant characteristic**
  d Regelstreckencharakteristik *f*
  f caractéristique *f* du système réglé
  r характеристика *f* регулируемого обекта

3273 **controlled plant identification**
  d Identifizierung *f* von Regelstrecken
  f identification *f* des systèmes de réglage
  r идентификация *f* объектов регулирования

*   **controlled plant with interacted
      parameters → 3269**

3274 **controlled power rectifier**
  d gesteuerter Leistungsgleichrichter *m*
  f redresseur *m* réglé de puissance
  r управляемый силовой выпрямитель *m*

3275 **controlled-speed generator**
  d Generator *m* mit gesteuerter Drehzahl
  f génératrice *f* à réglage de vitesse
  r генератор *m* с управляемым числом
       оборотов

3276 **controlled store**
  d gesteuerter Speicher *m*
  f mémoire *f* commandée
  r управляемое запоминающее устройство *n*

3277 **controlled system**
  d gesteuertes System *n*
  f système *m* réglé
  r управляемая система *f*

3278 **controlled system with transportation lag**
  d Regelstrecke *f* mit Totzeit
  f système *m* réglé à temps mort
  r регулируемая система *f* с запаздыванием

3279 **controlled temperature**
  d geregelte Temperatur *f*
  f température *f* contrôlée; température réglée
  r регулируемая температура *f*

3280 **controlled time interval**
  d gesteuertes Zeitintervall *n*
  f intervalle *m* de temps commandé
  r регулируемый временной интервал *m*

3281 **controlled value**
  d Regelgröße *f*
  f grandeur *f* réglée
  r регулируемая величина *f*

3282 **controlled value range**
  d Regelgrößenbereich *m*
  f étendue *f* de variation de la grandeur réglée
  r диапазон *m* изменения регулируемой
       величины

3283 **controlled variable**
  d gesteuerte Größe *f*; Steuergröße *f*;
       Steuerparameter *m*
  f variable *f* commandée; grandeur *f*
       commandée
  r управляемая величина *f*

3284 **controller; control device; control unit**
  d Kontroller *m*; Leitwerk *n*; Steuereinheit *f*;
       Regler *m*; Steuergerät *n*
  f contrôleur *m*; unité *f* de commande;
       combinateur *m*; régulateur *m*
  r контроллер *m*; управляющее
       устройство *n*; регулятор *m*

3285 **controller action**
  d Reglerfunktion *f*; Reglerwirkung *f*
  f action *f* du régulateur
  r воздействие *n* регулятора

3286 **controller resistance**
  d Reglerwiderstand *m*
  f résistance *f* de combinateur
  r сопротивление *n* управляющего звена

3287 **controller transfer function**
  d Reglerübertragungsfunktion *f*
  f fonction *f* de transfert du régulateur
  r функция *f* передачи регулятора

3288 **controller with locking device**
  d Fallbügelregler *m*
  f régulateur *m* à étrier
  r регулятор *m* со стопорным механизмом

3289 **control level**
  d Regelungspegel *m*
  f niveau *m* de commande
  r уровень *m* управления

3290 **control limits**
  d Regelgrenzen *fpl*
  f limites *fpl* de réglage
  r пределы *mpl* регулирования

3291 **control line**
  d Steuerleitung *f*

  f ligne *f* de commande
  r линия *f* управления

3292 controlling flow
  d Stellstrom *m*
  f flux *m* réglant; courant *m* réglant
  r регулирующий поток *m*

3293 controlling machine
  d regelnde Maschine *f*; steuernde Maschine
  f machine *f* réglante
  r управляющая машина *f*

3294 controlling motion
  d Steuerungsantrieb *m*; Steuerbewegung *f*
  f mouvement *m* de commande; calage *m*
  r управляющее движение *n*

3295 controlling power station
  d Regelkraftwerk *n*
  f usine-pilote *f* électrique
  r управляющая силовая станция *f*

  * **controlling system → 3230**

3296 controlling torque; driving torque
  d Steuerungsmoment *n*; Einstellmoment *n*; Antriebsmoment *n*
  f couple *m* directeur; couple moteur
  r управляющий момент *m*

3297 control loop
  d Regelkreis *m*; Regelschleife *f*
  f circuit *m* de réglage; boucle *f* de réglage
  r контур *m* регулирования

3298 control macro
  d Steuermakro *n*
  f macro *m* de commande
  r макрокоманда *f* управления

3299 control manipulator
  d Steuermanipulator *m*
  f manipulateur *m* de commande
  r управляющий манипулятор *m*

3300 control message; command message
  d Steuernachricht *f*; Befehlsnachricht *f*
  f message *m* de commande
  r управляющее сообщение *n*; командное сообщение

3301 control method
  d Regelverfahren *n*
  f mode *m* de réglage
  r метод *m* регулирования

3302 control mode

  d Steuermodus *m*
  f mode *m* de commande
  r режим *m* управления

3303 control model
  d Steuerungsmodell *m*
  f modèle *m* de commande
  r модель *f* управления

3304 control module
  d Steuerungsbaustein *m*
  f module *m* de commande
  r модуль *m* управления

  * **control needs → 3325**

3305 control of feedback system
  d Regelung *f* des Rückkopplungssystems
  f commande *f* du système asservi
  r управление *n* системой с обратной связью

  * **control of many-variable system → 8458**

3306 control of metal transfer
  d Steuern *n* des Werkstoffübergangs; Regelung *f* der Werkstoffübertragung
  f réglage *m* du transfert de métal; contrôle *m* de la transition du métal
  r регулирование *n* переноса металла; контроль *m* перехода металла

3307 control of operations
  d Operationssteuerung *f*
  f commande *f* des opérations
  r управление *n* операциями

3308 control of rotation velocity
  d Drehzahlregelung *f*
  f réglage *m* de la vitesse de rotation
  r регулирование *n* скорости вращения

3309 control of small flows
  d Regelung *f* kleiner Durchflussmengen
  f réglage *m* de petits débits
  r регулирование *n* малых расходов

3310 control optimization
  d Steuerungsoptimierung *f*
  f optimisation *f* de commande
  r оптимизация *f* управления

3311 control-oriented instruction set
  d steuerungsorientierter Befehlssatz *m*
  f jeu *m* d'instructions pour commandes
  r система *f* команд, ориентированная на решение задач управления

  * **control panel → 3192**

*   control path → 353

3312 control performance
d Regelgüte *f*; Regelverlauf *m*
f performance *f* de réglage
r качество *n* управления

3313 control process design
d Entwurf *m* des Regelvorganges
f projet *m* du processus réglé
r проект *m* процесса регулирования

3314 control process network
d Steuerverfahrensnetzwerk *n*
f réseau *m* de contrôle-commande
r схема *f* процесса управления

3315 control processor
d Steuerprozessor *m*
f processeur *m* de commande
r управляющий процессор *m*

3316 control program
d Steuerprogramm *n*; Lenkungsprogramm *n*
f programme *m* de commande
r управляющая программа *f*

3317 control program for object production
d Steuerprogramm *n* für Objektherstellung
f programme *m* pour production d'objet
r программа *f* управления производством изделий

3318 control protocol
d Steuerungsprotokoll *n*; Steuerungsprozedurvorschrift *f*
f procès-verbal *m* de commande
r протокол *m* управления

*   control pulse → 354

3319 control quantity
d Steuergröße *f*
f grandeur *f* de commande
r управляющая величина *f*

3320 control range
d Regelbereich *m*
f zone *f* de contrôle; étendu *f* de mesure de la grandeur réglée
r диапазон *m* регулирования

3321 control rate
d Regelgeschwindigkeit *f*; Stellgeschwindigkeit *f*
f vitesse *f* de réglage
r скорость *f* перестановки регулирующего органа

3322 control ratio
d Steuerverhältnis *n*
f rapport *m* de commande
r коэффициент *m* пропорционального регулятора

3323 control register
d Steuerungsregister *n*; Befehlszähler *m*
f registre *m* d'instructions
r регистр *m* управления

3324 control relay circuit
d Steuerungsrelaisstromkreis *m*
f circuit *m* de relais de commande
r схема *f* релейного управления

3325 control requirements; control needs
d Steuerungsanforderungen *fpl*; Steuerungserfordernisse *npl*
f exigences *fpl* de commande
r требования *npl* к управлению

3326 control rod calibration
d Eichung *f* des Regelstabes; Regelstabeichung *f*
f étalonnage *m* de la barre de réglage
r калибровка *f* регулирующего стержня

3327 control room
d Schaltwarte *f*; Überwachungsraum *m*
f salle *f* de commande
r зал *m* управления

*   control sensing → 12320

3328 control sensitive element
d Regelfühlglied *n*
f élément *m* sensible de réglage
r чувствительный элемент *m* регулирования

3329 control sequence
d Steuerfolge *f*; Regelungsfolge *f*; Steuerungsfolge *f*
f séquence *f* de contrôle; séquence de commande
r управляющая последовательность *f*

3330 control set
d Regelgarnitur *f*
f trousse *f* de réglage
r регулирующая установка *f*

*   control signal → 2639

3331 control signal generation
d Steuersignalerzeugung *f*
f génération *f* d'un signal de commande
r генерирование *n* сигнала управления

3332 control software
d Steuerungssoftware *f*
f software *m* de commande
r программное обеспечение *n* управления

* **control speed → 518**

3333 control stability
d Regelungssabilität *f*
f stabilité *f* de régulation
r устойчивость *f* регулирования

3334 control state
d Steuerzustand *m*
f état *m* de commande
r состояние *n* управления

3335 control statics
d Regelungsstatik *f*
f statique *f* du réglage
r статика *f* регулирования

3336 control station
d Kontrollstelle *f*; Wartestelle *f*;
Kontrollwarte *f*; Leitstation *f*
f station *f* directrice; poste *m* directeur
r станция *f* управления

3337 control step
d Regelungsstufe *f*
f pas *m* de réglage
r ступень *f* регулирования

3338 control store
d Steuerspeicher *m*
f mémoire *f* de commande
r накопитель *m* управляющего устройства

3339 control switch
d Steuerschalter *m*
f commutateur *m* de commande
r командный выключатель *m*

3340 control symbol
d Steuersymbol *n*
f symbole *m* de commande
r символ *m* управления

3341 control synchro
d Steuerdrehmelder *m*
f synchro-machine *f* de commande
r управляющий сельсин *m*

3342 control system analyzer
d Regelsystemanalysator *m*
f analyseur *m* du système de réglage
r анализатор *m* системы регулирования

3343 control system component
d Regelsystemkomponente *f*
f constituant *m* du circuit de commande
r блок *m* системы управления

3344 control system continuous simulator
d kontinuierliches Steuersystemmodell *n*
f simulateur *m* continu de système de
commande
r моделирующее устройство *n*
непрерывного действия системы
управления

**3345 control system with nonrational transfer
function**
d Regelkreis *m* mit nichtrationaler
Übertragungsfunktion
f système *m* asservi à fonction de transfert non
rationnelle
r система *f* регулирования с
нерациональными передаточными
функциями

3346 control system with time delay
d Regelungssystem *n* mit Totzeit
f système *m* asservi à retard
r система *f* регулирования с задержкой
времени

3347 control tape; pilot tape
d Steuerband *n*
f bande *f* de commande; bande pilote
r управляющая лента *f*

3348 control task
d Steuerungsaufgabe *f*
f tâche *f* de commande
r задача *f* по управлению

3349 control-technical connection
d steuerungstechnische Verknüpfung *f*
f combinaison *f* technique de commande
r контрольно-техническая связь *f*

**3350 control-technical coupling of machine
system**
d steuerungstechnische Kopplung *f* eines
Maschinensystems
f couplage *m* technique de la commande d'un
système de machine
r машинная система *f* со связью по технике
автоматического управления

3351 control-technical feature
(of manipulator generation)
d steuerungstechnisches Merkmal *n*
f caractéristique *f* technique de commande
r [характерный] параметр *m* системы
управления

3352 **control-technical possibilities**
   d steuerungstechnische Möglichkeiten *fpl*
   f possibilités *fpl* pilotetechniques
   r возможности *fpl* системы управления

3353 **control technique**
   d Regelungstechnik *f*
   f technique *f* de réglage
   r техника *f* регулирования

3354 **control technologies**
   d Steuerungstechnologien *fpl*
   f technologies *fpl* de commande
   r технология *f* управления

3355 **control technologies for automatons**
   d Steuerungstechnologien *fpl* für Automaten;
         Automatensteuerungstechnologien *fpl*
   f technologies *fpl* de commande pour
         automaten
   r технология *f* управления автоматами

3356 **control theory**
   d Regelungstheorie *f*; Kontrolltheorie *f*
   f théorie *f* de réglage; théorie de contrôle
   r теория *f* [автоматического] управления

3357 **control time**
   d Regelzeit *f*; Regeldauer *f*
   f temps *m* de réponse; durée *f* de réglage
   r время *n* регулирования;
         продолжительность *f* переходного
         процесса

3358 **control transfer; transfer of control**
   d Steuerübertragung *f*; Übertragung *f* dcr
         Steuerung
   f transfert *m* de contrôle
   r передача *f* управления

3359 **control transmitter**
   d Steuerungssender *m*; Steuersignalsender *m*
   f transmetteur *m* de commande
   r устройство *n* передачи управляющих
         сигналов

   *   **control unit → 3284**

   *   **control unit of industrial robot → 11082**

3360 **control valve for small flows**
   d Regelventil *n* für kleine Durchflussmengen
   f vanne *f* de réglage pour petits débits
   r регулирующий клапан *m* для малых
         расходов

   *   **control variable → 2642**

3361 **control vector**
   d Steuerungsvektor *m*
   f vecteur *m* de commande
   r управляющий вектор *m*

3362 **control voltage**
   d Steuerspannung *f*
   f tension *f* de commande
   r управляющее напряжение *n*

3363 **control winding**
   d Steuerwicklung *f*
   f bobinage *m* de commande; enroulement *m* de
         commande
   r управляющая обмотка *f*

   *   **control with fixed set point → 3061**

3364 **control word**
   d Steuerbefehl *m*
   f mot *m* de contrôle
   r управляющее слово *n*

3365 **control-word technique**
   d Steuerworttechnik *f*
   f technique *f* de mot de commande
   r метод m применения управляющих слов

3366 **convection current**
   d Konvektionsstrom *m*; Übertragungsstrom *m*
   f courant *m* de convection
   r конвекционный ток *m*

3367 **convectron**
   d Konvektron *n*
   f convectron *m*
   r конвектрон *m*

3368 **convention**
   d Konvention *f*; Vereinbarung *f*
   f convention *f*
   r соглашение *n*

3369 **conventional assembly sequence**
   d konventioneller Montageablauf *m*
   f séquence *f* d'assemblage traditionnelle
   r обыкновенная последовательность *f*
         сборки

3370 **conventional component**
   d konventionelles Bauelement *n*
   f composant *m* conventionnel
   r стандартный компонент *m*

3371 **conventional linear system**
   d herkömmliches lineares System *n*
   f système *m* linéaire conventionnel
   r обычная линейная система *f*

3372 **convergence**
 d Konvergenz *f*
 f convergence *f*
 r сходимость *f*

3373 **convergence acceleration**
 d Konvergenzbeschleunigung *f*
 f accélération *f* de convergence
 r ускорение *n* сходимости

3374 **convergence adjustment**
 d Konvergenzeinstellung *f*
 f ajustage *m* de la convergence
 r регулирование *n* сходимости

3375 **convergence algorithm**
 d Konvergenzalgorithmus *m*
 f algorithme *m* convergent
 r сходящийся алгоритм *m*

3376 **convergence attribute**
 d Konvergenzkriterium *n*
 f signe *m* de convergence
 r критерий *m* сходимости

3377 **convergence control**
 d Konvergenzregelung *f*
 f réglage *m* de la convergence
 r сходимость *f* процессов регулирования

3378 **convergence degree**
 d Konvergenzgrad *m*
 f degré *m* de convergence
 r степень *f* сходимости

3379 **convergence domain**
 d Konvergenzgebiet *n*
 f domaine *m* de convergence
 r облась *f* сходимости

3380 **convergence indicator**
 d Konvergenzmesser *m*
 f dispositif *m* de mesure de convergence
 r прибор *m* для измерения сходимости

3381 **convergence of series**
 d Reihenkonvergenz *f*
 f convergence *f* de la série
 r сходимость *f* ряда

3382 **convergence recorder**
 d Konvergenzschreiber *m*
 f enregistreur *m* de convergence
 r самописец *m* сходимости

3383 **convergent component**
 d Konvergenzelement *n*
 f élément *m* de convergence
 r собирательный элемент *m*

3384 **convergent control**
 d konvergente Regelung *f*
 f réglage *m* convergent
 r регулировака *f* сходимости

3385 **convergent oscillations; damped oscillations**
 d abnehmende Schwingungen *fpl*; abklingende Schwingungen; gedämpfte Schwingungen *fpl*
 f oscillations *fpl* convergentes; oscillations amorties
 r затухающие колебания *npl*

3386 **conversational communication**
 d Dialogverkehr *m*
 f communication *f* à dialogue
 r передача *f* [данных] в режиме диалога

3387 **conversational data entry**
 d Dialog-Dateneingabe *f*
 f entrée *f* de données par dialogue
 r ввод *m* данных в режиме диалога

3388 **conversational mode; dialogue mode; session mode**
 d Dialogbetrieb *m*; Dialogmodus *m*
 f mode *m* dialogué; mode conversationnel
 r режим *m* диалога

3389 **conversion; converting; transformation; transforming**
 d Konvertierung *f*; Umformung *f*; Umwandlung *f*; Transformation *f*
 f conversion *f*; inversion *f*; transformation *f*
 r конверсия *f*; преобразование *n*; превращение *n*

3390 **conversion accuracy**
 d Konversionsgenauigkeit *f*
 f précision *f* de conversion
 r точность *f* преобразования

3391 **conversion coefficient**
 d Konversionskoeffizient *m*
 f coefficient *m* de conversion
 r коэффициент *m* конверсии

3392 **conversion efficiency**
 d Umwandlungswirkungsgrad *m*
 f rendement *m* de conversion
 r эффективность *f* преобразования

3393 **conversion frequency**
 d Überlagerungsfrequenz *f*
 f fréquence *f* de conversion
 r частота *f* преобразования

3394 conversion loss
d Konversionsverlust *m*;
  Umwandlungwerlust *m*
f pertes *fpl* de conversion
r потери *fpl* на преобразование

3395 conversion time
d Umwandlungszeit *f*; Umsetzungszeit *f*
f temps *m* de conversion
r время *n* преобразования

3396 conversion transconductance
d Mischsteilheit *f*; Überlagerungssteilheit *f*
f pente *f* de conversion
r крутизна *f* преобразования

3397 conversion unit
d Umwandlungseinheit *f*
f unité *f* de conversion
r блок *m* преобразования

3398 convert *v*
d konvertieren; umwandeln; verwandeln
f convertir
r преобразовывать; превращать

3399 converted input signal
d umgeformtes Eingangssignal *n*
f signal *m* traduit d'entrée
r преобразованный входной сигнал *m*

3400 converted output signal
d umgeformtes Ausgangssignal *n*
f signal *m* traduit de sortie
r преобразованный выходной сигнал *m*

3401 converted variable
d umgeformte Größe *f*
f grandeur *f* traduite
r преобразованная переменная *f*

3402 converter; transducer; transductor
d Konverter *m*; Umwandler *m*; Wandler *m*;
  Geber *m*; Umsetzer *m*; Transduktor *m*
f convertisseur *m*; traducteur *m*
r конвертер *m*; преобразователь *m*

* **converter process** → 1830

3403 converter program
d Umwandlerprogramm *n*
f programme *m* de convertisseur
r программа *f* преобразователя

3404 converter protection
d Stromrichterschutz *m*
f protection *f* de convertisseur
r предохранение *n* преобразователей

3405 converter sensibility
d Umformerempfindlichkeit *f*
f sensibilité *f* de convertisseur
r чувствительность *f* преобразователя

* **converting** → 3389

3406 converting system
d Konvertiersystem *n*
f système *m* convertisseur
r система *f* преобразований

3407 conveying belt
  (for movement of party)
d Transportband *n*
f ruban *m* de transport
r конвейер *m*

3408 conveying error
d Beförderungsfehler *m*
f défaut *m* d'avancement
r ошибка *f* транспортировки

3409 conveyor chain
d Förderkette *f*
f chaîne *f* transporteuse
r передаточная цепь *f*

3410 conveyor device
d Fördereinrichtung *f*
f convoyeur *m*
r транспортное устройство *n*

3411 convolution of probability distribution
d Faltung *f* der Wahrscheinlichkeitsverteilung
f convolution *f* de ladistribution de probabilité
r свертка *f* распределения вероятностей

3412 convolution theorem
d Faltungssatz *m*
f théorème *m* de Duhamel
r теорема *f* свертывания

3413 cooling agent cycle
d Kühlmittelkreislauf *m*
f circulation *f* de l'agent réfrigérant
r циркуляция *f* охлаждающей среды

3414 cooling capacity
d Kühlungsleistung *f*
f puissance *f* de refroidissement
r производительность *f* охлаждающей
  установки

3415 cooperative multiaxis sensor
d kooperativer Mehrachsensensor *m*
f capteur *m* multiaxes coopératif; senseur *m*
  multiaxes coopératif
r совместный многоосевой сенсор *m*

3416 **cooperative structure**
   d kooperative Struktur *f*
   f structure *f* coopérative
   r объединенная структура *f*

3417 **coordinate assignment**
   d Koordinatenzuordnung *f*
   f assignation *f* de coordonnées
   r присваивание *n* координат

3418 **coordinated control system**
   d koordiniertes Steuerungssystem *n*
   f système *m* coordonné de commande
   r связанная система *f* управления

3419 **coordinate position**
   d Koordinatenlage *f*
   f position *f* de coordonnées
   r координатная позиция *f*

3420 **coordinate recorder; variable recorder**
   d Koordinatenschreiber *m*
   f enregistreur *m* de coordonnées
   r координатный самописец *m*

3421 **coordinate selector for data logging**
   d Koordinatenwähler *m* für Messzentralen
   f sélecteur *m* à coordonnées pour les centrales de mesure
   r координатный селектор *m* для регистрации данных

3422 **coordinate setting**
   d Punktsteuerung *f*; Koordinateneinstellung *f*
   f positionnement *m* par coordonnées
   r координатное регулирование *n*

3423 **coordinate store**
   d Koordinatenspeicher *m*
   f mémoire *f* à coordonnées
   r координатное запоминающее устройство *n*

3424 **coordinate system**
   d Koordinatensystem *n*
   f système *m* de coordonnées
   r система *f* координат

3425 **coordinate system of gear element**
   d Koordinatensystem *n* eines Getriebegliedes
   f système *m* de coordonnées d'un élément d'engrenage
   r координатная система *f* звена передачи

3426 **coordinate system origin**
   d Koordinatensystemursprung *m*; Ursprung *m* eines Koordinatensystems
   f origine *f* d'un système de coordonnées

   r начало *n* системы координат

3427 **coordinate transformation of space path**
   d Koordinatentransformation *f* einer Raumbahn
   f transformation *f* de coordonnées d'un chemin d'espace
   r преобразование *n* координат пространственной траектории

3428 **coordinate trasformation**
   d Koordinatentransformation *f*
   f transformation *f* de coordonnées
   r преобразование *n* координат

3429 **coordinate values**
   d Koordinatenwerte *mpl*; Koordinatendaten *pl*
   f valeurs *fpl* de coordonnées
   r координатные данные *pl*

3430 **coordinate values of movement points**
   d Koordinatenwerte *mpl* der Bewegungspunkte
   f valeurs *fpl* de coordonnées des points de mouvement
   r координаты *fpl* точек перемещения

3431 **coordinating element**
   d Koordinierungsglied *n*
   f organe *m* coordonnateur
   r координационное звено *n*

3432 **copy check**
   d Copy-Prüfung *f*
   f essai *f* copy
   r проверка *f* дублирования

   * **copy device → 3433**

3433 **copy[ing] device**
   d Kopiergerät *n*
   f appareil *m* à copier; copieur *m*
   r копировальное устройство *n*

3434 **correct** *v*
   d korrigieren
   f corriger
   r исправлять; корректировать

3435 **corrected actuating signal**
   d koorrigiertes Stellsignal *n*
   f signal *m* d'influence corrigé
   r корректированный управляющий сигнал *m*

3436 **corrected assembly process**
   d korrigierter Montagevorgang *m*
   f procédé *m* d'assemblage corigé
   r корректированный сборочный процесс *m*

3437 **corrected joint regulation**
   d   korrigierte Gelenkregelung *f*
   f   régulation *f* de joint corrigée
   r   корректированное регулирование *n*
       шарнира

3438 **corrected perturbance**
   d   ausgeregelte Störung *f*
   f   perturbation *f* surveillée
   r   скорректированное возмущение *n*

3439 **corrected program**
   d   korrigiertes Programm *n*
   f   programme *m* corrigé
   r   скорректированная программа *f*

3440 **correcting circuit**
   d   Korrekturkreis *m*
   f   circuit *m* correcteur
   r   корректирующая цепь *f*

3441 **correcting condition**
   d   Korrekturzustand *m*; Korrekturwert *m*
   f   condition *f* de correction; condition
       d'influence
   r   корректирующее воздействие *n*

3442 **correcting control; correction adjusting**
   d   Korrektursteuerung *f*; Korrekturregelung *f*
   f   commande *f* correctrice; réglage *m* du
       correction
   r   корректирующее регулирование *n*

3443 **correcting element; correction element;**
      correction term
   d   korrigierendes Element *n*; Korrekturglied *n*;
       Entzerrungselement *n*
   f   élément *m* correcteur; élément compensateur;
       terme *m* de correction
   r   корректирующий элемент *m*;
       корректирующее звено *n*

3444 **correcting feedback**
   d   Entzerrückkopplung *f*
   f   réaction *f* correctrice
   r   корректирующая обратная связь *f*

3445 **correcting feedforward**
   d   Entzerrvorkopplung *f*
   f   réaction *f* positive de correction
   r   корректирующая положительная обратная
       связь *f*

3446 **correcting filter**
   d   Korrektionsfilter *n*
   f   filtre *m* de correction
   r   корректирующий фильтр *m*

3447 **correcting function**
   d   Korrekturfunktion *f*
   f   fonction *f* de correction
   r   корректирующая функция *f*

3448 **correcting pulse**
   d   Korrekturimpuls *m*
   f   impulsion *f* de correction
   r   корректирующий импульс *m*; импульс
       добавления

3449 **correcting range**
   d   Stellbereich *m*; Korrekturbereich *m*
   f   étendue *f* de l'action correctrice; étendue de
       l'action réglante
   r   диапазон *m* коррекции

3450 **correcting unit**
   d   Korrekturblock *m*; Stellwerk *n*
   f   unité *f* de correction
   r   корректирующий блок *m*

3451 **correcting variable; manipulated variable**
   d   Stellgröße *f*
   f   grandeur *f* réglante; grandeur de commande
   r   регулирующая величина *f*

*   **correction adjusting** → 3442

3452 **correction data**
   d   Berichtigungsangaben *fpl*
   f   données *fpl* correctives
   r   корректирующие данные *pl*

*   **correction element** → 3443

3453 **correction factor**
   d   Korrekturfaktor *m*; Berichtigungsfaktor *m*
   f   facteur *m* de correction
   r   коэффициент *m* коррекции

3454 **correction lag; corrective lag**
   d   Korrekturverögerung *f*
   f   retard *m* de correction
   r   корректирующее запаздывание *n*;
       задержка *f* коррекции

3455 **correction member**
   d   Korrektionsglied *m*
   f   membre *m* de correction
   r   член *m* коррекции

*   **correction motion** → 3456

3456 **correction movement; correction motion**
   d   Korrekturbewegung *f*
   f   mouvement *m* de correction
   r   корректирующее движение *n*

3457 **correction of dynamic properties**
 d Korrektur *f* dynamischer Eigenschaften
 f correction *f* de propriétés dynamiques
 r коррекция *f* динамических свойств

3458 **correction rate**
 d Regelgeschwindigkeit *f*
 f vitesse *f* de régulation
 r скорость *f* корректирования

3459 **correction system**
 d Korrektursystem *n*
 f système *m* de correction
 r система *f* коррекции

3460 **correction table**
 d Korrektionstabelle *f*
 f tableau *m* de correction
 r таблица *f* поправок

 * **correction term** → 3443

3461 **corrective action**
 d Berichtigung *f*; Richtigstellung *f*;
   Stellgrößenänderung *f*; korrigierender
   Eingriff *m*
 f action *f* corrective
 r корректирующее воздействие *n*

3462 **corrective factor**
 d Korrekturfaktor *m*
 f coefficient *m* de correction
 r поправочный коэффициент *m*

 * **corrective lag** → 3454

3463 **corrective procedure**
 d Verbesserungsverfahren *n*
 f procédé *m* de correction
 r корректирующая процедура *f*

3464 **correlated controllers**
 d Korrelationsregler *mpl*; korrelierte
   Regler *mpl*
 f régulateurs *mpl* corrélés
 r связанные регуляторы *mpl*

3465 **correlation**
 d Korrelation *f*
 f corrélation *f*
 r корреляция *f*

3466 **correlation coefficient matrix**
 d Korrelationskoeffizientenmatrix *f*
 f matrice *f* du coefficient de corrélation
 r матрица *f* коэффициентов корреляции

3467 **correlation computer**

3467
 d Korrelationsrechner *m*
 f calculateur *m* de corrélation; calculatrice *f* de
   corrélation
 r машина *f* для вычисления корреляции

3468 **correlation degree**
 d Korrelationsgrad *m*
 f degré *m* de corrélation
 r степень *f* корреляции

3469 **correlation dependence**
 d Korrelationsabhängigkeit *f*
 f dépendance *f* de corrélation
 r корреляционная зависимость *f*

3470 **correlation electronics**
 d Korrelationselektronik *f*
 f électronique *f* de corrélation
 r корреляционная электроника *f*

3471 **correlation function**
 d Korrelationsfunktion *f*
 f fonction *f* de corrélation
 r функция *f* корреляции

3472 **correlation meter**
 d Korrelationsmesser *m*
 f appareil *m* de mesure de corrélation
 r коррелометр *m*

3473 **correlation method**
 d Korrelationsmethode *f*
 f méthode *f* de corrélation
 r метод *m* корреляции

3474 **correlation technique**
 d Korrelationsverfahren *n*;
   Korrelationstechnik *f*
 f technique *f* de corrélation
 r техника *f* корреляционного анализа

3475 **correlation theory**
 d Korrelationstheorie *f*
 f théorie *f* de corrélation
 r теория *f* корреляции

3476 **correlation tracking**
 d Korrelationsbahnverfolgung *f*
 f poursuite *f* par corrélation
 r корреляционное сопровождение *n*

3477 **correlation tracking system**
 d Korrelationsbahnverfolgungssystem *n*
 f système *m* de poursuite par corrélation
 r корреляционная система *f* слежения

3478 **correlation triangulation**
 d Korrelationstriangulation *f*

   f triangulation *f* par corrélation
   r корреляционная триангуляция *f*

3479 correlative compensation
   d Korrelationskompensation *f*
   f compensation *f* corrélative
   r корреляционная компенсация *f*

3480 correspondence rule
   d Übereinstimmungsgesetz *n*
   f loi *f* de correspondance
   r правило *n* соответствия

3481 corresponding value
   d korrespondierender Wert *m*
   f valeur *f* correspondante
   r соответствующее значение *n*

3482 cosine wave
   d Kosinuswelle *f*
   f onde *f* cosinusoïdale
   r косинуидальная волна *f*

   * **Coulomb friction** → 3483

3483 Coulomb friction [force]; dry friction
   d Coulombsche Reibung *f*; trockene Reibung;
        Störung *f* mit Coulombscher
        Reibungscharakteristik
   f friction *f* forcée de Coulomb; frottement *m*
        sec; frottement solide; frottement de
        Coulomb
   r сила *f* трения Кулона; сухое трение *n*

3484 countable set; enumerable set
   d abzählbare Menge *f*
   f ensemble *m* comptable
   r счётное множество *n*

3485 counter circuit
   d Zählkreis *m*
   f circuit *m* compteur
   r счётная цепь *f*

3486 counter connection; connection in oposition
   d Gegenschaltung *f*
   f montage *m* en opposition
   r противовключение *n*; встречное
        включение *n*

3487 counter control
   d Zählersteuerung *f*
   f commande *f* par compteur
   r контроль *m* с помощью счётчиков

   * **countercurrent principle** → 1546

3488 countercurrent system
   d Gegensinnschaltung *f*
   f montage *m* contre-courant
   r инверсная схема *f*

3489 counterflow apparatus
   d Gegenstromapparat *m*
   f appareil *m* contre-courant
   r противоточный аппарат *m*

3490 counterflow operation; counterflow process
   d Gegenstromprozess *m*;
        Gegenstromverfahren *n*
   f opération *f* contre-courant; procédé *m* à
        contre-courant
   r противоточный процесс *m*

   * **counterflow process** → 3490

3491 counter reset
   d Zählerrückstellung *f*
   f remise *f* à zéro du compteur
   r установка *f* счётчика в ноль

3492 counter terminal
   d Zählerendgerät *n*
   f terminal *m* de compteur
   r оконечный терминал *m* со счётчиком

3493 counter-timer unit
   d Zähler-Zeitgeber-Einheit *f*
   f unité *f* compteur-rythmeur
   r устройство *n* задания [интервала] времени

3494 counting apparatus
   d Zählapparatur *f*
   f appareil *m* de comptage
   r счётная аппаратура *f*

3495 counting assembly for absolute measurements
   d Zählanordnung *f* für Absolutmessungen
   f ensemble *m* de comptage pour des mesures
        absolues
   r счётный прибор *m* для абсолютных
        измерений

3496 counting circuit
   d Zählerschaltung *f*
   f circuit *m* compteur
   r счётная схема *f*; сумирующая цепь *f*

3497 counting decade
   d Zähldekade *f*
   f décade *f* de comptage
   r счётная декада *f*

3498 **counting decoder**
  d Zähldekodierer *m*
  f décodeur *m* à comptage
  r счётный дешифратор *m*

3499 **counting error**
  d Zählfehler *m*
  f erreur *f* de comptage
  r ошибка *f* счёта

3500 **counting pulse**
  d Zählimpuls *m*
  f impulsion *f* de comptage
  r счётный импульс *m*

3501 **counting relay**
  d Zählrelais *n*
  f relais *m* de comptage
  r релейный счётчик *m*

*   **counting stage → 406**

3502 **coupled circuits**
  d gekoppelte Stromkreise *mpl*
  f circuits *mpl* couplés
  r связанные контуры *mpl*

3503 **coupled control element combination**
  d Verbundregelung *f*
  f réglage *m* combiné
  r смешанное регулирование *n*

3504 **coupling adjustment**
  d Kopplungseinstellung *f*
  f réglage *m* de couplage
  r регулировка *f* соединения

3505 **coupling circuit**
  d Kopplungskette *f*
  f circuit *m* de couplage
  r схема *f* соединения

3506 **coupling component**
  d Kopplungskomponente *f*
  f composant *m* de couplage
  r элемент *m* соединения

3507 **coupling diagram of control**
  d Koppelplan *m* einer Steuerung
  f plan *m* de couplage de la commande
  r схема *f* связей в системе управления

3508 **coupling effect**
  d Kopplungseffekt *m*
  f effet *m* de couplage
  r эффект *m* связи

3509 **coupling factor**
  d Kopplungsfaktor *m*
  f coefficient *m* de couplage
  r коэффициент *m* связи

3510 **coupling function**
  d Kopplungsfunktion *f*
  f fonction *f* de couplage
  r функция *f* связи

3511 **coupling resistance; coupling resistor** (US)
  d Kopplungswiderstand *m*
  f résistance *f* de couplage
  r сопротивление *n* [в цепи] связи

*   **coupling resistor → 3511**

3512 **course correction**
  d Kursberichtigung *f*
  f correction *f* du parcours; correction de route
  r поправка *f* курса

3513 **covariance**
  d Kovarianz *f*
  f covariance *f*
  r ковариантность *f*

3514 **covariant element**
  d kovariantes Element *n*
  f élément *m* covariant
  r ковариантный элемент *m*

3515 **Cramer's rule**
  d Cramersche Regel *f*
  f règle *f* de Cramer
  r правило *n* Крамера

3516 **crest factor; peak factor**
  d Scheitelfaktor *m*
  f facteur *m* de crête; facteur de pointe
  r пикфактор *m*

3517 **crest value; peak value**
  d Spitzenwert *m*; Scheitelwert *m*;
     Höchstwert *m*
  f valeur *f* de crête
  r пиковое значение *n*

3518 **criterion**
  d Kriterium *n*
  f critère *m*
  r критерий *m*

3519 **criterion function**
  d Gütefunktion *f*
  f fonction *f* de qualité
  r функция *f* качества

3520 **criterion of middle losses**
  d Kriterium *n* der Verlustmittelwerte

f critérium *m* de pertes moyennes
r критерий *m* средных потерь

3521 criterion of non-interaction
d beeinflussungsfreies Kriterium *n*
f critère *m* d'autonomie
r критерий *m* автономности

3522 criterion of optimal modulus
d Kriterium *n* des optimalen Moduls
f critère *m* du module optimal
r критерий *m* оптимального модуля

3523 criterion of performance
d Leistungskriterium *n*
f critère *m* de puissance
r критерий *m* качества

3524 critical alignment
d Feinjustierung *f*
f alignement *m* critique
r прецизионное совмещение *n*

3525 critical attitude
d kritische Haltung *f*; kritische Stellung *f*
f attitude *f* critique
r критическое положение *n*

3526 critical constant
d kritische Konstante *f*
f constante *f* critique
r критическая константа *f*

3527 critical damping
d kritische Dämpfung *f*
f amortissement *m* critique
r критическое демпфирование *n*

3528 critical density
d kritische Dichte *f*
f densité *f* critique
r критическая плотность *f*

3529 critical distance
d kritische Entfernung *f*
f distance *f* critique
r критическое расстояние *n*

3530 critical error angle
d kritischer Verstimmungswinkel *m*
f angle *m* critique d'erreur
r критический угол *m* рассогласования

* **critical frequency → 3617**

3531 critical grid current
d kritischer Gitterstrom *m*
f courant *m* critique de grille

r критический сеточный ток *m*

3532 critical grid voltage
d kritische Gitterspannung *f*
f tension *f* critique de grille
r критическое сеточное напряжение *n*

3533 critical induction
d kritische Induktion *f*
f induction *f* critique
r критическое значение *n* индукции

3534 critical limit
d Abfallgrenze *f*
f limite *f* critique
r критический предел *m*

3535 critical matching
d kritische Anpassung *f*
f adaptation *f* critique
r согласование *n* по критическим параметрам

3536 critical point
d kritischer Punkt *m*
f point *m* critique
r критическая точка *f*

3537 critical potential
d kritisches Potential *n*
f potentiel *m* critique
r критический потенциал *m*

3538 critical pressure
d kritischer Druck *m*
f pression *f* critique
r критическое давление *n*

3539 critical resistance
d kritischer Widerstand *m*; Grenzwiderstand *m*
f résistance *f* critique
r критическое сопротивление *n*

3540 critical stability; stability limit; stability boundary
d Stabilitätsgrenze *f*
f limite *f* de stabilité; frontière *f* du domaine de stabilité
r граница *f* устойчивости

3541 critical state
d kritischer Zustand *m*
f état *m* critique
r критическое состояние *n*

3542 critical temperature
d kritische Temperatur *f*

f température *f* critique
r критическая температура *f*

3543 critical value
  d kritischer Wert *m*
  f valeur *f* critique
  r критическое значение *n*

3544 critical voltage difference
  d kritische Spannungsdifferenz *f*
  f différence *f* critique des tensions
  r критическая разность *f* напряжений

3545 critical volume
  d kritisches Volumen *n*
  f volume *m* critique
  r критический объём *m*

3546 critical wavelength
  d kritische Wellenlänge *f*
  f longueur *f* d'onde critique
  r критическая длина *f* волны

3547 cross-bar switch
  d Koordinatenschalter *m*
  f sélecteur *m* à coordonnées
  r координатный переключатель *m*

3548 cross correlation
  d Kreuzkorrelation *f*
  f corrélation *f* mutuelle
  r взаимная корреляция *f*

3549 cross-correlation function
  d Kreuzkorrelationsfunktion *f*
  f fonction *f* de corrélation mutuelle; fonction d'intercorrélation
  r функция *f* взаимной корреляции

3550 cross coupling
  d Kreuzkopplung *f*
  f couplage *m* parasite
  r перекрестная взаимная связь *f*

3551 cross-development system
  d Cross-Entwicklungssystem *n*
  f système *m* de développement cross
  r кросс-система *f* проектирования

3552 cross distortion
  d Kreuzverzerrung *f*; gegenseitige Modulationsverzerrung *f*
  f perturbance *f* de diaphonie
  r перекрестное искажение *n*

3553 cross modulation
  d Kreuzmodulation *f*
  f transmodulation *f*; intermodulation *f*

r перекрестная модуляция *f*

3554 cross noise
  d Übersprechstörungen *fpl*
  f diaphonie *f*
  r перекрестные помехи *fpl*

3555 cross-over frequency
  d Übergangsfrequenz *f*
  f fréquence *f* de coupure
  r частота *f* разделения

* cross point → 6838

3556 cross section
  d Querschnitt *m*
  f section *f* transversale
  r поперечное сечение *n*

3557 cross sensitivity
  d Querrichtungsempfindlichkeit *f*
  f sensibilité *f* transversale
  r поперечная чувствительность *f*

3558 cross simulator
  d Cross-Simulator *m*
  f simulateur *m* cross
  r моделирующая программа *f* операционной кросс-системы

3559 cross-spectral density
  d Kreuzspektraldichte *f*
  f densité *f* réciproque spectrale
  r взаимная спектральная плотность *f*

3560 cross-talk meter
  d Nebensprechmesser *m*
  f diaphonomètre *m*
  r измеритель *m* переходного затухания

3561 cryogenic data processor
  d Kryogenik-Datenverarbeitungsanlage *f*
  f appareil *m* cryogénique de traitement des données
  r криогенное устройство *n* обработки данных

3562 cryogenic elements
  d Tieftemperaturelemente *npl*
  f éléments *mpl* cryogéniques
  r криогенные элементы *mpl*

3563 cryogenic equipment
  d Kryogengerät *n*
  f équipement *m* cryogénique
  r криогенная аппаратура *f*

3564 cryogenic extraction process
  d Kälteextraktionsverfahren *n*

f procédé *m* d'extraction cryogénique
r метод *m* экстрагирования глубоким
  охлаждением

3565 cryogenic fractionation process
d Kältefraktionierungsverfahren *n*
f fractionnement *m* cryogénique; procédé *m* de
  séparation par cryogénique
r метод *m* фракционирования глубоким
  охлаждением

3566 cryogenics
d Kriogenik *f*; Tieftemperaturtechnik *f*;
  Technik *f* der Supraleiter
f cryogénie *f*
r криогеника *f*; криогенная техника *f*

3567 cryogenic store
d Kryogenspeicher *m*
f mémoire *f* cryogénique
r криогенный накопитель *m*

3568 cryogenic system
d Kryogensystem *n*
f système *m* cryogène
r криогенная система *f*

3569 crystal structure
d kristallische Struktur *f*; Kristallstruktur *f*
f structure *f* cristalline
r кристаллическая структура *f*

3570 cubical equation
d kubische Gleichung *f*
f équation *f* cubique
r кубическое уравнение *n*

3571 cumulative down-time
d kumulative Ausfallzeit *f*
f temps *m* cumulatif hors de service
r совокупное время *n* отказа

3572 cumulative operating time
d kumulative Betriebszeit *f*
f temps *m* de service cumulatif
r совокупное производственное время *n*

3573 cumulative spectral density
d kumulative spektrale Dichte *f*
f densité *f* spectrale cumulée
r кумулятивная спектральная плотность *f*

3574 current amplification factor
d Stromverstärkungsfaktor *m*
f facteur *m* d'amplification en courant
r коэффициент *m* усиления тока

3575 current amplifier

d Stromverstärker *m*
f amplificateur *m* de courant
r усилитель *m* тока

3576 current-carrying capacity
d Strombelastbarkeit *f*
f courant *m* de régime continu; capacité *f* de
  charge de courant
r пропускная способность *f* по току

3577 current circuit
d Stromkreis *m*
f circuit *m* de courant
r цепь *f* тока

3578 current control
d Stromregelung *f*
f réglage *m* de courant
r регулирование *n* тока

3579 current density
d Stromdichte *f*
f densité *f* de courant
r плотность *f* тока

3580 current direction
d Stromrichtung *f*
f direction *f* de courant
r направление *n* тока

3581 current-driven device
d stromgesteuertes Bauelement *n*
f module *m* à commande par courant
r возбуждаемое током устройство *n*

3582 current efficiency
d Stromausbeute *f*
f rendement *m* en courant
r коэффициент *m* использования тока

3583 current feedback
d Stromrückkopplung *f*
f réaction *f* de courant
r обратная связь *f* по току

3584 current gain
d Stromgewinn *m*
f gain *m* en courant
r усиление *n* тока

**3585 current impulse; current pulse; current
  surge**
d Stromstoß *m*
f impulsion *f* de courant
r импульс *m* напряжения тока

3586 current leakage effect
d Leckstromeffekt *m*

f effet *m* de fuite de courant
r эффект *m* тока утечки

3587 current limit
d Strombegrenzung *f*
f limitation *f* d'intensité de courant
r ограничение *n* тока

3588 current-limit control
d Strombegrenzungsregelung *f*
f réglage *m* par limiteur de courant
r управление *n* с ограничением тока

3589 current-limit starting
d stromgesteuertes Anlassen *n*
f démarrage *m* ampèremétrique
r пуск *m* в ход с ограничением тока

3590 current mode
d laufender Modus *m*
f mode *m* courant
r текущий режим *m*

3591 current peak
d Stromspitze *f*
f pointe *f* de courant
r пик *m* тока

3592 current protection
d Stromschutz *m*
f protection *f* ampèremétrique
r защита *f* от тока

*   **current pulse → 3585**

3593 current pulse amplifier
d Stromimpulsverstärker *m*
f amplificateur *m* à impulsion de courant
r усилитель *m* импульсов тока

3594 current regulator; current stabilizer
d Stromregler *m*; Stromstabilisator *m*
f stabilisateur *m* d'intensité; stabilisateur de courant
r регулятор *m* тока; стабилизатор *m* тока

3595 current sensitivity
d Stromempfindlichkeit *f*
f sensibilité *f* en courant
r чувствительность *f* по току

3596 current source
d Stromquelle *f*
f source *f* de courant
r источник *m* тока

3597 current spreading
d Stromaufweitung *f*

f étalement *m* du courant
r расходимость *f* тока

*   **current stabilizer → 3594**

*   **current surge → 3585**

3598 current systems
d derzeitige Systeme *npl*
f systèmes *mpl* actuels
r существующие системы *fpl*

3599 current telemeter
d Stromfernmessgerät *n*
f appareil *m* de télémesure à couplage par intensité
r токовый дистанционный измерительный прибор *m*

3600 current-time integral
d Zeitintegral *n* des Stromes; integrierter Strom *m*
f courant *m* intégré
r интеграл *m* тока во времени

3601 current-to-frequency conversion
d Strom-Frequenz-Umsetzung *f*; Strom-Frequenz-Wandlung *f*
f conversion *f* courant fréquence
r преобразование *n* ток-частота

3602 current transfer ratio
d Stromübertragungsverhältnis *n*
f rapport *m* de transfert de courant
r коэффициент *m* передачи тока

3603 current value
d gegenwärtiger Wert *m*
f valeur *f* actuelle
r текущее значение *n*

3604 current-voltage characteristic
d Strom-Spannungs-Charakteristik *f*
f caractéristique *f* tension-courant
r вольт-амперная характеристика *f*

3605 curve analyzer
d Kurvenanalysator *m*
f analyseur *m* de courbes
r прибор *m* для анализа кривых

3606 curve follower logic
d Kurvenverfolgungslogik *f*; Verfahren *n* für Kurvenverfolgung
f procédé *m* pour enregistrement des courbes
r логика *f* программного слежения

*   **curve plotter → 5971**

3607 curve point
 d Kurvenpunkt *m*
 f point *m* de courbes
 r точка *f* кривой

3608 curve scanner
 d Kurvenabtaster *m*; Kurvenabtastgerät *n*
 f dispositif *m* d'exploration de courbes
 r прибор *m* для снятия кривых

3609 curve tracing
 d Kurvendarstellung *f*
 f représentation *f* en forme de courbe
 r прочерчивание *n* кривой

3610 customer information control system
 d Kundeninformationssteuersystem *n*
 f système *m* de commande pour information
    de clients
 r система *f* управления информацией
    пользователей

3611 custom interface
 d Kunden-Interface *n*
 f interface *f* pour client
 r интерфейс *m* с нестандартными
    устройствами пользователя

3612 cut-off; cut-off switch
 d Abschalter *m*; Schalter *m*; Trennschalter *m*
 f interrupteur *m*
 r выключатель *m*

3613 cut off *v*
 d abschneiden; abschalten; trennen;
    unterbrechen
 f couper; séparer; déconnecter; disjoindre
 r выключать; отключать

3614 cut-off attenuation
 d Grenzdämpfung *f*
 f affaiblissement *m* critique
 r предельное ослабление *n*

**3615 cut-off characteristic of a current-limiting
      fuse**
 d Abschaltcharakteristik *f* einer
    Strombegrenzungssicherung
 f caractéristique *f* d'amplitude du courant
    coupé
 r характеристика *f* выключения
    токоограничителя

3616 cut-off condition
 d Abschaltungsbedingung *f*
 f condition *f* de coupure
 r условие *n* выключения

3617 cut-off frequency; critical frequency
 d Grenzfrequenz *f*; Trennfrequenz *f*;
    Schnittfrequenz *f*; kritische Frequenz *f*
 f fréquence *f* de coupure; fréquence critique
 r частота *f* среза; критическая частота

3618 cut-off signal
 d Sperrsignal *n*
 f signal *m* de coupure
 r сигнал *m* отсечки; запирающий сигнал

* **cut-off switch** → 3612

3619 cut-off time
 d Sperrzeit *f*; Ausschaltzeit *f*; Trennzeit *f*
 f temps *m* de coupure
 r момент *m* запирания

3620 cybernetic automaton
 d kybernetischer Automat *m*
 f automate *m* cybernétique
 r кибернетический автомат *m*

3621 cybernetic control; cybernetic direction
 d kybernetische Regelung *f*; kybernetische
    Leitung *f*; kybernetische Führung *f*
 f réglage *m* cybernétique; gestion *f*
    cybernétique
 r кибернетическое управление *n*

* **cybernetic direction** → 3621

* **cybernetic model** → 3624

3622 cybernetic representation
 d kybernetische Darstellung *f*
 f représentation *f* cybernétique
 r кибернетическое представление *n*

3623 cybernetics
 d Kybernetik *f*
 f cybernétique *f*
 r кибернетика *f*

3624 cybernetic simulator; cybernetic model
 d kybernetisches Modell *n*
 f simulateur *m* cybernétique; modèle *m*
    cybernétique
 r кибернетическая модель *f*

3625 cybernetic structure
 d kybernetische Struktur *f*
 f structure *f* cybernétique
 r кибернетическая структура *f*

3626 cybernetic system
 d kybernetisches System *n*

    f système *m* cybernétique
    r кибернетическая система *f*

3627 cycle; cyclic process; circulation
    d Kreisprozess *m*; Kreislauf *m*
    f cycle *m*; circulation *f*; circuit *m* fermé
    r циклический процесс *m*; циркуляция *f*

3628 cycle code; cyclic code
    d zyklischer Kode *m*
    f code *m* cyclique
    r циклический код *m*

3629 cycle control
    d Zyklussteuerung *f*
    f commande *f* à cycles
    r циклическое управление *n*

3630 cycle criterion
    d Zykluskriterium *n*
    f critère *m* de cycle
    r число *n* повторений цикла

3631 cycle delay
    d Zyklusverzögerung *f*
    f retard *m* de cycle
    r циклическая задержка *f* времени

3632 cycle duration measurement
    d Periodendauermessung *f*
    f mesure *f* de la durée de la periode
    r измерение *n* длительности цикла

3633 cycle index
    d Schleifenindex *m*
    f index *m* de cycles
    r число *n* цикла

3634 cycle principle
    d Kreislaufprinzip *n*
    f principe *m* de recirculation
    r принцип *m* циркуляции

3635 cycle progress
    d Zyklusverlauf *m*
    f déroulement *m* du cycle
    r ход *m* цикла

3636 cycle ratio
    d Kreislaufverhältnis *n*
    f taux *m* de recirculation
    r фактор *m* рециркуляции

3637 cycle signal generation
    d Taktsignalerzeugung *f*
    f génération *f* de signal de rythme
    r выработка *f* тактового сигнала

3638 cycle stage
    d Zyklusstufe *f*
    f échelon *m* du cycle
    r стадия *f* цикла

3639 cyclic admittance
    d zyklischer Leitwert *m*
    f admittance *f* cyclique
    r периодическая полная проводимость *f*

3640 cyclical binary code
    d zyklisch binärer Kode *m*
    f code *m* binaire cyclique
    r циклический двойчный код *m*

3641 cyclic algorithm
    d zyklischer Algorithmus *m*
    f algorithme *m* cyclique
    r циклический алгоритм *m*

3642 cyclically magnetized condition
    d zyklisch magnetisierter Zustand *m*
    f état *m* à magnétisation cyclique
    r условие *n* циклического намагничивания

3643 cyclic automatic manipulator
    d zyklischer automatischer Manipulator *m*
    f manipulateur *m* cyclique automatique
    r цикловой автоматический манипулятор *m*

3644 cyclic automaton
    d zyklischer Automat *m*
    f automate *m* cyclique
    r циклический автомат *m*

3645 cyclic check
    d zyklische Prüfung *f*
    f contrôle *m* cyclique
    r циклический контроль *m*

    * **cyclic code → 3628**

3646 cyclic control system
    d zyklisches Regelsystem *n*
    f système *m* de commande cyclique
    r циклическая система *f* регулирования

3647 cyclic graph
    d zyklischer Graf *m*
    f graphe *m* cyclique
    r циклический граф *m*

3648 cyclic mode
    d zyklische Arbeitsweise *f*; zyklische
       Betriebsart *f*
    f mode *m* cyclique
    r циклический режим *m*

3649 cyclic permutation
  d zyklische Permutation *f*
  f permutation *f* cyclique
  r циклическая перестановка *f*

  * **cyclic process → 3627**

3650 cyclic program
  d zyklisches Programm *n*
  f programme *m* cyclique
  r циклическая программа *f*

3651 cyclic shift
  d zyklische Verschiebung *f*
  f décalage *m* circulaire
  r циклический сдвиг *m*

3652 cyclic storage; cyclic store
  d zyklischer Speicher *m*
  f mémoire *f* cyclique
  r периодическая память *f*

  * **cyclic store → 3652**

3653 cyclic telemetering
  d zyklische Fernmessung *f*
  f télémétrie *f* cyclique
  r циклическая телеметрия *f*

3654 cyclization
  d Zyklisierung *f*
  f cyclisation *f*
  r циклизация *f*

3655 cymoscope; wave detector
  d Wellendetektor *m*; Wellenanzeiger *m*
  f détecteur *m* d'ondes; déceleur *m* d'ondes
  r индикатор *m* колебаний

D

3656 daisy-chain interrupt servicing
  d Daisy-Chain-Interruptbedienung *f*
  f service *m* d'interruption en chaîne
  r обслуживание *n* источников прерываний

3657 daisy-chain interrupt structure
  d Daisy-Chain-Interruptstruktur *f*
  f structure *f* d'interruption en chaîne
  r схема *f* формирования сигналов
       прерываний

3658 daisy-chain priority interrupt logic
  d Daisy-Chain-Interruptprioritätslogik *f*
  f chaîne *f* de priorités d'interruptions
  r логика *f* формирования сигналов
       прерываний

3659 daisy-chain structure
  d Daisy-Chain-Struktur *f*
  f structure *f* à signaux en chaîne
  r последовательно-приоритетная
       структура *f*

3660 damped frequency (US)**; damped natural
       frequency**
  d gedämpfte Eigenfrequenz *f*
  f fréquence *f* naturelle amortie
  r собственная частота *f* затухающих
       колебаний

  * **damped oscillations** → **3385**

3661 damped periodic element
  d periodisch gedämpftes Element *n*
  f élément *m* à amortissement périodique
  r подвижная часть *f* с периодическим
       демпфированием

3662 damped sinusoid
  d abklingende Sinusoide *f*; abklingende
       Sinusschwingung *f*
  f sinusoïde *f* amortie
  r затухающая синусоида *f*

3663 damper; damping device; antivibrator
  d Dämpfer *m*; Dämpfungsglied *n*;
       Dämpfungsvorrichtung *f*
  f amortisseur *m*
  r демпфирующее звено *n*; демпфер *m*;
       демпфирующее устройство *n*

3664 damping action
  d Dämpfung *f*; Dämpfwirkung *f*
  f amortissement *m*
  r затухающее действие *n*; демпфирующее
       действие

3665 damping adjustment
  d Dämpfungseinstellung *f*
  f réglage *m* de l'amortissement
  r регулировка *f* демпфирования

3666 damping circuit
  d Dämpfungskreis *m*
  f circuit *m* d'atténuation
  r демпфирующая схема *f*

  * **damping coefficient** → **1099**

3667 damping constant; decay constant
  d Dämpfungskonstante *f*; Zerfallskonstante *f*;
       Ausschwingkonstante *f*
  f constante *f* d'amortissement; constante
       d'évanouissement
  r постоянная *f* затухания

**3668 damping couple; damping moment;
       damping torque**
  d Dämpfungsmoment *n*
  f couple *m* d'amortissement; moment *m*
       d'affaiblissement
  r демпфирующий момент *m*

3669 damping curve
  d Abklingkurve *f*
  f courbe *f* d'amortissement
  r кривая *f* затухания

3670 damping decrement
  d Dämpfungsdekrement *n*;
       Dämpfungsabnahme *f*
  f décrément *m* d'amortissement
  r декремент *m* затухания

  * **damping device** → **3663**

3671 damping element
  d Dämpfungselement *n*
  f élément *m* d'amortissement
  r демпфирующий элемент *m*

3672 damping elimination
  d Entdämpfung *f*
  f élimination *f* d'amortissement
  r устранение *n* демпфирования

  * **damping factor** → **1099**

3673 damping magnet
  d Dämpfungsmagnet *m*

f aimant *m* amortisseur
r демпфирующий магнит *m*

* **damping moment → 3668**

3674 **damping of a signal**
d Signaldämpfung *f*
f atténuation *f* d'un signal
r затухание *n* сигнала

3675 **damping period**
d Abklingperiode *f*
f période *f* d'extinction
r период *m* затухания

3676 **damping resistance**
d Dämpfungswiderstand *m*
f résistance *f* d'amortissement
r демпфирующее сопротивление *n*

3677 **damping time**
d Dämpfungszeit *f*; Beruhigungszeit *f*
f durée *f* d'amortissement
r время *n* успокоения

* **damping torque → 3668**

3678 **dark current**
d Dunkelstrom *m*
f courant *m* d'obscurité
r темновой ток *m*

3679 **data**
d Daten *pl*
f données *fpl*
r данные *pl*

3680 **data acceptance**
d Datenannahme *f*; Datenübernahme *f*
f acceptation *f* de données; admission *f* de données
r приём *m* данных

3681 **data access**
d Datenzugriff *m*
f accès *m* de données
r доступ *m* к данным

3682 **data acquisition; data collection**
d Datenerfassung *f*; Messwerterfassung *f*
f collection *f* de données; acquisition *f* de données; saisie *f* de données
r сбор *m* данных

3683 **data-acquisition system**
d Datenerfassungssystem *n*
f système *m* d'acquisition de données
r система *f* сбора данных

3684 **data-acquisition unit**
d Datenerfassungseinheit *f*
f unité *f* d'acquisition de données
r устройство *n* сбора данных

3685 **data adapter unit**
d Prozessdatenadapter *m*;
Prozessdatenanschlusseinheit *f*
f adapteur *m* de données [de procédé]
r устройство *n* согласования данных, снятых с промышленного процесса

3686 **data address**
d Datenadresse *f*
f adresse *f* de données
r адрес *m* данных

3687 **data bank system**
d Datenbanksystem *n*
f système *m* de banque de données
r система *f* банка данных

3688 **data basis**
d Datenbasis *f*
f base *f* de données
r база *f* данных

3689 **data bit converter**
d Datenbitumsetzer *m*
f convertisseur *m* pour bits de données
r двоичный преобразователь *m* данных

3690 **data block**
d Datenblock *m*
f bloc *m* de données
r блок *m* данных

3691 **data bus**
d Datenbus *m*
f bus *m* de données
r шина *f* данных

3692 **data bus buffer**
d Datenwegpuffer *m*
f tampon *m* de trajet de données
r буфер *m* информационного канала

* **data byte → 2009**

3693 **data category**
d Datenkategorie *f*
f catégorie *f* de données
r категория *f* данных

3694 **data channel**
d Datenkanal *m*
f canal *m* de données; piste *f* de données
r канал *m* передачи данных

3695 **data channel attachment**
 d Datenkanalzusatzeinrichtung *f*; Steuerung *f* für Datenkanal
 f attachement *m* pour canal de données
 r вспомагательное устройство *n* информационного канала

3696 **data check indicator**
 d Datenprüfanzeiger *m*
 f indicateur *m* de test pour données
 r индикатор *m* контроля данных

3697 **data checking**
 d Datenprüfung *f*
 f vérification *f* de données
 r контроль *m* данных

3698 **data checking program**
 d Datenprüfprogramm *n*
 f programme *m* d'essai des données
 r программа *f* для контроля данных

3699 **data coding system**
 d Datenkodier[ungs]system *n*
 f système *m* de codification des données
 r система *f* кодирования данных

 * **data collection** → 3682

3700 **data communication control**
 d Datenverkehrssteuerung *f*; Datenübermittlungssteuerung *f*
 f commande *f* de communication de données
 r управление *n* передачей данных

3701 **data communication equipment**
 d Datenverkehrseinrichtung *f*; Datenübermittlungseinrichtung *f*
 f équipement *m* de communication de données; dispositif *m* de communication de données
 r аппаратура *f* передачи данных

3702 **data communication protocol**
 d Datenübertragungsprotokoll *n*
 f procès-verbal *m* de communication de données
 r протокол *m* обмена данных

3703 **data communication terminal**
 d Datenverkehrsterminal *n*; Datenkommunikationsendgerät *n*
 f terminal *m* de communication de données
 r терминал *m* для передачи данных

3704 **data control block**
 d Datensteuerblock *m*
 f bloc *m* de commande des données
 r блок *m* управления данными

3705 **data converter; data translator**
 d Datenwandler *m*; Datenumsetzer *m*
 f convertisseur *m* de données
 r преобразователь *m* данных

3706 **data-directed**
 d datengesteuert
 f commandé par données
 r управляемый данными

3707 **data display panel**
 d Datendarstellungstafel *f*
 f panneaum de représentation de données; tableau *m* d'affichage de données
 r панель *f* индикации данных

3708 **data dump; information loss**
 d Informationsverlust *m*
 f perte *f* d'information
 r потеря *f* информации

3709 **data encoding**
 d Datenkodierung *f*
 f codage *m* de données
 r кодирование *n* данных

3710 **data exchange**
 d Datenaustausch *m*
 f échange *m* de données
 r обмен *m* данными

3711 **data input**
 d Dateneingabe *f*
 f entrée *f* de données
 r ввод *m* данных

3712 **data input into analog computer**
 d Dateneingabe *f* in Analogrechner
 f introduction *f* de données dans une calculatrice analogique
 r ввод *m* данных в аналоговую вычислительную машину

3713 **data input into digital computer**
 d Dateneingabe *f* in einen Ziffernrechenautomaten
 f introduction *f* de données dans une calculatrice digitale
 r ввод *m* данных в цифровую вычислительную машину

3714 **data input technique**
 d Dateneingabetechnik *f*
 f technique *f* d'entrée des données
 r техника *f* ввода данных

3715 data link control
  d Datenleitungssteuerung *f*;
     Datenverbindungssteuerung *f*
  f commande *f* de liaison de données
  r управление *n* каналом передачи данных

**3716 data link system; data transmission
system**
  d Datenübertragungssystem *n*
  f système *m* de transmission de données
  r система *f* передачи данных

3717 data logger
  d Datenspeicher *m*; automatische
     Registriervorrichtung *f*
  f enregistreur *m* de données
  r устройство *n* для автоматической записи
     данных

3718 data logging
  d Datenerfassung *f*
  f consignation *f* des informations
  r запись *f* данных

3719 data logging machine
  d Datenerfassungsanlage *f*
  f dispositif *m* d'enregistrement de mesures
  r машина *f* для регистрации данных

3720 data management
  d Datenverwaltung *f*; Datenmanagement *n*
  f gestion *f* des données
  r управление *n* данными

3721 data manipulation
  d Datenmanipulation *f*
  f manipulation *f* de données
  r манипуляция *f* данными

3722 data modem
  d Datenmodem *n*
  f modem *m* de données
  r модем *m* данных

3723 data multiplexing
  d Datenmultiplexierung *f*
  f multiplexage *m* de données
  r мультиплексный режим *m* обработки
     данных

3724 data net control
  d Datennetzsteuerung *f*
  f commande *f* de réseau de données
  r управление *n* информационной сетью

3725 data of environment model
  d Umweltmodelldaten *pl*
  f données *fpl* d'un modèle d'environnement

  r данные *pl* модели окружающего
     пространства

3726 data of manipulator control
  d Manipulatorsteuerungsdaten *pl*; Daten *pl*
     einer Manipulatorsteuerung
  f données *fpl* d'une commande de
     manipulateur
  r параметры *mpl* системы управления
     манипулятором

3727 data output
  d Datenausgabe *f*
  f sortie *f* de données
  r вывод *m* данных

3728 data output rate
  d Datenausgaberate *f*
  f part *f* d'une sortie des données
  r интенсивность *f* вывода данных

3729 data packet switching system
  d Datenpaketvermittlungssystem *n*
  f système *m* de communication pour paquets
     de données
  r система *f* коммутации пакета данных

3730 data packet system
  d Datenpaketsystem *n*
  f système *m* de paquet de données
  r система *f* пакета данных

3731 data presentation
  d Datendarstellung *f*
  f présentation *f* de données
  r представление *n* данных

3732 data processing
  d Datenverarbeitung *f*
  f traitement *m* de données
  r обработка *f* данных

3733 data processing centre; computing centre
  d Rechenzentrum *n*; Rechenzentrale *f*;
     Datenverarbeitungszentrum *n*
  f centre *m* de calcul; centre de traitement de
     données
  r центр *m* обработки данных;
     вычислительный центр

3734 data processing machine
  d Datenverarbeitungsmaschine *f*
  f machine *f* à traiter l'information
  r машина *f* для обработки данных

3735 data processing rate
  d Datenverarbeitungsgeschwindigkeit *f*

f vitesse *f* de traitement de données
r скорость *f* обработки данных

3736 data processing system
d Datenverarbeitungssystem *n*
f système *m* de traitement de données
r система *f* обработки данных

3737 data processing theory
d Datenverarbeitungstheorie *f*; Theorie *f* der Datenverarbeitung
f théorie *f* du traitement des données
r теория *f* обработки данных

3738 data protection
d Datenschutz *m*
f protection *f* de données
r защита *f* данных

3739 data reader
d Datenleser *m*
f lecteur *m* des données
r устройство *n* считывания данных

3740 data ready signal
d Datenbereitschaftssignal *n*
f signal *m* de disposition de données
r сигнал *m* готовности данных

3741 data recorder
d Datenanzeiger *m*; Datenregistriergerät *n*
f appareil *m* enregistreur de données
r устройство *n* записи данных

3742 data reduction
d Datenverminderung *f*; Datenreduktion *f*
f réduction *f* des informations
r обработка *f* информации

3743 data remote processing
d Datenfernverarbeitung *f*
f traitement *m* de données à distance; informatique *f* à distance
r дистанционная обработка *f* данных

3744 data remote transfer
d Datenfernübertragung *f*
f transmission *f* de données à distance
r дистанционная передача *f* данных

3745 data set
d Datenmenge *f*
f ensemble *m* de données
r количество *n* данных

3746 data setup time
d Datenbereitstellzeit *f*
f temps *m* de mise à disposition de données

r время *n* установки данных

3747 data storage control
d Steuerung *f* von gespeicherten Daten
f commande *f* de données emmagasinées
r управление *n* накопленной информацией

3748 data storage device
d Datenspeicherungsvorrichtung *f*
f dispositif *m* d'emmagasinage de données
r устройство *n* накопления данных

3749 data strobe
d Datenstrobe *m*; Datenmarkierimpuls *m*
f impulsion *f* de marquage de données
r сигнал *m* стробирования

3750 data structure
d Datenstruktur *f*; Datenwort *n*
f structure *f* de données
r структура *f* данных

3751 data switching
d Datenvermittlung *f*
f communication *f* de données
r коммутация *f* данных

3752 data terminal
d Datenendstelle *f*; Datenterminal *n*
f terminal *m* de données
r терминал *m*

3753 data terminal system
d Datenendstellensystem *n*
f système *m* d'un terminal de données
r система *f* терминалов

3754 data transceiver
d Daten-Sender/Empfänger *m*
f émetteur-récepteur *m* de données
r приёмопередатчик *m* данных

3755 data transfering
d Informationsübertragung *f*
f transfert *m* de données
r перенос *m* данных

3756 data transfer operation
d Datenübertragungsoperation *f*
f opération *f* de transfert de données
r операция *f* по передаче данных

*   data translator → 3705

3757 data transmission link
d Datenübertragungsverbindung *f*
f lien *m* de transmission de données
r линия *f* передачи данных

*   **data transmission system** → 3716

3758 data transmission technique
 d Datenübertragungsverfahren *n*;
      Datenübertragungstechnik *f*
 f technique *f* de transmission de données;
      procédé *m* de transmission de données
 r техника *f* передачи данных

3759 data transmitter
 d Datenübermittler *m*
 f transmetteur *m* de données
 r передатчик *m* данных

3760 data type
 d Datentyp *m*; Datenart *f*
 f type *m* de données
 r тип *m* данных

3761 data type transformation
 d Datentypumwandlung *f*;
      Datentyptransformation *f*
 f transformation *f* de type de données
 r преобразование *n* типа данных

3762 data-valid signal
 d Datengültigkeitssignal *n*
 f signal *m* de validité de données
 r сигнал *m* истинности данных

**3763 dating pulse; master pulse;
       synchronization pulse**
 d Synchronisationsimpuls *m*
 f top *m* de synchronisation; impulsion *f* de
      synchronisation
 r синхронизирующий импульс *m*

3764 datum error
 d Messwertfehler *m*
 f erreur *f* de valeur indiquée
 r ошибка *f* индикации

*   **D-controller** → 10629

3765 dead band; dead zone
 d Unempfindlichkeitszone *f*;
      Unempfindlichkeitsbereich *m*; totes Band *n*
 f bande *f* d'insensibilité; zone *f* morte
 r зона *f* нечувствительности

*   **dead-beat** → 908

3766 dead-beat ammeter
 d gedämpftes Strommessgerät *n*;
      nichtperiodisches Amperemeter *n*
 f ampèremètre *m* apériodique
 r апериодический амперметр *m*

3767 dead-beat instrument
 d gedämpftes Messgerät *n*
 f mesureur *m* amorti; mesureur apériodique
 r апериодический измерительный прибор *m*

3768 deadlock handling
 d Systemblockierungsbehandlung *f*
 f traitement *m* de blocage [de système]
 r обработка *f* ситуаций взаимоблокировки

3769 dead stroke
 d toter Gang *m*
 f course *f* à vide
 r мертвый ход *m*

3770 dead time; idle time
 d Totzeit *f*
 f temps *m* mort
 r мертвое время *n*

3771 dead time correction
 d Totzeitkorrektur *f*
 f correction *f* du temps mort
 r поправка *f* на мертвое время; поправка на
      время запаздывания

*   **dead zone** → 3765

*   **deblock** *v* → 12696

3772 debug *v*
 d fehlerfrei machen
 f dépanner
 r отлаживать

*   **debug aids** → 3774

3773 debugging
 d Störbeseitigung *f*
 f dépannage *m*; déparasitage
 r отладка *f* программы

3774 debug[ging] aids
 d Fehlersuchhilfen *fpl*;
      Fehlerkorrekturmittel *npl*
 f aides *fpl* de dépannage; assistance *f* au
      débugage
 r [вспомагательные] средства *npl* отладки

3775 decade block
 d Dekadenblock *m*
 f bloc *m* à décades
 r декадный блок *m*

3776 decade capacitance box
 d dekadischer Kondensatorensatz *m*
 f boîte *f* à capacités à décades
 r декадный магазин *m* ёмкостей

3777 decade conductance box
d dekadische Konduktanz *f*
f boîte *f* de conductance à décades
r декадный магазин *m* проводимостей

3778 decade frequency divider
d dekadischer Frequenzteiler *m*
f diviseur *m* décadique de fréquence
r декадный делитель *m* частоты

3779 decade resistance box; decimal resistance
d Dekadenwiderstand *m*
f résistance *f* en décades
r декадный магазин *m* сопротивлений

3780 decade scaler
d Dekadenzähler *m*
f compteur *m* à échelle décimale
r декадный счётчик *m*

3781 decade switch
d Dekadenschalter *m*
f commutateur *m* de décades
r декадный переключатель *m*

* **decay constant → 3667**

3782 decay curve
d Abklingkurve *f*; Ausschwingcharakteristik *f*; Zerfallscharakteristik *f*
f courbe *f* d'extinction; courbe de désintégration
r кривая *f* спада

3783 decaying pulses
d abklingende Impulse *mpl*
f impulsions *fpl* décroissantes
r затухающие импульсы *mpl*

3784 decay of power
d Absinken *n* der Leistung
f chute *f* de puissance
r падение *n* мощности

3785 decay time
d Zerfallszeit *f*; Abklingzeit *f*; Ausschwingzeit *f*; Nachleuchtdauer *f*
f période *f* d'extinction; durée *f* d'évanouissement; durée de retour à zéro
r время *n* затухания

3786 decay time of sinusoidal oscillation
d Periode *f* des Sinuskurvenabklingens
f pseudo-période *f*
r время *n* затухания синусоидальных колебаний

3787 deceleration
d Verzögerung *f*; Geschwindigkeitsabnahme *f*
f ralentissement *m*; décélération *f*
r торможение *n*; замедление *n*; отрицательное ускорение *n*

3788 decelerometer
d Verzögerungsmesser *m*; Verzögerunsmessgerät *n*
f décéléromètre *m*
r децелерометр *m*

3789 decentral drive of gripper
d dezentraler Greiferantrieb *m*
f entraînement *m* décentralisé d'un grappin
r индивидуальный привод *m* захватного устройства

3790 decentralization of function
d Funktionsdezentralisierung *f*
f décentralisation *f* de fonction
r децентрализация *f* функций

3791 decentralized control
d dezentralisierte Steuerung *f*
f commande *f* décentralisée
r децентрализованное управление *n*

3792 decentralized data bank
d dezentrale Datenbank *f*
f banque *f* de données décentralisée
r децентрализованный банк *m* данных

3793 decentralized information acquisition
d dezentralisierte Informationserfassung *f*
f acquisition *f* décentralisée des informations
r децентрализованный сбор *m* информации

3794 decentralized information processing
d dezentralisierte Informationsverarbeitung *f*
f traitement *m* décentralisé des informations
r децентрализованная обработка *f* информации

3795 decentralized inquiry unit
d dezentrale Abfrageeinheit *f*
f unité *f* d'interrogation décentralisée
r блок *m* децентрализованного опроса

3796 decentralized process control
d dezentralisierte Prozesssteuerung *f*
f commande *f* de processus décentralisée
r децентрализованное управление *n* процессом

3797 decentral process guide system
d dezentrales Prozessleitsystem *n*
f système *m* de commande de processus décentralisé
r децентрализованная система *f* управления процессом

3798 **decibel-log-frequency**
d logarithmische
Amplitudenfrequenzcharakteristik *f*
f courbe *f* amplitude-fréquence logarithmique
r логарифмическая амплитудно-частотная
характеристика *f*

3799 **decibelmeter; noise test set**
d Dezibelmessgerät *n*; Phonmesser *m*;
Rauschmesser *m*
f décibelmètre *m*; hypsomètre *m*; sonomètre *m*
r децибелметр *m*

3800 **decimal add circuit**
d Dezimaladdierkreis *m*
f addeur *m* décimal; additionneur *m* décimal
r схема *f* десятичного сумматора

3801 **decimal notation**
d dezimale Schreibweise *f*; dezimale
Zahlendarstellung *f*
f notation *f* décimale; système *m* décimal
r представление *n* чисел в десятичной
системе

3802 **decimal processing**
d Dezimalverarbeitung *f*
f traitement *m* décimal
r обработка *f* десятичных чисел

* **decimal resistance** → 3779

3803 **decimal-to-binary conversion**
d Dezimal-Binär-Konvertierung *f*;
Dezimal-Dual-Umwandlung *f*
f conversion *f* décimale-binaire
r преобразование *n* десятичного счисления
в двоичное

3804 **decimal-to-binary converter**
d Dezimal-Dual-Umwandler *m*
f convertisseur *m* décimal-binaire
r преобразователь *m* из десятичного кода в
двойчный

3805 **decimal value**
d Dezimalwert *m*
f valeur *f* décimale
r десятичное значение *n*

3806 **decision criteria**
d Entscheidungskriterien *npl*
f critères *mpl* de décision
r критерии *mpl* принятия

3807 **decision element; logical element**
d Entscheidungsschaltung *f*; logisches
Element *n*; Verknüpfungsglied *n*

f circuit *m* de décision; circuit logique;
élément *m* logique
r логический элемент *m*

3808 **decision function**
d Entscheidungsfunktion *f*
f fonction *f* de décision
r функция *f* принятия решения

3809 **decision instruction**
d Entscheidungsbefehl *m*
f instruction *f* de décision
r инструкция *f* принятия решения

3810 **decision model**
d Entscheidungsmodell *n*
f modèle *m* de décision
r модель *f* принятия решения

3811 **decision-oriented information**
d entscheidungsorientierte Information *f*
f information *f* orientée sur décision
r информация f, ориентированная на
принятие решения

3812 **decision procedure**
d Entscheidungsverfahren *n*
f procédé *m* de décision
r разрешающая процедура *f*

3813 **decision rules**
d Entscheidungsregeln *fpl*
f règles *fpl* de décision
r правила *npl* принятия решения

3814 **decision theory**
d Entscheidungstheorie *f*
f théorie *f* de décision
r теория *f* принятия решений

3815 **decision threshold**
d Entscheiderschwelle *f*
f seuil *m* de décision
r порог *m* решения

3816 **decision value**
d Entscheidungswert *m*
f valeur *f* de décision
r решающее значение *n*

3817 **decode** *v*
d dekodieren; entschlüsseln
f décoder
r декодировать

3818 **decoder; decoding machine**
d Dekodierer *m*; Entschlüssler *m*;
Dekodieranlage *f*

f décodeur m; dispositif m décodeur
r декодирующее устройство n

3819 decoding
   d Dekodierung f; Entschlüsselung f
   f décodage m
   r декодирование n

   * **decoding automaton** → 1437

3820 decoding circuit
   d Dekodierschaltung f; Entschlüsslerkreis m
   f circuit m décodeur
   r декодирующая цепь f

   * **decoding machine** → 3818

3821 decomposer
   d Zersetzungsapparat m; Zersetzer m
   f appareil m de décomposition;
      décomposeur m
   r аппарат m для разложения

3822 decomposition
   d Dekomposition f
   f décomposition f
   r декомпозиция f

3823 decomposition of block diagrams
   d Zerlegung f von Blockschaltungen
   f décomposition f de diagrammes synoptiques
   r разложение n блок-схем

3824 decomposition of logical function
   d Zerlegung f logischer Funktionen
   f décomposition f de fonctions logiques
   r разложение n логических функций

3825 decomposition process
   d Abbauprozess m; Zerfallsprozess m;
      Dekompositionsverfahren n
   f procédé m de décomposition; processus m de
      désintégration
   r декомпозоционный процесс m

3826 decomposition voltage
   d Zersetzungsspannung f
   f tension f de décomposition
   r напряжение n разложения

3827 decoupling of multiloop control
   d Entkopplung f von Mehrfachregelungen
   f découplage m de réglages à plusieurs boucles
   r развязывающее устройство n
      многоконтурных систем регулирования

3828 decrease of intensity
   d Intensitätsrückgang m

f diminution f de l'intensité
r уменьшение n интенсивности

3829 decreasing time function
   d abnehmende Zeitfunktion f
   f fonction f décroissante du temps
   r убывающая функция f времени

3830 decrement
   d Dekrement n
   f décrément m
   r декремент m

3831 decremeter
   d Dämpfungsmesser m; Dekremeter n
   f décrémètre m
   r декреметр m

3832 dedicated
   d zugeordnet; zweckentsprechend;
      zweckorientiert
   f attribué; conforme
   r специализированный

3833 dedicated channel
   d Einzweckkanal m
   f canal m attribué
   r некоммутируемый канал m

3834 dedicated control
   d zugeordnete Steuerung f; zweckorientierte
      Steuerung
   f commande f attribuée
   r управление n выполнением специальных
      функций

3835 dedicated-function application
   d zweckorientierte Anwendung f; Einsatz m
      mit zugeschnittenem Funktionsspektrum
   f application f attribuée à une fonction
   r применение n для выполнения
      специальных действий

3836 dedicated microprocessor
   d zweckorientierter Mikroprozessor m
   f microprocesseur m attribué
   r специализированный микропроцессор m

3837 dedicated mode
   d Einzelverarbeitungsmodus m
   f mode m de traitement solitaire
   r специальный режим m [работы]

3838 dedicated structure
   d speziell zugeschnittene Struktur f
   f structure f attribuée
   r специализированная структура f

3839 **dedicated system**
d zweckbestimmtes System n
f système m attribué
r специализированная система f

3840 **deenergization**
d Aberregung f
f désexcitation f
r снятие n возбуждения

3841 **deenergize**
d entregen; beruhigen
f désexciter
r успокаивать; снимать возбуждение

3842 **default declaration** (US)
d Standardvereinbarung f
f déclaration f standard
r стандартное описание n

3843 **default vector**
d Standardvektor m
f vecteur m standard
r стандартный вектор m

3844 **defect localization**
d Defektlokalisierung f; Fehlerlokalisierung f
f localisation f de défaut
r обнаружение n дефектов

3845 **defectoscopy**
d Defektoskopie f
f contrôle m de défaut
r дефектоскопия f

3846 **defects structure**
d Defektstruktur f
f structure f des défauts
r структура f дефектов

3847 **defined coordinate orientation**
d definierte Koordinatenorientierung f
f orientation f définie de coordonnées
r определенная координатная ориентация f

3848 **defined coordinate position**
d definierte Koordinatenlage f
f position f définie de coordonnées
r определенная координатная позиция f

3849 **defined effector behaviour**
d definiertes Effektorverhalten n
f comportement m d'effecteur défini
r определенное поведение n эффектора

3850 **defined interface**
d definierte Schnittstelle f
f interface f défini

r определенный интерфейс m

3851 **defined movement**
d definierte Bewegung f
f mouvement m défini
r определенное движение n

3852 **defined working point**
d definierter Arbeitspunkt m
f point m de travail défini
r определенная рабочая точка f

3853 **defining equation**
d Bestimmungsgleichung f
f équation f de définition; équation déterminative
r определяющее уравнение n

3854 **definite automaton**
d definiter Automat m
f automate m défini
r определенный автомат m

3855 **definite data type**
d definierter Datentyp m
f type m de données défini
r определенный тип m данных

3856 **definite time lag; fixed time lag; permanent delay**
d unabhängige Zeitverzögerung f; Festzeitverzögerung f; unabhängige Verzögerung f; konstante Verzögerung f
f retard m indépendant; retard m permanent
r фиксированное запаздывание n во времени; постоянное запаздывание; фиксированная задержка f времени

3857 **definite time-lag circuit-breaker**
d unabhängig verzögerter Selbstunterbrecher m
f disjoncteur m à retard indépendant
r выключатель m с независимой задержкой во времени

3858 **definite time-lag over-current release**
d Überstromauslöser m mit unabhängiger Zeitverzögerung
f déclencheur m à maximum de courant à temporisation déterminée
r выключатель m свертка с независимой выдержкой времени

3859 **definite time-lag relay**
d unabhängiges Zeitrelais n; unabhängig verzögertes Relais n; Relais für bestimmte Zeitverzögerung
f relais m à retard constant
r реле n с независимой выдержкой времени

3860 definite time-limit release
  d unabhängig verzögerter Auslöser *m*
  f déclencheur *m* à retard indépendant
  r выключающее устройство *n* с
     независимой выдержкой времени

3861 definition of the accuracy of digital voltmeters
  d Genauigkeitsbestimmung *f* digitaler Voltmeter
  f définition *f* de la précision des voltmètres numériques
  r определение *n* погрешности цифровых вольтметров

3862 deflectability
  d Ablenkbarkeit *f*
  f déviabilité *f*
  r отклоняемость *f*

3863 deflected beam
  d abgelenkter Strahl *m*
  f faisceau *m* dévié
  r отклоненный луч *m*

3864 deflecting field
  d Ablenkfeld *n*
  f champ *m* de déviation
  r отклоняющее поле *n*

3865 deflecting torque
  d Ablenkungsmoment *n*
  f couple *m* de déviation; moment *m* de déviation
  r отклоняющий момент *m*

3866 deflecting voltage
  d Ablenkspannung *f*
  f tension *f* de déviation
  r отклоняющее напряжение *n*

3867 deflection aberration
  d Ablenkfehler *m*
  f aberration *f* de déviation
  r искажение *n* отклонения

3868 deflection action
  d Ablenkungseingriff *m*
  f action *f* par déviation
  r отклоняющее действие *n*

3869 deflection amplifier
  d Ablenkverstärker *m*
  f amplificateur *m* de déviation
  r усилитель *m* отклоняющего напряжения

3870 deflection coefficient; deflection factor
  d Ablenkungskoeffizient *m*;

Ausschlagfaktor *m*
  f coefficient *m* de déviation
  r коэффициент *m* отклонения

* deflection factor → 3870

3871 deflection modulation
  d Ablenkungsmodulation *f*
  f modulation *f* par déviation
  r модуляция *f* отклонением

3872 deflection potentiometer
  d Ausschlagpotentiometer *n*; Ablenkpotentiometer *n*
  f potentiomètre *m* de déviation
  r дефлекторный потенциометр *m*

3873 deflection sensitivity
  d Ablenkempfindlichkeit *f*
  f sensibilité *f* de déviation
  r чувствительность *f* отклонения

3874 deflection synchronization
  d Ablenkungssynchronisierung *f*
  f synchronisation *f* de la déviation
  r синхронизация *f* отклонения

3875 deflection system
  d Ablenksystem *n*
  f système *m* de déviation
  r отклоняющая система *f*

3876 defocusing
  d Defokussierung *f*
  f défocalisation *f*
  r дефокусировка *f*

3877 deformation potential
  d Deformationspotential *n*
  f potentiel *m* de déformation
  r потенциал *m* деформации

3878 degenerate energy level
  d entartetes Energieband *n*
  f niveau *m* dégénéré d'énergie
  r вырожденный энергетический уровень *m*

3879 degenerative amplifier
  d gegengekoppelter Verstärker *m*; Gegenkopplungsverstärker *m*
  f amplificateur *m* à contre-réaction
  r усилитель *m* с отрицательной обратной связью

3880 degenerative circuit
  d Gegenkopplungsschaltung *f*; gegengekoppelter Kreis *m*

f montage *m* à contre-réaction; circuit *m* à
contre-réaction
r схема *f* с отрицательной обратной связью

3881 degenerative electronic controller
d gegengekoppelter elektronischer Regler *m*
f régulateur *m* électronique à contre-réaction
r электронный регулятор с отрицательной
обратной связью

3882 degenerative feedback; negative feedback
d negative Rückführung *f*; Gegenkopplung *f*
f réaction *f* négative; contre-réaction *f*
r отрицательная обратная связь *f*

*   **degree of accuracy** → 203

3883 degree of approximation
d Annäherungsstufe *f*
f degré *m* d'approximation
r степень *f* приближения

*   **degree of attenuation** → 1098

**3884 degree of automatization; automation
degree**
d Automatisierungsgrad *m*
f degré *m* d'automatisation
r степень *f* автоматизации

3885 degree of degeneration
d Entartungsgrad *m*
f degré *m* de dégénération
r степень *f* вырождения

3886 degree of depolarization
d Depolarisationsgrad *m*;
Entpolarisierungsgrad *m*
f degré *m* de dépolarisation
r степень *f* деполяризации

3887 degree of disturbance
d Störungsgrad *m*
f degré *m* de perturbation
r степень *f* возмущений

3888 degree of freedom
d Freiheitsgrad *m*
f degré *m* de liberté
r степень *f* свободы

3889 degree of freedom of manipulator
d Manipulatorfreiheitsgrad *m*; Freiheitsgrad *m*
eines Manipulators
f liberté *f* de mouvement d'un manipulateur
r степень *f* свободы манипулятора

3890 degree of integration

d Integrationsgrad *m*
f degré *m* d'intégration
r степень *f* интеграции

3891 degree of irregularity
d Ungleichförmigkeitsgrad *m*
f degré *m* d'irrégularité
r степень *f* неравномерности

3892 degree of rationalization
d Rationalisierungsgrad *m*
f degré *m* de rationalisation
r степень *f* рационализации

3893 degree of reliability
d Zuverlässigkeitsgrad *m*
f degré *m* de fiabilité
r степень *f* надёжности

3894 degree of rigidity
d Steifigkeitsgrad *m*
f degré *m* de rigidité
r степень *f* жёсткости

3895 degree of saturation
d Sättigungsgrad *m*
f degré *m* de saturation
r степень *f* насыщения

3896 degree of sensitivity
d Empfindlichkeitsgrad *m*
f degré *m* de sensibilité
r степень *f* чувствительности

3897 degree of stability; stability degree
d Stabilitätsgrad *m*
f degré *m* de stabilité
r степень *f* устойчивости

3898 degree of thermal dissociation
d thermischer Dissoziationsgrad *m*
f degré *m* de dissociation thermique
r степень *f* термической диссоциации

3899 dehumidifying plant
d Entfeuchtungsanlage *f*
f installation *f* de déshydratation
r установка *f* для обезвоживания

3900 dehydration process
d Trocknungsprozess *m*;
Entfeuchtungsprozess *m*;
Entfeuchtungsvorgang *m*
f processus *m* de déshydratation; procédé *m* de
séchage
r процесс *m* обезвоживания

3901 dehydrogenation plant
d Dehydrierungsanlage *f*

f unité f de déshydrogénation; installation f de déshydrogénation
r установка f дегидрирования

3902 dehydrogenation process
d Dehydrierungsprozess m
f procédé m de déshydrogénation
r процесс m дегидрирования

3903 delay; lag; retardation
d Verzögerung f; Verspätung f
f retard m; retardement m
r задержка f; запаздывание n

3904 delay amplifier
d Verzögerungsverstärker m; Verstärker m mit Verzögerungsanordnung
f amplificateur m à ligne de retard
r усилитель m с задержкой

3905 delay basis operation
d Betrieb m mit Vorbereitung
f exploitation f avec attente
r заказная система f эксплуатации

3906 delay cable
d Verzögerungskabel n
f câble m de retard; câble retardeur
r кабель m задержки

3907 delay coincidence method
d Methode f der verzögerten Koinzidenz
f methode f de coïncidence retardée
r метод m замедленного совпадения

3908 delay correction network
d Laufzeitentzerrungsschaltung f; Verzögerungskorrekturschaltung f
f correcteur m de phase; réseau m correcteur de retard
r схема f задержки коррекции

  *   **delay distortion → 9584**

3909 delayed alarm
d verzögerte Alarmgabe f
f alarme f retardée
r задержанная сигнализация f

3910 delayed application
d verzögerte Anwendung f
f application f retardée
r замедленное применение n

3911 delayed automatic gain control
d verzögerte selbständige Verstärkungsregelung f
f réglage m automatique retardé de gain

r автоматическое регулирование n усиления с задержкой времени

3912 delayed carry
d verzögerter Übertrag m
f transfert m retardé
r задержанный перенос m

3913 delayed collector conduction
d verzögerte Kollektorleitung f
f conduction f retardée du collecteur
r задержанная проводимость f коллектора

3914 delayed control
d verzögerte Regelung f
f réglage m retardé
r замедленное регулирование n

3915 delayed feedback
d verzögerte Rückführung f
f réaction f retardée
r запаздывающая обратная связь f

3916 delayed ignition
d verzögerte Zündung f
f amorçage m retardé
r замедленное зажигание n

3917 delayed-line network chain
d Laufzeitkette f
f réseau m à retard
r схема f с линией задержки

3918 delayed manipulation error
d verzögerter Manipulationsfehler m
f défaut m de manipulation à retardement
r ошибка f манипулирования с выдержкой времени

3919 delayed reactivity
d verzögerte Reaktivität f
f reactivité f retardée
r замедленное восстановление n

3920 delayed scanning
d verzögerte Abtastung f
f balayage m retardé; exploration f retardée
r сканирование n с запаздыванием

3921 delay element
d Verzögerungselement n; Verzögerungsglied n
f élément m de retardement
r элемент m задержки

3922 delay feedback generator
d Verzögerungsrückkopplungsgenerator m

f oscillateur *m* à réaction retardée
r генератор *m* с задержанной обратной
   связью

3923 delaying member
d Verzögerungsglied *n*
f membre *m* de retard
r запаздывающее звено *n*

3924 delaying unit
d Verzögerungseinheit *f*
f unité *f* de retard
r блок *m* запаздывания

3925 delay line
d Verzögerungsleitung *f*;
   Verzögerungsstrecke *f*
f ligne *f* de retardement
r линия *f* задержки

3926 delay-line decoder
d Dekodierer *m* mit Verzögerungsleitung
f décodeur *m* à ligne de retard
r дешифратор *m* с линией задержки

3927 delay-line memory
d Verzögerungsleitungsspeicher *m*;
   Laufzeitspeicher *m*
f mémoire *f* à ligne de retard
r запоминающее устройство *n* на линиях
   задержки

3928 delay-line register
d Verzögerungsleitungsregister *n*
f registre *m* de ligne à retard
r регистр *m* на линиях задержки

3929 delay-line-shaped pulse
d durch Verzögerungsleitung geformter
   Impuls *m*
f impulsion *f* formée par une ligne de retard
r импульс *m* сформированный линией
   задержки

3930 delay of operation
d Wirkungsverzug *m*
f retard *m* du fonctionnement
r замедление *n* действия

3931 delay representation
d Verzögerungswiedergabe *f*
f reproduction *f* du retard
r воспроизведение *n* запаздывания

3932 delay system
d Verzögerungssystem *n*
f système *m* de retard
r система *f* задержки

3933 delay time
d Verzögerungszeit *f*
f temps *m* de retardement
r время *n* задержки

3934 delay-time characteristic
d Laufzeitcharakteristik *f*
f caractéristique *f* de temps de retard
r характеристика *f* запаздывания

3935 deliberate actuation
d gewollte Betätigung *f*
f manœuvre *f* volontaire
r выбранное действие *n*

* **delimiter → 2444**

3936 delink *v*
d entketten
f déchaîner
r разрывать [цепь]

3937 delta function; unit impulse function
d Deltafunktion *f*; Einheitsimpuls *m*;
   Einheitsimpulsfunktion *f*
f impulsion *f* en delta; fonction *f* delta;
   fonction impulsion unitaire
r дельта-функция *f*; единичная импульсная
   функция *f*

3938 delta noise
d Deltarauschen *n*
f bruit *m* delta
r дельта-шум *m*

3939 demagnetizing effect
d Entmagnetisierungseffekt *m*
f effet *m* de désaimantation
r эффект *m* размагничивания

3940 demand power
d Solleistung *f*; Leistungsbedarf *m*
f puissance *f* demandée
r необходимая мощность *f*

3941 demodulate *v*
d demodulieren
f démoduler
r демодулировать

3942 demodulator
d Demodulator *m*
f démodulateur *m*
r демодулятор *m*

3943 demonstration model
d Anschauungsmodell *n*; Mustermodell *n*;
   Demonstrationsmodell *n*

f modèle *m* de référence
r макет *m*; образец *m*

3944 demultiplexer
d Demultiplexer *m*
f démultiplexeur *m*
r демультиплексор *m*

3945 densitometry
d Densitometrie *f*
f densitométrie *f*
r денситометрия *f*

3946 density change
d Dichteänderung *f*
f variation *f* de densité
r изменение *n* плотности

3947 density control
d Dichteregelung *f*
f réglage *m* de densité
r регулирование *n* плотности

3948 density controller
d Dichteregler *m*
f régulateur *m* de densité
r регулятор *m* плотности

3949 density curve
d Dichtekurve *f*
f courbe *f* de densité
r кривая *f* плотности

3950 density distribution
d Dichteverteilung *f*
f distribution *f* de densité
r распределение *n* плотности

3951 density gradient
d Dichtegradient *m*
f gradient *m* de densité
r градиент *m* плотности

3952 density indicator
d Dichteanzeiger *m*
f indicateur *m* de densité
r плотномер *m*; индикатор *m* плотности

3953 density/moisture meter
d [kombiniertes] Dichte- und
   Feuchtemessgerät *n*
f densimètre/humidimètre *m*
r комбинированный плотномер-влагомер *m*

**3954 dependability; eqiupment dependability;
   equipment safety; reliability**
d Sicherheit *f*; Zuverlässigkeit *f*;
   Zuverlässigkeit *f* von Geräten

f fiabilité *f*; sécurité *f*; sûreté *f*; endurance *f*
r надёжность *f*

3955 dependent control
d Folgeregelung *f*
f réglage *m* lié
r связанное регулирование *n*

3956 dependent inverter
d netzabhängiger Wechselrichter *m*
f onduleur *m* alimenté à partir du réseau
r инвертор *m* питаемый от сети

3957 dependent variable
d abhängige Variable *f*
f variable *f* dépendante
r зависимая переменная *f*

3958 depolarization
d Depolarisation *f*
f dépolarisation *f*
r деполяризация *f*

3959 depressed equation
d reduzierte Gleichung *f*
f équation *f* réduite
r приведённое уравнение *n*

3960 depth pressure recorder
d Tiefendruckregistriergerät *n*
f enregistreur *m* de pression sous-marine
r самописец *m* глубинного давления

**3961 depth probe; depth sound; immersion
   probe**
d Tauchsonde *f*
f sonde *f* plongeante
r глубинный зонд *m*

* depth sound → 3961

3962 derivative
d Differentialquotient *m*; Ableitung *f*
f dérivée
r производная *f*

3963 derivative action
d Vorhaltverhalten *n*
f action *f* dérivée; comportement *m* à
   corrélation de dérivée
r воздействие *n* по производной

**3964 derivative action coefficient; derivative
   action factor**
d D-Einfluß-Koeffizient *m*;
   Differentiationsbeiwert *m*
f coefficient *m* d'action par dérivation
r коэффициент *m* дифференцирования

* derivative action factor → 3964

3965 derivative action time
d Vorhaltzeit *f* bei D-Wirkung
f temps *m* d'action dérivée
r время *n* упреждения

3966 derivative component
d Vorhalteglied *n*
f élément *m* de dérivation
r шунтирующая составляющая *f*

3967 derivative control
d Vorhalteregelung *f*
f réglage *m* à action dérivée
r регулирование *n* по производной

3968 derivative element; differentiating element; differentiator
d Differenzierglied *n*; Differentiator *m*; differenzierendes Glied *n*
f dérivateur *m*; organe *m* différentiateur; dispositif *m* de différentiation; différentiateur *m*
r дифференцирующее звено *n*; дифференциатор *m*

3969 derivative-proportional-integral control
d Proportional-Integral-Differential-Regelung *f*; PID-Regelung *f*
f réglage *m* à action proportionnelle-dérivée-intégrale
r регулирование *n* по координате, производной и интегралу

3970 derivative time constant
d Vorhaltzeitkonstante *f*; Derivationszeitkonstante *f*
f constante *f* de temps de l'action dérivée
r постоянная *f* времени дифференцирующего звена

3971 descending
d absteigend
f décroissant; descendant
r низходящий; убывающий

3972 describing function; equivalent admittance
d äquivalente Admittanz *f*; äquivalenter Verstärkungskoeffizient *m*; Beschreibungsfunktion *f*
f admittance *f* équivalente; gain *m* complexe équivalent
r эквивалентный адмитанц *m*; комплексный эквивалентный коэффициент *m* усиления

3973 describing function method; harmonic balance method
d Methode *f* der Funktionsbeschreibung
f méthode *f* de balance harmonique
r метод *m* гармонического баланса; метод описывающей функции

3974 descriptive model
d beschreibendes Modell *n*
f modèle *m* descriptif
r описательная модель *f*

3976 design *v*; construct *v*
d entwerfen; anlegen; konstruieren
f concevoir; construire; dessiner
r проектировать; конструировать

3975 design
d Entwurf *m*
f projet *m*; conception *f*
r разработка *f*; проект *m*

3977 design algorithm
d Entwurfalgorithmus *m*
f algorithme *m* de dessin
r алгоритм *m* проектирования

3978 design analysis
d Entwurfsanalyse *f*
f analyse *f* de projet; analyse de concept
r анализ *m* проекта

3979 design automation; automation of project
d Entwurfsautomatisierung *f*
f automatisation *f* de projet; automatisation de conception
r автоматизация *f* проектирования

3980 design conditions
d Entwurfsbedingungen *fpl*
f conditions *fpl* de projet
r проектные условия *npl*

3981 design criterion of gripper mechanism
d Entwurfskriterium *n* eines Greifermechanismus
f dessin *m* de critère d'un mécanisme de grappin
r критерий *m* разработки механизма захвата

3982 design goal
d Entwurfsziel *n*
f but *m* de concept
r цель *f* разработки

* designing → 10332

3983 design issues
d Entwurfsabkömmlinge *mpl*

f   dérivés *mpl* de concept
r   результаты *mpl* разработки

3984   design load
d   Entwurfsbelastung *f*
f   charge *f* de dessin
r   проектная нагрузка *f*

**3985   design meeting the demands of
          automation; automation-friendly design**
d   automatisierungsgerechte Gestaltung *f*;
     automatisierungsfreundlicher Entwurf *m*
f   projection *f* correspondant aux exigences de
     l'automatisation
r   оформление *n* процесса, удобное для
     автоматизации

3986   design method
d   Gestaltungsmethode *f*;
     Konstruktionsverfahren *n*;
     Entwurfsverfahren *n*
f   méthode *f* de projection; méthode de
     construction
r   метод *m* проектирования

3987   design model
d   Entwurfsmodell *n*; Auslegungsmodell *n*
f   modèle *m* de conception
r   модель *f* расчёта

3988   design of chemical engineering system
d   Entwurf *m* verfahrenstechnischer Systeme
f   conception *f* des systèmes du génie chimique
r   проектирование *n* химико-
     технологических систем

3989   design of experiments
d   Versuchsplanung *f*
f   planning *m* d'expériences
r   планирование *n* эксперимента

3990   design of switching circuits
d   Schaltkreisentwurf *m*; Entwurf *m* von
     Schaltkreisen
f   projet *m* des circuits de commutation
r   разработка *f* переключающих схем

3991   design optimization
d   Auslegungsoptimierung *f*
f   optimisation *f* de conception
r   оптимизация *f* основных параметров
     процесса или системы

3992   design parameter
d   Entwurfsparameter *n*
f   paramètre *m* de projet
r   проектный параметр *m*

3993   design phase

d   Entwurfsphase *f*
f   phase *f* de dessin
r   стадия *f* проектирования

3994   design principle
d   Entwurfsprinzip *n*; Entwurfsgrundsatz *m*
f   principe *m* de conception
r   принцип *m* проектирования

3995   design procedure
d   Entwurfsmethode *f*
f   méthode *f* de projet
r   процедура *f* расчёта

3996   design strategy
d   Entwurfsstrategie *f*
f   stratégie *f* de conception
r   стратегия *f* проектирования

3997   design tree
d   Entwurfsbaum *m*
f   arbre *m* de dessin
r   сеть *f* проекта

3998   desired portion; useful component
d   Nutzkomponente *f*
f   composante *f* utile
r   полезная составляющая *f*

*   **desired signal** → 12723

*   **desired value** → 346

3999   desired value change
d   Sollwertänderung *f*
f   changement *m* de la valeur désirée
r   изменение *n* регулируемой величины

4000   destroyed information
d   zerstörte Information *f*
f   information *f* détruite
r   разрушенная информация *f*

4001   destructive reading; destructive read-out
d   destruktives Lesen *n*
f   lecture *f* destructive
r   считывание *n* со стиранием; считывание с
     разрушением

*   **destructive read-out** → 4001

4002   destructive test
d   nicht zerstörungsfreie Prüfung *f*
f   essai *m* destructif
r   разрушающее испытание *n*

4003   detailed model
d   ausführliches Modell *n*

f modèle *m* détaillé
r подробная модель *f*

4004 detailled planning
d detaillierter Arbeitsablaufplan *m*
f planning *m* détaillé
r подробный график *m* работ

4005 detecting element
d Messfühler *m*; Messwertgeber *m*
f capteur *m* de mesure
r датчик *m* измеряемой величины

4006 detecting filter
d Suchfilter *n*
f filtre *m* de détection
r фильтр *m* поиска

4007 detecting threshold
d Nachweisschwelle *f*
f seuil *m* de détectabilité
r порог *m* детектирования

4008 detection limit
d Detektionsschwelle *f*; Detektionsgrenze *f*
f seuil *m* de détection
r порог *m* обнаружения

4009 detection of radiation
d Strahlennachweis *m*
f détection *f* de la radiation
r обнаружение *n* излучения

4010 detection of signal in noise
d Signalabsonderung *f* aus dem Rauschen
f séparation *f* du signal du bruit
r выделение *n* сигнала из шума

4011 detection range
d Detektorbereich *m*
f étendue *f* de détection
r область *f* детектирования

4012 detection sensibility
d Nachweisempfindlichkeit *f*
f sensibilité *f* de détection
r чувствительность *f* обнаружения

4013 detection subassembly
d Detektorblock *m*; Detektorbaugruppe *f*
f sous-ensemble *m* de détection
r блок *m* детектирования

4014 detection time
d Nachweiszeit *f*
f temps *m* de détection
r время *n* обнаружения

* detector → 4017

4015 detector analyzer
d Detektoranalysator *m*
f analyseur *m* détecteur
r детекторный анализатор *m*

4016 detector efficiency
d Ansprechwahrscheinlichkeit *f*
f rendement *m* d'un detecteur
r эффективность *f* детектора

4017 detector element; sensing element; sensor; detector
d Detektorelement *n*; Abtastglied *n*; Detektor *m*
f élément *m* détecteur; senseur *m*
r чувствительный элемент *m*; детектор *m*

4018 detector-noise limited
d begrenzt durch das Detektorrauschen
f limité par le bruit du détecteur
r ограниченный шумами детекторного каскада

4019 determinated machine; disciplined machine
d determinierte Maschine *f*
f machine *f* déterminée
r детерминированная машина *f*

4020 determinated sequence of motions
d determinierter Bewegungsablauf *m*
f séquence *f* de mouvement déterminée
r определенная последовательность *f* движений

4021 determination method
d Abgrenzungsmethode *f*
f méthode *f* de délimitation
r метод *m* разграничения

4022 determination of coarse position
d Grobpositionsermittlung *f*; Ermittlung *f* der Grobposition
f recherche *f* de la position grossière
r определение *n* глобальной позиции

4023 determination of errors
d Fehlerbestimmung *f*
f détermination *f* des erreurs
r определение *n* ошибки

4024 determination of position by angle encoder
d Positionsermittlung *f* durch Winkelkodierer
f recherche *f* de position avec codeur d'angle
r определение *n* позиции угловым кодирующим устройством

4025 **determination of position by**
      potentiometer
      d Positionsermittlung *f* durch Potentiometer
      f recherche *f* de position avec potentiomètre
      r определение *n* позиции с помощью
           потенциометров

4026 **deterministic**
      d deterministisch
      f déterministique
      r детерминированный

4027 **deterministic process**
      d deterministischer Prozess *m*; determinierter
           Prozess
      f procédé *m* déterministique
      r детерминированный процесс *m*

4028 **deterministic system**
      d deterministisches System *n*; determiniertes
           System
      f système *m* déterministique
      r детерминированная система *f*

4029 **development control memory**
      d Ablaufsteuerungsspeicher *m*; Speicher *m*
           einer Ablaufsteuerung
      f mémoire *f* d'une commande de déroulement
      r память *f* программного управления с
           обратной связью

4030 **development of integrated electronics**
      d Entwicklung *f* der integrierten Elektronik
      f développement *m* de l'électronique intégrée
      r развитие *n* интегральной электроники

4031 **development of sensor; sensor design**
      d Sensorentwicklung *f*
      f développement *m* de senseur; dessin *m* de
           senseur
      r разработка *f* сенсора

4032 **development point**
      (theory of automata)
      d Entwicklungspunkt *m*
      f point *m* de développement
      r точка *f* разложения

4033 **development potential**
      d Entwicklungspotential *n*
      f potentiel *m* de développement
      r потенциал *m* разработки

4034 **development program**
      d Ablaufprogramm *n*
      f programme *m* de déroulement
      r программа *f* последовательности

4035 **development project**
      d Entwicklungsprojekt *n*
      f projet *m* de développement
      r проект *m* разработки

4036 **development stage of control technique**
      d Entwicklungsstand *m* der Steuerungstechnik
      f stade *m* de développement de la technique de
           commande
      r уровень *m* разработки техники
           [автоматического] управления

4037 **development technique**
      d Entwicklungsverfahren *n*
      f procédé *m* de développement
      r методика *f* разработки

4038 **development time**
      d Entwicklungszeit *f*
      f temps *m* de développement
      r время *n* разработки

4039 **development trend**
      d Entwicklungstendenz *f*
      f tendance *f* de développement
      r тенденция *f* развития

4040 **development work; experimental work**
      d Entwicklungsarbeit *f*; experimentelle Arbeit *f*
      f mise *f* au point; travaux *mpl* de
           développement; travail *m* expérimental
      r экспериментальная работа *f*

4041 **deviation amplitude**
      d Abweichungsamplitude *f*
      f amplitude *f* de déviation
      r амплитуда *f* отклонения

4042 **deviation area**
      d Abweichungsfläche *f*
      f surface *f* des écarts
      r область *f* отклонения

4043 **deviation from nominal value**
      d Abweichung *f* vom Sollwert
      f écart *m* de consigne
      r отклонение *n* от номинального значения

4044 **deviation indicator**
      d Abweichungsanzeiger *m*
      f indicateur *m* de déviation
      r индикатор *m* отклонения

4045 **deviation measuring method**
      d Ausschlagmessmethode *f*
      f méthode *f* de déviation
      r измерение *n* методом отклонения;
           измерение методом рассогласования

4046 deviation ratio; offset ratio
d Abweichungsverhältnis n; dynamischer Regelfaktor m
f rapport m de déviation; facteur m dynamique de réglage; taux m de statisme
r отношение n отклонения; наклон m статической характеристики

4047 deviation value
d Abweichungsgröße f
f grandeur f d'écart
r величина f отклонения

* device → 7024

4048 device allocation; device assignment
d Gerätezuordnung f; Zuordnung f der Einheiten
f assignation f du dispositif; allocation f des dispositifs
r назначение n устройств

* device assignment → 4048

4049 device complex
d Gerätekomplex m
f ensemble m des dispositifs
r комплекс m устройств

4050 device control
d Gerätesteuerung f
f commande f d'appareil
r управление n устройством оборудования

4051 device driver program
d Gerätesteuerprogramm n
f programme m de commande d'appareil
r программа f управления устройством

4052 device end status
d Einrichtungsendzustand m
f état m de fin d'un dispositif
r конечное состояние n устройства

4053 device error
d Gerätefehler m
f erreur f de dispositif; défaut m d'appareil
r неисправность f прибора

4054 device for feeding and changing of tools
d Vorrichtung f zur Zuführung und Auswechselung von Werkzeugen
f dispositif m pour alimentation et échange d'outils
r устройство n для подвода и смены инструмента

4055 device for transport of parts
d Vorrichtung f zum Transport von Teilen
f dispositif m pour le transport de pièces
r устройство n для транспортировки деталей

4056 device independence
d Geräteunabhängigkeit f; Einheitenunabhängigkeit f; Einrichtungsunabhängigkeit f
f indépendance f de dispositif
r независимость f от устройств

4057 device of information winning
d Einrichtung f zur Informationsgewinnung
f dispositif m pour l'élaboration des informations
r устройство n получения информации

4058 device priority
d Gerätepriorität f
f priorité f d'appareil
r приоритет m устройства

4059 device selection
d Geräteauswahl f
f sélection f d'appareil
r выбор m устройства

4060 device specifications
d Geräteangaben fpl; Gerätedaten pl
f spécifications fpl d'appareils
r спецификации fpl устройств

4061 device status; unit status
d Gerätestatus m; Status m der Geräteeinheit
f état m d'appareil; état de l'unité
r состояние n устройства

4062 device test; unit test
d Gerätetest m
f test m de dispositif
r тест m устройства

4063 device type
d Gerätetyp m
f type m de dispositif; type d'appareil
r тип m устройства

4064 deviometer
d Abweichungsmesser m
f déviomètre m
r девиометр m

* diagnosis → 4065

4065 diagnostic; diagnosis
d Diagnose f
f diagnostic m
r диагностика f

4066 diagnostic bus
  d Diagnosebus *m*
  f bus *m* de diagnostic
  r диагностическая шина *f*

4067 diagnostic checking; diagnostic test
  d Diagnosetest *m*
  f test *m* de diagnostic
  r диагностический тест *m*

4068 diagnostic data; diagnostic information
  d Diagnosedaten *pl*; Diagnoseinformation *f*
  f données *fpl* de diagnostic; information *f* de diagnostic
  r диагностическая информация *f*

4069 diagnostic device
  d Diagnoseeinrichtung *f*
  f dispositif *m* de diagnostic
  r устройство *n* диагностики

  * **diagnostic information → 4068**

4070 diagnostic machine
  d Diagnosemaschine *f*
  f machine *f* diagnostique; machine de recherche d'erreur
  r диагностическая машина *f*

4071 diagnostic message
  d Diagnosemitteilung *f*; Diagnosenachricht *f*
  f message *m* de diagnostic
  r диагностическое сообщение *n*

4072 diagnostic procedure
  d Diagnoseverfahren *n*
  f procédé *m* de diagnostic
  r диагностическая процедура *f*

4073 diagnostic program; diagnostic routine
  d Diagnoseprogramm *n*
  f programme *m* diagnostique; programme de recherche d'erreur
  r программа *f* для диагностики; программа для выделения ошибок

  * **diagnostic routine → 4073**

4074 diagnostic signal
  d Diagnosesignal *n*
  f signal *m* diagnostique
  r сигнал *m* диагностирования

4075 diagnostic strategy
  d Diagnosestrategie *f*
  f stratégie *f* diagnostique
  r стратегия *f* диагностирования

  * **diagnostic test → 4067**

4076 diagnostic test system
  d Diagnosetestsystem *n*
  f système *m* de test de diagnostic
  r контрольно-диагностическая система *f*

4077 diagonalization of matrices
  d Diagonalisierung *f* von Matrizen
  f diagonalisation *f* des matrices
  r приведение *n* матриц к диагональному виду

4078 diagonal matrix
  d Diagonalmatrix *f*
  f matrice *f* diagonale
  r диагональная матрица *f*

4079 diagram of Euler and Venn
  d Euler-Venn-Diagramm *n*
  f diagramme *m* d'Euler et de Venn
  r диаграмма *f* Эйлера и Венна

4080 diagram of Karnaugh
  d Karnaugh-Plan *m*
  f table *f* de Karnaugh
  r таблица *f* Карно

4081 dialling pulse
  d Wählimpuls *m*
  f impulsion *f* d'appel
  r импульс *m* набора

4082 dialogue capability
  d Dialogfähigkeit *f*
  f faculté *f* de dialogue
  r диалоговая возможность *f*

4083 dialogue control
  d Dialogsteuerung *f*
  f commande *f* de dialogue
  r диалоговое управление *n*

  * **dialogue mode → 3388**

4084 dial telephone system
  d Selbstwählfernsprechsystem *n*
  f système *m* de téléphonie automatique
  r автоматическая телефонная система *f*

4085 diaphanometer
  d Diaphanometer *n*
  f diaphanomètre *m*
  r диафанометр *m*

4086 diaphragm
  d Membrane *f*
  f membrane *f*; diaphragme *m*
  r диафрагма *f*; мембрана *f*

4087 diaphragm actuator; diaphragm drive
d Membranantrieb m
f commande f à membrane
r мембранный привод m

* **diaphragm drive → 4087**

4088 diaphragm servomotor
d Membranstellmotor m
f servomoteur m à membrane
r мембранный сервомотор m

4089 diastimeter
d Diastimeter n; Distanzmesser m
f diastimomètre m
r дальномер m

4090 dielectric drying
d dielektrisches Trocknen n
f séchage m diélectrique
r диэлектрическая сушка f

4091 dielectric gradient
d dielektrischer Gradient m
f gradient m diélectrique
r диэлектрический градиент m

4092 dielectric heating
d dielektrische Heizung f
f chauffage m diélectrique
r нагрев m диэлектрика

4093 dielectric heating generator
d dielektrischer Heizungsgenerator m
f générateur m de chauffage diélectrique
r генератор m с диэлектрическим нагревом

4094 dielectric heating in cavity resonator
d dielektrische Erwärmung f im Hohlraumresonator
f chauffage m diélectrique dans la cavité résonnante
r нагрев m диэлектрика в объёмном резонаторе

4095 dielectric heating of thermoplastic materials
d dielektrische Erwärmung f von Thermoplasten
f chauffage m diélectrique de matériaux thermoplastiques
r диэлектрический нагрев m термопластических материалов

4096 dielectric interference filter
d dielektrisches Interferenzfilter n
f filtre m interférentiel diélectrique
r диэлектрический интерференциальный фильтр m

4097 dielectric leakage measurement
d Dielektrizitätsverlustmessung f
f mesure f des pertes diélectriques
r измерение n диэлектрических потерь

4098 dielectric losses
d dielektrische Verluste mpl
f pertes fpl diélectriques
r диэлектрические потери fpl

4099 dielectric stress
d dielektrische Beanspruchung f; dielektrische Belastung f
f contrainte f diélectrique
r диэлектрическое напряжение n

4100 difference between desired value and set value
d Differenz f zwischen Aufgabenwert und Sollwert
f écart m de statisme
r разность f между заданной и исходной величинами

4101 difference pulse; difference signal; differential pulse
d Differenzimpuls m; Differenzsignal n; Differentialimpuls m
f impulsion f différentielle
r разностный импульс m; разностный сигнал m; дифференциальный импульс m

* **difference signal → 4101**

4102 differentiable structure
d differenzierbare Struktur f
f structure f différentiable
r дифференцируемая структура f

4103 differential absorption method
d Differentialabsorptionsmethode f; Differentialabsorptionsverfahren n
f méthode f d'absorption différentielle
r метод m дифференциального поглощения

4104 differential address
d eigenrelative Adresse f
f adresse f différentielle
r разностный адрес m

4105 differential amplifier
d Differentialverstärker m
f amplificateur m différentiel
r дифференциальный усилитель m

4106 differential analyzer
d Differentialanalysator m

f analyseur *m* différentiel
r дифференциальный анализатор *m*

4107 differential booster
d Zusatzmaschine *f* mit Differentialerregung
f survolteur *m* différentiel
r дифференциальный бустер *m*

4108 differential bridge
d Differentialbrücke *f*
f pont *m* différentiel
r дифференциальный мост *m*

4109 differential calorimeter
d Differentialkalorimeter *n*
f calorimètre *m* différentiel
r дифференциальный калориметр *m*

4110 differential capacitance
d Differentialkapazität *f*
f capacité *f* différentielle
r дифференциальная ёмкость *f*

4111 differential circuit
d differenzierendes Netzwerk *n*; Differentialstromkreis *m*
f chaîne *f* de dérivation; circuit *m* de différentiation
r дифференциальная цепь *f*

4112 differential coefficient
d Differentialkoeffizient *m*
f coefficient *m* différentiel
r дифференциальный коэффициент *m*

4113 differential concatenation control
d differentiale Kaskadenregelung *f*; Gegenverbundkaskadenregelung *f*
f réglage *m* différentiel en cascade
r регулировка *f* скорости дифференциальным каскадным включением

4114 differential connexion
d Differentialschaltung *f*; Vergleichsschaltung *f*
f alimentation *f* différentielle
r дифференциальное соединение *n*; дифференциальное включение *n*

* **differential controller** → 10629

4115 differential cross-section
d differentieller Wirkungsquerschnitt *m*
f section *f* efficace différentielle
r дифференциальное [поперечное] сечение *n*

4116 differential current

d Differenzstrom *m*
f courant *m* différentiel
r дифференциальный ток *m*; разностный ток

4117 differential curve
d Differentialkurve *f*
f courbe *f* différentielle
r дифференциальная кривая *f*

4118 differential-difference equation
d Differential-Differenzgleichung *f*
f équation *f* de différence-différentielle
r дифференциально-разностное уравнение *n*

4119 differential element
d Differentialelement *n*; Differenzierglied *n*
f élément *m* différentiel
r дифференциальный элемент *m*

4120 differential equation with retarded argument
d Differentialgleichung *f* mit nacheilendem Argument
f équation *f* différentielle à l'argument retardé
r дифференциальное уравнение *n* с затухающим аргументом

4121 differential excitation
d Gegenverbunderregung *f*
f excitation *f* différentielle
r дифференциальное возбуждение *n*

4122 differential expression
d Differentialausdruck *m*
f expression *f* différentielle
r дифференциальное выражение *n*

4123 differential form
d Differentialform *f*
f forme *f* différentielle
r дифференциальная форма *f*

4124 differential gain control
d differentielle Gewinnregelung *f*; differentielle Gewinnsteuerung *f*
f réglage *m* différentiel du gain
r дифференциальное регулирование *n* усиления

4125 differential gear
d Ausgleichgetriebe *n*; Differential *n*
f engrenage *m* de différentiel
r дифференциал *m*

4126 differential-logarithmic pulse code modulation
d differential-logarithmische PCM-Modulation *f*

    f modulation *f* PCM par impulsions codées
       différentielle logarithmique
    r дифференциально-логарифмическая
       кодовоимпульсная модуляция *f*

**4127 differentially coherent transmission
      system**
    d differentiellkohärentes
       Übertragungssystem *n*
    f système *m* de transmission à cohérence
       différentielle
    r дифференциально-когерентная система *f*
       передачи

4128 differential measurement
    d Differentialmessmethode *f*
    f méthode *f* différentielle de mesure
    r дифференциальный метод *m* измерения;
       дифференциальное измерение *n*

4129 differential method
    d Differentialmethode *f*
    f méthode *f* différentielle
    r дифференциальный метод *m*

4130 differential modulation
    d Differentialmodulation *f*
    f modulation *f* différentielle
    r дифференциальная модуляция *f*

4131 differential pick-up
    d Differentialgeber *m*
    f capteur *m* différentiel
    r датчик *m* перепада; дифференциальный
       датчик

4132 differential pressure control
    d Druckdifferenzregelung *f*
    f réglage *m* de pression différentielle
    r регулирование *n* перепада давления

4133 differential pressure gauge
    d Differenzdruckmanometer *n*
    f manomètre *m* différentiel
    r дифференциальный манометр *m*

**4134 differential pressure indicator; pressure
      difference indicator**
    d Druckdifferenzanzeiger *m*;
       Differenzdruckwächter *m*
    f indicateur *m* de pression différentielle
    r указатель *m* перепада давления

4135 differential pressure measurement
    d Druckdifferenzmessung *f*
    f mesure *f* de pression différentielle
    r измерение *n* перепада давления

4136 differential pressure recorder
    d Druckdifferenzschreiber *m*
    f enregistreur *m* de pression différentielle
    r самописец *m* перепада давления

4137 differential pressure switch
    d Differentialdruckschalter *m*
    f commutateur *m* à différence de pression
    r дифференциальный манометрический
       выключатель *m*

**4138 differential pressure transducer;
      differential pressure transmitter**
    d Druckdifferenzgeber *m*
    f palpeur *m* de pression différentielle;
       transmetteur *m* de pression différentielle
    r датчик *m* дифференциального давления

* differential pressure transmitter → 4138

4139 differential protection
    d Differentialschutz *m*
    f protection *f* différentielle
    r дифференциальная защита *f*

* differential pulse → 4101

4140 differential receiver
    d Differentialempfänger *m*
    f récepteur *m* différentiel
    r дифференциальный приёмник *m*

4141 differential relay; balanced relay
    d Differentialrelais *n*
    f relais *m* différentiel
    r дифференциальное реле *n*; балансное реле

4142 differential resistance
    d differentieller Widerstand *m*
    f résistance *f* différentielle
    r дифференциальное сопротивление *n*;
       внутреннее сопротивление

4143 differential selsyn
    d differentiales Synchro *n*;
       Differentialsynchroübertrager *m*
    f selsyn *m* différentiel; synchro-transmetteur *m*
       différentiel
    r дифсельсин *m*; дифференциальный
       сельсин *m*

4144 differential servo
    d differentiales Folgeregelungssystem *n*
    f système *m* différentiel d'asservissement
    r дифференциальная следящая система *f*

4145 differential spectral sensitivity
    d differentielle Spektralempfindlichkeit *f*

 f sensibilité *f* spectrale différentielle
 r дифференциальная спектральная
      чувствительность *f*

4146 differential synchroreceiver
 d Differentialsynchroempfänger *m*
 f synchro-récepteur *m* différentiel
 r дифференциальный сельсин-приёмник *m*

4147 differential synchrotransmitter
 d Differentialsynchrosender *m*
 f synchro-transmetteur *m* différentiel
 r дифференциальный сельсин-датчик *m*

4148 differential telemeter transmitter
 d Differentialfernmesssender *m*
 f émetteur *m* de télémesure différentielle
 r передатчик *m* дифференциального
      телеметра

4149 differential temperature control
 d differentiale Temperaturregelung *f*;
      differentiale Temperatursteuerung *f*
 f réglage *m* différentiel de la température
 r регулирование *n* разности температур

4150 differential thermal analysis
 d Differentialthermoanalyse *f*
 f analyse *f* thermique différentielle
 r дифференциальный термический
      анализ *m*

4151 differential thermogravimetry
 d Differentialthermogravimetrie *f*
 f thermogravimétrie *f* différentielle
 r дифференциальная термогравиметрия *f*

4152 differential thermometer
 d Differentialthermometer *n*
 f thermomètre *m* différentiel
 r дифференциальный термометр *m*

4153 differentiating action
 d differenzierendes Verhalten *n*
 f action *f* dérivative
 r дифференцирующее воздействие *n*

4154 differentiating circuit
 d differenzierende Schaltung *f*
 f circuit *m* de différentiation
 r дифференцирующая схема *f*

 * **differentiating element** → 3968

4155 differentiating network
 d differenzierendes Netzwerk *n*
 f circuit *m* différentiateur; réseau *m* dérivateur
 r дифференцирующая цепь *f*

4156 differentiating network control system
 d differenzierendes
      Netzwerksteuerungssystem *n*
 f système *m* de commande de réseau de
      différentiation
 r дифференцирующая система *f* контурного
      управления

4157 differentiation symbol
 d Differentiationssymbol *n*
 f symbole *m* de la dérivation
 r символ *m* дифференцирования

 * **differentiator** → 3968

4158 differentiator time constant
 d Vorhaltezeit *f*; Differentiatorzeitkonstante *f*
 f durée *f* de l'action dérivée
 r постоянная *f* времени
      дифференцирующего звена

4159 differntial configuration
 d Differentialanordnung *f*
 f configuration *f* différentielle
 r дифференциальная схема *f*

4160 difficulty coefficient of assembly
 d Montageschwierigkeitsgrad *m*
 f coefficient *m* d'une difficulté d'assemblage
 r степень *f* сложности сборки

4161 diffraction of X-rays
 d Röntgenstrahlenbeugung *f*
 f diffraction *f* des rayons X
 r дифракция *f* ренгеновских лучей

4162 diffuser
 d Diffusionsapparat *m*
 f appareil *m* de diffusion
 r экстрактор *m*; диффузионный аппарат *m*

4163 diffuse radiation spectrum
 d diffuses Strahlungsspektrum *n*
 f spectre *m* diffusé de rayonnement
 r спектр *m* рассеянного излучения

4164 diffusion model
 d Diffusionsmodell *n*
 f modèle *m* de diffusion
 r диффузионная модель *f*

4165 diffusion process
 d Diffusionsprozess *m*
 f processus *m* de diffusion
 r процесс *m* диффузии

4166 digital-analog converter
 d Digital-Analog-Umsetzer *m*; Digital-Analog-
      Konverter *m*; Digital-Analog-Wandler *m*

   f convertisseur *m* digital analogique
   r преобразователь *m* дискретных данных в
      непрерывные

4167 digital angle measuring system
   d digitales Winkelmesssystem *n*
   f système *m* de goniométrie numérique
   r цифровая система *f* измерения углов

4168 digital automatic machine
   d digitale automatische Maschine *f*
   f machine *f* automatique digitale
   r цифровая автоматическая машина *f*

4169 digital automaton
   d digitaler Automat *m*
   f automate *m* digital
   r цифровой автомат *m*

4170 digital averager
   d Digital-Mittelwertsbildner *m*
   f producteur *m* numérique de moyenne
   r цифровой усреднитель *m*

4171 digital code
   d digitaler Kode *m*; Ziffernkode *m*
   f code *m* digital
   r цифровой код *m*

4172 digital coding of conceptions
   d digitale Kodierung *f* von Begriffen
   f codage *m* numérique de notions
   r числовое кодирование *n* понятий

4173 digital communication system
   d digitales Kommunikationssystem *n*
   f système *m* digital de communication
   r система *f* дискретной связи

4174 digital computer
   d digitale Rechenanlage *f*; Digitalrechner *m*
   f machine *f* à calculer numérique; calculateur *m* numérique; calculatrice *f* digitale
   r цифровая вычислительная машина *f*

4175 digital computer structure designing
   d Strukturentwurf *n* von Digitalrechnern
   f projet *m* de structure de calculatrices numériques
   r проектирование *n* цифровых вычислительных машин

4176 digital computer system of central control
   d Zentralsteuerungsanlage *f* eines Digitalrechners
   f commande *f* centrale par calculatrice numérique
   r система *f* цифровых вычислительных

устройств для центрального управления

4177 digital connection
   d Digitalverbindung *f*
   f connexion *f* numérique
   r цифровая связь *f*

4178 digital control; numerical control; discrete control
   d numerische Steuerung *f*; digitale Steuerung; diskrete Steuerung
   f commande *f* numérique; commande digitale; commande *f* discrète
   r цифровое управление *n*; дискретное управление

4179 digital control circuits theory
   d Theorie *f* der Digitalregelkreise
   f théorie *f* des circuits digitaux de réglage
   r теория *f* цифровых систем регулирования; теория дискретных систем регулирования

4180 digital control computer
   d digitaler Steuerrechner *m*
   f calculateur *m* numérique de commande
   r цифровое управляющее вычислительное устройство *n*

4181 digital control computer structure
   d Struktur *f* eines digitalen Steuerrechners
   f structure *f* de calculatrice numérique de commande
   r конструкция *f* цифрового управляющего вычислительного устройства

4182 digital control system
   d System *n* mit digitaler Steuerung; digitales Steuerungssystem *n*
   f système *m* de commande digitale
   r система *f* цифрового управления

4183 digital converter
   d Digitalumsetzer *m*; Digitalwandler *m*
   f convertisseur *m* numérique; traducteur *m* digital
   r цифровой преобразователь *m*

4184 digital data communication
   d numerische Datenübertragung *f*
   f transmission *f* numérique de données
   r цифровая связь *f* данных

4185 digital data processing
   d digitale Datenverarbeitung *f*
   f traitement *m* numérique de données
   r цифровая обработка *f* данных

  * **digital differential analyzer → 4196**

4186 digital display
d digitale Darstellung *f*
f représentation *f* digitale
r цифровая индикация *f*

4187 digital distribution frame
d Digitalsignalverteiler *m*
f répartiteur *m* numérique
r распределитель *m* цифровых сигналов

4188 digital encoder
d numerischer Kodierer *m*
f codeur *m* numérique
r цифровое кодирующее устройство *n*

4189 digital error
d Digitalfehler *m*
f erreur *f* numérique
r цифровая погрешность *f*

4190 digital filter
d digitales Filter *n*; Digitalfilter *n*; Filter mit digitalen Elementen
f filtre *m* digital; filtre avec des éléments digitaux
r цифровой фильтр *m*

4191 digital force sensor
d digitaler Kraftsensor *m*
f senseur *m* de force digital
r цифровой сенсор *m* усилий

4192 digital frequency meter
d numerischer Frequenzmesser *m*
f fréquencemètre *m* numérique
r цифровой частотомер *m*

4193 digital gripper control
d digitale Greifersteuerung *f*
f commande *f* digitale d'un grappin
r цифровое управление *n* схватом

4194 digital image processing
d digitale Bildverarbeitung *f*
f traitement *m* numérique d'images
r цифровая обработка *f* изображений

4195 digital-incremental odometry
d digital-inkrementale Wegmessung *f*
f odométrie *f* numérique différentielle
r цифровая дифференциальная одометрия *f*

4196 digital [integrating] differential analyzer
d digitaler Differentialanalysator *m*; Digitaldifferentiator *m*
f analyseur *m* différentiel digital; analyseur différentiel numérique
r цифровой дифференциальный анализатор *m*

4197 digital integrator
d digitaler Integrator *m*
f intégrateur *m* digital; intégrateur numérique
r цифровой интегратор *m*

4198 digital interpolator
d Zifferninterpolator *m*
f interpolateur *m* numérique
r цифровой интерполятор *m*

* **digitalization** → **4245**

4199 digitalized work piece
d digitalisiertes Werkstück *n*
f pièce *f* à usiner digitalisée
r обрабатываемая деталь f, преобразованная в цифровую форму

4200 digitalizing of graphical sheets
d Digitalisierung *f* grafischer Vorlagen
f conversion *f* digitale des documents graphiques
r цифровое преобразование *n* графических документации

4201 digital laser beam deflector
d digitale Laserstrahlablenkungseinrichtung *f*
f dispositif *m* de déviation digitale du faisceau de laser
r цифровое устройство *n* отклонения лазерного луча

4202 digital length measurement
d digitale Längenmessung *f*
f mesure *f* digitale de la longueur
r дискретное измерение *n* длин

4203 digital logic system
d digitales logisches System *n*
f système *m* logique digital
r цифровая логическая система *f*

4204 digitally controlled element
d digital gesteuertes Element *n*
f élément *m* à commande numérique
r элемент *m* с цифровым управлением

4205 digital main line
d Digitalhauptleitung *f*
f ligne *f* principale numérique
r цифровая магистраль *f*

4206 digital measuring device
d Digitalmessgerät *n*
f appareil *m* de mesure numérique
r цифровой измерительный прибор *m*

4207 digital ohmmeter with limit value checking
d Digitalohmmeter *n* mit Grenzwertkontrolle
f ohmmètre *m* numérique à contrôle de valeurs limites
r цифровой омметр *m* с контролем предельных значений

4208 digital optical processing
d digitale optische Signalverarbeitung *f*
f traitement *m* de signal optique numérique
r цифровая оптическая обработка *f* сигналов

4209 digital output
d Digitalausgabe *f*; Digitalausgang *m*; digitales Ausgangssignal *n*
f sortie *f* numérique
r цифровой выход *m*

4210 digital output transducer
d Messwertgeber *m* mit Digitalausgang
f transmetteur *m* à sortie numérique
r выходной цифровой преобразователь *m*

4211 digital phase meter
d Digitalphasenmesser *m*
f phasemètre *m* numérique
r цифровой измеритель *m* фазы

4212 digital position
d Ziffernstellung *f*
f position *f* digitale
r цифровая позиция *f*

4213 digital position control
d digitale Lagesteuerung *f*; Positionssteuerung *f*
f commande *f* digitale de position
r цифровое управление *n* положением

4214 digital position measurement
d digitale Lagemessung *f*
f mesure *f* digitale position
r измерение *n* цифровой позиции

4215 digital positionning system
d digitales Positionierungssystem *n*
f système *m* de positionnement digital
r цифровая система *f* позиционирования

4216 digital position servo
d digitaler Verstellservomechanismus *m*
f servomécanisme *m* positionneur digital
r цифровая позиционная следящая система *f*

4217 digital processing unit
d digitale Verarbeitungseinheit *f*
f unité *f* de traitement digitale
r цифровой процессорный блок *m*

4218 digital processor
d digitaler Prozessor *m*; Arrayprozessor *m*
f processeur *m* digital
r цифровой процессор *m*

4219 digital quantity
d Digitalgröße *f*; numerische Größe *f*
f quantité *f* digitale; quantité numérique
r цифровая величина *f*

4220 digital read-out
d Ziffernablesung *f*
f lecture *f* des chiffres
r цифровой отсчёт *m*

4221 digital recorder
d digitalgesteuerter Schreiber *m*
f enregistreur *m* digital; appareil *m* enregistreur
r цифровое устройство *n* записи

4222 digital recording
d Digitalaufzeichnung *f*
f enregistrement *m* digital
r цифровая запись *f*

4223 digital regulation algorithm
d digitaler Regelungsalgorithmus *m*
f algorithme *m* de régulation digital
r алгоритм *m* цифрового регулирования

4224 digital relay servosystem
d Ziffernrelaisfolgesystem *n*; digitales Relaisfolgesystem *n*
f système *m* asservi numérique à relais
r цифровая релейная следящая система *f*

4225 digital representation
d digitale Darstellung *f*
f représentation *f* digitale; représentation numérique
r цифровое изображение *n*; представление *n* в цифровой форме

4226 digital scientific computing system
d digitale wissenschaftliche Rechenanlage *f*
f système *m* de calcul scientifique digital
r цифровая вычислительная система *f* для научных расчётов

4227 digital sensor
d Digitalsensor *m*
f senseur *m* digital
r цифровой сенсор *m*

4228 **digital signal**
 d digitales Signal *n*; numerisches Signal *n*
 f signal *m* numérique; signal digital
 r цифровой сигнал *m*

4229 **digital simulation**
 d digitale Simulation *f*
 f simulation *f* digitale
 r дигитальное моделирование *n*; цифровое
   моделирование

4230 **digital simulator**
 d Digitalsimulator *m*
 f simulateur *m* numérique; simulateur digital
 r цифровое моделирующее устройство *n*

4231 **digital store**
 d Digitalspeicher *m*; Zifferenspeicher *m*
 f mémoire *f* numérique
 r цифровой накопитель *m*

4232 **digital system for production**
 d digitales System *n* für die Produktion
 f système *m* digital pour production
 r цифровая система *f* управления
   производством

4233 **digital system realization**
 d Realisierung *f* eines digitalen Systems
 f réalisation *f* d'un système digital
 r реализация *f* цифровой системы

 * **digital technique** → 8862

4234 **digital telemetering**
 d digitales Fernmessverfahren *n*
 f télémétrie *f* digitale
 r цифровая телеметрия *f*

4235 **digital transient analyzer**
 d digitaler Analysator *m* von
   Übergangserscheinungen; digitales
   Einschwingmodell *n*
 f analyseur *m* digital de processus transitoires
 r цифровой анализатор *m* переходных
   процессов

4236 **digital transmission**
 d Digitalübertragung *f*
 f transmission *f* numérique
 r передача *f* цифровой информации

4237 **digital unit**
 d Digitaleinheit *f*
 f unité *f* digitale
 r цифровой блок *m*

4238 **digit impulse**

 d Ziffernimpuls *m*
 f impulsion *f* digitale
 r цифровой импульс *m*; дискретный
   импульс

 * **digitized device** → 4244

4239 **digitized drawing**
 d digitalisierte Zeichnung *f*
 f dessin *m* numérique
 r преобразованная в цифровую форму
   графическая информация *f*

4240 **digitized image**
 d digitalisiertes Bild *n*
 f image *f* digitalisée
 r цифровое отображение *n*

4241 **digitized spark chamber**
 d digital dargestellte Funkenkammer *f*
 f chambre *f* d'étincelle numérique
 r цифровая искровая камера *f*

4242 **digitized structure**
   (of information)
 d digitalisierte Struktur *f*
 f structure *f* digitalisée
 r дискретная структура *f*

4243 **digitized thermocouple compensation**
 d digitalisierte Thermoelementkompensation *f*
 f compensation *f* de thermocouple digitalisée
 r цифровая конпенсация *f* с помощью
   термоэлемента

4244 **digitizer; digitized device**
 d Ziffernddarstellung *f*; Digitalgeber *m*;
   Digitalisierer *m*
 f codeur *m* numérique; dispositif *m* digitalisé;
   digitaliseur *m*
 r квантущюее устройство *n*; дигитайзер *m*

4245 **digitizing; digitalization**
 d Digitalisierung *f*
 f digitalisation *f*
 r преобразование *n* в цифровую форму

4246 **digit scanning**
 d Ziffernabtastung *f*
 f exploration *f* des chiffres
 r цифровая развёртка *f*

4247 **digit selector**
 d Zahlenverteiler *m*; Zahlenwähler *m*
 f sélecteur *m* digit
 r цифровой селектор *n*

4248 **dimension**
 d Dimension *f*

   f dimension *f*
   r размерность *f*

4249 dimensional analysis
   d Dimensionsanalyse *f*
   f analyse *f* dimensionelle
   r анализ *m* размерностей

4250 3-dimensional phase space
   d dreidimensionaler Phasenraum *m*
   f espace *m* de phase à trois dimensions
   r трехмерное фазовое пространство *n*

4251 dimensional stability
   d Dimensionsstabilität *f*
   f stabilité *f* de dimension
   r устойчивость *f* размерностей

4252 dimension control
   d Dimensionssteuerung *f*
   f commande *f* des dimensions
   r контрол *m* размеров

4253 dimensionless coefficient; dimensionless value; non-dimensional value
   d dimensionsloser Koeffizient *m*; dimensionslose Größe *f*; bezogene Größe
   f coefficient *m* sans dimension; grandeur *f* sans dimension; grandeur non dimensionnelle
   r безмерный коэффициент *m*; безразмерная величина *f*

   * **dimensionless value** → 4253

4254 dimensionless variable; reduced variable; non-dimensional variable
   d dimensionslose Variable *f*
   f variable *f* sans dimensions
   r безразмерная переменная *f*

4255 diode circuit
   d Diodenschaltung *f*
   f montage *m* à diode
   r диодная схема *f*

4256 diode counter
   d Diodenzähler *m*
   f compteur *m* à diodes
   r диодный счётчик *m*

4257 diode current limiter
   d Diodenstrombegrenzer *m*
   f limiteur *m* de courant à diode
   r диодный ограничитель *m* тока

4258 diode detection
   d Diodengleichrichtung *f*
   f détection *f* à diode

   r диодное детектирование *n*

4259 diode function generator
   d Diodenfunktionsgenerator *m*
   f générateur *m* de fonctions à diodes
   r диодный генератор *m* функций

4260 diode multiplier
   d Diodenvervielfacher *m*
   f multiplicateur *m* à diode
   r диодный умножитель *m*

4261 Dirac delta-function
   d Dirac-Funktion *f*; Diracsche Funktion *f*
   f fonction *f* impulsive; fonction de Dirac; fonction d'impulsion unitaire
   r функция *f* Дирака

4262 direct access method
   d Direktzugriffsmethode *f*
   f méthode *f* à accès direct
   r метод *m* прямого доступа

4263 direct acting controller; direct operating controller
   d Regler *m* ohne Hilfsenergie; direkt wirkender Regler; Direktregler *m*
   f régulateur *m* à action directe; régulateur *m* direct
   r регулятор *m* прямого действия

4264 direct acting recording instrument
   d direkt betätigtes Registrierinstrument *n*; direkt betätigter Schreiber *m*
   f appareil *m* enregistreur à action directe
   r регистрирующий прибор *m* с прямой записью

4265 direct action circuit
   d Stromkreis *m* mit direkter Wirkungsrichtung; Kette *f* direkter Einwirkung
   f circuit *m* d'action dirigée
   r цепь *f* направленного действия

4266 direct code
   d direkter Kode *m*
   f code *m* direct
   r прямой код *m*

4267 direct concatenation control
   d direkte Kaskadenregelung *f*
   f réglage *m* direct en cascade
   r регулирование *n* прямым каскадным включением

4268 direct control
   d unmittelbare Regelung *f*
   f régulation *f* directe
   r непосредственное регулирование *n*

*   **direct control checking** → 4271

4269   direct controlled regulation valve
   d direkt gesteuertes Regelventil *n*
   f soupape *f* de réglage à commande directe
   r регулирующий клапан *m* с
      непосредственным управлением

4270   direct control of tools
   d direkte Steuerung *f* von Werkzeugen
   f commande *f* directe des outils
   r непосредственное управление *n*
      инструментами

4271   direct control supervision; direct control checking
   d direkte Steuerungsüberwachung *f*
   f inspection *f* de commande directe
   r прямое супервизорное управление *n*

4272   direct control system
   d direkt wirkendes System *n*;
      Regelungssystem *n* ohne Hilfsenergie
   f système *m* de commande à action directe
   r система *f* прямого регулирования

4273   direct cooling system
   d Direktkühlsystem *n*
   f système *m* de refroidissement direct
   r система *f* непосредственного охлаждения

4274   direct current amplifier
   d Gleichstromverstärker *m*
   f amplificateur *m* de courant continu
   r усилитель *m* постоянного тока

*   **direct current bias** → 4275

4275   direct current bias [voltage]
   d Gleichstromvorspannung *f*
   f polarisation *f* de courant continu
   r напряжение *n* смещения

4276   direct current converter
   d Gleichspannungswandler *m*
   f convertisseur *m* de tension continue
   r преобразователь *m* постоянного тока

4277   direct current dump
   d Gleichstromversorgungsausfall *m*
   f manque *m* de courant continu; panne *f* de
      courant continu
   r внезапное отключение *n* постоянного
      напряжения

4278   direct current input voltage
   d Eingangsgleichspannung *f*
   f tension *f* continue d'entrée

   r постоянная составляющая *f* входного
      напряжения

4279   direct current level
   d Gleichspannungspegel *m*
   f niveau *m* de courant continu
   r уровень *m* постоянной составляющей

4280   direct current output voltage
   d Ausgangsgleichspannung *f*
   f tension *f* continue de sortie
   r постоянная составляющая *f* выходного
      напряжения

4281   direct current [power] supply
   d Gleichstromversorgung *f*
   f alimentation *f* en courant continu
   r источник *m* постоянного напряжения

4282   direct current signalling
   d Gleichstromübertragung *f*
   f transmission *f* à courant continu
   r связь *f* по постоянному току

*   **direct current supply** → 4281

4283   direct data change
   d direkter Datenaustausch *m*
   f changement *m* direct des données
   r прямой обмен *m* данными

4284   direct data channel
   d direkter Datenkanal *m*
   f canal *m* direct des données
   r прямой информационный канал *m*

4285   direct data processing
   d direkte Datenverarbeitung *f*
   f traitement *m* direct de données
   r управляемая обработка *f* данных

4286   direct data traffic
   d direkter Datenverkehr *m*
   f trafic *m* direct des données
   r непосредственный обмен *m* данными

4287   direct deflection measuring method
   d Ausschlagmessmethode *f*
   f méthode *f* de mesure de déviation
   r метод *m* прямого измерения отклонения

4288   direct digital control
   d direkte digitale Regelung *f*
   f contrôle *m* digital direct
   r прямое цифровое управление *n*

4289   direct digital controller
   d direkter digitaler Mehrkanalregler *m*; direkter
      digitaler Regler *m*

f régulateur *m* numérique direct
r контроллер *m* прямого цифрового
   управления

4290 direct drive
  d direkter Antrieb *m*
  f entraînement *m* direct
  r прямой привод *m*

4291 direct execution
  d Direktausführung *f*
  f exécution *f* directe
  r непосредственое исполнение *n*

4292 direct force destination
  d direkte Kraftbestimmung *f*
  f destination *f* de force directe
  r непосредственное определение *n* усилий

4293 direct frequency modulation
  d direkte Frequenzmodulation *f*
  f modulation *f* directe de fréquence
  r непосредственная частотная модуляция *f*

4294 direct information exchange
  d direkter Informationsaustausch *m*
  f échange *m* direct d'informations
  r непосредственный обмен *m* информации

4295 direct input
  d direkter Eingang *m*; direkte Eingabe *f*
  f entrée *f* directe
  r прямой вход *m*

4296 directional diagram
  d Richtungsdiagramm *n*
  f diagramme *m* directionnel
  r диаграмма *f* направлености

4297 directional operation
  d richtungsabhängiges Arbeiten *n*
  f fonctionnement *m* directionnel
  r направленное действие *n*

4298 directional power protection
  d gerichteter Leistungsschutz *m*
  f dispositif *m* de protection directionnel
      wattmétrique
  r направленная защита *f* мощности

4299 directional relay
  d Richtrelais *n*; Richtungsrelais *n*
  f relais *m* protecteur directionnel
  r направленное реле *n*

4300 directional scintillation counter
  d gerichteter Szintillationszähler *m*
  f compteur *m* directif à scintillation

r направленный сцинтилляционный
   счётчик *m*

4301 directional transmission
  d gerichtete Übertragung *f*
  f transmission *f* directionnelle
  r направленная передача *f*

4302 direction for use
  d Bedienungsvorschrift *f*;
      Gebrauchsanweisung *f*
  f instruction *f* de service; mode *m* d'emploi
  r инструкция *f* эксплуатации

4303 direction of flow
  d Strömungsrichtung *f*
  f direction *f* du courant
  r направление *n* потока

4304 direction of magnetization
  d Magnetisierungsrichtung *f*
  f direction *f* d'aimantation
  r направление *n* намагничености

4305 direction of polarization
  d Polrisationsrichtung *f*
  f direction *f* de polarisation
  r направление *n* поляризации

4306 direction of propagation
  d Ausbreitungsrichtung *f*
  f direction *f* de propagation
  r направление *n* распространения

**4307 direction of transfer; transmission
       direction**
  d Übertragungsrichtung *f*
  f direction *f* de transfert; direction de
      transmission
  r направление *n* передачи

4308 direction switching condition
  d Schaltrichtungsbedingung *f*
  f condition *f* de direction de commutation
  r условие *n* переключения направления

4309 direct line
  d Direktleitung *f*; Standleitung *f*
  f ligne *f* directe
  r прямая линия *f*

4310 directly operated valve
  d direkt betätigtes Ventil *n*; Direktregler *m*
  f soupape *f* à action directe
  r вентиль *m* прямого действия

4311 direct manipulator control
  d direkte Manipulatorsteuerung *f*

    f commande *f* directe de manipulateur
    r непосредственное управление *n* манипулятором

4312 direct method of Liapunov
    d direkte Methode *f* von Ljapunow
    f méthode *f* directe de Ljapunow
    r прямой метод *m* Ляпунова

4313 direct moment destination
    d direkte Momentenbestimmung *f*
    f destination *f* directe de moments
    r непосредственное определение *n* моментов

4314 direct numerical control
    d direkte numerische Steuerung *f*; direkte NC-Steuerung *f*; DNC
    f commande *f* numérique directe
    r система *f* непосредственного числового программного управления

    * **direct operating controller** → 4263

4315 direct processing
    d Drektverarbeitung *f*
    f traitement *m* direct
    r прямая обработка *f*

4316 direct reading dosimeter
    d direkt anzeigendes Dosimeter *n*
    f dosimètre *m* à indication directe
    r дозиметр *m* с непосредственным отсчётом

4317 direct reading instrument
    d Instrument *n* mit unmittelbarer Ablesung
    f appareil *m* à lecture directe
    r прибор *m* с непосредственным отсчётом

4318 direct reading transmission measuring set
    d Pegelzeiger *m* mit unmittelbarer Ablesung
    f décibelmètre à lecture directe
    r измеритель *m* уровня передачи с непосредственным отсчётом

4319 direct relation telemeter
    d Directbeziehungsfernmesser *m*
    f télémètre *m* à lecture proportionnelle
    r прямозависящая телеметрическая система *f*

4320 direct routing system
    d System *n* mit unmittelbarer gesteuerter Wahl; System mit direkter Steuerung
    f système *m* à commande directe
    r устройство *n* прямого программирования

4321 direct series trip
    d Direktserienauslöser *m*; Primärserienauslöser *m*
    f déclencheur *m* à série direct
    r непосредственный последовательный разъединитель *m*

4322 direct short-circuit interruption
    d Direktkurzschlußabschaltung *f*
    f coupure *f* directe du courant de court-circuit
    r непосредственное выключение *n* короткого замыкания

4323 direct signal control
    d direkte Signalsteuerung *f*
    f commande *f* directe au moyen de la signalisation
    r прямое сигнальное управление *n*

4324 direct simulation
    d direkte Simulation *f*
    f simulation *f* directe
    r непосредственное моделирование *n*

4325 direct vision spectroscope
    d Geradsichtspektroskop *n*
    f spectroscope *m* à vision directe
    r спектроскоп *m* с прямым наблюдением

4326 direct visual supervision
    d direkte visuelle Überwachung *f*
    f inspection *f* visuelle directe
    r прямой супервизорный контроль *m*

4327 disabling of the interrupt system
    d Sperrung *f* des Interruptsystems
    f verrouillage *m* du système d'interruption
    r блокировка *f* системы прерываний

    * **disabling pulse** → 1848

    * **disabling signal** → 6611

4328 disaster propagation
    d Störungsausbreitung *f*; Störungsfortpflanzung *f*
    f propagation *f* de perturbations
    r распространение *n* аварии

    * **disciplined machine** → 4019

4329 disconnect *v*
    d abschalten
    f déconnecter; couper
    r разъединять; отсоединять

4330 disconnect indication
    d Abtrennanzeige *f*

f indication *f* de déconnexion
r индикация *f* разъединения

4331 disconnecting lever for mechanism
d Abstellhebel *m* für Automatik
f levier *m* de suppression des automatismes
r рычаг *m* выключения автоматики

4332 disconnection element
d Abschaltelement *n*
f élément *m* de coupure
r отключающее устройство *n*

4333 disconnection time delay
d Ausschaltzeitverzögerung *f*
f retard *m* du temps d'interruption
r запаздывание *n* при отключении

4334 discontinuous action
d diskontinuierliche Wirkung *f*; unstetige
Wirkung
f action *f* discontinue; fonctionnement *m*
discontinu
r прерывистое воздействие *n*

4335 discontinuous-action controller
d unstetiger Regler *m*; diskontinuierlich
wirkender Regler
f régulateur *m* à action discontinue
r регулятор *m* импульсного действия

4336 discontinuous-action servomechanism
d diskontinuierlicher Servomechanismus *m*
f servomécanisme *m* à action intermittente
r севомеханизм *m* прерывистого действия

4337 discontinuous control; intermittent control
d unstetige Regelung *f*; diskontinuierliche
Regelung
f réglage *m* intermittent; contrôle *m* discontinu
r прерывистое регулирование *n*

**4338 discontinuous controller; discrete action
controller; intermittent controller**
d unstetiger Regler *m*
f régulateur *m* discontinu; régulateur à action
intermittente
r регулятор *m* дискретного действия;
регулятор прерывистого действия

4339 discontinuous function
d unstetige Funktion *f*
f fonction *f* discontinue
r прерывистая функция *f*

4340 discontinuous integration gearing
d unstetiges Integrationsgetriebe *n*
f mécanisme *m* d'intégration discontinue

r прерывистый интегрирующий
механизм *m*

4341 discontinuous production
d diskontinuierliche Produktion *f*
f production *f* non continue
r прерывное производство *n*

4342 discontinuous signal; intermittent signal
d unstetiges Signal *n*; diskontinuierliches
Signal
f signal *m* discontinu; signal intermittent
r прерывистый сигнал *m*

* **discontinuous system → 6903**

4343 discontinuous term
d unstetiges Glied *n*
f terme *m* discontinu
r неоднородный член *m*

4344 discontinuous variable
d unstetige Größe *f*
f grandeur *f* discontinue
r прерывистая переменная *f*

* **discrete action controller → 4338**

4345 discrete automaton
d diskreter Automat *m*
f automate *m* discret
r дискретный автомат *m*

4346 discrete circuit
d diskrete Schaltung *f*
f circuit *m* discret
r дискретная схема *f*

4347 discrete component
d diskretes Bauelement *n*
f composant *m* discret
r дискретный компонент *m*

4348 discrete-continuous system
d diskrete und stetiges System *n*
f système *m* discret et continu
r дискретно-непрерывная система *f*

* **discrete control → 4178**

4349 discrete distribution
d diskrete Verteilung *f*
f répartition *f* discrète
r дискретное распределение *n*

4350 discrete filter
d diskretes Filter *n*
f filtre *m* discret
r дискретный фильтр *m*

4351 discrete input; intermittent input
  d diskrete Einwirkung *f*
  f action *f* discrète; action intermittente
  r дискретное воздействие *n*

4352 discrete logic
  d diskrete Logik *f*
  f logique *f* discrète
  r дискретная логика *f*

4353 discrete maximum principle
  d diskretes Maximumprinzip *n*
  f principe *m* de maximum discret
  r дискретный принцип *m* максимума

4354 discrete optimizing system
  d diskretes Optimisierungssystem *n*
  f système *m* d'optimisation à action discontinue
  r система *f* оптимизации дискретного действия

4355 discrete path state
  d diskreter Bahnzustand *m*
  f état *m* de chemin discret
  r дискретное состояние *n* траектории

4356 discrete process
  d diskreter Prozess *m*
  f procédé *m* discret
  r дискретный процесс *m*

4357 discrete programming
  d diskrete Optimierung *f*
  f programmation *f* discrète; optimisation *f* discrète
  r дискретное программирование *n*

4358 discrete pulses
  d Einzelimpulse *mpl*; diskrete Impulse *mpl*
  f impulsions *fpl* discrètes
  r дискретные импульсы *mpl*

4359 discrete quantity
  d diskrete Größe *f*
  f quantité *f* discrète; grandeur *f* discrète
  r дискретная величина *f*

4360 discrete representation
  d diskrete Darstellung *f*
  f représentation *f* discrète
  r дискретное представление *n*

4361 discrete-signal distance transmission
  d Fernübertragung *f* mit quantisiertem Signal
  f transmission *f* à distance de signaux discrets
  r дискретная дистанционная передача *f* сигнала

4362 discrete stochastic multistage decision process
  d diskreter stochastischer Mehrstufenentscheidungsprozess *m*
  f procédé *m* décisif à plusieurs étages discret et aléatoire
  r дискретный случайный многошаговый процесс *m* решения; дискретный стохастическии многошаговый процесс решения

4363 discretization of continuous-time system
  d Diskretisierung *f* eines stetigen System
  f discrétisation *f* d'un système continu
  r квантование *n* непрерывной системы

4364 discret-time system
  d diskretes System *n*
  f système *m* discret de temps
  r дискретная система *f*

4365 discriminating element
  d Diskriminatorglied *n*
  f élément *m* discriminateur
  r избирательный элемент *m*

4366 discriminating protective system
  d Selektivschutzsystem *n*
  f système *m* de protection sélectif
  r различающая защитная система *f*

4367 discriminating relay
  d Selektivschütz *n*; Selektivschutzrelais *n*
  f relais *m* sélecteur
  r резонансное реле *n*

4368 discrimination
  d Unterscheidungsvermögen *n*; Trennungsvermögen *n*
  f discrimination *f*
  r избирательность *f*

4369 disengaging zero position
  d Nullstellungsausschaltung *f*
  f débrayage *m* de position de zéro
  r выключение *n* нулевых положений

4370 disintegration energy
  d Zersetzungsenergie *f*; Zerfallsenergie *f*
  f énergie *f* de décomposition; énergie de désintégration
  r энергия *f* распада

4371 disjoint convex region
  d getrenntes konvexes Gebiet *n*
  f région *f* convexe disjoint
  r изолированная выпуклая область *f*

4372 **disjunction**
d Disjunktion f
f disjonction f
r дизъюнкция f

4373 **disjunctive normal form**
d disjunktive Normalform f
f forme f normale disjonctive
r дизъюнктивная нормальная форма f

4374 **disk coder**
d Plattenkodierer m
f codeur m à disque
r дисковое кодирующее устройство n

4375 **disk controller**
d Plattenspeicher-Steuergerät n
f appareil m de commande de disque
r контроллер m дискового запоминающего
   устройства

4376 **disk operating system; DOS**
d Plattenbetriebssystem n; DOS
f système m d'exploitation sur disques; DOS
r дисковая операционная система f: ДОС

4377 **disk system software**
d Plattensystemsoftware f; Software f eines
   Plattensystems
f software f d'un système de disques
r программное обеспечение n системы на
   дисках

4378 **dislocation density**
d Verschiebungsdichte f; Versetzungsdichte f
f densité f des dislocations
r плотность f дислокации

4379 **disordered handling object**
d ungeordnetes Handhabungsobjekt n
f objet m de manutention désordonné
r неориентированный объект m
   манипулирования

4380 **dispatching desk**
d Dispatcherpult n
f pupitre m de dispatching
r диспетчерский пульт m

4381 **dispatching object**
d Zuteilungsobjekt n
f objet m de dispatching
r объект m диспетчеризации

4382 **dispatching priority**
d Zuteilungsvorrang m
f priorité f d'expédition
r приоритет m диспетчеризации

4383 **dispatching system; supervisory control
system**
d Dispatchersystem n
f système m de télécontrôle
r система f диспетчеризации

4384 **dispersion coefficient**
d Dispersionskoeffizient m
f coefficient m de dispersion
r коэффициент m дисперсии

4385 **dispersion index**
d Dispersionsindex m
f index m de dispersion
r показатель m рассеяния

4386 **displacement constant**
d Verschiebungskonstante f
f constante f de déplacement
r постоянная f смещения

4387 **displacement controller**
d Verschiebungsregler m
f régulateur m du déplacement
r регулятор m перемещений

4388 **displacement indicator**
d Verschiebungsgeber m
f indicateur m de déplacement
r датчик m для измерения смещений

4389 **displacement modulation; phase-pulse
modulation**
d Phasenimpulsmodulation f
f modulation f par déplacement d'impulsion;
   modulation d'espacement d'impulsion
r фазово-импульсная модуляция f

4390 **displacement sensor**
d Wegsensor m
f senseur m de route
r сенсор m перемещений

4391 **displacement travel**
d Verschiebeweg m
f course f de coulissement
r траектория f перемещения

4392 **display console**
d Bildschirmkonsole f
f console f à écran de visualisation
r консоль f с дисплеем

4393 **display control**
d Anzeigesteuerung f
f commande f de l'affichage
r управление n отображением

4394 display control logic
 d Displaysteuerlogik *f*
 f logique *f* de commande d'affichage
 r логический блок *m* управления дисплеем

4395 display device; display unit
 d Anzeigevorrichtung *f*; Bildschirmeinheit *f*; Sichtgerät *n*
 f dispositif *m* d'indicateur; dispositif d'affichage; unité *f* d'affichage
 r показывающее устройство *n*; модуль *m* визуального выхода; устройство визуального изображения; дисплей *m*

4396 display error
 d Darstellungsfehler *m*
 f erreur *f* de représentation
 r ошибка *f* индикации

 * **display for non-validation** → 216

4397 display modes
 d Anzeigebetriebsarten *fpl*
 f modes *mpl* d'affichage
 r режимы *mpl* отображения

4398 display multiplexor
 d Anzeigemultiplexer *m*
 f multiplexer *m* pour indication
 r мультиплексор *m* системы отображения

4399 display position
 d Anzeigestelle *f*; Anzeigeposition *f*
 f position *f* d'affichage
 r позиция *f* на экране дисплея

4400 display technique
 d Anzeigeverfahren *n*; Anzeigetechnik *f*
 f procédé *m* d'affichage
 r техника *f* отображения

 * **display unit** → 4395

 * **disregard** *v* → 8567

4401 disruptive voltage; breakdown voltage
 d Durchschlagsspannung *f*
 f tension *f* disruptive
 r пробивное напряжение *n*

4402 dissipation
 d Dissipation *f*
 f dissipation *f*
 r диссипация *f*

4403 dissipation effect
 d Dissipationseinwirkung *f*
 f action *f* dissipative
 r диссипативное [воз]действие *n*

4404 dissipation function
 d dissipative Funktion *f*
 f fonction *f* dissipative
 r диссипативная функция *f*

4405 distance control; distant control; remote control
 d Fernsteuerung *f*; Fernbedienung *f*; Fernschaltung *f*; Fernlenkung *f*; Fernbetätigung *f*
 f commande *f* à distance; télécommande *f*; contrôle *m* à distance; téléguidage *m*
 r дистанционное управление *n*; телерегулирование *n*

4406 distance difference measurement
 d Ablagemessung *f*
 f mesure *f* de la différence de distance
 r измерение *n* разности расстояний

4407 distance intervention
 d Ferneingriff *m*
 f intervention *f* à distance
 r дистанционное вмешательство *n*

 * **distance measurement** → 10938

4408 distance protection
 d Distanzschutz *m*
 f protection *f* de distance
 r дистанционная защита *f*

4409 distance recognition system
 d Entfernungserkennungssystem *n*
 f système *m* d'identification à distance
 r система *f* определения расстояния

4410 distance sensor
 d Abstandssensor *m*; Distanzsensor *m*; Entfernungssensor *m*
 f senseur *m* de distance
 r сенсор *m* расстояния

 * **distant control** → 4405

4411 distant reading
 d Fernablesung *f*
 f lecture *f* à distance
 r дистанционное считывание *n*

4412 distortion
 d Verzerrung *f*
 f distorsion *f*
 r искажение *n*

4413 distortion analyzer
 d Verzerrungsanalysator *m*

  f analyseur *m* de distorsions
  r анализатор *m* искажений

4414 distortion bridge
  d Klirrfaktormessbrücke *f*;
     Verzerrungsmessbrücke *f*
  f pont *m* de distorsion
  r мост *m* для измерения искажений

4415 distortion coefficient; distortion factor
  d Klirrfaktor *m*; Verzerrungsfaktor *m*
  f coefficient *m* de distorsion
  r коэффициент *m* искажения

4416 distortion due to feedback
  d Rückkopplungsverzerrung *f*
  f distorsion *f* par réaction
  r искажение *n* от обратной связи

**4417 distortion elimination; compensation of
     distortion**
  d Entzerrung *f*; Verzerrungskompensation *f*
  f correction *f* de distorsion; compensation *f* de
     distorsion
  r устранение *n* искажения

  * **distortion factor** → 4415

4418 distortion limited operation
  d verzerrungsbegrenzter Betrieb *m*
  f fonctionnement *m* limité par la distorsion
  r работа f, ограничиваемая искажением

4419 distortion meter
  d Verzerrungsmesser *m*
  f distorsiomètre *m*
  r измеритель *m* искажений

4420 distortion transmission impairment
  d Minderung *f* der Übertragungsgüte durch
     Verzerrung
  f réduction *f* de qualité de transmission due à
     distorsion
  r ухудшение *n* качества передачи
     вследствие искажения

4421 distribute *v*
  d verteilen
  f distribuer
  r распределять

4422 distributed
  d verteilt
  f distribué; réparti
  r распределенный; дистрибутивный

4423 distributed capacity
  d verteilte Kapazität *f*

  f capacité *f* distribuée; capacité répartie
  r распределенная ёмкость *f*

4424 distributed computer system
  d verteiltes Rechnersystem *n*
  f système *m* de calculateurs réparti
  r распределенная вычислительная система *f*

4425 distributed constants
  d Verteilungskonstanten *fpl*
  f constantes *fpl* réparties
  r распределенные параметры *mpl*

4426 distributed control
  d verteilte Steuerung *f*
  f commande *f* répartie
  r распределенное управление *n*

4427 distributed data entry
  d verteilte Dateneingabe *f*
  f entrée *f* de données répartie
  r децентрализованный ввод *m* данных

4428 distributed induction
  d verteilte Induktivität *f*
  f inductance *f* distribuée
  r распределенная индуктивность *f*

**4429 distributed-intelligence microprocessor
     system**
  d Mikroprozessorsystem *n* mit verteilter
     Intelligenz
  f système *m* de microprocesseurs à intelligence
     répartie
  r микропроцессорная система *f* с
     распределенной обработкой

4430 distributed logic
  d verteilte Logik *f*
  f logique *f* répartie
  r распределенная логика *f*

4431 distributed network
  d verteilte Netzwerke *fpl*
  f réseaux *mpl* distribués
  r дистрибутивные сети *fpl*; распределенные
     сети

4432 distributed parameter amplifier
  d Verstärker *m* mit verteilten Parametern
  f amplificateur *m* à paramètres répartis
  r усилитель *m* с распределенными
     параметрами

4433 distributed parameter system
  d System *n* mit verteilten Parametern
  f système *m* à paramètres répartis
  r система *f* с распределенными
     параметрами

4434 distributed processing system
  d verteiltes Datenverarbeitungssystem *n*
  f système *m* informatique réparti
  r распределенная система *f* обработки
       данных

4435 distributed random number
  d distributive Zufallszahl *f*
  f nombre *m* au hasard distribué
  r дистрибутивное случайное число *n*

4436 distributed redundancy
  d verteilte Redundanz *f*
  f redondance *f* distribuée
  r распределенная избыточность *f*

4437 distributed switching system
  d System *n* mit verteilter Vermittlung
  f système *m* de commutation réparti
  r система *f* с распределенной коммутацией

4438 distribution
  d Verteilung[sfunktion] *f*
  f distribution *f*; fonction *f* de distribution
  r распределение *n*

4439 distribution circuit
  d Verteilungsschaltung *f*
  f circuit *m* de distribution
  r распределительная схема *f*

4440 distribution code
  d Verteilungskode *m*
  f code *m* de distribution
  r код *m* распределения

   * **distribution coefficient** → 9454

4441 distribution function
  d Verteilungsfunktion *f*
  f fonction *f* de distribution
  r функция *f* распределения

4442 distribution law
  d Verteilungsgesetz *n*
  f loi *f* de distribution
  r закон *m* распределения

4443 distribution model
  d Verteilungsmodell *n*
  f modèle *m* de distribution
  r модель *f* распределения

4444 distribution of control
  d Verteilung *f* der Steuerung
  f distribution *f* de commande
  r распределение *n* управления

4445 distribution register
  d Verteilerregister *n*
  f registre *m* distributeur
  r регистр *m* распределения

4446 distribution variance
  d Verteilungsdispersion *f*
  f dispersion *f* de distribution
  r дисперсия *f* распределения

4447 distributor
  d Verteiler *m*; Aufgeber *m*;
       Aufgabevorrichtung *f*; Verteilanlage *f*
  f distributeur *m*; installation *f* de distribution
  r распределитель *m*

4448 disturbance; fault breakdown;
       interference
  d Störung *f*
  f perturbation *f*; défaut *m*; interférence *f*
  r помеха *f*; возмущение *n*

4449 disturbance analysis
  d Störungsauswertung *f*; Störungsanalyse *f*;
       Störungsaufklärung *f*
  f analyse *f* des dérangements
  r анализ *m* причин отказа

4450 disturbance band; disturbance range
  d Störbereich *m*
  f domaine *m* de perturbation
  r область *f* помех

4451 disturbance error function
  d Störungsfehlerfunktion *f*
  f fonction *f* d'erreur de perturbation
  r функция *f* распределения вероятности
       ошибок от помех

4452 disturbance feedforward; disturbance
       superposition
  d Störungskompensierung *f*
  f compensation *f* de perturbation
  r компенсация *f* возмущающего
       воздействия

4453 disturbance level
  d Störpegel *m*
  f niveau *m* de perturbations
  r уровень *m* помех; уровень возмущений

   * **disturbance quantity** → 4458

   * **disturbance range** → 4450

4454 disturbance registration
  d Störungserfassung *f*

f enregistrement *m* de perturbations
r регистрация *f* возмущения

4455 disturbance response
d Störverhalten *n*
f comportement *m* au dérangement
r реакция *f* на возмущение

4456 disturbance storage
d Anhäufung *f*
f emmagasinage *m*; stockage *m*
r накопление *n* возмущений

* **disturbance superposition → 4452**

4457 disturbance time
d Störungsdauer *f*
f durée *f* de dérangement; période *f* de
   perturbation
r продолжительность *f* возмущения

* **disturbance upset → 4458**

**4458 disturbance variable; disturbing variable;
        disturbance upset; disturbance quantity**
d Störgröße *f*; störende Veränderliche *f*;
   Störvorgang *m*; Störeinwirkung *f*
f grandeur *f* perturbatrice; variable *f*
   perturbatrice; action *f* perturbatrice
r возмущающая переменная *f*;
   возмущающая величина *f*; возмущающее
   воздействие *n*

4459 disturbance variable compensation
d Störgrößenaufschaltung *f*;
   Störungskompensation *f*
f compensation *f* de perturbation
r компенсация *f* возмущений

4460 disturbed motion
d gestörte Bewegung *f*
f mouvement *m* perturbé
r возмущенное движение *n*

4461 disturbed state
d gestörter Zustand *m*
f état *m* perturbé
r возмущенное состояние *n*

4462 disturbed value
d gestörter Wert *m*
f valeur *f* perturbée
r искаженное заначение *n*

4463 disturbed-zero output
d gestörtes Nullausgangssignal *n*
f signal *m* zéro de sortie perturbé
r искаженный нулевой выходной сигнал *m*;

выходной сигнал разрушенного нуля

4464 disturbing force
d störende Kraft *f*; Störkraft *f*
f force *f* perturbatrice
r искажающая сила *f*

4465 disturb[ing] pulse; noise pulse
d Störimpuls *m*
f impulsion *f* perturbatrice; impulsion de bruit
r искажающий импульс *m*; шумовой
   импульс

* **disturbing variable → 4458**

* **disturb pulse → 4465**

4466 diverge *v*
d divergieren
f diverger
r расходиться

4467 divergence of series
d Reihendivergenz *f*
f divergence *f* de la série
r расходимость *f* ряда

4468 divergent
d divergent
f divergent
r расходящий[ся]

4469 divergent oscillations
d Divergenzschwingungen *fpl*
f oscillations *fpl* divergentes
r расходящиеся колебания *npl*

4470 diversity factor
d Verschiedenheitsfaktor *m*
f facteur *m* de diversité
r коэффициент *m* разновременности

4471 dividing circuit; division circuit
d Dividierkreis *m*; Divisionsschaltung *f*
f circuit *m* diviseur; circuit de division
r схема *f* деления

4472 dividing device
d Teilgerät *n*
f appareil *m* diviseur
r делительное устройство *n*

* **division algorithm → 647**

* **division circuit → 4471**

4473 document handling
d Belegbearbeitung *f*

f travail *m* de documents
r [предварительная] подготовка *f* документов

4474 domain of definition
d Definitionsbereich *m*
f intervalle *m* de définition; domaine *m* de définition
r область *f* определения

* **DOS → 4376**

* **dosimeter → 1701**

4475 dosimetry probe
d Strahlenmesskopf *m*; Strahlenmesssonde *f*
f sonde *f* dosimétrique
r дозиметрический зонд *m*

4476 double amplification circuit; reflex circuit
d doppelt verstärkende Schaltung *f*; Schaltung mit Doppelverstärkung; Reflexschaltung *f*
f circuit *m* réflexe; circuit à double amplification
r схема *f* двойного усиления; рефлексная схема

4477 double amplitude peak; peak-to-peak amplitude
d Spitze-zu-Spitze-Amplitude *f*
f amplitude *f* crête à crête
r амплитуда *f* суммарного колебания; двойная амплитуда

4478 double computer system
d Doppelrechnersystem *n*
f système *m* à computer double
r дублированная компьютерная система *f*

4479 double error-correcting code
d Kode *m* mit doppelter Fehlerkorrektur
f code *m* avec correction double de l'erreur
r код *m* с исправлением двойных ошибок

4480 double-loop servomechanism; two-loop servomechanism
d Doppelschleifenservomechanismus *m*
f servomécanisme *m* à double boucle
r двухконтурная следящая система *f*

4481 double phantom circuit; superphantom circuit
d Achterkreis *m*; Achterstromkreis *m*
f circuit *m* superfantôme; circuit fantôme double
r двойная фантомная схема *f*; суперфантомная схема

4482 double pulse generator
d Doppelimpulsgenerator *m*
f générateur *m* à impulsions doubles
r генератор *m* двойных импульсов

4483 double pulse modulation
d Doppelimpulsmodulation *f*
f modulation *f* impulsive double
r двойная импульсная модуляция *f*

4484 double-range recording flowmeter
d registrierender Doppelbereich-Durchflussmesser *m*
f débitmètre *m* enregistreur à double gamme
r регистрирующий расходомер *m* с двумя пределами измерений

4485 double sideband transmission
d Zweiseitenbandübertragung *f*
f transmission *f* sur deux bandes latérales
r передача *f* с двумя боковыми полосами

4486 double stage process
d Zweistufenprozess *m*
f procédé *m* à deux étages
r двухступенчатый процесс *m*

4487 double-throw contact; two-way contact
d Umschaltkontakt *m*
f contact *m* à commutateur
r переключающий контакт *m*

4488 double-throw contact with neutral position
d Umschaltkontakt *m* mit neutraler Stellung
f contact *m* à commutateur à position neutre
r контакт *m* двухстороннего действия

4489 doughnut-type transformer
d Ringtransformator *m*
f transformateur *m* à tore; transformateur toroïdal
r кольцеобразный трансформатор *m*

4490 downtime
d Ausfallzeit *f*
f temps *m* d'arrêt
r время *n* остановки; продолжительность *f* выключения

4491 drawing machine
d Zeichenmaschine *f*
f machine *f* à dessiner
r чертёжная машина *f*

* **drawing work automation → 1424**

4492 draw-off mode
d Entnahmebetrieb *m*

f régime *m* avec soutirage
r режим *m* с отбором

* **dressing → 10022**

4493 drift
d Drift *f*; Abweichung *f*
f dérive *m*; glissement *m*; écart *m*
r дрейф *m*; смещение *n*; отклонение *n*

4494 drift-corrected amplifier
d driftkompensierter Verstärker *m*
f amplificateur *m* à compensation de dérive
r усилитель *m* с коррекцией дрейфа

4495 drift correction
d Driftkorrektur *f*
f correction *f* de dérive
r коррекция *f* отклонения

4496 drift error
d Driftfehler *m*
f erreur *f* de dérive
r ошибка *f* дрейфа

4497 drift factor
d Driftfaktor *m*
f facteur *m* de dérive
r коэффициент *m* смещения

4498 driftmeter
d Driftmesser *m*
f dérivomètre *m*
r измеритель *m* угла сноса

4499 drive *v*
d treiben; speisen; lenken
f entraîner; alimenter
r приводить в действие; вести

4500 drive current
d Treiberstrom *m*; Steuerstrom *m*
f courant *m* de basculeur
r ток *m* возбуждения

4501 drive dimensioning
d Antriebsdimensionierung *f*
f dimensionnement *m* d'entraînement
r выбор *m* размеров [и параметров] привода

4502 drive element
d Antriebselement *n*
f élément *m* d'entraînement
r элемент *m* привода

4503 drive element dynamics
d Antriebselementedynamik *f*
f dynamique *f* d'élément d'entraînement

r динамика *f* элементов привода

4504 drive element number; element number of manipulator drive
d Antriebsgliederzahl *f*; Gliederzahl *f* eines Manipulatorantriebes
f nombre *m* d'éléments d'un entraînement de manipulateur
r число *n* звеньев привода манипулятора

* **drive gear → 8992**

4505 drive moment
d Antriebsmoment *n*
f moment *m* d'entraînement; couple *m* d'entraînement
r приводной момент *m*

4506 drive motor
d Antriebsmotor *m*
f moteur *m* de commande; moteur *m* d'entraînement
r приводной двигатель *m*

4507 drive movement
d Antriebsbewegung *f*
f mouvement *m* d'entraînement
r приводное движение *n*

4508 drive movement number
d Anzahl *f* der Antriebsbewegungen
f nombre *m* des mouvements d'entraînement
r количество *n* приводных движений

4509 drive of regulated unit
d Antrieb *m* des Stellgliedes
f commande *f* de l'organe de réglage final
r привод *m* исполнительного механизма

4510 drive of rotation angle
d Drehwinkelantrieb *m*
f entraînement *m* de l'angle de rotation
r сельсин-привод *m*

* **driver → 4512**

4511 drive regulator
d Antriebsregler *m*
f régulateur *m* d'entraînement
r регулятор *m* привода

4512 driver stage; driving stage; driver
d Treiberstufe *f*
f étage *m* driver; étage d'attaque
r задающий каскад *m*; предконечный каскад *m*

4513 drive system
d Antriebssystem *n*

f système *m* d'entraînement
r система *f* привода

* **driving circuit** → 3202

4514 driving power
 d Steuerleistung *f*; Antriebskraft *f*
 f puissance *f* de commande
 r управляющая мощность *f*

* **driving pulse** → 354

* **driving stage** → 4512

* **driving torque** → 3296

4515 droop
 d bleibende Regelabweichung *f*
 f écart *m* résiduel permanent
 r статистическое отклонение *n*

4516 droop correction
 d Abfallberichtigung *f*
 f correction *f* de la chute
 r поправка *f* на падение

* **drop analysis** → 11755

4517 drop indicator relay
 d Fallklappenrelais *n*
 f lapin *m* indicateur
 r указательное реле *n*; блинкер *m*

4518 drop in pressure
 d Druckverlust *m*; Druckabfall *m*
 f perte *f* de pression; chute *f* de pression
 r потеря *f* давления

4519 drum controller
 d Trommelkontroller *m*;
   Trommelfahrschalter *m*;
   Walzenfahrschalter *m*
 f combinateur *m* à tambour
 r барабанный контроллер *m*

4520 drum recorder
 d Trommelschreiber *m*
 f appareil *m* enregistreur à tambour
 r барабанный самописец *m*

4521 drum store
 d Trommelspeicher *m*
 f mémoire *f* tambour
 r барабанный накопитель *m*

* **dry friction** → 3483

4522 dry system

 d Trockensystem *n*; Trocknungssystem *n*
 f système *m* de séchage
 r сухая система *f*

4523 dual channel microprogram
 d Dualkanal-Mikroprogramm *n*
 f piste *f* duale pour microprogramme
 r микропрограмма *f* канала передачи
   двоичной информации

**4524 dual-coded number; binary-coded
     number**
 d binär kodierte Zahl *f*; dual kodierte Zahl
 f nombre *m* codé binaire; nombre codé dual
 r двоично-кодированное число *n*

4525 dual component
 d Dualkomponente *f*
 f composante *f* duale
 r двойная составляющая *f*

* **dual control** → 4529

4526 duality
 d Dualität *f*
 f dualité *f*
 r двойственность *f*

4527 duality theory
 d Dualitätstheorie *f*
 f théorie *f* de dualité
 r теория *f* двойственности

4528 dual measuring instrument
 d Doppelmessgerät *n*; Zweifachmessgerät *n*
 f appareil *m* de mesure à deux lectures
 r двойной измерительный прибор *m*

4529 dual[-mode] control
 d duale Steuerung *f*; Dualsteuerung *f*;
   Doppelsteuerung *f*; Zweipunktregelung *f*
 f commande *f* double; commande *f* à deux
   modes de fonctionnement
 r дуальное управление *n*

4530 dual operation
 d Doppelbetätigung *f*; Doppelwirkungsweise *f*
 f fonctionnement *m* à double effet
 r двойное действие *n*

4531 dual spectral method
 d duales Spektralverfahren *n*
 f procédé *m* dual-spectral
 r двоичный спектральный метод *m*

4532 dual speed synchro system
 d Selsyn *n* mit zwei Geschwindigkeiten;
   Drehmelder *m* mit zwei Geschwindigkeiten

  f selsyn *m* à deux vitesses
  r двухскоростная следящая система *f*

4533 dual state space
  d dualer Zustandsraum *m*
  f espace *m* d'état dual
  r дуальное пространство *n* состояний

4534 dual switch board
  d Doppelschalttafel *f*
  f panneau *m* double de commutation
  r двойной коммутатор *m*

4535 dual theorem
  (operations research)
  d Dualitätstheorem *n*
  f théorème *m* dual
  r теорема *f* дуализма

4536 ductilimeter
  d Dehnbarkeitsmesser *m*; Dehnungsmesser *m*
  f ductilimètre *m*; mesureur *m* de ductilité
  r дуктилометр *m*

4537 Duhamel integral
  d Duhamelsches Integral *n*
  f intégrale *f* de Duhamel
  r интеграл *m* Дюамеля

4538 dummy address
  d Scheinadresse *f*
  f adresse *f* fictive
  r фиктивный адрес *m*

4539 dummy information
  d belanglose Information *f*
  f information *f* fictive
  r фиктивная информация *f*

4540 dump
  d Speisungsunterbrechung *f*
  f arrêt *m* d'alimentation; disjonction *f*
  r прекращение *n* питания

4541 dump check
  d Übertragungskontrolle *f*
  f essai *m* de transfert; contrôle *m* de transfert
  r проверка *f* передачи

4542 duplex adding machine
  d Duplex-Addiermaschine *f*
  f machine *f* à additionner duplex
  r дуплексная суммирующая машина *f*

4543 duplex channel
  d Duplexkanal *m*; Zweiwegkanal *m*
  f canal *m* duplex
  r дуплексный канал *m*

4544 duplex circuit
  d Duplexleitung *f*
  f circuit *m* duplex
  r дуплексная линия *f*; двухсторонняя линия

4545 duplex mode
  d Duplexbetriebsweise *f*; Zweiwegmodus *m*
  f mode *m* duplex
  r дуплексный режим *m*

4546 duplicating device
  d Wiederholungseinrichtung *f*
  f dispositif *m* de duplication
  r дублирующее устройство *n*

4547 duplication check
  d Zwillingskontrolle *f*
  f vérification *f* par double opération
  r двойной контроль *m*

4548 duplication factor
  d Vielfaktor *m*
  f facteur *m* de duplication
  r коэффициент *m* дублирования

4549 duration of action; duration of effect
  d Einwirkungsdauer *f*; Wirkungsdauer *f*
  f durée *f* de l'attaque; durée d'action
  r продолжительность *f* воздействия

4550 duration of cycle
  d Zyklusdauer *f*; Zykluszeit *f*
  f durée *f* de cycle
  r продолжительность *f* цикла

  *   **duration of effect** → 4549

4551 duration of self-regulation
  d Ausgleichszeit *f*
  f durée *f* d'auto-équilibrage
  r длительность *f* саморегулирования

4552 duration of the impulse
  d Impulsdauer *f*
  f durée *f* de l'impulsion
  r длительность *f* импульса

4553 dust removal plant
  d Entstaubungsanlage *f*
  f installation *f* de dépoussiérage
  r пылеуловительная установка *f*

4554 duty cycle; work cycle
  d Arbeitsphase *f*; Arbeitszyklus *m*
  f cycle *m* de travail; cycle opératoire
  r рабочий цикл *m*

4555 duty factor
  d Betriebsfaktor *m*

f facteur *m* de charge
r коэффициент *m* заполнения

* **duty ratio** → 1934

4556 dwell phase of manipulator system
d Ruhephase *f* eines Manipulatorsystems
f phase *f* de repos d'un système de manipulation
r нерабочая фаза *f* манипуляционной системы

4557 dwell time
d Verweilzeit *f*
f temps *m* d'arrêt
r программируемая временная задержка *f*

4558 dynamic
d dynamisch
f dynamique
r динамический

4559 dynamic accuracy; dynamic precision
d dynamische Genauigkeit *f*
f précision *f* dynamique
r динамическая точность *f*

4560 dynamic allocation
d dynamische Zuordnung *f*
f allocation *f* dynamique
r динамическое распределение *n*

4561 dynamic analysis
d dynamische Analyse *f*
f analyse *f* dynamique
r динамический анализ *m*

4562 dynamic balance; dynamic equilibrium
d dynamisches Gleichgewicht *n*; dynamischer Gleichgewichtszustand *m*
f équilibre *m* dynamique
r динамическое равновесие *n*

4563 dynamic behaviour
d dynamisches Verhalten *n*
f tenue *f* dynamique; allure *f* dynamique; comportement *m* dynamique
r динамический режим *m* работы

* **dynamic binding** → 4583

4564 dynamic boundary condition
d dynamische Randbedingung *f*
f condition *f* marginale dynamique
r динамическое краевое условие *n*

4565 dynamic calculation
d dynamische Berechnung *f*

f calcul *m* dynamique
r динамический расчёт *m*

4566 dynamic characteristic
d dynamische Arbeitskurve *f*
f caractéristique *f* dynamique
r динамическая характеристика *f*

4567 dynamic circuit
d dynamische Schaltung *f*
f circuit *m* dynamique
r динамическая схема *f*

4568 dynamic control system
d dynamisches Regelsystem *n*; dynamisches Steuersystem *n*
f système *m* de réglage dynamique; réglage *m* dynamique
r динамическая система *f* управления

4569 dynamic damper
d dinamischer Dämpfer *m*; dinamischer Stoßdämpfer *m*
f amortisseur *m* dynamique
r динамический демпфер *m*

4570 dynamic data
d dynamische Daten *pl*
f données *fpl* dynamiques
r динамические данные *pl*

4571 dynamic definition
d dynamische Definition *f*
f définition *f* dynamique
r динамическое определение *n*

4572 dynamic deviation of mechanical subassemblies
d dynamische Abweichung *f* mechanischer Baugruppen
f écart *m* dynamique d'unités mécaniques
r динамическое отклонение *m* механических узлов

4573 dynamic deviation of simultaneous movement
d dynamische Abweichung *f* von Simultanbewegungen
f écart *m* dynamique de mouvements simultanés
r динамическое отклонение *n* синхронных движений

4574 dynamic device
d dynamisches Gerät *n*
f dispositif *m* dynamique
r динамическое устройство *n*

4575 dynamic dump
  d dynamischer Speicherauszug *m*
  f extrait *m* dynamique de mémoire
  r динамическая выдача *f*

4576 dynamic element
  d dynamisches Element *n*
  f élément *m* dynamique
  r динамический элемент *m*

  * **dynamic equilibrium → 4562**

4577 dynamic error
  d dynamischer Fehler *m*; vorübergehende Regelabweichung *f*
  f erreur *f* dynamique; erreur de réglage
  r динамическая погрешность *f*; динамическая ошибка *f*

4578 dynamic extension measurement
  d Messung *f* dynamischer Dehnungsvorgänge
  f mesure *f* de procédés d'allongement dynamiques
  r измерение *n* динамических процессов растяжения

4579 dynamic fidelity
  d dynamische Wiedergabegenauigkeit *f*
  f fidélité *f* dynamique de reproduction
  r динамическая точность *f* воспроизведения

4580 dynamic friction
  d dynamische Reibung *f*
  f friction *f* dynamique
  r динамическое трение *n*

4581 dynamic generator characteristic
  d dynamische Generatorkennlinie *f*
  f caractéristique *f* dynamique de générateur
  r динамическая характеристика *f* генератора

4582 dynamic lag
  d dynamische Verzögerung *f*
  f retard *m* dynamique
  r динамическое запаздывание *n*

  * **dynamic link → 4583**

4583 dynamic link[ing]; dynamic binding
  d dynamische Verknüpfung *f*
  f liaison *f* dynamique; lien *m* dynamique
  r динамическая связь *f*

4584 dynamic magnetic storage technique
  d dynamische Magnetspeichertechnik *f*
  f technique *f* d'emmagasinage magnétique dynamique

  r техника *f* динамического магнитного запоминания

4585 dynamic manipulator property
  d dynamische Manipulatoreigenschaft *f*; dynamische Eigenschaft *f* eines Manipulators
  f propriété *f* dynamique d'un manipulateur
  r динамическая характеристика *f* манипулятора

4586 dynamic mass-spectrometer
  d dynamische Massenspektrometer *n*
  f spectromètre *m* dynamique de masse
  r динамический массспектрометр *m*

  * **dynamic memory → 4608**

4587 dynamic operation
  d dynamische Operation *f*
  f opération *f* dynamique
  r динамическая операция *f*

4588 dynamic operational behaviour
  d dynamisches Betriebsverhalten *n*
  f comportement *m* opérationnel dynamique
  r динамическая рабочая характеристика *f*

4589 dynamic optimization
  d dynamische Optimierung *f*
  f optimisation *f* dynamique
  r динамическая оптимизация *f*

4590 dynamic parallel program structure
  d dynamische parallele Programmstruktur *f*
  f structure *f* parallèle dynamique du programme
  r программа *f* с динамической параллельной структурой

  * **dynamic precision → 4559**

4591 dynamic problem checking
  d dynamische Problemprüfung *f*
  f essai *m* de problème dynamique
  r динамический контроль *m* проблемы

4592 dynamic programming
  d dynamische Programmierung *f*
  f programmation *f* dynamique
  r динамическое программирование *n*

4593 dynamic program structure
  d dynamische Programmstruktur *f*
  f structure *f* de programme dynamique
  r динамическая структура *f* программы

4594 dynamic properties correction
  d Korrektur *f* der dynamischen Eigenschaften

f correction *f* des propriétés dynamiques
r коррекция *f* динамических свойств

4595 dynamic range
d Dynamikbereich *m*
f gamme *f* dynamique; plage *f* dynamique
r динамический диапазон *m*

4596 dynamic regime
d dynamischer Betrieb *m*
f régime *m* dynamique
r динамический режим *m*

4597 dynamic relocation
d dynamische Verschiebung *f*
f relocation *f* dynamique
r динамическое смещение *n*

4598 dynamic resistance
d dynamischer Widerstand *m*
f résistance *f* dynamique
r динамическое сопротивление *n*

4599 dynamic responses of automatic measurement means
d dynamische Kennlinien *fpl* von automatischen Messgliedern
f caractéristiques *fpl* dynamiques d'appareils mesureurs automatiques
r динамические характеристики *fpl* автоматических измерительных приборов

4600 dynamic sensitivity
d dynamische Empfindlichkeit *f*
f sensibilité *f* dynamique
r динамическая чувствительность *f*

4601 dynamic sequential control
d Programmablaufänderung *f*
f passage *m* de programme dynamique
r динамическое программное управление *n*

4602 dynamic simulation
d dynamische Simulation *f*
f simulation *f* dynamique
r динамическая симуляция *f*

4603 dynamic single-mode laser
d dynamischer Einmodenlaser *m*
f laser *m* unimodal dynamique
r динамический одномодовый лазер *m*

4604 dynamics of interconnected steam systems
d Dynamik *f* vermaschter Dampfsysteme
f dynamique *f* de systèmes interconnectés à vapeur
r динамика *f* сопряженных паровых систем

4605 dynamics of ramified control circuits
d Dynamik *f* verzweigter Regelkreise
f dynamique *f* des circuits ramifiés de réglage
r динамика *f* разветвленных систем регулирования

4606 dynamics of the operation of the automaton
d Dynamic *f* der Automatenoperation
f dynamique *f* du fonctionnement d'automate
r динамика *f* функционирования автомата

4607 dynamic state
d dynamischer Zustand *m*
f état *m* dynamique
r динамическое состояние *n*

4608 dynamic storage; dynamic memory
d dynamischer Speicher *m*
f mémoire *f* dynamique
r динамическая память *f*; запоминающее устройство *n* динамического типа

4609 dynamic subroutine
d dynamisches Unterprogramm *n*
f sous-routine *f* dynamique
r динамическая подпрограмма *f*

4610 dynamic system
d dynamisches System *n*
f système *m* dynamique
r динамическая система *f*

4611 dynamic test
(of manipulation system)
d dynamische Probe *f*
f essai *m* dynamique
r динамический контроль *m*

4612 dynamic unit
d dynamische Einheit *f*
f unité *f* dynamique
r динамическое звено *n*

4613 dynamic wavemeter
d dynamischer Wellenmesser *m*
f ondemètre *m* dynamique
r динамический волномер *m*

4614 dynamo-governor
d Dynamoregler *m*
f régulateur *m* de dynamo
r динамо-регулятор *m*

E

4615 earth-fault
d Erdschluß m
f défaut m à laterre
r замыкание n на землю

4616 earth-fault protection
d Erdfehlerschutz m
f protection f contre les défauts à la terre
r защита f от замыкания на землю

4617 earthing contact
d Erdungskontakt m
f contact m de mise à la terre
r контакт m заземления

4618 earth leakage current
d Erdschlußstrom m
f courant m de fuite à la terre
r ток m утечки [в землю]

4619 earth leakage current breaker
d Erdschlußstromunterbrecher m;
   Erdschlußstromschutzschalter m
f disjoncteur m de protection à courant de
   perte à la terre
r выключатель m тока замыкания на землю

4620 earth terminal
d Erdungsanschluss m
f borne f de mise à la terre
r зажим m заземления

4621 earth tester
d Erdungsmesser m; Erdleitungsprüfer m
f tellurohmmètre m
r измеритель m заземления; меггер m

4622 echo altimeter
d akustischer Höhenmesser m; Echolot n
f altimètre m acoustique; altimètre à écho
r отражательный высотомер m

4623 echo checking; echo testing
d Echoprüfung f; Prüfung f durch
   Rückübertragung
f contrôle m par écho; essai m d'écho; contrôle
   d'écho
r испытание n на эхо

4624 echo pulse
d Echoimpulse m

f impulsion f réfléchie
r отраженный [волновой] импульс m

4625 echo signal
d Echosignal n
f signal m d'écho
r эхо-сигнал m

* **echo testing → 4623**

4626 economical cybernetics
d ökonomische Kybernetik f; Kybernetik in der
   Ökonomie
f cybernétique f économique
r экономическая кибернетика f

4627 eddy-current brake
d Wirbelstrombremse f
f frein m à courants parasites; frei à courants
   de Foucault
r тормоз m на вихревых токах

4628 eddy-currents measurement method
d Wirbelstrommessverfahren n
f méthode f de mesure au moyen de courants
   de Foucault
r измерение n возбуждением вихревых
   токов

4629 edge-frequency
d Grenzfrequenz f
f fréquence f limite
r граничная частота f

4630 editing equipment
d Redigiereinrichtung f
f équipement m d'arrangement
r редактирующее устройство n

* **EDP → 4881**

* **EDP system → 4882**

4631 educational data system
d Datensystem n für Ausbildungszwecke
f système m de données pour éducation
r информационная система f для обучения

4632 effective
d effektiv; wirksam; tatsächlich
f effectif; efficace
r эффективный; действующий

* **effective address → 326**

4633 effective area
d Effektivfläche f
f surface f effective
r эффективная поверхность f

4634 **effective driving-current density amplitude**
   d Amplitude *f* der effektiven Erregerstromdichte
   f amplitude *f* de la densité efficace du courant d'excitation
   r амплитуда *f* эффективной плотности управляющего тока

4635 **effective input admittance**
   d wirksame Eingangsadmittanz *f*
   f admittance *f* effective d'entrée
   r эффективная входная [полная] проводимость *f*

4636 **effective input capacitance**
   d wirksame Eingangskapazität *f*
   f capacité *f* effective d'entrée
   r эффективная входная [полная] ёмкость *f*

4637 **effective input impedance**
   d wirksame Eingangsimpedanz *f*
   f impédance *f* effective d'entrée
   r эффективное входное [полное] сопротивление *n*

4638 **effective mass**
   d wirksame Masse *f*
   f masse *f* effective
   r эффективная масса *f*

4639 **effective part of scale**
   d Wirkteil *m* der Skale
   f partie *f* efficace du cadran
   r эффективная часть *f* шкалы

   * **effective power** → 12722

4640 **effective range of measurement**
   d effektiver Messbereich *m*
   f étendue *f* effective de mesure
   r эффективный диапазон *m* измерений

4641 **effective resistance**
   d effektiver Widerstand *m*
   f résistance *f* effective
   r эффективное сопротивление *n*

4642 **effective transmission equivalent**
   d Bezugsdämpfung *f* eines Übertragungssystems; Nutzdämpfung *f*
   f équivalent *m* effective de transmission
   r эквивалент *m* затухания передачи

4643 **effective value; virtual value**
   d Effektivwert *m*
   f valeur *f* efficace; valeur effective
   r эффективное значение *n*

   * **effector** → 351

4644 **effector computer**
   d Effektorrechner *m*
   f ordinateur *m* d'effecteur
   r компьютер *m* исполнительного органа

4645 **effector coordinates**
   d Effektorkoordinaten *fpl*
   f coordonnées *fpl* d'effecteur
   r эффекторные координаты *fpl*

4646 **effector exchange**
   d Effektoraustausch *m*
   f échange *m* d'effecteur
   r эффекторный обмен *m*

   * **effector field of application** → 4647

4647 **effector field of use; effector field of application**
   d Effektoranwendungsgebiet *n*; Anwendungsgebiet *n* für Effektoren
   f domaine *m* d'application pour effecteurs
   r область *f* применения эффекторов

4648 **effector-fixed reference point**
   d effektorfester Referenzpunkt *m*
   f point *m* de référence à effecteur fixe
   r фиксированная опорная точка *f* исполнительного органа

4649 **effector force measurement**
   d Effektorkraftmessung *f*
   f mesurage *m* de force d'effecteur
   r измерение *n* эффекторных усилий

4650 **effector function**
   d Effektorfunktion *f*
   f fonction *f* d'effecteur
   r функция *f* эффектора; функция исполнительного органа

4651 **effector geometry**
   d Effektorgeometrie *f*
   f géométrie *f* d'effecteur
   r геометрия *f* эффектора

4652 **effector movement**
   d Effektorbewegung *f*
   f mouvement *m* d'effecteur
   r движение *n* исполнительного органа

4653 **effector movement guiding**
   d Effektorbewegungsführung *f*
   f guidage *m* de mouvement d'effecteur
   r направление *n* движения исполнительного органа

4654   effector orientation data
   d Effektororientierungsdaten pl
   f données fpl d'orientation d'effecteur
   r данные pl ориентации эффектора

4655   effector place in base coordinates
   d Effektorort m in Basiskoordinaten
   f place f d'effecteur en coordonnées de base
   r положение n эффектора в базовых
      координатах

4656   effector position
   d Effektorposition f
   f position f d'effecteur
   r позиция f эффектора

4657   effector reference system
   d Effektorbezugssystem n
   f système m de référence d'effecteur
   r базовая система f эффектора

4658   effector regulation
   d Effektorregelung f
   f régulation f d'effecteur
   r эффекторное регулирование n

4659   effector state
   d Effektorzustand m
   f état m d'effecteur
   r эффекторное состояние n

4660   efficiency of the system; system efficiency
   d Wirksamkeit f des Systems;
      Systemwirksamkeit f
   f efficacité f du système
   r эффективность f системы

4661   efficiency theorem
   d Effizienztheorem n
   f théorème m de l'efficacité
   r теорема f эффективности

4662   efficient algorithm
   d wirkungsvoller Algorithmus m
   f algorithme m efficace
   r эффективный алгоритм m

4663   efficient programming system
   d effizientes Programmiersystem n
   f système m de programmation efficient
   r эффективная система f программирования

4664   efficient pulse transmission
   d leistungsfähige Impulsübertragung f
   f transmission f d'impulsion efficace
   r эффективная импульсная передача f

4665   eigenfunction

   d Eigenfunktion f; Eigenlösung f
   f fonction f propre
   r собственная функция f

4666   eigenfunction expansion
   d Entwicklung f nach Eigenwerten
   f développement m en fonctions propres
   r разложение n по собственным функциям

4667   eigenvalue equation
   d Eigenwertgleichung f; charakteristische
      Gleichung
   f équation f de valeur propre
   r характеристическое уравнение n

4668   eigenvalue problem
   d Eigenwertproblem n; Eigenwertaufgabe f
   f problème m de valeur propre
   r задача f собственных значений

4669   eigenvector
   d Eigenvektor m
   f vecteur m propre
   r собственный вектор m

   *   EIS → 5084

4670   elastic feedback; variable feedback
   d variable Rückführung f; elastische
      Rückführung
   f contre-réaction f fléchissante; contre-réaction
      non proportionnelle; retour m élastique
   r гибкая обратная связь f

4671   elastic feedback controller; variable
       feedback controller
   d Regeleinrichtung f mit nachgebender
      Rückführung
   f régulateur m à réaction non proportionnelle;
      régulateur à réaction variable
   r регулятор m с гибкой обратной связью

   *   electric → 4675

4672   electric acquisition sensor
   d elektrischer Erfassungssensor m
   f senseur m d'acquisition électrique
   r электрический датчик m сбора

4673   electric actuator; electric servomotor
   d elektrischer Steuermotor m; elektrischer
      Stellmotor m; elektrischer Servomotor m
   f servomoteur m électrique; organe m
      électrique de réglage
   r электрический исполнительный
      механизм m

4674   electric adapting
   d elektrische Anpassung f

     f adaptation *f* électrique
     r электрическое сопряжение *n*

4675 electric[al]
     d elektrisch
     f électrique
     r электрический

4676 electrical analogy
     d elektrische Analogie *f*
     f analogie *f* électrique
     r электрическая аналогия *f*

4677 electrical analyzer
     d elektrischer Analysator *m*
     f analyseur *m* électrique
     r электрический анализатор *m*

4678 electrical balance
     d elektrisches Gleichgewicht *n*
     f équilibre *m* électrique
     r электрический баланс *m*

4679 electrical characteristics
     d elektrische Kenndaten *pl*
     f caractéristiques *fpl* électriques
     r электрические характеристика *fpl*

4680 electrical conductance; electric conductivity
     d elektrische Leitfähigkeit *f*
     f conductivité *f* électrique; conductibilité *f* électrique
     r электропроводимость *f*

4681 electrical contact controller
     d elektrischer Kontaktregler *m*
     f régulateur *m* électrique à contacts
     r регулятор *m* с электрическими контактами

4682 electrical correction
     d elektrische Korrektur *f*
     f correction *f* électrique
     r электрокорректировка *f*

4683 electrical dilatometer; electric strain gauge
     d elektrischer Dehnungsmesser *m*
     f extensomètre *m* électrique; jauge *f* électrique de contrainte
     r электрический тензометр *m*

4684 electrical engineering
     d Elektrotechnik *f*
     f électrotechnique
     r электротехника *f*

4685 electrical Fourier's analysis
     d elektrische Fourier-Analyse *f*
     f analyse *f* électrique de Fourier
     r электрический Фурье-анализ *m*

4686 electrical heat generator
     d Generator *m* für Dielektrikheizung
     f générateur *m* pour chauffage diélectrique
     r диэлектрический генератор *m* для нагрева

4687 electrical installation
     d elektrische Anlage *f*; Elektroinstallation *f*
     f installation *f* électrique
     r электроустановка *f*; электрооборудование *n*

4688 electrical insulation
     d elektrische Isolierung *f*
     f isolation *f* électrique
     r электрическая изоляция *f*

4689 electrically operated
     d elektrisch gesteuert
     f commandé par électricité
     r электроуправляемый

4690 electrically operated control
     d elektrische Regelung *f*
     f réglage *m* électrique
     r электрическое регулирование *n*

4691 electrically operated controller; electric controller; electric regulator
     d elektrischer Regler *m*; Elektroregler *m*; elektrische Regelvorrichtung *f*
     f régulateur *m* électrique; combinateur *m* électrique
     r регулятор *m* с электрическим приводом; электрический регулятор

4692 electrically operated drive; electric drive; electric propulsion
     d elektrischer Antrieb *m*; Elektroantrieb *m*
     f actionnement *m* électrique; commande *f* électrique
     r электрический привод *m*

4693 electrical measuring and test equipment
     d elektrische Mess- und Prüfeinrichtung *f*
     f équipement *m* de mesure et de test électrique
     r электроизмерительные приборы *mpl* и испытательное оборудование *n*

4694 electrical pressure measuring converter
     d elektrischer Druck-Messumformer *m*
     f convertisseur *m* mesureur électrique de pression
     r электрический измерительный преобразователь *m* давления

4695 electrical recorder
d elektrischer Schreiber *m*
f enregistreur *m* électrique
r электрический самопишущий прибор *m*

4696 electrical relay element
d elektrisches Relaiselement *n*
f élément *m* électrique de relais
r электрический релейный элемент *m*

4697 electrical scanner
d elektrisches Abtastgerät *n*
f dispositif *m* électrique de balayage
r электрическое развёртывающее
   устройство *n*; электрическое
   сканирующее устройство

4698 electrical sensing
d elektrische Abfühlung *f*
f lecture *f* électrique
r электрическое восприятие *n* сигнала
   датчиком

4699 electrical separation
d elektrische Abscheidung *f*
f électroséparation *f*
r электросепарация *f*

4700 electric[al] signal evaluation
d elektrische Signalauswertung *f*
f évaluation *f* de signal électrique
r обработка *f* электрических сигналов

4701 electrical specifications
d Kenndatenblatt *n* elektrischer Parameter
f spécifications *fpl* électriques
r требования *npl* к электрическим
   параметрам

4702 electrical zero
d elektrische Nullstellung *f*
f zéro *m* électrique
r электрический нуль *m*

4703 electric circuit
d Stromkreis *m*
f circuit *m* électrique
r электрический контур *m*; электрическая
   схема *f*

* **electric conductivity → 4680**

4704 electric connection
d elektrische Verbindung *f*
f connexion *f* électrique
r электрическое соединение *n*

4705 electric control

d elektrische Regelung *f*
f réglage *m* électrique
r электроуправление *n*

4706 electric control gear
d elektrisches Steuerungsgerät *n*; elektrischer
   Steuerungsantrieb *m*
f appareil *m* de commande électrique
r электрический управляющий механизм *m*

* **electric controller → 4691**

4707 electric data output
d elektrische Datenausgabe *f*
f sortie *f* de données électrique
r электрический вывод *m* данных

4708 electric data scanning
d elektrische Datenabtastung *f*
f exploration *f* électrique des données
r электрическое сканирование *n*

4709 electric data storage
d elektrische Datenspeicherung *f*
f emmagasinage *m* électrique des données
r электрическое накопление *n* данных

4710 electric data transmission
d elektrische Datenübertragung *f*
f transmission *f* électrique des données
r электрическая передача *f* данных

4711 electric digital reading
d elektrische Ziffernlesung *f*
f lecture *f* de chiffres électrique
r электрическое считывание *n* цифр

**4712 electric discharge vacuum gauge; vacuum
     electric discharge gauge**
d elektrischer Entladungsvakuummesser *m*;
   Elektroentladungsvakuummesser *m*
f vacuomètre *m* à décharge électrique
r электроразрядный вакуумметр *m*

4713 electric displacement
d elektrische Verschiebung *f*
f déplacement *m* électrique
r электрическое смещение *n*

* **electric drive → 4692**

4714 electric drive power coefficient
d Leistungsfaktor *m* eines Elektroantriebes
f facteur *m* de puissance d'une commande
   électrique
r коэффициент *m* мощности
   электропривода

4715 **electric drives blocking**
   d Elektroantriebsblockierung *f*
   f blocage *m* de commandes électriques
   r блокировка *f* электроприводов

4716 **electric drive with progressive movement;**
      electric drive with rectilinear motion
   d Elektroantrieb *m* mit geradliniger Bewegung
   f traction *f* électrique à mouvement rectiligne
   r электропривод *m* с поступательным
      движением

   * **electric drive with rectilinear**
      motion → 4716

4717 **electric feedback**
   d elektrische Rückführung *f*
   f réaction *f* électrique; action *f* en retour
      électrique
   r електрическая обратная связь *f*

4718 **electric field gradient**
   d Gradient *m* elektrischer Feldstärke
   f gradient *m* de champ électrique
   r градиент *m* электрического поля

4719 **electric field intensity**
   d elektrische Feldstärke *f*
   f intensité *f* de champ électrique
   r напряженность *f* электрического поля

4720 **electric final control elements**
   d elektrische Stellglieder *npl*
   f organes *mpl* de commande
   r исполнительные электрические
      органы *mpl*

4721 **electric information storage**
   d elektrische Informationsspeicherung *f*
   f emmagasinage *m* d'information électrique
   r электрическое накопление *n* информации

4722 **electric information technique**
   d elektrische Informationstechnik *f*
   f technique *f* électrique d'information
   r электрическая техника *f* [обработки]
      информации

4723 **electric interface**
   d elektrisches Interface *n*
   f interface *f* électrique
   r электрический интерфейс *m*

4724 **electric interface condition**
   d elektrische Interfacebedingung *f*
   f condition *f* d'interface électrique
   r электрическое условие *n* интерфейса

4725 **electric leakage tester**
   d Leckstrommesser *m*
   f indicateur *m* de courant de fuite
   r измеритель *m* утечки электрического тока

4726 **electric limit fuse**
   d elektrische Endlagensicherung *f*
   f fusible *m* électrique de position finale
   r электрический предохранитель *m*
      конечного положения

4727 **electric load lift device**
   d elektrische Lasthebeeinrichtung *f*
   f élévateur *m* de charge électrique
   r электрическое грузоподъемное
      устройство *n*

4728 **electric machine compounding**
   d Compoundierung *f* von Elektromaschinen
   f compoundage *m* de machines électriques
   r компаундирование *n* электрических
      машин

4729 **electric motor assembly**
   d Elektromotorenmontage *f*
   f assemblage *m* de moteur électrique
   r процесс *m* сборки электродвигателей

4730 **electric motor mechanical characteristics**
   d mechanische Elektromotorenkennlinien *fpl*
   f caractéristiques *fpl* mécaniques de moteurs
      électriques
   r механические характеристики *fpl*
      электродвигателей

4731 **electric-operated linear unit**
   d elektrisch angetriebene Lineareinheit *f*
   f unité *f* linéaire entraînée électriquement
   r блок *m* линейных перемещений с
      электрическим приводом

4732 **electric-pneumatic switch**
   d elektropneumatischer Schalter *m*
   f commutateur *m* électropneumatique
   r электро-пневматический выключатель *m*

4733 **electric power system cybernetics**
   d Kybernetik *f* der
      Elektroenergiversorgungssysteme
   f cybernétique *f* de grands réseaux électriques
   r кибернетика *f* электрической
      энергосистемы

   * **electric propulsion** → 4692

   * **electric regulator** → 4691

4734 **electric remote control**
   d elektrische Fernsteuerung *f*

f télécommande f électrique; commande f
électrique à distance
r электрическое дистанционное
управление n

4735 electric remote transmission
d elektrische Fernübertragung f
f transmission f électrique à distance
r электрическая дистанционная передача f

4736 electric resistance
d elektrischer Widerstand m
f résistance f électrique
r электрическое сопротивление n

4737 electric resistance thermometer
d elektrisches Widerstandsthermometer n
f thermomètre m à résistance électrique
r электрорезистивный термометр m

4738 electric resonance relay
d elektrisches Resonanzrelais n
f relais m électrique à résonance
r электрическое резонансное реле n

4739 electric sensor property
d elektrische Sensoreigenschaft
f propriété f de senseur électrique
r электрическое свойство n сенсора

*   **electric servomotor → 4673**

4740 electric signal
d elektrisches Signal n
f signal m électrique
r электрический сигнал m

*   **electric signal evaluation → 4700**

4741 electric store register
d elektrisches Speicherregister n
f registre m de mémoire électrique
r электрический запоминающий регистр m

*   **electric strain gauge → 4683**

4742 electric structure of bus
d elektrische Busstruktur f
f structure f électrique du bus
r электрическая структура f шины

4743 electric telemeter
d elektrischer Entfernungsmesser m;
elektrisches Fernmessgerät n
f appareil m électrique de télémesure
r электрическое телеметрическое
устройство n

4744 electric telemetering system

d elektrisches Fernmesssystem n
f système m électrique de mesure
r электрическая система f телеизмерений

*   **electric test → 4745**

4745 electric test[ing]
d elektrische Prüfung f; elektrische Kontrolle f
f essai m électrique
r испытание n электрических свойств;
электрическое испытание; электрический
контроль m

4746 electric time schedule transmitter
d elektrischer Zeitplangeber m
f transmetteur m électrique du plan temporaire
r электрический передатчик m расписания

**4747 electric transmitter of differential
pressure**
d elektrischer Differenzdruckgeber m
f transmetteur m électrique de la différence de
pression
r электрический датчик m перепада
давлений

4748 electric transmitter of mechanical values
d elektrischer Geber m mechanischer Größen
f transmetteur m électrique de grandeurs
mécaniques
r электрический датчик m механических
величин

4749 electric verifier
d elektrischer Lochprüfer m; elektrischer
Prüfer m
f vérificatrice f électrique
r электрическое контрольное устройство n

4750 electroacoustic effct
d elektroakustischer Effekt m
f effet m électro-acoustique
r электроакустический эффект m

4751 electroacoustic transducer
d elektroakustischer Wandler m
f transucteur m électro-acoustique
r электроакустический преобразователь m

4752 electroanalysis
d Elektroanalyse f
f électro-analyse f; analyse f électrique
r электроанализ m

4753 electroautomatic power control
d elektroautomatische Leistungsregelung f
f réglage m électro-automatique de puissance
r электрическое автоматическое
регулирование n мощности

4754 electrochemical process
 d elektrochemisches Verfahren *f*
 f processus *m* électrochimique
 r электрохимический процесс *m*

4755 electrocontact machining
 d Elektrokontaktbearbeitung *f*
 f traitement *m* par contacts électrique
 r электроконтактная обработка *f*

4756 electrodrive dynamic braking
 d Gegenstrombremsung *f* von Elektroantrieben
 f freinage *m* dynamique de commande
   électrique
 r динамическое торможение *n*
   электропривода

4757 electrodynamic analogy
 d elektrodynamische Analogie *f*
 f analogie *f* électrodynamique
 r электродинамическая аналогия *f*

4758 electrodynamic flowmeter
 d electrodynamischer Durchflussmesser *m*
 f débitmètre *m* électrodynamique
 r электродинамический расходомер *m*

4759 electrodynamic instrument
 d elektrodynamisches Instrument *n*
 f appareil *m* électrodynamique
 r электродинамический прибор *m*

4760 electrodynamic radiator
 d elektrodynamischer Strahler *m*
 f radiateur *m* électrodynamique
 r электродинамический излучател *m*

4761 electrodynamic relay
 d elektrodynamisches Relais *n*
 f relais *m* électrodynamique
 r электродинамическое реле *n*

4762 electrodynamic vibration pick-up
 d elektrodynamischer Schwingungsgeber *m*
 f capteur *m* électrodynamique de vibrations
 r электродинамический вибродатчик *m*

4763 electroerosion treatment
 d elektroerosive Bearbeitung *f*
 f usinage *m* électro-érosif
 r электроэрозионная обработка *f*

4764 electrographic recording technique
 d elektrografische Aufzeichnungstechnik *f*
 f technique *f* d'enregistrement
   électrographique
 r техника *f* электрографической записи

4765 electrohydraulically operated

 d elektrohydraulisch angetriben
 f à commande électrohydraulique
 r с электрогидравлическим приводом

4766 electrohydraulic control circuit
 d elektrohydraulischer Regelkreis *m*
 f circuit *m* de régulation électrohydraulique
 r электрогидравлический контур *m*
   регулирования

4767 electrohydraulic controller
 d elektrohydraulischer Regler *m*
 f régulateur *m* électrohydraulique
 r электрогидравлический регулятор *m*

4768 electrohydraulic control system
 d elektrohydraulisches Regelungssystem *n*
 f système *m* de réglage électrohydraulique
 r электрогидравлическая система *f*
   регулирования

4769 electrohydraulic converter
 d elektrohydraulischer Umformer *m*
 f convertisseur *m* électrohydraulique
 r электрогидравлический
   преобразователь *m*

4770 electrohydraulic effect
 d elektrohydraulischer Effekt *m*
 f effet *m* électrohydraulique
 r электрогидравлический эффект *m*

4771 electrohydraulic gripper drive
 d elektrohydraulischer Greiferantrieb *m*
 f entraînement *m* de grappin
   électrohydraulique
 r электрогидравлический привод *m* схвата

4772 electrohydraulic servomechanism
 d elektrohydraulischer Servomechanismus *m*
 f servomécanisme *m* électrohydraulique
 r электрогидравлический сервомеханизм *m*

4773 electrohydraulic servovalve
 d elektrohydraulisches Servoventil *n*
 f servo-soupape *f* électrohydraulique
 r электрогидравлический сервоклапан *m*

4774 electrokinetic potential
 d elektrokinetisches Potential *n*
 f potentiel *m* électrocinétique
 r электрокинетический потенциал *m*

4775 electroluminescence sensor
 d Elektrolumineszenzgeber *m*;
   Elektrolumineszenzsensor *m*
 f capteur *m* électroluminescent; senseur *m*
   électroluminescent
 r электролюминесцентный сенсор *m*

4776 **electroluminescence sensor for manipulator control**
   d Elektrolumineszenzsensor *m* für Manipulatorsteuerung
   f capteur *m* électroluminescent pour la commande de manipulateurs
   r электролюминесцентный сенсор *m* для системы управления манипулятором

4777 **electroluminescent display panel**
   d Elektrolumineszenzdarstellungsschirm *m*
   f tableau *m* électroluminescent de display
   r электролюминесцентная индикаторная панель *f*

4778 **electroluminescent element**
   d Elektrolumineszenzelement *n*
   f élément *m* électroluminescent
   r электролюминесцентный элемент *m*

4779 **electroluminescent screen**
   d Elektrolumineszenzschirm *m*
   f panneau *m* électroluminescent
   r электролюминесцентный экран *m*

4780 **electrolytic capacitor**
   d Elektrolytkondensator *m*
   f condensateur *m* électrolytique
   r электролитический конденсатор *m*

4781 **electrolytic hygrometer**
   d elektrolytisches Hygrometer *n*
   f hygromètre *m* électrolytique
   r электролитический гигрометр *m*

4782 **electrolytic polarization**
   d elektrolytische Polarisation *f*
   f polarisation *f* électrolytique
   r электролитическая поляризация *f*

4783 **electrolytic solution pressure**
   d elektrolytischer Lösungsdruck *m*
   f tension *f* électrolytique
   r давление *n* раствора электролита

4784 **electrolytic store**
   d elektrolytischer Speicher *m*
   f mémoire *f* électrolytique
   r электролитический накопитель *m*

4785 **electrolytic tank**
   d elektrolytische Wanne *f*
   f cuve *f* électrolytique
   r электролитическая ванна *f*

4786 **electrolytic timer**
   d elektrolytischer Zeitschalter *m*
   f minuterie *f* électrolytique

   r электролитическое реле *n* времени

4787 **electromagnetic compensation**
   d elektromagnetische Kompensation *f*
   f compensation *f* électromagnétique
   r электромагнитная компенсация *f*

4788 **electromagnetic constant**
   d elektromagnetische Konstante *f*
   f constante *f* électromagnétique
   r электромагнитная постоянная *f*

4789 **electromagnetic contactless relay**
   d elektromagnetisches kontaktloses Relais *n*
   f relais *m* statomagnétique
   r электромагнитное бесконтактное реле *n*

4790 **electromagnetic contactor**
   d elektromagnetisches Schütz *n*
   f contacteur *m* électromagnétique
   r электромагнитный контактор *m*

4791 **electromagnetic control**
   d elektromagnetische Steuerung *f*
   f commande *f* électromagnétique
   r электромагнитное управление *n*

4792 **electromagnetic copying**
   d elektromagnetisches Kopieren *n*
   f copiage *m* électromagnétique
   r электромагнитное копирование *n*

4793 **electromagnetic counter**
   d elektromagnetischer Zähler *m*
   f compteur *m* électromagnétique
   r электромагнитный счётчик *m*

4794 **electromagnetic coupling**
   d elektromagnetische Kopplung *f*
   f accouplement *m* électromagnétique
   r электромагнитная связь *f*

4795 **electromagnetic criteria**
   d elektromagnetische Kriterien *npl*
   f critères *mpl* électromagnétiques
   r электромагнитный критерий *m*

4796 **electromagnetic damping**
   d elektromagnetische Dämpfung *f*
   f amortissement *m* électromagnétique
   r электромагнитное демпфирование *n*

4797 **electromagnetic deflection**
   d elektromagnetische Auslenkung *f*; elektromagnetische Ablenkung *f*
   f déviation *f* électromagnétique
   r электромагнитное отклонение *n*

4798 **electromagnetic delay line**
 d elektromagnetische Verzögerungsleitung *f*
 f ligne *f* de retardement électromagnétique
 r электромагнитная линия *f* задержки

4799 **electromagnetic flowmeter**
 d elektromagnetischer Strömungsmesser *m*
 f débitmètre *m* électromagnétique
 r электромагнитный расходомер *m*

4800 **electromagnetic gripper equipment**
 d elektromagnetische Greifereinrichtung *f*
 f équipement *m* de grappin électromagnétique
 r электромагнитное захватное устройство *n*

4801 **electromagnetic lens**
 d elektromagnetische Linse *f*
 f lentille *f* électromagnétique
 r электромагнитная линза *f*

4802 **electromagnetic locking**
 d elektromagnetische Verriegelung *f*
 f verrouillage *m* électromagnétique
 r электромагнитная блокировка *f*

4803 **electromagnetic order device**
 d elektromagnetischer Ordner *m*
 f dispositif *m* d'ordre électromagnétique
 r электромагнитное устройство *n*
  упорядочения

4804 **electromagnetic oscillations**
 d elektromagnetische Schwingungen *fpl*
 f oscillations *fpl* électromagnétiques
 r электромагнитные колебания *npl*

4805 **electromagnetic perceptive system**
 d elektromagnetisches
  Wahrnehmungssystem *n*
 f système *m* de perception électromagnétique
 r электромагнитная система *f* восприятия

4806 **electromagnetic pump**
 d elektromagnetische Pumpe *f*
 f pompe *f* électromagnétique
 r электромагнитный насос *m*

4807 **electromagnetic release**
 d elektromagnetische Auslösung *f*
 f déclenchement *m* électromagnétique
 r электромагнитное размыкание *n*

4808 **electromagnetic screening;**
  electromagnetic shielding
 d elektromagnetische Abschirmung *f*
 f blindage *m* électromagnétique
 r электромагнитное экранирование *n*

* **electromagnetic shielding** → 4808

4809 **electromagnetic thickness measurement of**
  layers
 d elektromagnetische Schichtdickenmessung *f*
 f mesure *f* électromagnétique de couches
 r электромагнитное измерение *n* толщины
  слоев

4810 **electromagnetic transducer**
 d elektromagnetischer Wandler *m*
 f transducteur *m* électromagnétique
 r электромагнитный преобразователь *m*

4811 **electromagnetic transmission line storage**
 d elektromagnetische Speicherleitung *f*
 f mémoire *f* à ligne électromagnétique
 r запоминающее устройство *n* на
  электромагнитных линиях

4812 **electromagnetic unit**
 d elektromagnetische Einheit *f*
 f unité *f* électromagnétique
 r электромагнитная единица *f*

4813 **electromagnetic valve**
 d elektromagnetisches Ventil *n*
 f soupape *f* électromagnétique
 r электромагнитный вентиль *m*

4814 **electromagnetic vibration-type buncer**
 d elektromagnetischer Vibrationsbunker *m*
 f soute *f* électromagnétique à vibration
 r электромагнитный вибрационный
  бункер *m*

4815 **electromechanical amplifier**
 d elektromechanischer Verstärker *m*
 f amplificateur *m* électromécanique
 r электромеханический усилитель *m*

4816 **electromechanical change-over switch**
 d elektromechanischer Umschalter *m*
 f commutateur *m* électromécanique
 r электромеханический переключатель *m*

4817 **electromechanical controller**
 d elektromechanischer Regler *m*
 f régulateur *m* électromécanique
 r электромеханический регулятор *m*

4818 **electromechanical differential analyzer**
 d elektromechanisches
  Differentialanalysiergerät *n*
 f analyseur *m* différentiel électromécanique
 r электромеханический дифференциальный
  анализатор *m*

4819 electromechanical dimension tranducer
  d Messwandler *m* geometrischer Größen
  f capteur *m* électromécanique de dimension
  r электромеханический датчик *m* размеров

4820 electromechanical drive
  d elektromechanischer Antrieb *m*
  f entraînement *m* électromécanique
  r электромеханический привод *m*

4821 electromechanical impulse recorder
  d elektromechanischer Impulsschreiber *m*
  f compteur-enregistreur *m* d'impulsions
     électromécanique
  r электромеханический импульсный
     регистратор *m*

4822 electromechanical interlock
  d elektromechanische Verriegelung *f*;
     elektromechanische Blockierung *f*
  f blocage *m* électromécanique
  r электромеханическая блокировка *f*

4823 electromechanical low-frequency control
      oscillator
  d elektromechanischer
     Niederfrequenzoszillator *m*
  f oscillateur *m* électromécanique basse
     fréquence
  r электромеханический низкочастотный
     генератор *m*

4824 electromechanical manipulator
  d elektromechanischer Manipulator *m*
  f manipulator *m* électromécanique
  r электромеханический манипулятор *m*

4825 electromechanical metering relay
  d elektromechanisches Zählrelais *n*
  f relais *m* compteur électromécanique
  r электромеханический реле-счётчик *m*

4826 electromechanical scanner
  d elektromechanischer Abtaster *m*
  f explorateur *m* électromécanique
  r электромеханическое развёртывающее
     устройство *n*

4827 electromechanical system
  d elektromechanisches System *n*
  f système *m* électromécanique
  r электромеханическая система *f*

4828 electrometric amplifier
  d elektrometrischer Verstärker *m*
  f amplificateur *m* électrométrique
  r электрометрический усилитель *m*

4829 electron
  d Elektron *n*
  f électron *m*
  r электрон *m*

4830 electron band spectrum
  d Elektronenbandspektrum *n*
  f spectre *m* de bande électronique
  r электронный полосовой спектр *m*

4831 electron beam
  d Elektronenstrahl *m*
  f faisceau *m* d'électrons
  r электронный луч *m*

4832 electron-beam distributor
  d Elektronenstrahlverteiler *m*
  f distributeur *m* à faisceau électronique
  r электроннолучевой распределитель *m*

4833 electron-beam magnetometer
  d Elektronenstrahlmagnetometer *n*
  f magnétomètre *m* à faisceau électronique
  r электроннолучевой магнитометр *m*

4834 electron-beam oscillograph
  d Elektronenstrahloszillograf *m*
  f oscillographe *m* à faisceau électronique
  r электроннолучевой осциллограф *m*

4835 electron-beam parametric amplifier
  d parametrischer Elektronenstrahlverstärker *m*
  f amplificateur *m* paramétrique à faisceau
     électronique
  r электроннолучевой параметрический
     усилитель *m*

4836 electron beam remelting process
  d Elektronenstrahlumschmelzverfahren *n*
  f procédé *m* de fusion par faisceau d'électrons;
     procédé de fusion par bombardement
     électronique
  r способ *m* переплава электронным лучом;
     способ электроннолучевого переплава

4837 electron-beam technology
  d Elektronenstrahltechnologie *f*
  f technologie *f* des rayons électoniques
  r электроннолучевая технология *f*

4838 electron-beam treatment
  d Elektronenstrahlbehandlung *f*
  f traitement *m* par faisceau électronique
  r электроннолучевая обработка *f*

4839 electron beam weld[ed] joint
  d Elektronenstrahlschweißverbindung *f*;
     elektronenstrahlgeschweißte Verbindung *f*

f jonction *f* de soudage par faisceau
d'électrons; jonction de soudage par
bombardement électronique
r соединение n, выполненное сваркой
электронным лучом; соединение,
выполненное электроннолучевой сваркой

4840 electron beam welding parameters
d Elektronenstrahlschweißparameter *mpl*
f paramètres *mpl* de soudage par faisceau
d'électrons; paramètres de soudage par
bombardement électronique
r параметры *mpl* сварки электронным
лучом; режим *m* электроннолучевой
сварки

4841 electron beam welding technique
d Elektronenstrahlschweißtechnik *f*
f technique *f* de soudage par bombardement
électronique; technique *f* de soudage par
faisceau d'électrons
r техника *f* сварки электронным лучом;
техника электроннолучевой сварки

4842 electron beam welding technology
d Elektronenstrahlschweißtechnologie *f*
f technologie *f* de soudage par faisceau
d'électrons
r технология *f* электроннолучевой сварки

* **electron beam weld joint** → 4839

4843 electron cascade
d Elektronenkaskade *f*
f cascade *f* d'électrons
r электронный каскад *m*

4844 electron conductivity
d Elektronenleitfähigkeit *f*
f conductibilité *f* électronique
r электронная проводимость *f*

4845 electron coupling
d Elektronenkopplung *f*; elektronische
Kopplung *f*
f couplage *m* électronique
r электронная связь *f*

4846 electron current
d Elektronenstrom *m*
f courant *m* électronique
r электронный ток *m*

4847 electron emission
d Elektronenemission *f*
f émission *f* électronique
r электронная эмиссия *f*

4848 electronic
d electronic
f électronique
r электронный

4849 electronic acquisition sensor
d elektronischer Erfassungssensor *m*
f capteur *m* d'acquisition électronique
r электронный датчик *m* сбора

4850 electronically controlled
d elektronisch gesteuert
f à commande électronique
r управляемый электронными средствами

4851 electronically controlled converter
d elektronisch gesteuerter Umformer *m*
f convertisseur *m* à commande électronique
r преобразователь *m* с электронным
управлением

**4852 electronically controlled power supslay
unit**
d elektronisch geregeltes
Stromversorgungsgerät *n*
f poste *m* d'alimentation à réglage électronique
r блок *m* электроснабжения с электронным
управлением

**4853 electronically operated control; electronic
control**
d elektronische Regelung *f*; elektronische
Steuerung *f*
f réglage *m* électronique; commande *f*
électronique
r электронное регулирование *n*;
электронное управление *n*

**4854 electronically operated controller;
electronic controller; electronic regulator**
d elektronischer Regler *m*
f régulateur *m* électronique
r электронный регулятор *m*

4855 electronically scanned optical tracker
d optische Nachführanlage *f* mit elektronischer
Abtastung
f dispositif *m* optique de poursuite à
exploration électronique; traceur *m* optique à
balayage électronique
r оптическое устройство *n* сопровождения с
электронным сканированием

4856 electronical simulating
d elektronisches Simulieren *n*
f simulation *f* par moyens électronique
r электронное моделирование *n*

4857 **[electronic] analog computer**
  d [elektronischer] Analogrechner *m*;
      Analogierechenmaschine *f*
  f calculateur *m* analogique [électronique];
      machine *f* à calcul analogique; calculatrice *f*
      analogique
  r аналоговая вычислителная машина *f*

4858 **electronic application**
  d Elektronikanwendung *f*
  f application *f* de l'électronique
  r применение *n* в электронике

4859 **electronic automatic swith**
  d elektronischer Schaltautomat *m*;
      elektronischer automatischer Schalter *m*
  f commutateur *m* électronique
  r электронный автоматический
      выключатель *m*

4860 **electronic automation**
  d elektronische Automatisierung *f*
  f automatisation *f* électronique
  r электронная автоматизация *f*

4861 **electronic balancing**
  d elektronische Auswuchtung *f*
  f équilibrage *m* électronique
  r электронное уравновешивание *n*

  * **electronic building bloc** → 4862

4862 **electronic building brick; electronic
      building bloc**
  d elektronischer Baustein *m*; elektronische
      Baugruppe *f*
  f élément *m* électronique de construction
  r электронный конструктивный блок *m*

4863 **electronic calculator device**
  d elektronisches Rechengerät *n*
  f dispositif *m* de calcul électronique
  r электронное вычислительное
      устройство *n*

4864 **electronic character recognition**
  d elektronische Zeichenerkennung *f*
  f reconnaissance *f* électronique des caractères
  r электронное распознавание *n* символов

4865 **electronic circuit**
  d elektronischer Kreis *m*
  f circuit *m* électronique
  r электронный контур *m*; электронная
      схема *f*

4866 **electronic circuit technique**
  d elektronische Schaltungstechnik *f*

  f technique *f* des circuits électroniques
  r электронная схемотехника *f*

4867 **electronic classifying instrument**
  d elektronisches Klassiergerät *n*
  f classeuse *f* électronique
  r электронный прибор *m* для
      классификация

4868 **electronic clock with coded digital signal**
  d elektronische Uhr *f* mit kodiertem
      Digitalsignal
  f horloge *f* électronique à signal digital code
  r электронные часы *pl* с кодированным
      цифровым сигналом

4869 **electronic compensation teletransmitter**
  d elektronischer Kompensationsferngeber *m*
  f télétransmetteur *m* de compensation
      électronique
  r электронно-компенсированный
      телепередатчик *m*

4870 **electronic component; electronic element**
  d elektronisches Bauteil *n*; elektronisches
      Bauelement *n*
  f composant *m* électronique; élément *m*
      électronique
  r электронный субблок *m*; электронный
      элемент *m*

4871 **electronic computing automaton**
  d elektronischer Rechenautomat *m*
  f automate *m* de calcul électronique
  r электронный вычислительный автомат *m*

4872 **electronic conductivity control**
  d elektronische Regelung *f* der Leitfähigkeit
  f réglage *m* électronique de la conductivité
  r электронное регулирование *n*
      проводимости

  * **electronic control** → 4853

4873 **electronic control desk**
  d elektronisches Steuerpult *n*
  f pupitre *m* de commande électronique
  r электронный пульт *m* управления

4874 **electronic-controlled final assembly**
  d elektronisch gesteuerte Endmontage *f*
  f assemblage *m* final à commande électronique
  r конечная сборка *f* с электронным
      управлением

  * **electronic controller** → 4854

4875 **electronic controller of vapour turbines**
  d elektronischer Dampfturbinenregler *m*

f régulateur *m* électronique de turbine à vapeur
r электронный регулятор *m* паровой
турбины

4876 electronic control system
d elektronisches Regelsystem *n*
f système *m* de réglage électronique
r электронная система *f* регулирования

4877 electronic coordinate setting
d elektronische Koordinateneinstellung *f*
f réglage *m* électronique de coordonnées
r электронная координатная установка *f*

4878 electronic couting circuit
d elektronische Zählschaltung *f*
f circuit *m* de comptage électronique
r электронная счётная схема *f*

4879 electronic data bank
d elektronische Datenbank *f*
f banque *f* de données électronique
r электронный банк *m* данных

4880 electronic data collection
d elektronische Datenerfassung *f*; elektronische
Datensammlung *f*; elektronisches
Datensammeln *n*
f acquisition *f* électronique des données;
collection *f* électronique des données
r электронный сбор *m* данных

4881 electronic data processing; EDP
d elektronische Datenverarbeitung *f*
f traitement *m* électronique de données
r электронная обработка *f* данных

**4882 electronic data processing system; EDP
system**
d elektronisches Datenverarbeitungssystem *n*;
EDV-System *n*
f système *m* de traitement électronique des
données; ensemble *m* électronique pour
traitement des données; système EDP
r система *f* электронной обработки данных

4883 electronic data station
d elektronische Datenstation *f*
f station *f* de données électronique
r электронное информационное
устройство *n*

4884 electronic decade counter
d elektronischer Dekadenzähler *m*
f compteur *m* électronique à décades
r электронный декадный счётчик *m*

4885 electronic device

d elektronisches Gerät *n*
f appareil *m* électronique
r электронный прибор *m*; электронное
устройство *n*

4886 electronic digit reading
d elektronische Ziffernlesung *f*
f lecture *f* électronique des chiffres
r электронное считывание *n* цифр

4887 electronic direction
d elektronische Leitung *f*; elektronische
Führung *f*
f gestion *f* électronique
r электронная управляющая система *f*

4888 electronic display viewing area
d elektronischer Anzeigesichtbereich *m*
f zone *f* visible de l'affichage électronique
r область *f* визуального обзора
электронного дисплея

4889 electronic drive
d elektronischer Antrieb *m*
f entraînement *m* électronique; propulsion *f*
électronique
r электронный привод *m*

4890 electronic drive regulator
d elektronischer Antriebsregler *m*
f régulateur *m* d'entraînement électronique
r электронный регулятор *m* привода

* electronic element → 4870

4891 electronic error detector
d elektronischer Fehlerdetektor *m*
f détecteur *m* électronique de défauts
r электронный чувствительный элемент *m*
следящей системы

4892 electronic exposure time indicator
d elektronischer Belichtungszeitmesser *m*
f posemètre *m* électronique
r электронный индикатор *m* времени
экспозиции

**4893 electronic generator of very low
frequencies**
d elektronischer Generator *m* sehr niedriger
Frequenzen
f générateur *m* électronique de très basses
fréquences
r электронный генератор *m* очень низких
частот

4894 electronic guidance equipment
d elektronisches Lenkungsgerät *n*

f dispositif *m* électronique de guidage
r электронное направляющее устройство *n*

4895 electronic hygrometer
d elektronischer Feuchtigkeitsmesser *m*
f hygromètre *m* électronique
r электронный гигрометр *m*

4896 electronic impact
d Elektronenstoß *m*
f choc *m* électronique
r электронный удар *m*

4897 electronic impulse regulator
d elektronischer Impulsregler *m*
f régulator *m* électronique à impulsions
r электронный импульсный регулятор *m*

4898 electronic information processing system
d elektronisches Informationsverarbeitungssystem *n*
f système *m* de traitement électronique des informations
r электронная система *f* обработки информации

4899 electronic instrumentation reliability
d Zuverlässigkeit *f* elektronischer Apparatur
f fiabilité *f* d'appareillage électronique
r надёжность *f* электронной аппаратуры

4900 electronic-integrated processing
d elektronisch integrierter Arbeitsablauf *m*
f procédé *m* électronique de traitement intégré
r электронная интегрированная обработка *f*

4901 electronic level control
d elektronische Niveauregelung *f*
f régulation *f* électronique du niveau
r электронное регулирование *n* уровня

4902 electronic limiting value indicator
d elektronischer Grenzwertmelder *m*
f indicateur *m* électronique de la valeur limite
r электронный индикатор *m* предельного значения

4903 electronic logging of measured values
d elektronische Messwerterfassung *f*
f enregistrement *m* électronique des valeurs mesurées
r электронный сбор *m* измерительных данных

4904 electronic magnetic stabilizer
d elektronisch-magnetischer Stabilisator *m*
f stabiliseur *m* électronique-magnétique
r электронно-магнитный стабилизатор *m*

4905 electronic measurement of revolutions
d elektronische Drehzahlmessung *f*
f mesurage *m* électronique du nombre de tours
r электронный метод *m* измерения числа оборотов

4906 electronic measuring result compensation
d elektronische Messergebniskompensation *f*
f compensation *f* électronique des résultats de mesure
r электронная компенсация *f* результатов измерения

4907 electronic microanalyzer
d Elektronenmikroanalysator *m*
f microanalyseur *m* électronique
r электронный микроанализатор *m*

4908 electronic model
d elektronisches Modell *n*
f simulateur *m* électronique
r электронная модель *f*

4909 electronic modular system
d elektronisches Bausteinsystem *n*
f système *m* modulaire électronique
r система *f* электронных модулей

4910 electronic multichannel analyzer
d elektronischer Vielkanalanalysator *m*
f analyseur *m* électronique à canaux multiples
r электронный многоканальный анализатор *m*

4911 electronic multiplier
d elektronischer Multiplikator *m*
f multiplicateur *m* électronique
r электронный умножитель *m*

4912 electronic overload detector
d elektronischer Überlastungsdetektor *m*
f détecteur *m* électronique de surcharge
r электронный индикатор *m* перегрузок

4913 electronic parallel digital computer
d elektronischer Paralleldigitalrechner *m*
f calculatrice *f* électronique digitale parallèle
r электронная цифровая вычислительная машина *f* паралельного действия

* **electronic pen** → **7256**

4914 electronic-pneumatic controller
d elektropneumatischer Regler *m*
f régulateur *m* électronique-pneumatique
r электроннопневматический регулятор *m*

4915 electronic polarimeter
d elektronisches Polarimeter *n*

f polarimètre *m* électronique
r электронный поляриметр *m*

4916 electronic pressure gauge
d elektronisches Manometer *n*
f manomètre *m* électronique
r электронный манометр *m*

4917 electronic process
d elektronisches Verfahren *n*
f procédé *m* électronique
r электронный метод *m*

4918 electronic profile projector
d elektronischer Profilprojektor *m*
f projecteur *m* électronique de profil
r электронный профильный проектор *m*

4919 electronic programming device
d elektronisches Programmiergerät *n*
f appareil *m* de programmation électronique
r электронное программное устройство *n*

4920 electronic quenching circuit
d elektronischer Löschkreis *m*; elektronische Löschschaltung *f*
f circuit *m* de coupure électronique
r гасящая электронная схема *f*

4921 electronic random number generator
d elektronischer Zufallszahlengenerator *m*
f générateur *m* électronique de nombres aléatoires
r электронный генератор *m* случайных чисел

4922 electronic reader; electronic reading device
d elektronische Leseeinrichtung *f*; Fotolekteur *m*; elektronischer Leser *m*
f dispositif *m* de lecture électronique; photolecteur *m*; lecteur *m* électronique
r электронное устройство *n* считывания

* **electronic reading device** → **4922**

* **electronic regulator** → **4854**

4923 electronic relay
d elektronisches Relais *n*
f relais *m* électronique
r электронное реле *n*

4924 electronic remote control
d elektronische Fernsteuerung *f*
f commande *f* à distance électronique
r электронное дистанционное управление *n*

4925 electronic reverse current controller

d elektronischer Rückstromregler *m*
f régulateur *m* du courant inverse automatique
r электронный регулятор *m* обратного тока

4926 electronic select automaton
d elektronischer Wählautomat *m*
f sélecteur *m* automatique
r электронный избирательный автомат *m*

4927 electronics industry
d Elektronikindustrie *f*
f industrie *f* électronique
r электронная промышленность *f*

4928 electronic slip measuring device
d elektronische Schlupfmesseinrichtung *f*
f appareil *m* de mesure du glissement électronique
r электронное устройство *n* для измерения скольжения

4929 electronic spatial thermostat
d elektronischer Raumthermostat *m*
f thermostat *m* spatial électronique
r электронный пространственный термостат *m*

4930 electronic speed controller
d elektronisches Geschwindigkeitssteuergerät *n*; elektronischer Drehzahlregler *m*
f variateur *m* de vitesse électronique
r электронный регулятор *m* скорости [вращения]

4931 electronic step-by-step system
d elektronisches Schrittgebersystem *n*
f système *m* électronique pas à pas
r электронная шаговая система *f*; электронная медленнодействующая система

4932 electronic storage device
d elektronische Speichereinheit *f*
f dispositif *m* électronique de mémoire
r электронный накопитель *m*

4933 electronic subsystem
d elektronisches Untersystem *n*
f sous-système *m* électronique
r электронная подсистема *f*

4934 electronic switch of marginal speed
d elektronischer Drehzahl-Grenzwertschalter *m*
f commtateur *m* électronique de limite de vitesse
r электронный выключатель *m* предельной скорости

4935 electronic system for temperature control
  d elektronisches System n für
     Temperaturregelung
  f système m électronique pour le réglage de la
     température
  r электронная система f для регулирования
     температуры

4936 electronic tachometer
  d elektronischer Drehzahlmesser m
  f tachymètre m électronique
  r электронный тахометр m

4937 electronic timer
  d elektronischer Zeitauslöser m; elektronisches
     Zeitrelais n
  f minuterie f électronique; relais m temporisé
     électronique
  r электронный хронизатор m

4938 electronic time sequence control unit
  d elektronisches Zeitfolge-Steuergerät n
  f dispositif m électronique de temporisation
     séquentielle
  r электронное устройство n для управления
     в определенной последовательности

4939 electronic trajectory deviation indicator
  d electronisches
     Bahnabweichungsanzeigegerät n
  f indicateur m électronique d'erreur de
     trajectoire
  r электронный индикатор m отклонения от
     расчётной траектории

4940 electronic tuning range
  d elektronischer Abstimmbereich m
  f gamme f d'accord électronique
  r диапазон m электронной настройки

4941 electronic tuning sensitivity
  d elektronische Abstimmempfindlichkeit f
  f sensibilité f d'accord électronique
  r чувствительность f электронной
     настройки

4942 electronic voltage controller; electronic
     voltage regulator
  d elektronische
     Spannungskontrolleinrichtung f;
     elektronischer Spannungsregler m
  f régulateur m électronique de tension;
     appareil m électronique de tension
  r электронный контроллер m напряжения;
     электронный регулятор m напряжения

  * electronic voltage regulator → 4942

4943 electronic warning signal device

  d elektronisches Warnsignalgerät n
  f appareil m signal avertisseur électronique
  r электронное сигнальное
     предупреждающее устройство n

4944 electronic writing automaton
  d elektronischer Schreibautomat m
  f automate m d'inscription électronique;
     automate d'écriture électronique
  r электронный пишущий автомат m

4945 electron-optical image converter
  d elektronenoptischer Bildwandler m
  f convertisseur m électronique-optique d'image
  r электронно-оптический
     преобразователь m изображения

4946 electron-optical input device
  d elektronische optische Eingabeeinrichtung f
  f dispositif m d'entrée optique électronique
  r электронно-оптическое устройство n
     ввода

4947 electron optics
  d Elektronenoptik f
  f optique f électronique
  r электронная оптика f

4948 electrooptical approximation sensor
  d elektro-optischer Annäherungssensor m
  f capteur m électro-optique de proximité
  r электро-оптический сенсор приближения

4949 electrooptical coupling
  d elektrooptische Kopplung f
  f couplage m électro-optique
  r электрооптическая связь f

4950 electrooptical function generator
  d elektrooptischer Funktionsgenerator m
  f générateur m électro-optique de fonctions
  r электрооптический генератор m функций

4951 electrooptical imaging and storage
  d elektrooptische Abbildung f und
     Speicherung f
  f reproduction f et emmagasinage m électro-
     optique
  r электрооптическое воспроизведение n и
     хранение n изображения

4952 electrooptically tuned laser
  d elektrooptisch abstimmbarer Laser m
  f laser m à accord électro-optique
  r лазер m с электрооптической настройкой

4953 electrooptical sensor
  d elektro-optischer Geber m; elektro-optischer
     Sensor m

f capteur *m* électro-optique; senseur *m* électro-
optique
r электронно-оптический датчик *m*

4954 electropneumatic
d elektropneumatisch
f électropneumatique
r электропневматический

4955 electropneumatic actuator
d elektropneumatischer Effektor *m*;
Wirkglied *n*
f élément *m* de commande électropneumatique
r электропневматический привод *m*

4956 electropneumatically controlled feed slide
d elektropneumatisch gesteuerter
Zuführapparat *m*
f système *m* d'avance à tiroir à commande
électropneumatique
r подающее устройство *n* с
электропневматическим управлением

4957 electropneumatic control device
d elektropneumatische Steuervorrichtung *f*
f dispositif *m* de commande
électropneumatique
r электропневматическое управляющее
устройство *n*

4958 electropneumatic controller
d elektropneumatischer Fahrschalter *m*
f combinateur *m* électropneumatique
r электропневматический контроллер *m*

4959 electropneumatic high-pressure converter
d elektropneumatischer
Hochdruckumformer *m*
f convertisseur *m* électropneumatique à haute
pression
r электропневматический
преобразователь *m* высокого давления

4960 electropneumatic interlock
d elektropneumatische Sperre *f*
f blocage *m* électropneumatique
r электропневматическая блокировка *f*

4961 electropneumatic level controller
d elektropneumatischer Pegelregler *m*
f régulateur *m* électropneumatique de niveau
r электропневматический регулятор *m*
уровня

4962 electropneumatic position governor
d elektropneumatischer Stellungsregler *m*
f régulateur *m* électropneumatique des
positions

r электропневматический регулятор *m*
положения

4963 electropneumatic sequential control
d elektropneumatische Folgesteuerung *f*
f commande *f* successive électropneumatique
r электропневматическое последовательное
регулирование *n*

4964 electropneumatic valve
d elektropneumatisches Ventil *n*
f valve *f* électropneumatique
r электропневматический клапан *m*

4965 electrostatic attraction
d elektrostatische Anziehung *f*
f attraction *f* électrostatique
r электростатическое притяжение *n*

**4966 electrostatic beaming; electrostatic
focusing**
d elektrostatische Bündelung *f*; elektrostatische
Fokussierung *f*
f concentration *f* électrostatique; focalisation *f*
électrostatique
r электростатическое фокусирование *n*

4967 electrostatic deflection
d elektrostatische Ablenkung *f*
f déviation *f* électrostatique
r электростатическое отклонение *n*

* electrostatic focusing → 4966

4968 electrostatic process
d elektrostatische Operation *f*
f processus *m* électrostatique
r электростатическая технология *f*

4969 electrostatic repulsion
d elektrostatische Abstoßung *f*
f répulsion *f* électrostatique
r электростатическое отталкивание *n*

4970 electrostatic scanning
d elektrostatische Abtastung *f*
f balayage *m* électrostatique; exploration *f*
électrostatique
r электростатическое разложение *n*

4971 electrostatic screen
d elektrostatischer Schirm *m*; elektrostatische
Abschirmung *f*
f écran *m* électrostatique; blindage *m*
électrostatique
r электростатический экран *m*

4972 electrostatic sensing
d elektrostatisches Abfühlen *n*

f palpation *f* électrostatique
r электростатическое считывание *n* сигнала

4973 electrostatic technique
d elektrostatische Technik *f*
f technique *f* électrostatique
r электростатическая техника *f*

4974 electrothermal
d elektrothermisch
f électrothermique
r электротермический

4975 electrothermal machining
d elektrothermische Bearbeitung *f*
f usinage *m* électrothermique
r электротермическая обработка *f*

*   **electrothermic printer → 12352**

*   **element → 8025**

4976 elementary algorithm; primary algorithm
d Elementaralgorithmus *m*
f algorithme *m* élémentaire
r элементарный алгоритм *m*

4977 elementary analysis
d Elementaranalyse *f*
f analyse *f* élémentaire
r элементарный анализ *m*

4978 elementary automaton
d elementarer Automat *m*
f automate *m* élémentaire
r элементарный автомат *m*

4979 elementary building block
d Elementarbaustein *m*
f bloc *m* fonctionel élémentaire
r элементарный блок *m*

4980 elementary function
d Elementarfunktion *f*
f fonction *f* élémentaire
r элементарная функция *f*

4981 elementary information
d elementare Information *f*
f information *f* élémentaire
r элементарная информация *f*

4982 elementary iteration procedure
d elementares Iterationsverfahren *n*
f méthode *f* d'itération élémentaire
r метод *m* элементарных итераций

4983 elementary logical connections

d elementare logische Verknüpfungen *fpl*
f connexions *fpl* logiques élémentaires
r элементарные логические связи *fpl*

4984 elementary logical theorem
d elementarer logischer Satz *m*
f théorème *m* logique élémentaire
r элементарная логическая теорема *f*

4985 elementary member
d Elementarglied *n*
f membre *m* élémentaire
r элементарное звено *n*

4986 elementary operation
d Elementaroperation *f*
f opération *f* élémentaire
r элементарная операция *f*

*   **elementary process → 1686**

4987 elementary reaction
d Elementarreaktion *f*
f réaction *f* élémentaire
r элементарная реакция *f*

4988 elementary sensor; simple sensor
d Elementarsensor *m*; elementarer Sensor *m*
f senseur *m* élémentaire; capteur *m* élémentaire
r простой чувствительный элемент *m*

4989 elementary system subprogram
d elementares Systemunterprogramm *n*
f sous-programme *m* élémentaire du système
r элементарная системная подпрограмма *f*

4990 element efficiency
d Elementwirkungsgrad *m*
f rendement *m* de l'élément; rendement
    élémentaire
r коэффициент *m* полезного действия
    элемента

4991 element for digital automation
d Element *n* für digitale Automatisierung
f élément *m* pour l'automatisation digitale
r элемент *m* [для] цифровой автоматизации

*   **element number of manipulator
     drive → 4504**

4992 element variable
d Elementvariable *f*
f variable *f* de l'élément
r переменная *f* элемента

4993 element with distributed parameters
d Element *n* mit verteilten Parametern

f élément m à constantes réparties
r элемент m с распределенными
параметрами

4994 element with lumped parameters
d Glied n mit konzentrierten Parametern
f élément m à paramètres localisés
r элемент m с сосредоточенными
параметрами

4995 element without inertia
d Element n ohne Trägheit
f élément m sans inertie
r безинерционный элемент m

4996 element without selfregulation
d Glied n ohne Ausgleich
f élément m sans autoréglage; élément sans
autorégulation
r звено n без компенсации; элемент m без
компенсации

4997 element with time delay
d Element n mit Totzeit
f élément m à retard
r элемент m с запаздыванием

4998 elevated temperature
d erhöhte Temperatur f
f température f élevée
r повышенная температура f

4999 elliptical equation
d elliptische Gleichung f
f équation f elliptique
r эллиптическое уравнение n

5000 elliptical system
d elliptisches System n
f système m elliptique
r эллиптическая система f

5001 elliptic function
d elliptische Funktion f
f fonction f elliptique
r эллиптическая функция f

5002 embedded temperature detector
d eingebauter Temperaturfühler m
f détecteur m encasté de température
r встроенный температурный детектор m

5003 emergency
d Not f
f urgence f
r непредвиденный случай m

5004 emergency control

d Noteregelung f; Ersatzregelung f;
Reserveregelung f
f régulation f de veille; régulation de secours
r аварийное управление n

5005 emergency cutout
d Notausschalter m
f disjoncteur m de secours
r аварийный выключатель m

5006 emergency measure
d Sofortmaßnahme f
f mesure f d'urgence
r неотложное мероприятие n

5007 emergency movement
d Ausweichbewegung f
f mouvement m d'évitement
r аварийное движение n

5008 emergency power supply
d Notstromversorgung f
f alimentation f de secours
r аварийный источник m питания

5009 emergency shut-down
d Notabschaltung f
f déclenchement m d'urgence
r аварийное выключение n; аварийная
остановка f

5010 emergency stop-button
d Notausschaltknopf m
f bouton m d'arrêt d'urgence
r аварийная кнопка f

5011 emission analysis
d Emissionsanalyse f
f analyse f par émission
r эмиссионный анализ m

5012 emission characteristic
d Emissionskennlinie f
f caractéristique f d'émission
r эмиссионная характеристика f

5013 emission control
d Emissionssteuerung f
f contrôle m d'émission
r управление n излучением

5014 emission measurement technology
d Emissionsmesstechnik f
f technique f de mesure d'émission
r эмиссионная измерительная техника f

5015 emission probability
d Emissionswahrscheinlichkeit f

f probabilité *f* d'émission
r вероятность *f* испускания

5016 emission pulse
d emittierter Impuls *m*; Emissionsimpuls *m*
f impulsion *f* d'émission
r импульс *m* излучения

5017 emission rate
d Emissionshäufigkeit *f*
f intensité *f* d'émission
r интенсивность *f* излучения;
интенсивность испускания

5018 emission spectral analysis
d Emissionsspektralanalyse *f*
f analyse *f* spectrale d'émission
r спектральный анализ *m* эмиссии

5019 emissivity measurement
d Emissionsstärkemessung *f*
f mesurage *m* du pouvoir émissif
r измерение *n* излучательной способности

* **emphasize** *v* → **139**

5020 empirical model
d empirisches Modell *n*
f modèle *m* empirique
r эмпирическая модель *f*

5021 empirical procedure
d empirisches Verfahren *n*
f procédé *m* empirique
r эмпирический процесс *m*

5022 empirical value
d empirischer Wert *m*
f valeur *f* empirique
r эмпирическая величина *f*

* **employment** → **923**

5023 employment case
d Einsatzfall *m*
f cas *m* d'application
r обстоятельство n, связанное с
применением

* **emptiness check** → **5024**

5024 emptiness check[ing]
d Leerkontrolle *f*
f contrôle *m* du manque
r холостой контроль *m*

5025 empty
d leer; abgearbeitet

f vide
r незанятый; пустой

5026 emulation system
d Emulationssystem *n*
f système *m* d'émulation
r эмуляционная система *f*

5027 emulsifier
d Emulgiermaschine *f*
f émulsionneuse *f*; émulseur *m*
r аппарат *m* для эмульгирования

5028 enabled interruption
d gestattete Unterbrechung *f*
f interruption *f* accessible
r разрешенное прерывание *n*

5029 enable signal
d Freigabesignal *n*
f signal *m* de libération
r сигнал *m* разрешения

5030 encipher *v*; **encode** *v*
d kodieren; verschlüsseln
f coder; codifier
r кодировать; зашифровать

* **encode** *v* → **5030**

* **encoder** → **2530**

5031 end-around carry
d Endübertrag *m*; Rücklaufübertrag *m*;
Einerrücklauf *m*
f retenue *f* en arrière
r циклический перенос *m*

5032 end cycle control and return to zero
d Befehl *m* zur Zyklusbeendigung und
Rückstellung auf Null
f commande *f* fin de cycle et remise à zéro
r команда *f* для окончания цикла и
возвращения в нулевое положение

5033 endless installation
d Endlosanlage *f*
f installation *f* sans fin
r бесконечная установка *f*

5034 end of cycle
d Ende *n* des Zyklus
f fin *f* de cycle
r конец *m* цикла

5035 end position; final position
d Endposition *f*

f   position f finale
r   конечная позиция f; конечное
    положение n

5036 end status
d   Endzustand m
f   état m de fin
r   состояние n завершения

**5037 energetic balance method; method of
        energetic balance; power balance method**
d   Methode f des energetischen Gleichgewichts;
    Energiebalancemethode f
f   méthode f d'équilibre énergétique; méthode
    de balance énergétique
r   метод m энергетического баланса

5038 energetic efficiency
d   energetischer Wirkungsgrad m
f   rendement m énergétique
r   энергетический коэффициент m полезного
    действия

5039 energetic method
d   energetische Methode f
f   méthode f énergétique
r   энергртический метод m

5040 energy
d   Energie f
f   énergie f
r   энергия f

5041 energy analyzer
d   Energieanalysator m
f   analyseur m d'énergie
r   энергетический анализатор m

5042 energy balance
d   Energiegleichgewicht n; Energiebilanz f
f   balance f d'énergie; équilibre m énergétique
r   энергетический баланс m

5043 energy band
d   Energieband n
f   bande f d'énergie
r   энергетическая зона f

5044 energy barrier
d   Energieschranke f; Energieschwelle f
f   seuil m énergétique
r   энергетический барьер m

5045 energy component
d   Wirksstromkomponente f; Wattkomponente f
f   composante f wattée
r   активная составляющая f

5046 energy consumption
d   Energieverbrauch m; Energieaufwand m
f   consommation f d'énergie; dépense f
    d'énergie
r   потребление n энергии

5047 energy converter
d   Energieumformer m
f   transformateur m d'énergie; convertisseur m
    d'énergie
r   преобразователь m энергии

5048 energy coupling of gripper
d   Greiferenergiekopplung f
f   couplage m d'énergie d'un grappin
r   обратная связь f захватного устройства по
    энергии

5049 energy decrement
d   Energiedekrement n; Energieverlust m;
    Energieabnahme f
f   décrément m d'énergie
r   декремент m энергии

5050 energy density
d   Energiedichte f
f   densité f énergétique
r   плотность f энергии

5051 energy-economical system
d   energieökonomisches System n
f   système m pour l'économie de l'énergie
r   энергоэкономная система f

5052 energy-independent store
d   energieunabhängiger Speicher m
f   mémoire f indépendante d'énergie
r   энергонезависимое запоминающее
    устройство n

5053 energy input; addition of energy
d   Energieeingang m; Energieeinspeisung f
f   entrée f de puissance; alimentation f d'énergie
r   подвод m энергии

5054 energy level
d   Energieniveau n
f   niveau m d'énergie
r   энергетический уровень m; уровень
    энергии

5055 energy-level analysis
d   Termanalyse f
f   analyse f des niveaux d'énergie
r   анализ m энергетических уровней

5056 energy-level change value
d   Energieniveauänderungswert m

f valeur f de changement du niveau
énergétique
r величина f изменения энергетического
уровня

5057 energy-level diagram
d Energiepegeldiagramm n; Energieschema n
f diagramme m de niveaux énergétiques
r схема f энергетических уровней;
диаграмма f энергетических уровней

5058 energy-level spacing
d Energiestufenabstand m;
Energieniveauabstand m
f espacement m des niveaux énergétiques;
distance f des niveaux énergétiques
r разность f энергетических уровней;
интервал m энергетических уровней

5059 energy model
d Energiemodell n; energetisches Modell n
f modèle m énergétique
r энергетическая модель f

5060 energy of absolute zero
d Energie f am absoluten Nullpunkt
f énergie f au zéro absolu
r энергия f абсолютного нуля; нулевая
энергия

5061 energy-optimal control
d energieoptimale Steuerung f
f commande f optimale d'énergie
r оптимальное управление n по энергии

5062 energy-saving technology
d energiesparende Technologie f
f technologie f économisatrice d'énergie
r энергосберегающая технология f

5063 energy sensitiveness
d Energieempfindlichkeit f
f sensibilité f à l'énergie
r чувствительность f к изменению энергии

5064 energy spectrum
d Energiespektrum n
f spectre m énergétique
r энергетический спектр m

5065 energy state
d Energiezustand m; energetischer Zustand m
f état m énergétique
r энергетическое состояние n

5066 energy storage unit
d Energiespeichereinheit f
f unité f d'accumulation d'énergie

r блок m накопления энергии

5067 energy theorem
d Energiesatz m; Energieerhaltungssatz m
f loi f de la conservation d'énergie
r закон m сохранения энергии

**5068 energy transfer; energy transmission;
power transmission**
d Energieübertragung f; Energiefortleitung f
f transmission f de l'énergie
r передача f энергии; перенос m энергии

5069 energy transfer coefficient
d Energieübertragungskoeffizient m;
Energieumsatz m
f coefficient m de transfert d'énergie
r коэффициент m передачи энергии

5070 energy transfer mechanics
d Mechanik f der Energieübertragung
f mécanique f de la transmission d'énergie
r механизм m передачи энергии

* **energy transmission** → 5068

* **engine control** → 8356

5071 engineering
d Konstruktionswesen n; Projektierung f
f construction f; conception f; technique f
r проектирование n; конструирование n

5072 engineering approximation
d technische Approximierung f
f approximation f technique
r инженерное приближение n

5073 engineering concept
d Konstruktionsprinzip n
f concept m constructif
r технический принцип m

5074 engineering console
d Wartungskonsole f
f console f d'entretien
r инженерный пульт m

5075 engineering constraints
d technische Grenzen fpl; technische
Restriktionen fpl
f limites fpl techniques; contraintes fpl
techniques
r технические ограничения npl

* **engineering cybernetics** → 2661

5076 engineering data
d technische Daten pl

f données *fpl* techniques
r технические данные *pl*

5077 engineering design
d technischer Entwurf *m*
f projet *m* technique; conception *f* technique
r техническое проектирование *n*

5078 engineering development
d technische Entwicklung *f*
f développement *m* technique
r техническая разработка *f*

5079 engineering flow sheet
d Betriebsschema *n*; technologisches Fließbild *n*
f schéma *m* de procédé; flow-sheet *m* technologique
r технологическая схема *f*

5080 engineering level
d Konstruktionsstand *m*
f niveau *m* de construction
r технический уровень *m*

5081 engineering solution
d technische Lösung *f*
f solution *f* technique
r техническое решение *n*

5082 engineering working method
d ingenieurtechnische Arbeitsmethode *f*
f méthode *f* de travail génietechnique
r инженерно-технический метод *m* работы

5083 enquire drawing
d Anfragezeichnung *f*
f dessin *m* de projet; dessin de demande
r техническая документация *f* по требованию

5084 enterprise information system; EIS
d Enterprise-Informationssystem *n*
f système *m* d'information d'entreprise; SIE
r информационная система *f* предприятия

* **enterprise investigation → 9025**

5085 entire function
d ganze Funktion *f*
f fonction *f* entière
r целевая функция *f*

5086 entire system
d vollständiges System *n*
f système *m* entier
r завершенная система *f*

5087 entrainment of frequency

d Frequenzmitzieheffekt *m*
f entraînement *m* de fréquence
r захватывание *n* частоты

5088 entropic stability
d Entropiestabilität *f*
f stabilité *f* entropique
r энтропийная устойчивость *f*

5089 entropy
d Entropie *f*
f entropie *f*
r энтропия *f*

* **enumerable set → 3484**

5090 envelope
d Einhüllende *f*
f enveloppe *f*
r огибающая *f*

* **envelope delay → 5091**

* **envelope delay meter → 5092**

5091 envelope delay [time]
d Gruppenlaufzeit *f*
f temps *m* de transit de groupe
r время *n* запаздывания огибающей

5092 envelope delay [time] meter
d Gruppenlaufzeitmesser *m*
f appareil *m* de mesure de temps de propagation de groupe
r прибор *m* для измерения группового времени распространения

5093 environment; environs
d Milieu *n*; Umwelt *f*; Umgebung *f*
f environs *mpl*; milieu *m*; environnement *m*
r среда *f*

5094 environment acquisition
d Umwelterfassung *f*
f acquisition *f* d'environnement
r сбор *m* данных об окружающей среде

5095 environmental condition
d Umgebungsbedingung *f*
f condition *f* ambiante; condition de l'ambiance
r условие *n* окружающей среды

5096 environmental disturbance variable
d Umweltstörgröße *f*
f grandeur *f* perturbatrice d'environnement
r возмущающее воздействие *n* окружающей среды

5097 environmental requirements
    d Umgebungsanforderungen *fpl*
    f exigences *fpl* à l'environnement
    r требования *npl* окружающей обстановки

5098 environmental stability
    d Umweltstabilität *f*
    f stabilité *f* d'ambiance
    r устойчивость *f* к внешним воздействиям

5099 environment model
    d Umweltmodell *n*
    f modèle *m* d'environnement
    r модель *f* окружающего пространства

5100 environment simulation
    d Umgebungssimulation *f*; Umweltsimulation *f*
    f simulation *f* de l'environnement
    r моделирование *n* внешних условий

    * environs → 5093

    * eqiupment dependability → 3954

5101 equality relation
    d Gleichheitsrelation *f*; Äquivalenzrelation *f*
    f relation *f* équivalente
    r соотношение *n* эквивалентности

5102 equalization; compensation
    d Ausgleich *m*; Kompensation *f*
    f compensation *f*
    r компенсация *f*

5103 equalizer
    d Ausgleichsglied *n*; Korrekturglied *n*;
       Entzerrer *m*
    f égalisateur *m*; compensateur *m*
    r выравниватель *m*

5104 equalizing pulse
    d Ausgleichsimpuls *m*
    f impulsion *f* d'égalisation
    r выравнивающий импульс *m*

5105 equals module
    d Vergleichsmodul *m*
    f module *m* de comparaison
    r модуль *m* сравнения

5106 equation in relative variables
    d Gleichung *f* in relativen Variablen
    f équation *f* aux variables relatives
    r уравнение *n* с относительными
       переменными

5107 equation of continuity
    d Kontinuitätsgleichung *f*

    f équation *f* de continuité
    r уравнение *n* непрерывности

5108 equation of controlled system
    d Gleichung *f* der Regelstrecke
    f équation *f* du système réglé
    r уравнение *n* регулируемой системы

5109 equation of decay
    d Zerfallsgleichung *f*
    f équation *f* de décomposition
    r уравнение *n* распада

5110 equation of free oscillations
    d Gleichung *f* freier Schwingungen
    f équation *f* d'oscillations libres
    r уравнение *n* свободных колебаний

5111 equation of halfperiods
    d Halbperiodengleichung *f*
    f équation *f* de demi-périodes
    r уравнение *n* полупериода

5112 equation of motion
    d Bewegungsgleichung *f*
    f équation *f* de mouvement
    r уравнение *n* движения

5113 equation of static control circuit
    d Gleichung *f* des ststischen Regelkreises
    f équation *f* du circuit de réglage statique
    r уравнение *n* статического контура
       регулирования

5114 equation solver
    d Gleichungslöser *m*
    f machine *f* à résoudre des équations
    r устройство *n* для решения уравнений

5115 equation system
    d Gleichungssystem *n*
    f système *m* d'équations
    r система *f* уравнений

5116 equidirectional derivative
    d gleichsinnige Ableitung *f*
    f dérivation *f* concordante
    r эквинаправленная производная *f*

5117 equidistant code
    d äquidistanter Kode *m*
    f code *m* équidistant
    r эквидистантный код *m*

    * equilibrium → 1585

5118 equilibrium conditions; balance conditions
    d Gleichgewichtsbedingungen *fpl*

f conditions *fpl* d'équilibre
r условия *npl* равновесия

5119 equilibrium data
d Gleichgewichtsdaten *pl*
f données *fpl* d'équilibre
r равновесные данные *pl*

5120 equilibrium energy
d Gleichgewichtsenergie *f*
f énergie *f* d'équilibre
r энергия *f* равновесия

5121 equilibrium gain
d Gleichgewichtsgewinn *m*
f gain *m* à l'équilibre; gain d'équilibre
r усиление *n* в состоянии равновесия

5122 equilibrium point
d Gleichgewichtspunkt *m*
f point *m* d'équilibre
r точка *f* равновесия

5123 equilibrium position; equilibrium state
d Gleichgewichtszustand *m*
f état *m* d'équilibre
r состояние *n* равновесия

5124 equilibrium power level
d stabiles Leistungsniveau *n*
f niveau *m* de puissance équilibré
r равновесный уровень *m* мощности

* **equilibrium state** → 5123

5125 equipment; fitting out
d Ausrüstung *f*; Anlage *f*
f équipement *m*; matériel *m*; installation *f*
r оборудование *n*; оснащение *n*

5126 equipment arrangement
d Ausrüstungsanordnung *f*
f disposition *f* d'équipement
r компоновка *f* оборудования

5127 equipment block
d Ausrüstungsblock *m*
f bloc *m* d'équipement
r блок *m* оборудования

5128 equipment compatibility
d Gerätekompatibilität *f*
f compatibilité *f* d'équipement
r совместимость *f* оборудования

5129 equipment engineering
d Bauweise *f*
f mode *m* de construction

r разработка *f* оборудования

5130 equipment failure
d Fehler *m* der Geräte
f panne *f* d'équipement
r отказ *m* в работе устройства

5131 equipment group
d Ausrüstungsgruppe *f*
f groupe *m* d'équipement
r группа *f* оборудования

5132 equipment innovation
d Ausrüstungserneuerung *f*
f renouvellement *m* d'équipement
r новшество *n* в оборудовании

* **equipment layout plan** → 6710

5133 equipment reliability; safety of operation
d Betriebssicherheit *f*
f sécurité *f* de fonctionnement; sûreté *f* de
   service
r надёжность *f* оборудования; надёжность
   эксплуатации

* **equipment safety** → 3954

5134 equipment system
d Anlagensystem *n*
f système *m* de l'équipement
r система *f* оборудования

5135 equipment trouble
d Gerätestörung *f*
f perturbation *f* d'appareil
r неисправность *f* оборудования

**5136 equipment with numerical program
         control**
d Ausrüstung *f* mit Ziffernprogrammsteuerung
f équipement *m* à commande de programme de
   chiffres
r оборудование *n* с числовым программным
   управлением

5137 equipotent
d gleichmächtig
f de même puissance
r равномощный

5138 equipotential line
d Äquipotentiallinie *f*
f ligne *f* équipotentielle
r эквипотенциальная линия *f*

5139 equispaced pulses
d abstandsgetreue Impulse *mpl*

f impulsions *fpl* équidistantes
r равноудаленные импульсы *mpl*

5140 equivalent action
d äquivalente Einwirkung *f*;
Äquivalenteinwirkung *f*
f action *f* équivalente
r эквивалентное воздействие *n*

* **equivalent admittance → 3972**

5141 equivalent binary digits
d äquivalente Binärstellenzahlen *fpl*
f nombres *mpl* de bits équivalents
r эквивалентные двоичные числа *npl*

* **equivalent circuit → 5145**

5142 equivalent damping
d Ersatzdämpfung *f*
f amortissement *m* équivalent
r эквивалентное затухание *n*

5143 equivalent impedance of a non-linear element
d äquivalente Impedanz *f* eines nichtlinearen Gliedes
f impédance *f* équivalente d'élément non linéaire
r эквивалентный импеданс *m* нелинейного элемента

5144 equivalent load
d äquivalente Belastung *f*
f charge *f* équivalente
r эквивалентная нагрузка *f*

5145 equivalent network; equivalent circuit
d Ersatzschaltung *f*
f circuit *m* d'équivalent
r эквивалентная схема *f*

5146 equivalent parameter
d Ersatzgröße *f*
f paramètre *m* équivalent
r эквивалентный параметр *m*

5147 equivalent power of noise; noise equivalent power
d äquivalente Rauschleistung *f*
f puissance *f* équivalente de bruit
r эквивалентная энергия *f* шума; эквивалентная мощность *f* шума

5148 equivalent state
d äquivalenter Zustand *m*
f état *m* équivalent
r эквивалентное состояние *n*

5149 equivalent structure transformation
d äquivalente Strukturwandlung *f*
f transformation *f* équivalente de structure
r эквивалентное преобразование *n* структуры

5150 erasable store
d löschbarer Speicher *m*
f mémoire *f* effaçable
r стираемая память *f*

5151 erase driver
d Löschtreiber *n*
f driver *m* d'effacement
r стирающий усилитель-формирователь *m*

5152 eraser switch
d Löschschalter *m*
f interrupteur *m* d'effacemrnt
r стирающий выключатель *m*

5153 erasure of information
d Löschen *n* der Information
f effacement *m* de l'information
r стирание *n* информации

5154 erecting tool
d Montageeinrichtung *f*
f installation *f* de montage; dispositif *m* de montage
r монтажное приспособление *n*; устройство *n* для монтажа

5155 erection drawing
d Montagedokumentation *f*
f documentation *f* de montage
r монтажная техническая документация *f*

5156 erection project
d Montageprojekt *n*
f projet *m* de montage
r монтажный проект *m*

5157 ergodic hypothesis
d Ergodenhypothese *f*; ergodische Vermutung *f*
f hypothèse *f* ergodique
r эргодическая гипотеза *f*

5158 ergodic property
d ergodische Eigenschaft *f*
f propriété *f* ergodique
r эргодическое свойство *n*

5159 ergometer
d Ergometer *n*
f ergomètre *m*
r эргометр *m*

5160 error-actuated system
  d fehlergesteuerte Anlage *f*; fehlerbetätigtes
       System *n*
  f système *m* commandé par signal erreur
  r система *f* действующая от
       рассогласования

5161 error and balance attenuation
  d Fehler- und Symmetriedämpfung *f*
  f affaiblissement *m* d'équilibrage et de
       symétrie
  r затухание *n* вследствие рассогласования и
       симметрии

5162 error-checking capability
  d Fehlerprüffähigkeit *f*;
       Fehlerprüfungstauglichkeit *f*
  f faculté *f* de vérification d'erreurs
  r возможность *f* контроля ошибок

5163 error compensation of the resolver
  d Fehlerkompensation *f* des Resolvers
  f compensation *f* d'erreurs de résolveur
  r компенсация *f* погрешностей решающего
       устройства

5164 error compensation of the tachogenerator
  d Fehlerkompensation *f* des Tachogenerators
  f compensation *f* des erreurs du générateur
       tachymétrique
  r компенсация *f* погрешностей
       тахогенератора

5165 error condition
  d Fehlerbedingung *f*
  f condition *f* d'erreurs
  r условие *n* ошибки

5166 error control
  d Fehlerkorrektur *f*; Fehlerüberwachung *f*;
       Fehlerkontrolle *f*
  f contrôle *m* d'erreurs; correction *f* d'erreurs
  r контроль *m* ошибок; нахождение *n*
       ошибок

5167 error-controlled code
  d fehlerkontrollierender Kode *m*
  f code *m* à surveillance d'erreur
  r помехоустойчивый код *m*

5168 error-correcting code
  d selbstkorrigierender Kode *m*
  f code *m* autocorrecteur
  r код *m* с исправлением ошибки

5169 error-correcting device
  d Fehlerkorrektureinrichtung *f*
  f dispositif *m* de correction des erreurs

  r устройство *n* исправления ошибок

5170 error-correcting program
  d Fehlerkorrekturprogramm *n*
  f programme *m* de correction d'erreurs
  r программа *f* для исправления ошибок

5171 error correction
  d Fehlerkorrektur *f*
  f correction *f* d'erreur[s]
  r исправление *n* ошибок

5172 error-correction circuit
  d Kreis *m* für Abweichungskorrektur
  f circuit *m* de correction d'écart
  r схема *f* исправления ошибок

5173 error criterion
  d Fehlerkriterium *n*
  f critère *f* d'erreurs; critérium *m* d'erreurs
  r критерий *m* ошибок

5174 error-detecting code; check code
  d fehlererkennender Kode *m*; Prüfkode *m*
  f code *m* détecteur; code d'essai
  r код *m* с самопроверкой

**5175 error-detection circuitry; fault-detection
       circuitry**
  d Fehlererkennungsschaltung *f*
  f circuit *m* de reconnaissance d'erreurs
  r схема *f* обнаружения ошибок

5176 error diagnostic
  d Fehlerdiagnose *f*; Fehlerbestimmung *f*
  f diagnostic *m* d'erreur
  r диагностика *f* ошибок

5177 error diagnostic signal
  d Fehlererkennungssignal *n*
  f signal *m* de détection d'une erreur
  r сигнал *m* распознавания ошибок

5178 error distance
  d Fehlerabstand *m*
  f distance *f* des erreurs
  r ошибка *f* интервала

5179 error estimation
  d Fehlerabschätzung *f*
  f estimation *f* d'erreur
  r оценка *f* ошибки

5180 error-free fall
  d fehlerfreier Fall *m*
  f chute *f* sans faute
  r безошибочный спад *m*

5181 error-free operation
   d fehlerfreier Betrieb *m*
   f régime *m* sans défaut
   r режим *m* [работы] без ошибок

5182 error-free transmission
   d fehlerfreie Übertragung *f*
   f transmission *f* sans défaut
   r передача *f* [данных] без ошибок

5183 error function
   d Fehlerfunktion *f*
   f fonction *f* d'erreur
   r функция *f* ошибки

5184 error handling routine
   d Fehlerbehandlungsroutine *f*
   f routine *f* de maniement d'erreurs
   r программа *f* обработки ошибок

5185 error-indicating circuit
   d Kreis *m* für Abweichungsfeststellung
   f circuit *m* détecteur d'écart
   r схема *f* индикации рассогласования

5186 error indicator
   d Fehleranzeiger *m*
   f indicateur *m* d'erreur
   r индикатор *m* ошибки

5187 error in the simulation
   d Fehler *m* der Simulation
   f erreur *f* de simulation
   r ошибка *f* моделирования

5188 error limit; limit of error
     (by object registration)
   d Fehlergrenze *f*
   f limite *f* d'erreur
   r предел *m* ошибки

5189 error localization; fault localization
   d Fehlerlokalisierung *f*
   f localisation *f* d'erreur
   r локализация *f* неисправности

5190 error localization of test routine
   d Fehlerlokalisierung *f* einer Testroutine
   f localisation *f* d'erreur d'une routine de test
   r обнаружение *n* неисправностей
     программы контроля

5191 error-mesuring device
   d Vergleichsglied *n*
   f comparateur *m*
   r блок *m* сравнения

5192 error of approximation

   d Approximationsgüte *f*;
     Approximationsfehler *m*; Verfahrensfehler *m*
   f erreur *m* de valeur approchée; erreur
     d'approximation
   r ошибка *f* аппроксимации

5193 error of construction
   d Konstruktionsfehler *m*
   f défaut *m* de construction
   r ошибка *f* конструкции

5194 error of reading
   d Ablesefehler *m*
   f erreur *f* de lecture
   r ошибка *f* отсчёта

5195 error protection procedure
   d Fehlerschutzverfahren *n*
   f procédure *f* de protection d'erreur
   r метод *m* защиты от ошибок

5196 error pulse
   d Fehlerimpuls *m*
   f impulsion *f* d'erreur
   r импульс *m* ошибочного сигнала

5197 error rate
   d Fehlerrate *f*; Fehlerhäufigkeit *f*
   f taux *m* d'erreurs
   r интенсивность *f* ошибок

5198 error routine
   d Fehlermaßnahmeprogramm *n*
   f programme *m* de mesure contre erreurs
   r программа *f* контроля ошибок

5199 error-sampled control system
   d Impulssystem *n* mit intermittierendem Signal
     der Regelabweichung
   f système *m* échantillonné à signal d'erreur
     intermittent
   r импульсная система *f* с сигналами
     рассогласования

5200 error-sensing device
   d Fehleraufspürgerät *n*
   f détecteur *m* d'erreurs
   r устройство *n* для приёма сигналов ошибок

   *   error source → 11701

5201 error theory; theory of error
   d Fehlertheorie *f*
   f théorie *f* d'erreurs
   r теория *f* ошибок

5202 error transfer function
   d Fehlerübertragungsfunktion *f*

f fonction *f* de transfert d'erreur
r передаточная функция *f* рассогласования

5203 estimated value
d Schätzwert *m*
f valeur *f* estimée
r оценочное значение *n*

5204 estimation; identification
d Kennwertermittlung *f*; Modellermittlung *f*; Identifikation *f*
f évaluation *f* de l'indice; estimation *f* de l'indice; identification *f* de l'indice
r идентификация *f*; опознание *n*; оценка *f*

5205 estimation of quality
d Güteabschätzung *f*; Gütekriterium *n*
f estimation *f* de qualité
r оценка *f* качества

5206 estimation of stability; stability estimation
d Stabilitätsabschätzung *f*; Stabilitätskriterium *n*
f estimation *f* de stabilité
r оценка *f* устойчивости

5207 estimation unit; evaluation unit
d Auswerteeinheit *f*
f bloc *m* d'estimation; organe *m* d'estimation
r блок *m* оценки

5208 Euler's differential equation
d Eulersche Differentialgleichung *f*
f équation *f* différentielle d'Euler
r дифференциальное уравнение *n* Эйлера

5209 evaluate *v*
d auswerten; erproben
f évaluer
r оценивать

* **evaluation → 942**

5210 evaluation control
d Bewertungssteuerung *f*
f commande *f* d'évaluation
r управление *n* оценкой

5211 evaluation factor
d Bewertungsfaktor *m*
f facteur *m* d'évaluation
r оценочный фактор *m*

5212 evaluation methods
d Bewertungsmethoden *fpl*
f méthodes *fpl* d'évaluation
r методы *mpl* оценки

5213 evaluation model
d Bewertungsmodell *n*
f modèle *m* d'évaluation
r оценочная модель *f*

5214 evaluation of integrals
d Berechnung *f* von Integralen
f calcul *m* des intégrales
r вычисление *n* интегралов

5215 evaluation of sensor information
d Auswertung *f* der Sensorinformation
f évaluation *f* de l'information de senseur
r анализ *m* сенсорной информации

5216 evaluation quantity
d Bewertungsgröße *f*
f grandeur *f* d'évaluation
r величина *f* оценки

5217 evaluation time
d Auswertezeit *f*
f temps *m* d'évaluation
r время *n* обработки

* **evaluation unit → 5207**

* **Evans method → 8074**

5218 even function; parity function
d gerade Funktion *f*; Paritätsfunktion *f*
f fonction *f* paire; fonction *f* de parité
r чётная функция *f*

5219 even harmonics
d geradzahlige Harmonische *f*
f harmonique *f* paire
r чётная гармоника *f*

5220 evenly divided scale; uniform scale
d gleichmäßige Skale *f*
f échelle *f* uniforme
r равномерная шкала *f*

5221 evenly spaced energy levels
d gleichmäßig verteilte Energieniveaus *npl*
f niveaux *mpl* énergétiques à distribution uniforme
r равномерно распределенные энергетические уровни *mpl*

5222 event; occurrence
d Ereignis *n*; Vorgang *m*
f événement *m*; occurrence *f*
r событие *n*

5223 event control block
d Ereignissteuerblock *m*

    f bloc *m* de commande d'événement
    r блок *m* управления событиями

5224 event identification
    d Ereigniskennzeichnung *f*
    f identification *f* d'événement
    r идентификация *f* события

5225 event representation
    d Erscheinungsdarstellung *f*; Darstellung *f*
       einer Erscheinung
    f représentation *f* d'événement
    r воспроизведение *n* события

5226 event synchronization
    d Ereignissynchronisation *f*
    f synchronisation *f* d'événement
    r синхронизация *f* событий

5227 exact approximation
    d geaue Annäherung *f*
    f approximation *f* exacte
    r точное приближение *n*

    * **exactitude** → 199

    * **exactness** → 199

5228 exact regulation algorithm
    d exakter Regelalgorithmus *m*
    f algorithme *m* de régulation exact
    r точный алгоритм *m* регулирования

5229 exception condition
    d Ausnahmebedingung *f*
    f condition *f* exceptionnelle
    r исключающее условие *n*

5230 exception handling
    d Bearbeitung *f* von Ausnahmebedingungen
    f traitement *m* d'exceptions
    r обработка *f* особых случаев

5231 excess loss
    d Zusatzverlust *m*
    f perte *f* en excès
    r дополнительная потеря *f*

5232 excess noise technique
    d Exzess-Geräuschtechnik *f*
    f méthode *f* du bruit redondant
    r метод *m* избыточного шума

5233 excess-six-code
    d Sechsexzesskode *m*; Sechsüberschußkode *m*;
       Plussechskode *m*
    f code *m* plus six
    r код *m* с избытком шесть

5234 excess-three-code
    d Dreiexzesskode *m*;
       Dreiexzessverschlüsselung *f*;
       Dreiüberschußkode *m*; Plusdreikode *m*
    f code *m* plus trois
    r код *m* с избытком три

    * **exchangeable** → 6839

5235 exchangeable control bar
    d auswechselbare Steuerschiene *f*
    f réglette *f* de commande échangeable
    r заменяемая управляющая шина *f*

5236 excitation and inhibition
    d Erregung *f* und Hemmung *f*
    f excitation *f* et inhibition *f*
    r возбуждение *n* и торможение *n*

5237 excitation circuit
    d Erregerkreis *m*
    f circuit *m* d'xcitation; chaîne *f* d'excitation;
       boucle *f* d'excitation
    r цепь *f* возбуждения

5238 excitation condition
    d Anregungsbedingung *f*
    f condition *f* d'excitation
    r условие *n* возбуждения

5239 excitation cross-section
    d Erregungsquerschnitt *m*
    f section *f* efficace d'excitation
    r поперечное сечение *n* возбуждения

5240 excitation curve
    d Anregungskurve *f*
    f courbe *f* d'excitation
    r кривая *f* возбуждения

5241 excitation density
    d Erregungsdichte *f*
    f densité *f* d'excitation
    r плотность *f* возбуждения

5242 excitation flow
    d Erregungsfluss *m*; Erregungsströmung *f*
    f courant *m* d'excitation
    r поток *m* возбуждения

5243 excitation frequency
    d Erregungsfrequenz *f*
    f fréquence *f* d'excitation
    r частота *f* возбуждения

5244 excitation function
    d Anregungsfunktion *f*
    f fonction *f* d'excitation
    r функция *f* возбуждения

5245 excitation level
d Anregungsniveau *n*; angeregtes Niveau *n*
f niveau *m* excité; niveau d'excitation
r уровень *m* возбуждения

5246 excitation potential; excitation voltage
d Erregungsspannung *f*
f tension *f* d'excitation
r напряжение *n* возбуждения; потенциал *m* возбуждения

5247 excitation pulse
d Erregungsimpuls *m*
f impulsion *f* d'excitation
r импульс *m* возбуждения

5248 excitation system
d Erregungssystem *n*
f système *m* d'excitation
r система *f* возбуждения

5249 excitation time
d Erregungszeit *f*
f temps *m* d'excitation
r время *n* возбуждения

5250 excitation transfer
d Erregungsübertragung *f*
f transfert *m* d'excitation
r передача *f* возбуждения

* **excitation voltage → 5246**

5251 excitation winding
d Erregungswicklung *f*; Magnetisierungswicklung *f*
f enroulement *m* d'excitation
r обмотка *f* возбуждения

5252 excited state spectrum
d Spektrum *n* der angeregten Zustände
f spectre *m* des états excités
r спектр *m* возбужденных состояний

5253 exciter
d Erreger *m*; Erregermaschine *f*
f excitatrice *f*; excitateur *m*; dispositif *m* d'excitation
r возбудитель *m*; задающий генератор *m*

5254 exciter force
d Erregerkraft *f*
f force *f* d'excitation
r возбуждающая сила *f*

5255 exciter set
d Erregersatz *m*; Erregeranlage *f*
f groupe *m* d'excitation
r возбуждающий агрегат *m*

5256 exciting voltage pulse shape
d Impulsform *f* der Erregungsspannung
f forme *f* de l'impulsion du voltage excitateur
r форма *f* возбуждающего импульса напряжения

5257 exclusion
d Exklusion *f*; [logische] Ausschließung *f*
f exclusion *f*
r исключение *n*

5258 exclusive access
d alleiniger Zugriff *m*
f accès *m* exclusif
r монопольный доступ *m*

5259 exclusive control
d exklusive Steuerung *f*
f commande *f* exclusive
r монопольное управление *n*

5260 EXCLUSIVE-OR gate; non-equivalence gate; modulo 2 sum gate
d EXKLUSIV-ODER-Gatter *n*; Inäquivalenzgatter *n*; Modulo-2-Addition-Gatter *n*
f porte *f* OU EXCLUSIF; porte de non-équivalence; porte d'addition en module 2
r схема *f* исключающее ИЛИ; схема неравнозначности; схема сложения по модулю 2

5261 EXCLUSIVE-OR operation; XOR
d EXKLUSIV-ODER-Operation *f*; EXOR; XOR
f opération *f* OU EXCLUSIF
r операция *f* исключающее ИЛИ

5262 executable
d ausführbar
f exécutable
r выполняемый; исполнительный

5263 executable program
d aufrufbares Programm *n*
f programme *m* d'appel
r готовая программа *f*

5264 execute *v*
d ausführen
f exécuter
r исполнять; выполнять

5265 execution cycle
d Ausführungszyklus *m*

 f cycle m d'exécution
 r цикл m выполнения [инструкции]

5266 execution time
 d Ausführungszeit f
 f temps m d'exécution
 r время n выполнения [инструкции]

5267 executive control
 d Exekutivsteuerung f ; Ausführungssteuerung f
 f commande f d'exécution
 r управление n исполнением

5268 executive device
 d Stellglied n ; Stelleinheit f
 f organe m de commande; organe d'exécution; positionneur m
 r исполнительный элемент m ; исполнительное звено n

5269 executive program
 d Exekutivprogramm n ; Ausführungssteuerprogramm n
 f programme m exécutif
 r исполнительная программа f

5270 executive system
 d Ausführungsorganisationssystem n
 f système m exécutif
 r исполнительная система f

5271 exergetic element efficiency
 d exergetischer Elementwirkungsgrad m
 f rendement m élémentaire exergétique
 r эксергетический коэффициент m полезного действия элемента

5272 exergetic system efficiency
 d exergetischer Systemwirkungsgrad m
 f rendement m exergétique du système
 r эксергетический коэффициент m полезного действия системы

5273 exhaust gas plant
 d Abgasanlage f
 f installation f d'exhaustion
 r система f выпуска газа

5274 existence conditions
 d Existenzbedingungen fpl
 f conditions fpl d'existence
 r условия npl существования

5275 existence theorem
 d Existenzsatz m
 f théorème m d'existence
 r теорема f существования

5276 exit instruction
 d Ausgangsbefehl m
 f instruction f de sortie
 r выходная команда f

5277 exit losses
 d Austrittsverluste mpl
 f pertes fpl de sortie
 r потери fpl на выходе

5278 exothermic process
 d exothermer Prozess m
 f processus m exothermique
 r экзотермический процесс m

5279 expandability; expansibility; extendibility; extensibility
 d Erweiterungsfähigkeit f ; Ausbaufähigkeit f
 f expansibilité f ; extensibilité f
 r расширяемость f

5280 expandable; extensible
 d erweiterbar
 f expansible; extensible
 r расширяемый

5281 expander
 d Erweiterungsbaustein m ; Erweiterungsmodul m
 f module m d'élargissement
 r модуль m расширения

5282 expander chip
 d Erweiterungsschaltkreis m
 f circuit m d'élargissement; puce f d'élargissement
 r [микро]схема f расширения

 $*$ **expansibility → 5279**

5283 expansion; extension
 d Erweiterung f
 f extension f ; élargissement m
 r расширение n

5284 expected average value
 d erwarteter Mittelwert m ; Mittelerwartungswert m
 f valeur f moyenne expectée
 r среднее ожидаемое значение n ; среднее вероятное значение

5285 experiment
 d Experiment n
 f expérience f
 r эксперимент m

5286 experimental assembly
 d Experimentalaufbau m ; Versuchsaufbau m

    f assemblage *m* expérimental
    r экспериментальное устройство *n*

5287 experimental conditions
    d experimentelle Bedingungen *fpl*
    f conditions *fpl* expérimentales
    r экспериментальные условия *npl*

5288 experimental data
    d experimentelle Daten *pl*
    f données *fpl* expérimentales
    r экспериментальные данные *pl*

5289 experimental facilities
    d Versuchseinrichtung *f*
    f installation *f* d'expérimentation
    r оборудование *n* для экспериментов

5290 experimental fluid dynamics
    d experimentelle Strömungsdynamik *f*
    f dynamique *f* des fluides expérimentale
    r динамика *f* экспериментального обтекания

5291 experimental identification of systems
    d experimentelle Identifizierung *f* von Regelstrecken
    f identification *f* expérimentale des systèmes
    r экспериментальная идентификация *f* систем

5292 experimental model
    d Experimentalmodell *n*
    f modèle *m* expérimental
    r экспериментальная модель *f*

5293 experimental model of simple production line
    d Experimentalmodell *n* einer einfachen Fertigungsstraße
    f modèle *m* expérimental d'une ligne de production simple
    r экспериментальная модель *f* простой производственной линии

    * **experimental result → 12322**

5294 experimental systems
    d Experimentalsysteme *npl*
    f systèmes *mpl* expérimentaux
    r экспериментальная система *f*

    * **experimental unit → 9718**

5295 experimental value
    d experimenteller Wert *m*
    f valeur *f* expérimentale
    r экспериментальный результат *m*

    * **experimental work → 4040**

5296 expiry date
    d Ablaufzeitpunkt *m*
    f date *f* d'expiration
    r дата *f* истечения срока

5297 explicit
    d explizit; ausdrücklich
    f explicite
    r явный

5298 explicit function
    d explizite Funktion *f*
    f fonction *f* explicite
    r явная функция *f*

    * **exploration → 1627**

5299 exponent; power exponent
    d Exponent *m*
    f exposant *m* [de puissance]
    r показатель *m*; экспонент *m*

5300 exponential absorption
    d exponentielle Absorption *f*
    f absorption *f* exponentielle
    r поглощение *n* по экспоненциальному закону

5301 exponential amplifier
    d exponentieller Verstärker *m*
    f amplificateur *m* exponentiel
    r экспоненциальный усилитель *m*

5302 exponential approximation
    d exponentielle Näherung *f*
    f approximation *f* exponentielle
    r экспоненциальное приближение *n*

5303 exponential curve
    d Exponentialkurve *f*
    f courbe *f* exponentielle
    r экспоненциальная кривая *f*

5304 exponential damping
    d Exponentialabklingen *n*
    f amortissement *m* exponentiel; évanouissement *m* exponentiel
    r экспоненциальное затухание *n*

5305 exponential decay
    d exponentieller Abfall *m*
    f décroissance *f* exponentielle
    r экспоненциальный распад *m*; экспоненциальное спадание *n*

5306 exponential decay time constant
    d Zeitkonstante *f* des exponentiellen Zerfalls

f constante *f* de temps du traînage exponentiel;
constante de temps de la décroissance
exponentielle
r постоянная *f* времени экспоненциального
распада

5307 exponential distortion
d exponentielle Verzerrung *f*;
Exponentialverzerrung *f*
f distorsion *f* exponentielle
r экспоненциальное искажение *n*

5308 exponential distribution
d Exponentialverteilung *f*
f distribution *f* exponentielle
r экспоненциальное распределение *n*

5309 exponential equation
d Exponentialgleichung *f*
f équation *f* exponentielle
r экспоненциальное уравнение *n*;
показательное уравнение

5310 exponential lag
d exponentielle Verzögerung *f*
f retard *m* exponentiel
r замедление *n* по экспоненциальному
закону

5311 exponentially increasing amplification
d exponentiell zunehmende Verstärkung *f*
f amplification *f* à croissance exponentielle
r экспоненциально возрастающее
усиление *n*

5312 exponential process
d Exponentialvorgang *m*
f processus *m* exponentiel
r экспоненциальный процесс *m*

5313 exponential unit
d exponentielle Einheit *f*
f unité *f* exponentielle
r экспоненциальный элемент *m*

5314 exponent of the root; root exponent
d Wurzelexponent *m*
f exposant *m* de la racine
r показатель *m* корня

5315 extended mnemonic code
d erweiterter mnemonischer Kode *m*
f code *m* mnémonique étendu
r расширенный мнемокод *m*

* **extendibility** → 5279

* **extensibility** → 5279

* **extensible** → 5280

5316 extensible measuring central
d ausdehnbare Messzentrale *f*
f centrale *f* de mesure extensible
r расширяемое измерительное устройство *n*

* **extension** → 5283

5317 extension of object zone
d Objektraumerweiterung *f*
f extension *f* d'une zone d'objet
r расширение *n* зоны объекта

5318 external action
d äußere Einwirkung *f*
f action *f* extérieure
r внешнее воздействие *n*

5319 external clock enable
d externe Taktfreigabe *f*
f libération *f* de rythme externe
r разрешение *n* внешней синхронизации

5320 external clocking
d externe Taktierung *f*
f rythme *m* externe
r внешняя синхронизация *f*

5321 external control
d externe Regelung *f*; äußere Regelung
f régulation *f* externe; conduite *f* externe
r внешний контроль *m*; наружный контроль

5322 external data
d externe Daten *pl*
f données *fpl* externes
r внешные данные *pl*

5323 external device
d Externgerät *n*
f appareil *m* externe
r внешнее устройство *n*

5324 external disturbance
d äußere Störung *f*
f perturbation *f* extérieure
r внешняя помеха *f*; внешнее возмущение *n*

5325 external drive
d äußere Erregung *f*
f excitation *f* externe; excitation extérieure
r внешний привод *m*

5326 external event
d externes Ereignis *n*
f événement *m* externe
r внешнее событие *n*

5327 **external excitation**
  d Fremderregung *f*
  f excitation *f* indépendante
  r внешнее возбуждение *n*

5328 **external feedback signal**
  d Außenrückkopplungssignal *n*
  f signal *m* de réaction extérieure
  r сигнал *m* внешней обратной связи

5329 **external function**
  d externe Funktion *f*
  f fonction *f* externe
  r внешняя функция *f*

5330 **external geometry description**
  d externe Geometriebeschreibung *f*
  f description *f* de géométrie externe
  r внешнее геометрическое описание *n*

5331 **external interpolator**
  d Außeninterpolator *m*
  f interpolateur *m* externe
  r внешний интерполятор *m*

5332 **external interrupt**
  d externe Unterbrechung *f*
  f interruption *f* externe
  r внешнее прерывание *n*

5333 **external load**
  d Außenlastwiderstand *m*
  f charge *f* externe
  r внешняя нагрузка *f*

5334 **external logic**
  d äußere Logik *f*
  f logique *f* extérieure
  r внешняя логика *f*

5335 **external magnetic field**
  d äußeres Magnetfeld *n*
  f champ *m* magnétique externe
  r внешнее магнитное поле *n*

5336 **external optical modulation**
  d äußere optische Modulation *f*
  f modulation *f* optique externe
  r внешняя оптическая модуляция *f*

5337 **external program**
  d externes Programm *n*
  f programme *m* extérieur
  r внешняя программа *f*

5338 **external sensor**
  d externer Geber *m*; externer Sensor *m*
  f capteur *m* externe; senseur *m* externe

  r датчик *m* внешней информации

5339 **external shielding**
  d äußere Abschirmung *f*
  f blindage *m* extérieur
  r внешнее экранирование *n*

5340 **external support logic**
  d externe Hilfslogikschaltung *f*
  f logique *f* auxiliaire externe
  r внешняя вспомагательная логика *f*

5341 **external voltage**
  d äußere Spannung *f*
  f tension *f* extérieure
  r внешнее напряжение *n*

5342 **extinction pulse**
  d Löschimpuls *m*
  f impulsion *f* d'extinction
  r гасящий импульс *m*

5343 **extract instruction**
  d Substitutionsbefehl *m*
  f instruction *f* d'extraction
  r команда *f* извлечения

5344 **extraction apparatus**
  d Extraktionsapparat *m*
  f appareil *m* d'extraction
  r экстакционный аппарат *m*

5345 **extraction of mark**
  d Merkmalsextraktion *f*
  f extraction *f* de marque
  r выделение *n* [характерных] признаков

5346 **extraction process**
  d Extraktionsprozess *m*;
     Extraktionsverfahren *n*
  f procédé *m* d'extraction
  r процесс *m* экстрагирования

5347 **extraction system**
  d Extraktionssystem *n*
  f système *m* d'extraction
  r экстракционная система *f*

5348 **extrapolation**
  d Extrapolation *f*
  f extrapolation *f*
  r экстаполация *f*

5349 **extremal controller; extremum controller;
      optimizing controller; peakholding
      controller**
  d Extremalwertregler *m*

  f régulateur *m* extrémal; régulateur à
   extrémum
  r экстремальный регулятор *m*

**5350 extremal system with storage of the
   extremum**
  d Extremalsystem *n* mit
   Extremwertspeicherung *f*
  f système *m* extrémal à mémorisation de
   valeur extrême
  r экстремальная система *f* с запоминанием
   экстрмума

5351 extremum conditions
  d Extremalbedingungen *fpl*
  f conditions *fpl* extrêmes
  r условия *npl* экстремума

**5352 extremum control; optimal control;
   peakholding control; optimum control;
   optimizing control**
  d optimale Regelung *f*; Optimalregelung *f*;
   Extremwertregelung *f*
  f réglage *m* optimal; réglage extrémal
  r оптимальное регулирование *n*;
   экстремальное регулирование

  * **extremum controller → 5349**

5353 extrinsic conduction
  d Störstellenleitung *f*
  f conduction *f* extrinsèque
  r несобственная проводимость *f*

F

5354 fabricating technique
d Herstellungsverfahren *n*
f procédé *m* de fabrication
r способ *m* изготовления

* **facility → 7024**

5355 factor analysis system
d Faktoranalysensystem *n*
f système *m* d'analyse des facteurs
r система *f* факторного анализа

5356 factor of demagnetization
d Entmagnetisierungsfaktor *m*
f facteur *m* de désaimantation
r коэффициент *m* размагничивания

5357 factor of safety
d Sicherheitsfaktor *m*
f coefficient *m* de sécurité
r коэффициент *m* безопасности

5358 factor value
d Faktorenwert *m*; Größe *f* des Faktors
f valeur *f* de facteur
r величина *f* фактора

5359 fading by polarization
d Polarisationsschwund *m*
f évanouissement *m* par polarisation
r поляризационное замирание *n*

5360 fading control
d Schwundregelung *f*
f réglage *m* du fading
r регулирование *n* замирания;
регулирование затухания

5361 fail-safe
d ausfallsicher; störsicher
f sûr contre perturbations
r защищенный от отказов

5362 fail-safe system
d ausfallsicheres System *n*; ausfallgeschütztes
System
f système *m* sûr contre défaillance
r защищенная от отказов система *f*

5363 fail-soft system
d fehlertolerantes System *n*; ausfallweiches

System
f système *m* flexible contre perturbations
r система *f* с постепенным отказом

5364 failure; breakdown; malfunction
d Ausfall *m*; Funktionsstörung *f*;
Fehlfunktion *f*
f défaillance *f*; panne *f*; arrêt *m*
r отказ *m*; разрушение *n*; повреждение *n*;
нарушение *n* работоспособности; сбой *m*

**5365 failure behaviour; behaviour during
breakdown**
d Ausfallverhalten *n*
f comportement *m* de défaillance
r характеристика *f* потери
работоспособности

5366 failure condition
d Störungszustand *m*
f état *m* de défaillance
r состояние *n* отказов

5367 failure detection; trouble shooting
d Fehlersuche *f*; Störungssuche *f*
f dépistage *m* des pannes
r обнаружение *n* неисправностей;
обнаружение повреждений

5368 failure probability
d Ausfallwahrscheinlichkeit *f*
f probabilité *f* de défaillance
r вероятность *f* отказа

5369 failure probability density distribution
d Ausfallwahrscheinlichkeitsdichteverteilung *f*
f distribution *f* probable des défaillances
r распределение *n* плотности вероятности
отказа

5370 failure rate
d Ausfallrate *f*
f taux *m* de défaillance; taux d'avarie
r степень *f* неисправности

5371 failure rate prediction
d Vorausberechnung *f* der Ausfallrate
f prévision *f* du taux de défaillance
r предварительный расчёт *m* интенсивности
отказов

5372 failure reason
d Ausfallursache *f*
f cause *f* de panne; origine *f* de panne
r причина *f* отказов

5373 failure test
d Ausfalltest *m*

f test *m* de défaillance
r тест *m* отказов

5374 **failure time; fault time**
d Ausfallzeit *f*
f temps *m* de panne
r время *n* простоя

5375 **fallback**
d Ersatzfunktion *f*
f fonction *f* de réserve
r запасная функция *f*

5376 **falling characteristic**
d fallende Kennlinie *f*
f caractéristique *f* décroissante
r нисходящая характеристика *f*

5377 **falling out of step**
d Außertrittfallen *n*
f décrochage *m*
r выпадение *n* из синхронизма

5378 **false control input signal**
d Falsch-Steuereingangssignal *n*
f signal *m* d'entrée de commande faux
r ложный управляющий входной сигнал *m*

5379 **false information input**
d falscher Informationseingang *m*
f entrée *f* d'information fausse
r ложный информационный вход *m*

5380 **false output signal**
d falsches Ausgangssignal *n*
f signal *m* de sortie faux
r ложный выходной сигнал *m*

5381 **false trip**
d Falschschaltung *f*; Falschbetätigung *f*;
   Falschauslösung *f*
f fausse manœuvre *f*; déclenchement *m* faux
r ложный ход *m*; ошибочное действие *n*

5382 **family of characteristics**
d Kennlinienschar *f*
f famille *f* de caractéristiques
r семейство *n* характеристик

*   **fan system** → 618

5383 **Faraday rotation automatic polarimeter**
d automatisches Polarimeter *n* mit magnetooptischer Drehung
f polarimètre *m* automatique à rotation
   magnétooptique
r автоматический поляриметр *m* с
   магнитно-оптическим вращением

5384 **far-end cross-talk attenuation**
d Gegennebensprechdämpfung *f*
f affaiblissement *m* télédiaphonique
r переходное затухание *n* на приёмном
   конце

5385 **far-field analyzer**
d Fernfeldanalysator *m*
f analyseur *m* du champ lointain
r анализатор *m* поля в дальней зоне

5386 **far-field interference pattern**
d Interferenzmuster *n* im Fernfeld
f figure *f* d'interférence dans le champ lointain
r интерференционная картина *f* в дальней
   зоне

5387 **fast-acting relay**
d schnellwirkendes Relais *n*;
   schnellansprechendes Relais;
   Schnellschaltrelais *n*
f relais *m* rapide
r быстродействующее реле *n*

5388 **fast chopper**
d schneller Zeracker *m*; Schnellzerhacker *m*
f interrupteur *m* rapide
r быстродействующий прерыватель *m*

5389 **fast coincidence circuit**
d schneller Koinzidenzkreis *m*;
   schnellwirkender Koinzidenzkreis
f circuit *m* rapide de coïncidence
r быстродействующая схема *f* совпадений;
   схема совпадения с высокой
   разрешающей способностью

5390 **fast data transmission; high-data rate;
      high-speed data transmission**
d schnelle Datenübertragung *f*
f transmission *f* rapide de données
r передача *f* данных с высокой скоростью

5391 **fast laser pulse**
d kurzer Laserimpuls *m*
f impulsion *f* courte du laser
r короткий импульс *m* лазера

5392 **fast operation speed; high-speed of
      operation**
d hohe Operationsgeschwindigkeit *f*
f rapidité *f* de fonctionnement grande; vitesse *f*
   d'opération grande
r высокое быстродействие *n*

5393 **fast response flowmeter**
d schneller Durchflussmesser *m*;
   Durchflussmesser mit kurzer Ansprechzeit

f débitmètre *m* à réponse rapide
r быстродействующий расходомер *m*;
расходомер с малой постоянной времени

5394 fast-response laser detection system
d Laserdetektionssystem *n* mit großer
Ansprechgeschwindigkeit
f système *m* laser de détection à réponse rapide
r система *f* индикации лазерного излучения
с малой инерционностью

5395 fast response laser receiver
d schnellansprechender
Laserstrahlenempfänger *m*
f récepteur *m* à laser réponse rapide
r безинерционный приёмник *m* лазерного
излучения

5396 fast response regulation algorithm
d reaktionsschneller Regelungsalgorithmus *m*
f algorithme *m* de régulation à temps de
réponse
r быстродействующий алгоритм *m*
регулирования

5397 fast signal
d schnelles Signal *n*
f signal *m* rapide
r кратковременный сигнал *m*

5398 fast stop; rapid stop
d Schnellhalt *m*
f arrêt *m* rapide
r быстрый останов *m*

5399 fast switching speed
d hohe Schaltgeschwindigkeit *f*
f vitesse *f* de commutation élevée
r высокая скорость *f* переключения

5400 fatigue
d Ermüdung *f*
f fatigue *f*
r усталость *f*; старение *n*

5401 fatigue test; fatigue trial
d Ermüdungstest *m*; Ermüdungsversuch *m*
f essai *m* de fatigue
r испытание *n* на усталость

* **fatigue trial** → 5401

* **fault breakdown** → 4448

5402 fault communication
d Fehlerübermittlung *f*
f communication *f* de défaut
r неисправная связь *f*

5403 fault-current relay protection
d Fehlerstromschutzschaltung *f*
f protection *f* par relais à courant de défaut
r релейная защита *f* от повреждений

* **fault-detection circuitry** → 5175

5404 faulted line
d gestörte Leitung *f*
f ligne *f* perturbée
r неисправная линия *f* связи

5405 fault finder
d Fehlersucher *m*; Störungssucher *m*
f détecteur *m* de pannes
r прибор *m* для отыскания повреждений

5406 fault-free hardware
d fehlerfreie Hardware *f*; fehlerfreie Technik *f*
f technique *f* sans défaut
r исправная аппаратура *f*

5407 fault indication
d Störungsmeldung *f*
f signalisation *f* de panne
r сигнализация *f* неисправности

5408 fault liability; susceptibility of failure
d Störanfälligkeit *f*
f sensibilité *f* aux perturbations; susceptibilité *f*
aux perturbations
r подверженность *f* воздействию помех

* **fault localization** → 5189

5409 fault-location instrument
d Fehlerortungsgerät *n*
f appareil *m* de dépistage de défauts
r прибор *m* для определения места
повреждения

* **fault time** → 5374

5410 fault tolerance
d Fehlertoleranz *f*; Fehlerzulässigkeit *f*
f tolérance *f* de défauts
r нечувствительность *f* к отказам

5411 fault tolerant computer system
d fehlertolerantes Rechnersystem *n*
f système *m* de calculateur tolérant des défauts
r нечувствительная к отказам
вычислительная система *f*

5412 faulty element
d fehlerhaftes Element *n*; gestörtes Element
f élément *m* défectueux; élément perturbé
r неисправный элемент *m*

5413 **faulty manipulation**
 d fehlerhafte Handhabung f
 f manipulation f défectueuse
 r ошибочная манипуляция f

5414 **faulty module**
 d fehlerhafter Modul m; schadhafter Modul
 f module m défectueux
 r неисправный модуль m

 * **faulty operator intervention → 8983**

5415 **faulty signal**
 d fehlerhaftes Signal n
 f signal m fautif
 r ошибочный сигнал m

5416 **feasibility conditions**
 d Bedingungen fpl der Realisierbarkeit;
    Realisierbarkeitsbedingungen fpl
 f conditions fpl de réalisation
 r условия npl осуществимости

5417 **feasible basis; admissible basis**
 d zulässige Basis f
 f base f admissible; base acceptable
 r допустимая база f

 * **feasible solution → 532**

5418 **feasible system**
 d realisierbares System n
 f système m réslisable
 r осуществимая система f

5419 **feature**
 d Merkmal n; wesentliche Eigenschaft f
 f propriété f; caractéristique f
 r особенность f; признак m

5420 **feature v**
 d aufweisen; kennzeichnen
 f caractériser; présenter
 r характеризовать

 * **feature of the computer → 2896**

 * **feed → 5447**

5421 **feed apparatus**
 d Aufgabeapparat m; Speiser m
 f appareil m alimentateur
 r механизм m подачи; питатель m

 * **feedback → 1541**

5422 **feedback adjustment**
 d Rückkopplungseinstellung f

 f mise f au point de la réaction
 r регулировка f обратной связи

5423 **feedback amplifier**
 d Gegenkopplungsverstärker m
 f amplificateur m à contre-réaction
 r усилитель m с обратной связью

5424 **feedback capacitor**
 d Rückkopplungskondensator m
 f condensateur m de réaction
 r конденсатор m обратной связи

5425 **feedback channel**
 d Rück[kopplungs]kanal m
 f canal m de réaction
 r канал m обратной связи

5426 **feedback circuit**
 d Rückkopplungsschaltung f
 f circuit m de réaction
 r схема f с обратной связью

5427 **feedback collector capacitance**
 d Rückwirkungskapazität f
 f capacité f de réaction
 r ёмкость f обратной связи

5428 **feedback control**
 d Rückkopplungsregelung f
 f commande f en boucle fermée; réglage m à
    réaction
 r регулирование n по замкнотому циклу

5429 **feedback controller**
 d Rückkopplungsregler m
 f régulateur m à reaction
 r регулятор m с обратными связями

5430 **feedback control system**
 d Rückkopplungsregleranlage f;
    Rückkopplungssteuerungssystem n
 f système m de réglage en boucle fermée
 r система f управления с обратной связью

5431 **feedback elements**
 d Rückführungselemente npl
 f éléments mpl de réaction
 r элементы mpl обратной связи

5432 **feedback factor**
 d Rückführungskoeffizient m
 f taux m de réaction
 r коэффициент m обратной связи

5433 **feedback lag**
 d Rückkopplungsverzögerung f
 f retard m de réaction
 r запаздывание n в цепи обратной связи

5434 feedback limiter
   d Rückkopplungsbegrenzer *m*
   f limiteur *m* de réaction
   r ограничитель *m* обратной связи

5435 feedback-regulated rectifier
   d rückkopplungsstabilisierter Gleichrichter *m*
   f redresseur *m* à stabilisation automatique par réaction
   r выпрямитель *m* с обратной связью

5436 feedback sensor
   d Rückkopplungssensor *m*
   f capteur *m* à réaction
   r датчик *m* обратной связи

5437 feedback shift register
   d rückgekoppeltes Schieberegister *n*
   f registre *m* de décalage à réaction
   r сдвигающий регистр *m* с обратной связью

5438 feedback-stabilized amplifier
   d rückkopplungsstabilisierter Verstärker *m*
   f amplificateur *m* stabilisé à réaction
   r стабилизированный усилитель *m* с обратной связью

5439 feedback-system transient response
   d Übertragungscharakteristik *f* des Rückführungssystems
   f réponse *f* transitoire du système à l'asservissement
   r переходная характеристика *f* системы обратной связи

5440 feedback transfer function
   d Übertragungsfunktion *f* Regelkreises
   f fonction *f* de transfer de réaction
   r передаточная функция *f* замкнутой системы [с обратной связью]

5441 feedback voltage ratio
   d Rückkopplungsspannungsverhältnis *n*
   f gain *m* inverse en tension
   r соотношение *n* обратной связи и напряжения

5442 feed change
   d Vorschubwechsel *m*
   f changement *m* de l'avance
   r изменение *n* [скорости] подачи

5443 feed check
   d Transportprüfung *f*
   f vérification *f* de transport
   r контроль *m* подачи

5444 feed controller

**   *d* Speiseregler *m*; Vorschubregler *m***
   f régulateur *m* d'alimentation; régulateur d'avance
   r регулятор *m* питания; регулятор подачи

5445 feed device of work pieces
   d Werkstückzubringeenrichtung *f*
   f dispositif *m* d'apport de pièces à usiner; dispositif d'alimentation de pièces à usiner
   r устройство *n* подачи обрабатываемых деталей

5446 feedforward
   d Vorwärtswirkung *f*
   f action *f* directe
   r прямая связь *f*

5447 feed[ing]
   d Speisung *f*; Vorschub *m*
   f alimentation *f*; avance *m*
   r питание *n*; подача *f*

5448 feeding circuit
   d Speiseschaltung *f*
   f circuit *m* d'alimentation
   r питающая схема *f*

5449 feeding device
   d Zuführungsvorrichtung *f*
   f dispositif *m* d'alimentation
   r загрузочное устройство *n*

5450 feeding equipment
   d Beschickungsvorrichtung *f*
   f dispositif *m* de chargage
   r загрузочно-разгрузочное устройство *n*

* feeding of parts → 663

5451 feeding station
   d Zuführungsstation *f*
   f station *f* d'alimentation
   r питающая станция *f*

5452 feed[ing] system
   d Beschickungssystem *n*; Speisevorrichtung *f*
   f système *m* de charge; dispositif *m* d'alimentation
   r система *f* подачи

5453 feed interlock
   d Zuführungssperre *f*
   f blocage *m* d'alimentation
   r блокировка *f* питания

* feed system → 5452

5454 ferrite core transformer
   d Transformator *m* mit Ferritkern

f transformateur *m* à noyau de ferrite
r трансформатор *m* с ферритовым
   сердечником

5455 ferrite-Hall-generator
d Ferrit-Hallgenerator *m*
f générateur *m* de Hall à ferrite
r ферритовый генератор *m* Холла

5456 ferrite transfluxor
d Ferrittransfluxor *m*
f transfluxor *m* en ferrite
r ферритовый трансфлюксор *m*

5457 ferrodynamic relay
d ferrodynamisches Relais *n*
f relais *m* ferrodynamique
r ферродинамическое реле *n*

5458 ferroresonant computing circuit
d Ferroresonanzrechenschaltung *f*
f circuit *m* calculateur à ferrorésonance
r феррорезонансная вычислительная
   схема *f*

5459 ferroresonant operation
d Ferroresonanzbetrieb *m*;
   Ferroresonanzwirkung *f*
f opération *f* ferroresonnante
r феррорезонансное действие *n*

5460 fiber optic computer interface
d Lichtwellenleiter-Rechner-Interface *n*
f interface *f* fibre optique-ordinateur
r световодный интерфейс *m* компьютера

5461 fiber optic data transmission system
d optisches Datenübertragungssystem *n*
f système *m* de transmission de données par
   fibres optiques
r оптическая система *f* передачи данных

5462 fibre optics application
d Lichtleiteranwendung *f*
f application *f* du conduit de lumière
r применение *n* светопровода

**5463 Fick's diffusion law; Fick's law of
       diffusion**
d Ficksches Gesetz *n*; Ficksches
   Diffusionsgesetz *n*
f loi *f* [de diffusion] de Fick
r закон *m* [диффузии] Фикка

* **Fick's law of diffusion → 5463**

5464 fictitious load; phantom load
d fiktive Belastung *f*

f charge *f* fictive
r фиктивная нагрузка *f*

5465 fidelity of information transmission
d Genauigkeit *f* der Informationsübertragung
f fidélité *f* de transmission de l'information
r точность *f* передачи информации

5466 fidelity of reproduction
d Wiedergabetreue *f*
f fidélité *f* de reproduction
r точность *f* воспроизведения

5467 field boundary
d Feldbegrenzung *f*
f délimitation *f* de la zone
r граница *f* поля

5468 field break switch
d Feldunterbrechungsschalter *m*
f interrupteur *m* d'excitation
r переключатель *m* возбуждения на
   шунтирующее сопротивление

5469 field definition
d Feldbestimmung *f*
f définition *f* de la zone
r определение *n* напряженности поля

5470 field pole
d Feldpol *m*
f pôle *m* du champ
r полюс *m* возбуждения

5471 field reduction
d Abschwächung *f* des Feldes
f réduction *f* de champ
r ослабление *n* поля

5472 field service
d Außendienst *m*; Kundendienst *m* im
   Einsatzbereich
f service *m* après-vente
r эксплуатационное обслуживание *n*

5473 field simulation
d Feldmodellierung *f*
f simulation *f* de champ
r моделирование *n* поля

5474 field test
d Einsatztest *m*; Einsatzerprobung *f*
f test *m* de mise en œuvre
r испытание *n* в условиях эксплуатации

5475 figurative constant
d vorgesehene Konstante *f*; figurative
   Konstante

f constante *f* figurative
r символьная константа *f*

* **figure of merit** → 10474

5476 file
d Datei *f*; File *n*
f fichier *m*
r файл *m*

5477 file control
d Dateisteuerung *f*
f commande *f* par le fichier
r управление *n* файлом

5478 file description attribute
d Dateibeschreibungsattribut *n*
f attribut *m* de description d'un fichier
r описатель *m* файла

5479 file format
d Dateiformat *n*
f format *m* de fichier
r формат *m* файла

5480 file management system
d Dateiverwaltungssystem *n*
f système-management *m* de fichier
r система *f* управления файлами

5481 file protection function
d Dateischutzfunktion *f*
f fonction *f* de protection d'un fichier
r функция *f* защиты файла

5482 film dosimetry
d Filmdosimetrie *f*
f dosimétrie *f* de film
r плёночная дозиметрия *f*

5483 filter attenuation
d Filterdämpfung *f*
f affaiblissement *m* de filtre
r затухание *n* фильтра

5484 filter characteristic
d Filtercharakteristik *f*
f réponse *f* de filtre; caractéristique *f* de filtre
r характеристика *f* фильтра

5485 filter circuit
d Filterkreis *m*
f circuit *m* de filtre
r фильтрующий контур *m*

5486 filtering
d Siebung *f*
f filtrage *m*
r фильтрация *f*

5487 filter of finite memory; finite-memory filter
d Filter *n* mit begrenztem Gedächtnis
f filtre *m* avec mémoire limitée
r фильтр *m* с ограниченной памятью

5488 filter of infinite memory; infinite-memory filter
d Filter *n* mit unbegrenztem Gedächtnis
f filtre *m* avec mémoire indéfinie
r фильтр *m* с неограниченной памятью

5489 filter photometer
d Filterfotometer *n*
f photomètre *m* à filtre
r фильтр-фотометр *m*

5490 filter range
d Filterbereich *m*
f bande *f* du filtre; gamme *f* du filtre
r диапазон *m* пропускания фильтра

5491 filter synthesis
d Filtersynthese *f*
f synthèse *f* de filtre
r синтез *m* фильтра

5492 filter synthesis program
d Filtersyntheseprogramm *n*; Programm *n* zur Filtersynthese
f programme *m* de synthèse de filtre
r программа *f* для синтеза фильтров

5493 filter with time delay
d Verzögerungsfilter *n*
f filtre *m* à retard
r фильтр *m* с запаздыванием

5494 filtration plant
d Filteranlage *f*
f installation *f* de filtrage
r установка *f* для фильтрации

5495 final adjustment
d Endeinstellung *f*; Fertigeinstellung *f*
f mise *f* au point finale; réglage *m* final
r окончательная настройка *f*

5496 final control condition
d Bedingung *f* der Endregelung; Zustand *m* der Endregelung; Sollwert *m* der Regelgröße
f condition *f* de réglage final
r установившееся значение *n* регулируемого параметра

5497 final control element
d Stelleinrichtung *f*; Stellwerk *n*; Stellorgan *n*; Steller *m*; Stellzeug *n*

f organe *m* de réglage final; organe exécutif
r исполнительный орган *m*; выходной
управляющий элемент *m*;
исполнительный элемент
системы регулирования

5498 final controlled variable
d Endregelgröße *f*
f grandeur *f* réglée finale
r установившееся значение *n* регулируемой
переменной

5499 final dog indicator
d Endanschlaggeber *m*; Endlagengeber *m*
f capteur *m* de butée de fin de course
r датчик *m* концевого положения

5500 final inspection
d Endprüfung *f*
f inspection *f* finale
r сдаточное испытание *n*

* **final position** → 5035

5501 final position initiators
d Endlageninitiatoren *mpl*
f initiateurs *mpl* de positions finales
r инициаторы *mpl* положения

5502 final position monitoring
d Endlagenüberwachung *f*
f contrôle *m* de fin de course
r контроль *m* конечного положения

5503 final quantity
d Endgröße *f*
f quantité *f* finale
r конечная величина *f*

5504 final regulation element
d Endregelunsorgan *n*
f organe *m* de réglage final
r предельный регулятор *m*

5505 final stage
d Endstufe *f*
f étage *m* final
r оконечная ступень *f*; оконечный каскад *m*

5506 final value
d Endwert *m*
f valeur *f* finale
r конечное значение *n*

5507 final value of amplification
d Endwert *m* der Verstärkung
f valeur *f* finale de l'amplification
r предельное значение *n* усиления

5508 final value theorem
d Endwertsatz *m*
f théorème *m* de valeur finale
r теорема *f* о конечном значении

* **finding circuit** → 11230

5509 fine adjustment
d Feinregelung *f*; Feineinstellung *f*
f réglage *m* précis
r точное регулирование *n*

5510 fine adjustment of manipulator system
d Feineinstellung *f* des Manipulatorsystems
f réglage *m* de précision de système de
manipulateur
r точная настройка *f* системы манипулятора

5511 fine positioning of effector
d Effektorfeinpositionierung *f*;
Feinpositionierung eines Effektors
f positionnement *m* fin d'un effecteur
r точное позиционирование *n* эффектора

* **finish** *v* → 2746

5512 finite
d endlich
f fini
r конечный

5513 finite automaton
d endlicher Automat *m*
f automate *m* fini
r конечный автомат *m*

5514 finite degree of stability
d endlicher Stabilitätsgrad *m*
f degré *m* fini de stabilité
r конечная степень *f* устойчивости

5515 finite differences
d endliche Differenzen *fpl*
f différences *fpl* finies
r конечные разности *fpl*

5516 finite-dimensional
d endlich-dimensional
f à dimension finie
r конечно размерный

5517 finite dimensional vector space
d endlich-dimensionaler Vektorraum *m*
f espace *f* vectorielle à dimension finie
r конечномерное векторное пространство *n*

5518 finite induction
d vollständige Induktion *f*

f induction f complète
r полная индукция f

* finite-memory filter → 5487

5519 finite pulse width
d endliche Impulsbreite f; endliche Breite f des Impulses
f durée f finie d'impulsion; largeur f finie d'impulsion
r конечная длительность f импульса

5520 finite sequence
d endliche Folge f
f séquence f finie
r конечная последовательность f

5521 finite time instant
d endliches Zeitmoment n
f moment m fini de temps
r конечный момент m времени

5522 firing pulse
d Auslöseimpuls m; Zündimpuls m
f impulsion f d'amorçage; impulsion d'excitation
r пусковой импульс m

5523 firmware
d Firmware f
f firmware m
r специализированное программное обеспечение n

5524 first approximation
d erste Annäherung f
f première approximation f
r первое приближение n

5525 first approximation method
d Methode f der ersten Annäherung
f méthode f de première approximation
r метод m первого приближения

5526 first clock pulse
d erster Taktimpuls m
f première impulsion f de rythme
r первый тактовый импульс m

5527 first harmonic; fundamental harmonic
d erste Harmonische f
f harmonique m fondamentale
r основная гармоника f; первая гармоника

5528 first order system
d System n erster Ordnung
f système m de premier ordre
r система f первого порядка

5529 first visual system generation
d erste Sichtsystemgeneration f; erste Generation f von Sichtsystemen
f génération f première d'un système visuel
r первое поколение n визуальных систем

5530 fission pulse
d Spaltungsimpuls m
f impulsion f de comptage due à la fission
r импульс m деления

* fitting out → 5125

5531 fixed command control
d Festwertregelung f
f réglage m de maintien; régulation f de maintien
r стабилизирующее управление n; регулирование n с фиксированным сигналом управления

5532 fixed cycle operation
d Taktgeberbetrieb m; Zeitgeberbetrieb m
f opération f à durée définie
r работа f с постоянным циклом

5533 fixed focus pyrometer
d Pyrometer n mit konstanter Brennweite
f pyromètre m à foyer fixe
r пирометр m с постоянным фокусом; пирометр Фостера

5534 fixed memory; permanent memory
d Totspeicher m; Fest[wert]speicher m; Permanentspeicher m
f mémoire f morte; mémoire fixe; mémoire permanente
r постоянная память f

5535 fixed memory control
d Festwertspeichersteuerung f
f commande f à mémoire fixe
r управление n постоянным запоминающим устройством

5536 fixed period
d konstante Periode f
f période f constante
r постоянный период m

5537 fixed program control
d Festprogrammsteuerung f
f commande f à programme fixe
r жёсткопрограммированная система f управления

5538 fixed range
d fester Bereich m; festgesetzter Bereich

f gamme *f* fixée
r постоянно установленный диапазон *m*

5539 fixed set point regulation; regulation with fixed set point
d Regelung *f* mit festem Sollwert
f réglage *m* à valeur constante
r регулирование *n* для стабилизации параметра

5540 fixed structure
d feste Struktur *f*
f structure *f* fixe
r постоянная сруктура *f*

5541 fixed terminal state
d fester Endzustand *m*
f état *m* final fixe; état final stable
r закрепленное конечное состояние *n*

* fixed time lag → 3856

5542 fixed trip; locked trip
d gesperrte Auslösung *f*
f déclenchement *m* verrouillé
r замкнутое расцепляющее устройство *n*

5543 fixed value
d fester Wert *m*
f valeur *f* fixe
r стационарное значение *n*

5544 fixed vertex
d fester Knoten *m*
f nœud *m* fixe
r постоянный узел *m*

5545 fixed wiring
d feste Verdrahtung *f*
f câblage *m* fixe
r жёсткий монтаж *m*

* fixture → 7024

5546 flame photometer
d Flammenfotometer *n*; Flammenlichtstärkemesser *m*
f photomètre *m* à flamme
r пламенный фотометр *m*

5547 flamestat control
d Flammenregelung *f*
f réglage *m* de flamme
r регулирование *n* пламени

5548 flashing signal
d Blinksignal *n*; Flackerzeichen *n*
f signal *m* à éclats

r проблесковый сигнал *m*

5549 flat pulse; flat-topped pulse
d flacher Impuls *m*
f impulsion *f* plate
r импульс *m* с плоской вершиной; плоский импульс

5550 flat switching circuit structure
d Flachstruktur *f* des Schaltstromkreises
f structure *f* plate du circuit de relais
r структура *f* плоской схемы переключения

* flat-topped pulse → 5549

5551 flexible assembly process
d flexibler Montageprozess *m*
f processus *m* de montage flexible
r гибкий сборочный процесс *m*

5552 flexible assembly system
d flexibles Montagesystem *n*
f système *m* de montage flexible
r гибкая сборочная система *f*

5553 flexible automatic assembly cell
d flexible automatische Montagezelle *f*
f cellule *f* d'assemblage automatique flexible
r гибкая ячейка *f* автоматической сборки

5554 flexible automatic assembly system
d flexibles automatisches Montagesystem *n*
f système *m* d'assemblage automatique flexible; système de montage automatique flexible
r гибкая система *f* автоматической сборки

5555 flexible automation
d flexible Automatisierung *f*
f automatisation *f* flexible
r гибкая автоматизация *f*

5556 flexible automation medium
d flexibles Automatisierungsmittel *n*
f moyen *m* d'automatisation flexible
r гибкое средство *n* автоматизации

5557 flexible fabrication system
d flexibles Fertigungssystem *n*
f système *m* de fabrication fexible
r гибкая система *f* производства

5558 flexible feed technique
d flexible Zuführungstechnik *f*
f technique *f* d'acheminement flexible
r гибкая техника *f* подачи

5559 flexible flow process
d flexible Fertigungslinie *f*

f ligne *f* de production flexible
r гибкая производственная линия *f*

5560 flexible free-programmable handling device
d flexibles freiprogrammierbares Handhabungsgerät *n*
f dispositif *m* de manutention programmable librement flexible
r свободное программируемое манипуляционное устройство *n*

5561 flexible gripper system
d flexibles Greifersystem *n*
f système *m* de grappin flexible
r гибкое захватное устройство *n*

5562 flexible handling device
d flexible Handhabeeinrichtung *f*
f dispositif *m* de manutention flexible
r гибкое устройство *n* манипулирования

5563 flexible handling system
d flexibles Handhabungssystem *n*
f système *m* de manutention fexible
r гибкая манипуляционная система *f*

5564 flexible handling technique
d flexible Handhabetechnik *f*
f technique *f* de manutention flexible
r гибкая техника *f* манипулирования

5565 flexible instruction system
d flexibles Befehlssystem *n*
f système *m* flexible de commandes
r гибкая система *f* команд

5566 flexible integrated tool system
d anpassungsfähiges integriertes Werkzeugsystem *n*
f système *m* d'outil intégré fexible
r гибкая совмещенная система *f* инструментов

5567 flexible linkage
d flexible Verkettung *f*
f enchaînement *m* flexible
r гибкая связь *f*

5568 flexible multilogic
d flexible Multilogik *f*
f logique *f* multiple flexible
r гибкая многозначная логика *f*

5569 flexible production complex
d flexibler Produktionskomplex *m*
f complexe *m* de production flexible
r гибкий производственный комплекс *m*

5570 flexible production module
d flexibler Produktionsmodul *m*
f module *m* de production flexible
r гибкий производственный модуль *m*

5571 flexible programming
d flexibles Programmieren *n*
f programmation *f* flexible
r гибкое программирование *n*

5572 flicker effect
d Flackereffekt *m*; Flimmereffekt *m*; Funkeleffekt *m*
f scintillation *f*; papillotement *m*
r фликер-эффект *m*

5573 flicker frequency
d Flimmerfrequenz *f*
f fréquence *f* de papillotement
r частота *f* мигания

5574 floating action
d Integralwirkung *f*; Nachstellwirkung *f*
f action *f* intégrale
r астатическое действие *n*

* **floating-action controller → 1052**

5575 floating component
d Integralkomponente *f*
f composante *f* intégrale
r астатический элемент *m*

5576 floating control; null offset
d Schwimmregelung *f*; astatische Regelung *f*
f réglage *m* flottant; régulation *f* astatique
r астатическое регулирование *n*

5577 floating-point instruction
d Gleitkommabefehl *m*
f instruction *f* à virgule flottante
r команда *f* выполнения операции с плавающей запятой

5578 floating-point method
d Gleitkommamethode *f*; Gleitkommaverfahren *n*
f méthode *f* de virgule flottante
r метод *m* плавающей запятой

5579 floating-point representation
d Gleitkommadarstellung *f*; halblogarithmische Zahlendarstellung *f*
f notation *f* à virgule flottante
r представление *n* чисел в системе с плавающей запятой

5580 floating potential
d Schwimmspannung *f*

f potentiel *m* flottant
r свободный потенциал *m*

5581 floating rate; floating speed
d Stellgeschwindigkeit *f*
f vitesse *f* d'ajustage
r скорость *f* астатического действия

*   **floating speed → 5581**

5582 floating state; high-impedance state
d Floating-Zustand *m*; inaktiver Zustand *m*
f état *m* flottant; état à impédance élevée
r высокоимпедансное состояние *n*

5583 floatless liquid-level controller
d schwimmerloser Niveauregler *m*
f régulateur *m* de niveau sans flotteur
r беспоплавковый регулятор *m* уровня

5584 float level gauge
d Flüssigkeitsniveaumesswandler *m*;
   Schwimmerhöheregelungsgeber *m*
f capteur *m* à flotteur de niveau
r поплавковый уровнемер *m*

5585 float-operated flowmeter
d Schwimmerdurchflussmesser *m*
f débitmètre *m* à flotteur
r поплавковый расходомер *m*

5586 float-operated pressure gauge
d Schwimmermanometer *n*;
   Schwimmerdruckmesser *m*
f manomètre *m* à flotteur
r поплаковый манометр *m*

5587 flow calorimetry
d Durchflusskalorimetrie *f*
f calorimétrie *f* à débit
r проточная калориметрия *f*

5588 flow capacity
d Durchflusskapazität *f*
f pouvoir *m* de débit
r расход *m*

5589 flow coefficient
d Durchflusszahl *f*
f coefficient *m* de débit
r коэффициент *m* расхода

*   **flow control→ 5603**

5590 flow controller
d Durchflussmengenmesser *m*
f régulateur *m* du débit
r регулятор *m* расхода

5591 flow density
d Durchflussdichte *f*; Strömungsdichte *f*
f débit *m* spécifique
r удельный расход *m*

5592 flow element
d Durchflusselement *n*
f élément *m* de débit
r элемент *m* расхода

5593 flow gauge
d Durchflussmessgerät *n*; Durchflussmesser *m*
f débitmètre *m*
r расходомер *m*; реометр *m*

5594 flow guard
d Durchflusswächter *m*
f garde *m* de débit
r прибор *m* для контроля расхода

5595 flow indicator
d Durchflussmengenanzeiger *m*
f indicateur *m* de débit
r индикатор *m* расхода; указатель *m* расхода

5596 flow measuring instrument
d Durchflussmengenmessgerät *n*
f appareil *m* de mesure de débit
r прибор *m* для измерения расхода
   жидкости

5597 flow meter for liquid metals
d Durchflussmengenmesser *m* für flüssige
   Metalle
f débitmètre *m* pour les métaux liquides
r расходомер *m* для жидких металлов

5598 flow meter with pneumatic transmitter
d Durchflussmengenmesser *m* mit
   pneumatischem Geber
f débitmètre *m* à transmetteur pneumatique
r расходомер *m* с пневматическим
   датчиком

5599 flow of information
d Informationsfluss *m*
f flux *m* d'informations
r поток *m* информации

5600 flow process
d Flussprozess *m*
f procédé *m* de flux
r процесс *m* потока

5601 flow proportional counter
d Durchflussproportionalzähler *m*
f compteur *m* proportionnel au flux
r поточный пропорциональный счётчик *m*

5602 flow-pulsation damping system
  d Dämpfer *m* der Flusspulsierung
  f système *m* amortisseur des pulsations de flux de courant; système amortisseur des oscillations pulsatoires
  r пульсирующая расходомерная система *f* с затуханием

5603 flow [rate] control; basic rate system
  d Durchflussmengenregelung *f*
  f réglage *m* de débit; contrôle *m* de débit
  r регулирование *n* расхода

5604 flow rate profile
  d Durchsatzprofil *n*
  f profil *m* du débit
  r профиль *m* расхода

5605 flow ratio control
  d Verhältnisregelung *f*; Durchflussverhältnisregelung *f*
  f commande *f* du rapport de débits
  r регулирование *n* коэффициента расхода

5606 flow regulator
  d Durchflussregler *m*
  f régulateur *m* du débit
  r регулятор *m* потока

5607 flow sensor
  d Strömungssensor *m*
  f senseur *m* d'écoulement
  r струйный сенсор *m*

  * **flow sheet → 10166**

5608 flow-through transmitter
  d Durchströmungsgeber *m*
  f transmetteur *m* de passage
  r датчик *m* протекания

5609 flow value
  d Durchflussgröße *f*; Durchflusswert *m*
  f valeur *f* du courant
  r величина *f* расхода

5610 fluctuating signal
  d schwankendes Signal *n*
  f signal *m* fluctuant; signal variable
  r пульсирующий сигнал *m*

5611 fluctuation of amplitude
  d Amplitudenschwankung *f*
  f fluctuation *f* d'amplitude
  r колебание *n* амплитуды

5612 fluctuation of density
  d Dichteschwankung *f*

  f fluctuation *f* de densité
  r флуктуация *f* плотности

5613 flue-gas analyzer
  d Rauchgasprüfer *m*
  f analyseur *m* des gaz d'échappment
  r анализатор *m* дымовых газов

5614 fluid distance sensor
  d fluidischer Abstandssensor *m*
  f senseur *m* de distance fluide
  r гидравлический сенсор *m* расстояния

5615 fluid-flow control
  d Flüssigkeitsstromsteuerung *f*
  f commande *f* du flux de fluide
  r регулирование *n* потока жидкости

5616 fluid-logic system
  d fluidisches Verknüpfungssystem *n*
  f système *m* logique fluidiquer
  r логическая система *f* на струйных элементах

5617 fluid network analyzer
  d Flüssigkeitsnetzwerkanalysator *m*
  f analysateur *m* à circuits fluidiques
  r анализатор *m* жидкостных и газовых сетей

5618 fluid pressure transducer
  d Flüssigkeitsdruckgeber *m*
  f capteur *m* de pression de fluide
  r датчик *m* давления жидкости

5619 fluid systems
  d hydraulische Regelsysteme *npl*
  f systèmes *mpl* de réglage hydrauliques
  r гидравлические системы *fpl* регулирования

5620 fluorometer
  d Fluorometer *n*; Fluoreszenzmesser *m*
  f fluoromètre *m*
  r флуорометр *m*

5621 flutter effect
  d Flattereffekt *m*
  f flutter-effet *m*; effet *m* vibratoire
  r вибрация *f*; флаттер *m*

5622 flyback pulses
  d Rücklaufimpulse *mpl*
  f impulsions *fpl* de retour du spot
  r импульсы *mpl* от обратного хода развёртки

5623 flywheel synchronization
  d Schwungradsynchronisation *f*; Kompensationssynchronisation *f*

f synchronisation *f* à volant électronique
r инерционная синхронизация *f*

5624 focusing acoustic system
   d Fokussierungsschallsystem *n*
   f système *m* acoustique focalisant
   r фокусирующая акустическая система *f*

5625 follower
   d Folge[r]stufe *f*
   f suiveur *m*
   r повторитель *m*; следящий элемент *m*

5626 follower controller
   d Folgeregler *m*
   f régulateur *m* de poursuite
   r контроллер *m* следящего механизма

5627 follow-on automatic mechanism
   d Folgeautomatik *f*
   f mécanisme *m* automatique à galet
   r автоматическое следящее звено *n*

5628 follow-up control
   d Folgeregelung *f*
   f réglage *m* en cascade
   r следящее регулирование *n*; каскадное регулирование

5629 follow-up controller; servofollower; servoregulator
   d Folgeregler *m*; Servoregler *m*
   f régulateur *m* en cascade; servorégulateur *m*
   r следящий регулятор *m*

5630 follow-up system; servosystem
   d Nachlaufregelungssystem *n*; Folgeregelungssystem *n*; Servosystem *n*
   f système *m* de servocommande; système suiveur; servosystème *m*
   r следящая система *f*; сервомеханизм *m*; сервосистема *f*

5631 forbidden band
   d verbotenes Energieband *n*
   f bande *f* interdite
   r запрещенный уровень *m*

5632 forbidden-combination check
   d Überprüfung *f* auf verbotene Kombinationen
   f essai *m* de combinaisons interdites
   r проверка *f* на появление недопустимых кодовых операций

5633 forbidden increment
   d verbotenes Inkrement *n*
   f incrément *m* défendu
   r недопустимое приращение *n*

5634 force-balanced potentiometer
   d Potentiometer *n* mit Kräfteausgleich; Kräfteausgleichpotentiometer *n*
   f potentiomètre *m* à équilibrage forcé; potentiomètre d'opposition
   r компенсированный потенциометр *m*; потенциометр равновесия сил

5635 force-balance pressure gauge
   d Kompensationsdruckgeber *m*
   f capteur *m* de pression à compensation
   r компенсированный датчик *m* давления

5636 force-balance regulator
   d Kräftvergleichsregler *m*
   f régulateur *m* de compensation
   r регулятор *m* силового равновесия

5637 force-balance transducer
   d Kräftegleichgewichtgeber *m*
   f trasducteur *m* d'équilibre de forces; capteur *m* d'opposition
   r датчик *m* с уравновешенным динамометрическим элементом

5638 forced-air cooling system
   d Kühlsystem *n* mit Zwangsbelüftung
   f système *m* de refroidissement à ventilation forcée
   r система *f* с принудительным воздушным охлаждением

5639 forced component
   d erzwungener Anteil *m*
   f réponse *f* forcée; composante *f* forcée
   r вынужденная составляющая *f*

5640 forced cooling
   d Zwangskühlung *f*
   f refroidissement *m* forcé
   r принудительное охлаждение *n*

5641 forced linearization
   d erzwungene Linearisierung *f*
   f linéarisation *f* forcée
   r принудительная линеаризация *f*

5642 forced oscillations
   d erzwungene Schwingungen *fpl*
   f oscillations *fpl* forcées
   r вынужденные колебания *npl*

*   **forced regime** → 5644

5643 forced response
   d erzwungene Reaktion *f*
   f réaction *f* forcée
   r вынужденная реакция *f*

5644 forced state; forced regime
d erzwungener Zustand *m*
f régime *m* forcé
r вынужденный режим *m*

5645 forced ventilation cooling
d Druckluftkühlung *f*
f refroidissement *m* par ventilation forcée
r принудительное воздушное охлаждение *n*

5646 force feedback
d Kraftrückführung *f*
f réaction *f* de force
r силовая обратная связь *f*

5647 force measurement device; force measurement installation
d Kraftmesseinrichtung *f*
f installation *f* de mesurage de force
r силоизмерительное устройство *n*

* force measurement installation → 5647

5648 force of inertia; mass force
d Trägheitskraft *f*; Massenkraft *f*
f force *f* d'inertie; force de masse
r сила *f* инерции

5649 force sensor system
d Kraftsensorsystem *n*
f système *m* de senseur de force
r система *f* датчиков усилия

5650 forcing function
d Störfunktion *f*
f fonction *f* de perturbations
r мешающая функция *f*

5651 foreground processing
d Vordergrundverarbeitung *f*
f traitement *m* en premier plan
r приоритетная обработка *f*

5652 fork control; tuning fork control
d Stimmgabelsteuerung *f*
f pilotage *m* par diapason
r камертонная стабилизация *f*

5653 formal logic
d formale Logik *f*
f logique *f* formelle
r формальная логика *f*

5654 formal parameter
d Formalparameter *m*
f paramètre *m* formel
r формальный параметр *m*

5655 formal system
d formales System *n*
f système *m* formel
r формальная система *f*

5656 forming unit
d Formungselement *n*
f élément *m* formateur
r формирующее звено *n*

5657 formula translator
d Formelübersetzer *m*
f traducteur *m* de formules
r преобразователь *m* формул

5658 fortuitous distortion; irregular distortion
d Zufallsverzerrung *f*
f distorsion *f* fortuite; distorsion irrégulière; distorsion accidentelle; distorsion aléatoire
r случайное искажение *n*

5659 forward channel; forward path
d Vorwärtspfad *m*
f chaîne *f* d'action
r канал *m* прямой связи

5660 forward controlling element
d Vorwärtssteuerungsglied *n*
f organe *m* de réglage direct
r управляющий элемент *m* прямой связи

* forward path → 5659

5661 forward recovery time
d Durchlasserholungszeit *f*
f temps *m* de recouvrement en sens direct
r начальное время *n* восстановления

5662 forward resistance
d Durchlasswiderstand *m*; Fließwiderstand *m*
f résistance *f* directe
r прямое сопротивление *n*

5663 forward signal
d Regelbefehl *m*; Anregungsgröße *f*
f signal *m* d'action
r сигнал *m* прямой цепи воздействия

5664 forward voltage
d Spannung *f* in Flußrichtung
f tension *f* directe
r прямое напряжение *n*

5665 Fourier analyzer; harmonic analyzer
d harmonischer Analysator *m*; Oberwellenanalysator *m*
f analyseur *m* harmonique
r гармонический анализатор *m*

5666 **Fourier expansion**
  d Fouriersche Reihenentwicklung *f*
  f développement *m* en série de Fourier
  r разложение *n* в ряд Фурье

5667 **Fourier transformation**
  d Fouriersche Transformation *f*
  f transformation *f* de Fourier
  r преобразование *n* Фурье

5668 **four level generator**
  d Vierpegelgenerator *m*
  f générateur *m* à quatre niveaux
  r четырёхуровневый генератор *m*

5669 **four level scheme**
  d Vierpegelanordnung *f*; Viertermschema *n*
  f disposition *f* à quatre niveaux
  r четырёхуровневая схема *f*

5670 **fractional rational function**
  d gebrochene rationale Funktion *f*
  f fonction *f* rationelle à fraction
  r дробно-рациональная функция *f*

5671 **frame impulse**
  d Bildimpuls *m*
  f impulsion *f* d'image
  r кадровый импульс *m*

5672 **frame leakage protection**
  d Rahmenschlußschutz *m*
  f protection *f* contre contact à la masse
  r защита *f* с заземляющей шиной

5673 **free automaton**
  d freier Automat *m*
  f automate *m* libre
  r независимый автомат *m*

5674 **free component**
  d freie Komponente *f*
  f composante *f* libre
  r свободная составляющая *f*

5675 **freedom of defects; absence of failures**
  d Fehlerfreiheit *f*; Ausfallfreiheit *f*
  f liberté *f* de défauts; absence *f* de défaillances
  r безошибочность *f*; безотказность *f*

5676 **free dynamic system**
  d ungestörtes dynamisches System *n*
  f système *m* dynamique non perturbé
  r свободная динамическая система *f*

5677 **free of distorion**
  d verzerrungsfrei
  f sans distorsion
  r без искажений

5678 **free of disturbances**
  d störungsfrei
  f exempt de perturbations
  r без помех

* **free oscillation** → 1449

5679 **free-oscillation regime**
  d freier Schwingungszustand *m*
  f régime *m* des oscillations libres
  r режим *m* свободных колебаний

5680 **free-programmable automatic assemblage**
  d freiprogrammierbare automatische Montage *f*
  f assemblage *m* automatique programmable librement
  r свободнопрограммируемая автоматическая сборка *f*

5681 **free-programmable miniature manipulator**
  d freiprogrammierbarer Kleinmanipulator *m*
  f manipulateur *m* miniature programmable librement
  r свободнопрограммируемый малогабаритный манипулятор *m*

5682 **free-running circuit; free-swinging circuit**
  d freischwingende Schaltung *f*
  f circuit *m* auto-oscillateur
  r несинхронизованная схема *f*

5683 **free-space attenuation**
  d Freiraumdämpfung *f*
  f affaiblissement *m* dans l'espace libre
  r пространственное затухание *n*

* **free-swinging circuit** → 5682

5684 **frequency adjustment**
  d Frequenzeinstellung *f*
  f réglage *m* de la fréquence
  r настройка *f* частоты; установка *f* частоты

5685 **frequency analysis**
  d Frequenzanalyse *f*
  f analyse *f* harmonique; analyse des fréquences
  r частотный анализ *m*

5686 **frequency band**
  d Frequenzband *n*
  f bande *f* de fréquences
  r полоса *f* частот

* **frequency-band limited** → 1631

5687 frequency bandwidth
  d Frequenzbandbreite *f*
  f largeur *f* de la bande de fréquence
  r ширина *f* полосы [частот]

5688 frequency changer
  d Frequenzwandler *m*
  f changeur *m* de fréquence; convertisseur *m* de fréquence
  r преобразователь *m* частоты

5689 frequency code
  d Frequenzkode *m*; frequenzmodulierter Kode *m*
  f code *m* à modulation de fréquence
  r частотный код *m*

5690 frequency compensation
  d Frequenzausgleich *m*
  f compensation *f* en fréquence
  r частотная коррекция *f*

5691 frequency controller; frequency regulator
  d Frequenzregler *m*
  f régulateur *m* de fréquence
  r регулятор *m* частоты

5692 frequency control of motors
  d Frequenzsteuerung *f* von Motoren
  f commande *f* en fréquence de moteurs
  r регулирование *n* частоты двигателей

5693 frequency converter characteristic slope
  d Kennliniensteilheit *f* des Frequenzwandlers
  f pente *f* de la caractéristique du convertisseur de fréquence
  r крутизна *f* характеристики преобразователя частоты

  * **frequency counter** → 5705

5694 frequency detector
  d Frequenzdetektor *m*
  f détecteur *m* de fréquence
  r частотный детектор *m*

5695 frequency deviation
  d Frequenzabweichung *f*
  f déviation *f* de fréquence
  r отклонение *n* частоты

5696 frequency discriminator
  d Frequenzdiskriminator *m*
  f discriminateur *m* de fréquence
  r частотный дискриминатор *m*

5697 frequency distortions
  d Frequenzverzerrungen *fpl*

  f distorsions *fpl* de fréquence
  r частотные искажения *npl*

5698 frequency division multiplex
  d Frequenzmultiplex *m*
  f multiplexage *m* par répartition en fréquence
  r частотное уплотнение *n*

5699 frequency division of channels
  d Frequenzteilung *f* von Kanälen
  f séparation *f* fréquentielle de canaux
  r частотное разделение *n* каналов

5700 frequency domain; frequency range
  d Frequenzbereich *m*; Bandbereich *m*
  f domaine *m* fréquentiel; bande *f* de fréquences; gamme *f* de fréquences
  r частотная область *f*; диапазон *m* частот

5701 frequency doubler
  d Frequenzverdoppler *m*
  f doubleur *m* de fréquence
  r удвоитель *m* частоты

5702 frequency error limits
  d Frequenzfehlergrenzen *fpl*
  f limite *f* des défauts de fréquence
  r пределы *mpl* погрешности частоты; пределы частотных ошибок

5703 frequency filter
  d Frequenzsieb *n*
  f filtre *m* de fréquence
  r частотный фильтр *m*

5704 frequency function
  d Häufigkeitsfunktion *f*
  f fonction *f* de fréquence
  r функция *f* частоты

5705 frequency meter; frequency counter
  d Frequenzmesser *m*
  f fréquencemètre *m*
  r волономер *m*; частотомер *m*

5706 frequency method
  d Frequenzmethode *f*
  f méthode *f* harmonique
  r частотный метод *m*

5707 frequency-modulated telecontrol system generator
  d frequenzmodulierter Generator *m* des Fernwirksystems
  f générateur *m* à modulation de fréquence pour système de télécommande
  r частотномодулированный генератор *m* системы телеуправления

5708 **frequency modulation**
 d Frequenzmodulation *f*; FM-Modulation *f*
 f modulation *f* des fréquences; modulation en
    fréquence
 r частотная модуляция *f*

5709 **frequency modulator**
 d Frequenzmodulator *m*
 f modulateur *m* de fréquence
 r частотный модулятор *m*

5710 **frequency monitor**
 d Frequenzkontrolleinrichtung *f*
 f moniteur *m* de fréquence
 r контроллер *m* частоты

5711 **frequency multiplier**
 d Frequenzvervielfacher *m*
 f multiplicateur *m* de fréquence
 r умножитель *m* частоты

5712 **frequency offset transponder**
 d Antwortsender *m* mit Frequenzversetzung
 f répondeur *m* à décalage de fréquence
 r импульсный повторитель *m* со смешением
    частоты сигнала

5713 **frequency output transducer**
 d Geber *m* mit Frequenzausgang;
    Frequenzausgangsgeber *m*
 f capteur *m* à sortie frequentielle
 r датчик *m* с частотным выходом

5714 **frequency-phase characteristic**
 d Frequenz-Phasen-Kennlinie *f*
 f caractéristique *f* fréquence-phase
 r частотно-фазовая характеристика *f*

5715 **frequency protection**
 d Frequenzschutz *m*
 f dispositif *m* de protection de fréquence
 r частотная защита *f*

 * **frequency range** → 5700

5716 **frequency range of a transmission system**
 d Frequenzbereich *m* eines
    Übertragungssystems
 f bande *f* de fréquences transmises par un
    système de transmission
 r полоса *f* частот передающей системы

 * **frequency regulator** → 5691

5717 **frequency relay**
 d Frequenzrelais *n*
 f rélais *m* de fréquence
 r резонансное реле *n*; частотное реле

5718 **frequency resolution constant**
 d Konstante *f* des
    Frequenzauflösungsvermögens; Beiwert *m*
    des Frequenzauflösungsvermögens
 f constante *f* de distinction de fréquence
 r постоянная *f* разрешения частоты

5719 **frequency response analyzer**
 d Frequenzganganalysator *m*
 f analysateur *m* de fréquences
 r частотный анализатор *m*

5720 **frequency-selective element**
 d frequenzselektives Element *n*
 f élément *m* sélectif de fréquence
 r частотно-избирательный элемент *m*

5721 **frequency-shift keying**
 d Frequenzumtastung *f*
 f modulation *f* par déplacement de fréquence
 r манипуляция *f* сдвигом частоты

5722 **frequency spectrum**
 d Frequenzspektrum *n*
 f spectre *m* de fréquences
 r спектр *m* частоты

5723 **frequency stability criterion**
 d Frequenzkriterium *n* der Stabilität
 f critérium *m* fréquentiel de stabilité
 r критерий *m* устойчивости частоты

5724 **frequency-stabilized**
 d frequenzstabilisiert; frequenzkonstant
 f stabilisé en fréquence
 r частотно-стабилизированный

5725 **frequency standard**
 d Frequenznormal *n*
 f étalon *m* de fréquence
 r эталон *m* частоты

5726 **frequency telemeter**
 d Frequenztelemeter *n*;
    Frequenzfernmessgerät *n*;
    Frequenzfernanzeiger *m*
 f appareil *m* télémesureur de fréquence
 r частотный телеметр *m*

5727 **frequency-telemetering system**
 d Frequenzfernmesssystem *n*
 f système *m* télémétrique fréquentiel
 r телеметрическая частотная система *f*

5728 **frequency tolerance**
 d Frequenztoleranz *f*
 f tolérance *f* de fréquence
 r допуск *m* по частоте

5729 **frequency value**
  d Häufigkeitswert *m*
  f valeur *f* de fréquence
  r величина *f* частоты

5730 **frequency variation relay**
  d Frequenzschwankungsrelais *n*
  f relais *m* à variation de fréquence
  r реле *n* вариации частот

5731 **friction adjuster**
  d Friktionsregler *m*
  f régulateur *m* à friction
  r регулятор *m* трения

5732 **frictional damping**
  d Reibungsdämpfung *f*
  f amortissement *m* par frottement
  r фрикционное демпфирование *n*

5733 **friction coefficient**
  d Reibungskoeffizient *m*
  f coefficient *m* du frottement
  r коэффициент *m* трения

5734 **front of logic pulse**
  d Front *f* des logischen Impulses
  f front *m* d'impulsion logique
  r фронт *m* логического импульса

5735 **front panel**
  d Frontplatte *f*; Frontbedienfeld *n*
  f panneau *m* frontal
  r передняя панель *f*

5736 **full-automatic analysis**
  d vollautomatische Analyse *f*
  f analyse *f* entièrement automatique
  r [полностью] автоматический анализ *m*

5737 **full-automatic checkig**
  d vollautomatische Prüfung *f*
  f vérification *f* complètement automatique
  r [полностью] автоматическая проверка *f*

5738 **full-automatic equipment**
  d vollautomatische Einrichtung *f*
  f dispositif *m* entièrement automatique
  r полностью автоматизированное
    оборудование *n*

5739 **full-automatic error correcting**
  d vollautomatische Fehlerkorrektur *f*
  f correction *f* d'erreur complètement
    automatique
  r [полностью] автоматическое
    исправление *n* ошибок

5740 **full-automatic processing**

  d vollautomatische Verarbeitung *f*
  f traitement *m* complètement automatique
  r полностью автоматизированная
    обработка *f*

5741 **full-compatible processor**
  d vollkompatibler Prozessor *m*
  f processeur *m* entièrement compatible
  r полностью совместимый процессор *m*

5742 **full-duplex communication line**
  d Vollduplexübertragungsleitung *f*
  f ligne *f* de communication duplex
  r дуплексная линия *f* связи

5743 **full-duplex transmission system**
  d Vollduplexübertragungssystem *n*
  f système *m* de transmission duplex
  r дуплексная система *f* передачи [данных]

5744 **full duty**
  d Volltrieb *m*
  f régime *m* en pleine activité
  r полный режим *m*

5745 **full home position**
  d Endlage *f*
  f position *f* fin de course
  r нулевое положение *n*; положение покоя

5746 **full-integrated production equipment**
  d vollintegrierte Fertigungsanlage *f*
  f installation *f* de fabrication complètement
    intégrée
  r интегральное производственное
    оборудование *n*

5747 **full load**
  d Vollast *f*
  f pleine charge *f*
  r полная нагрузка *f*

5748 **full-magnetic controller**
  d vollmagnetischer Fahrschalter *m*
  f combinateur *m* tout à fait magnétique
  r магнитный контроллер *m*

  * **full mechanization** → 2755

5749 **full-programmable manipulator control**
  d vollprogrammierbare Manipulatorsteuerung *f*
  f commande *f* de manipulateur complètement
    programmable
  r полносью программируемая система *f*
    управления манипулятором

5750 **full-scale value**
  d Skalenendwert *m*

f valeur *m* de déviation maximale
r максимальное значение *n* считываемое на
шкале

5751 full-time circuit
d permanente Schaltung *f*
f liaison *f* permanente
r контур *m* с полным рабочим циклом

5752 full-wave rectifier
d Zweiweggleichrichter *m*
f redresseur *m* push-pull; redresseur biphasé
r двухполупериодный выпрямитель *m*

5753 full-wave voltage impulse
d volle Stoßwelle *f*
f onde *f* de tension pleine de choc
r двухполупериодный импульс *m*
напряжения; полная ударная волна *f*

**5754 fully automatic coordinated trafic
        regulation**
d vollautomatische koordinierte
Verkehrsregelung *f*
f régulation *f* coordonnée de circulation
complètement automatique
r полностью автоматизированное
координированное регулирование *n*
движением

5755 fully automatic diaphragm
d vollautomatische Blende *f*
f diaphragme *m* entièrement automatique
r полностью автоматизированная
диафрагма *f*

**5756 fully automatic Diesel emergency power
        supply unit**
d vollautomatische Dieselnotstromanlage *f*
f générateur *m* Diesel de secours
complètement automatique
r полностью автоматизированный запасной
дизель-агрегат *m*

**5757 fully automatic equipment; fully
        automatic machine**
d Vollautomat *m*
f équipement *m* complètement automatique
r автоматическое оборудование *n*

* **fully automatic machine** → 5757

5758 functional
d Funktional *n*
f fonctionnelle *f*
r функционал *m*

5759 functional analysis

d Funktionalanalysis *f*
f analyse *f* fonctionnelle
r функциональный анализ *m*

* **functional block diagram** → 1838

5760 functional capabilities
d funktionelle Möglichkeiten *fpl*
f capacités *fpl* fonctionnelles
r функциональные возможности *fpl*

5761 functional characteristics
d Funktionskennwerte *mpl*
f caractéristiques *fpl* fonctionnelles
r функциональные характеристики *fpl*

5762 functional circuit of machine
d Maschinenfunktionsschaltung *f*
f circuit *m* fonctionnel de machine
r функциональная схема *f* машины

5763 functional classification
d funktionelle Klassifizierung *f*
f classification *f* fonctionnelle
r функциональная классификация *f*

5764 functional decomposition
d funktionelle Dekomposition *f*
f décomposition *f* fonctionnelle
r функциональная декомпозиция *f*

5765 functional definition
d Funktionsdefinition *f*
f définition *f* des fonctions
r функциональное определение *n*

5766 functional dependence
d funktionelle Abhängigkeit *f*
f dépendance *f* fonctionnelle
r функциональная зависимость *f*

5767 functional description
d Funktionsbeschreibung *f*
f description *f* de fonctionnement
r описание *n* функционирования

5768 functional design
d [logisch-]funktioneller Entwurf *m*
f conception *f* fonctionnelle
r функциональное проектирование *n*

5769 functional determinant
d Funktionaldeterminante *f*
f déterminant *m* fonctionnel
r функциональный определитель *m*

5770 functional electronic block
d elektronischer Funktionsblock *m*

   f bloc *m* fonctionnel électronique
   r электронный функциональный блок *m*

5771 functional interleaving
   d funktionelle Verschachtelung *f*
   f encastrement *m* fonctionnel
   r функциональное чередование *n*

5772 functionality
   d Zweckmäßigkeit *f*
   f convenance *f*; utilité *f*
   r функциональность *f*

5773 functionally complete base
   d funktionell vollständige Basis *f*
   f base *f* fonctionnellement complète
   r функционально полный базис *m*

5774 functional module
   d Funktionsmodul *m*
   f module *m* fonctionnel
   r функциональный модуль *m*

5775 functional monitoring
   d Funktionsüberwachung *f*
   f surveillance *f* fonctionnelle
   r функциональный контроль *m*

5776 functional optionality
   d funktionelle Auswahlmöglichkeit *f*
   f optionalité *f* fonctionnelle; possibilité *f* fonctionnelle supplémentaire
   r дополнительные функциональные возможности *fpl*

5777 functional organization
   d Funktionsorganisation *f*
   f organisation *f* fonctionnelle
   r функциональная организация *f*

5778 functional overview
   d Funktionsübersicht *f*
   f vue *f* synoptique de fonctionnement
   r обзор *m* функциональных возможностей

5779 functional relationship
   d funktionale Beziehung *f*
   f rapport *m* fonctionnel
   r функциональное соотношение *n*

   * **functional scheme** → 12088

5780 functional signal
   d Funktionssignal *n*
   f signal *m* de fonction
   r функциональный сигнал *m*

5781 functional structure

   d Funktionsstruktur *f*
   f structure *f* fonctionnelle
   r функциональная структура *f*

5782 functional symbol
   d funktionelles Symbol *n*
   f symbole *m* fonctionnel
   r функциональный символ *m*

5783 functional test
   d Funktionstest *m*
   f test *m* fonctionnel
   r проверка *f* работоспособности

5784 function checking of relay circuits
   d Funktionskontrolle *f* von Relaiskreisen
   f contrôle *m* de fonction des circuits à relais
   r контроль *m* действия релейных контуров

5785 function distribution analyzer
   d Funktionsverteilungsanalysator *m*
   f analyseur *m* de fonctions de distribution
   r анализатор *m* функции распределения

5786 function element of controller
   d Funktionsglied *n* des Reglers; Regelglied *n*
   f élément *m* fonctionnel du régulateur
   r функциональное звено *n* регулятора

5787 function generator
   d Functionsgenerator *m*; Functionsumformer *m*; Funktionsgeber *m*
   f générateur *m* de fonctions; tranducteur *m* de fonctions
   r генератор *m* функций

5788 function instruction
   d Funktionsbefehl *m*
   f instruction *f* de fonction
   r рабочая инструкция *f*

5789 function key
   d Funktionstaste *f*
   f touche *f* de fonction
   r ключ *m* функции

   * **function of state** → 11841

   * **function of time** → 12425

5790 function-oriented programming
   d funktionsorientierte Programmierung *f*
   f programmation *f* orientée sur les fonctions
   r функционально-ориентированное программирование *n*

5791 function oscillation
   d Schwankung *f* einer Funktion; Funktionsschwankung *f*

f oscillation *f* d'une fonction
r колебание *n* функции

5792 function potentiometer
  d Funktionspotentiometer *n*
  f potentiomètre *m* fonctionnel
  r функциональный потенциометр *m*

5793 function simulator
  d Funktionsmodell *n*
  f simulateur *m* fonctionnel; modèle *m*
    fonctionnel
  r функциональная модель *f*

5794 function table program
  d Funktionstabellenprogramm *n*
  f programme *m* de la table de fonctions
  r функциональная табличная программа *f*

5795 function unit
  d Funktionseinheit *f*
  f unité *f* fonctionnelle
  r функциональный блок *m*

5796 fundamental component of the current variations
  d Grundkomponente *f* der Stomänderungen
  f composante *f* fondamentale des variations de courant
  r основная составляющая *f* изменений тока

  * **fundamental equation** → 1655

  * **fundamental harmonic** → 5527

5797 fundamental interval
  d Fundamentalabstand *m*
  f intervalle *m* fondamental
  r основной интервал *m*

5798 fundamental law
  d Grundgesetz *n*
  f loi *f* fondamentale
  r основной закон *m*

5799 fundamental system
  d Fundamentalsystem *n*; System *n* linear unabhängiger Lösungen
  f système *m* fondamental; système des solutions linéaires indépendantes
  r фундаментальная система *f*

5800 fundamental theorem of information transmission
  d Fundamentaltheorem *n* der Informationsübertragung
  f théorème *m* fondamental de transmission d'informations

  r фундаментальная теорема *f* передачи информации

5801 fundamental vibration mode
  d Grundschwingungstyp *f*
  f mode *m* fondamental de vibration
  r основной режим *m* колебаний

5802 fusion frequency
  d Flimmerfrequenz *f*
  f fréquence *f* de fusion rétinienne
  r частота *f* слияния

5803 fuzzy filtration system
  d unklares Filtrierungssystem *n*
  f système *m* de filtrage flou
  r нечёткая система *f* фильтрации

G

5804 gain
  d Gewinn *m*; Zuwachs *m*
  f gain *m*
  r усиление *n*

5805 gain adjustment
  d Verstärkungsnachstellung *f*
  f ajustement *m* de gain
  r подстройка *f* усиления

5806 gain characteristic
  d Verstärkungskennlinie *f*
  f caractéristique *f* d'amplification
  r характеристика *f* усиления

  * **gain coefficient** → 733

5807 gain control
  d Verstärkungsregelung *f*
  f réglage *m* de gain
  r регулировка *f* усиления

5808 gain cross-over frequency
  d Amplitudenschnittfrequenz *f*
  f fréquence *f* de coupure de gain
  r частота *f* раделения по коэффициенту усиления

5809 gain-guided
  d gewinngeführt
  f à guidage par le gain
  r управляемый усилением

5810 gain-guided laser
  d gewinngeführter Laser *m*
  f laser *m* à guidage par le gain
  r лазер m, направленный эффектом усиления

5811 gain level
  d Gewinnpegel *m*
  f niveau *m* de gain
  r уровень *m* усиления

5812 gain margin
  d Verstärkungsgrenze *f*
  f marge *m* de gain; plage *m* de sécurité
  r предел *m* усиления; граница *f* усиления

5813 gain mechanism
  d Wirkungsweise *f* des Gewinns;
Mechanismus *m* des Gewinns
  f mécanisme *m* de gain
  r механизм *m* усиления

5814 gain-phase characteristic
  d Amplituden-Phasen-Charakteristik *f*
  f diagramme *m* amplitude-phase; lieu *m* de transfert
  r амплитудно-фазовая характеристика *f*

5815 gain range
  d Verstärkungsbereich *m*
  f champ *m* d'amplification
  r диапазон *m* усиления

5816 gain set
  d Verstärkungsmesseinrichtung *f*; Verstärkungsmesser *m*
  f kerdomètre *m*
  r измеритель *m* усиления; указатель *m* усиления

5817 gain stabilization
  d Verstärkungsstabilisierung *f*
  f stabilisation *f* d'amplification
  r стабилизация *f* усиления

5818 galvanic coupling
  d galvanische Kopplung *f*
  f couplage *m* conductif
  r гальваническая связь *f*

5819 galvanometer constant
  d Galvanometerkonstante *f*
  f constante *f* de galvanomètre
  r постоянная *f* гальванометра

5820 galvanometer recorder
  d registrierendes Element *n* eines Galvanometers
  f élément *m* enregistreur d'un galvanomètre
  r регистрирующий элемент *m* гальванометра

  * **game theory** → 12331

5821 gamma control
  d Gammaeinstellung *f*
  f réglage *m* de gamma
  r гамма-коррекция *f*

5822 gamma function
  d Gammafunktion *f*
  f fonction *f* gamma
  r гамма-функция *f*

5823 gamma radiation absorption
  d Gammastrahlungsabsorption *f*

    f absorption *f* du rayonnement gamma
    r поглощение *n* гамма-излучения

5824 gamma radiometer
    d Gammaradiometer *n*
    f radiomètre *m* gamma
    r гаммарадиометр *m*

5825 gamma[-ray] spectrometry
    d Gammaspektrometrie *f*
    f spectrométrie *f* gamma
    r спектрометрия *f* гамма-излучения; гамма-
       спектрометрия *f*

  * **gamma spectrometry** → 5825

5826 gang *v*
    d kuppeln
    f coupler; jumeler
    r сопрягать; сочленять; спаривать

5827 gang circuit
    d Gleichgangstromkreis *m*
    f circuit *m* à commande unique
    r сгруппированная схема *f*

5828 gang control
    d Gleichgangsteuerung *f*
    f commande *f* unique
    r групповое управление *n*

5829 gang switch
    d Gangschalter *m*
    f commutateur *m* jumelé
    r двойной выключатель *m*

5830 gap adjustment
    d Spaltbreiteneinstellung *f*
    f réglage *m* de l'entrefer
    r регулировка *n* зазора

5831 garbled information
    d verstümmelte Information *f*
    f information *f* mutilée
    r искаженная информация *f*

5832 garbled signal
    d verstümmeltes Signal *n*; verzerrtes Signal
    f signal *m* distordu
    r искаженный сигнал *m*

5833 gas analysis
    d Gasanalyse *f*
    f analyse *m* du gaz
    r газовый анализ *m*

5834 gas analyzer
    d Gasanalysator *m*

    f analyseur *m* de gaz
    r газоанализатор *m*

5835 gas calorimeter
    d Gaskalorimeter *n*
    f calorimètre *m* analyseur de gaz
    r газовый калориметр *m*

5836 gaschromatographic analysis
    d gaschromatografische Analyse *f*
    f analyse *f* chromatographique en phase
       gazeuse
    r газохроматографический анализ *m*

5837 gas chromatography
    d Gaschromatografie *f*
    f chromatographie *f* gazeuse
    r газовая хроматография *f*

5838 gas cleaning process
    d Gasreinigungsprozess *m*
    f procédé *m* d'épuration de gaz
    r процесс *m* газоочистки

5839 gas cleaning unit
    d Gasreinigungsanlage *f*
    f unité *f* d'épuration de gaz
    r газоочистительная установка *f*

5840 gas constant
    d Gaskonstante *f*
    f constante *f* de gaz
    r газовая постоянная *f*

5841 gas-cooled reactor
    d gasgekühlter Reaktor *m*
    f réacteur *m* à refroidissement au gaz
    r реактор *m* с газовым охлаждением

5842 gas detector
    d Gasanzeiger *m*
    f détecteur *m* de gaz
    r детектор *m* газа

5843 gas discharge display
    d Gasentladungsanzeige *f*
    f affichage *m* à décharge au gaz
    r газоразрядный индикатор *m*

**5844 gas-discharge relay; gas-filled relay; ionic
       relay**
    d Gas[entladungs]relais *n*; Ionenrelais *n*
    f relais *m* électronique; relais ionique
    r газонаполненное реле *n*

5845 gas distribution automatic control
    d selbsttätige Regelung *f* der Gasverteilung

   f réglage *m* automatique de distribution de gaz
   r автоматическое регулирование *n* распределения газов

5846 gas dynamics
   d Gasdynamik *f*
   f dynamique *f* de gaz
   r газовая динамика *f*

5847 gas expansion refrigerating system
   d Gasexpansionskühlsystem *n*
   f système *m* frigorifique à détente de gaz
   r холодильная система *f* газового расширения

5848 gas-filled phototube
   d gasgefüllte Fotozelle *f*
   f cellule *f* photoélectrique à gaz
   r газонаполненный фотоэлемент *m*

   * **gas-filled relay** → **5844**

5849 gas-flow meter
   d Gasmesser *m*
   f gazomètre *m*
   r газомер *m*

5850 gas-flow recorder
   d Gasflussschreiber *m*
   f enregistreur *m* de débit du gaz
   r регистратор *m* расхода газа

5851 gas fractionation plant
   d Gastrennanlage *f*
   f unité *f* de fractionnement de gaz
   r установка *f* фракционирования газа

5852 gasification plant
   d Vergasungsanlage *f*
   f unité *f* de gazéification
   r установка *f* газификации

5853 gasification process
   d Vergasungsverfahren *n*; Vergasungsprozess *m*
   f procédé *m* de gazéification
   r процесс *m* газификации

5854 gasification reactor
   d Vergasungsreaktor *m*
   f réacteur *m* de gazéification
   r реактор *m* газификации

5855 gas liquefaction plant
   d Gasverflüssigungsanlage *f*
   f installation *f* de liquéfaction de gaz
   r установка *f* для сжижения газа

5856 gas mixture lens
   d Mischgaslinse *f*
   f lentille *f* à gaz mixte
   r линза *f* для газовых смесей

5857 gas moisture measurement
   d Gasfeuchtemessung *f*
   f mesure *f* de l'humidité de gaz
   r измерение *n* влажности газа

5858 gasoline pressure gauge
   d Benzindruckmesser *m*
   f indicateur *m* de pression d'essence; indicateur de pression de carburant
   r манометр *m* для жидкого топлива

5859 gas operated
   d gasbetätigt
   f commandé par gaz
   r с газовым исполнительным механизмом

5860 gas pressure regulator
   d Gasdruckregler *m*
   f régulateur *m* de pression de gaz
   r регулятор *m* давления газа

5861 gas producting plant
   d Gaserzeugungsanlage *f*
   f installation *f* de production de gaz
   r газогенераторная установка *f*

5862 gas recovery process
   d Gasrückgewinnungsverfahren *n*
   f procédé *m* de récupération de gaz
   r процесс *m* газоулавливания

5863 gas reforming plant
   d Gasspaltanlage *f*
   f unité *f* de réformation de gaz
   r установка *f* риформинга газа

5864 gas traces recording device
   d Gasspurenschreiber *m*
   f enregistreur *m* traceur de gaz
   r прибор *m* регистрирующий присуствие газа

5865 gas turbine reactor
   d Gasturbinenreaktor *m*
   f réacteur *m* à turbine à gaz
   r реактор *m* с газовой турбиной; газотурбинный реактор *m*

   * **gate** → **2564**

   * **gate circuit** → **2564**

5866 gate circuit logic
   d Torschaltungslogik *f*; Gatterschaltungslogik *f*

f logique f des circuits de porte
r логика f на схемах совпадения

5867 gated automatic gain control
d getastete automatische
Verstärkungsregelung f
f antifading m à déclenchement périodique
r автоматическое регулирование n усиления
с ручной манипуляцией

5868 gate pulse; gating pulse
d Strobimpuls m; Auftastimpuls m;
Gatterimpuls m
f impulsion f de fixation; créneau m; impulsion
de porte
r стробирующий импульс m

* **gating circuit** → 2564

* **gating pulse** → 5868

5869 gating stage
d Torschaltung f; Toröffnungsstufe f
f étage m à commande de passage
r каскад m стробирования

5870 gating switch
d Sperrschalter m; Blockschalter m
f interrupteur m cyclique
r коммутирующий выключатель m

5871 gating system; selector system
d Auswahlsystem n; Selektorsystem n
f système m sélecteur
r селекторная система f

5872 gauge v; calibrate v
d kalibrieren; eichen
f calibrer; étalonner
r калибровать; градуировать

5873 gauge transformation
d Eichtransformation f
f transformation f d'étalonnage
r эталонное преобразование n;
калибровочное преобразование

5874 Gaussian curve
d Gausssche Kurve f
f courbe f gaussienne; courbe de Gauss
r кривая f Гаусса

**5875 Gaussian distribution; normal
distribution**
d Gausssche Verteilung f; Normalverteilung f
f distribution f de Gauss; répartition f normale
r распределение n Гаусса; нормальное
распределение

5876 Gaussian low pass filter
d Gauss-Tiefpass m
f filtre m passe-bas de Gauss
r гауссовский низкочастотный фильтр m

5877 Gaussian random process
d Gaussscher Zufallsprozess m
f processus m stochastique Gaussien
r случайный процесс m Гаусса

5878 Gay-Lussac's law
d Gay-Lussac-Gesetz n
f loi f de Gay-Lussac
r закон m Гей-Люсака

5879 gear control
d Getriebesteuerung f
f commande m à engrenage
r управление n зубчатой передачей

5880 gear for servomechanisms
d Getriebe n für Servomechanismen
f transmission f pour les servomécanismes
r передача f для сервомеханизмов

5881 gear ratio; transfer ratio; transfer number
d Getriebeübersetzungsverhältnis n;
Übertragungsverhältnis n;
Übertragungszahl f
f rapport m de denture; rapport de réduction;
démultiplication f; taux m de transfert;
nombre m de transfert
r передаточное число n

5882 Geiger counter
d Geigerzähler m
f compteur m Geiger
r счётчик m Гейгера

5883 general application in automation
d allgemeine Anwendung f in der
Automatisierung
f application f générale en automatisation
r общее применение n в автоматизации

5884 general chart
d allgemeines Diagramm n
f diagramme m d'ensemble
r общая схема f

5885 generalized active element
d verallgemeinertes aktives Element n
f élément m actif généralisé
r обобщённый активный элемент m

5886 generalized automaton
d verallgemeinerter Automat m
f automate m généralisé
r обобщённый автомат m

5887 generalized coordinates
  d verallgemeinerte Koordinaten *fpl*
  f coordonnées *fpl* généralisées
  r обобщённые координаты *fpl*

5888 generalized frequency response
  d verallgemeinerte Frequenzcharakteristik *f*
  f caractéristique *f* fréquentielle de généralisation
  r обобщённая частотная характеристика *f*

5889 generalized model
  d verallgemeinertes Modell *n*
  f modèle *m* généralisé
  r обобщённая модель *f*

5890 generalized parameter
  d verallgemeinerter Parameter *m*
  f paramètre *m* généralisé
  r обобщённый параметр *m*

5891 generalized transfer function
  d verallgemeinerte Übertragungsfunktion *f*
  f transmittance *f* généralisée
  r обобщённая передаточная функция *f*

5892 general machine program
  d allgemeines Maschinenprogramm *n*
  f programme *m* de machine général
  r общая машинная программа *f*

5893 general monitor checking routine
  d allgemeines Überwachungsprogramm *n*
  f programme *m* d'analyse général
  r общая программа *f* контрольных испытаний

  * **general plan** → **7205**

  * **general program** → **5900**

5894 general-purpose application
  d Universalanwendung *f*
  f application *f* universelle
  r универсальное применение *n*

  * **general-purpose automatic machine** → **5895**

5895 general-purpose automatic machine [for assembly]
  d Mehrzweckautomat *m* [für Montage]
  f automate *m* à usage multiple [pour assemblage]; machine *f* automatique à usage multiple [pour assemblage]
  r универсальный [сборочный] автомат *m*

5896 general-purpose computer
  d Universalrechner *m*; Allzweckrechner *m*
  f ordinateur *m* universel
  r универсальный компьютер *m*

  * **general-purpose interface** → **12681**

5897 general-purpose radar
  d Mehrzweckradar *n*
  f radar *m* universel
  r радиолокатор *m* общего назначения

5898 general reactor equation
  d allgemeine Reaktorgleichung *f*
  f équation *f* du réacteur générale
  r общее уравнение *n* реактора

5899 general-register architecture
  d Universalregisterstruktur *f*
  f structure *f* à registres généraux
  r архитектура f, базирующаяся на регистрах общего назначения

5900 general routine; general program
  d allgemeines Programm *n*
  f programme *m* général; routine *f* générale
  r универсальная программа *f*

5901 general solution
  d allgemeine Lösung *f*
  f solution *f* générale
  r общее решение *n*

5902 general store
  d Hauptspeicher *m*
  f mémoire *f* générale
  r универсальный накопитель *m*; запоминающее устройство *n*

5903 generated assembly data
  d generierte Montagedaten *pl*
  f données *fpl* d'assemblage engendrées
  r выработанные данные *pl* для сборки

5904 generated checking data
  d generierte Prüfdaten *pl*
  f données *fpl* d'essai engendrées
  r выработанные данные *pl* для контроля

5905 generating equation
  d erzeugende Gleichung *f*
  f équation *f* génératrice
  r генерирующее уравнение *n*

5906 generating frequency
  d Erzeugungsfrequenz *f*; erzeugende Frequenz *f*
  f fréquence *f* génératrice
  r генерирующая частота *f*

5907 generating function
d Erzeugungsfunktion *f*; erzeugende Funktion *f*
f fonction *f* génératrice
r генерирующая функция *f*

5908 generating routine
d erzeugendes Programm *n*
f autoprogrammeur *m*
r программирующая программа *f*

5909 generation of automatons
d Automatengeneration *f*
f génération *f* d'automates
r поколение *n* автоматов

5910 generation of drive moment
d Antriebsmomenterzeugung *f*
f génération *f* de moment d'entraînement
r выработка *f* приводящего момента

5911 generation of geometric structures
d Erzeugung *f* geometrischer Strukturen
f génération *f* des structures géométriques
r генерирование *n* геометрических структур

5912 generation of random numbers
d Zufallszahlengewinnung *f*
f obtention *f* de nombres aléatoires
r генерирование *n* случайных чисел

5913 generation of sensors
d Sensorgeneration *f*
f génération *f* de capteurs
r поколение *n* сенсоров

5914 generator of random signals
d Generator *m* von Zufallssignalen
f générateur *m* de signaux aléatoires
r генератор *m* случайных сигналов

5915 generator pulse regime
d Generatorimpulsbetrieb *m*
f régime *m* impulsionnel du générateur
r импульсный режим *m* генератора

5916 generator voltage controller
d Generatorspannungsregler *m*
f régulateur *m* de tension de générateur
r регулятор *m* напряжения генератора

5917 generator zero pulse time
d Nullzeitimpuls *m*
f impulsion *f* d'un générateur déterminant le temps zéro
r начало *n* отсчёта времени импульса генератора

5918 geometric analog data

d geometrische Analogdaten *pl*
f données *fpl* analogiques géométriques
r геометрические аналоговые данные *pl*

5919 geometric information
d geometrische Information *f*
f information *f* géométrique
r геометрическая информация *f*

5920 geometric model
d geometrisches Modell *n*
f modèle *m* géométrique
r геометрическая модель *f*

5921 geometric object
d geometrisches Objekt *n*
f objet *m* géométrique
r геометрический объект *m*

5922 geometric structure
d geometrische Struktur *f*
f structure *f* géométrique
r геометрическая структура *f*

5923 geometric structure analysis
d geometrische Strukturanalyse *f*
f analyse *f* des structures géométriques
r анализ *m* геометрических структур

5924 geometric structure input
d geometrische Struktureingabe *f*; Eingabe *f* geometrischer Strukturen
f entrée *f* des structures géométriques
r ввод *m* геометрических структур

5925 geometric structure output
d geometrische Strukturausgabe *f*
f sortie *f* des structures géométriques
r вывод *m* геометрических структур

5926 geometry description
d Geometriebeschreibung *f*
f description *f* de géométrie
r геометрическое описание *n*

5927 geometry treatment
d Geometrieverarbeitung *f*
f traitement *m* de géométrie
r обработка *f* геометрических данных

5928 geothermometer
d Geothermometer *n*
f géothermomètre *m*
r геотермометр *m*

5929 global automation system
d globales Automatisierungssystem *n*
f système *m* d'automation globale
r глобальная автоматизированная система *f*

5930   global-controlled effector state
  d global geregelter Effektorzustand *m*
  f état *m* d'effecteur global réglé
  r эффекторное состояние *n* с глобальным
     регулированием

  * global reactivity → 11863

  * global stability → 11770

5931   goal-seeking system
  d Selbstanpassungssystem *n*
  f système *m* auto-adaptatif
  r самоприспосабливающая система *f*

  * governor valve → 615

5932   grade
  d Güteklasse *f*
  f classe *f* de qualité; degré *m* de qualité
  r класс *m*; степень *f*

5933   graded potentiometer
  d abgestufter Spannungsteiler *m*
  f potentiomètre *m* gradué
  r нелинейный потенциометр *m*

5934   graded time-lag relay
  d Zeitrelais *n* mit abgestufter Verzögerung
  f relais *m* de temporisation à retard gradué
  r регулируемое реле *n* со ступенчатой
     выдержкой

5935   gradient
  d Gradient *m*
  f gradient *m*
  r градиент *m*

5936   grading automatic sorter
  d Klassifizierautomat *m*
  f trieur *m* automatique
  r автомат *m* для сортировки

5937   gradiometer
  d Gradiometer *n*
  f gradomètre *m*; indicateur *m* de pente
  r градиометр *m*; измеритель *m* уклонов

5938   Gram determinant
  d Gramsche Determinante *f*
  f déterminant *m* de Gram
  r детерминант *m* Грама

5939   graph chart
  d grafische Darstellung *f*; Schautafel *f*
  f représentation *f* graphique
  r график *m*

5940   graph data
  d grafische Angaben *fpl*
  f données *fpl* graphiques
  r графические данные *pl*

5941   graphic access method
  d grafische Zugriffsmethode *f*
  f méthode *f* d'accès graphique
  r метод *m* графического обращения

5942   graphic acquisition unit
  d grafische Erfassungseinheit *f*
  f unité *f* d'acquisition graphique
  r блок *m* графического распознавания

5943   graphic addition
  d grafische Addition *f*
  f addition *f* graphique
  r графическое сложение *n*

5944   graphical analysis
  d grafische Analyse *f*
  f analyse *f* graphique
  r графический анализ *m*

5945   graphical data input technique
  d grafische Dateneingabetechnik *f*
  f technique *f* d'entrée graphique des données
  r техника *f* графического ввода данных

5946   graphical determination
  d grafische Bestimmung *f*
  f détermination *f* graphique
  r графическое определение *n*

5947   graphical interactive interface
  d grafisches interaktives Interface *n*
  f interface *f* graphique interactive
  r графический интерактивный интерфейс *m*

5948   graphical interactive program system
  d grafisches interaktives Programmsystem *n*
  f système *m* de programme interactif
     graphique
  r графическая интерактивная система *f*
     программ

5949   graphical record converting into electric
       voltage wave
  d Umwandlung *f* grafischer Darstellung in
     elektrische Spannungswellen
  f conversion *f* d'enregistrement graphique en
     onde de tension électrique
  r преобразование *n* графической записи в
     волну электрического напряжения

5950   graphic attention control block
  d grafischer Achtungssteuerblock *m*

f bloc *m* de commande d'attention graphique
r блок *m* управления графическим устройством

5951 graphic basic element
d grafisches Grundelement *n*
f élément *m* de base graphique
r графический базовый элемент *m*

5952 graphic code
d Formkode *m*
f code *m* graphique
r графический код *m*

5953 graphic communication
d grafischer Informationsaustausch *m*
f communication *f* graphique
r передача *f* графической информации

5954 graphic data processing
d grafische Datenverarbeitung *f*; Verarbeitung *f* grafischer Informationen
f traitement *m* d'informations graphiques
r обработка *f* графической информации

5955 graphic data structure
d grafische Datenstruktur *f*
f structure *f* graphique de données
r графическая структура *f* данных

5956 graphic design
d grafischer Entwurf *m*
f dessin *m* graphique
r графическое проектирование *n*

5957 graphic display
d grafisches Display *n*; grafisches Sichtanzeigegerät *n*
f appareil *m* de visualisation graphique
r графический дисплей *m*

5958 graphic display program
d grafisches Bildschirmprogramm *n*
f programme *m* de l'affichage graphique
r графическая программа *f* дисплея

5959 graphic information
d grafische Information *f*
f information *f* graphique
r графическая информация *f*

5960 graphic interpretation
d grafische Interpretation *f*
f interprétation *f* graphique
r графическая интерпретация *f*

5961 graphic man-machine communication
d grafische

Mensch-Maschine-Kommunikation *f*
f communication *f* graphique entre homme-machine
r графическая связь *f* между человеком и машиной

5962 graphic method
d grafische Methode *f*
f méthode *f* graphique
r графический метод *m*

5963 graphic noise
d grafisches Rauschen *n*
f bruit *m* graphique
r искажение *n* графика

5964 graphic panel
d grafisches Paneel *n*
f panneau *m* graphique
r панель *f* индикации графических данных

5965 graphic processor
d Grafikprozessor *m*
f processeur *m* graphique
r графический процессор *m*

5966 graphic quantity
d grafische Größe *f*
f quantité *f* graphique
r величина *f* в графической форме

5967 graphic representation
d grafische Darstellung *f*
f représentation *f* graphique
r графическое представление *n*

5968 graphics hardware
d Grafikhardware *f*
f hardware *m* graphique
r графические аппаратные средства *npl*

5969 graphic structure
d grafische Struktur *f*
f structure *f* graphique
r графическая структура *f*

5970 graphic terminal
d grafisches Endgerät *n*
f terminal *m* graphique
r графический терминал *m*

5971 graph plotter; curve plotter
d Kurvenschreiber *m*; Plotter *m*
f enregistreur *m* graphique; traceur *m* de courbes
r графопостроитель *m*; прибор *m* для записи кривых

5972 graph theory
 d grafische Darstellungstheorie *f*;
      Grafentheorie *f*
 f théorie *f* des graphes
 r теория *f* графов

5973 grating spectroscope
 d Gitterspektroskop *n*
 f spectroscope *m* à grille
 r дифракционный спектроскоп *m*

5974 gravitational acceleration; gravity acceleration
 d Schwerkraftbeschleunigung *f*;
      Erdbeschleunigung *f*
 f accélération *f* gravitationnelle; accélération
      de la pesanteur
 r ускорение *n* силы тяжести; ускорение
      свободного падения

 * **gravity acceleration → 5974**

5975 gravity-controlled instrument
 d Instrument *n* mit gegenwirkendem Gewicht
 f appareil *m* de mesure à contrepoids
 r измерительное устройство *n* с
      противодействующей массой

5976 gravity conveyor
 d Schwerkraftförderer *m*
 f transporteur *m* à gravité
 r гравитационный транспортер *m*;
      гравитационный конвейер *m*

5977 gravity correction
 d Schwerkraftberichtigung *f*
 f correction *f* de gravitation
 r гравитационная поправка *f*

 * **gravity regulator → 2227**

5978 Green's function
 d Greensche Funktion *f*
 f fonction *f* de Green
 r функция *f* Грина

5979 grid bias modulation
 d Gittervorspannungsmodulation *f*
 f modulation *f* par variation de polarisation de
      grille
 r модуляция *f* [сеточным] смещением

5980 grid circuit
 d Gitterkreis *m*
 f circuit *m* de grille
 r сеточный контур *m*; цепь *f* сетки

5981 grid control

 d Gittersteuerung *f*
 f commande *f* par grille
 r сеточное управление *n*

5982 grid current characteristic
 d Gitterstromkennlinie *f*
 f caractéristique *f* de courant de grille
 r характеристика *f* сеточного тока

5983 grid cut-off voltage
 d Steuergittereinsatzspannung *f*
 f tension *f* de blocage
 r запирающий потенциал *m* сетки

5984 grid driving power
 d Gittersteuerleistung *f*
 f puissance *f* d'excitation de grille
 r мощность *f* возбуждения цепи

5985 grid leak
 d Gitterverlustwiderstand *m*;
      Gitterableitwiderstand *m*
 f résistance *f* de fuite de grille
 r сопротивление *n* сеточной утечки

5986 grip aids
 d Greifhilfen *fpl*
 f aides *fpl* de grippion
 r дополнительные устройства *npl* для
      захватывания

5987 grip characteristic
 d Greifcharakteristik *f*
 f caractéristique *f* de grippion
 r характеристика *f* захватывания

5988 grip cycle
 d Greifzyklus *m*
 f cycle *m* de grippion; cycle de préhension
 r цикл *m* захватывания

5989 grip device; gripper equipment
 d Greifvorrichtung *f*
 f dispositif *m* de grippion; équipement *m* de
      grappin
 r захватное устройство *n*

5990 grip force survey
 d Greifkraftüberwachung *f*
 f surveillance *f* de force de grippion
 r контроль *m* захватывающего усилия

5991 grip of handling objects
 d Greifen *n* von Handhabungsobjekten
 f grippion *f* d'objets de manutention
 r захватывание *n* манипулируемых
      объектов

5992 **grip organ drive**
  d Greiferorganantrieb m
  f entraînement m d'un organe de grippion
  r привод m захватного органа

5993 **gripper change device**
  d Greiferwechseleinrichtung f
  f dispositif m de changement de grappin
  r устройство n смены схвата

5994 **gripper change system**
  d Greiferwechselsystem n
  f système m de changement de grappin
  r система f смены схвата

5995 **gripper control**
  d Greifersteuerung f
  f commande f d'un grappin
  r управление n схватом

5996 **gripper energy line**
  d Greiferenergieleitung f
  f ligne f d'énergie d'un grappin
  r линия f передачи энергии схвата

  * **gripper equipment** → 5989

5997 **gripper gear**
  d Greifergetriebe n
  f engrenage m de grappin
  r передаточный механизм m схвата

5998 **gripper gear elements**
  d Greifergetriebeglieder npl;
     Getriebeglieder npl eines Greifers
  f éléments mpl d'engrenage d'un grappin
  r звенья npl передаточного механизма
     схвата

5999 **gripper program**
  d Greiferprogramm n
  f programme m de grappin
  r программа f управления захватным
     устройством

6000 **gripper review**
  d Greiferüberblick m
  f vue f d'ensemble de grappin
  r обзор m захватных устройств

6001 **gripper sensing element**
  d Greiferfühlelement n
  f élément m de palpage de grappin
  r чувствительный элемент m захвата

6002 **gripper technique**
  d Greifertechnik f
  f technique f de grappin

  r техника f захвата

6003 **gripper trace control**
  d Greiferablaufsteuerung f
  f commande f de poursuite de grappin
  r управление n слежением захватного
     устройства

6004 **gross error; total error**
  d Gesamtfehler m
  f erreur f totale
  r общая ошибка f

6005 **ground detector**
  d Erdschlußprüfer m
  f détecteur m de masse
  r указатель m заземления

6006 **grounded-grid circuit**
  d Gitterbasisschaltung f
  f montage m avec grille à la masse
  r схема f с заземленной сеткой

6007 **ground state**
  d Grundzustand m
  f état m de base
  r основное состояние n

6008 **group theory**
  d Gruppentheorie f
  f théorie f des groupes
  r теория f групп

  * **guard relay** → 1852

6009 **guidance; guide**
  d Steuerung f; Führung f; Lenkung f
  f guidage m; guide m
  r наведение n; управление n

6010 **guidance computer**
  d Lenkungsrechner m
  f ordinateur m de guidage
  r наводящий компьютер m

6011 **guidance phase**
  d Lenkungsphase f
  f phase f de guidage
  r фаза f наведения

  * **guide** → 6009

6012 **guided transmission**
  d Wellenleiterübertragung f;
     Wellenleiterverbindung f
  f transmission f par guide d'ondes
  r управляемый привод m

6013 guide numbers calculation
   d Leitzahlenberechnung *f*
   f calculation *f* de nombres-guides
   r расчёт *m* управляемых элементов

6014 gyro-control
   d Kreiselkompaßregelung *f*
   f réglage *m* à compas gyroscopique
   r гироскопическое регулирование *n*

6015 gyro-frequency
   d Gyrofrequenz *f*; Drehfrequenz *f*
   f fréquence *f* de rotation; gyrofréquence
   r гирочастота *f*; частота *f* вращения

H

6016 half life
d Halbwertszeit *f*
f demi-vie *f*; période *f* radioactive
r период *m* полураспада

6017 half period; semicycle
d Halbperiode *f*; Halbwelle *f*
f demi-période *f*; demi-onde *f*; alternance *f*
r полупериод *m*

6018 half-process automation
d Halbbetriebautomatisierung *f*
f automatisation *f* de demi-exploitation
r полуавтоматизация *f* процессов

6019 half-stable limit cycle
d halbstabiler Grenzzyklus *m*
f cycle *m* limite demistable
r полуустойчивый предельный цикл *m*

6020 half-wave rectifier
d Halbweggleichrichter *m*;
   Einweggleichrichter *m*
f redresseur *m* demi-onde
r однополупериодный выпрямитель *m*

6021 half-write pulse
d Rückschreibimpuls *m*
f impulsion *f* de réinscription
r импульс *m* полузаписи

6022 Hall effect
d Hall-Effekt *m*
f effet *m* Hall
r эффект *m* Холла

6023 Hamilton-Jacobi equation
d Hamilton-Jacobische Gleichung *f*
f équation *f* de Hamilton-Jacobi
r уравнение *n* Гамильтона-Якоби

6024 Hamilton's principe
d Hamiltonsches Prinzip *n*
f principe *m* de Hamilton
r принцип *m* Гамильтона

6025 handle *v*; manipulate *v*
d behandeln; handhaben; manipulieren
f traiter; manipuler
r обрабатывать; манипулировать

* **handler** → 6035

* **handler checking** → 7790

6026 handler packet
d Handlerpaket *n*
f paquet *m* de handler
r пакет *m* данных для манипулятора

6027 handling
d Handhaben *n*; Handhabung *f*
f manutention *f*
r манипулирование *n*

**6028 handling accuracy of mechanical
      components**
d Genauigkeit *f* bei der Handhabung
   mechanischer Bauteile
f exactitude *f* de manipulation de pièces
   mécaniques
r точность *f* манипулирования
   механическими элементами

6029 handling aides
d Handhabungshilfen *fpl*
f aides *fpl* de manipulation
r средства *npl* манипулирования

**6030 handling assembly unit; manipulation
      assembly unit**
d Handhabebaukasten *m*
f boîte *f* de construction de manipulation
r система *f* манипулирования агрегатного
   типа

6031 handling automation
d Handhabungsautomatisierung *f*
f automatisation *f* de manutention
r автоматизация *f* процесса
   манипулирования

6032 handling automaton
d Handhabungsautomat *m*
f automate *m* de manutention
r манипуляционный автомат *m*

6033 handling control device
d Handhabesteuergerät *n*
f dispositif *m* de commande de manutention
r прибор *m* управления манипуляцией

6034 handling cycle; manipulation cycle
d Handhabezyklus *m*
f cycle *m* de manutention
r цикл *m* манипулирования

* **handling defect** → 6038

6035 handling device; manipulation device; manipulator; handler
d Handhabegerät *n*
f appareil *m* de manipulation
r манипулятор *m*

6036 handling device with computer coupling; manipulation device with computer coupling
d Handhabegerät *n* mit Rechnerkopplung
f appareil *m* d'opération à couplage de calculatrice
r манипулятор m, подключенный к вычислительной машине

6037 handling equipment; manipulation equipment
d Handhabeeinrichtung *f*
f équipement *m* de manutention
r манипуляционное устройство *n*

6038 handling error; handling defect; manipulation error; manipulation defect
d Handhabefehler *m*
f défaut *m* de manipulation
r ошибка *f* манипулирования

6039 handling function
d Handhabefunktion *f*
f fonction *f* de manutention
r функция *f* манипулирования

6040 handling object
d Handhabungsobjekt *n*; Handhabeobjekt *n*
f objet *m* de manutention
r объект *m* манипулирования

6041 handling of assembly parts; manipulation of assembly parts
d Montageteilhandhabung *f*; Handhabung *f* von Montageteilen
f manutention *f* de parts d'assemblage
r манипулирование *n* собираемыми деталями

6042 handling of prism parts
d Handhabung *f* von Prismateilen
f manutention *f* de parts d'un prisme
r манипулирование *n* призматическими изделиями

6043 handling of rotation parts
d Handhabung *f* von Rotationsteilen
f manutention *f* de parts de rotation
r манипулирование *n* вращающими деталями

6044 handling operation

d Handhabungsoperation *f*; Handhabeoperation *f*
f opération *f* de manutention
r операция *f* манипулирования

6045 handling process; manipulation process
d Handhabungsvorgang *m*; Handhabevorgang *m*
f procédé *m* manutention; processus *m* de manipulation
r процесс *m* манипулирования

6046 handling process data
d Daten *pl* eines Handhabeprozesses
f données *fpl* d'un procédé de manutention
r параметры *mpl* процесса манипулирования

6047 handling program
d Handhabeprogramm *n*
f programme *m* de manutention
r программа *f* манипулирования

6048 handling requisition
d Handhabungsanforderung *f*
f demande *f* de manutention
r требование *n* к манипулированию

6049 handling sequence
d Handhabungssequenz *f*
f suite *f* de manutention
r последовательность *f* манипуляционных операций

6050 handling system
d Handhabungssystem *n*
f système *m* de manipulation
r манипуляционная система *f*

6051 handling task
d Handhabeaufgabe *f*
f problème *m* de manipulation
r манипуляционная задача *f*

6052 handling technique; manipulation technique
d Handhabungstechnik *f*; Handhabetechnik *f*
f technique *f* de manutention; technique de manipulation
r техника *f* манипулирования

6053 handling technique building block
d Handhabungstechnikbaukasten *m*; Baukasten *m* für Handhabungstechnik
f construction *f* par blocs pour technique de manutention
r модульная система *f* манипулирования; агрегатная система манипулирования

6054 handling technology
d Handhabungstechnologie *f*
f technologie *f* de manutention
r технология *f* манипулирования

6055 hand operation; manual control; manual setting
d Handregelung *f*; Handsteuerung *m*; Handbetätigung *f*
f commande *f* manuelle; réglage *m* à la main
r ручное регулирование *n*; ручная установка *f*

6056 hardness tester
d Härteprüfer *m*; Härteprüfgerät *n*
f duromètre *m*
r твердомер *m*; измеритель *m* твердости

6057 hardness-testing automaton
d Härteprüfautomat *m*
f automate *m* d'essai de dureté
r автомат *m* для испытания на твердость

6058 hardware
d Bausteine *mpl*; Bauelemente *npl*; Hardware *f*
f pièces *fpl* constructives; hardware *m*
r конструктивные элементы *mpl*; техническое оборудование *n*

6059 hardware area
d Hardwarebereich *m*
f domaine *m* d'hardware
r область *f* аппаратного обеспечения

6060 hardware compatibility
d Hardware-Kompatibilität *f*
f compatibilité *f* de matériel
r аппаратурная совместимость *f*

6061 hardware configuration
d Hardware-Konfiguration *f*
f configuration *f* de matériel
r конфигурация *f* аппаратных средств

6062 hardware control
d Hardware-Steuerung *f*; Steuerung *f* der Gerätetechnik
f commande *f* par hardware
r управление *n* аппаратурой

6063 hardware error
d Hardwarefehler *m*
f erreur *f* de l'hardware
r аппаратная ошибка *f*

6064 hardware failure
d Hardware-Ausfall *m*; Gerätetechnikausfall *m*
f défaillance *f* de matériel

r отказ *m* аппаратуры

6065 hardware fault
d Hardware-Fehler *m*; Gerätetechnikfehler *m*
f défaut *m* de matériel
r аппаратурная ошибка *f*

6066 hardware identification
d Hardware-Identifikation *f*; Identifikation *f* der Gerätetechnik
f identification-hardware *f*
r идентификация *f* оборудования

6067 hardware-implemented
d gerätetechnisch realisiert
f réalisé par matériel
r аппаратно реализованный

6068 hardware interface
d Hardware-Schnittstelle *f*; Gerätetechnikschnittstelle *f*
f interface *f* de matériel
r аппаратный интерфейс *m*

6069 hardware-interrupt
d Hardware Unterbrechung *f*; gerätetechnisch ausgelöste Unterbrechung *f*
f interruption *f* par matériel
r прерывание *n* от технических средств

6070 hardware priority interrupts
d Interruptprioritätsstaffelung *f* durch Hardware; verdrahtetes Interruptprioritätssystem *n*
f système *m* d'interruptions à priorité par matériel
r система *f* прерываний с аппаратно реализуемыми приоритетами

6071 hardware redundancy testing
d Testung *f* mittels Hardware-Redundanz
f test *m* à l'aide de redondance de matériel
r метод *m* контроля с помощью введенной аппаратурной избыточности

6072 hardware signal
d Hardwaresignal *n*
f signal *m* d'hardware
r сигнал *m* аппаратных средств

6073 hardware standardization
d Hardwarestandardisierung *f*
f standardisation *f* de l'hardware
r стандартизация *f* аппаратного обеспечения

6074 harmonic action
d harmonischer Eingriff *m*

f action *f* harmonique
r гармоническое воздействие *n*

6075 harmonic analysis
d harmonische Analyse *f*
f analyse *f* harmonique
r гармонический анализ *m*

* **harmonic analyzer** → 5665

6076 harmonic balance
d harmonische Balance *f*
f balance *f* harmonique; approximation *f* du premier harmonique
r гармонический баланс *m*

* **harmonic balance method** → 3973

6077 harmonic components
d harmonische Komponenten *fpl*; harmonische Teilschwingungen *fpl*
f composantes *fpl* harmoniques
r гармонические составляющие *fpl*

* **harmonic distortion** → 760

6078 harmonic distortion measuring set
d Klirrfaktormessgerät *n*
f mesureur *m* de distorsion non linéaire
r устройство *n* для измерения нелинейного искажения

6079 harmonic drive
d Harmonic Drive *n*; Wellgetriebe *n*
f entraînement *m* harmonique
r волновая передача *f*

6080 harmonic drive system
d Harmonic-Drive-System *n*; Wellgetriebesystem *n*
f système *m* harmonic drive
r система *f* волновых передач

6081 harmonic filter
d harmonisches Filter *n*
f filtre *m* harmonique
r фильтр *m* [для подавления] гармоник

6082 harmonic frequency converter
d harmonischer Frequenzwandler *m*; harmonischer Frequenzteiler *m*
f changeur *m* harmonique de fréquence
r преобразователь *m* частоты гармоник

6083 harmonic function of time
d harmonische Zeitfunktion *f*
f fonction *f* harmonique du temps
r гармоническая функция *f* времени

6084 harmonic linearization
d harmonische Linearisierung *f*
f linéarisation *f* harmonique
r гармоническая линеаризация *f*

6085 harmonic motion
d harmonische Bewegung *f*
f mouvement *m* harmonique
r гармоническое движение *n*

6086 harmonic oscillator
d harmonischer Oszillator *m*
f oscillateur *m* harmonique
r генератор *m* гармонических колебаний

6087 harmonics
d Harmonische *f*
f harmonique *f*
r гармоника *f*

6088 harmonic spectrum of signal
d harmonisches Signalspektrum *n*
f spectre *m* harmonique de signal
r гармонический спектр *m* сигнала

6089 harmonic synthesizer
d harmonischer Synthetisator *m*
f appareil *m* de synthèse d'harmoniques
r гармонический синтезатор *m*

6090 hazard analysis
d Gefährdungsanalyse *f*
f analyse *f* des dangers
r аварийный анализ *m*

6091 head amplifier; preamplifier
d Vorverstärker *m*
f préamplificateur *m*
r предусилитель *m*

6092 heat capacity
d Wärmekapazität *f*
f capacité *f* calorifique
r теплоёмкость *f*

6093 heat conductivity
d Wärmeleitfähigkeit *f*
f conductibilité *f* thermique
r теплопроводность *f*

6094 heat controller; heating controller; thermoregulator
d Wärmeregler *m*; Heizungsregler *m*; Thermoregler *m*
f régulateur *m* thermique; régulateur d'échauffement; thermorégulateur *m*
r терморегулятор *m*

6095 heat converter
  d Temperaturwandler *m*
  f convertisseur *m* thermique
  r термопреобразователь *m*

6096 heat effect
  d Wärmewirkungsgrad *m*; thermischer
     Wirkungsgrad *m*
  f rendement *m* thermique
  r тепловой коэффициент *m* полезного
     действия

6097 heat engineering
  d Wäremtechnik *f*; Heizungstechnik *f*
  f technique *f* thermique; technique du
     chauffage
  r теплотехника *f*

6098 heater power circuit
  d Heizenergiekreis *m*
  f circuit *m* d'incandescence; circuit d'énergie
     de chauffage
  r цепь *f* нагрева

6099 heat flow remote measurement
  d Wärmeflussfernmessung *f*
  f télémesure *f* du débit de chaleur
  r дистанционное измерение *n* теплового
     потока

6100 heat homing guidance
  d Zielansteuerung *f* durch Wärmewirkung
  f autoguidage *m* par chaleur
  r самонаведение *n* по тепловому излучению

  * **heating controller → 6094**

6101 heating effect
  d Erwärmungseffekt *m*
  f effet *m* de chauffage
  r тепловой эффект *m*

6102 heating plant
  d Heizanlage *f*
  f installation *f* de chauffage
  r отопительная установка *f*

6103 heating system
  d Heizsystem *n*
  f système *m* de chauffage
  r система *f* нагрева

  * **heating value → 2061**

6104 heat-limited servomechanism
  d Folgesystem *n* mit Erwärmungsbegrenzung;
     erwärmungsbegrenztes Folgesystem
  f système *m* de poursuite à limite
     d'échauffement

  r сервомеханизм *m* с ограничением по
     нагреву

6105 heat loss
  d Wärmeverlust *m*
  f pertes *fpl* par échauffement
  r тепловые потери *fpl*

6106 heat-oriented sensor
  d wärmeorientierter Sensor *m*
  f senseur *m* orienté à la chaleur
  r теплоориентированный сенсор *m*

6107 heat-sensitive sensor
  d wärmeempfindlicher Sensor *m*
  f senseur *m* sensible à la chaleur
  r термочувствительный сенсор *m*

6108 heat supervision
  d Wärmekontrolle *f*
  f contrôle *m* de chaleur
  r тепловой контроль *m*

6109 heat transfer
  d Wärmedurchgang *m*
  f passge *m* de chaleur; transmission *f* de
     chaleur; écoulement *m* thermique
  r теплопередача *f*

6110 heat transfer coefficient
  d Wärmeübergangsbeiwert *m*
  f coefficient *m* de transmission de chaleur
  r коэффициент *m* теплопередачи

6111 heat transmission counter flow principle
  d Gegenstromwärmeübertragungsprinzip *n*
  f principe *m* de transfert de chaleur à contre-
     courant
  r принцип *m* теплопередачи противотоком

6112 height adjustment
  d Höheneinstellung *f*
  f réglage *m* en hauteur; réglage d'altitude
  r регулировка *f* частоты

6113 height control
  d Höhenregelung *f*
  f regulation *f* de hauteur
  r управление *n* высотой; стабилизация *f*
     высоты

6114 height gauge
  d Höhenmesser *m*
  f altimètre *m*
  r высотомер *m*; альтиметр *m*

6115 helical scanning
  d Wendelabtastung *f*;
     Schraubenlinienabtastung *f*

f balayage *m* hélicoïdal
r винтовое развёртывание *n*

6116 helium-neon laser
d Helium-Neon-Laser *m*
f laser *m* à néon-hélium
r гелий-неоновый лазер *m*

6117 hermetic sealing of electronic devices
d hermetischer Verschluß *m* elektronischer
Apparaturen
f étanchéité *f* des appareils électroniques;
ecellement *m* des appareils électronique
r герметизация *f* электронной аппаратуры

6118 heterodyne signal
d Heterodynsignal *n*
f signal *m* hétérodyne
r гетеродинный сихнал *m*

6119 heterodyne wavemeter
d Interferenzwellenmesser *m*
f ondemètre *m* hétérodyne
r гетеродинный волномер *m*

6120 heterogeneous equation; inhomogeneous equation
d inhomogene Gleichung *f*
f équation *f* hétérogène
r неоднородное уравнение *n*

6121 heterogeneous reactor model
d heterogenes Reaktormodell *n*
f modèle *m* de réacteur hétérogène
r гетерогенная модель *f* реактора

6122 heteropolar bond
d heteropolare Bindung *f*
f liaison *f* hétéropolaire
r гетерополярная связь *f*

6123 heterostatic circuit
d heterostatische Schaltung *f*;
Quadrantenschaltung *f*
f circuit *m* hétérostatique
r гетеростатический контур *m*

6124 heterostatic instrument
d heterostatisches Messgerät *n*
f appareil *m* de mesure hétérostatique
r гетеростатический прибор *m*

6125 heuristic design method
d heuristische Entwurfsmethode *f*
f méthode *f* heuristique de projection
r эвристический метод *m* проектирования

6126 heuristic input data

d heuristische Eingabedaten *pl*
f données *fpl* d'entrée heuristiques
r эвристические вводные данные *pl*

6127 heuristic methods
d heuristische Methoden *fpl*
f méthodes *fpl* heuristiques
r эвристические методы *mpl*

6128 heuristic solution method
d heuristische Lösungsmethode *f*
f méthode *f* de solution heuristique
r эвристический метод *m* решения

* **HF-generator → 6150**

6129 hidden response
d latente Reaktion *f*
f réaction *f* latente
r скрытая реакция *f*

6130 hierarchically structured design
d hierarchisch strukturierter Entwurf *m*
f conception *f* à structure hiérarchique
r иерархически структурированное проектирование *n*

6131 hierarchical multicomputer system
d hierarchisches Mehrrechnersystem *n*
f système *m* hiérarchique de calculateurs multiples
r иерархическая многомашинная система *f*

6132 hierarchical structure for data managements
d hierarchische Struktur *f* für Datenmanagements
f structure *f* hiérarchique pour le management des données
r иерархическая структура *f* управления данными

6133 hierarchical system
d hierarchisch aufgebautes System *n*;
Hierarchiesystem *n*
f système *m* hiérarchique
r иерархическая система *f*

6134 hierarchy
d Rangordnung *f*; Hierarchie *f*
f hiérarchie *f*
r иерархия *f*

6135 high-active signal
d high-aktives Signal *n*; obenaktives Signal
f signal *m* actif en état supérieur
r сигнал m, активный при высоком уровне
[напряжения]

* **high-breaking-capacity contactor** → 6183

6136 high-capacity sensor
 d leistungsfähiger Sensor m
 f capteur m capable
 r эффективный сенсор m; мощный сенсор

* **high-data rate** → 5390

6137 high-definition lidar
 d Hochauflösungslidar m; Lidar m mit hohem
    Auflösungsvermögen
 f lidar m à résolution élevée
 r лазерный локатор m с высокой
    разрешающей способностью

6138 high-definition rangefinder
 d Entfernungsmesser m mit hoher Auflösung
 f télémètre m à grand pouvoir résolvant;
    télémètre à résolution élevée
 r дальномер m с высокой разрешающей
    способностью

6139 high-degree of automation
 d hoher Automatisierungsgrad m
 f grand degré m d'automation
 r высокая степень f автоматизации

6140 high-density circuit technique
 d hochdichte Schaltungstechnik f
 f technique f de circuits à haute densité
 r технология f схем с высокой плотностью
    упаковки

**6141 high-detectivity detector; high-sensitivity
    detector**
 d Detektor m mit hoher Empfindlichkeit;
    Detektor mit hohem Auflösungsvermögen
 f détecteur m à sensibilité élevée
 r детектор m с высокой [пороговой]
    чувствительностью

6142 high-energy laser
 d Hochleistungslaser m
 f laser m à haute énergie
 r лазер m с высокой энергией

6143 higher geometric programming language
 d höhere geometrische Programmiersprache f
 f language m de programmation géométrique
    supérieur
 r геометрический язык m
    программирования на высоком уровне

* **higher harmonic** → 9333

* **high frequency** → 10565

**6144 high-frequency alternator for ultrasonic
    transducer**
 d Ultraschallhochfrequenzmaschine f
 f alternateur m haute fréquence pour
    transducteur ultrasonique
 r генератор m высокой частоты для
    ультразвукового преобразователя

6145 high-frequency amplifier
 d Hochfrequenzverstärker m
 f amplificateur m haute fréquence
 r высокочастотный усилитель m

**6146 high-frequency analytical measuring
    method**
 d analytisches Hochfrequenzmessverfahren n
 f méthode f de mesure analytique à haute
    fréquence
 r высокочастотный аналитический метод m
    измерения

6147 high-frequency communication channel
 d Hochfrequenznachrichtenkanal m
 f canal m de télécommunication à haute
    fréquence
 r высокочастотный канал m связи

6148 high-frequency distortion
 d Hochfrequenzverzerrung f
 f distorsion f de haute fréquence
 r высокочастотное искажение n

6149 high-frequency filter
 d Hochfrequenzfilter n
 f filtre m haute fréquence
 r фильтр m высокой частоты

**6150 high-frequency generator; HF-generator;
    radio-frequency alternator**
 d Hochfrequenzgenerator m; HF-Generator m
 f générateur m haute fréquence; génératrice f à
    haute fréquence
 r высокочастотный генератор m; генератор
    высокой частоты

6151 high-frequency induction heating
 d Hochfrequenzinduktionsheizung f
 f chauffage m par induction haute fréquence
 r индукционный нагрев m токами высокой
    частоты

6152 high-frequency measuring technique
 d Hochfrequenzmesstechnik f
 f technique f de mesure haute fréquence
 r техника f высокочастотных измерений

**6153 high-frequency remote signalling
    apparatus**
 d Hochfrequenzfernmeldeapparat m

f appareil *m* de télécommunication par courant haute fréquence
r дистанционный сигнальный аппарат *m* высокой частоты

6154 high-frequency spectroscopy
d Hochfrequenzspektroskopie *f*
f spectroscopie *f* haute fréquence
r высокочастотная спектроскопия *f*

6155 high-frequency telemetering system
d Hochfrequenzfernmesssystem *n*
f système *m* haute fréquence de télémesure
r телеметрическая система *f* высокой частоты

6156 high-frequency variable
d schnell veränderliche Variable *f*
f variable *f* rapide changeante
r быстроменяющаяся переменная *f*

6157 high-gain amplifier
d Verstärker *m* mit hohem Gewinn
f amplificateur *m* à gain élevé
r усилитель *m* с большим коэффициентом усиления

6158 high-gain transition
d Hochgewinnübergang *m*
f transition *f* à gain élevé
r переход *m* с большим усилением

6159 high-grade component
d hochwertiges Bauteil *n*
f composant *m* de degré élevé
r высокоуровневый компонент *m*

6160 high-impedance output
d hochohmiger Ausgang *m*
f sortie *f* à impédance élevée
r высокоимпедансный выход *m*

* **high-impedance state → 5582**

6161 high-inverse-voltage rectifier
d Gleichrichter *m* für hohe Sperrspannungen
f redresseur *m* à haute tension inverse
r выпрямитель *m* обратного высокого напряжения

6162 high-level logic
d Großpegel-Logik *f*
f logique *f* de niveau élevé
r логическая схема *f* с высокими уровнями переключения

6163 high-level signal
d Signal *n* mit hohem Pegel

f signal *m* de niveau élevé
r сигнал *m* высокого уровня

6164 high-low control
d Stark-Schwach-Steuerung *f*; Stark-Schwach-Regelung *f*
f commande *f* tout ou peu; réglage *m* à deux états
r управление *n* типа разгон-торможение

6165 high-low level control
d Zweipunktniveauregelung *f*
f réglage *m* tout ou peu de niveau
r двухпозиционное регулирование *n* уровня

6166 high-order add circuit
d Addierkreis *m* löherer Ordnung
f circuit *m* sommateur additionneur d'ordres élevés
r схема *f* сложения старших разрядов

6167 high-order digit
d Zahl *f* hoher Ordnung
f chiffre *m* d'ordre élevé
r старший разряд *m*

6168 high-pass filter
d Hochpass *m*
f filtre *m* passe-haut
r фильтр *m* высокого пропускания

6169 high-performance system
d Hochleistungssystem *n*
f système *m* de grande puissance; système de haute performance
r система *f* высокой производительности

6170 high-power laser radiation
d Hochleistungslaserstrahlung *f*
f rayonnement *m* cohérent de grande puissance; rayonnement de laser de grande puissance
r высокоинтенсивное излучение *n* лазера

6171 high-power laser system
d Lasersystem *n* hoher Leistung
f système *m* laser à puissance élevée
r система *f* мощных лазеров

6172 high-pressure facility
d Hochdruckanlage *f*
f unité *f* à haute pression
r установка *f* высокого давления

6173 high-pressure process
d Hochdruckprozess *m*; Hochdruckverfahren *m*

f procédé *m* à haute pression; processus *m* à
haute pression
r процесс *m* при высоком давлении

6174 high-pressure pulse
d Hochdruckstoß *m*
f choc *m* de haute pression
r импульс *m* высокого давления

6175 high-pressure technology
d Hochdrucktechnologie *f*
f technologie *f* haute pression
r технология *f* высокого давления

6176 high-pressure transport
d Hochdruckförderung *f*
f transport *m* sous haute pression
r высоконапорный транспорт *m*

6177 high-priority event
d Ereignis *n* hoher Priorität
f événement *m* de haute priorité
r событие *n* с высоким приоритетом

6178 high-resolution detector
d Detektor *m* mit hohem Auflösungsvermögen
f détecteur *m* à pouvoir résolvant élevé;
détecteur à résolution élevée
r детектор *m* с высокой разрешающей
способностью

6179 high-sensitive laser detection system
d hochempfindliches Laserdetektionssystem *n*
f système *m* laser de détection à sensibilité
élevée
r высокочувствительная лазерная система *f*
детектирования

6180 high-sensitivity
d hochempfindlich
f à haute sensibilité
r высокочувствительный

*   **high-sensitivity detector** → 6141

*   **high-speed action controller** → 10533

6181 high-speed circuit
d Hochgeschwindigkeitsschaltung *f*;
Hochgeschwindigkeitsschaltkreis *m*
f circuit *m* à grande vitesse
r быстродействующая схема *f*

6182 high-speed computer
d Schnellrechner *m*
f calculatrice *f* à grande vitesse
r быстродействующий компьютер *m*

**6183 high-speed contactor; high-breaking-
        capacity contactor**
d Schnellschütz *n*
f contacteur *m* rapide;
contacteur-disjoncteur *m*
r быстродействующий контактор *m*

6184 high-speed control
d Schnellsteuerung *f*; Schnellregelung *f*
f commande *f* rapide
r скоростное регулирование *n*

*   **high-speed data transmission** → 5390

6185 high-speed logic
d Hochgeschwindigkeitslogik *f*
f logique *f* à grande vitesse
r быстродействующая логика *f*

*   **high-speed of operation** → 5392

**6186 high-speed response of remote control
        systems**
d schnelles Ansprechen *n* von
Fernsteuerungssystemen
f réponse *f* rapide de systèmes de
télécommande
r высокая скорость *f* реакции систем
дистанционного управления

6187 high-speed scientific computing unit
d wissenschaftliche Schnellrecheneinheit *f*
f unité *f* de calcul scientifique à grande vitesse
r быстродействующий процессор *m* для
научных расчётов

6188 high-speed servomechanism
d Schnellfolgesystem *n*
f système *m* de poursuite à action rapide
r быстродействующий сервомеханизм *m*

6189 high-temperature thermocouples
d Thermoelemente *npl* für hohe Temperaturen
f couples *mpl* thermo-électriques pour
températures élevées
r термопары *fpl* для высоких температур

6190 high-velocity scanning
d Abtastung *f* mit hoher Geschwindigkeit
f balayage *m* à vitesse élevée
r высокоскоростное сканирование *n*

6191 high-voltage accelerator
d Hochspannungsbeschleuniger *m*;
Hochvoltbeschleuniger *m*
f accélérateur *m* à haute tension
r высоковольтный ускоритель *m*

6192 **Hilbert transform**
  d Hilbert-Transformation *f*
  f transformation *f* d'Hilbert
  r преобразование *n* Гильберта

6193 **hill climbing**
  d Suchschrittmethode *f*
  f recherche *f* de l'extremum
  r поиск *m* экстремума

  * **hold circuit → 6196**

6194 **hold condition**
  d Haltebedingung *f*; Haltbedingung *f*
  f condition *f* de maintien
  r условие *n* блокировки

6195 **holding action**
  d Haltewirkung *f*
  f action *f* de maintien
  r воздействие *n* с запоминанием

6196 **hold[ing] circuit**
  d Haltekreis *m*; Haltestromkreis *m*;
      Halteschaltung *f*
  f circuit *m* de maintien
  r схема *f* блокировки

6197 **holding element**
  d Halteglied *n*
  f organe *m* de maintien
  r элемент *m* памяти

6198 **holding key**
  d Sperrschalter *m*; Halteschalter *m*;
      Halteschlüssel *m*
  f clé *m* de garde
  r ключ *m* блокировки

6199 **hold time**
  d Haltezeit *f*
  f temps *m* de halte
  r время *n* удержания

6200 **homing information**
  d Zielsuchdaten *npl*
  f données *fpl* du système d'autoguidage
  r данные *pl* вырабатываемые системой
      самонаведения

6201 **homing phase**
  d Zielauflugphase *f*
  f phase *f* d'autoguidage
  r этап *m* самонаведения

6202 **homing receiver**
  d Zielflugempfänger *m*; Zielpeilempfänger *m*
  f récepteur *m* de guidage automatique

  r приёмник *m* системы самонаведения

6203 **homing sensor**
  d Zielanflugfühler *m*
  f organe *m* sensible du système d'autoguidage
  r датчик *m* системы самонаведения

6204 **homogeneous automaton**
  d homogener Automat *m*
  f automate *m* homogène
  r однородный автомат *m*

6205 **homogeneous equation system**
  d homogenes Gleichungssystem *n*
  f système *m* d'équations homogènes
  r однородная система *f* уравнений

6206 **homogeneous system**
  d homogenes System *n*
  f système *m* homogène
  r однородная система *f*

6207 **homomorphic[al] model**
  d homomorphes Modell *n*
  f modèle *m* homomorphique
  r гомоморфная модель *f*

  * **homomorphic model → 6207**

6208 **homomorphic system**
  d homomorphes System *n*
  f système *m* homomorphique
  r гомоморфная система *f*

6209 **homopolar sequence power**
  d homöopolare Leistung *f*
  f puissance *f* homopolaire
  r гомополярная мощность *f*

6210 **homostructure**
  d Homostruktur *f*
  f homostructure *f*
  r гомоструктура *f*

6211 **homostructure laser**
  d Homostrukturlaser *m*
  f laser *m* homostructure
  r лазер *m* с гомоструктурой

6212 **Hopkinson's leakage coefficient**
  d Hopkinsonscher Streufaktor *m*
  f coefficient *m* de Hopkinson de dispersion
  r коэффициент *m* магнитной утечки
      Хопкинсона

6213 **horizontal-deflection amplifier**
  d Horizontalablenkverstärker *m*
  f amplificateur *m* de déviation horizontale
  r усилитель *m* строчной развёртки

6214 **horizontal-deflection circuit; horizontal-scanning circuit**
  d horizontale Ablenkschaltung *f*
  f circuit *m* de déviation horizontale
  r схема *f* горизонтальной развёртки

6215 **horizontal part of pulse**
  d Impulsdach *n*
  f partie *f* horizontale de l'impulsion
  r горизонтальная часть *f* импульса

6216 **horizontal polarization**
  d Horizontalpolarisation *f*; horizontale Polarisation *f*
  f polarisation *f* horizontale
  r горизонтальная поляризация *f*

  *   **horizontal-scanning circuit** → 6214

6217 **hot-strip instrument**
  d Hitzmessstreifengerät *n*
  f appareil *m* mesureur thermique à bande
  r тепловой измерительный прибор *m* с лентой

6218 **hot-wire instrument**
  d Hitzdrahtinstrument *n*
  f appareil *m* thermique à fil chaud
  r тепловой измерительный прибор *m*

6219 **humidification efficiency**
  d Befeuchtungswirkungsgrad *m*
  f rendement *m* d'humectation
  r коэффициент *m* полезного действия увлажнения

6220 **humidification system**
  d Befeuchtungssystem *n*
  f système *m* d'humidification
  r система *f* увлажнения

6221 **humidifying machinery**
  d Befeuchtungsausrüstung *f*
  f machinerie *f* d'humidification; installation *f* d'humectation
  r оборудование *n* для увлажнения

6222 **humidity control**
  d Feuchtigkeitsregelung *f*
  f réglage *m* d'humidité
  r регулирование *n* влажности

6223 **humidity controller**
  d Feuchtigkeitsregler *m*
  f régulateur *m* de l'humidité
  r регулятор *m* влажности

6224 **humidity feeler resistant to compression**
  d druckfester Feuchtefühler *m*
  f tâteur *m* d'humidité résistant à la compression
  r датчик *m* влажности, устойчивый против давления

6225 **humidity measuring instrument**
  d Feuchtigkeitsmessgerät *n*
  f appareil *m* de mesure de l'humidité
  r прибор *m* для измерения влажности; влагомер *m*

6226 **humidity meter of the gas under pressure**
  d Druckgasfeuchtigkeitsmesser *m*
  f hygromètre *m* du gaz sous pression
  r влагомер *m* [для] сжатого газа

6227 **humidity range**
  d Feuchtigkeitsbereich *m*
  f gamme *f* d'humidité
  r диапазон *m* влажности

6228 **humidity-sensitive element**
  d Feuchtigkeitsgeber *m*; Feuchtigkeitsabnehmer *m*
  f capteur *m* de l'humidité
  r влагочувствительный элемент *m*

6229 **hunting; parasitic oscillations**
  d Pendelung *f*; Regelschwankung *f*; Selbstausgleich *m*; parasitäre Schwingungen *fpl*
  f auto-équilibrage *m*; instabilité *f*; oscillations *fpl* parasitaires
  r рыскание *n*; паразитные колебания *npl*

6230 **hunting detection**
  d Feststellung *f* parasitärer Schwingungen
  f dépistage *m* des oscillations parasites
  r обнаружение *n* паразитных колебаний

6231 **hunting period**
  d Nachlaufperiode *f*; Wendeperiode *f*; Wendezeitraum *m*
  f période *f* de pompage
  r период *m* рыскания

6232 **hunt mode**
  d Fangbetrieb *m*
  f mode *m* de saisie
  r режим *m* активного ожидания

6233 **Hurwitz stability criterion**
  d Stabilitätskriterium *n* von Hurwitz
  f critère *m* de stabilité de Hurwitz
  r критерий *m* устойчивости Гурвица

6234 **hybrid circuit**
  d Hybridschaltung *f*

    f circuit *m* hybride
    r гибридная схема *f*

6235 hybrid computation technique application
    d Anwendung *f* der Hybridrechentechnik
    f application *f* de la technique de calcul
       hybride
    r применение *n* гибридной вычислительной
       техники

6236 hybrid computer
    d Hybridrechner *m*; gekoppelter Analog- und
       Digitalrechner *m*; hybrides Rechnersystem *n*
    f calculateur *m* hybride; calculatrice *f* hybride
    r гибридный компьютер *m*

6237 hybrid computer system evolution
    d Entwicklung *f* der Hybridrechnersysteme
    f développement *m* de systèmes de calculatrice
       hybride
    r разработка *f* гибридных вычислительных
       систем

6238 hybrid converter
    d hybrider Umsetzer *m*
    f convertisseur *m* hybride
    r гибридный преобразователь *m*

6239 hybrid digital-analog circuit
    d gemischte Digital-Analog-Schaltung *f*
    f circuit *m* numérique-analogique hybride
    r гибридная цифро-аналоговая схема *f*

    * **hybrid IC** → 6240

6240 hybrid integrated circuit; hybrid IC
    d Hybridschaltkreis *m*
    f circuit *m* intégré hybride; CI *m* hybride
    r гибридная интегральная схема *f*

6241 hybrid models
    d hybride Modelle *npl*
    f modèles *mpl* hybrides
    r гибридные модели *fpl*

6242 hybrid operation
    d hybride Arbeitsweise *f*
    f fonctionnement *m* hybride
    r гибридный режим *m* работы

6243 hybrid optimization
    d hybride Optimierung *f*; Optimierung mittels
       Hybridrechner
    f optimisation *f* hybride
    r гибридная оптимизация *f*

6244 hybrid program
    d hybrides Programm *n*; Programm zur

       Simulation hybrider Schaltungen
    f programme *m* hybride
    r гибридная программа *f*

6245 hybrid radar-infrared system
    d hybrides Radar- und
       Infrarotdetektionssystem *n*
    f système *m* hybride de radar et de détection à
       rayons infrarouges
    r гибридная инфракрасная
       радиолокационная система *f*

6246 hybrid technique
    d Hybridtechnik *f*
    f technique *f* hybride
    r гибридная техника *f*

6247 hydration plant
    d Hydratationsanlage *f*
    f unité *f* d'hydratation
    r гидратационная установка *f*

6248 hydration process
    d Hydratationsvorgang *m*
    f procédé *m* d'hydratation
    r процесс *m* гидратации

6249 hydraulic actuator
    d hydraulischer Stellantrieb *m*
    f servomoteur *m* hydraulique
    r гидравлический сервомотор *m*

6250 hydraulic amplifier
    d hydraulischer Verstärker *m*
    f amplificateur *m* hydraulique
    r гидравлический усилитель *m*

6251 hydraulic analogy
    d hydraulische Analogie *f*
    f analogie *f* hydraulique
    r гидравлическая аналогия *f*

6252 hydraulic circuit
    d hydraulischer Kreis *m*
    f circuit *m* hydraulique
    r гидравлическая схема *f*; гидросистема *f*

6253 hydraulic controller; hydraulic regulator;
       oil-operated controller
    d hydraulische Regelvorrichtung *f*;
       hydraulischer Regler *m*
    f régulateur *m* hydraulique
    r гидравлический регулятор *m*

6254 hydraulic differential analyzer
    d hydraulische Integrieranlage *f*; hydraulischer
       Differentialanalysator *m*

f analyseur *m* différentiel hydraulique
r гидравлический дифференциальный анализатор *m*

6255 hydraulic digital controls
d hydraulische Digitalsteuerungen *fpl*
f commandes *fpl* hydrauliques numériques
r цифровое управление *n* на гидравлических элементах

* **hydraulic drive → 8915**

6256 hydraulic-driven
d hydraulisch angetrieben
f entraîné hydrauliquement
r с гидравлическим приводом

6257 hydraulic drive of rotation angle
d hydraulischer Drehwinkelantrieb *m*
f entraînement *m* hydraulique de l'angle de rotation
r гидравлический сельсин-привод *m*

6258 hydraulic drive system
d hydraulisches Antriebssystem *n*
f système *m* d'entraînement hydraulique
r система *f* гидравлического привода

6259 hydraulic efficiency
d hydraulischer Wirkungsgrad *m*
f rendement *m* hydraulique
r гидравлический коэффициент *m* полезного действия

6260 hydraulic integrator
d hydraulischer Integrator *m*
f intégrateur *m* hydraulique
r гидравлический интегратор *m*

6261 hydraulic logic
d hydraulische Logik *f*
f logique *f* hydraulique
r гидравличрская логика *f*

6262 hydraulic positional servomechanism
d hydraulischer Positionierungsservomechanismus *m*; hydraulischer Verstellservomechanismus *m*; hydraulischer Folgesteuerungsmechanismus *m*
f servomécanisme *m* positionneur hydraulique
r гидравлический позиционный сервомеханизм *m*

* **hydraulic regulator → 6253**

6263 hydraulic remote transmission
d hydraulische Fernübertragung *f*

f transmission *f* hydraulique à distance
r гидравлическая дистанционная система *f* передачи

6264 hydraulic servosystem
d hydraulischer Regelkreis *m*
f circuit *m* de réglage hydraulique
r гидравлическая система *f* регулирования

6265 hydraulic tracer control
d hydraulische Fühlersteuerung *f*
f commande *f* à palpeur hydraulique
r управление *n* при помощи гидравлического следящего устройства

6266 hydrodynamic torque transformer
d hydrodynamischer Drehmomentwandler *m*
f convertisseur *m* hydrodynamique de couple
r гидродинамический преобразователь *m* крутящего момента

6267 hydroelectronic program control
d hydroelektronische Programmsteuerung *f*
f commande *f* de programme hydroélectronique
r гидроэлектронное программное управление *n*

6268 hydrogenation process
d Hydrierprozess *m*
f procédé *m* d'hydrogénation
r процесс *m* гидрирования

6269 hydropneumatic
d hydropneumatisch
f hydropneumatique
r гидропневматический

6270 hydrorefining plant
d Hydroraffinationsanlage *f*
f unité *f* d'hydroraffinage; unité d'hydrofining
r установка *f* гидроочистки

6271 hydrostatic drive
d hydrostatischer Antrieb *m*
f entraînement *m* hydrostatique
r гидростатический привод *m*

6272 hygroscope
d Feuchtigkeitsanzeiger *m*; Luftfeuchtigkeitsanzeiger *m*
f hygroscope *m*
r гигроскоп *m*

6273 hyperbolic function
d hyperbolische Funktion *f*
f fonction *f* hyperbolique
r гиперболическая функция *f*

6274 **hyperbolic navigation**
   d hyperbolische Navigation *f*
   f navigation *f* hyperbolique
   r гиперболическая система *f* навигации

6275 **hyperbolic velocity**
   d hyperbolische Geschwindigkeit *f*
   f vitesse *f* hyperbolique
   r скорость *f* при движении по
      гиперболической траектории

6276 **hypergeometric[al] probability**
   d hypergeometrische Wahrscheinlichkeit *f*
   f probabilité *f* hypergéométrique
   r гипергеометрическая вероятность *f*

   * **hypergeometric probability → 6276**

6277 **hypothesis**
   d Hypothese *f*
   f hypothèse *m*; supposition *f*
   r гипотеза *f*

6278 **hysteresigraph**
   d Hysterese[schleifen]schreiber *m*
   f hystérésigraphe *m*
   r гистерезиграф *m*

6279 **hysteresis characteristic**
   d Hysteresecharakteristik *f*
   f caractéristique *f* d'hystérésis
   r гистерезисная характеристика *f*

6280 **hysteresis constant**
   d Hysteresekonstante *f*
   f constante *f* d'hystérésis
   r постоянная *f* гистерезиса

6281 **hysteresis cycle**
   d Hysteresisschleife *f*
   f boucle *f* d'hystérésis
   r гистерезисный цикл *m*

6282 **hysteresis error**
   d Hysteresefehler *m*
   f erreur *f* due à l'hystérésis
   r погрешность f, вносимая гистерезисом

6283 **hysteresis losses**
   d Hystereseverluste *mpl*
   f pertes *fpl* par hystérésis
   r потери *fpl* на гистерезис

6284 **hysteresis meter**
   d Hysteresemesser *m*; Hysteresismesser *m*
   f hystérésimètre *m*
   r гистерезиметр *m*

6285 **hysteresis motor**
   d Hysteresemotor *m*
   f moteur *m* à hystérésis
   r гистерезисный двигатель *m*

6286 **hysteresis non-linearity**
   d Hysteresenichtlinearität *f*
   f non-linéarité *f* d'hystérésis
   r нелинейность f, вносимая гистерезисом

6287 **hysteresis phenomenon**
   d Hystereseerscheinung *f*
   f phénomène *m* d'hystérésis
   r явление *n* гистерезиса

I

* **I-action** → 6760

* **IC** → 6782

6288 idealized system
  d idealisiertes System *n*
  f système *m* idéalisé
  r идеализированная система *f*

6289 ideal pulse
  d Idealimpuls *m*
  f impulsion *f* idéale
  r идеальный импульс *m*

6290 ideal theory
  d ideale Theorie *f*
  f théorie *f* idéale
  r идеализированная теория *f*

6291 ideal value
  d Idealwert *m*
  f valeur *f* idéale
  r идеальное значение *n* величины

6292 identical equation
  d identische Gleichung *f*
  f équation *f* identique
  r идентичное уравнение *n*

6293 identical sensor
  d identischer Geber *m*
  f capteur *m* identique
  r идентичный сенсор *m*

6294 identifiable
  d identifizierbar; erkennbar; nachweisbar
  f possible à identifier
  r идентифицируемый

* **identification** → 5204

6295 identification algorithm
  d Erkennungsalgorithmus *m*
  f algorithme *m* d'identification
  r алгоритм *m* идентификации

6296 identification block
  d Identifikationsblock *m*
  f bloc *m* d'identification
  r идентификационный блок *m*

6297 identification check
  d Identifizierungsprüfung *f*
  f vérification *f* d'identification
  r идентификационный контроль *m*

6298 identification code
  d Kennungskode *m*; kodierte Kennung *f*
  f code *m* identificateur; code d'identification
  r идентификационный код *m*

* **identification label** → 6307

6299 identification method
  d Identifizierungsmethode *f*
  f méthode *f* d'identification
  r метод *m* идентификации

6300 identification procedure
  d Identifikationsprozedur *f*
  f procédure *f* d'identification
  r процедура *f* идентификации

6301 identification process
  d Identifizierungsverfahren *n*
  f procédé *m* d'identification
  r процесс *m* идентификации

6302 identification pulse
  d Kennungsimpuls *m*
  f impulsion *f* d'identification
  r импульс *m* идентификации

* **identification sensor** → 9506

6303 identification software
  d Erkennungssoftware *f*
  f software *m* d'identification
  r программное обеспечение *n* системы распознавания

6304 identified component
  d identifiziertes Bauelement *n*
  f composant *m* identifié
  r идентифицированное изделие *n*

6305 identified variable
  d identifizierte Variable *f*
  f variable *f* identifiée
  r идентифицированная переменная *f*

6306 identified work piece
  d identifiziertes Werkstück *n*
  f pièce *f* à travailler identifiée
  r распознанная обрабатываемая деталь *f*

6307 identifier; identification label
  d Identifizierer *m*;
      Identifizierungskennzeichen *n*

f identificateur *m*; label *m* d'identification
r идентификатор *m*; метка *f* идентификации

6308 identify *v*
d identifizieren; bezeichnen
f dentifier
r идентифицировать

6309 identifying code
d Erkennungskode *m*
f code *m* de reconnaissance
r опознавательный код *m*

6310 identifying of linear continuous systems
d Identifizierung *f* linearer stetiger Systeme
f identification *f* de systèmes continus linéaires
r идентификация *f* линейных непрерывных систем

6311 identifying signal
d Kennsignal *n*
f signal *m* d'identification
r сигнал *m* идентификации

6312 identity
d Identität *f*
f identité *f*
r идентичность *f*

6313 identity element
d Einselement *n*
f élément *m* unité
r единичный элемент *m*

6314 identity matrix
d Einheitsmatrix *f*
f matrice *f* unitaire
r единичная матрица *f*

6315 idiostatic circuit
d idiostatischer Stromkreis *m*; idiostatische Schaltung *f*
f montage *m* idiostatique
r идиостатическая схема *f*; идиостатический контур *m*

6316 idle current
d Leerlaufstrom *m*
f courant *m* à vide
r ток *m* холостого хода

* **idle cycle → 1827**

6317 idle mode
d Leerlaufbetriebsart *f*
f mode *m* à vide
r режим *m* простоя

6318 idle speed adjustment
d Leerlaufeinstellung *f*; Leerlaufgeschwindigkeitseinstellung *f*
f ajustage *m* de la vitesse à vide
r регулировка *f* скорости холостого хода

6319 idle state
d leerer Zustand *m*; operationsloser Zustand; Leerzustand *m*
f état *m* à vide; état sans opération
r состояние *m* простоя

* **idle time → 3770**

6320 idling appliance
d Leerlaufeinrichtung *f*
f dispositif *m* de ralenti; dispositif de marche à vide
r система *f* холостого хода

6321 idling conditions
d Leerlaufzustand *m*
f conditions *fpl* d'inactivité; état *m* de marche à vide
r режим *m* холостого хода

6322 idling cycle
d Leerlaufgang *m*
f cycle *m* à vide
r холостой цикл *m*

6323 idling switching-off automatic unit
d Leerlauf-Abschaltautomatik *f*
f automatisme *m* de disjonction pour marche à vide
r автоматика *f* останова при холостом ходе

* **IFAC → 6916**

6324 ignitron control
d Ignitronregelung *f*
f commande *f* par ignitron
r игнитронная регулировка *f*

6325 illegal operation
d unerlaubte Operation *f*
f opération *f* illégale
r запрещенная операция *f*

6326 illegal value
d unzulässiger Wert *m*
f valeur *f* illégale
r недопустимое значение *n*

6327 illogical logic
d verbotene Logik *f*
f logique *f* illogique
r запрещенная логика *f*

6328 illumination level
 d Beleuchtungspegel m
 f niveau m d'éclairage
 r уровень m освещенности

6329 image
 d Bild n
 f image f
 r изображение n

6330 image acquisition
 d Bilderfassung f
 f acquisition f d'image
 r сбор m данных отображения

6331 image analysis
 d Bildanalyse f
 f analyse f d'image
 r анализ m изображений

6332 image analysis software
 d Bildanalysesoftware f; Software f für
    Bildanalyse
 f software m pour analyse d'image
 r программное обеспечение n анализа
    изображений

6333 image converter tube
 d Bildwandlerröhre f
 f tube-convertisseur m
 r электронно-оптический
    преобразователь m

6334 image-forming infrared system
 d infrarotes Abbildungssystem n
 f système m infrarouge de formation d'image
 r инфракрасная система f формирования
    изображения

6335 image input interface
 d Bildeingabeinterface n
 f interface f d'entrée d'image
 r интерфейс m ввода изображения

 * image measuring → 9668

 * image of object → 8872

6336 image processing module
 d Bildverarbeitungsmodul m
 f module m de traitement d'image
 r модуль m обработки изображений

6337 image processing system
 d Bildverarbeitungssystem n
 f système m de traitement d'images
 r система f обработки изображений

6338 image sensor
 d Bildsensor m
 f senseur m d'image
 r сенсор m отображения

6339 image signal processing
 d Bildsignalverarbeitung f
 f traitement m de signal d'image
 r обработка f сигнала изображения

6340 image system
 d Abbildungssystem n
 f système m d'image
 r система f изображения

6341 image transducer system
 d Bildwandlersystem n
 f système m de transducteur d'image
 r система f преобразования оптического
    изображения в электронное

6342 image translator
 d Bildübertrager m
 f translateur m d'image
 r передатчик m изображения

6343 imaginary characteristic of an non-linear
      element
 d Imaginärcharakteristik f eines nichtlinearen
    Elementes
 f caractéristique f imaginaire d'élément non
    linéaire
 r мнимая характеристика f нелинейного
    элемента

6344 imaginary frequency response
 d imaginärer Frequenzgang m
 f caractéristique f imaginaire en fréquence
 r мнимая частотная характеристика f

6345 immediate access; instantaneous access
 d direkter Zugriff m; sofortiger Zugriff
 f accès m instantané; accès direct
 r немедленный доступ m

6346 immediate image system
 d Sofortbildsystem n
 f système m à images instantanées
 r система f с непосредственным
    отображением

6347 immersion gain
 d Immersionsgewinn m
 f gain m par immersion
 r коэффициент m усиления в результате
    иммерсии

 * immersion probe → 3961

6348 **immersion thermostat**
  d Tauchthermostat *m*
  f thermostat *m* à tige plongeante
  r иммерсионный термостат *m* для
     погружения

6349 **immiscibility analysis**
  d Nichtmischbarkeitsanalyse *f*
  f analyse *f* de l'immiscibilité
  r анализ *m* несмешаемости

6350 **immunity from electromagnetic
     interference**
  d Immunität *f* gegenüber elektromagnetischen
     Störungen
  f insensibilité *f* aux parasites
     électromagnétiques
  r нечувствительность *f* к электромагнитным
     помехам

6351 **impact acceleration**
  d Stoßbeschleunigung *f*
  f accélération *f* par choc
  r импульсное ускорение *n*

6352 **impedance**
  d Impedanz *f*; Scheinwiderstand *m*
  f impédance *f*
  r импеданс *m*; полное сопротивление *n*

6353 **impedance balancing block; impedance
     balancing unit**
  d Impedanzausgleichsglied *n*
  f équilibreur *m* d'impédance
  r блок *m* балансировки импедансов

  * **impedance balancing unit** → 6353

6354 **impedance coil**
  d Drosselspule *f*
  f bobine *f* d'impédance
  r дроссель *m*; реактивная катушка *f*

6355 **impedance comparator**
  d Impedanzkomparator *m*
  f comparateur *m* d'impédances
  r импедансный компаратор *m*

6356 **impedance control**
  d Impedanzsteuerung *f*
  f commande *f* d'impédance
  r импедансное управление *n*

6357 **impedance converter**
  d Impedanzwandler *m*
  f convertisseur *m* d'impédance
  r преобразователь *m* импеданса

6358 **impedance corrector**
  d Impedanzkorrektor *m*
  f correcteur *m* d'impédance
  r импедансный корректор *m*

6359 **impedance matching**
  d Scheinwiderstandsanpassung *f*;
     Impedanzanpassung *f*
  f adaptation *f* d'impédance
  r согласование *n* импедансов

6360 **impedance matching transformer**
  d Impedanzwandler *m*
  f transformateur *m* d'adaptation d'impédance
  r трансформатор *m* для согласования
     сопротивлений

6361 **impedance measurement**
  d Impedanzmessung *f*
  f mesure *f* d'impédance
  r импедансное измерение *n*

6362 **impedance protection**
  d Impedanzschutz *m*
  f dispositif *m* de protection à impédance
  r импедансная защита *f*

6363 **impedance transfer**
  d Impedanzübertragung *f*
  f transfert *m* à impédance
  r взаимный импеданс *m*

6364 **implemented programming system**
  d implementiertes Programmiersystem *n*
  f système *m* de programmation implementé
  r реализованная система *f*
     программирования

6365 **implication**
  d Implikation *f*
  f implication *f*
  r импликация *f*

6366 **implicit; implied**
  d implizit; inbegriffen; impliziert;
     miteinbegriffen
  f implicite; impliqué
  r неявный; подразумеваемый

6367 **implicit finite difference method**
  d implizites Differenzenverfahren *n*
  f méthode *f* des différences finies implicites
  r неявный метод *m* конечных разностей

6368 **implicit function**
  d implizite Funktion *f*
  f fonction *f* implicite
  r неявная функция *f*

*   implied → 6366

6369 improved circuit technique
 d verbesserte Schaltkreistechnik *f*
 f technique *f* améliorée de circuits
 r усовершенствованная схемотехника *f*

6370 improved precision; improvement of control quality
 d verbesserte Präzision *f*;
 Regelgüteverbesserung *f*
 f précision *f* meilleure; perfectionnement *m* de qualité de réglage
 r повышение *n* качества управления

6371 improved precision control
 d verbesserte Präzisionssteuerung *f*
 f commande *f* de précision meilleure
 r улучшенное прецизионное управление *n*

6372 improvement
 d Verbesserung *f*
 f amélioration *f*
 r улучшение *n*; усовершенствование *n*

*   **improvement of control quality** → 6370

6373 impulse; pulse
 d Impuls *m*
 f impulsion *f*
 r импульс *m*

6374 impulse accumulator; impulse store
 d Impulsspeicher *m*
 f accumulateur *m* des impulsions
 r импульсный накопитель *m*

6375 impulse approximation
 d Impulsannäherung *f*
 f approximation *f* impulsionnelle
 r импульсное приближение *n*

6376 impulse circuit
 d Impulsstromkreis *m*
 f circuit *m* d'impulsion
 r импульсный контур *m*; импульсная цепь *f*

6377 impulse code
 d Impulskode *m*
 f code *m* d'impulsions
 r импульсный код *m*

6378 impulse corrector
 d Impulskorrektor *m*
 f filtre *m* correcteur d'impulsions
 r корректор *m* импульсов

6379 impulse current withstand test
 d Stoßstromhalteprüfung *f*
 f essai *m* de tenue au courant de choc
 r испытание *n* импульсным током

6380 impulse differential counter
 d Impulsdifferenzzähler *m*
 f compteur *m* d'impulsions différentiel
 r дифференциальный счётчик *m* импульсов

6381 impulse element
 d Impulselement *n*; Impulsglied *n*; periodischer Taster *m*
 f élément *m* impulsionnel
 r импульсный элемент *m*

6382 impulse flashover voltage
 d Stoßüberschlagsspannung *f*
 f tension *f* d'éclatement au choc
 r импульсное пробивное напряжение *n*

6383 impulse frequency; pulse-recurrence frequence
 d Impulsfolgefrequenz *f*
 f fréquence *f* de répétition d'impulsions
 r частота *f* следования импульсов

6384 impulse-frequency telemetering
 d Impulsfrequenzfernmessung *f*
 f télémesure *f* à fréquence d'impulsions
 r телеизмерение *n* частотой [следования] импульсов

6385 impulse front
 d Impulsflanke *f*
 f flanc *m* d'impulsion; front *m* d'impulsion
 r фронт *m* импульса

6386 impulse function; pulse function
 d Impulsfunktion *f*
 f fonction *f* impulsionnelle
 r импульсная функция *f*

6387 impulse period
 d Impulsperiode *f*; Impulsdauer *f*
 f période *f* d'impulsions
 r период *m* импульсов

6388 impulse preselection
 d Impulsvorwahl *f*
 f présélection *f* d'impulsion
 r предварительный выбор *m* импульса

6389 impulse ratio modulation; pulse width modulation
 d Pulslängenmodulation *f*
 f modulation *f* de largeur d'impulsions
 r широтно-импульсная модуляция *f*

6390 impulse recorder
 d Impulsschreiber *m*; schreibender
  Impulszähler *m*
 f enregistreur *m* d'impulsions
 r регистратор *m* числа импульсов

6391 impulse response; pulse response
 d Impulsantwort *f*
 f réponse *f* d'impulsion; réaction *f* d'impulsion
 r реакция *f* на импульс

6392 impulse signal; pulse signal
 d Impulssignal *n*
 f signal *m* à impulsion
 r импульсный сигнал *m*

6393 impulse signalling
 d Impulssignalisieren *n*
 f signalisation *f* impulsionnelle
 r импульсная сигнализация *f*

 * **impulse store** → 6374

6394 impulse telemeter
 d Impulsfernmesser *m*
 f télémètre *m* à couplage par impulsions
 r телеметрическое устройство *n*
  импульсного типа

6395 impulse train
 d Impulsfolge *f*
 f train *m* d'impulsions
 r серия *f* импульсов

6396 impulse transmission in pneumatic lines
 d Impulsübertragung *f* in pneumatischen
  Leitungen
 f transmission *f* d'impulsions dans les lignes
  pneumatiques
 r передача *f* импульсов по пневматическим
  линиям

6397 impulse-type modulator
 d Impulsmodulator *m*
 f modulateur *m* d'impulsion
 r импульсный модулятор *m*

6398 impulse-type multiplier
  (for analog systems)
 d Impulsmultiplikationsgerät *n*
 f multiplicateur *m* à impulsion
 r импульсное умножающее устройство *n*

6399 impulse-type output amplifier
 d Impulsausgangsverstärker *m*
 f amplificateur *m* de sortie impulsionnel
 r импульсный выходной усилитель *m*

6400 impulse-voltage protection
 d Stoßspannungsschutz *m*
 f protection *f* au choc de tension
 r защита *f* от импульсного напряжения

6401 impulse wave
 d Stoßwelle *f*
 f onde *f* de choc électrique
 r импульсная волна *f*

6402 impulse wavetail
 d Stoßwellenrücken *m*
 f queue *f* de l'onde de choc
 r конец *m* импульса

6403 impurity analysis
 d Analyse *f* von Verunreinigungen
 f analyse *f* des impuretés
 r анализ *m* загрязнения

6404 impurity conductivity
 d Störstellenleitfähigkeit *f*
 f conductibilité *f* par impuretés
 r примесная проводимость *f*

6405 inaccessible value
 d gesperrter Wert *m*
 f valeur *f* inaccessible
 r недоступное значение *n*; недоступная
  величина *f*

 * **inactivation of a memory** → 8042

6406 inadmissible state
 d unzulässiger Zustand *m*
 f état *m* inadmissible
 r недопустимое состояние *n*

 * **inching** → 6408

6407 inching control
 d Tippbetriebsteuerung *f*
 f commande *f* par impulsions
 r управление *n* короткими включениями

6408 inching [service]; jogging service
 d Tippbetrieb *m*; Tastbetrieb *m*
 f commande *f* par fermetures successives
  rapides d'un circuit
 r медленное перемещение *n* в импульсном
  режиме

6409 in-circuit
 d schaltungsintern
 f interne au circuit
 r внутрисхемный

6410 in-circuit check
 d schaltungsinterne Prüfung *f*; Prüfung
  innerhalb einer Schaltung

f vérification *f* interne au circuit
r внутрисхемный контроль *m*

6411 in-circuit emulation
   d schaltungsinterne Emulation *f*
   f émulation *f* interne au circuit
   r внутрисхемная эмуляция *f*

6412 in-circuit emulator
   d schaltungsinnerer Emulator *m*; einsteckbarer
      Mikroprozessor-Emulator *m*
   f émulateur *m* interne au circuit
   r внутрисхемный эмулятор *m*

6413 inclusion
   d Inklusion *f*; [logische] Einschließung *f*
   f inclusion *f*
   r включение *n*

6414 inclusion relation
   d Inklusion *f*; Enthaltensein-Relation *f*
   f relation *f* d'inclusion
   r отношение *n* включения

6415 incoherence
   d Inkohärenz *f*
   f incohérence *f*
   r некогерентность *f*

6416 incoherent
   d inkohärent
   f incohérent
   r некогерентный

6417 incoherent analog modulation
   d unkohärente Analogmodulation *f*
   f modulation *f* analogue non cohérente
   r некогерентная аналоговая модуляция *f*

6418 incoherent detection
   d nichtkohärente Detektion *f*
   f détection *f* non cohérente
   r некогерентное детектирование *n*

6419 incoherent reception system
   d nichtkohärentes Empfängersystem *n*
   f système *m* récepteur non cohérent
   r некогерентная приёмная система *f*

6420 incoherent signal
   d nichtkohärentes Signal *n*
   f signal *m* incohérent
   r некогерентный сигнал *m*

6421 incoming-material control
   d Eingangskontrolle *f*;
      Eingangswarenkontrolle *f*
   f contrôle à l'entrée

r контроль *m* на входе

6422 incoming stream; input stream
   d Eintrittsstrom *m*; eintretender Strom *m*
   f courant *m* d'entrée; fux *m* entrant
   r поступающий поток *m*

6423 incompatibility
   d Inkompatibilität *f*; Nichtkompatibilität *f*;
      Unverträglichkeit *f*
   f incompatibilité *f*
   r несовместимость *f*

6424 incompatible microprocessor
   d nichtkompatibler Mikroprozessor *m*
   f microprocesseur *m* incompatible
   r несовместимый микропроцессор *m*

6425 incomplete convergence
   d Teilkonvergenz *f*
   f convergence *f* partielle
   r частичное схождение *n*

6426 inconsistent
   d überbestimmt
   f inconsistant
   r переопределенный

6427 inconsistent system of equation
   d unverträgliches System *n* von Gleichungen
   f système *m* inconstant; système disconvenant
   r несовместимая система *f* уравнений

6428 inconstancy
   d Unbeständigkeit *f*; Instabilität *f*
   f instabilité *f*
   r нестабильность *f*

6429 incorrect operation
   d unrichtige Operation *f*
   f fonctionnement *m* incorrect
   r неправильное срабатывание *n*

6430 increased productivity
   d Produktivitätssteigerung *f*
   f augmentation *f* de la productivité
   r повышение *n* производительности

6431 increasing oscillation
   d aufklingende Schwingung *f*
   f oscillation *f* croissante
   r возрастающее колебание *n*

6432 increment
   d Zuwachs *m*
   f incrément *m*
   r приращение *n*

6433 **increment** v
  d erhöhen; aufzählen
  f incrémenter; hausser; compter
  r инкрементировать; увеличивать

6434 **incremental angle sensor**
  d inkrementaler Winkelsensor m
  f capteur m angulaire incrémental
  r датчик m угловых приращений

6435 **incremental computer**
  d Inkrementrechner m
  f calculateur m incrémental
  r инкрементное вычислительное
    устройство n

6436 **incremental control**
  d Inkrementalregelung f
  f réglage m différentiel
  r инкрементальное регулирование n

6437 **incremental digital control**
  d inkrementale digitale Steuerung f
  f commande f digitale incrémentale
  r инкрементное цифровое управление n

6438 **incremental equivalent circuit**
  d Abgleichschaltung f; Abgleichvorrichtung f
  f circuit m incrémentiel équivalent
  r инкрементальная эквивалентная схема f

6439 **incremental indicator; incremental sensor**
  d Inkrementalgeber m
  f transmetteur m incrémental
  r датчик m приращения

6440 **incremental permeability**
  d zusätzliche Permeabilität f
  f perméabilité f additionnelle
  r дифференциальная магнитная
    проницаемость f

  * **incremental sensor** → 6439

6441 **incremental vector**
  d Inkrementvektor m
  f vecteur m d'incrément
  r инкрементный вектор m

6442 **incrementation parameter**
  d Schrittweitenparameter m
  f paramètre m d'incrémentation
  r параметр m величины шага

6443 **indefinite integral**
  d unbestimmtes Integral n
  f intégrale f indéfinie
  r неопределенный интеграл m

6444 **independent control**
  d unabhängige Regelung f
  f régulation f autonome
  r независимое регулирование n;
    независимое управление n

6445 **independent-controlled drive movement**
  d unabhängig gesteuerte Antriebsbewegung f
  f mouvement m d'entraînement commandé
    indépandant
  r приводное движение n с независимым
    управлением

6446 **independent control system**
  d autonomes Regelsystem n
  f système de réglage sans interactions
  r независимая система f регулирования;
    независимая система управления

6447 **independent manual operation**
  d unabhängiger Handbetrieb m
  f service m manuel indépendant
  r независимая ручная операция f

6448 **independent of the time; time-independent**
  d zeitunabhängig
  f indépendant de temps
  r независимы от времени

  * **independent portable gamma-ray
    spectrometer** → 1446

6449 **independent time-lag relay**
  d unabhängiges Zeitrelais n; konstantes
    Zeitverzögerungsrelais n
  f relais m à retard constant
  r реле n с независимой задержкой времени

6450 **independent variable**
  d unabhängige Veränderliche f
  f variable f indépendante
  r независимая переменная f

6451 **indeterminacy**
  d Indeterminiertheit f
  f indéterminisme m
  r неопределенность f

  * **index accumulator** → 1660

6452 **index of oscillation**
  d Schwingindex m
  f indice m d'oscillation
  r индекс m колебательности; показатель m
    колебательности

6453 **index of quality; quality index**
  d Güteparameter m

   f index *m* de qualité
   r показатель *m* качества

6454 index register
   d Adresseninkrementregister *n*
   f registre *m* d'accroissement d'adresse
   r регистр *m* адреса

6455 index value
   d Sollwert *m*
   f valeur *f* de consigne
   r заданное значение *n*

6456 indicated angle
   d angezeigter Winkel *m*
   f angle *m* affiché
   r индикаторный угол *m*; отмеченный угол

6457 indicated work
   d induzierte Arbeit *f*
   f travail *m* induit
   r индикаторная работа *f*

6458 indicating circuit
   d Indikatorstromkreis *m*
   f circuit *m* d'indication
   r индикаторный контур *m*

6459 indicating controller
   d anzeigender Regler *m*
   f régulateur *m* à indication
   r шкальный регулятор *m*; индикаторный контроллер *m*

6460 indicating instrument
   d Ablesegerät *n*; Anzeigegerät *n*
   f appareil *m* de mesure; indicateur *m*
   r показывающий прибор *m*; индикатор *m*

6461 indicating lamp
   d Signallampe *f*; Anzeigelampe *f*
   f lampe *f* indicatrice; lampe de signalisation
   r индикаторная лампа *f*; сигнальная лампа

6462 indicating range; indication range; range of indication
   d Anzeigebereich *m*
   f gamme *f* d'indication
   r диапазон *m* показаний

6463 indicating relay; indicator relay; annunciator relay
   d Anzeigerelais *n*; Melderelais *n*
   f relais *m* de signalisation
   r сигнальное реле *n*

6464 indicating self-balancing potentiometer
   d anzeigendes und selbsttätig abgleichendes

      Potentiometer *n*
   f potentiomètre *m* à auto-équilibrage et lecture directe
   r самобалансирующийся потенциометр *m* с индикатором

6465 indicating selsin
   d anzeigender Selsyn *m*
   f selsyn *m* indicateur
   r сельсин-индикатор *m*

6466 indication
   d Anzeige *f*; Hinweis *m*
   f indication *f*; affichage *m*
   r индикация *f*; указание *n*

   * **indication range** → 6462

6467 indication summation in telemetering
   d Fernmessanzeigensummierung *f*
   f sommation *f* d'indications en télémesure
   r суммирование *n* показаний при телеизмерении

6468 indicator diagram
   d Indikatordiagramm *n*
   f diagramme *m* d'indicateur
   r индикаторная диаграмма *f*

6469 indicator gate pulse
   d Gatterimpuls *m*
   f impulsion *f* de déclenchement
   r индикаторный обратный импульс *m*

   * **indicator relay** → 6463

6470 indirect-acting recording instrument
   d indirekt arbeitendes Registriergerät *n*
   f appareil *m* enregistreur à action indirecte
   r регистрирующий прибор *m* косвенного действия

6471 indirect action controller
   d Regler *m* mit Hilfsenergie; Hilfsenergieregler *m*
   f régulateur *m* à action indirecte
   r регулятор *m* непрямого действия

6472 indirect control
   d indirekte Steuerung *f*
   f commande *f* indirecte
   r непрямое управление *n*

6473 indirect control checking; indirect control supervision
   d indirekte Steuerungsüberwachung *f*
   f inspection *f* de commande indirecte
   r непрямое супервизорное управление *n*

*   indirect control supervision → 6473

6474 indirect control system; system with power amplification
d indirekt wirkendes System *n*; Regelungssystem *n* mit Hilfsenergie
f système *m* à action indirecte; système de commande à amplification
r система f, управляемая по косвенным параметрам

6475 indirect drive
d indirekter Antrieb *m*
f entraînement *m* indirect
r непрямой привод *m*

6476 indirect visual supervision
d indirekte visuelle Überwachung *f*
f inspection *f* visuelle indirecte
r непрямой супервизорный контроль *m*

6477 individual pulse
d Einzelimpuls *m*
f impulsion *f* individuelle; impulsion solitaire
r одиночный импульс *m*

6478 inductance
d Induktivität *f*
f inductance *f*
r индуктивность *f*

6479 inductance-capacitance delay line
d Reaktanz-Verzögerungslinie *f*; induktiv-kapazitive Verzögerungslinie *f*
f ligne *f* de retard à inductance et capacité
r реактивная линия *f* задержки; индуктивно-ёмкостная линия задержки

6480 inductance potential divider
d induktiver Potentialteiler *m*
f potentiomètre *m* inductif
r индуктивный делитель *m* напряжения

6481 inductance strain gauge
d Induktionsdehnungsmessstreifen *m*
f extensomètre *m* à fil; jauge *f* de contrainte à induction
r индуктивный тензометр *m*

6482 induction balance
d Induktionswaage *f*
f balance *f* d'induction
r индукционная уравновешенная схема *f*

6483 induction coupling; inductive coupling; magnetic coupling
d induktive Kopplung *f*; magnetische Kopplung

f couplage *m* inductif; accouplement *m* inductif
r индуктивная связь *f*

6484 induction flowmeter
d Induktionsdurchflussmesser *m*
f débitmètre *m* inductif
r индукционный расходомер *m*

6485 induction guard of liquid flow
d Induktionswächter *m* des Flüssigkeitsdurchflusses
f garde *f* à induction du débit de liquide
r индукционный ограничитель *m* расхода жидкости

6486 induction heater
d induktives Heizgerät *n*
f inducteur *m* de chauffage
r индукционный нагревательный прибор *m*

6487 induction-heating current frequency
d Induktionsheizungswechselstromfrequenz *f*
f fréquence *f* de courant pour chauffage par induction
r частота *f* тока индукционного нагрева

6488 induction motor
d Induktionsmotor *m*
f moteur *m* asynchrone; moteur à induction
r индукционный двигатель *m*

6489 induction tacho-generator
d Induktionsdrehzahlgeber *m*
f génératrice *f* tachymétrique asynchrone
r индукционный тахогенератор *m*

6490 induction voltage regulator
d Induktionsspannungsregler *m*; Drehumwandler *m*
f régulateur *m* de tension d'induction
r индукционный регулятор *m* напряжения

6491 induction wattmeter
d Drehfeldleistungsmesser *m*
f wattmètre *m* à champ tournant; wattmètre d'induction
r индукционный ваттметр *m*

*   inductive coupling → 6483

6492 inductive implication
d induktive Implikation *f*
f implication *f* inductive
r индуктивная импликация *f*

6493 inductive pick-off
d induktiver Abgriff *m*

    f capteur *m* inductif
    r индуктивный датчик *m*

6494 inductor
    d Drosselspule *f*; Induktivität *f*
    f inducteur *m*
    r катушка *f* индукции

6495 inductosyn scale
    d Induktosynmaßstab *m*
    f échelle *f* d'inductosyne
    r шкала *f* индуктосина

6496 industrial acquisition terminal
    d industrielles Erfassungsterminal *n*
    f terminal *m* de saisie industriel
    r промышленный терминал *m* сбора данных

6497 industrial application of manipulators
    d industrielle Manipulatoranwendung *f*
    f application *f* industrielle de manipulateurs
    r применение *n* манипуляторов в
       промышлености

6498 industrial automation
    d industrielle Automatisierung *f*
    f automatisation *f* industrielle
    r промышленная автоматизация *f*

6499 industrial control
    d Regelung *f* von Produktionsprozessen;
       selbsttätige Fertigungssteuerung *f*
    f réglage *m* des processus industriels;
       contrôle *m* industriel
    r регулирование *n* производственных
       процессов

6500 industrial control centre
    d Industriemesszentrale *f*
    f centrale *f* de mesure industrielle
    r промышленный контрольно-
       измерительный пункт *m*

6501 industrial control system
    d industrielles Steuerungssystem *n*
    f système *m* de commande industriel
    r промышленная система *f* управления

6502 industrial ecology
    d Industrieökologie *f*
    f écologie *f* industrielle
    r промышленная экология *f*

6503 industrial electronics
    d industrielle Elektronik *f*
    f électronique *f* industrielle
    r промышленная электроника *f*

6504 industrial handling
    d industrielle Handhabung *f*
    f manutention *f* industrielle
    r промышленное манипулирование *n*

6505 industrial handling technique
    d industrielle Handhabetechnik *f*
    f technique *f* de manutention industrielle
    r промышленная манипуляционная
       техника *f*

6506 industrial instrument
    d Betriebsinstrument *n*
    f appareil *m* industriel
    r технический прибор *m*

6507 industrial interface card
    d industrielle Interfacekarte *f*
    f carte *f* d'interface industrielle
    r промышленная интерфейсная карта *f*

6508 industrially usable indicator
    d industriell nutzbarer Geber *m*
    f transmetteur *m* d'usage industriel
    r датчик *m* промышленного применения

6509 industrially usable power source
    d industriell nutzbare Stromquelle *f*
    f source *f* de courant d'usage industriel
    r источник *m* питания для промышленного
       применения

6510 industrially usable sensor
    d industriell nutzbarer Sensor *m*
    f capteur *m* d'usage industriel
    r сенсор *m* для промышленного применения

6511 industrial modular check system
    d industrielles modulares Kontrollsystem *n*
    f système *m* modulaire de contrôle industriel
    r промышленная контрольная система *f*
       модульного типа

6512 industrial process control
    d Industrieprozesssteuerung *f*
    f commande *f* de procédés industriels
    r управление *n* производственным
       процессом

6513 industrial process simulation
    d Modellierung *f* von Produktionsvorgängen
    f simulation *f* de processus industriels
    r моделирование *n* производственных
       процессов

6514 industrial production
    d Industrieproduktion *f*
    f production *f* industrielle
    r промышленное производство *n*

6515 industrial remote signalling
  d industrielle Fernsignalisierung *f*;
    Betriebsfernmeldung *f*
  f télésignalisation *f* industrielle
  r производственная дистанционная
    сигнализация *f*

6516 industrial research
  d Betriebsforschung *f*; Industrieforschung *f*
  f recherche *f* industrielle
  r промышленное исследование *n*

6517 [industrial] robot
  d Roboter *m*; Industrieroboter *m*
  f robot *m*; robot industriel
  r робот *m*; промышленный робот

6518 [industrial] robot drive
  d Roboterantrieb *m*; Antrieb *m* eines
    Industrieroboters
  f entraînement *m* de robor [industriel]
  r привод *m* [промышленного] робота

6519 [industrial] robot hardware solution
  d Roboterhardwarelösung *f*; Industrieroboter-
    Hardwarelösung *f*
  f solution *f* d'hardware pour les robots
    industriels
  r [выбранное] решение *n* технических
    средств [промышленного] робота

  * industrial robot operation → 936

6520 [industrial] robot regulating system
  d Roboterregelsystem *n*; Regelsystem *n* eines
    Industrieroboters
  f système *m* de régulation de robot industriel
  r система *f* регулирования
    [промышленного] робота

6521 [industrial] robot regulator
  d Roboterregler *m*; Industrieroboterregler *m*
  f régulateur *m* d'un robot [industriel]
  r регулятор *m* [промышленного] робота

  * industrial robots technique → 11087

  * industrial robot technique → 11087

6522 industrial safety
  d Arbeitssicherheit *f*
  f sécurité *f* des travailleurs
  r безопасность *f* труда

6523 industrial telemetering system
  d industrielles Fernmesssystem *n*
  f système *m* industriel de télémesure
  r промышленная телеметрическая система *f*

6524 industry specifications; industry standard
  d Industrienorm *f*; Industriestandard *m*
  f norme *f* industrielle
  r промышленные требования *npl*;
    промышленные нормы *fpl*;
    промышленный стандарт *m*

  * industry standard → 6524

6525 inert gas circuit
  d Inertgaskreislauf *m*
  f circuit *m* de gaz inerte
  r схема *f* циркуляции инертного газа

6526 inert gas plant
  d Inertgasanlage *f*
  f unité *f* de gaz inerte
  r установка *f* по производству инертного
    газа

6527 inertia
  d Beharrungsvermögen *n*; Trägheit *f*
  f inertie *f*; conservation *f* des forces vives
  r инерционность *f*; инерция *f*

6528 inertia constant
  d Trägheitskonstante *f*
  f constante *f* d'inertie
  r постоянная *f* инерции

6529 inertial guidance
  d Trägheitslenkung *f*
  f guidage *m* inertiel
  r инерциальное наведение *n*; инерциальное
    управление *n*

6530 infinite
  d unendlich; infinit
  f infini; sans fin
  r бесконечный

6531 infinite automaton
  d unendlicher Automat *m*
  f automate *m* infini
  r бесконечный автомат *m*

6532 infinite degree of stability
  d unendlicher Stabilitätsgrad *m*
  f degré *m* infini de stabilité
  r неограниченная степень *f* устойчивости

  * infinitely fine control → 3130

6533 infinitely variable speed gearing
  d stufenloses Regelgetriebe *n*
  f mécanisme *m* continu réglable
  r зубчатая передача *f* для непрерывного
    регулирования скорости

* **infinite-memory filter** → 5488

6534 influence of disturbance variable
d Störgrößeneinfluß *m*
f influence *f* d'un grandeur perturbatrice
r влияние *n* возмущающего воздействия

6535 influence size
d Einflußgröße *f*
f grandeur *f* d'influence
r параметр *m* влияния

6536 influence size of work piece
d Werkstückeinflußgröße *f*
f grandeur *f* d'influence de pièce à usiner
r характерный параметр *m*
⌈обрабатываемой⌉ детали

6537 influencing function
d Eingriffsfunktion *f*
f fonction *f* d'influence
r функция *f* воздействия

6538 influencing value
d Einflußgröße *f*
f grandeur *f* d'influence
r воздействующая величина *f*

* **influencing variable** → 359

6539 information
d Information *f*
f information *f*
r информация *f*

6540 information acquisition sensor
d Informationserfassungssensor *m*
f senseur *m* d'acquisition d'information
r сенсор *m* сбора информации

6541 information and control system
d Informations- und Steuerungssystem *n*
f système *m* d'information et de commande
r информационная и управляющая
система *f*

* **information bit** → 6579

6542 information build-up
d Informationsaufbau *m*; Nachrichtenaufbau *m*
f établissement *m* d'information
r информационная структура *f*

6543 information carrier
d Informationsträger *m*
f porteur *m* d'information
r носитель *m* информации

6544 information channel
d Informationskanal *m*; Datenkanal *m*
f canal *m* d'information; canal de
communication
r информационный канал *m*

6545 information checking
d Informationskontrolle *f*
f examen *m* d'information
r контроль *m* информации

6546 information circuit
d Informationskreis *m*
f circuit *m* d'information
r информационная цепь *f*

6547 information coding levels
d Informationskodierungsniveaus *npl*
f niveaux *mpl* de codage d'information
r уровни *mpl* кодирования информации

6548 information content
d Informationsinhalt *m*
f contenu *m* en informations
r содержание *n* информации

6549 information cycle
d Informationszyklus *m*
f cycle *m* d'information
r цикл *m* информации

6550 information density
d Informationsdichte *f*
f densité *f* d'information
r плотность *f* информации

6551 information engineering
d Informationstechnik *f*
f technique *f* d'information
r информационная техника *f*

6552 information exchange
d Informationsaustausch *m*
f échange *m* d'informations
r информационный обмен *m*

6553 information extraction
d Informationsextraktion *f*
f extraction *f* d'information
r извлечение *n* информации

6554 information flow
d Informationsstrom *m*
f flux *m* d'information
r поток *m* информации

6555 information format
d Informationsformat *n*

f format *m* d'information
r формат *m* информации

*   **information loss** → **3708**

6556   **information lossless automaton**
d Automat *m* ohne Informationsverluste
f automate *m* sans pertes d'information
r автомат *m* без потери информации

6557   **information management system**
d Informationsmanagementsystem *n*
f système *m* pour le management
   d'information
r информационно-управляющая система *f*

6558   **information parameter**
d Informationsparameter *m*
f paramètre *m* d'information
r параметр *m* информации

6559   **information pattern**
d Informationsspektrum *n*
f image *f* d'information; spectre *m*
   d'information
r спектр *m* информации

6560   **information perceptive system**
d Informationswahrnehmungssystem *n*
f système *m* de perception pour prendre des
   informations
r система *f* восприятия информации

6561   **information processing**
d Informationsverarbeitung *f*
f traitement *m* d'information
r обработка *f* информации

6562   **information processing system**
d Informationsverarbeitungssystem *n*
f système *m* de traitement d'information
r система *f* обработки информации

6563   **information processing theory**
d Informationsverarbeitungstheorie *f*
f théorie *f* du traitement d'information
r теория *f* обработки информации

6564   **information quantity**
d Informationsmenge *f*
f quantité *f* d'information
r количество *n* информации

6565   **information reception**
d Informationsaufnahme *f*
f réception *f* d'information
r приём *m* информации

6566   **information representation**
d Informationsdarstellung *f*
f représentation *f* d'information
r представление *n* информации

6567   **information retrieval**
d Informationserschließung *f*
f recouvrement *m* des informations
r восстановление *n* информации

6568   **information selection**
d Informationsauswahl *f*
f sélection *f* d'informations
r информационная селекция *f*

6569   **information signal**
d Informationssignal *n*
f signal *m* d'information
r информационный сигнал *m*

6570   **information system**
d Informationssystem *n*
f système *m* d'information
r информационная система *f*

6571   **information system planning**
d Informationssystemplanung *f*
f planification *f* d'un système d'information
r проектирование *n* информационных
   систем

6572   **information-technical coupling of machine
       system**
d informationstechnische Kopplung *f* eines
   Maschinensystems
f couplage *m* technique de l'information d'un
   système de machine
r машинная система *f* с информационной
   связью

6573   **information technology**
d Informationstechnologie *f*
f technologie *f* d'information
r информационная технология *f*

6574   **information terminal**
d Informationsterminal *n*
f terminal *m* des informations
r информационный терминал *m*

6575   **information theory**
d Informationstheorie *f*; Theorie *f* der
   Informations- und Nachrichtenübertragung
f théorie *f* de l'information
r теория *f* информации

*   **information transfer** → **6576**

6576 information transmission; information transfer
d Informationsübertragung *f*
f transmission *f* d'information
r передача *f* информации

6577 information transmission rate
d Informationsübertragungsgeschwindigkeit *f*
f vitesse *f* de transmission d'information
r скорость *f* передачи информации

6578 information transmitter
d Nachrichtenquelle *f*; Informationssender *m*
f transmetteur *m* d'information
r передатчик *m* информации

6579 information unit; information bit
d Nachrichtenelement *n*; Nachrichteneinheit *f*;
   Informationsbit *n*
f unité *f* d'information; bit *m* d'information
r единица *f* информации; бит *m*

6580 information winning
d Informationsgewinnung *f*
f élaboration *f* des informations
r получение *n* информации

6581 informatized control system
d informatisiertes Steuersystem *n*
f système *m* de commande informatisé
r информационная система *f* управления

6582 infrared analyzer of gases
d infrarotes Analysiergerät *n* für Gase
f analyseur *m* infrarouge de gaz
r инфракрасный анализатор *m* газов

6583 infrared communication equipment
d Infrarotstrahlenverbindung *f*
f appareillage *m* de télécommunications à
   rayons infrarouges
r аппаратура *f* инфракрасной системы связи

6584 infrared emissing ability
d Infrarotstrahlungsvermögen *n*
f capacité *f* d'émission en infrarouge; aptitude *f*
   d'émission en infrarouge; pouvoir *m*
   d'émission en infrarouge
r инфракрасная излучательная
   способность *f*

6585 infrared equipment
d Infrarotanlage *f*; Infrarotsystem *n*
f système *m* à lumière infrarouge
r инфракрасная установка *f*

6586 infrared Fourier transform spectrometry
d Infrarotstrahlenspektrometrie *f* mit Fourier-
   Transformation
f spectrométrie *f* infrarouge à transformée de
   Fourier
r инфракрасная спектроскопия *f* с
   использованием преобразования Фурье

6587 infrared gas analyzer
d Infrarotgasanalysator *m*
f analyseur *m* infrarouge des gaz
r инфракрасный газоанализатор *m*

6588 infrared homing action
d Infrarotziellenkung *f*
f autoguidage *m* à rayons infrarouges
r самонаведение *n* по информационному
   излучению

6589 infrared identification
d Infrarotstrahlenkennung *f*; Kennung *f* durch
   Infrarotstrahlen
f identification *f* à rayons infrarouges
r идентификация *f* по инфракрасному
   излучению

6590 infrared input flow
d Eingangsinfrarotstrahl *m*
f flux *m* d'entrée de rayons infrarouges
r поток *m* инфракрасного излучения

6591 infrared laser radiation
d Laserinfrarotstrahlung *f*
f rayonnement *m* infrarouge de laser
r инфракрасное излучение *n* лазера

6592 infrared modulation
d Infrarotmodulation *f*
f modulation *f* infrarouge
r модуляция *f* инфракрасного излучения

6593 infrared pulse modulation system
d impulsmoduliertes Infrarotsystem *n*
f système *m* infrarouge à modulation
   d'impulsions
r инфракрасная система *f* с импульсной
   модуляцией

6594 infrared radiation detecting system
d Infrarotstrahlungserfassung *f*;
   Infrarotstrahlungsauffindung *f*
f système *m* de détection du rayonnement
   infrarouge
r система *f* обнаружения инфракрасного
   излучения

6595 infrared scanning device
d Infrarotstrahlenabtastgerät *n*
f dispositif *m* de balayage à rayons infrarouges
r инфракрасное сканирующее устройство *n*

6596 infrared search system
d Infrarotzielsuchsystem *n*
f système *m* infrarouge de recherche; système infrarouge d'exploration
r инфракрасная поисковая система *f*

6597 infrared sensing element
d infrarotempfindliches Element *n*
f palpeur *m* à rayons infrarouges
r инфракрасный чувствительный элемент *m*

6598 infrared sensor
d Infrarotsensor *m*
f senseur *m* à infrarouges; capteur *m* infrarouge
r инфракрасный сенсор *m*

6599 infrared signal entropy
d Infrarotsignalentropie *f*
f entropie *f* du signal infrarouge
r энтропия *f* инфракрасного сигнала

6600 infrared spectrometer detector
d Infrarotspektrometerdetektor *m*
f détecteur *m* du spectromètre à l'infrarouge
r детектор *m* инфракрасного спектрометра

6601 infrared spectroscopic examination of samples
d infrarotspektroskopische Probenuntersuchung *f*
f examination *f* des échantillons par spectroscopie infrarouge
r испытание *n* образцов методом инфракрасной спектроскопии

6602 infrared surveillance system
d Infrarotstrahlenüberwachungssystem *n*
f système *m* de surveillance à rayons infrarouges
r обзорная инфракрасная система *f*

6603 infrasonic frequency
d Infraschallfrequenz *f*
f fréquence *f* infrasonore; fréquence infra-acoustique
r инфразвуковая частота *f*

6604 inherent feedback
d innere Rückkopplung *f*
f autoréaction *f*; réaction *f* propre
r внутренная обратная связь *f*

6605 inherent regulation rate
d Selbstregulierungsgeschwindigkeit *f*
f vitesse *f* d'autorégulation
r скорость *f* саморегулирования

6606 inherent stability
d Eigenstabilität *f*
f stabilité *f* propre
r собственная устойчивость *f*

6607 inherited error
d mitgeschleppter Fehler *m*
f erreur *f* entraînée
r привнесенная ошибка *f*

6608 inhibit *v*
d sperren; verhindern
f inhiber
r запрещать

6609 inhibiting circuit; time delay circuit
d Sperrschaltung *f*; Verzögerungskreis *m*
f circuit *m* inhibiteur; circuit bloqueur; circuit de retard
r схема *f* задержки; задерживающая цепь *f*

6610 inhibiting input
d Verbotssignal *n*
f signal *m* d'inhibition
r запрещающий вход *m*

6611 inhibiting signal; disabling signal
d Sperrsignal *n*; Blockiersignal *n*
f signal *m* inhibiteur
r сигнал *m* запрета

6612 inhibit pulse
d Sperrimpuls *m*
f impulsion *f* d'inhibition
r импульс *m* запрета

* **inhomogeneous equation** → 6120

6613 inhomogeneous magnetic field; non-homogeneous magnetic field
d inhomogenes Magnetfeld *n*; nichthomogenes Magnetfeld
f champ *m* magnétique inhomogène; champ magnétique non homogène
r неоднородное магнитное поле *n*

6614 initial address
d Anfangsadresse *f*
f adresse *f* initiale
r начальный адрес *m*

6615 initial adjustment
d Anfangseinstellung *f*
f mise *f* au point initiale
r начальная установка *f*; предварительная установка

6616 initial automaton
d Anfangsautomat *m*

f automate *m* initial
r начальный автомат *m*

6617 initial conditions; starting conditions
d Anfangsbedingungen *fpl*; Anfangswerte *mpl*;
Startbedingungen *fpl*
f conditions *fpl* initiales; conditions de départ
r начальные условия *npl*; исходные условия

6618 initial data
d Anfangsdaten *pl*
f données *fpl* initiales
r начальные данные *pl*

6619 initial displacement
d Anfangsstörung *f*
f trouble *m* initial; dérangement *m* initial
r начальное воздействие *n*; начальное
возмущение *n*

6620 initial failure
d Frühausfall *m*
f erreur *f* initiale
r начальное повреждение *n*

*   **initial internal pressure → 10025**

6621 initialization; initializing
d Initialisierung *f*; Initialisieren *n*; Herstellen *n*
von Anfangsbedingungen
f initialisation *f*
r инициализация *f*; установка *f* в начальное
состояние

*   **initializing → 6621**

6622 initial position
d Ausgangsposition *f*
f position *f* initiale
r исходное положение *n*

6623 initial rate; initial speed
d Anfangsgeschwindigkeit *f*
f vitesse *f* initiale
r начальная скорость *f*

6624 initial selection sequence
d Anfangsauswahlfolge *f*
f séquence *f* initiale de sélection
r начальная последовательность *f* выбора

*   **initial speed → 6623**

6625 initial stability
d Anfangsstabilität *f*
f stabilité *f* initiale
r начальная устойчивость *f*

6626 initial state
d Anfangszustand *m*
f état *m* initial; situation *f* initiale
r начальное состояние *n*

6627 initial susceptibility
d Anfangsaufnahmefähigkeit *f*;
Anfangssuszeptibilität *f*
f susceptibilité *f* initiale
r начальная [магнитная] восприимчивость *f*

6628 initial system
d Ausgangssystem *n*
f système *m* initial
r исходная система *f*

6629 initial test routine
d Anfangstestprogramm *n*
f programme *m* de test initial
r начальная программа *f* контроля

6630 initial value; start value
d Anfangswert *m*; Ausgangswert *m*;
Startwert *m*
f valeur *f* initiale
r начальное значение *n*

6631 initial-value problem
d Anfangswertproblem *n*
f problème *m* de valeur initiale
r начальная задача *f*; задача Коши

6632 initial value theorem
d Lehrsatz *m* vom Anfangswert
f théorème *m* de valeur initiale
r теорема *f* о начальном значении

6633 initiating fuse element
d Steuerschmelzleiter *m*
f conducteur *m* fusible de commande
r запускающий плавкий элемент *m*;
регулирующий плавкий элемент

6634 initiating pulse; release pulse
d Startimpuls *m*; Auslöseimpuls *m*
f impulsion *f* de déclenchement
r пусковой импульс *m*; размыкающий
импульс

6635 injection contact
d Injektionskontakt *m*
f contact *m* d'injection
r инъекцирующий контакт *m*

6636 inlet system
d Einlaßsystem *n*
f système *m* d'admission
r приёмная система *f*; впускная система

6637 **inlet temperature control**
  d Eintrittstemperaturregelung *f*
  f réglage *m* de température d'entrée
  r регулирование *n* температуры на входе

6638 **inner loop**
  d Innenschleife *f*
  f boucle *f* intérieure
  r внутренный контур *m*; побочный контур

6639 **inopportune operation**
  d ungelegene Betätigung *f*; unpassendes
     Ansprechen *n*
  f fonctionnement *m* intempestif
  r несвоевременное действие *n*

6640 **inphase amplitude detection**
  d Phasengleichheitsdetektion *f*;
     Inphasedetektion *f*; Detektion *f* des
     gleichphasigen Zustandes
  f détection *f* de cophasage; détection de
     synchronisme
  r амплитудное детектирование *n*
     синфазного сигнала

6641 **inphase detector**
  d Inphasedetektor *m*; Gleichphasendetektor *m*
  f détecteur *m* de synchronisme
  r синфазный детектор *m*

6642 **in-phase signal**
  d gleichphasiges Signal *n*
  f signal *m* en phase
  r синфазный сигнал *m*

6643 **input action**
  d Eingabeverfahren *n*
  f action *f* d'entrée
  r входное воздействие *n*

6644 **input amplifier**
  d Eingangsverstärker *m*
  f amplificateur *m* d'entrée
  r входной усилитель *m*

6645 **input amplitude range**
  d Eingangsamplitudenbereich *m*
  f champ *m* d'amplitude d'entrée
  r амплитудный диапазон *m* входных
     сигналов

6646 **input area**
  d Eingabebereich *m*
  f domaine *m* d'entrée
  r область *f* ввода

6647 **input bus type**
  d Eingabe-Bustyp *m*

  f type *m* de bus d'entrée
  r входная шина *f*

6648 **input channel**
  d Eingabekanal *m*
  f canal *m* d'entrée
  r входной канал *m*

6649 **input control unit**
  d Eingabesteuereinheit *f*
  f unité *f* de commande d'entrée
  r устройство *n* управления вводом

6650 **input coordinate**
  d Eingangskoordinate *f*
  f coordonnée *f* d'entrée
  r входная координата *f*

6651 **input current**
  d Eingangsstrom *m*
  f courant *m* d'entrée
  r входной ток *m*

6652 **input data**
  d Eingangsdaten *pl*; Eingabedaten *pl*
  f données *fpl* d'entrée
  r входные данные *pl*

6653 **input device; input element**
  d Eingangsvorrichtung *f*
  f dispositif *m* d'entrée
  r вводное устройство *n*; входное устройство

  * **input element → 6653**

6654 **input enable signal**
  d Eingabe-Freigabesignal *n*; Eingabetor-
     Aktivierungssignal *n*
  f signal *m* de libération d'entrée
  r сигнал *m* разрешения ввода

6655 **input equipment**
  d Eingabegerät *n*
  f appareillage *m* d'entrée
  r входная аппаратура *f*

6656 **input filter**
  d Eingangsfilter *m*
  f filtre *m* d'entrée
  r входный фильтр *m*

6657 **input flow**
  d Eingabefluss *m*; Eingabestrom *m*
  f flux *m* entrant
  r входный поток *m*

6658 **input function**
  d Eingabefunktion *f*

f fonction f d'entrée
r входная функция f

6659 input grid capacity
d Eingangsgitterkapazität f
f capacité f d'entrée de la grille
r входная ёмкость f сетки

6660 input high voltage
d Eingangsspannung f oberer Pegel; Eingangs-
Highpegelspannung f
f tension f d'entrée de niveau supérieur
r входное напряжение n высокого уровня

6661 input impedance
d Eingangsscheinwiderstand m
f impédance f d'entrée
r входное полное сопротивление n;
комплексное сопротивление

6662 input information
d Eingangsinformation f
f information f d'entrée
r входная информация f

6663 input information duration
d Eingangsinformationsdauer f
f durée f d'information d'entrée
r длительность f [поступления] входной
информации

6664 input low voltage
d Eingangsspannung f unterer Pegel; Eingangs-
Lowpegelspannung f
f tension f d'entrée de niveau inférieur
r входное напряжение n низкого уровня

6665 input machine
d Eingabemaschine f; Eingabemechanismus m
f machine f d'entrée
r устройство n загрузки

6666 input of control informations
d Eingabe f von Steuerinformationen
f entrée f d'informations de commande
r ввод m управляющей информации

6667 input of measured value
d Messwerteingabe f
f entrée f des valeurs de mesure
r ввод измеряемых величин m

6668 input-output buffer store
d Eingabe-Ausgabe-Pufferspeicher m
f mémoire-tampon f entrée-sortie
r буферное запоминающее устройство на
входе-выходе

6669 input-output control
d Ein- und Ausgangssteuerung f
f commande f d'entrée et de sortie
r управление n по входу и выходу

6670 input-output control system
d Eingabe-Ausgabe-Steuersystem n
f système m de commande d'entrée-sortie
r система f управления вводом-выводом

6671 input-output cycle
d Eingabe-Ausgabe-Zyklus m
f cycle m d'entrée-sortie
r цикл m ввода-вывода

6672 input-output device; input-output unit
d Eingabe-Ausgabe-Gerät n;
Eingabe-Ausgabe-Einheit f
f dispositif m entrée-sortie; unité f entrée-sortie
r устройство n ввода-вывода

* **input-output devices control table → 6677**

6673 input-output interface
d Eingabe-Ausgabe-Interface n
f interface m d'entrée-sortie
r интерфейс m ввода-вывода

6674 input-output limited system
d von Ein- und Ausgabe abhängiges System n
f système m limité par entrée et sortie
r система f с ограничением по входу и
выходу

6675 input-output model
d Eingabe-Ausgabe-Modell n
f modèle m d'entrée-sortie
r модель f ввода-вывода

6676 input-output monitoring
d Eingabe-Ausgabe-Uberwachung f
f surveillance f d'entrée-sortie
r контроль m над вводом-выводом

* **input-output unit → 6672**

**6677 input-output units control table; input-
output devices control table**
d Gerätesteuertabelle f für Ein- und Ausgabe
f table f de commande des dispositifs pour
entrée-sortie
r таблица f управления устройствами ввода-
вывода

6678 input perturbation
d Eingangsstörung f
f trouble m d'entrée
r входное возмущение n

6679 input power
  d Eingangsleistung *f*
  f puissance *f* d'entrée
  r входная мощность *f*

6680 input [power] winding; supply winding
  d Eingangswicklung *f*
  f enroulement *m* d'alimentation; enroulement d'entrée
  r входная обмотка *f*

6681 input pressure
  d Eintrittsdruck *m*
  f pression *f* d'entrée
  r входное давление *n*

6682 input pulse
  d Eingangsimpuls *m*
  f impulsion *f* d'entrée
  r входной импульс *m*

6683 input quantity
  d Eingangsgröße *f*
  f grandeur *f* d'entrée
  r входная величина *f*

6684 input quantity of the system
  d Eingangsgröße *f* des Systems
  f grandeur *f* d'entrée du système
  r входной параметр *m* системы

6685 input sensitivity
  d Eingangsempfindlichkeit *f*
  f sensibilité *f* à l'entrée
  r чувствительность *f* на входе

6686 input signal
  d Eingangssignal *n*
  f signal *m* d'entrée; signal incident; signal d'attaque
  r входной сигнал *m*

6687 input state
  d Eingangszustand *m*
  f état *m* d'entrée
  r входное состояние *n*

6688 input store
  d Eingabespeicher *m*; Eingangsspeicher *m*
  f bloc *m* d'entrée; mémoire *f* d'entrée
  r входной накопитель *m*

  * **input stream → 6422**

6689 input time constant
  d Eingangszeitkonstante *f*
  f constante *f* de temps d'entrée
  r постоянная *f* времени на входе

6690 input transformer
  d Eingangstransformator *m*
  f transformateur *m* d'entrée
  r входной трансформатор *m*

6691 input unit; sensing unit
  d Eingabegerät *n*; Aufnahmegerät *n*
  f organe *m* d'entrée
  r входной блок *m*

6692 input variable
  d Eingangsvariable *f*
  f variable *m* d'entrée
  r входная переменная *f*

  * **input winding → 6680**

6693 inquiry
  d Untersuchung *f*; Anfrage *f*
  f exploration *f*; étude *f*; examen *m*; inspection *f*; demande *f*
  r исследование *n*; изучение *n*; запрос *m*

6694 inquiry control
  d Abfragesteuerung *f*
  f commande *f* d'interrogation
  r управление *n* запросами

6695 inquiry of measuring points
  d Abfrage *f* von Messstellen
  f demande *f* de points de mesure
  r система *f* опроса измерительных пунктов

6696 insensitive
  d unempfindlich
  f insensible
  r нечувствительный

6697 insensitive sensor type
  d unempfindlicher Sensortyp *m*
  f type *m* de senseur insensible
  r нечувствительный тип *m* сенсора

6698 insensitivity; non-sensitivity
  d Unempfindlichkeit *f*
  f insensibilité *f*
  r нечувствительность *f*

6699 insertion of data
  d Dateneinführung *f*
  f introduction *f* de données; alimentation *f* en données
  r включение *n* данных; ввод *m* данных

6700 insolubility degree
  d Unlösbarkeitsgrad *m*
  f degré *m* de l'insolubilité
  r степень *f* неразрешимости

6701 **instability**
   d Unstabilität *f*
   f instabilité *f*
   r неустойчивость *f*

6702 **instability region**
   d Instabilitätsbereich *m*
   f domaine *m* d'instabilité
   r область *f* неустойчивости

6703 **instable state; unstable state**
   d instabiler Zustand *m*
   f état *m* instable
   r неустойчивое состояние *n*

6704 **instalation of manipulator**
   d Manipulatorinstallation *f*
   f installation *f* de manipulateur
   r установка *f* манипулятора

6705 **install** *v*; **setup** *v*
   d einbauen; installieren; aufstellen
   f installer
   r инсталировать; располагать

6706 **installation**
   d Installicrung *f*; Anlage *f*; Montage *f*
   f installation *f*; montage *m*
   r инсталляция *f*; установка *f*; монтаж *m*

6707 **installation diagram; installation lay-out**
   d Montageschema *n*
   f plan *m* de montage
   r монтажная схема *f*

   * **installation lay-out** → 6707

6708 **installation of industrial manipulator**
   d Industriemanipulatorinstallierung *f*
   f installation *f* de manipulateur industriel
   r установка *f* промышленных манипуляторов

6709 **installation optimozation**
   d Aufstellungsoptimierung *f*
   f optimisation *f* de disposition
   r оптимизация *f* размещения оборудования

6710 **installation schedule; equipment layout plan**
   d Aufstellungsplan *m*
   f plan *m* de disposition [d'équipement]
   r план *m* расстановки; план расположения [оборудования]

   * **instantaneous access** → 6345

6711 **instantaneous action detector**
   d unverzögerter Detektor *m*
   f détecteur *m* instantané
   r безынерционный детектор *m*

6712 **instantaneous contact**
   d Schnellkontakt *m*; Springkontakt *m*
   f contact *m* instantané
   r быстродействующий контакт *m*

6713 **instantaneous deviation of controlled variable**
   d augenblickliche Regelabweichung *f*
   f écart *m* instantané de réglage
   r мгновенное значение *n* регулируемого параметра величины

6714 **instantaneous electromagnetic release**
   d Schnellauslöser *m*
   f déclencheur *m* électromagnétique à action instantanée
   r быстродействующий электромагнитный размыкающий механизм *m*

6715 **instantaneous error**
   d Momentanfehler *m*
   f erreur *f* instantanée
   r мгновенная ошибка *f*

6716 **instantaneous frequency**
   d Momentanfrequenz *f*; Augenblicksfrequenz *f*
   f fréquence *f* instantanée
   r мгновенная частота *f*

6717 **instantaneous magnetic relay**
   d magnetisches Momentanrelais *n*; magnetische Schnellauslösung *f*
   f relais *m* magnétique à action instantanée
   r быстродействующее магнитное реле *m*

6718 **instantaneous power**
   d Momentanleistung *f*; Augenblicksleistung *f*
   f puissance *f* instantanée; puissance momentanée
   r мгновенная мощность *f*

6719 **instantaneous reading**
   d Momentanablesung *f*
   f lecture *f* instantanée
   r мгновенный отсчёт *m*

6720 **instantaneous short-circuit current**
   d Stoßkurzschlußstrom *m*
   f courant *m* instantané de court-circuit
   r ударный ток *m* короткого замыкания

6721 **instantaneous value**
   d Momentanwert *m*; Augenblickswert *m*
   f valeur *f* instantanée
   r мгновенное значение *n*

6722 instantaneous value converter
d Momentanwertumsetzer *m*
f convertisseur *m* de valeur instantanée
r преобразователь *m* мгновенного значения

6723 instantaneous voltage
d Augenblicksspannung *f*
f tension *f* instantanée
r мгновенное напряжение *n*

6724 instant of time
d Zeitmoment *n*; Zeitpunkt *m*
f moment *m* de temps
r момент *m* времени

6725 instructed carry
d anbefohlener Übertrag *m*; gesteuerter Übertrag
f report *m* commandé; transfert *m* commandé
r управляемая передача *f*; управляемый перенос *m*

6726 instruction; command
d Instruktion *f*; Befehl *m*
f instruction *f*; commande *f*
r инструкция *f*; команда *f*

6727 instruction address change
d Befehlsadressenänderung *f*
f changement *m* d'adresse d'instruction
r переадресация *f* команд

6728 instruction algorithm
d Belehrungsalgorithmus *m*
f algorithme *m* d'instruction
r алгоритм *m* обучения

6729 instruction array
d Befehlsfeld *n*; Insruktionsfeld *n*
f zone *f* d'instruction
r распределение *n* команд

6730 instruction block; block of instruction
d Befehlsblock *m*; Gruppe *f* von Befehlen
f bloc *m* des instructions
r блок *m* команд

6731 instruction by remote control
d Fernbefehl *m*
f transmission *f* d'ordres à distance; télétransmission *f* d'ordres
r передача *f* команд дистанционного управления

6732 instruction classification
d Befehlsklassifizierung *f*
f classification *f* d'instructions
r классификация *f* команд

6733 instruction code
d Befehlskode *m*
f code *m* d'instructions
r код *m* команд

* instruction counter → 3215

6734 instruction decoder; command decoder
d Befehlsdekodierer *m*
f décodeur *m* d'instructions
r дешифратор *m* команд

6735 instruction distribution channel
d Befehlsverteilerkanal *m*
f canal *m* de distribution d'instructions
r канал *m* распределения команд

6736 instruction element
d Befehlselement *n*
f élément *m* d'instruction
r элемент *m* команды

6737 instruction forming
d Instruktionsformierung *f*
f formage *m* d'instructions
r формирование *n* команды

6738 instruction machine
d Kommandogerät *n*
f machine *f* à instructions
r командная машина *f*; управляющая машина

6739 instruction main line
d Befehlshauptleitung *f*
f ligne *f* principale d'instructions
r командная магистраль *f*

6740 instruction manual
d Bedienungsanweisung *f*
f réglement *m* de service
r регламент *m*

6741 instruction modification; command modification
d Befehlsmodifikation *f*; Modifikation *f* der Befehle
f modification *f* d'instruction; modification de commandes
r модификация *f* команды

6742 instruction processing unit
d Befehls[verarbeitungs]einheit *f*
f unité *f* de traitement des instructions
r устройство *n* обработки команд

6743 instruction register
d Befehlsregister *n*

    f registre *m* d'instructions
    r регистр *m* [запоминания] команд

6744 instruction sequence
    d Befehlsfolge *f*
    f séquence *f* de commande
    r последовательность *f* команд

6745 instruction structure; command structure
    d Befehlsstruktur *f*; Struktur *f* des
        Befehlswortes
    f structure *f* d'instruction; structure de
        commande
    r структура *f* команды

6746 instruction system; command system
    d Befehlssystem *n*
    f système *m* d'instructions; système de
        commandes
    r система *f* команд

6747 instrumental error
    d Instrumentenfehler *m*
    f erreur *f* due à l'instrument
    r ошибка *f* прибора; инструментальная
        погрешность *f*

6748 instrumentation correcting
    d Messgerätekorrektur *f*
    f correction *f* d'instrument [de mesure]
    r инструментальная поправка *f*

6749 instrument for selsyn zeroing
    d Selsynnullstellungsgerät *n*
    f appareil *m* de remise à zéro de selsyns
    r устройство *n* для установки нуля
        сельсинов

6750 instrument manufacture
    d Gerätebau *m*
    f production *f* d'instruments
    r приборостроение *n*

6751 instrument programmability
    d Messgeräteprogrammierbarkeit *f*
    f programmabilité *f* d'instrument [de mesure]
    r программируемость *f* прибора

6752 instrument range
    d Messbereich *m*
    f gamme *m* de mesure
    r пределы *mpl* измерений прибора

6753 instrument table; test stand; test board; test table
    d Instrumententisch *m*; Messtisch *m*;
        Teststand *m*; Messtafel *f*; Prüftisch *m*;
        Eichplatz *m*
    f table *f* de mesure; table d'essai; table

        d'étalonnage
    r измерительный стол *m*; испытательный
        стенд *m*

6754 instrument transformer
    d Messwandler *m*
    f transformateur *m* de mesure
    r измерительный трансформатор *m*

6755 instrument with magnetic screening
    d magnetisch abgeschirmtes Instrument *n*
    f appareil *m* à écran magnétique
    r прибор *m* с магнитным экраном

6756 insulate *v*; isolate *v*
    d isolieren
    f isoler
    r изолитовать

6757 insulation meter
    d Isolationsmesser *m*
    f mesureur *m* d'isolations
    r прибор *m* для измерения изоляции

6758 insulation testing unit
    d Isolationsprüfer *m*
    f contrôleur *m* d'isolation
    r прибор *m* для измерения сопротивления
        изоляции

6759 insulator material
    d Isoliermaterial *n*
    f matériau *m* d'isolation
    r изоляционный материал *m*

6760 integral action; I-action
    d Integralwirkung *f*; I-Wirkung *f*
    f action *f* par intégration; action I
    r интегральное действие *n*

6761 integral action coefficient
    d Integralwirkungskoeffizient *m*
    f coefficient *m* d'action par intégration
    r коэффициент *m* интегрального
        воздействия

6762 integral action rate
    d Integralwirkungsmaß *n*
    f taux *m* d'action par intégration
    r степень *f* интегрального воздействия

6763 integral [action] time constant
    d Integralzeitkonstante *f* des Reglers;
        Zeitkonstante *f* des integrierenden Gliedes
    f constante *f* de temps de l'action intégrale;
        durée *f* de flottement; durée de l'action
        intégrale
    r интегральная постоянная *f* времени;
        постоянная времени изодрома

6764 **integral characteristic**
   d Integralkennlinie *f*
   f réponse *f* intégrale; caractéristique *f* intégrale
   r интегральная характеристика *f*

   * **integral circuit** → 6808

6765 **integral compensation**
   d Kompensation *f* durch integrierendes Glied
   f compensation *f* par réseau intégrateur
   r компенсация *f* с использованием
     интегрального управления

6766 **integral constraint**
   d Integralnebenbedingung *f*
   f limitation *f* intégrale; condition *f* aux limites
     intégrales
   r интегральная связь *f*

6767 **integral control**
   d I-Regelung *f*; integralwirkende Regelung *f*
   f réglage *m* intégral; régulation *f* à corrélation
     d'intégrale
   r регулирование *n* с интегральным
     поведением

   * **integral controller** → 1052

6768 **integral dependence**
   d Integralbeziehung *f*
   f dépendance *f* intégrale
   r интегральная зависимость *f*

6769 **integral error method**
   d Methode *f* der Integralfehler
   f méthode *f* d'erreurs intégrales
   r интегральный метод *m* подбора

6770 **integral estimation**
   d integrale Bewertung *f*
   f estimation *f* intégrale
   r интегральная оценка *f*

6771 **integral formula**
   d Integralformel *f*
   f formule *f* intégrale
   r интегральная формула *f*

6772 **integral linear estimation**
   d lineares Integralkriterium *n*
   f estimation *f* linéaire intégrale
   r интегральная линейная оценка *f*

6773 **integral method**
   d Integralverfahren *n*; Integralmethode *f*
   f méthode *f* intégrale
   r интегральный метод *m*

6774 **integral performance criterion**
   d integrales Qualitätskriterium *n*
   f critère *m* intégral de qualité
   r интегральный критерий *m* качества

6775 **integral performance index**
   d Integralkennwert *m* der Güte
   f indice *m* intégral de performance
   r интегральный показатель *m* качества

6776 **integral representation**
   d Inegraldarstellung *f*
   f représentation *f* intégrale
   r интегральное представление *n*

6777 **integral square estimation**
   d quadratisches Integralkriterium *n*
   f estimation *f* quadratique intégrale
   r квадратичная интегральная оценка *f*

   * **integral time constant** → 6763

6778 **integral transformation**
   d Integraltransformation *f*
   f transformation *f* intégrale
   r интегральное преобразование *n*

6779 **integral transistorized amplifier**
   d transistorisierter Integralverstärker *m*
   f amplificateur *m* intégral transistorisé
   r транзисторный интегральный
     усилитель *m*

6780 **integrate** *v*
   d integrieren; zusammenfassen
   f intégrer
   r интегрировать; объединять

6781 **integrated automation; complex**
   automati[zati]on
   d integrierte Automatisierung *f*; komplexe
     Automatisierung *f*
   f automati[sati]on *f* intégrée; automatisation
     complexe
   r интегральная автоматизация *f*;
     комплексная автоматизация *f*

6782 **integrated circuit; IC**
   d integrierte Schaltung *f*
   f circuit *m* intégré; CI *m*
   r интегральная схема *f*; ИС *f*

6783 **integrated circuit analyzer**
   d Analysator *m* für integrierte Schaltungen
   f analyseur *m* pour circuits intégrés
   r анализатор *m* интегральных схем

6784 **integrated circuit test system**
   d Testsystem *n* für integrierte Schaltkreise

f système *m* de test de circuits intégrés
r контрольно-измерительная система *f* для
проверки интегральных схем

6785 integrated component
  d integriertes Bauteil *n*; integrierter Baustein *m*
  f composant *m* intégré
  r интегрированный компонент *m*

6786 integrated control
  d integrierte Steuerung *f*
  f commande *f* intégrée
  r интегрированное управление *n*

6787 integrated data processing
  d integriertes Datenverarbeitungszentrum *n*
  f centre *m* de traitement intégré des données
  r центр *m* интегрированной обработки
     данных

6788 integrated data processing
  d integrierte Datenverarbeitung *f*
  f exploitation *f* de données par intégration;
     opération *f* intégrée de données
  r интегрированная обработка *f* данных

6789 integrated electronics
  d integrierte Elektronik *f*
  f électronique *f* intégrée
  r интегральная электроника *f*

6790 integrated force sensor
  d integrierter Kraftsensor *m*
  f senseur *m* de force intégré
  r объединенный датчик *m* усилия

6791 integrated graphic microsystem
  d integriertes grafisches Mikrosystem *n*
  f microsystème *m* graphique intégré
  r интегрированная графическая
     микросистема *f*

6792 integrated image system
  d integriertes Abbildunssystem *n*
  f système *m* d'image intégré
  r интегральная система *f* отображения

6793 integrated information system
  d integriertes Informationssystem *n*
  f système *m* intégré des informations
  r интегрированная информационная
     система *f*

6794 integrated machine manipulator
  d maschinenintegrierter Manipulator *m*
  f manipulator *m* intégré de machine
  r машинно-интегрированный
     манипулятор *m*

6795 integrated manipulation system
  d integriertes Handhabungssystem *n*
  f système *m* de manutention intégré
  r интегрированная манипуляционная
     система *f*

6796 integrated manipulator
  d integrierter Manipulator *m*
  f manipulateur *m* intégré
  r интегральный манипулятор *m*

6797 integrated optical circuit
  d integrierte optische Schaltung *f*
  f circuit *m* intégré optique
  r оптическая интегральная схема *f*

6798 integrated optical directional coupler
  d integrierter optischer Richtkoppler *m*
  f coupleur *m* directif en optique intégrée
  r оптический интегрированный
     направленный ответвитель *m*

6799 integrated optical spectrum analyzer
  d integrierter optischer Spektrumanalysator *m*
  f analyseur *m* de spectres optiques intégré
  r интегральный оптический
     спектроанализатор *m*

6800 integrated single-chip computer
  d integrierter Einchiprechner *m*
  f ordinateur *m* intégré à un chip
  r интегрированный однокристальный
     микрокомпьютер *m*

6801 integrated structure
  d Vereinigungsstruktur *f*
  f structure *f* intégrée
  r интегрально-гипотетическая структура *f*

6802 integrating circuit introduction
  d Integriergliedeinführung *f*
  f compensation *f* par contrôle intégrant
  r введение *n* интегрирующего звена

6803 integrating counter circuit
  d integrierender Zählkreis *m*
  f circuit *m* intégrateur de comptage
  r счётная интегрирующая схема *f*

6804 integrating digital voltmeter
  d digitales Integralvoltmeter *n*
  f voltmètre *m* digital intégral
  r интегрирующий цифровой вольтметр *m*

6805 integrating element
  d Integrationsglied *n*; I-Glied *n*
  f organe *m* d'action intégrale
  r интегрирующий элемент *m*;
     интегрирующее звено *n*

6806 **integrating frequency meter; master frequency meter**
   d integrierender Frequenzmesser *m*
   f fréquencemètre *m* intégrateur; fréquencemètre de référence
   r интегрирующий частотомер *m*

6807 **integrating input**
   d Integrationseingang *m*
   f entrée *f* intégrale
   r интегрирующий вход *m*

6808 **integrating network; integral circuit**
   d Integriernetzwerk *n*
   f circuit *m* intégrateur
   r интегрирующий контур *m*; интегрирующая цепь *f*

6809 **integrating part**
   (of a system)
   d integrierender Bestandteil *m*; integrierter Teil *m*
   f partie *f* intégrante
   r интегрирующая составная часть *f*

   * **integrating relay → 1521**

6810 **integration**
   d Integration *f*
   f intégration *f*
   r интегрирование *n*

6811 **integration constant; constant of integration**
   d Integrationskonstante *f*
   f constante *f* d'intégration
   r постоянная *f* интегрирования

6812 **integration of electric signals**
   d Integration *f* elektrischer Signale
   f intégration *f* de signaux électriques
   r интегрирование *n* электрических сигналов

6813 **integration of pulses**
   d Integration *f* von Impulsen
   f intégration *f* d'impulsions
   r сложение *n* импульсов

6814 **integration path**
   d Integrationsweg *m*
   f chemin *m* d'intégration
   r траектория *f* интегрирования

6815 **integration step**
   d Integrationsschritt *m*
   f pas *m* d'intégration
   r шаг *m* интегрирования

6816 **integration theorem**
   d Integrationssatz *m*
   f théorème *m* d'intégration
   r теорема *f* интегрирования

6817 **integrator**
   d Integrator *m*; Integrierer *m*
   f intégrateur *m*
   r интегратор *m*

6818 **integrity**
   d Ganzheit *f*
   f intégrité *f*
   r целостность *f*

6819 **integro-differentiating network; lead-lag network**
   d Integrations-Differentiations-Netzwerk *n*
   f circuit *m* intégrant de différentiation
   r интегрально-дифференцирующий контур *m*

6820 **intelligent measuring system**
   d intelligentes Messsystem *n*
   f système *m* de mesure intelligent
   r интеллектуальная измерительная система *f*

6821 **intelligent sensor**
   d intelligenter Sensor *m*
   f senseur *m* intelligent
   r интеллектуальный сенсор *m*

6822 **intelligent sensorics**
   d intelligente Sensorik *f*; intelligente Sensortechnik *f*
   f sensorique *f* intelligente
   r интеллектуальная сенсорика *f*

6823 **intelligent sensor visual system**
   d intelligentes Sensorsichtsystem *m*
   f système *m* intelligent de senseur visuel
   r интеллектуальная визуальная сенсорная система *f*

6824 **intensity maximum**
   d Intensitätsmaximum *n*
   f maximum *m* d'intensité
   r максимум *m* интенсивности

6825 **intensity modulation**
   d Intensitätsmodulation *f*; Leistungsmodulation *f*
   f modulation *f* d'intensité
   r модуляция *f* интенсивности

6826 **intensity modulator**
   d Intensitätsmodulator *m*

    f modulateur *m* d'intensité
    r модулятор *m* интенсивности

6827 intensity telemetering system
    d Intensitätsfernmesssystem *n*
    f système *m* de télémesure d'intensité
    r система *f* телеизмерения интенсивности

6828 intentional non-linearity
    d Nebennichtlinearität *f*; zusätzliche Nichtlinearität *f*
    f non-linéarité *f* intentionnelle
    r дополнительная нелинейность *f*; намеренно вводимая нелинейность

6829 interact *v*
    d zusammenwirken; wechselwirken
    f collaborer; agir mutuellement
    r взаимодействовать

6830 interacting control; multivariable control
    d gekoppelte Selbstregelung *f*; vermaschte Regelung *f*
    f régulation *f* multiple
    r взаимосвязанное автоматическое регулирование *n*

6831 interaction automatic control system
    d vermaschtes Regelungssystem *n*; vermaschter Regelkreis *m*
    f système *m* asservi à plusieurs variables; système de réglage à plusieurs variables
    r взаимосвязанная система *f* автоматического регулирования

6832 interaction time
    d Wechselwirkungszeit *f*
    f temps *m* d'interaction
    r время *n* взаимодействия

6833 interactive feature
    d Dialogeigenschaft *f*
    f propriété *f* de dialogue
    r возможность *f* для диалога

6834 interactive graphical technique
    d interaktive grafische Technik *f*
    f techniques *fpl* graphiques interactives
    r интерактивная графическая техника *f*

6835 interactive mode
    d Dialogarbeitsweise *f* Bediener-Rechner; interaktive Arbeitsweise *f*
    f mode *m* à dialogue; mode interactif
    r интерактивный режим *m*

6836 interactive processing
    d Dialogverarbeitung *f*

    f traitement *m* interactif
    r интерактивная обработка *f*

6837 interactive terminal
    d Dialogterminal *n*; interaktives Terminal *n*
    f terminal *m* de dialogue; terminal interactif
    r интерактивный терминал *m*

6838 intercept point; cross point
    d Schnittpunkt *m*; Kreuzungspunkt *m*
    f point *m* d'intersection; point de croisement
    r точка *f* пересечения

6839 interchangeable; exchangeable
    d austauschbar; auswechselbar
    f interchangeable; échangeable
    r [взаимо]заменяемый

6840 interconnected controls
    d verkoppelte Steuerungen *fpl*
    f commandes *fpl* réunies; commandes reliées
    r сопряженные органы *mpl* управления

6841 interconnected network
    d Verbundnetz *n*
    f réseau *m* interconnecté
    r связанная сеть *f*

   *   **interconnected system** → **2837**

6842 interconnecting cable
    d Zusammenschaltkabel *n*
    f câble *m* d'interconnexions
    r [внутренний] соединительный кабель *m*

6843 interconnection
    d Verbundbetrieb *m*
    f marche *f* interconnectée
    r работа *f* в системе

6844 interconnection line
    d Zusammenschaltleitung *f*; Koppelleitung *f*
    f ligne *f* d'interconnexions
    r [внутренная] соединительная линия *f*

6845 interconnection test system
    d Zusammenschalt-Testsystem *n*
    f système *m* de test d'interconnexions
    r система *f* контроля [внутренных] соединений

6846 interface
    d Interface *n*; Schnittstelle *f*; Anschlussstelle *f*; Anschlussbild *n*
    f interface *f*
    r интерфейс *m*; устройство *n* сопряжения

6847 interface channel
    d Schnittstellenkanal *m*; Anschlusskanal *m*

    f canal *m* d'interface
    r канал *n* сопряжения

6848 interface circuit
    d Schnittstellenschaltung *f*;
    Anschlussschaltung *f*
    f circuit *m* d'interface
    r схема *f* сопряжения

6849 interface control
    d Schnittstellensteuerung *f*
    f commande *f* d'interface
    r управление *n* интерфейсом

6850 interface controller
    d Schnittstellen-Steuerbaustein *m*;
    Anschlusssteuereinheit *f*
    f unité *f* de commande d'interface
    r контроллер *m* интерфейса

6851 interface electronics
    d Interface-Elektronik *f*
    f électronique *f* interface
    r электроника *f* интерфейса

6852 interface logic
    d Schnittstellen-Logikschaltung *f*
    f logique *f* d'interface
    r интерфейсная логика *f*

6853 interface module
    d Schnittstellenmodul *m*;
    Schnittstellenbaustein *m*
    f module *m* d'interface
    r интерфейсный модуль *m*

6854 interface signal sequence
    d Interface-Signalfolge *f*
    f séquence *f* de signal d'interface
    r последовательность *f* сигналов
    интерфейса

6855 interface standard
    d Schnittstellenstandard *m*;
    Interface-Standard *m*; Anschlussstandard *m*
    f norme *f* d'interface; standard *m* d'interface
    r стандарт *m* интерфейса

6856 interface timing
    d Interface-Zeitteilung *f*; Zeitberechnung *f*
    einer Schnittstelle
    f distribution *f* de temps d'interface
    r синхронизация *f* интерфейса

6857 interfacial tensiometer
    d Grenzschichtspannungsmesser *m*
    f appareil *m* de mesure de la tension
    interfaciale

    r прибор *m* для измерения напряжения
    поверхностного слоя

6858 interfacing characteristics
    d Schnittstellenkenndaten *pl*
    f caractéristiques *fpl* d'interface
    r характеристики *fpl* интерфейса

6859 interfacing kit
    d Schnittstellenbausatz *m*
    f jeu *m* de construction d'interfaces
    r набор *m* компонентов для интерфейса

6860 interfacing technique
    d Schnittstellentechnik *f*; Anschlusstechnik *f*
    f technique *f* d'interface
    r техника *f* сопряжения

    * **interference → 4448**

6861 interference area
    d Störungsgebiet *n*
    f zone *f* de brouillage
    r зона *f* помех; область *f* интерференции

6862 interference comparator
    d Interferenzkomparator *m*
    f comparateur *m* interférentiel
    r интерференционный компаратор *m*

6863 interference effect
    d Interferenz *f*; Interferenzeffekt *m*; Störung *f*;
    Störungseffekt *m*
    f interférence *f*; brouillage *m*; effet *m* de
    brouillage
    r влияние *n* помех

6864 interference elimination measuring
    d Entstörungsmessung *f*
    f mesure *f* de perturbation résiduelle
    r измерение *n* с подавлением помех

**6865 interference field strength measuring
instrument**
    d Störfeldstärkemessgerät *n*
    f appareil *m* de mesure de l'intensité du champ
    parasite
    r измеритель *m* силы поля помех

6866 interference limiter
    d Störbegrenzer *m*
    f limiteur *m* de brouillage; limiteur de
    parasites
    r ограничитель *m* помех

**6867 interference lines for measuring material
strain**
    d Interferenzlinien *fpl* zur
    Materialspannungsmessung

f lignes *fpl* d'interférence pour mesurer la
tension des matériaux
r измерение *n* напряжения [материала]
интерференционными методами

6868 interference measuring apparatus
d Störungsmessgerät *n*
f appareil *m* à mesurer le brouillage
r измеритель *m* помех

6869 interference microscope
d Interferenzmikroskop *n*
f microscope *m* à interférence
r интерференционный микроскоп *m*

6870 interference noise
d Interferenzrauschen *n*
f bruit *m* interférentiel
r шум *m* вследствие интерференции

6871 interference peak
d Störspitze *f*
f pointe *f* de brouillage; crête *f* de brouillage
r пик *m* помехи

6872 interference pulse
d Störimpuls *m*
f impulsion *f* parasite
r интерференционный импульс *m*

6873 interference refractometer
d Interferenzrefraktometer *n*
f réfractomètre *m* à interférence
r интерференционный рефрактометр *m*

6874 interference relay
d Störungsrelais *n*
f relais *m* à perturbation
r интерференционное реле *n*

6875 interference wavelength
d interferentielle Wellenlänge *f*
f longueur *f* d'ondes dues à l'interférence
r длина *f* волн интерференции

6876 interferometric control
d interferometrische Kontrolle *f*
f contrôle *m* par interféromètre
r интерферометрический контроль *m*

6877 interferometric measurement
d interferometrische Messung *f*
f mesure *f* interférométrique
r интерферометрическое измерение *n*

6878 interferometric modulator
d interferometrischer Modulator *m*
f modulateur *m* interférométrique

r интерферометрический модулятор *m*

6879 interferometric sensor
d interferometrischer Sensor *m*
f capteur *m* interférométrique
r интерферометрический сенсор *m*

6880 interferometric system
d interferometrisches System *n*
f système *m* interférométrique
r интерферометрическая система *f*

6881 interferometry
d Interferometrie *f*
f interférométrie *f*
r интерферометрия *f*

6882 interlinked transfer line
d verkettete Transferstraße *f*
f voie-transfert *f* enchaînée
r сопряженная автоматическая линия *f*

*   **interlock** → **1840**

*   **interlock circuit** → **1843**

6883 interlocked operation
d verriegelter Betrieb *m*
f opération *f* verrouillée
r блокировка *f* функционирования

*   **interlocking** → **1840**

6884 interlocking device
d Verriegelungseinrichtung *f*;
Verblockungssystem *n*
f dispositif *m* de verrouillage
r устройство *n* для блокировки

*   **interlocking relay** → **1852**

6885 interlock time
d Verriegelungszeit *f*
f temps *m* de verrouillage
r время *n* блокировки

6886 intermediate amplifier
d Zwischenverstärker *m*; Mittelverstärker *m*
f amplificateur *m* intermédiaire
r промеждуточный усилитель *m*

*   **intermediate check** → **6887**

6887 intermediate check[ing]
d Zwischenkontrolle *f*
f contrôle *m* intermédiaire
r промеждуточный контроль *m*

6888 intermediate component
  d Zwischenglied *n*
  f composant *m* intermédiaire
  r промежуточное звено *n*

6889 intermediate control
  d Hauptgruppensteuerung *f*
  f commande *f* à l'étage intermédiaire
  r промежуточное управление *n*

6890 intermediate frequency filter
  d Zwischenfrequenzfilter *n*
  f filtre *m* moyenne fréquence
  r фильтр *m* промежуточной частоты

6891 intermediate means
  d Zwischenorgane *npl*
  f dispositifs *mpl* intermédiaires
  r промежуточные элементы *mpl*;
      вспомогательные средства *npl*

6892 intermediate memory; intermediate store
  d Zwischenspeicher *m*
  f mémoire *f* intermédiaire
  r промежуточная память *f*;
      промежуточное запоминающее
      устройство *n*

6893 intermediate position
  d Mittelstellung *f*
  f position *f* intermédiaire
  r промежуточное положение *n*

6894 intermediate process changing mechanism
  d Zwischenprozessaustauschmechanismus *m*
  f mécanisme *m* d'échange interprocessus
  r взаимозаменяемый механизм *m* для
      промежуточного процесса

6895 intermediate quantity
  d Zwischengröße *f*
  f quantité *f* intermédiaire
  r промежуточная величина *f*

  *   **intermediate repeater** → 10812

6896 intermediate result
  d Zwischenergebnis *n*
  f résultat *m* intermédiaire
  r промежуточный результат *m*

6897 intermediate stop
  d Zwischenstillsetzung *f*
  f arrêt *m* intermédiaire
  r промежуточная остановка *f*

  *   **intermediate store** → 6892

6898 intermeshed circuit
  d vermaschte Schaltung *f*
  f montage *m* maillé
  r многоконтурная схема *f*

6899 intermitted feed
  d schrittweiser Vorschub *m*;
      Sprungvorschub *m*
  f avance *f* intermittente
  r прерывистое питание *n*

6900 intermittent
  d unstetig; unterbrochen
  f discontinu; intermittent
  r прерывистый

  *   **intermittent control** → 4337

  *   **intermittent controller** → 4338

6901 intermittent duty
  d Aussetzbetriebsertrag *m*
  f service *m* intermittent; régime *m* intermittent
  r прерывистый режим *m* работы

6902 intermittent fault
  d intermittierend auftretende Störung *f*
  f défaut *m* intermittent
  r перемежающаяся неисправность *f*;
      неустойчивое положение *n*

  *   **intermittent input** → 4351

  *   **intermittent signal** → 4342

6903 intermittent system; discontinuous system
  d unstetiges System *n*; diskontinuierliches
      System
  f système *m* discontinu
  r прерывистая система *f*

6904 intermodulation
  d Zwischenmodulation *f*; gegenseitige
      Modulation *f*
  f intermodulation *f*
  r взаимная модуляция *f*

6905 internal circuit
  d innerer Stromkreis *m*
  f circuit *m* intérieur
  r внутренняя цепь *f*

6906 internal control
  d innere Regelung *f*
  f réglage *m* interne
  r внутреннее регулирование *n*

6907 **internal element of the system**
 d inneres Element *n* des Systems
 f élément *m* intérieur du système
 r внутренний элемент системы

6908 **internal energy**
 d innere Energie *f*
 f énergie *f* interne
 r внутренняя энергия *f*

6909 **internal idle time**
 d innere Leerlaufzeit *f*
 f temps *m* interne inactif
 r внутренее холостое время *n*

6910 **internal logic**
 d interne Logik *f*
 f logique *f* interne
 r внутренная логика *f*

6911 **internal losses**
 d Eigenverluste *mpl*
 f pertes *fpl* propres
 r собственные потери *fpl*

6912 **internal model**
 d inneres Modell *n*
 f modèle *m* interne
 r внутренная модель *f*

6913 **internal operating ratio**
 d innere Ausbeute *f*
 f rendement *m* interne
 r внутренний коэффициент *m*
    использования

6914 **internal photoelectric effect**
 d innerer Fotoeffekt *m*
 f effet *m* photo-électrique interne
 r внутренний фотоэлектрический эффект *m*

6915 **internal state**
 d Innenzustand *m*
 f état *m* interne
 r внутреннее состояние *n*

6916 **International Federation of Automatic
       Control; IFAC**
 d Internationale Föderation *f* für automatische
    Steuerung
 f Fédération *f* internationale de commande
    automatique
 r Международная федерация *f* по
    автоматическому управлению; ИФАК

6917 **interpolate** *v*
 d interpolieren
 f interpoler

 r интерполировать

6918 **interpolating function**
 d Interpolatiosfunktion *f*
 f fonction *f* d'interpolation
 r интерполирующая функция *f*

6919 **interpolation**
 d Interpolation *f*
 f interpolation *f*
 r интерполяция *f*

6920 **interpolator**
 d Interpolator *m*
 f appareil *m* d'interpolation
 r интерполятор *m*

6921 **interpret** *v*
 d interpretieren
 f interpréter
 r интерпретировать

6922 **interpretation method**
 d Interpretationsmethode *f*
 f méthode *f* d'interprétation
 r метод *m* интерпретации

6923 **interpretative simulation**
 d interpretierende Simulation *f*
 f simulation *f* interprétative
 r интерпретационное моделирование *n*

6924 **interpreter**
 d interpretierendes Organ *n*
 f dispositif *m* interprète
 r интерпретирующее устройство *n*

6925 **interpreter code**
 d Zuordnerkode *m*
 f code *m* interprète
 r интерпретирующий код *m*

6926 **interpreting routine**
 d interpretierendes Programm *n*
 f routine *f* d'interprétation
 r интерпретирующая программа *f*

6927 **interprocess communication**
 d Kommunikation *f* zwischen Einzelprozessen
 f communication *f* entre processus différents
 r взаимодействие *n* между процессами

6928 **interprocess coupling**
 d Zwischenprozesskopplung *f*
 f couplage *m* interprocessus
 r связь *f* между процессами

6929 **interrogation frequency**
 d Abfragefrequenz *f*

f fréquence *f* d'interrogation
r частота *f* запроса

6930 interrogation pulse
  d Abfrageimpuls *m*
  f impulsion *f* de demande; impulsion
     d'interrogation
  r импульс *m* запроса

6931 interrogation register
  d Abfrageregister *n*
  f registre *m* d'interrogation
  r регистр *m* запроса

6932 interrogation system
  d Abfragesystem *n*
  f système *m* d'interrogation
  r система *f* опроса

6933 interrupt *v*; break *v*; abort *v*
  d unterbrechen; abbrechen
  f interrompre
  r прерывать; обрывать

6934 interrupt acknowledge signal
  d Interruptbestätigengssignal *n*;
     Unterbrechungsanerkennungssignal *n*
  f signal *m* de confirmation d'interruption;
     signal d'acceptation d'interruption
  r сигнал *m* подтверждения прерывания

6935 interrupt-based system
  d System *n* auf Interruptbasis
  f système *m* basé sur interruptions
  r система *f* обмена на основе прерываний

6936 interrupt call
  d Unterbrechungsaufforderung *f*
  f appel *m* d'interruption
  r вызов *m* по прерыванию

6937 interrupt channel
  d Interruptkanal *m*; Unterbrechungskanal *m*
  f canal *m* d'interruption; piste *f* d'interruption
  r канал *m* прерываний

6938 interrupt-controlled microprocessor
  d interruptgesteuerter Mikroprozessor *m*
  f microprocesseur *m* commandé par
     interruption
  r микропроцессор *m* с контролем
     прерываний

6939 interrupt controller
  d Interruptsteuerungsgerät *n*;
     Unterbrechungssteuerungsgerät *n*
  f appareil *m* de commande d'interruption
  r контроллер *m* прерываний

6940 interrupt control unit
  d Interruptsteuereinheit *f*;
     Unterbrechungssteuereinheit *f*
  f unité *f* de commande d'interruption
  r блок *m* обработки прерываний

6941 interrupt disarm
  d Interruptquellenblockierung *f*;
     Unterbrechungsquellensperrung *f*
  f blocage *m* de source d'interruption
  r блокировка *f* источника прерываний

6942 interrupt-driven environment
  d interruptbetriebene Systemumgebung *f*
  f environnement *m* actionné par interruption
  r оборудование n, приводимое в действие
     сигналом прерывания

6943 interrupt-driven system
  d interruptgesteuertes System *n*;
     unterbrechungsgesteuertes System
  f système *m* à commande par interruptions
  r система f, управляемая прерываниями

6944 interrupted autooscillations
  d diskrete Selbstschwingungen *fpl*
  f auto-oscillations *fpl* discontinues
  r прерывистые автоколебания *npl*

6945 interrupt identification
  d Interruptkennung *f*;
     Unterbrechungsidentifizierung *f*
  f identification *f* d'interruption
  r идентификация *f* прерывания

6946 interrupt[ing] signal; break signal
  d Interruptsignal *n*; Eingriffsignal *n*;
     Breaksignal *n*
  f signal *m* d'interruption
  r сигнал *m* прерывания

6947 interrupting time
  d Unterbrechungszeit *f*
  f retard *m* de coupure
  r время *n* прерывания

6948 interruption; break[ing]; abort
  d Interrupt *m*; Unterbrechung *f*; Abbruch *m*
  f interruption *f*; rupture *f*; coupure *f*
  r прерывание *n*; обрывание *n*

6949 interruption request block
  d Anforderungsblock *m* für
     Unterbrechungs[sbehandlung]
  f bloc *m* de requête pour interruption
  r блок *m* запросов на прерывание

6950 interruption unit
  d Unterbrechungseinheit *f*

f   unité f d'interruption
r   блок m прерывания

6951 interrupt priority system
d   Interruptvorrangsystem n;
    Unterbrechungsprioritätssystem n
f   système m de priorité d'interruption
r   система f приоритетов прерываний

6952 interrupt register
d   Unterbrechungsregister n
f   registre m d'interruption
r   регистр m прерывания

6953 interrupt request reset signal
d   Interruptanforderungs-Rücksetzsignal n
f   signal m de mise à zéro de demandes
    d'interruption
r   сигнал m установки в нуль запроса
    прерывания

6954 interrupt request synchronization register
d   Interruptanforderungs-
    Synchronisierungsregister n
f   registre m de synchronisation de demandes
    d'interruption
r   регистр m синхронизации запросов
    прерывания

*   **interrupt signal → 6946**

6955 interrupt source
d   Interruptquelle f; Unterbrechungsquelle f
f   source f d'interruption
r   источник m прерывания

6956 interrupt tripping
d   Interruptauslösung f
f   déclenchement m d'interruption
r   пуск m прерывания

*   **intersection → 8615**

6957 interstage
d   Zwischenstufe f; Zwischenzustand m
f   état m intermédiaire
r   промеждуточная ступень f

6958 interstage transformer
d   Zwischentransformator m
f   transformateur m intermédiaire
r   межкаскадный трансформатор m

6959 interval
d   Intervall n
f   intervalle m
r   интервал m

6960 interval of high frequencies

d   Hochfrequenzintervall n
f   intervalle m de hautes fréquences
r   интервал m высоких частот

6961 interval of low frequencies
d   Niederfrequenzintervall n
f   intervalle m de fréquences basses
r   интервал m низких частот

6962 intrinsic kinetics
d   innere Kinetik f; Eigenkinetik f
f   cinétique f propre; cinétique intrinsèque
r   внутренная кинетика f; собственная
    кинетика

6963 invariable
d   unveränderlich; invariabel
f   invariable
r   неизменный

6964 invariance
d   Invarianz f; Unveränderlichkeit f
f   invariance f
r   инвариантность f

6965 invariance of a cybernetic[al] system
d   Invarianz f eines kybernetischen Systems
f   invariance f d'un système cybernétique
r   инвариантность f кибернетической
    системы

*   **invariance of a cybernetic system → 6965**

6966 invariance principe
d   Konstanthaltungsprinzip n;
    Invarianzprinzip n
f   principe m d'invariance
r   принцип m инвариантности

6967 invariant
d   invariant
f   invariable
r   инвариантный

6968 invariant control system
d   invariantes Regelsystem n
f   système m invariant de réglage
r   инвариантная система f регулирования

*   **inverse coupling → 1541**

6969 inverse element
d   inverses Element n
f   élément m inverse
r   обратный элемент m

6970 inverse Fourier transform
d   inverse Fourier-Transformation f;
    Fourier-Rücktransformation f

f transformation f inverse de Fourier
r обратное преобразование n Фурье

6971 inverse function
d Umkehrfunktion f; inverse Funktion f
f fonction f inverse
r обратная функция f

6972 inverse integrator
d Umkehrintegrator m
f intégrateur-inverseur m
r обратный интегратор m

6973 inverse Laplace transformation
d Laplace-Rücktransformation f
f transformation f inverse de Laplace
r обратное преобразование n Лапласа

6974 inverse manipulator model
d inverses Manipulatormodell n
f modèle m de manipulateur inverse
r инверсная модель f манипулятора

* **inverse-parallel connexion** → 906

6975 inverse phase-amplitude characteristic
d umgekehrte
   Amplituden-Phasen-Charakteristik f
f réponse f de transfert inverse
r обратная амплитудно-фазовая
   характеристика f

**6976 inverse relation telemeter; inversion
        telemeter**
d Invertentfernungsmesser m
f télémètre m à inversion
r обратнозависимое телеметрическое
   устройство n

6977 inverse sequence
d umgekehrte Folge f
f séquence f inverse
r обратная последовательность f

6978 inverse structure
d Inversionsstruktur f
f structure f d'inversion
r инверсная структура f

6979 inverse time-lag circuit-breaker
d abhängig verzögerter Selbstauslöser m
f disjoncteur m à retard dépendant
r выключатель m цепи с обратным
   запаздыванием

**6980 inverse transfer function; return transfer
        function**
d inverse Übertragungsfunktion f;

Rückführübertragungsfunktion f
f transmittance f inverse; transmittance de la
   chaîne de réaction
r обратная передаточная функция f

6981 inversion
d Inversion f; Invertierung f
f inversion f
r инверсия f

6982 inversion level
d Inversionspegel m
f niveau m d'inversion
r уровень m инверсии

6983 inversion scanning
d Abtastung f nicht markierter Stellen
f exploration f de positions non marquées
r развёртка f с обращением

* **inversion telemeter** → 6976

6984 invert v
d invertieren; umkehren
f invertir; inverser
r инвертировать; обращать

6985 inverted mode
d umgekehrte Betriebsweise f
f mode m inverti
r инверсный режим m работы

6986 inverter stage
d Inverterstufe f; Umkehrstufe f
f étage m d'inversion
r каскад m преобразователя

6987 invertible automaton
d invertibler Automat m
f automate m invertible
r инвертируемый автомат m

6988 inverting circuit
d invertierende Schaltung f
f circuit m d'inversion
r инвертирующая схема f

* **ionic relay** → 5844

6989 irrecoverable; unrecoverable
d unverbesserlich; irreparabel
f irrémédiable; irréparable
r неисправимый; невосстановимый

6990 irreducible system
d irreduzibles System n
f système m irréductible
r неприводимая система f

6991 **irregular**
  d irregulär; unregelmäßig
  f irrégulier
  r нерегулярный; неравномерный

6992 **irregular code**
  d unregelmäßiger Kode *m*
  f code *m* irrégulier
  r нерегулярный код *m*

* **irregular distortion** → 5658

6993 **irregularity coefficient**
  d Unregelmäßigkeitsfaktor *m*;
     Ungleichförmigkeitsgrad *m*
  f coefficient *m* d'irrégularité
  r коэффициент *m* неравномерности

6994 **irregular part of the function**
  d regelloser Teil *m* der Funktion
  f partie *f* irrégulière de la fonction
  r нерегулярная часть *f* функции

6995 **irrelevant data**
  d unwesentliche Daten *pl*
  f données *fpl* sans importance
  r несущественные данные *pl*

6996 **isochromate**
  d Isochromate *f*
  f ligne *f* isochrome; ligne isochromatique
  r изохромата *f*

6997 **isochrone region**
  d Isochronbereich *m*
  f région *f* isochrone
  r изохронная область *f*

6998 **isoelectronic row**
  d isoelektronische Reihe *f*
  f série *f* isoélectronique
  r изоэлектронный ряд *m*

* **isolate** *v* → 6756

6999 **isolating circuit**
  d Entkopplungsschaltung *f*
  f circuit *m* de découplage
  r отключающая цепь *f*

7000 **isomorphism**
     (of automata)
  d Isomorphismus *m*
  f isomorphisme *m*
  r изоморфизм *m*

7001 **isoperimetric problem**
  d isoperimetrisches Problem *n*

  f problème *m* isopérimétrique
  r изопериметрическая проблема *f*

7002 **item counter**
  d Postenzähler *m*
  f compteur *m* de postes; compteur de ventes
  r счётчик *m* операции

7003 **iteration**
  d Iteration *f*
  f itération *f*
  r итерация *f*

7004 **iteration cycle**
  d Iterationszyklus *m*
  f cycle *m* d'itération
  r цикл *m* итерации

7005 **iteration method; iteration procedure**
  d Iterationsverfahren *n*
  f méthode *f* d'itération
  r итерационный метод *m*; итерационная
     процедура *f*

7006 **iteration number**
  d Anzahl *f* der auszuführenden Schritte;
     Iterations[an]zahl *f*
  f nombre *m* d'itérations
  r число *n* итераций

* **iteration procedure** → 7005

7007 **iteration variable**
  d Iterationsvariable *f*
  f variable *f* itérative
  r итерационная переменная *f*

7008 **iterative**
  d iterativ
  f itératif
  r итеративный; итерационный

7009 **iterative addition**
  d iteratives Addieren *n*; schrittweise Addition *f*
  f addition *f* itérative
  r итеративное сложение *n*

7010 **iterative analog computer**
  d iterativ arbeitender Analogrechner *m*
  f ordinateur *m* analogique itératif
  r итерационная аналоговая вычислительная
     машина *f*

7011 **iterative attenuation**
  d Kettendämpfung *f*
  f affaiblissement *m* itératif
  r повторное затухание *n*

7012 iterative computation
  d iterative Berechnung *f*
  f calcul *m* itératif
  r итеративный расчёт *m*

7013 iterative decomposition
  d iterative Dekomposition *f*
  f décomposition *f* itérative
  r итеративная декомпозиция *f*

7014 iterative design
  d iterative Entwurf *m*; Entwurf durch Iteration
  f conception *f* itérative
  r итеративная разработка *f*

7015 iterative simulation
  d iterative Simulation *f*
  f simulation *f* itérative
  r итерационное моделирование *n*

7016 iterative solution
  d iterative Lösung *f*
  f solution *f* itérative
  r итерационное решение *n*

7017 iterative technique
  d iterative Technik *f*
  f technique *f* itérative
  r метод *m* итераций

7018 iterator
  d Iterator *m*
  f itérateur *m*
  r повторитель *m*

J

7019 jack panel
  d Buchsenfeld *n*
  f panneau *m* à broches
  r коммутационная панель *f*

7020 jamming signal
  d Störsignal *n*
  f signal *m* brouilleur
  r мешающий сигнал *m*

7021 jerkmeter
  d Ruckmesser *m*
  f suraccéléromètre *m*
  r измеритель *m* скорости изменения
     ускорения

7022 jet-pipe oil-operated control
  d hydraulische Strahlrohrregelung *f*
  f réglage *m* hydraulique à tuyau oscillant
  r гидравлическое регулирование *n*
     струйного типа

7023 jet-pipe oil-operated controller
  d hydraulischer Strahlstromregler *m*
  f régulateur *m* hydraulique à tuyau oscillant
  r гидравлический струйный регулятор *m*

7024 jig; fixture; appliance; device; facility
  d Vorrichtung *f*; Apparatur *f*
  f disposition *f*; dispositif *m*; appareil *m*
  r устройство *n*; приспособление *n*

  * **job control → 7025**

7025 job management; job control
  d Jobverwaltung *f*; Auftragsverwaltung *f*
  f gestion *f* de travaux
  r управление *n* заданиями

7026 job-oriented terminal
  d joborientiertes Terminal *n*
  f terminal *m* orienté vers problème
  r проблемно-ориентированный терминал *m*

7027 job processing
  d Auftragsverarbeitung *f*
  f traitement *m* de travail
  r обработка *f* задания

  * **jogging service → 6408**

  * **join** *v* **→ 1040**

7028 join distribution
  d gemeinsame Verteilung *f*
  f répartition *f* commune
  r совместное распределение *n*

7029 joining element
  d Verbindungselement *n*
  f élément *m* de jonction; élément de
     connexion; raccordement *m*
  r соединяющий элемент *m*

  * **joining of the states → 2614**

  * **joint aid→ 7030**

7030 joint|ing| aid
  d Fügehilfe *f*
  f aide *f* de jointage
  r вспомогательное устройство *n* для
     процессов сопряжения

7031 jointing system
  d Fügesystem *n*
  f système *m* de jointage
  r система *f* сопряжения

7032 jointing theory
  d Fügetheorie *f*
  f théorie *f* de jointage
  r теория *f* сопряжения

7033 jointing unit
  d Fügeeinheit *f*
  f unité *f* de jointage
  r блок *m* сопряжения

7034 joint kinematics
  d Gelenkkinematik *f*
  f cinématique *f* de joint
  r кинематика *f* шарнирного соединения

7035 joint manipulator
  d Gelenkmanipulator *m*
  f manipulateur *m* articulé
  r манипулятор *m* шарнирного типа

7036 joint regulation
  d Gelenkregelung *f*
  f régulation *f* de joint
  r регулирование *n* шарниров

7037 Joulean effect
  d Erwärmungsverlust *m*; Joule-Effekt *m*
  f effet *m* de Joule
  r эффект *m* Джоуля

7038 jump
d Sprung *m*
f saut *m*
r перепад *m*; передача *f* управления

7039 jump function; step function
d Sprungfunktion *f*
f fonction *f* échelon; fonction de saut
r ступенчатая функция *f*

7040 jump instruction
d Sprungbefehl *m*
f instruction *f* de saut
r команда *f* перехода

7041 jump resonance
d Resonanz *f* mit sprungförmiger
  Charakteristik; Resonanz mit Hysterese
f résonance *f* de saut
r скачкообразный резонанс *m*

* **junction → 8615**

7042 junction rectifier
d Flächengleichrichter *m*;
  Sperrschichtgleichrichter *m*
f redresseur *m* à jonction
r контактный выпрямитель *m*

K

7043 Karnaugh map
d Karnaugh-Karte *f*
f diagramme *m* de Karnaugh
r диаграмма *f* Карно

* **KB → 7047**

* **KBPS → 7048**

7044 keyboard-controlled
d tastaturgesteuert
f commandé à clavier
r управляемый клавиатурой

7045 keyboard programming unit
d Programmsteuereinrichtung *f* mit Drucktasten
f programmateur *m* à clavier
r клавиатурное програмирующее устройство *n*

7046 key relay
d Tasterrelais *n*
f relais *m* de manipulation
r манипуляторное реле *n*

* **kHz → 7049**

7047 kilobyte; KB
d Kilobyte *n*; KB
f kilo-octet *m*; Ko
r килобайт *m*; Кб

7048 kilobyte per second; KBPS
d Kilobytes *npl* pro Sekunde
f kilo-octets *mpl* par seconde
r килобайт *mpl* в секунду

* **kilocycles per second → 7049**

7049 kilohertz; kHz; kilocycles per second
d Kilohertz *n*
f kilohertz *m*; kilocycles *mpl* par seconde
r килогерц *m*; кгц

7050 kinematic analysis
(of gripper gear)
d kinematische Analyse *f*
f analyse *f* cinématique
r кинематический анализ *m*

7051 kinematic chain
d kinematische Kette *f*
f chaîne *f* cinématique
r кинематическая цепь *f*

7052 kinematic manipulator structure
d kinematische Manipulatorstruktur *f*
f structure *f* de manipulateur cinématique
r кинематическая структура *f* манипулятора

7053 kinematic output
d kinematische Ausgabe *f*
f extraction *f* cinématique
r кинематический вывод *m*

* **kinematic schedule → 7054**

7054 kinematic scheme; kinematic schedule
d kinematisches Schema *n*
f schéma *m* cinématique
r кинематическая схема *f*

7055 kinematic state
d kinematischer Zustand *m*
f état *m* cinématique
r кинематическое состояние *n*

7056 kinetic control system; positional servosystem (US)
d Positionsfolgessystem *n*
f asservissement *m* de position
r позиционная следящая система *f*

7057 kinetic energy of thermal motion
d kinetische Energie *f* der Wärmebewegung
f énergie *f* cinétique du mouvement thermique
r кинетическая энергия *f* теплового движения

7058 kinetic measuring data
d kinetische Messdaten *pl*
f valeurs *fpl* mesurées cinétiques
r кинетические опытные данные *pl*

7059 kinetic parameter
d kinetischer Parameter *m*
f paramètre cinétique
r кинетический параметр *m*

7060 kinetic regulation system
d kinetisches Regelungssystem *n*
f système *m* de réglage cinétique
r кинетическая система *f* регулирования

7061 king class
d radioelektrisches Fernsteuerungssystem *n*
f système *m* de télécommande radioélectrique
r радиоэлектрическая телеуправляемая система *f*

7062 knee process unit; console process unit
   d Konsolverfahrenseinheit *f*
   f unité *f* de procédé de console
   r консольный блок *m* перемещений

7063 knife-edge relay
   d Schneidenankerrelais *n*
   f relais *m* à tranchant
   r реле *n* с якорем на призматической опоре

  * **knowbot → 7066**

7064 knowledge
   d Kenntnis *f*; Wissenschaft *f*
   f savoir *m*; science *f*; connaissance *f*
   r знание *n*; познание *n*

7065 knowledge engineering
   d Wissenstechnologie *f*
   f ingénierie *f* de connaissance
   r технология *f* представления и обработки знаний

7066 knowledge robot; knowbot
   d Wissensroboter *m*; Knowbot *m*
   f robot *m* bibliothécaire; robot de nature logicielle
   r робот *m* знаний

L

7067 label coding
 d Kodermarkierung *f*
 f indication *f* de code
 r маркерное кодирование *n*

7068 laboratory automat
 d Laboratoriumsautomat *m*
 f automate *m* de laboratoire
 r лабораторная установка *f*

7069 laboratory equipment
 d Laboreinrichtung *f*
 f équipement *m* de laboratoire
 r лабораторное оборудование *n*

7070 laboratory logic testing
 d Labor-Logikprüfung *f*; Laborprüfung *f* der Logik
 f examen *m* logique au laboratoire
 r лабораторная проверка *f* [функционирования] логики

7071 laboratory measuring instruments
 d Laboratoriumsmessgeräte *npl*
 f instruments *mpl* de mesure de laboratoire
 r лабораторные измерительные приборы *mpl*

7072 laboratory model
 d Labormodell *n*
 f modèle *m* de laboratoire
 r лабораторная модель *f*

7073 laboratory test
 d Laborversuch *m*
 f essai *m* de laboratoire
 r лабораторное испытание *n*

7074 laboratory use
 d Laboreinsatz *m*
 f usage *m* au laboratoire
 r лабораторное применение *n*

 * **lag** → 3903

7075 lag angle
 d Nacheilungswinkel *m*
 f angle *m* de retard
 r угол *m* отставания

7076 lag coefficient
 d Verzögerungsfaktor *m*
 f facteur *m* de retard
 r коэффициент *m* запаздывания

7077 lag curve
 d Verzögerungskennlinie *f*
 f courbe *f* de délai
 r характеристика *f* инерционности

7078 lag element
 d Verzögerungsglied *n*
 f organe *m* de retard
 r блок *m* задержки

7079 lagging phase
 d Verzögerungsphase *f*
 f phase *f* en retard
 r запаздывающая фаза *f*

7080 lag network
 d Stromkreis *m* mit Phasenverzögerung
 f circuit *m* à phase retardée
 r контур *m* с запаздыванием по фазе

7081 Lagrangian function
 d Lagrangesche Funktion *f*
 f fonction *f* de Lagrange
 r функция *f* Лагранжа

7082 Lagrangian multiplier method
 d Lagrangesche Multiplikatorenmethode *f*
 f méthode *f* des paramètres de Lagrange
 r метод *m* множителей Лагранжа

7083 lag representation
 d Nacheilungsdarstellung *f*
 f présentation *f* du retard
 r воспроизведение *n* запаздывания

7084 lag theorem
 d Lehrsatz *m* von der Phasennacheilung
 f théorème *m* du retard
 r теорема *f* запаздывания

7085 lamp signalling
 d Lampensignalisierung *f*
 f signalisation *f* par lampe
 r световая сигнализация *f*

7086 lamp signalling switchboard
 d Schalttafel *f* für Glühlampensignalanlage
 f tableau *m* de commutation de signaux lumineux
 r панель *f* с сигнализацией

7087 language analysis
 d Sprachanalyse *f*

f analyse *f* de langage
r анализ *m* языка

7088 language for automatic mechanical assembly
 d Sprache *f* für automatische mechanische Montage
 f langage *m* pour assemblage automatique mécanique
 r язык *m* [программирования] автоматической механической сборки

7089 Laplace transformation
 d Laplacesche Transformation *f*; Laplacesche Umformung *f*
 f transformation *f* de Laplace
 r преобразование *n* Лапласа

7090 large-area proportional counter
 d Großflächenproportionalzähler *m*
 f compteur *m* proportionnel à grande surface
 r пропорциональный счётчик *m* большой плоскости

7091 large-irradiation plant
 d große Bestrahlungsanlage *f*
 f installation *f* d'irradiation à grande capacité
 r мощная радиационная установка *f*

7092 large-scale computing system
 d Großrechnersystem *n*; große EDV-Anlage *f*
 f système *m* calculateur à grande échelle
 r большая вычислительная система *f*

7093 large-scale integration
 d hohe Integration *f*; Großbereichsintegration *f*
 f intégration *f* élevée
 r высокая степень *f* интеграции

7094 large-signal analysis
 d Großsignalanalyse *f*
 f analyse *f* en régime signal fort
 r анализ *m* большого сигнала

7095 large-signal region
 d Großsignalbereich *m*
 f domaine *m* de signaux grands
 r область *f* большого сигнала

7096 large transient
 d große Transiente *f*
 f grand transitoire *m*
 r большой переходный процесс *m*

7097 laser accelerometer
 d Laserbeschleunigungsmesser *m*
 f accéléromètre *m* à laser
 r лазерный акселерометр *m*

7098 laser activity
 d Lasertätigkeit *f*
 f action *f* du laser; travail *m* du laser
 r действие *n* лазера

7099 laser aiming
 d Laseranzielen *n*; Laserlenkung *f*
 f visée *f* à laser; guidage *m* à laser
 r наведение *n* лазера

7100 laser alignment
 d Lasereinstellung *f*
 f ajustage *m* du laser; ajustage à l'aide du laser
 r регулировка *f* положения при помощи лазера

7101 laser altimeter; laser altitude gauge
 d Laserhöhenmesser *m*
 f altimètre *m* à laser
 r лазерный высотомер *m*

* **laser altitude gauge** → 7101

7102 laser amplifier
 d Laserverstärker *m*
 f amplificateur *m* à laser
 r лазерный усилитель *m*

7103 laser amplifier bandwidth
 d Laserverstärkerbandbreite *f*
 f largeur *f* de bande de l'amplificateur à laser
 r полоса *f* пропускания лазерного усилителя

7104 laser amplifier based integrated optical circuit
 d integrierte optische Schaltung *f* auf der Basis von Laserverstärkern
 f circuit *m* d'optique intégrée basé sur des lasers amplificateurs
 r интегральная оптическая схема *f* на основе лазерных усилителей

7105 laser aperture
 d Laserapertur *f*; Laseröffnung *f*
 f ouverture *f* du laser
 r апертюра *f* лазера

7106 laser application
 d Laseranwendung *f*
 f application *f* du laser
 r применение *n* лазера

7107 laser arrangement
 d Lasergerät *n*; Laseranordnung *f*
 f dispositif *m* à laser; appareil *m* à laser
 r лазерная установка *f*

7108 **laser-beam danger**
  d Laserstrahlgefahr *f*
  f danger *m* du faisceau laser
  r опасность *f* поражения лазерным лучом

7109 **laser-beam deflecting circuit**
  d Laserstrahlablenkschaltung *f*
  f circuit *m* de déviation du faisceau laser
  r схема *f* отклонения лазерного луча

7110 **laser-beam focusing**
  d Laserstrahlfokussierung *f*;
       Laserstrahlbündelung *f*
  f focalisation *f* du faisceau laser
  r фокусировка *f* лазерного луча

7111 **laser-beam machining device**
  d Laserstrahlbearbeitungsmaschine *f*
  f dispositif *m* d'usinage à laser
  r установка *f* для механической обработки
       лазерным лучом

7112 **laser-beam modulator**
  d Laserstrahlmodulator *m*
  f modulateur *m* de faisceau laser
  r модулятор *m* лазерного луча

7113 **laser[-beam] printer**
  d Laserdrucker *m*
  f imprimante *f* [à] laser
  r лазерный принтер *m*

7114 **laser-beam sensor**
  d Laserstrahlsensor *m*; Sensor *m* mit
       Laserstrahlen
  f senseur *m* à faisceau laser
  r сенсор *m* на лазерном излучении

7115 **laser-beam welding**
  d Laserstrahlschweißen *n*
  f soudage *m* par laser; soudage à faisceau de
       laser
  r сварка *f* лазерным лучом

7116 **laser bistable device**
  d bistabiles Lasergerät *n*
  f dispositif *m* bistable à laser
  r бистабильное лазерное устройство *n*

7117 **laser cascade connection**
  d Laserkaskadenverbindung *f*
  f connexion *f* en cascade des lasers
  r каскадное включение *n* лазеров

7118 **laser channel capacity**
  d Laserkanalkapazität *f*;
       Laserkanalübertragungsfähigkeit *f*
  f capacité *f* du canal laser

  r пропускная способность *f* лазерного
       канала [связи]

7119 **laser circuit**
  d Laserkreis *m*
  f circuit *m* du laser
  r лазерная схема *f*; лазерный контур *m*

7120 **laser coherence**
  d Laserstrahlkohärenz *f*
  f cohérence *f* des rayons du laser
  r когерентность *f* излучения лазера

7121 **laser communication**
  d Laserverbindung *f*
  f communication *f* à laser
  r лазерная связь *f*

7122 **laser communication engineering**
  d Lasernachrichtentechnik *f*
  f technique *f* des communications à laser
  r лазерная техника *f* связи

7123 **laser communication equipment**
  d Laserfernmeldeausrüstung *f*;
       Laserfernmeldeeinrichtung *f*;
       Laserfernmeldevorrichtung *f*
  f appareillage *m* de communication à laser
  r оборудование *n* лазерной системы связи

7124 **laser communication system**
  d Laser-Nachrichtenübertragungssystem *n*
  f système *m* de communication par laser
  r лазерная система *f* передачи

7125 **laser cut**
  d Laserschnitt *m*
  f échancrure *f* laser
  r вырез m, сформированный лазерным
       лучом; лазерный вырез

7126 **laser data display equipment**
  d Laserdatendarstellungsgerät *n*
  f appareillage *m* à laser pour la reproduction
       de données
  r лазерная аппаратура *f* индикации данных

7127 **laser data processing equipment**
  d Laserdatenverarbeitungsanlage *f*
  f ordinateur *m* à laser
  r лазерная аппаратура *f* обработки данных

7128 **laser data transmission line**
  d Laser-Datenübertragungsleitung *f*
  f ligne *f* de transmission à laser de données
  r лазерная линия *f* передачи данных

7129 **laser detection system**
  d Lasererfassungssystem *n*

f système *m* de détection à laser
r лазерная система *f* обнаружения

7130 laser determination of the trajectory
d Flugbahnbestimmung *f* durch Laser
f détermination *f* de la trajectoire à l'aide du laser
r определение *n* траектории при помощи лазерных средств

7131 laser direct imaging
d direkte Abbildung *f* mit Laserstrahlen
f imagerie *f* par faisceau laser
r формирование *n* изображений лазерным лучом

* **laser display → 7178**

7132 laser display system
d Laserdarstellungssystem *n*
f système *m* de représentation à laser
r лазерная система *f* индикации

7133 laser effect
d Lasereffekt *m*
f effet *m* laser
r лазерный эффект *m*

7134 laser emission
d Laseremission *f*; Laserausstrahlung *f*
f rayonnement *m* de laser; émission *f* du laser
r излучение *n* лазера

7135 laser energy
d Laserenergie *f*
f énergie *f* de laser
r энергия *f* [излучения] лазера

7136 laser enrichment
d Laseranreicherung *f*
f enrichissement *m* laser
r лазерное обогощение *n*

7137 laser excitation
d Lasererregung *f*
f excitation *f* du laser
r возбуждение *n* лазера

7138 laser flowmeter
d Laserdurchflussmesser *m*
f débitmètre *m* à laser
r лазерный расходомер *m*

7139 laser fluctuations
d Laserleistungsschwankungen *fpl*
f fluctuations *fpl* du laser
r флуктуации *fpl* излучения лазера

7140 laser focus

d Laserfokus *m*
f foyer *m* laser
r лазерный фокус *m*

7141 laser frequency
d Laserfrequenz *f*
f fréquence *f* du laser
r частота *f* излучения лазера

7142 laser frequency stability
d Laserfrequenzstabilität *f*
f stabilité *f* de la fréquence du laser
r частотная стабильность *f* излучения лазера

7143 laser gain
d Lasergewinn *m*; Gewinn *m* des Lasers
f gain *m* du laser
r коэффициент *m* усиления лазера

7144 laser generation
d Laserschwingungserzeugung *f*
f génération *f* d'oscillement par laser
r лазерная генерация *f*

7145 laser generator
d Lasergenerator *m*
f générateur *m* laser
r лазерный генератор *m*

7146 laser gyroscope
d Lasergyroskop *n*
f gyroscope *m* à laser
r лазерный гироскоп *m*

7147 laser harmonic
d Harmonische *f* des Lasers
f harmonique *f* du laser
r гармоника *f* излучения лазера

7148 laser-induced defect
d durch Laserstrahlen verursachter Fehler *m*
f défaut *m* crée par radiation laser
r дефект m, вызванный лазерным излучением

7149 laser-induced heating
d Laserstrahlerwärmung *f*; Laserstrahlerhitzung *f*
f chauffage *m* par laser
r нагревание *n* лучом лазера

7150 laser information display system
d Informationsdarstellungslasersystem *n*; Laserinformationsdarstellungssystem *n*
f système *m* à laser de représentation de l'information
r лазерная система *f* отображения информации

7151 laser interferometric alignment
  d laserinterferometrische Justierung *f*
  f alignement *m* à l'aide d'interféromètre laser
  r совмещение *n* с помощью лазерного
     интерферометра

7152 laser irradiation
  d Laserbestrahlung *f*; Lasereinstrahlung *f*
  f irradiation *f* par laser
  r лазерное облучение *n*

7153 laser level
  d Laserpegel *m*; Laserniveau *n*
  f niveau *m* du laser
  r энергетический уровень *m* лазера

7154 laser light; laser radiation
  d Laserlicht *n*; Laserstrahlung *f*
  f lumière *f* du laser; rayonnement *m* laser
  r видимое излучение *n* лазера; лазерное
     излучение

7155 laser linewidth
  d Laserlinienbreite *f*
  f largeur *f* de raie d'un laser
  r ширина *f* линии лазерного излучения

7156 laser-linewidth determining mechanism
  d Mechanismus m, der dir Laserlinienbreite
     bestimmt
  f mécanisme *m* déterminant la largeur de raie
     du laser
  r механизм *m* для определения ширины
     спектральной линии лазера

7157 laser machine
  d Lasermaschine *f*
  f machine *f* à laser
  r лазер *m*

  * **laser medium → 299**

7158 laser microspectroanalyzer
  d Lasermikrospektroanalysator *m*
  f microspectroanalyseur *m* à laser
  r лазерный микроспектроанализатор *m*

7159 laser modulation
  d Lasermodulation *f*
  f modulation *f* d'un laser
  r модуляция *f* лазера

7160 laser operation
  d Laserbetrieb *m*
  f opération *f* du laser; fonctionnement *m* du
     laser
  r действие *n* лазера

7161 laser oscillator
  d Laseroszillator *m*
  f oscillateur *m* laser
  r лазер-генератор *m*

7162 laser output
  d Laserausgangsleistung *f*
  f puissance *f* de sortie de laser
  r выходная мощность *f* лазера

7163 laser output characteristic
  d Laserausgangscharakteristik *f*
  f caractéristique *f* de sortie du laser
  r характеристика *f* выходной мощности
     лазера

7164 laser output spectrum
  d Laserausgangsspektrum *n*
  f spectre *m* de sortie du laser
  r спектр *m* выходного излучения лазера

7165 laser penetration
  d Laserstrahlendurchgriff *m*
  f pénétration *f* des rayons du laser
  r проникновение *n* лазерного излучения

7166 laser pick-off unit
  d Empfänger *m* für kohärente Strahlen
  f récepteur *m* d'ondes cohérentes
  r чувствительный элемент *m* лазера

7167 laser piercing power
  d Laserdurchschlagsvermögen *n*
  f pouvoir *m* de perçage du laser
  r пронизивающая способность *f* лазера

7168 laser power source
  d Laserenergiequelle *f*
  f source *f* d'alimentation du laser
  r источник *m* питания лазера

7169 laser pressure gauge
  d Laserdruckmesser *m*
  f gauge *f* de pression à laser
  r лазерный датчик *m* давления

  * **laser printer → 7113**

7170 laser probe mass spectrography
  d Massenspektrografie *f* mit Lasersonde
  f spectrographie *f* de masse à sonde laser
  r масс-спектрография *f* с лазерным зондом

7171 laser pulse compression
  d Laserimpulskompression *f*
  f compression *f* d'impulsions lasers
  r сжатие *n* лазерного импульса

7172 **laser pulse control**
 d Laserimpulssteuerung *f*
 f commande *f* impulsionnelle du laser
 r импульсное управление *n* с помощью
  лазера

7173 **laser radar technique**
 d Laserradartechnik *f*
 f technique *f* du radar à laser
 r лазерная радиолокационная техника *f*

 * **laser radiation** → 7154

7174 **laser Raman system**
 d Laser-Raman-System *n*
 f système *m* laser Raman
 r лазерная система *f* Рамана

7175 **laser rangefinder; laser ranging device;
  laser ranging equipment**
 d Laserentfernungsmesser *m*
 f télémètre *m* à laser
 r лазерный дальномер *m*

 * **laser ranging device** → 7175

 * **laser ranging equipment** → 7175

7176 **laser receving station**
 d Laserempfängerstation *f*
 f station *f* réceptrice des rayons de laser
 r лазерная приёмная станция *f*

7177 **laser rounding**
 d Laserabrunden *n*
 f arrondi *m* par laser
 r лазерное сглаживание *n*

7178 **laser[-scan] display**
 d Laserbildschirm *m*
 f écran *m* à [rayon] laser
 r лазерный экран *m*

7179 **laser scanner**
 d Laserscanner *m*; Laserlesegerät *n*
 f scanner *m* [à] laser
 r лазерный сканер *m*

7180 **laser scanning microscope**
 d Laserrastermokroskop *n*
 f microscope *m* laser à balayage
 r лазерный растровый микроскоп *m*

7181 **laser search apparatus**
 d Laseraufklärungsgerät *n*; Lasersuchgerät *n*
 f dispositif *m* laser de surveillance; dispositif
  exploreur à laser
 r лазерная поисковая аппаратура *f*

 * **laser sending station** → 7190

7182 **laser signal**
 d Lasersignal *n*
 f signal *m* du laser
 r лазерный сигнал *m*

7183 **laser spectroscopy**
 d Laserspektroskopie *f*
 f spectroscopie *f* par laser
 r лазерная спектроскопия *f*

7184 **laser superheterodyne receiver**
 d Laserüberlagerungsempfänger *m*
 f récepteur *m* superhétérodyne à laser
 r лазерный супергетеродинный приёмник *m*

7185 **laser surveillance**
 d Laserüberwachung *f*
 f surveillance *f* à laser
 r обзорная лазерная установка *f*

7186 **laser switch**
 d Laserschalter *m*
 f interrupteur *m* à laser
 r лазерный переключатель *m*

7187 **laser technology**
 d Lasertechnik *f*
 f technique *f* du laser
 r лазерная технология *f*

7188 **laser threshold**
 d Laserschwelle *f*
 f seuil *m* d'effet laser
 r порог *m* лазерного эффекта

7189 **laser tracking data**
 d Lasernachlaufdaten *pl*;
  Lasernachlaufangaben *fpl*
 f données *fpl* de poursuite à laser
 r данные *pl* системы лазерного
  сопровождения

7190 **laser transmitting station; laser sending
  station**
 d Lasersendestation *f*
 f station *f* émettrice à laser; station émettrice
  des rayons de laser
 r лазерная передающая станция *f*

7191 **lasing time**
 d Laserbetriebszeit *f*; Laserbetriebsintervall *n*
 f temps *m* de travail du laser; période *f* de
  travail du laser
 r время *n* генерации лазера

7192 **last-event estimator**
 d Absorptionspunktschätzung *f*

f estimateur *m* du dernier événement
r оценка *f* последнего события

7193 latch *v*
d selbsthaltend aufschalten; auffangen
f commuter à verrouillage
r фиксировать

7194 latched relay
d Sperrklinkenrelais *n*
f relais *m* à cliquet
r реле *n* с механической блокировкой

7195 latch register
d Latchregister *n*; Auffangregister *n*;
Informationsfangregister *n*
f registre *m* de tampon d'informations
r регистр-фиксатор *m*

7196 latency time
d Latenzzeit *f*
f temps *m* de latence
r время *n* ожидания

7197 lateral stability
d Querstabilität *f*
f stabilité *f* transversale
r поперечная устойчивость *f*

7198 launch control
d Startkontrolle *f*
f contrôle *m* du lancement
r управление *n* стартом

7199 law of regulating action
d Stellgrößengesetz *n*
f loi *f* d'action réglante
r правило *n* регулирующего воздействия

7200 law of similarity
d Ähnlichkeitsregel *f*; Ähnlichkeitstheorie *f*
f principe *m* de similitude; règle *f* de similitude
r теория *f* подобия; принцип *m* подобия

7201 layer interface
d Zwischeninterface *n*
f interface *f* intermédiaire
r межуровневый интерфейс *m*

7202 layout
d Anordnung *f*; Layout *n*
f disposition *f*; layout *m*
r схема *f* размещения

7203 layout of erection site
d Montageplatzlayout *n*; Layout *n* eines
Montageplatzes
f layout *m* de poste d'assemblage

r схема *f* сборочного места

7204 layout of work pieces
d Werkstückanordnung *f*
f groupement *m* des pièces à usiner
r схема *f* расположения обрабатываемых
деталей

7205 layout plan; general plan
d Übersichtsschema *n*; Gesamtplan *m*
f plan *m* [de disposition] d'ensemble
r обзорная схема *f*; общий план *m*

7206 lead capacitance
d Zuleitungskapazität *f*
f capacité *f* de conducteur
r ёмкость *f* подводящего провода

7207 leading circuit
d Überholungsstromkreis *m*;
Überholungsschleife *f*
f chaîne *f* d'anticipation
r цепь *f* опережения

7208 leading decision
d Grundsatzentscheidung *f*
f décision *f* principale
r принципиальное решение *n*;
фундаментальное решение

7209 leading element
d Voreilglied *n*; Vorhaltelement *n*
f élément *m* d'anticipation
r звено *n* опережения

* **lead-lag network** → 6819

7210 leakage path
d Kriechweg *m*; Kriechstrecke *f*
f voie *f* de fuite
r путь *m* утечки

7211 leak-tight encapsulation
d lecksichere Verkappung *f*
f encapsulation *f* à étanchéité
r герметизация *f*

7212 leak tightness
d Lecksicherheit *f*
f herméticité *f*; étanchéité *f*
r герметичность *f*

7213 leaky package
d undichtes Gehäuse *n*
f boîtier *m* non étanche
r негерметичный корпус *m*

* **learning automaton** → 7215

7214 learning classification system
d lernendes Klassifikationssystem *n*
f système *m* de classification à apprentissage
r система *f* классификации с обучением

7215 learning machine; learning automaton
d Lernmaschine *f*; lernender Rechenautomat *m*; lernender Automat *m*
f machine *f* élève; automate *m* élève [autodidacte]; machine enseignante
r самообучающаяся машина *f*; самообучающийся автомат *m*

7216 learning model
d Lernmodell *n*
f modèle *m* à apprentissage
r модель *f* обучения

7217 learning phase
d Lernphase *f*
f phase *f* à apprentissage; phase d'auto-apprentissage
r фаза *f* обучения

7218 learning program
d Lernprogramm *n*
f programme *m* d'apprentissage
r обучающая программа *f*

7219 leased line
d Mietleitung *f*
f ligne *f* louée
r выделенная линия *f*

7220 level change value
d Schaltpunkt *m*
f point *m* de commutation; valeur *f* de commutation
r пороговая величина *f*

7221 level controller
d Niveauregler *m*
f régulateur *m* de niveau
r регулятор *m* уровня

7222 level crossing
d Niveaukreuzung *f*
f passage *m* à niveau
r переход *m* на одинаковом уровне

7223 level detector
d Niveaudetektor *m*; Pegelprüfer *m*
f détecteur *m* de niveau
r уровнемер *m*

7224 level indicator
d Niveauanzeiger *m*
f indicateur *m* de niveau

r указатель *m* уровня

7225 level measurement
d Pegelmessung *f*
f mesure *f* de niveau
r измерение *n* уровня

* **level of activity** → 323

7226 level of capacity
d Leistungsniveau *n*
f niveau *m* de puissance
r уровень *m* мощности

7227 level of interpretation
d Interpretierungsniveau *n*
f niveau *m* d'interprétation
r уровень *m* интерпретации

* **level of process** → 323

7228 level of significance
d Signifikanzniveau *n*
f seuil *m* de probabilité
r уровень *m* значимости

7229 level power cylinder
d Stellmotor *m* mit Kurbelantrieb; Servomotor *m* mit Kurbelantrieb
f servomoteur *m* à levier
r кривошипный сервомотор *m*

7230 level regulation
d Pegelregelung *f*
f réglage *m* de niveau
r регулирование *n* уровня

7231 level safety valve
d Sicherheitsventil *n* mit Gewichtshebel
f soupape *f* de sûreté à levier
r рычажный предохранительный клапан *m*

7232 level setting
d Pegeleinstellung *f*
f ajustement *m* de niveau
r установка *f* уровня

7233 level structure
d Termstruktur *f*
f structure *f* de terme
r структура *f* уровня

7234 level teleindicator; remote level indicator
d Niveaufernanzeiger *m*; Pegelfernanzeiger *m*
f indicateur *m* de niveau à distance
r дистанционный уровнемер *m*

7235 level transmitter
d Niveaugeber *m*

f transmetteur *m* de niveau
r датчик *m* уровня

7236 level-triggered
d pegelgettriggert; pegelgeschalter
f basculé par niveau
r переключаемый уровнем

7237 Liapunov function
d Ljapunow-Funktion *f*
f fonction *f* de Ljapunow
r функция *f* Ляпунова

7238 life curve
d Lebensdauerkurve *f*
f courbe *f* de longévité
r кривая *f* срока службы

7239 life test
d Lebensdauerprüfung *f*
f essai *m* de durée de vie
r испытание *n* для определения срока службы

7240 lifting force
d Hubvermögen *n*; Hubkraft *f*
f force *f* de levage
r подъёмная способность *f*

7241 lifting magnet
d Hubmagnet *m*; Hebemagnet *m*
f électro-aimant *m* d'ascension; électro-aimant de levage
r подъёмный электромагнит *m*

7242 lifting method
d Hebeverfahren *n*; Hubverfahren *n*
f méthode *f* de levage
r метод *m* подъёма

7243 light barrier checking
d Lichtschrankenkontrolle *f*
f contrôle *m* de barrière photo-électrique
r контроль *m* с помощью фотоэлектрических устройств

7244 light barrier sensor
d Lichtschrankensensor *m*
f senseur *m* d'une barrière photo-électrique
r сенсор *m* с фотоэлектрическим устройством

7245 light conductor technology; optical fibres technology
d Lichtleitertechnologie *f*
f technologie *f* par fibre optique
r технология *f* световодов

7246 light-dependent control element
d lichtabhängiges Steuerglied *n*
f élément *m* de contrôle dépendant de lumière
r светочувствительный управляющий элемент *m*

7247 light detecting component
d optisches Empfangsbauelement *n*
f composant *m* récepteur de lumière
r оптический приемочный схемный элемент *m*

7248 light detector
d Lichtdetektor *m*
f détecteur *m* de la lumière
r детектор *m* света

7249 light-gap regulator
d Lichtspaltregler *m*
f régulateur *m* de fente lumineuse
r регулятор *m* световой щели

7250 light-gap testing equipment
d Lichtspaltprüfeinrichtung *f*
f appareil *m* de contrôle de la fente lumineuse
r оборудование *n* для испытания методом световой щели

7251 light-homing guidance
d automatisches Zielsuchen *n* durch Lichtstrahlen
f giudage *m* automatique par faisceaux lumineux
r оптическое самонаведение *n*

7252 light impulse
d Lichtimpuls *m*
f impulsion *f* lumineuse
r световой импульс *m*

7253 lighting control
d Beleuchtungsregulierung *f*
f réglage *m* d'éclairage
r регулирование *n* освещения

7254 lighting installation
d Lichtanlage *f*; Beleuchtungsanlage *f*
f installation *f* d'éclairage
r осветительная установка *f*

7255 light-intensity modulation
d Modulation *f* durch Lichtintensität
f modulation *f* en intensité de flux lumineux
r модуляция *f* интенсивности света

7256 light pen; light stylos; beam pen; electronic pen
d Lichtgriffel *m*; Lichtstift *m*; elektronischer Stift *m*

    f crayon *m* lumineux; crayon électronique;
      photostyle *m*
    r световое перо *n*; световой карандаш *m*

7257 light-pen-controlled program
    d lichtstiftgesteuertes Programm *n*
    f programme *m* contrôlé par crayon lumineux
    r управляемая световым пером программа *f*

7258 light pen display
    d Lichtstiftanzeige *f*
    f affichage *m* de crayon lumineux
    r дисплей *m* со световым пером

7259 light pen sensitivity
    d Lichtstiftempfindlichkeit *f*
    f sensibilité *f* d'un crayon lumineux
    r чувствительность *f* светового пера

7260 light relay
    d Lichtrelais *n*
    f relais *m* lumineux
    r фотореле *n*

7261 light signal; luminous signal
    d Lichtsignal *n*; Leuchtsignal *n*
    f signal *m* lumineux
    r световой сигнал *m*

    * **light stylos** → 7256

7262 light-wave device
    d Lichtwellengerät *n*; oproelektronisches
      Gerät *n*
    f dispositif *m* opto-électronique
    r оптоэлектронный прибор *m*

7263 light-wave measuring unit
    d Lichtwellenlängemaßeinheit *f*
    f unité *f* de mesure de longueur d'onde
      lumineuse
    r устройство *n* для измерения [длины]
      световой волны

    * **limit** → 1891

7264 limitation of the consequences
    d Folgenbegrenzung *f*; Begrenzung *f* der
      Folgen
    f limitation *f* de conséquences
    r ограничение *n* последствий

7265 limit case
    d Grenzfall *m*
    f cas *m* limite
    r предельный случай *m*

7266 limit comparator

    d Grenzwertvergleicher *m*
    f comparateur *m* de valeur limite
    r устройство *n* сравнения предельных
      значений

7267 limit contact
    d Grenzkontakt *m*
    f contact *m* limite
    r предельный контакт *m*

7268 limit cycle
    d Grenzzyklus *m*
    f cycle *m* de limite
    r предельный цикл *m*

7269 limited action
    d begrenzte Einwirkung *f*
    f action *f* limiteé
    r ограниченное воздействие *n*

7270 limited power
    d begrenzte Leistung *f*
    f puissance *f* limiteé
    r ограниченная мощность *f*

7271 limited quantity
    d begrenzte Größe *f*
    f quantité *f* limiteé
    r ограничивающая величина *f*

7272 limited variable
    d begrenzte Variable *f*
    f variable *f* bornée
    r ограниченная переменная *f*

    * **limiter** → 2444

7273 limiter characteristic
    d Begrenzerkennlinie *f*
    f caractéristique *f* du limiteur
    r характеристика *f* ограничителя

7274 limiter circuit
    d Begrenzerschaltung *f*
    f circuit *m* limiteur
    r контур *m* ограничителя

7275 limit force
    d Grenzkraft *f*
    f force *f* limite
    r предельное усилие *n*

7276 limit force-controlled jointing movement
    d grenzkraftgesteuerte Fügebewegung *f*
    f mouvement *m* de jointage commandé par
      force limite
    r движение *n* сопряжения, управляемое от
      предельного усилия

7277 limiting characteristic function
  d charakteristische Limitfunktion *f*
  f fonction *f* caractéristique de limite
  r предельная характеристическая функция *f*

7278 limiting condidion
  d Grenzbedingung *f*; begrenzende Bedingung *f*
  f condition *f* limitative
  r ограничивающее условие *n*

7279 limiting controller
  d Grenzwertregler *m*
  f régulateur-limiteur *m*
  r ограничивающий регулятор *m*

7280 limiting current
  d Grenzstrom *m*
  f courant *m* limite
  r предельный ток *m*

7281 limiting device
  d Begrenzungsvorrichtung *f*
  f dispositif *m* limiteur
  r ограничительное устройство *n*

7282 limiting feedback
  d begrenzende Rückkopplung *f*
  f réaction *f* limitante
  r ограничивающая обратная связь *f*

7283 limiting feed forward
  d begrenzende Vorwärtswirkung *f*
  f action *f* limitante
  r ограничивающая прямая связь *f*

7284 limiting frequency
  d Grenzfrequenz *f*
  f fréquence *f* de coupure
  r предельная частота *f*

7285 limiting process
  d Grenzwertprozess *m*;
     Grenzwertbetrachtung *f*
  f procédé *m* limite
  r предельный процесс *m*

7286 limiting resistance
  d Begrenzungswiderstand *m*
  f résistance *f* limitante
  r [токо]ограничивающее сопротивление *n*

7287 limiting value switch
  d Grenzwertschalter *m*
  f commutateur *m* de valeur limitée
  r переключатель *m* предельных значений

7288 limit moment
  d Grenzmoment *n*

  f moment *m* limite
  r предельный момент *m*

7289 limit of capacity
  d Leistungsgrenze *f*
  f limite *f* de puissance
  r предел *m* мощности

* **limit of error → 5188**

7290 limit of measurement
  d Messgrenze *f*
  f limite *f* de mesure
  r предел *m* измерения

* **limit signal → 7297**

7291 limits of integration
  d Integrationsbereich *m*;
     Integrationsgrenzen *fpl*
  f domaine *m* d'intégration; limites *fpl*
     d'intégration
  r пределы *mpl* интегрирования

7292 limit stability
  d Grenzstabilität *f*
  f stabilité *f* de limite
  r предельная устойчивость *f*

7293 limit switch of manipulator
  d Manipulatorendschalter *m*
  f interrupteur *m* limite d'un manipulateur
  r концевой выключатель *m* манипулятора

7294 limit temperature
  d Grenztemperatur *f*
  f température *f* limite
  r предельная температура *f*

7295 limit tolerance
  d Grenztoleranz *f*
  f tolérance *f* limite
  r предельный допуск *m*

7296 limit value
  d Grenzwert *m*
  f valeur *f* limite
  r предельное значение *n*

7297 limit [value] signal
  d Grenzwertsignal *n*
  f signal *m* limite
  r сигнал *m* превышения предельного
     значения

7298 line adapter
  d Leitungsadapter *m*; Leitungsanpassungsteil *n*

f adapteur *m* de ligne
r линейный адаптер *m*

7299 line adaption
  d Leitungsanpassung *f*
  f adaptation *f* de ligne
  r адаптация *f* линии

7300 linear accelerator
  d Linearbeschleuniger *m*
  f accélérateur *m* linéaire
  r линейный ускоритель *m*

  * **linear acting element** → 7354

7301 linear activity
  d Linienaktivität *f*
  f activité *f* linéaire
  r линейная активность *f*

7302 linear actuator
  d linearer Effektor *m*
  f élément *m* linéaire de commande
  r линейный исполнительный механизм *m*

7303 linear adjusting
  d lineares Einstellen *n*
  f ajustage *m* linéaire
  r линейная юстировка *f*

7304 linear algebraic equation
  d lineare algebraische Gleichung *f*
  f équation *f* algébrique linéaire
  r линейное [алгебраическое] уравнение *n*

7305 linear amplifier
  d Linearverstärker *m*; Proportionalverstärker *m*
  f amplificateur *m* linéaire
  r линейный усилитель *m*

7306 linear approximation
  d lineare Annäherung *f*
  f approximation *f* linéaire
  r линейное приближение *n*

7307 linear attenuation
  d lineare Dämpfung *f*
  f affaiblissement *m* linéaire
  r линейное затухание *n*

7308 linear automaton
  d linearer Automat *m*
  f automate *m* linéaire
  r линейный автомат *m*

  * **linear block** → 7328

7309 linear channel

d Linearkanal *m*
f canal *m* linéaire
r линейный канал *m*

7310 linear circuit; linear network
  d Linearstromkreis *m*
  f circuit *m* linéaire
  r линейная схема *f*

7311 linear code
  d linearer Kode *m*
  f code *m* linéaire
  r линейный код *m*

7312 linear combination of control-loop elements
  d Serienschaltung *f* der Regelkreisglieder; Serienschaltung der Glieder im Regelkreis
  f combinaison *f* en série des éléments du système asservi
  r последовательное соединение *n* звеньев в цепи регулирования

7313 linear constraints
  d lineare Beschränkungen *fpl*
  f contraintes *fpl* linéaires
  r линейные ограничения *npl*

7314 linear control
  d lineare Regelung *f*; Linearregelung *f*
  f réglage *m* linéaire
  r линейное регулирование *n*; линейное управление *n*

7315 linear control electromechanism
  d linearer Stellantrieb *m*
  f électromécanisme *m* de commande linéaire
  r электромеханическое устройство *n* линейного перемещения регулирующих элементов

7316 linear damping
  d Lineardämpfung *f*
  f amortissement *m* linéaire
  r линейное демпфирование *n*

7317 linear degree
  d linearer Grad *m*
  f degré *m* linéaire
  r степень *f* линейности

7318 linear density
  d Liniendichte *f*; lineare Dichte *f*
  f densité *f* linéique; densité linéaire
  r линейная плотность *f*

7319 linear dependence
  d lineare Abhängigkeit *f*

    f relation *f* linéaire
    r линейная зависимость *f*

7320 linear detection
    d lineare Gleichrichtung *f*
    f détection *f* linéaire
    r линейное детектирование *n*

7321 linear dimension
    d lineare Abmessung *f*; Linearabmessung *f*
    f dimension *f* linéaire
    r линейный размер *m*

7322 linear discrete-time system
    d lineares diskretes System *n*; lineares Tastsystem *n*
    f système *m* discret linéaire
    r линейная дискретная система *f*

7323 linear dispersion
    d Lineardispersion *f*; lineare Dispersion *f*
    f dispersion *f* linéaire
    r линейная дисперсия *f*

7324 linear displacement
    d lineare Verschiebung *f*; Linearverschiebung *f*
    f déplacement *m* linéaire
    r линейное перемещение *n*

7325 linear distortion
    d lineare Verzerrung *f*
    f distorsion *f* linéaire
    r линейное искажение *n*

7326 linear drive
    d Linearantrieb *m*
    f entraînement *m* linéaire
    r привод *m* линейных перемещений

7327 linear-elastic fracture mechanics
    d linear-elastische Bruchmechanik *f*
    f mécanique *f* de fracture linéaire-élastique
    r линейно-упругая механика *f* разрушения

7328 linear element; linear block
    d lineares Glied *n*
    f élément *m* linéaire
    r линейное звено *n*; линейный элемент *m*

7329 linear equation system
    d lineares Gleichungssystem *n*
    f système *m* des équations linéaires
    r система *f* линейных уравнений

7330 linear error
    d linearer Fehler *m*
    f erreur *f* linéaire
    r линейная погрешность *f*

7331 linear extrapolation
    d lineare Extrapolation *f*
    f extrapolation *f* linéaire
    r линейная экстраполяция *f*

7332 linear extrapolation distance
    d lineare Extrapolationslänge *f*; linearer Extrapolationsabstand *m*
    f distance *f* d'extrapolation linéaire
    r линейная длина *f* экстраполяции

7333 linear filter
    d lineares Filter *n*; Linearfilter *n*
    f filtre *m* linéaire
    r линейный фильтр *m*

7334 linear frequency spectrum
    d Linearfrequenzspektrum *n*
    f spectre *m* linéaire de fréquences
    r линейный частотный спектр *m*

7335 linear function
    d lineare Funktion *f*; Linearfunktion *f*
    f fonction *f* linéaire
    r линейная функция *f*

7336 linear-guided component
    d linear geführtes Bauelement *n*
    f composant *m* guidé linéaire
    r элемент *m* с линейными направляющими

7337 linear heat rate
    d lineare Wärmebelastung *f*; Längenbelastung *f*; Stablängenbelastung *f*
    f charge *f* thermique linéique
    r линейная тепловая нагрузка *f*

7338 linear independence
    d lineare Unabhängigkeit *f*
    f indépendance *f* linéaire
    r линейная независимость *f*

7339 linear interpolation
    d lineare Interpolation *f*
    f interpolation *f* linéaire
    r линейная интерполяция *f*

7340 linearity
    d Linearität *f*
    f linéarité
    r линейность *f*

7341 linearity control
    d Linearitätsregelung *f*
    f réglage *m* de linéarité
    r регулировка *f* линейности

7342 linearity error
    d Linearitätsfehler *m*

f défaut *m* de linéarité
r ошибка *f* линеаризации

7343 linearity in amplitude
d Amplitudenlinearität *f*
f linéarité *f* d'amplitude
r линейность *f* по амплитуде

7344 linearity of capacitive micrometers
d Linearität *f* von kapazitiven Mikrometern
f linéarité *f* de micromètres capacitifs
r линейность *f* ёмкостных микрометров

7345 linearity of radiation receivers
d Linearität *f* von Strahlungsempfängern
f linéarité *f* de récepteurs de radiation
r линейность *f* приёмников излучения

7346 linearity of signal
d Signallinearität *f*
f linéarité *f* d'un signal
r линейность *f* сигнала

7347 linearity theorem
d Satz *m* über die Linearität; Lehrsatz *m* von der Linearität
f théorème *m* de linéarité
r теорема *f* линейности

7348 linearization
d Linearisierung *f*
f linéarisation *f*
r линеаризация *f*

7349 linearization by method of small oscillations
d Linearisierung *f* durch die Methode kleiner Schwingungen
f linéarisation *f* par méthode de petites oscillations
r линеаризация *f* методом малых колебаний

7350 linearization of relay systems
d Linearisierung *f* der Relaissysteme
f linéarisation *f* de systèmes à relais
r линеаризация *f* релейных систем

7351 linearization range
d Linearisierungsbereich *m*
f plage *f* de linéarisation
r диапазон *m* линеаризации

7352 linearize *v*
d linearisieren
f linéariser
r линеаризовать

7353 linearized system model

7354 linear joint; linear acting element
d linearisiertes Systemmodell *n*
f modèle *m* de système linéarisé
r линеаризованная модель *f* системы

7354 linear joint; linear acting element
d linearer Akteur *m*; lineares Bewegungsglied *n*
f actionneur *m* linéaire
r линейное активное звено *n*

7355 linearly independent
d linear unabhängig
f indépendant linéaire[ment]
r линейно независимый

7356 linear model
d lineares Modell *n*
f modèle *m* linéaire
r линейная модель *f*

7357 linear-mode region
d linearer Arbeitsbereich *m*
f champ *m* de travail linéaire
r линейный диапазон *m* [работы]

7358 linear momentum resolution
d Impulsauflösung *f*
f résolution *f* en impulsion
r разрешение *n* по импульсу

7359 linear motor
d Linearmotor *m*
f moteur *m* linéaire
r электродвигатель *m* с прямолинейным полем

* linear network → 7310

7360 linear optimal systems
d linearte Optimalsysteme *npl*
f systèmes *mpl* optimaux linéaires
r линейные оптимальные системы *fpl*

7361 linear optimization
d Linearoptimierung *f*
f optimisation *f* linéaire
r линейная оптимизация *f*

7362 linear optimization program
d lineares Optimierungsprogramm *n*
f programme *m* d'optimisation linéaire
r программа *f* линейной оптимизации

7363 linear polarization
d lineare Polarisation *f*
f polarisation *f* rectiligne
r линейная поляризация *f*

7364 linear position control
 d lineare Stellungsregelung *f*
 f régulation *f* de position linéaire
 r линейное регулирование *n* положения

7365 linear potentiometer
 d Linearpotentiometer *n*
 f potentiomètre *m* linéaire
 r линейный потенциометр *m*

7366 linear programming
 d lineare Programmierung *f*
 f programmation *f* linéaire
 r линейное программирование *n*

7367 linear program part
 d gerades Programmstück *n*
 f partie *f* rectiligne de programme
 r линейная часть *f* программы

7368 linear pulse amplifier
 d linearer Impulsverstärker *m*
 f amplificateur *m* d'impulsion linéaire
 r линейный импульсный усилитель *m*

7369 linear range
 d linearer Bereich *m*
 f gamme *f* linéaire
 r линейная область *f*

7370 linear receiver
 d linearer Empfänger *m*
 f récepteur *m* linéaire
 r линейный приёмник *m*

7371 linear resistance flowmeter
 d Linearwiderstandsdurchflussmesser *m*
 f débitmètre *m* à résistance linéaire
 r расходомер *m* с линейным
    сопротивлением

7372 linear rising signal
 d linear ansteigendes Signal *n*; Rampensignal *n*
 f signal *m* rampe
 r линейно нарастающий сигнал *m*

7373 linear scale
 d lineare Skale *f*
 f échelle *f* linéaire
 r равномерная шкала *f*

7374 line[ar] scanning
 d lineare Zerlegung *f*; Zeilenabtastung *f*;
    zeilenweise Abtastung *f*
 f balayage *m* linéaire
 r линейная развёртка *f*; строчное
    сканирование *n*

7375 linear servosystem dynamics
 d Dynamik *f* des linearen Servosystems
 f dynamique *f* du servomécanisme linéaire
 r динамика *f* линейной следящей системы

7376 linear single-loop control system
 d lineares einschleifiges Regelungssystem *n*
 f système *m* de réglage linéaire à boucle
    unique
 r линейная одноконтурная система *f*
    регулирования

7377 linear-slope delay filter
 d Verzögerungsfilter *n* mit linearer
    Kennliniensteilheit
 f filtre *m* à retard à pente linéaire
 r фильтр *m* задержки с линейным спадом
    характеристики

7378 linear-slope group delay characteristic
 d Gruppenlaufzeitcharakteristik mit konstanter
    Steilheit
 f caractéristique *f* de délai de groupe à pente
    linéaire
 r характеристика *f* групповой задержки с
    линейным спадом

7379 linear store
 d Linearspeicher *m*
 f mémoire *f* linéaire
 r линейная память *f*

7380 linear sweep generator
 d Linearzeitablenkgenerator *m*
 f générateur *m* à base de temps linéaire
 r линейный генератор *m* качающейся
    частоты

7381 linear switching circuit
 d linearer Schaltkreis *m*
 f circuit *m* de commutation linéaire
 r линейная переключательная схема *f*

7382 linear system
 d lineares System *n*
 f système *m* linéaire
 r линейная система *f*

 * **linear system of constant
    coefficients → 3042**

 * **linear system of variable
    coefficients → 12757**

7383 linear system parts
 d lineare Systemanteile *mpl*
 f parts *fpl* de système linéaire
 r линейные составляющие *fpl* системы

7384 **linear system stability inverstigation**
   d Stabilitätsuntersuchung *f* linearer Systeme
   f vérification *f* de stabilité des systèmes
      linéaires
   r определение *n* стабильности линейных
      систем

7385 **linear system theory**
   d lineare Systemtheorie *f*
   f théorie *f* des systèmes linéaire
   r линейная теория *f* систем

7386 **linear system with variable parameters**
   d Linearsystem *n* mit variablen Parametern
   f système *m* linéaire à paramètres variables
   r линейная система *f* с переменными
      параметрами

7387 **linear time-invariant system**
   d zeitlich unveränderliches lineares System *n*
   f système *m* linéaire invariant dans le temps
   r линейная система f, не зависящая от
      времени

7388 **linear-to-log converter**
   d linear-logarithmischer Umsetzer *m*
   f convertisseur *m* linéaire-logarithmique
   r линейно-логарифмический
      преобразователь *m*

7389 **linear trajectory interpolation**
   d lineare Bahninterpolation *f*
   f interpolation *f* trajectoire linéaire
   r линейная интерполяция *f* траектории

7390 **linear transducer**
   d linearer Wandler *m*
   f transducteur *m* linéaire
   r линейный преобразователь *m*

7391 **linear transfer circuit**
   d lineares Übertragungsglied *n*
   f organe *m* linéaire
   r линейное передаточное звено *n*

7392 **linear transfer function**
   d lineare Übertragungsfunktion *f*
   f fonction *f* de transfert linéaire
   r линейная передаточная функция *f*

7393 **linear transformation of coordinates**
   d lineare Koordinatenumformung *f*
   f transformation *f* linéaire de coordonnées
   r линейная трансформация *f* координат

7394 **linear unit**
   d Lineareinheit *f*
   f unité *f* linéaire

   r блок *m* линейных перемещений

7395 **linear unit structure**
   d Lineareinheitenaufbau *m*; Aufbau *m* einer
      Lineareinheit
   f structure *f* d'unité linéaire
   r структура *f* линейных перемещений

7396 **linear variable resistor**
   d linearveränderlicher Widerstand *m*
   f résistance *f* à variation linéaire
   r линейное сопротивление *n*

7397 **line assembly work**
   d Gleitmontageprinzip *n*
   f montage *m* à la chaîne
   r сборка *f* на поточной линии

* **line clock** → 7398

7398 **line clock [pulse]**
   d Leitungstakt *m*
   f horloge *f* de ligne
   r тактовый импульс *m* канала

7399 **line concentrator**
   d Leitungskonzentrator *m*
   f concentrateur *m* de ligne
   r линейный концентратор *m*

7400 **line control**
   d Leitungssteuerung *f*
   f commande *f* de ligne
   r управление *n* линией

7401 **line-descriptive element**
   d linienbeschreibendes Element *n*
   f élément *m* descriptif d'une ligne
   r элемент m, описывающий линии

7402 **line dialling**
   d Leitungsanwahl *f*
   f sélection *f* de ligne
   r линейный вызов *m*

7403 **line driver**
   d Leitungstreiber *m*
   f basculeur *m* de ligne
   r линейный формирователь *m*

7404 **line equipment**
   d Leitungseinrichtung *f*
   f équipement *m* de ligne
   r линейный прибор *m*

7405 **line impedance**
   d Leitungsimpedanz *f*

f impédance *f* de ligne
r импеданс *m* линии

7406 line interface
d Linieninterface *n*
f interface *f* de ligne
r интерфейс *m* линии

7407 line noise
d Leitungsrauschen *n*
f bruit *m* de lignes
r шум *m* в линии передачи

7408 line of stability; stability line
d Stabilitätslinie *f*
f ligne *f* de stabilité
r линия *f* устойчивости

7409 line of stop
d Anschlaglinie *f*
f ligne *f* de butée
r линия *f* упора

7410 line pattern
d Linienstruktur *f*
f structure *f* linéaire
r линейная структура *f*

7411 line position register
d Zeilenstellungsregister *n*
f registre *m* de position des lignes
r строчный регистр *m* положения

7412 line pulse
d Zeilenimpuls *m*
f impulsion *f* de ligne
r строчный [ведущий] импульс *m*

7413 line relay
d Linienrelais *n*; Leitungsrelais *n*;
    Anrufrelais *n*
f relais *m* de ligne; relais d'un circuit
r линейное реле *n*

7414 line resistance compensation
d Ausgleich *m* des Leitungswiderstandes
f compensation *f* de la résistance de la ligne
r компенсация *f* линейного сопротивления

*   **line scanning → 7374**

7415 line sensor
d Zeilensensor *m*
f capteur *m* de ligne; senseur *m* de ligne
r строка *f* фотодатчика

7416 line-sharing system
d Leitungsteilnehmersystem *n*;

Leitungsteilungssystem *n*
f système *m* de lignes à abonnés
r система *f* с разделением линий

7417 line signal
d Leitungssignal *n*
f signal *m* de ligne
r линейный сигнал *m*

7418 lines of interruption bus
d Leitungen *fpl* des Unterbrechungsbusses
f lignes *fpl* du bus d'interruption
r линии *fpl* шины прерывания

7419 line spectrum
d Linienspektrum *n*
f spectre *m* de raies
r линейный спектр *m*

*   **line variation → 7422**

7420 line voltage
d Netzspannung *f*
f tension *f* du secteur
r сетевое напряжение *n*

7421 line voltage fluctuations
d Netzspannungsschwankungen *fpl*
f variation *f* de tension du secteur
r флуктуации *fpl* сетевого напряжения

7422 line voltage regulator; line variation
d Netzspannungsregler *m*
f régulateur *m* de tension du secteur
r регулятор *m* сетевого напряжения

7423 link *v*
d verbinden; verketten
f lier; enchaîner
r связывать

7424 linkage control table
d Verbindungssteuertabelle *f*
f table *f* de commande de liaison
r таблица *f* управления связями

7425 linkage convention
d Verbindungsrichtlinie *f*
f convention *f* de liaison
r соглашение *n* о связи

7426 linkage register
d Verbindungsregister *n*
f registre *m* de liaison
r регистр *m* связи

*   **linking energy → 1789**

7427 linking sequence
d Verbindungsfolge f
f séquence f de liaison
r последовательность f связи

7428 links synchronization
d Verbindung-Synchronisation f
f synchronisation f de liaisons
r синхронизация f связей

7429 Lipschitz condition
d Lipschitz-Bedingung f
f condition f de Lipschitz
r условие n Липшица

7430 liquid chromatography
d Flüssigkeitschromatografie f
f chromatographie f en phase liquide
r жидкостная хроматография f

7431 liquid cooled reactor
d flüssigkeitsgekühlter Reaktor m
f réacteur m refroidi par liquide; réacteur au refroidissement liquide
r реактор m с жидкостным охлаждением

7432 liquid flow
d Flüssigkeitsströmung f; Strömung f von Flüssigkeiten
f écoulement m liquide
r течение n жидкостей; поток m жидкостей

7433 liquid fuel homogeneous reactor
d flüssig-homogener Reaktor m; homogener Reaktor mit flüssigem Brennstoff
f réacteur m homogène [à combustible] liquide
r гомогенный реактор m на жидком топливе

7434 liquid infrared analyzer
d Ultrarotflüssigkeitsanalysator m
f analysateur m de liquides infrarouge
r инфракрасный анализатор m жидкости

7435 liquid level control
d Flüssigkeitsstandsregelung f
f réglage m de niveau du liquide
r регулирование n уровня жидкости

7436 liquid metal coolant circuit
d Flüssigmetall-Kühlkreislauf m; Flüssigmetallkühlkreis m
f circuit m de refroidissement par métal liquide
r контур m жидкометаллического охлаждения

7437 liquid recirculating system
d Flüssigkeitsumlaufsystem n
f cycle m de liquide; système m de recirculation de liquide
r система f рециркуляции жидкости

7438 liquid shutdown system
d Flüssig[keits]abschaltsystem n
f système m d'arrêt liquide
r система f жидкостного выключения

7439 load action
d Belastungseinwirkung f; Belastungseingriff m; Belastungseinfluß m
f effet m de charge; action f de charge
r воздействие n по нагрузке

7440 load chamber
d Ladekammer f
f chambre f de chargement
r загрузочная камера f

7441 load change
d Belastungsänderung f
f changement m de charge
r изменение n нагрузки

7442 load characteristic curve
d Belastungskennlinie f
f ligne f caractéristique en charge
r кривая f нагрузки; нагрузочная характеристика f

7443 load circuit
d Belastungsstromkreis m; Verbraucherstromkreis m
f circuit m de charge; circuit d'utilisation
r схема f нагрузки

7444 load controller
d Belastungsregler m
f régulateur m de charge
r регулятор m нагрузки

7445 load curve
d Belastungskurve f
f ligne f de charge
r кривая f воздействия нагрузки

7446 loaded request block
(of programs)
d Anforderungsblock m für geladene Programme
f bloc m de requête pour programmes chargés
r блок m запросов на загруженные программы

7447 load factor
d Beschickungsfaktor m; Belastungsfaktor m; Belastungsgrad m
f facteur m de charge
r коэффициент m нагрузки

7448 **load impedance**
  d Abschlußimpedanz *f*
  f impédance *f* de charge
  r полное сопротивление *n* нагрузки

7449 **loading capacity**
  d Belastungsvermögen *n*; zulässige Belastung *f*
  f capacité *f* admissible de charge
  r предельная допустимая нагрузка *f*

  * **loading handler** → 2316

7450 **loading handling technique**
  d Einlege-Handhabetechnik *f*
  f technique *f* de manutention de chargement
  r техника *f* манипулирования [процессом] загрузки

7451 **loading line**
  d Belastungslinie *f*
  f ligne *f* de charge
  r нагрузочная линия *f*

7452 **loading pattern**
  d Beschickungsschema *n*; Beladungsplan *m*
  f modèle *m* de chargement
  r схема *f* загрузки

7453 **load life**
  d Lebensdauer *f* bei Belastung
  f durée *f* de vie à pleine charge
  r срок *m* службы при полной нагрузке

7454 **load limiting resistor**
  d Belastungsbegrenzungswiderstand *m*
  f résistance *f* limiteuse de charge
  r резистор m, ограничивающий нагрузку

7455 **load point**
  d Ladepunkt *m*; Belastungspunkt *m*
  f point *m* de charge
  r точка *f* приложения нагрузки

7456 **load regulation**
  d Belastungsregelung *f*
  f réglage *m* de charge
  r регулирование *n* нагрузки

7457 **load sharing mode**
  d Lastteilverfahren *n*
  f mode *m* de partage de charge
  r режим *m* распределения нагрузки

7458 **load test**
  d Belastungstest *m*
  f test *m* de charge
  r контроль *m* нагрузки

7459 **load vector**
  d Belastungsvektor *m*
  f vecteur *m* de charge
  r вектор *m* нагрузки

7460 **load voltage**
  d Lastspannung *f*; Verbraucherspannung *f*
  f tension *f* de charge
  r напряжение *n* загрузки

7461 **local control**
  d lokale Steuerung *f*
  f commande *f* locale
  r локальное управление *n*

7462 **local data collection**
  d lokale Datenerfassung *f*; Datenerfassung am Entstehungsort
  f acquisition *f* locale de données
  r локальный сбор *m* данных

7463 **local exchange**
  d Ortsvermittlungsstelle *f*
  f communication *f* locale
  r локальный коммутатор *m*

7464 **local feedback**
  d lokale Rückführung *f*; örtliche Rückkopplung *f*
  f réaction *f* locale
  r локальная обратная связь *f*

7465 **local iteration block**
  d lokaler Iterationsblock *m*
  f bloc *m* d'itération local
  r локальный итерационный блок *m*

7466 **localization**
  d Lokalisierung *f*; Ortung *f*; Eingrenzung *f*
  f localisation *f*
  r локализация *f*

7467 **localization of defect causes**
  d Fehlerursachenlokalisierung *f*
  f localisation *f* d'origines de défaut
  r локализация *f* причин неисправностей

7468 **localization system**
  d Lokalisierungssystem *n*
  f système *m* de localisation
  r система *f* локализации

7469 **localize** *v*
  d lokalisieren
  f localiser
  r локализовывать

7470 **localized component**
  d lokalisiertes Bauelement *n*

  f composant *m* localisé
  r локализованный компонент *m*

7471 locally continuous function
  d stückweise stetige Funktion *f*
  f fonction *f* continue à pièces
  r локально-непрерывная функция *f*

7472 local management interface
  d lokales Management-Interface *n*; lokale Managementschnittstelle *f*
  f interface *f* de gestion locale
  r интерфейс *m* локального управления

7473 local mode
  d lokale Betriebsweise *f*; abgegrenzte Betriebsweise
  f mode *m* local
  r локальный режим *m*

7474 local network
  d lokales Netz *n*
  f réseau *m* local
  r локальная сеть *f*

7475 local processing
  d lokale Verarbeitung *f*; Vor-Ort-Verarbeitung *f*
  f traitement *m* local
  r локальная обработка *f*

  * **local stability** → 11771

7476 local structure
  d lokale Struktur *f*
  f structure *f* locale
  r локальная структура *f*

7477 local unit
  d lokale Einheit *f*
  f unité *f* locale
  r локальное устройство *n*

7478 locate function
  d Suchfunktion *f*
  f fonction *f* de recherche
  r функция *f* поиска

7479 locate mode
  d Zeigermodus *m*
  f mode *m* d'aiguille
  r режим *m* локализации

7480 lock *v*
  d Ziel erfassen und Spur einhalten
  f accrocher et suivre
  r захватывать

  * **lock circuit** → 7483

  * **locked trip** → 5542

7481 lock-in amplifier
  d Blockierverstärker *m*
  f amplificateur *m* de blocage
  r блокирующий усилитель *m*

7482 lock-in detector
  d Blockierdetektor *m*
  f détecteur *m* de blocage
  r блокирующий детектор *m*

7483 lock[ing] circuit
  d Synchronisierkreis *m*; Sperrschaltung *f*; Haltestromschaltung *f*
  f circuit *m* de synchronisation; circuit de verrouillage
  r синхронизирующая схема *f*

7484 locking member
  d Verriegelungselement *n*; Verregelungsglied *n*
  f organe *m* de verrouillage
  r элемент *m* блокировки

7485 logarithmic additivity
  d logarithmische Additivität *f*
  f additivité *f* logarithmique
  r логарифмическая аддитивность *f*

7486 logarithmic amplifier
  d logarithmischer Verstärker *m*
  f amplificateur *m* logarithmique
  r логарифмический усилитель *m*; усилитель с логарифмической характеристикой

7487 logarithmic amplitude characteristic
  d logarithmische Amplitudencharakteristik *f*
  f réponse *f* logarithmique en amplitude
  r логарифмическая амплитудная характеристика *f*

7488 logarithmic attenuator
  d logarithmisches Dämpfungsglied *n*
  f atténuateur *m* logarithmique
  r логарифмический ослабитель *m*

7489 logarithmic computing circuit
  d logarithmischer Rechenstromkreis *m*
  f circuit *m* de calcul logarithmique
  r логарифмическая вычислительная схема *f*

7490 logarithmic register
  d logarithmisches Register *n*
  f registre *m* logarithmique
  r логарифмический регистр *m*

7491 logarithmic search method
d logarithmisches Suchverfahren n
f méthode f de recherche logarithmique
r метод m логарифмического поиска

7492 logarithmic work principle
d logarithmisches Arbeitsprinzip n
f principe m de travail logarithmique
r логарифмический принцип m действия

7493 logger
d Messwerterfassungsgerät n
f enregistreur m automatique
r устройство n регистации данных

7494 logical adder
d logischer Adder m
f addeur m logique
r логический сумматор m

7495 logic[al] analysis
d Logikanalyse f; logische Analyse f
f analyse f logique
r логический анализ m

7496 logical AND circuit
d logische UND-Schaltung f;
UND-Verknüpfungsglied n
f circuit m logique ET
r логическая схема f И

7497 logical automaton
d logischer Automat m
f automate m logique
r логический автомат m

7498 logical block
d logischer Block m
f bloc m logique
r логический блок m

7499 logical channel
d logischer Kanal m
f canal m logique
r логический канал m

7500 logical circuits equivalence
d Äquivalenz f logischer Schaltungen
f équivalence f de circuits logiques
r равноценность f логических схем

7501 logical comparison
d logischer Vergleich m
f comparaison f logique
r логическое сравнение n

7502 logical component
d logische Komponente f; logisches Glied n

f composante f logique
r логическая составляющая f

7503 logical connection
d logische Verknüpfung f
f liaison f logique
r логическая связка f

7504 logical control signal
d logisches Steuersignal n
f signal m de commande logique
r логический сигнал m управления

7505 logical data processing
d logische Datenverarbeitung f
f traitement m des données logique
r логическая обработка f данных

7506 logical decision function
d logische Entscheidungsfunktion f
f fonction f de décision logique
r функция f проверки логического условия

7507 logical design
d logischer Entwurf m
f projet m logique; dessin m logique
r логическая разработка f, логический синтез m

7508 logical design of switching circuits
d logischer Entwurf m von Schaltkreisen
f projet m logique des circuits de commutation
r логическое проектирование n переключающих схем

7509 logical diagram
d logisches Diagramm n;
Verknüpfungsdiagramm n
f diagramme m logique
r логическая диаграмма f

7510 logical direction
d logische Leitung f; logische Führung f
f gestion f logique
r логическое управление n

* logical element → 3807

7511 logical error
d logischer Fehler m
f erreur f logique
r логическая ошибка f

7512 logical expression
d logischer Ausdruck m
f expression f logique
r логическое выражение n

7513 logical function minimal member
  d Minimalglied *n* einer logischen Funktion
  f membre *m* minimal de fonction logique
  r минимальное звено *n* регулирования
      логической функции

7514 logical gate element
  d logisches Gatter *n*
  f porte *f* logique
  r логический элемент *m* схемы совпадения

7515 logical interface condition
  d logische Interfacebedingung *f*
  f condition *f* d'interface logique
  r логическое условие *n* интерфейса

7516 logical method for the analysis
  d Methode *f* zur logischen Analyse
  f méthode *f* d'analyse logique
  r метод *m* логического анализа

7517 logical multiplication
  d logische Multiplikation *f*
  f multiplication *f* logique
  r логическая мултипликация *f*

7518 logical net
  d logisches Netz *n*
  f réseau *m* logique
  r логическая сеть *f*

7519 logical noise
  d logisches Rauschen *n*
  f bruit *m* logique
  r логический шум *m*

7520 logical NOT circuit
  d logische NICHT-Schaltung *f*
  f circuit *m* logique NON
  r логическая схема *f* типа НЕТ

7521 logical operation of control system
  d logische Operation *f* des Steuersystems
  f opération *f* logique de système de commande
  r логическая операция *f* системы
      управления

7522 logical OR circuit
  d logische ODER-Schaltung *f*
  f circuit *m* logique OU
  r логическая схема *f* ИЛИ

7523 logical program scheme
  d logisches Programmschema *n*
  f schéma *m* logique de programme
  r логическая схема *f* программы

7524 logical shift

  d logische Verschiebung *f*
  f décalage *m* logique
  r логический сдвиг *m*

7525 logical signal
  d logisches Signal *n*
  f signal *m* logique
  r логический сигнал *m*

7526 logical switching element
  d logisches Schaltelement *n*
  f élément *m* de commutation logique
  r логический переключательный элемент *m*

7527 logical system
  d logisches System *n*
  f système *m* logique
  r логическая система *f*

  * **logic analysis** → 7495

  * **logic analyzer** → 7540

7528 logic base circuit
  d logische Grundschaltung *f*
  f circuit *m* logique de base
  r основная логическая цепь *f*

7529 logic device
  d Logikbauteil *n*
  f module *m* logique
  r логическое устройство *n*

7530 logic display system
  d logisches Anzeigesystem *n*
  f système *m* d'affichage logique
  r система *f* отображения данных на
      логическом уровне

  * **logic function** → 1761

7531 logic high level
  d oberes logisches Niveau *n*;
      High-Logikpegel *m*
  f niveau *m* logique supérieur
  r высокий логический уровень *m*

7532 logic implication
  d logische Implikation *f*
  f implication *f* logique
  r логическая импликация *f*

7533 logic instruction
  d logischer Befehl *m*
  f instruction *f* logique
  r логическая команда *f*; логическая
      инструкция *f*

7534 **logic low level**
  d unteres logisches Niveau *n*;
  Low-Logikpegel *m*
  f niveau *m* logique inférieur
  r низкий логический уровень *m*

7535 **logic of the computer**
  d Schaltkreislogik *f* des Rechners
  f logique *f* de la calculatrice; logique du
  calculateur
  r логическая схема *f* компьютера

7536 **logic operation**
  d logische Operation *f*; logischer Vorgang *m*
  f opération *f* logique
  r логическая операция *f*

7537 **logic pulse**
  d logischer Impuls *m*
  f impulsion *f* logique
  r логический импульс *m*

7538 **logic sequential control**
  d logische Folgesteuerung *f*
  f commande *f* séquentielle logique
  r последовательный контроль *m* с
  применением логических операций

7539 **logic simulation**
  d Logiksimulation *f*
  f simulation *f* logique
  r логическое симулирование *n*

7540 **logic[-state] analyzer**
  d Logikanalysator *m*
  f analuseur *m* logique
  r логический анализатор *m*

7541 **logic switch**
  d Logikschalter *m*
  f commutateur *m* logique
  r логическая переключательная схема *f*

7542 **logic tester**
  d Logiktester *m*
  f appareil *m* de test logique
  r логический тестер *m*

7543 **logistics**
  d Logistik *f*
  f logistique *f*
  r логистика *f*

7544 **long-chain**
  d langkettig
  f à chaîne longue
  r длинноцепный

7545 **long-chain branching**
  d langkettige Verzweigung *f*
  f ramification *f* à chaînes longues
  r длинноцепное разветвление *n*

7546 **long-distance heating installation
  controller**
  d Regler *m* für Fernheizungsanlage
  f régulateur *m* des installations de chauffage à
  distance
  r дистанционный регулятор *m* установки
  нагрева

7547 **long-distance wave-guided transmission**
  d Wellenleiterweitverbindung *f*;
  Wellenleiterfernübertragung *f*
  f transmission *f* à grande distance par guide
  d'ondes
  r управляемая передача *f* на дальние
  растояния

7548 **longitudinal differential protection**
  d Longitudinaldifferentialabschirmung *f*;
  längsgerichtete Schützeinrichtung *f*
  f protection *f* différentielle longitudinale
  r осевое дифференциальное защитное
  устройство *n*

7549 **longitudinal multipoint sensor**
  d Längsmehrpunktsensor *m*
  f senseur *m* multipoint longitudinal; capteur *m*
  multipoint longitudinal
  r продольный многоточечный сенсор *m*

7550 **longitudinal stability**
  d Längsstabilität *f*
  f stabilité *f* longitudinale
  r продольная устойчивость *f*

7551 **long-pulse laser**
  d Laser *m* für lange Impulse
  f laser *m* à longue impulsion de sortie
  r лазер *m* длинных импульсов

7552 **long-range laser communication**
  d Laserweitverkehrsverbindung *f*
  f communication *f* par laser à grande portée
  r дальняя лазерная связь *f*

7553 **long-run test; long-time test**
  d Langzeittest *m*; Dauertest *m*
  f test *m* d'endurance
  r длительное испытание *n*

7554 **long-term continuous duty; long-term
  continuous operation**
  d kontinuierliche Fahrweise *f*

f opération *f* continue
r непрерывный режим *m* работы

* **long-term continuous operation** → 7554

7555 long-term effect
  d Langzeitwirkung *f*
  f effet *m* à long terme
  r длительный эффект *m*

7556 long-term speed variation
  d langsame Geschwindigkeitsänderung *f*
  f variation *f* de vitesse lente
  r медленное изменение *n* скорости

7557 long-term stability
  d Langzeitstabilität *f*
  f stabilité *f* à long terme
  r долговременная стабильность *f*

* **long-time test** → 7553

* **loop circuit** → 7559

7558 loop dialling system
  d Schleifensystem *n*
  f système *m* à boucle
  r шлейф-система *f*

7559 loop[ed] circuit
  d Schleife *f*; Doppelleitung *f*
  f circuit *m* fermé; circuit bouclé
  r кольцевая цепь *f*; двухпроводная цепь

7560 loop element
  d Regelkreisglied *n*; Regelkreiselement *n*
  f organe *m* de boucle
  r элемент *m* контура

7561 loop network
  d Schleifennetz *n*
  f réseau *m* en boucle
  r шлейфовая сеть *f*

7562 loop resistance
  d Schleifenwiderstand *m*;
     Doppelleitungswiderstand *m*
  f résistance *f* du bouclage
  r сопротивление *n* шлейфа

7563 loop resolution
  d Schleifenzerlegung *f*;
     Doppelleitungszerlegung *f*;
     Doppelleitungsspaltung *f*
  f insensibilité *f* en boucle ouverte
  r разрешающая способность *f* шлейфа

7564 loop rule

d Schleifenregel *f*
f règle *m* de boucle
r правило *n* для разрыва циклов

7565 loop structure
  d Schleifenstruktur *f*; Schleifenaufbau *m*
  f structure *f* de boucle
  r структура *f* цикла

7566 loose coupling; weak coupling
  d schwache Kopplung *f*
  f accouplement *m* lâche; couplage *m* faible
  r слабая связь *f*

7567 loss factor
  d Verlustfaktor *m*
  f facteur *m* de perte
  r коэффициент *m* потери

7568 loss modulus
  d Verlustmodul *m*
  f module *m* de perte
  r модуль *m* потери

7569 loss of accuracy
  d Genauigkeitsverlust *m*
  f perte *f* de précision
  r потеря *f* точности

7570 loss of cycle
  d Gangverlust *m*
  f perte *f* de cycle
  r ослабление *n* циклического процесса

7571 loss resistance
  d Verlustwiderstand *m*
  f résistance *f* des pertes
  r сопротивление *n* потерь

7572 low-frequency demodulator
  d Niederfrequenzdemodulator *m*
  f démodulateur *m* basse fréquence
  r низкочастотный демодулятор *m*

7573 low-frequency distortion
  d Niederfrequenzverzerrung *f*
  f distorsion *f* de basse fréquence
  r низкочастотное искажение *n*

7574 low-frequency filter
  d Niederfrequenzfilter *n*
  f filtre *m* basse fréquence; filtre BF
  r фильтр *m* низкой частоты

7575 low-frequency induction heating
  d Niederfrequenzinduktionsheizung *f*
  f chauffage *m* basse fréquence par induction
  r низкочастотный индукционный нагрев *m*

* **low-high transition** → 7609

7576 low-level amplifier
d Kleinsignalverstärker *m*
f amplificateur *m* pour signaux faibles
r малосигнальный усилитель *m*

7577 low-level circuit
d Niedrigpegelschaltung *f*
f circuit *m* à niveau inférieur
r малосигнальная схема *f*

7578 low-level counter
d Zähler *m* für schwache Intensität
f compteur *m* de bas niveau
r счётчик *m* малой мощности

7579 low-level modulation
d Vorstufenmodulation *f*
f modulation *f* à faible niveau
r модуляция *f* на малой мощности

7580 low-level signal
d Signal *n* mit niedrigem Pegel
f signal *m* de niveau inférieur
r сигнал *m* с малой амплитудой

7581 low-level stage
d Stufe *f* mit niedrigem Pegel
f étage *m* à faible niveau
r маломощный каскад *m*

7582 low-loss
d verlustarm
f à faible perte
r с малыми потерями

7583 low-noise amplifier
d rauscharmer Verstärker *m*
f amplificateur *m* à faible bruit
r малошумящий усилитель *m*

7584 low-noise parametric amplifier
d rauscharmer parametrischer Verstärker *m*
f amplificateur *m* paramétrique à bruit faible
r параметрический усилитель *m* с малыми шумами

7585 low of probability; probability low
d Wahrscheinlichkeitsgesetz *n*
f loi *f* de probabilité
r закон *m* вероятности

7586 low-pass filter
d Tiefpass *m*; Tiefpassfilter *n*
f passe-bas *m*; filtre *m* passe-bas
r фильтр *m* нижних частот

7587 low-pass filters in control-loops
d Tiefpassfilter *npl* in Regelkreisen
f filtres *mpl* passe-bas dans les systèmes asservis
r низкочастотные фильтры *mpl* в контурах регулирования

7588 low-power application
d Niederleistungsanwendung *f*
f application *f* à faible puissance
r применение n, требующее низкой потребляемой мощности

7589 low-power consumption
d geringe Leistungsaufnahme *f*
f faible consommation *f*
r низкое потребление *n* мощности

7590 low-power drain
d geringe Stromaufnahme *f*
f consommation *f* de courant faible
r низкое энергопотребление *n*

7591 low-power source
d geringe Stromabgabe *f*; Quelle *f* kleiner Leistung
f source *f* à faible puissance
r маломощный источник *m* питания

7592 low-power system
d Low-Power-System *n*; System *n* niedriger Leistungsaufnahme
f système *m* à consommation réduite
r система *f* с низким потреблением мощности

7593 low-pressure circuit
d Niederdruckkreislauf *m*
f circuit *m* de basse pression
r схема *f* низкого давления

7594 low-pressure process
d Niederdruckprozess *m*
f processus *m* à basse pression
r процесс *m* низкого давления

7595 low-pressure range
d Niederdruckbereich *m*
f domaine *m* de basse pression
r диапазон *m* низких давлений

7596 low-pressure recording flowmeter
d registrierendes Niederdruckdurchflussmesser *m*
f débitmètre *m* enregistreur basse pression
r регистрирующий расходомер *m* для малых перепадов

7597 low-pressure subsystem
  d Niederdruckteil n
  f partie f à basse pression
  r подсистема f низкого давления

7598 low-priority event
  d Ereignis n niederer Priorität
  f événement m de basse priorité
  r событие n с низким приоритетом

7599 low-resolution detector
  d Detektor m mit niedrigem
      Auflösungsvermögen
  f détecteur m à pouvoir résolvant réduit
  r детектор m с низкой разрешающей
      способностью

7600 low state
  d Low-Zustand m; niedriger Zustand m
  f état m inférieur
  r состояние n с низким уровнем сигнала

7601 low-temperature bolometer
  d Niedertemperaturbolometer n
  f bolomètre m à basse température
  r низкотемпературный болометр m

7602 low-temperature demodulator
  d Niedertemperaturdemodulator m
  f démodulateur m à basse température
  r низкотемпературный демодулятор m

7603 low-temperature detector
  d Niedertemperaturdetektor m
  f détecteur m à températures basses
  r низкотемпературный детектор m

7604 low-temperature field
  d Tieftemperaturbereich m
  f champ m de basse température; domaine m
      de basse température
  r низкотемпературный диапазон m

7605 low-temperature operation
  d Tieftemperaturbetrieb m
  f opération f à basse température
  r низкотемпературный режим m

7606 low-temperature plant
  d Tieftemperaturanlage f
  f installation f frigorifique
  r установка f низких температур

7607 low-temperature processing
  d Tieftemperaturbehandlung f
  f traitement m à basse température
  r низкотемпературная обработка f

7608 low-threshold current density
  d niedrige Schwellenstromdichte f
  f densité f de courant de seuil faible
  r низкая плотность f порогового тока

7609 low[-to]-high transition
  d Low-High-Übergang m; Übergang m vom
      unteren zum oberen Signalpegel
  f transition f du niveau bas au niveau haut
  r переход m из состояния с низким в
      состояние с высоким уровнем напряжения

7610 low value
  d unterer Wert m
  f valeur f inférieure
  r нижнее значение n

7611 low-voltage circuit
  d Niederspannungskreis m
  f circuit m à basse tension
  r низковольтная цепь f

  *   luminance control → 1945

  *   luminous signal → 7261

7612 lumped capacity
  d konzentrierte Kapazität f
  f capacité f concentrée
  r сосредоточенная ёмкость f

7613 lumped characteristic
  d konzentrierte Charakteristik f
  f caractéristique f composée
  r сосредоточенная характеристика f

7614 lumped parameter
  d konzentrierter Parameter m
  f paramètre m localisé
  r сосредоточенный параметр m

7615 lumped parameter system
  d System n mit konzentrierten Parametern
  f système m à paramètres localisès
  r система f с сосредоточенными
      параметрами

M

7616 machine allowance
d Maschinentoleranz *f*
f tolérance *f* de machine
r машинная погрешность *f*

7617 machine automation
d Maschinenautomatisierung *f*
f automatisation *f* de machines
r автоматизация *f* работы машин

7618 machine-available time
d Maschinenwirkzeit *f*; Rechnerwirkzeit *f*
f temps *m* d'exploitation
r рабочее время *n* машины

7619 machine check
d Maschineprüfung *f*
f vérification *f* de machine
r машинный контроль *m*

7620 machine-code compatible instruction set
d maschinenkodekompatibler Befehlssatz *m*;
    kompatibler Befehlssatz auf
    Maschinenbefehlsebene
f jeu *m* d'instructions compatible en code
    machine
r система *f* команд, совместимая на уровне
    машинного кода

7621 machine-computing technique
d maschinelle Rechentechnik *f*
f technique *f* de calcul mécanique; techniques
    fpl de calcul mécaniques
r вычислительная техника *f*

7622 machine condition
d Maschinenbedingung *f*
f condition *f* de la machine
r машинное условие *n*

7623 machine configuration
d Maschinenkonfiguration *f*
f configuration *f* de machine
r машинная конфигурация *f*

7624 machine coordinate system
d Maschinenkoordinatensystem *n*
f système *m* de coordonnées de la machine
r машинная система *f* координат

7625 machine cycle

d Maschinenperiode *f*; Rechnerperiode *f*
f cycle *m* de machine
r машинный цикл *m*

7626 machine defect analysis
d Maschinendefektanalyse *f*; Analyse *f* von
    Maschinendefekten
f analyse *f* de défaut de machine
r анализ *m* неисправностей машин

7627 machine defect effect
d Maschinendefektauswirkung *f*; Auswirkung *f*
    von Maschinendefekten
f effet *m* de défaut de machine
r последствия *npl* неисправностей машин

7628 machine down-time
d Maschinenstillstandszeit *f*
f temps *m* de repos de machine
r время *n* простоя машины

7629 machine equation; computer equation
d Rechnergleichung *f*; Maschinengleichung *f*
f équation *f* de machine; équation *f* de
    calculateur
r машинное уравнение *n*

7630 machine failure
d Maschinenausfall *m*; Anlagenausfall *m*
f panne *f* de machine; panne d'installation
r отказ *m* машины

7631 machine feed control
d Vorschubregelung *f*
f contrôle *m* de l'avancement d'une machine
r управление *n* подачей машины

7632 machine-independent interface
d maschinenunabhängiges Interface *n*;
    maschinentypunabhängige Schnittstelle *f*
f interface *f* indépendante de machine
r машинно-зависимый интерфейс *m*

7633 machine instruction
d Maschinenbefehl *m*
f instruction *f* de machine
r машинная команда *f*

**7634 machine-integrated field of
instrumentation**
d maschinenintegrierte Gerätetechnik *f*
f technique *f* de l'instrumentation intégrée de
    machine
r машинно-интегрированные контрольно-
    измерительные приборы *mpl*

7635 machine language
d Rechnersprache *f*

    f langue *f* de machine
    r машинный язык *m*

7636 machine-limited system
    d durch Rechnergeschwindigkeit begrenztes
       System *n*
    f système *m* limité par vitesse de machine
    r система f, ограниченная быстродействием
       вычислительной машины

7637 machine load
    d Maschinenbelastung *f*;
       Maschinenauslastung *f*
    f charge *f* de machine
    r загрузка *f* машины

7638 machine logic
    d Maschinenlogik *f*
    f logique *f* de machine
    r машинная логика *f*

7639 machine logic design
    d Maschinenlogikentwurf *m*; Logikentwurf *m*
       einer Rechenmaschine
    f conception *f* logique de machine
    r проектирование *n* логической структуры
       вычислительной машины

7640 machine operation
    d Maschinenoperation *f*
    f opération *f* de machine
    r машинная операция *f*

7641 machine-operation synchronizing
    d Synchronisierung *f* der Maschinenarbeit
    f synchronisation *f* du fonctionnement de la
       machine
    r синхронизация *f* работы машины

7642 machine-oriented
    d maschinenorientiert
    f orienté machine
    r машинно-ориентированный

7643 machine-oriented geometry language
    d maschinenorientierte Geometriesprache *f*
    f langage *m* de géométrie orienté sur la
       machine
    r машинно-ориентированный
       геометрический язык *m*

7644 machine-oriented testing service program
    d maschinenorientiertes Testhilfsprogramm *n*
    f programme *m* de service de test orienté sur la
       mashine
    r машинно-ориентированная
       дополнительная программа *f* контроля

7645 machine program; machine routine
    d Maschinenprogramm *n*
    f programme *m* de machine
    r машинная программа *f*

7646 machine reliability
    d Zuverlässigkeit *f* der Maschine
    f fiabilité *f* de la machine
    r надёжность *f* машины

7647 machine representation
    d Maschinendarstellung *f*; maschineninterne
       Darstellung *f*; rechnerinterne Darstellung
    f représentation *f* de machine
    r машинное представление *n*

  * **machine routine** → 7645

7648 machinery diagram
    d konstruktives Fließbild *n*
    f flow-sheet *m* constructif
    r конструктивная схема *f*

7649 machine set-up time
    d Maschinenrüstzeit *f*
    f temps *m* de préparation de machine
    r время *n* наладки оборудования

7650 machine signal
    d Maschinensignal *n*
    f signal *m* de machine
    r машинный сигнал *m*

7651 machine system
    d Maschinensystem *n*
    f système *m* de machine
    r машинная система *f*

  * **machine time** → 2947

7652 machine-tool industry
    d Werkzeugmaschinenbau *m*
    f construction *f* de machinesoutils
    r станкостроение *n*

7653 machine-tool linkage
    d Werkzeugmaschinenverkettung *f*
    f enchaînement *m* de machineoutil
    r связь *f* между станками

7654 machine translation
    d automatische Übersetzung *f*
    f traduction *f* par machine
    r автоматический перевод *m*

7655 machine word
    d Maschinenwort *n*; Rechnerwort *n*

f   mot *m* de machine
r   машинное слово *n*

7656   machine zero point
   d   Maschinennullpunkt *m*
   f   point *m* d'origine de machine
   r   машинный нуль *m*

7657   machining
   d   maschinelle Bearbeitung *f*; spanende
      Bearbeitung
   f   usinage *m*
   r   [механическая] обработка *f*

7658   machining step
   d   Bearbeitungsstufe *f*
   f   degré *m* d'usinage; phase *m* d'usinage
   r   ступень *f* обработки

7659   Maclaurin expansion
   d   Maclaurinsche Reihenentwicklung *f*
   f   développement *m* de Maclaurin
   r   разложение *n* в ряд Лорана

   *   **macro → 7664**

7660   macroblock
   d   Makroblock *m*
   f   macrobloc *m*
   r   макроблок *m*

   *   **macrocode → 7664**

   *   **macrocommand → 7664**

7661   macrodefinition
   d   Makrodefinition *f*;
      Makrooperationsspezifikation *f*
   f   définition *f* de macro
   r   макроопределение *n*

7662   macrofacility
   d   Makroeinrichtung *f*
   f   dispositif *m* à macros
   r   макросредство *n*

**7663   macrogenerating program;
        macrogenerator**
   d   Makrogenerierprogramm *n*;
      Makrogenerator *m*
   f   programme *m* de génération macro;
      générateur *m* macro
   r   макрогенератор *m*

   *   **macrogenerator → 7663**

7664   macroinstruction; macrocommand;

macrocode; macro[s]
   d   Makrobefehl *m*; Makroinstruktion *f*;
      Makros *n*
   f   macro-instruction *f*; macrocommande *f*;
      macros *m*
   r   макроинструкция *f*; макрокоманда *f*;
      макрос *m*

7665   macrooperation
   d   Makrooperation *f*
   f   macro-opération *f*
   r   макрооперация *f*

7666   macroprocessor
   d   Makroprozessor *m*
   f   macroprocesseur *m*
   r   макропроцессор *m*

7667   macroprogramming
   d   Makroprogrammierung *f*;
      Maschinenprogrammierung *f* über
      Makrobefehle
   f   programmation *f* macro
   r   макропрограммирование *n*

   *   **macros → 7664**

   *   **macrostructure → 2501**

7668   magnetically recorded program
   d   magnetisch aufgezeichnetes Programm *n*
   f   programme *m* enregistré par voie magnétique
   r   магнитная запись *f* программы

7669   magnetic amplifier electric drive control
   d   Magnetverstärkersteuerung *f* elektrischer
      Getriebe
   f   commande *f* de moteurs électriques par
      amplificateur magnétique
   r   управление *n* электроприводом при
      помощи магнитного усилителя

7670   magnetic amplifier servosystem
   d   Servosystem *n* mit magnetischem Verstärker
   f   système *m* asservi à amplificateur
      magnétique
   r   следящая система *f* с магнитным
      усилителем

7671   magnetic amplifier temperature controller
   d   Transduktortemperaturregler *m*
   f   régulateur *m* de température à transducteur
   r   терморегулятор *m* с магнитным
      усилителем

7672   magnetic analog-digital converter
   d   magnetischer Analog-Digital-Umsetzer *m*

f convertisseur *m* magnétique analogique-digital
r магнитный аналого-цифровой преобразователь *m*

7673 magnetic analyzer
d Magnetanalysator *m*; magnetischer Analysator *m*
f analyseur *m* magnétique
r магнитный анализатор *m*

7674 magnetic attenuator
d magnetischer Abschwächer *m*
f atténuateur *m* magnétique
r магнитный аттенюатор *m*

7675 magnetic attractive force
d magnetische Anziehungskraft *f*
f force *f* d'atraction magnétique
r сила *f* магнитного притяжения

7676 magnetic carrier
d magnetischer Träger *m*
f porteur *m* magnétique
r магнитный носитель *m*

7677 magnetic characteristics
d magnetische Eigenschaften *fpl*
f caractéristiques *fpl* magnétiques
r магнитные свойства *npl*

7678 magnetic circuit
d magnetischer Kreis *m*
f circuit *m* magnétique
r схема *f* на магнитных элементах

7679 magnetic controlling equipment
d magnetische Steuereinrichtung *f*
f appareil *m* de commande magnétique
r установка *f* для магнитного контроля

* **magnetic coupling → 6483**

7680 magnetic current
d magnetischer Fluss *m*
f flux *m* magnétique
r магнитный поток *m*

7681 magnetic damping
d magnetische Dämpfung *f*
f amortissement *m* magnétique
r электромагнитное демпфирование *n*

7682 magnetic deflection
d magnetische Ablenkung *f*
f déviation *f* magnétique
r магнитное отклонение *n*

7683 magnetic delay line
d magnetische Verzögerungsleitung *f*; magnetisches Laufzeitglied *n*
f ligne *f* magnétique à retard
r магнитная линия *f* задержки

7684 magnetic demodulation
d magnetische Demodulation *f*
f démodulation *f* magnétique
r магнитная демодуляция *f*

7685 magnetic electron spectrometer
d magnetisches Elektronenspektrometer *n*
f spectromètre *m* électronique magnétique
r магнитный электронный спектрометр *m*

7686 magnetic field intensity
d magnetische Feldstärke *f*
f intensité *f* de champ magnétique
r напряженность *f* магнитного поля

7687 magnetic field sensitivity
d Empfindlichkeit *f* gegenüber elektromagnetischen Feldern
f sensibilité *f* au champ magnétique
r чувствительность *f* к электромагнитным полям

7688 magnetic field sensor
d Magnetfeldsensor *m*
f senseur *m* de champ magnétique
r сенсор *m* магнитного поля

7689 magnetic field stabilization
d Magnetfeldstabilisierung *f*
f stabilisation *f* du champ magnétique
r стабилизация *f* магнитного поля

7690 magnetic field strength meter
d magnetischer Feldstärkenmesser *m*
f magnétomètre *m*
r измеритель *m* напряженности магнитного поля

7691 magnetic float-type level transmitter
d magnetischer Schwimmerniveaugeber *m*
f transmetteur *m* magnétique de niveau à flotteur
r магнитный поплавковый датчик *m* уровня

7692 magnetic flux density
d Magnetflussdichte *f*
f densité *f* de flux magnétique
r плотность *f* магнитного потока

7693 magnetic fluxmeter
d Magnetflussmesser *m*

f fluxmètre *m* magnétique
r измеритель *m* магнитного потока

7694 magnetic flux stabilizer
d Magnetflussstabilisator *m*
f stabilisateur *m* de flux magnétique
r стабилизатор *m* магнитного потока

7695 magnetic gas analyzer
d magnetischer Gasanalysator *m*
f analyseur *m* magnétique de gaz
r магнитный газоанализатор *m*

7696 magnetic hysteresis; magnetic lag
d magnetische Hysterese *f*
f hystérésis *f* magnétique
r магнитный гистерезис *m*

7697 magnetic induction
d magnetische Induktion *f*
f induction *f* magnétique
r магнитная индукция *f*

* **magnetic lag** → 7696

7698 magnetic leakage
d magnetische Streuung *f*
f dispersion *f* magnétique
r магнитное рассеяние *n*

7699 magnetic logical element
d magnetisches Verknüpfungsglied *n*
f élément *m* logique magnétique
r магнитный логический элемент *m*

7700 magnetic loss
d magnetische Verluste *mpl*
f pertes *fpl* magnétiques
r магнитные потери *fpl*

7701 magnetic micropulsation
d magnetische Mikropulsation *f*
f micropulsation *f* magnétique
r магнитная микропульсация *f*

7702 magnetic nuclear resonance
d magnetische Nuklearresonanz *f*
f résonance *f* magnétique nucléaire
r магнитный ядерный резонанс *m*

7703 magnetic nuclear resonance spectrometry
d magnetische Kernresonanzspektrometrie *f*
f spectrométrie *f* à résonance nucléaire magnétique
r магнитная спектрометрия *f* ядерного резонанса

7704 magnetic object

d magnetisches Objekt *n*
f objet *m* magnétique
r магнитный объект *m*

7705 magnetic permeability
d magnetische Permeabilität *f*
f perméabilité *f* magnétique
r магнитная проницаемость *f*

7706 magnetic perturbation
d magnetische Störung *f*
f perturbation *f* magnétique
r магнитное возмущение *n*

7707 magnetic polarization
d magnetische Polarisation *f*
f polarisation *f* magnétique
r магнитная поляризация *f*

7708 magnetic potential
d magnetisches Potential *n*
f potentiel *m* magnétique
r магнитный потенциал *m*

7709 magnetic recording technique
d Magnetaufzeichnungstechnik *f*
f technique *f* d'enregistrement magnétique
r техника *f* магнитной записи

7710 magnetic saturation
d magnetische Sättigung *f*
f saturation *f* magnétique
r магнитное насыщение *n*

7711 magnetic-sensitive sensor
d magnetempfindlicher Sensor *m*
f senseur *m* magnétique sensible
r магниточувствительный сенсор *m*

7712 magnetic shield
d magnetische Schirmung *f*
f écran *m* magnétique
r магнитный экран *m*

7713 magnetic spectrograph
d magnetischer Spektrograf *m*
f spectrographe *m* magnétique
r магнитный спектрограф *m*

7714 magnetic switching
d magnetisches Schalten *n*
f commutation *f* magnétique
r магнитная коммутация *f*

7715 mag[netic] tape
d Magnetband *n*
f bande *f* magnétique
r магнитная лента *f*

7716 **magnetic tape device; tape device; tape unit**
  d Magnetbandgerät n; Magnetbandeinheit f
  f dispositif m à bande [magnétique]; unité f à bande
  r магнитно-лентовое устройство n

7717 **[magnetic] tape mode**
  d Magnetbandmodus m
  f mode m de bande [magnétique]
  r режим m работы [магнитной] ленты

7718 **magnetic tester**
  d Magnetprüfgerät n
  f perméamètre m
  r магнитный тестер m

7719 **magnetic thermal relay**
  d magnetisches Thermorelais n
  f relais m magnéto-thermique
  r термо-магнитное реле n

7720 **magnetic time relay**
  d magnetisches Zeitrelais n
  f relais m magnétique temporisé
  r магнитное реле n с выдержкой времени

7721 **magnetic variometer**
  d magnetisches Variometer n
  f variomètre m magnétique
  r магнитный вариометр m

7722 **magnetic voltage controller**
  d magnetischer Spannungsregler m
  f régulateur m magnétique de la tension
  r магнитный регулятор m напряжения

7723 **magnetization curve**
  d Magnetisierungskurve f
  f courbe f d'aimantation
  r кривая f намагничивания

7724 **magnetization vector**
  d Magnetisierungsvektor m
  f vecteur m de magnétisation
  r вектор m намагниченности

7725 **magnetoacoustic delay line**
  d magnetoakustische Verzögerungsleitung f
  f ligne f de retard magnéto-acoustique
  r магнитоакустическая линия f задержки

7726 **magnetoelectric transducer**
  d magnetoelektrischer Wandler m
  f transducteur m magnétoélectrique
  r магнитоэлектрический преобразователь m

7727 **magneto-fluid dynamics**

  d Magnetohydrodynamik f; Plasmadynamik f
  f magnétohydrodynamique f
  r магнитогидродинамика f

7728 **magnetomechanical damping**
  d magnetomechanische Dämpfung f
  f amortissement m magnétomécanique
  r магнитомеханическое демпфирование n

7729 **magnetomechanical gas analyzer**
  d magnetomechanischer Gasanalysator m
  f analyseur m magnétomécanique de gaz
  r магнитомеханический газоанализатор m

7730 **magnetooptical diplay**
  d magnetooptische Anzeige f
  f affichage m magnétooptique
  r магнитно-оптическая индикация f

7731 **magnetooptical effect**
  d magnetooptischer Effekt m
  f effet m magnétooptique
  r магнитооптический эффект m

7732 **magnetostriction**
  d Magnetostriktion f
  f magnétostriction f
  r магнитострикция f

7733 **magnetostriction control**
  d Magnetostriktionsregelung f
  f réglage m à magnétostriction
  r магнитострикционное регулирование n

7734 **magnetostriction delay line; magnetostrictive delay line**
  d magnetostriktive Verzögerungsleitung f
  f ligne f à retard à magnétostriction
  r магнитострикционная линия f задержки

7735 **magnetostriction oscillator; magnetostrictor**
  d magnetostriktiver Oszillator m; Magnetostriktionsgenerator m
  f oscillateur m à magnétostriction
  r магнитострикционный генератор m

* **magnetostrictive delay line** → 7734

7736 **magnetostrictive filter**
  d Magnetostriktionsfilter n
  f filtre m à magnétostriction
  r магнитострикционный фильтр m

* **magnetostrictor** → 7735

7737 **magnistor**
  d Magnistor m

f magnistor *m*
r магнистор *m*

*   **mag type** → 7715

7738 main comparison unit
d Hauptvergleicher *m*;
   Hauptvergleichereinheit *f*
f unité *f* principale de comparaison
r главное устройство *n* сравнения

7739 main component
d Hauptkomponente *f*
f constituant *m* essentiel
r главный компонент *m*

*   **main control board** → 2202

7740 main coupling
d Hauptkopplung *f*
f couplage *m* principal
r основная связь *f*

7741 main expansion joint
d Haupttrennfuge *f*
f joint *m* de dilatation principal
r главный разъём *m*

7742 main [fabrication] process
d Haupt[fertigungs]prozess *m*
f procédé *m* [de fabrication] principal
r основной [производственный] процесс *m*

7743 main gear
d Hauptantrieb *m*
f commande *f* principale
r главный привод *m*

7744 main group
d Hauptgruppe *f*
f groupe *m* principal
r главная группа *f*; главный узел *m*

7745 main laser
d Hauptlaser *m*
f laser *m* principal
r основной лазер *m*

7746 main line
d Hauptleitung *f*
f conduite *f* principale
r магистраль *f*; магистральная линия *f*

7747 main line switch
d Hauptschalter *m*
f interrupteur *m* principal
r главный коммутатор *m* каналов

7748 main logic board
d Zentral-Steckeinheit *f*; zentrale Logik-
   Leiterplatte *f*
f unité *f* centrale enfichable
r главный логический модуль *m*

7749 main plant
d Hauptanlage *f*
f installation *f* principale
r главная установка *f*

*   **main process** → 7742

7750 main processor
d Hauptprozessor *m*
f processeur *m* principal
r основной процессор *m*

7751 main resonance
d Hauptresonanz[stelle] *f*
f résonance *f* principale
r основной резонанс *m*

7752 main stage
d Hauptstufe *f*
f étage *m* principal
r основная ступень *f*

7753 main steam maximum pressure controller
d Frischdampfmaximaldruckregler *m*;
   Dampfabwurfregler *m*
f régulateur *m* de la pression maximale de
   vapeur fraîche; régulateur de la pression
   maximale vapeur
r регулятор *m* максимального давления
   свежего пара

7754 main steam minimum pressure limiter
d Frischdampfminimaldruckbegrenzer *m*
f limiteur *m* de la pression minimale vapeur
r ограничитель *m* минимального давления
   свежего пара

7755 maintain *v*
d instand halten; warten
f maintenir
r обслуживать

7756 maintainability
d Reparaturfähigkeit *f*
f maintenabilité *f*
r ремонтопригодность *f*

7757 main technologies of manipulators
d Manipulatorhaupttechnologien *fpl*
f technologies *fpl* principales de manipulateurs
r основные технологические процессы *mpl*
   манипулятора

7758 **maintenance aids**
  d Wartungshilfen *fpl*
  f aides *fpl* de maintenance
  r вспомагательные устройства *npl* технического обслуживания

7759 **maintenance equipment; servicing equipment**
  d Instandhaltungseinrichtungen *fpl*; Wartungseinrichtungen *fpl*
  f équipement *m* d'entretien
  r обслуживающие устройства *npl*

7760 **maintenance error**
  d Instandhaltungsfehler *m*
  f erreur *f* d'entretien
  r ошибка *f* в [тех]обслуживании

7761 **maintenance function test**
  d Wartungsfunktionstest *m*; Funktionstest *m* bei Wartung
  f test *m* de fonctionnement d'entretien
  r функциональный тест *m* системы обслуживания

7762 **maintenance part**
  d Ersatzteil *n*; Wartungsteil *n*
  f pièce *f* d'entretien
  r запасная часть *f*; запасная деталь *f*

7763 **maintenance period**
  d Instandhaltungsperiode *f*
  f période *f* d'entretien
  r период *m* технического обслуживания

7764 **maintenance process**
  d Instandhaltungsprozess *n*
  f processus *m* d'entretien
  r процесс *m* технического обслуживания

7765 **maintenance time**
  d Instandhaltungszeit *f*
  f temps *m* de maintenance
  r время *n* технического обслуживания

7766 **majorant**
  d Majorante *f*
  f majorante *f*
  r мажоранта *f*

7767 **major control**
  d Übergruppenkontrolle *f*
  f contrôle *m* majeur
  r мажоритарный контроль *m*

7768 **major cycle**
  d Hauptperiode *f*; Hauptzyklus *m*
  f cycle *m* majeur; cycle principal

  r основной цикл *m*

7769 **major failure**
  d großer Ausfall *m*; großer Schaden *m*
  f défaillance *f* majeure
  r значительный отказ *m*

7770 **major feedback**
  d Hauptrückführung *f*; Hauptrückkopplung *f*
  f réaction *f* principale
  r основная обратная связь *f*

7771 **majority element**
  d Majoritätselement *n*
  f élément *m* à majorité
  r мажоритарный элемент *m*

7772 **majority function**
  d Majoritätsfunktion *f*
  f fonction *f* majoritaire
  r мажоритарная функция *f*

7773 **majority logic**
  d Mehrheitslogik *f*
  f logique *f* majoritaire
  r мажоритарная логика *f*

7774 **major loop**
  d Hauptschleife *f*
  f boucle *f* principale
  r основной контур *m*

7775 **make contact**
  d Schließkontakt *m*
  f contact *m* à fermeture
  r замыкающий контакт *m*

7776 **make critical** *v*
  d kritisch machen
  f rendre critique
  r сделать критическим

7777 **make impulse**
  d Schließungsstromstoß *m*; Schließungsimpuls *m*
  f impulsion *f* de fermeture
  r импульс *m* замыкания

  * **make-up** → 7778

7778 **make-up [feed]**
  d Zuspeisung *f*; Zusatz *m*; Nachspeisung *f*
  f appoint *m*; suralimentation *f*
  r подпитка *f*; подпитывание *n*

7779 **making alive**
  d Unterspannungsetzen *n*

f mise *f* sous tension
r подача *f* напряжения

7780 maladjustment
d Falscheinstellung *f*
f déréglage *m*; dérangement *m*
r плохая регулировка *f*

7781 maldistribution
d Fehlverteilung *f*; falsche Verteilung *f*
f fausse distribution *f*
r неправильное распределение *n*;
нарушение *n* распределения

* **malfunction → 5364**

7782 management
d Leitung *f*; Management *n*
f gestion *f*; direction *f*; management *m*
r управление *n*; организация *f*

7783 management information
d Verwaltungsinformation *f*
f information *f* de gestion
r информация *f* управления

7784 management information base
d Management-Informationsbasis *f*
f base *f* d'information de gestion
r информационная база *f* для управления

7785 management science
d Managementlehre *f*; Betriebswissenschaft *f*
f théorie *f* de gestion; science *f* de gestion
r теория *f* управления

7786 manipulated object
d manipuliertes Objekt *n*
f objet *m* manipulé
r манипулированный объект *m*

* **manipulated variable → 3451**

* **manipulate *v* → 6025**

* **manipulation assembly unit → 6030**

* **manipulation cycle → 6034**

* **manipulation defect → 6038**

* **manipulation device → 6035**

* **manipulation device with computer coupling → 6036**

* **manipulation equipment → 6037**

* **manipulation error → 6038**

* **manipulation of assembly parts → 6041**

* **manipulation process → 6045**

* **manipulation technique → 6052**

* **manipulator → 6035**

7787 manipulator application condition
d Manipulatoreinsatzbedingung *f*;
Einsatzbedingung *f* eines Manipulators
f condition *f* d'application de manipulateur
r условия *npl* применения манипулятора

7788 manipulator arrangement
d Manipulatoranordnung *f*
f arrangement *m* de manipulateur
r схема *f* расположения манипулятора

7789 manipulator building-block system
d Manipulatorbaukastensystem *n*
f système *m* de construction par blocs d'un manipulateur
r модульная система *f* построения манипулятора

7790 manipulator checking; handler checking
d Manipulatorkontrolle *f*
f contrôle *m* d'un manipulateur
r контроль *m* манипулятора

7791 manipulator computer control
d Manipulatorrechnersteuerung *f*;
Rechnersteuerung *f* eines Manipulators
f commande *f* de calculateur d'un manipulateur
r управление *n* манипулятором с использованием компьютера

* **manipulator condition → 7805**

7792 manipulator control impulsion
d Manipulatorsteuerimpuls *m*; Steuerimpuls *m* eines Manipulators
f impulsion *f* de commande de manipulateur
r импульс *m* управления манипулятором

7793 manipulator control system
d Manipulatorregelungssystem *n*
f système *m* de régulation de manipulateur
r регулирующая система *f* манипулятора

* **manipulator data → 7800**

7794 manipulator defect analysis
d Manipulatordefektanalyse *f*; Analyse *f* von Manipulatordefekten

f analyse *f* de défaut de manipulateur
r анализ *m* неисправностей манипулятора

7795 manipulator drive variant
d Manipulatorantriebsvariante *f*
f variante *f* d'entraînement de manipulateur
r вид *m* привода манипулятора

7796 manipulator dynamics
d Manipulatordynamik *f*
f dynamique *f* de manipulateur
r динамика *f* манипулятора

7797 manipulator generation
d Manipulatorgeneration *f*
f génération *f* de manipulateur
r поколение *n* манипуляторов

7798 manipulator memory
d Manipulatorspeicher *m*
f mémoire *f* de manipulateur
r память *f* манипулятора

7799 manipulator model
d Manipulatormodell *n*
f modèle *m* de manipulateur
r модель *f* манипулятора

7800 manipulator parameters; manipulator data
d Manipulatorparameter *mpl*;
Manipulatordaten *pl*
f paramètres *mpl* de manipulateur; données *fpl* de manipulateur
r технические данные *pl* манипулятора

7801 manipulator point control
d Manipulatorpunktsteuerung *f*;
Punktsteuerung *f* eines Manipulators
f commande *f* par points d'un manipulateur
r точечное управление *n* манипулятором

7802 manipulator precision
d Manipulatorpräzision *f*
f précision *f* de manipulateur
r точность *f* манипулятора

7803 manipulator regulator distance
d Manipulatorreglerstrecke *f*
f distance *f* de régulateur de manipulateur
r линия *f* регулятора манипулятора

7804 manipulator speed
d Manipulatorgeschwindigkeit *f*
f vitesse *f* de manipulateur
r скорость *f* манипулятора

7805 manipulator state; manipulator condition

d Manipulatorzustand *m*
f état *m* de manipulateur
r состояние *n* манипулятора

7806 manipulator system
d Manipulatorsystem *n*
f système *m* de manipulateur
r система *f* манипулятора

7807 manipulator technique
d Manipulatortechnik *f*
f technique *f* de manipulateur
r манипуляционная техника *f*

7808 manipulator track control
d Manipulatorbahnsteuerung *f*;
Bahnsteuerung *f* eines Manipulators
f commande *f* de chemin d'un manipulateur
r система *f* контурного управления манипулятором

7809 manipulator type
d Manipulatortyp *m*
f type *m* de manipulateur
r тип *m* манипулятора

7810 manipulator utilization
d Manipulatorauslastung *f*; Auslastung *f* von Manipulatoren
f utilisation *f* de manipulateurs
r коэффициент *m* использования манипуляторов

7811 manipulator with numerical structure
d Manipulator *m* mit numerischer Struktur
f manipulateur *m* à structure numérique
r манипулятор *m* с программной структурой

7812 manipulator with position control; position controlled manipulator
d Manipulator *m* mit Positionsregelung
f télémanipulateur *m* à système de positionnement
r манипулятор *m* с регулированием по положению

7813 man-machine interface
d Mensch-Maschine-Schnittstelle *f*
f interface *f* homme-machine
r человеко-машинный интерфейс *m*

7814 man-machine system
d Mensch-Maschine-System *n*
f système *m* homme-machine
r человеко-машинная система *f*

7815 manometer test press
d Manometerprüfpresse *f*

f presse *f* d'essai pour manomètres
r пресс *m* для испытания манометров

7816 manual closed-loop control system; manual-monitored control system
d Handregelsystem *n*
f système *m* de commande manuelle à asservissement
r замкнутая система *f* с ручным управлением

* **manual control** → 6055

7817 manual control unit
d Handbetätigungseinheit *f*
f unité *f* de manœuvre manuelle
r блок *m* ручного управления

7818 manual guided effector
d manuell geführter Effektor *m*
f effecteur *m* guidé manuel
r эффектор *m* с ручным управлением

* **manual-monitored control system** → 7816

7819 manual operation of manipulator
d Handbetrieb *m* eines Manipulators
f opération *f* manuelle d'un manipulateur
r ручной режим *m* работы манипулятора

7820 manual reset adjustment
d Handrückstellung *f*; Rückstellung *f* von Hand
f rétablissement *m* à main; réenclenchement *m* manuel; remise *f* à zéro manuelle
r ручная установка *f* зоны регулирования

* **manual setting** → 6055

7821 manufacturability
d Fertigungsmöglichkeit *f*; Herstellbarkeit *f*
f manufacturabilité *f*
r технологичность *f*

7822 manufacturing control
d Fertigungssteuerung *f*
f commande *f* de fabrication
r управление *n* производством

7823 manufacturing data
d Fabrikationsdaten *pl*
f données *fpl* de fabrication
r производственные данные *pl*

7824 manufacturing mass memory
d Fertigungsmassenspeicher *m*; Massenspeicher *m* für Fertigung
f mémoire *f* de grande capacité pour la fabrication

r массовое запоминающее устройство *n* производства

7825 manufacturing process
d Fertigungsprozess *m*
f procédé *m* de fabrication
r производственный процесс *m*

7826 manufacturing program memory
d Fertigungsprogrammspeicher *m*; Speicher *m* für Fertigungsprogramm
f mémoire *f* pour le programme de fabrication
r память *f* производственной программы

7827 manufacturing system
d Fertigungssystem *n*
f système *m* de fabrication
r производственная система *f*

7828 manufacturing task
d Fertigungsaufgabe *f*
f problème *m* de fabrication
r производственное задание *n*

7829 manufacturing technique
d Herstellungstechnik *f*; Herstellungstechnologie *f*
f technique *f* de fabrication
r технические средства *npl* производства

7830 manufacturing time data
d Fertigungszeitdaten *pl*; Daten *pl* für Fertigungszeiten
f données *fpl* de temps de production
r данные *pl* производственного времени

7831 many-element laser
d Mehrelementlaser *m*
f laser *m* à éléments multiples
r многоэлементный лазер *m*

7832 many-stage counter
d Vielstufenzähler *m*
f compteur *m* à plusieurs étages
r многокаскадный счётчик *m*

7833 many-valued
d mehrwertig
f polyvalent
r многозначный

7834 many-valued function
d vielwertige Funktion *f*
f fonction *f* polyvalente
r многозначная функция *f*

* **many-valued logic** → 8468

7835 **many-variable system**
  d Mehrvariablensystem *n*
  f système *m* à variables multiples
  r система *f* со многими переменными

7836 **mapping**
  d Einteilung *f*; Abbildung *f*
  f représentation *f*
  r отображение *n*

7837 **mapping system**
  d Einteilungssystem *n*
  f système *m* de classement
  r система *f* отображения

7838 **marginal checking**
  d Grenzwertprüfung *f*
  f essai *m* marginal
  r проверка *f* на надёжность; граничное
     испытание *n*

7839 **marginal service life**
  d Grenznutzungsdauer *f*
  f vie *f* utile marginale; vie utile limite
  r предельная длительность *f* использования

7840 **marginal stability**
  d Randstabilität *f*
  f stabilité *f* marginale
  r краевая устойчивость *f*

  * **marginal utility theory → 1899**

7841 **marginal voltage check**
  d Spannungsgrenzwerttest *m*; Test *m* bei
     reduzierter Betriebsspannung
  f test *m* à tension limite
  r испытание *n* при предельном отклонении
     напряжения

7842 **marker impulse**
  d Markierimpuls *m*
  f impulsion *f* de marquage
  r маркерный импульс *m*

7843 **marking circuit**
  d Markierungsschaltung *f*
  f circuit *m* marqueur
  r схема *f* маркировки

7844 **Markovian process**
  d Markowscher Prozess *m*
  f procédé *m* de Markov
  r марковский процесс *m*

7845 **Markovian-type process**
  d Prozess *m* vom Markowschen Typ
  f processus *m* de Markov; processus du type
     Markov

  r процесс *m* марковского типа

7846 **maskable interrupt**
  d maskierbare Unterbrechung *f*
  f interruption *f* masquable
  r замаскированное прерывание *n*

7847 **mass flowmeter**
  d Mengenmesser *m*;
     Durchflussmengenmesser *m*
  f débitmètre *m* massique
  r объёмный расходомер *m*

7848 **mass flow-sheet**
  d Mengenfließschema *n*
  f flow-sheet *m* quantitatif
  r технологическая схема *f* материальных
     потоков

  * **mass force → 5648**

7849 **mass scanning**
  d Massenabtastung *f*
  f balayage *m* de masse
  r масс-сканирование *n*

7850 **mass spectrometer with vacuum lock**
  d Massenspektrometer *n* mit Vakuumschleuse
  f spectromètre *m* de masse avec sas à vide
  r масс-спектрометр *m* с вакуумным
     затвором

7851 **mass spectrometric analysis**
  d massenspektrometrische Analyse *f*
  f analyse *f* au spectrographe de masse
  r масс-спектрометрический анализ *m*

  * **master → 7855**

7852 **master algorithm**
  d Hauptalgorithmus *m*
  f algorithme *m* principal
  r основной алгоритм *m*

7853 **master command routine**
  d Kommandohauptroutine *f*
  f routine *f* principale de commandes
  r главная управляющая программа *f*

7854 **master controller**
  d Hauptregler *m*; Hauptschalter *m*
  f commutateur *m* principal; organe *m* directeur
  r главный контроллер *m*

  * **master control panel → 2202**

7855 **master device; master**
  d Hauptgerät *n*; Mastergerät *n*;
     steuerungsführendes Gerät *n*

f appareil *m* maître; maître *m*
r ведущее устройство *n*

*   **master frequency meter** → 6806

7856 master mode
d steuerungsführende Betriebsart *f*
f mode *m* maître
r основной режим *m*

7857 master program
d Steuerprogramm *n*; organisatorisches Programm *n*
f programme *m* directeur
r главная программа *f*

*   **master pulse** → 3763

7858 master reset signal
d Hauptrücksetzsignal *n*; allgemeines Rücksetzsignal *n*
f signal *m* principal de remise; signal général de remise
r сигнал *m* общего сброса

7859 master synchronizing pulse
d Mastersynchronimpuls *m*
f impulsion *f* de synchronisation pilote
r главный имульс *m* синхронизации

7860 master unit
d Haupteinheit *f*
f unité *f* principale
r основной блок *m*

7861 match *v*
d anpassen; abgleichen
f conformer; coïncider
r согласовывать; сопоставлять; совпадать

7862 matched impedance
d angepasste Impedanz *f*
f impédance *f* adaptée
r согласованный импеданс *m*

7863 matched load
d angepasste Belastung *f*
f charge *f* accordée
r согласованная нагрузка *f*

*   **matching** → 370

7864 matching condition
d Anpassungsbedingung *f*
f condition *f* d'adaptation
r условие *n* согласования

7865 matching device; matching equipment

d Anpassungseintrichtung *f*
f dispositif *m* d'adaptation; équipement *m* d'adaptation
r согласующее устройство *n*

*   **matching equipment** → 7865

7866 matching error
d Anpassungsfehler *m*
f erreur *f* d'adaptation
r ошибка *f* согласования

7867 matching impedance
d Anpassungswiderstand *m*
f impédance *f* d'adaptation
r согласующее сопротивление *n*

7868 matching priority
d Vergleichspriorität *f*
f priorité *f* de comparaison
r приоритет *m* сравнения

7869 matching value
d Vergleichswert *m*
f valeur *f* de comparaison
r величина *f* для сопоставления

7870 mathematical algorithm
d mathematischer Algorithmus *m*
f algorithme *m* mathématique
r математический алгоритм *m*

7871 mathematical approximation
d mathematische Annäherung *f*; mathematische Approximation *f*
f approximation *f* mathématique
r математическое приближение *n*

7872 mathematical calculations
d mathematische Berechnungen *fpl*
f calculs *mpl* mathématiques
r математические расчёты *mpl*

*   **mathematical check** → 980

*   **mathematical checking** → 980

7873 mathematical expectation
d mathematische Erwartung *f*
f espérance *f* mathématique
r математическое ожидание *n*

7874 mathematical fundamental
d mathematische Grundlage *f*
f base *f* mathématique; considération *f* mathématique fondamentale
r математический принцип *m*

7875 **mathematical logic**
   d symbolische Logik *f*; mathematische Logik
   f logique *f* mathématique
   r математическая логика *f*; символическая
        логика

7876 **mathematical machine**
   d mathematische Maschine *f*
   f machine *f* mathématique
   r математическая машина *f*

7877 **mathematical manipulator model**
   d mathematisches Manipulatormodell *n*
   f modèle *m* de manipulateur mathématique
   r математическая модель *f* манипулятора

7878 **mathematical model**
   d mathematisches Modell *n*
   f modèle *m* mathématique
   r математическая модель *f*

7879 **mathematical operation with pneumatic
        signals**
   d mathematische Operation *f* mit
        pneumatischen Signalen
   f opération *f* mathématique aux signaux
        pneumatiques
   r математическая операция *f* с
        пневматическими сигналами

7880 **mathematical simulation**
   d mathematische Modellierung *f*
   f simulation *f* mathématique
   r математическое моделирование *n*

7881 **mathematical statistics**
   d mathematische Statistik *f*
   f statistique *f* mathématique
   r математическая статистика *f*

7882 **mathematical structure**
   d mathematische Struktur *f*
   f structure *f* mathématique
   r математическая структура *f*

7883 **mathematical subroutines**
   d mathematische Subroutinen *fpl*;
        mathematische Teilprogramme *npl*
   f sous-routines *fpl* mathématiques
   r математические подпрограммы *fpl*

7884 **matrix**
   d Matrix *f*
   f matrice *f*
   r матрица *f*

7885 **matrix analysis**
   d Matrixanalysis *f*

   f analyse *f* de matrice
   r матричный анализ *m*

7886 **matrix arithmetic**
   d Matrixarithmetik *f*
   f arithmétique *f* matricielle
   r матричная арифметика *f*

7887 **matrix arrangement**
   d Matrixanordnung *f*
   f disposition *f* matricielle
   r матричная конструкция *f*

7888 **matrix decoder**
   d Matrixentzifferer *m*
   f déchiffreur *m* matriciel
   r матричный дешифратор *m*

7889 **matrix element**
   d Matrixelement *n*; Element *n* der Matrix
   f élément *m* d'une matrice
   r матричный элемент *m*

7890 **matrix encoder**
   d Matrixkodierschaltung *f*
   f circuit *m* de codage à matrice
   r матричный кодер *m*

7891 **matrix exponential function**
   d Matrizenexponentialfunktion *f*
   f fonction *f* matricielle exponentielle
   r матричная экспоненциальная функция *f*

   *   **matrix memory** → 7894

7892 **matrix method**
   d Matrixmethode *f*
   f méthode *f* de matrice
   r матричный метод *m*

   *   **matrix notation** → 7893

7893 **matrix representation; matrix notation**
   d Matrixdarstellung *f*; Matrixschreibweise *f*
   f représentation *f* matricielle; notation *f*
        matricielle
   r представление *n* в матричной форме

7894 **matrix store; matrix memory**
   d Matrixspeicher *m*
   f mémoire *f* matricielle
   r матричная память *f*

7895 **matrix telemetering system**
   d Matrixzenfermesssystem *n*
   f système *m* télémétrique à matrice
   r матричная телеметрическая система *f*

7896 matrix transformation
d Matrixtransformation *f*
f transformation *f* de matrice
r матричное преобразование *n*

7897 maximize *v*
d maximieren
f maximaliser
r максимизировать

7898 maximum
d Maximum *n*
f maximum *m*
r максимум *m*

7899 maximum counting speed
d maximale Zählgeschwindigkeit *f*
f vitesse *f* maximum de comptage
r максимальная скорость *f* счёта

7900 maximum current setting of starting relay
d maximaler Einstellstrom *m* des Einschaltrelais
f réglage *m* maximum d'intensité de relais de démarrage
r регулирование *n* максимального тока пускового реле

7901 maximum cut-out
d Maximalausschalter *m*; Höchstausschalter *m*
f disjoncteur *m* à maxima; interrupteur *m* à maximum
r максимальный разъединитель *m*

7902 maximum deflection
d Ablenkamplitude *f*
f déviation *f* maximum
r максимальное отклонение *n*

7903 maximum demand indicator
d Maximumverbrauchszähler *m*; Spitzenzähler *m*; Höchstlastanzeiger *m*; Zähler *m* mit Höchstverbrauchsangabe
f compteur *m* à indicateur de maximum
r счётчик *m* с указателем максимума

7904 maximum demand recorder
d Zähler *m* mit schreibendem Höchstverbrauchsanzeiger
f compteur *m* à enregistreur de maximum
r регистратор *m* максимального потребления

7905 maximum error diagnostic
d maximale Fehlerdiagnostik *f*
f diagnostic *m* maximum d'erreurs
r диагностика *f* максимальных ошибок

7906 maximum frequency of oscillation
d Schwinggrenzfrequenz *f*
f fréquence *f* maximale d'oscillation
r предельная частота *f* колебаний

7907 maximum modulation frequency
d höchste Modulationsfrequenz *f*
f fréquence *f* de modulation maximale
r максимальная частота *f* модуляции

7908 maximum number of command
d maximale Kommandozahl *f*
f nombre *m* d'ordre maximal
r максимальное число *n* команд

7909 maximum output
d Höchstleistung *f*; Spitzenleistung *f*; maximale Ausgangsleistung *f*
f puissance *f* maximale de sortie
r максимальная мощность *f*

7910 maximum overshoot
d Überschwingweite *f*
f élongation *f* de dépassement; taux *m* de dépassement
r максимальное перерегулирование *n*; максимальный заброс *m*

7911 maximum permeability
d maximale Permeabilität *f*
f perméabilité *f* maximale
r максимальная проницаемость *f*

7912 maximum permissible exposure values
d maximal zulässige Bestrahlungswerte *mpl*
f valeurs *fpl* d'exposition maximale permises
r максимально допустимые величины *fpl* облучения

7913 maximum principle; Pontrjagin maximum principle
d [Pontrjaginsches] Maximumprinzip *n*; Maximalprinzip *n*
f principe *m* du maximum; principe maximum [de Pontrjagin]
r принцип *m* максимума [Понтрягина]

7914 maximum rating
d Grenzdaten *pl*
f valeurs *fpl* limite
r предельные параметры *mpl*

7915 maximum recording attachment
d schreibendes Maximumwerk *n*
f enregistreur *m* de maximum
r устройство *n* регистрации максимума

7916 maximum scale value
d Maximalskalenwert *m*

f valeur *f* maximale d'echelle
r максимальное показание *n* шкалы

7917 Maxwell element
d Maxwellsches Element *n*;
Maxwell-Körper *m*
f élément *m* de Maxwell
r элемент *m* Максвелла

7918 Maxwell equation
d Maxwellsche Gleichung *f*
f équation *f* de Maxwell
r уравнение *n* Максвелла

* **MB → 8024**

7919 mean angular velocity
d mittlere Winkelgeschwindigkeit *f*
f vitesse *f* angulaire moyenne
r среднее угловое перемещение *n*

7920 mean deviation
d mittlere Abweichung *f*
f écart *m* moyen; déviation *f* moyenne;
variation *f* moyenne
r разброс *m*; рассеяние *n*

* **mean fuel rating → 1507**

7921 mean power; average power
d mittlere Leistung *f*
f puissance *f* moyenne
r средняя мощность *f*

7922 mean probability
d mittlere Wahrscheinlichkeit *f*
f probabilité *f* moyenne
r средняя вероятность *f*

* **mean-pulse indicator → 1518**

7923 mean repair time
d mittlere Reparaturzeit *f*
f temps *m* moyen de réparation
r среднее время *n* востановления

7924 means of rationalization
d Rationalisierungsmittel *n*
f équipement *m* de rationalisation
r средство *n* рационализации

7925 mean square deviation; standard deviation
d mittlere quadratische Abweichung *f*
f écart *m* moyen quadratique
r среднеквадратичное отклонение *n*;
стандартное отклонение

7926 mean square error
d mittlerer quadratischer Fehler *m*
f erreur *f* quadratique moyenne
r среднеквадратичная ошибка *f*

7927 mean square error minimum
d Minimum *n* des mittleren quadratischen
Fehlers
f erreur *f* quadratique moyenne minimum
r минимум *m* среднеквадратичной ошибки

7928 mean square error moment
d mittleres quadratisches Fehlermoment *n*
f moment *m* d'erreur quadratique moyenne
r момент *m* среднеквадратичной ошибки

7929 mean square estimation
d quadratisches Mittelwertkriterium *n*
f estimation *f* quadratique moyenne
r среднеквадратичная оценка *f*

7930 mean square of intensity fluctuation
d mittleres Quadrat *n* der
Intensitätsschwankung
f valeur *f* moyenne quadratique des
fluctuations d'intensité
r среднеквадратичное значение *n*
интенсивности флуктуации

7931 mean time between failures
d mittlerer Ausfallabstand *m*; mittlere
störungsfreie Zeit *f*
f temps *m* moyen entre défaillances
r среднее время *n* безотказной работы

7932 mean time to maintain
d mittlere Wartungszeit *f*
f temps *m* moyen de maintenance
r среднее время *n* обслуживания

7933 measurability
d Messbarkeit *f*
f mesurabilité *f*
r измеримость *f*

7934 measurable
d messbar
f mesurable
r измеримый

7935 measured operating value
d gemessener Betriebswert *m*
f valeur *f* de fonctionnement mesurée
r измеренное значение *n* параметра

7936 measured quantity
d Messgröße *f*

f grandeur *f* mesurée
r измеряемая величина *f*

7937 measured response
d gemessene Reaktion *f*
f réaction *f* mesurée
r измеренная реакция *f*

7938 measured signal
d Messsignal *n*
f signal *m* de mesure
r измеряемый сигнал *m*

**7939 measured value acquisition; measured
value logging**
d Messwerterfassung *f*
f acquisition *f* des valeurs mesurées
r сбор *m* измеренных значений

*   **measured value logging** → 7939

7940 measurement
d Messung *f*
f mesure *f*; mesurage *m*
r измерение *n*

7941 measurement of gripper force
d Greiferkraftmessung *f*
f mesure *f* de force de grappin
r измерение *n* усилия захвата

7942 measurement of load and extension
d Last- und Dehnungsmessung *f*
f mesure *f* d'allongement sous charge
r измерение *n* нагрузки и удлинения

7943 measurement of mechanical stress
d Messung *f* der mechanischen Beanspruchung
f mesure *f* des efforts mécaniques
r измерение *n* механической
напряжённости

**7944 measurement precision; measuring
accuracy; accuracy of measurement**
d Messgenauigkeit *f*
f précision *f* de mesure
r точность *f* измерения

7945 measurement setup
d Messplatz *m*
f place *f* de mesure
r установка *f* для измерения

*   **measuring accuracy** → 7944

*   **measuring apparatus** → 7956

7946 measuring apparatus constant

d Messgerätekonstante *f*
f constante *f* d'appareil de mesure
r постоянная *f* измерительного прибора

7947 measuring block
d Messblock *m*
f bloc *m* de mesure
r измерительный блок *m*

7948 measuring bridge
d Messbrücke *f*
f pont *m* de mesure
r измерительный мост *m*

7949 measuring circuit
d Messkreis *m*
f circuit *m* de mesure
r измерительная цепь *f*

7950 measuring device
d Messeinrichtung *f*
f dispositif *m* de mesurage
r измерительное устройство *n*

7951 measuring diaphragm; orifice plate
d Messblende *f*
f diaphragme *m* de mesure
r измерительная диафрагма *f*

7952 measuring equipment
d Messausrüstung *f*; Messgeräteausstattung *f*
f équipement *m* de mesure
r измерительное оборудование *n*

7953 measuring error
d Messfehler *m*
f erreur *f* de mesure
r ошибка *f* измерения

7954 measuring impulse
d Messimpuls *m*
f impulsion *f* de mesure
r измерительный импульс *m*

7955 measuring installation
d Messanlage *f*
f installation *f* de mesure
r измерительная установка *f*

**7956 measuring instrument; measuring means;
measuring apparatus**
d Messorgan *n*; Messinstrument *n*;
Messapparat *m*; Messvorrichtung *f*
f appareil *m* de mesure; appareil mesureur
r измерительный прибор *m*

7957 measuring instrument calibration
d Eichung *f* von Messgeräten

f étalonnage *m* d'appareils de mesure
r калибровка *f* измерительных приборов

7958 measuring instrument classification
  d Klasseneinteilung *f* für Messgeräte
  f classification *f* d'appareils de mesure
  r классификация *f* измерительных приборов

7959 measuring instrument guide
  d Messwerkzeugführung *f*
  f guidage *m* d'outil de mesure
  r управление *n* измерительным
     инструментом

7960 measuring instruments remote reading
  d Fernablesung *f* von Messinstrumenten
  f lecture *f* à distance des instruments de
     mesure
  r дистанционный отсчёт *m* показаний
     измерительного прибора

**7961 measuring instrument with digital
     indication**
  d Messgerät *n* mit digitaler Anzeige
  f appareil *m* mesureur à indication digitale
  r измерительный прибор *m* с цифровой
     индикацией

7962 measuring magnitude acquisition
  d Messgrößenerfassung *f*
  f acquisition *f* d'une grandeur de mesure
  r сбор *m* данных о нескольких параметрах

  * **measuring means** → 7956

7963 measuring minicentral
  d Messminizentrale *f*
  f mini-centrale *f* de mesure
  r измерительное миниатюрное устройство *n*

7964 measuring of the correlation functions
  d Korrelationsfunktionsmessung *f*
  f mesure *f* des fonctions de corrélation
  r измерение *n* корреляционных функции

**7965 measuring of the dispersion by
     refractometer**
  d Dispersionsmessung *f* mit Refraktometer
  f mesurage *m* de dispersion par réfractomètre
  r измерение *n* рассеяния при помощи
     рефрактометра

7966 measuring of the phase angle changes
  d Messen *n* der Phasenwinkelschwankungen
  f mesurage *m* des déviations d'angle de phase
  r измерение *n* [изменений] сдвига фаз угла

7967 measuring panel for testing installations

  d Messtafel *f* für Prüfanlagen
  f tableau *m* de mesure pour installations d'essai
  r измерительная панель *f* испытательных
     установок

7968 measuring point
  d Messpunkt *m*; Messstelle *f*
  f point *m* de mesure
  r точка *f* измерения

7969 measuring potentiometer
  d Messpotentiometer *n*
  f potentiomètre *m* de mesure
  r измерительный потенциометр *m*

7970 measuring precision
  d Messgenauigkeit *f*
  f précision *f* de mesure
  r точность *f* измерений

7971 measuring procedure
  d Messverfahren *n*
  f procédé *m* de mesure
  r процедура *f* измерения

7972 measuring range
  d Messbereich *m*
  f domaine *m* de mesure; étendue *f* de mesurage
  r диапазон *m* измерения

7973 measuring safety
  d Messsicherheit *f*
  f sécurité *f* de mesure
  r надёжность *f* измерения

7974 measuring sensor
  d messender Sensor *m*
  f capteur *m* mesurable
  r измерительный сенсор *m*

7975 measuring signal
  d Messsignal *n*
  f signal *m* de mesure
  r измерительный сигнал *m*

7976 measuring system
  d Messsystem *n*
  f système *m* de mesure
  r измерительная система *f*

7977 measuring technique
  d Messtechnik *f*
  f technique *f* de mesure
  r измерительная техника *f*

7978 measuring transducer
  d Messumformer *m*

f traducteur *m* de mesure; convertisseur *m* de
mesure
r преобразователь *m* измеряемой величины

7979 measuring transducer for gas analyzers
d Messumformer *m* für Gasanalysatoren
f mesureur *m* transmetteur pour analyseurs de
gaz
r измерительный преобразователь *m* для
газоанализаторов

7980 measuring transmitter
d Messumformer *m*
f convertisseur *m* de mesure
r измерительный датчик *m*

7981 measuring value
d Messwert *m*
f valeur *f* de mesure
r измеренное значение *n*; измеренная
величина *f*

7982 measuring values transmission system
d Messwertübertragungssystem *n*
f système *m* de transmission de valeurs de
mesure
r система *f* передачи измерительных
величин

7983 mechanical amplifier
d mechanischer Verstärker *m*
f amplificateur *m* mécanique
r механический усилитель *m*

7984 mechanical analyzer
d mechanischer Analysator *m*
f analyseur *m* mécanique
r механический анализатор *m*

7985 mechanical assembly
d mechanische Baugruppe *f*
f assemblage *m* mécanique; module *m*
mécanique
r механический узел *m*

* **mechanical check → 7986**

7986 mechanical check[ing]
d mechanische Prüfung *f*
f contrôle *m* mécanique
r контроль *m* с помощью механических
устройств

**7987 mechanical data acquisition; mechanical
data collection**
d maschinelle Datenerfassung *f*
f saisie *f* mécanique des données; collection *f*
mécanique des données

r машинизированный сбор *m* данных

* **mechanical data collection → 7987**

7988 mechanical design
d mechanischer Aufbau *m*
f structure *f* mécanique
r механическая конструкция *f*

7989 mechanical digital clock display system
d mechanisches Digitaluhranzeigesystem *n*
f système *m* mécanique de lecture numérique
de montres
r механическая система *f* цифровой
индикации времени

* **mechanical drawing → 12233**

7990 mechanical drive
d mechanischer Antrieb *m*
f commande *f* mécanique
r механический привод *m*

7991 mechanical efficiency
d mechanischer Wirkungsgrad *m*
f rendement *m* mécanique
r механический коэффициент *m* полезного
действия

7992 mechanical-electrical sensor
d mechanisch-elektrischer Wandler *m*
f capteur *m* mécanique-électrique
r электромеханический датчик *m*

7993 mechanical full-automatic production
d mechanische vollautomatische Produktion *f*
f production *f* entièrement automatique
mécanique
r полностью автоматизированное
механическое производство *n*

7994 mechanical gripper expense
d mechanischer Greiferaufwand *m*
f frais *mpl* de grappin mécanique
r механические затраты *fpl* на захватное
устройство *n*

7995 mechanical gripper flexibility
d mechanische Greiferflexibilität *f*
f flexibilité *f* de grappin mécanique
r механическая гибкость *f* захвата

7996 mechanical gripper parameter
d mechanischer Greiferparameter *m*
f paramètre *m* de grappin mécanique
r механический параметр *m* захватного
устройства

7997 **mechanical interlocking**
  d mechanische Sperrung *f*; mechanische
     Verriegelung *f*
  f verrouillage *m* mécanique
  r механическая блокировка *f*

7998 **mechanically coupled**
  d mechanisch gekoppelt
  f à couplage mécanique
  r с механической связью

7999 **mechanical manipulator energy**
  d mechanische Manipulatorenergie *f*
  f énergie *f* mécanique d'un manipulateur
  r механическая энергия *f* манипулятора

8000 **mechanical measurement**
  d mechanische Messung *f*
  f mesure *f* mécanique
  r механическое измерение *n*

  *   **mechanical motion unit → 8001**

8001 **mechanical movement unit; mechanical
     motion unit**
  d mechanische Bewegungseinheit *f*
  f unité *f* de mouvement mécanique
  r механический блок *m* перемещений

8002 **mechanical network**
  d mechanisches Netzwerk *n*
  f réseau *m* mécanique
  r механическая сеть *f*

8003 **mechanical optical switch**
  d mechanischer optischer Schalter *m*
  f commutateur *m* optique à commande
     mécanique
  r оптико-механический переключатель *m*

8004 **mechanical order principle**
  d mechanisches Ordnungsprinzip *n*
  f principe *m* d'ordre mécanique
  r механический принцип *m* упорядочения

8005 **mechanical overload protection**
  d mechanischer Überlastschutz *m*
  f protection *f* de surcharge mécanique
  r механическое устройство *n* защиты от
     перегрузки

8006 **mechanical parts machining**
  d mechanische Teilebearbeitung *f*
  f usinage *m* de pièces mécanique
  r механическая обработка *f* деталей

8007 **mechanical production engineering**
  d mechanische Produktionstechnik *f*

  f technique *f* de production mécanique
  r механическое производственное
     оборудование *n*

8008 **mechanical programmer**
  d mechanischer Programmierer *m*
  f programmateur *m* mécanique
  r механическое программирующее
     устройство *n*

8009 **mechanical remote control**
  d mechanische Fernsteuerung *f*; mechanische
     Fernlenkung *f*
  f commande *f* mécanique à distance;
     télécommande *f* mécanique
  r механическое дистанционное
     управление *n*

8010 **mechanical sensor**
  d mechanischer Sensor *m*
  f capteur *m* mécanique
  r механический датчик *m*

8011 **mechanical servosystem**
  d mechanisches Folgesystem *n*
  f servosystème *m* mécanique
  r механическая сервосистема *f*

8012 **mechanical specifications**
  d mechanische Spezifikation *f*;
     Kenndatenblatt *n* mechanischer Parameter
  f spécification *f* mécanique
  r требования *npl* к механическим
     параметрам

8013 **mechanical strength**
  d mechanische Festigkeit *f*
  f résistance *f* mécanique
  r механическая прочность *f*

8014 **mechanical switch arrangement**
  d mechanische Schalteranordnung *f*
  f arrangement *m* d'interrupteur mécanique
  r механическое расположение *n*
     выключателей

8015 **mechanical zero setting**
  d mechanische Nulleinstellung *f*
  f réglage *m* mécanique de zéro
  r отметка *f* механического нуля

8016 **mechanization**
  d Mechanisierung *f*
  f mécanisation *f*
  r механизация *f*

8017 **mechanized direction**
  d mechanisierte Verwaltung *f*; mechanisierte
     Leitung *f*

f gestion f mécanisée
r механизированное управление n

8018 mechanochemical effect
d mechanochemische Einwirkung f;
  mechanochemischer Effekt m
f effet m mécano-chimique
r механохимическое воздействие n

8019 medium complexity
d mittlere Komplexität f
f intégration f à moyenne échelle
r средняя степень f интеграции

8020 medium infrared
d mittleres Infrarot n
f infrarouge m moyen
r средний инфракрасный диапазон m

8021 medium-speed data transmission
d mittelschnelle Datenübertragung f
f transmission f de données à vitesse moyenne
r передача f данных со средней скоростью

8022 medium-speed line
d mittelschnelle Leitung f
f ligne f à vitesse moyenne
r линия f среднего быстродействия

8023 megabit
d Megabit n
f mégabit m
r мегабит m

8024 megabyte; MB
d Megabyte n; MB
f méga-octet m; Mo
r мегабайт m; Мб

8025 member; element
d Member n; Element n
f membre m; élément m
r член m; элемент m

8026 membrane valve
d Membranventil n
f soupape f à membrane
r мембранный клапан m

8027 memistor
d Memistor m
f memistor m
r мемистор m

8028 memory
d Gedächtnis n
f mémoire f; organe m de mémoire;
  dispositif m de mémoire

r память f

8029 memory access protection
d Speicherzugriffsschutz m
f protection f d'accès à la mémoire
r защита f от несанкционированного
  доступа к памяти

* **memory area** → 8052

8030 memory assembly; store assembly
d Speicheraufbau m
f montage m de mémoire
r структура f памяти

8031 memory block
d Speicherblock m
f bloc m de mémoire
r блок m памяти

8032 memory capacity
d Speicherkapazität f
f capacité f de mémoire
r ёмкость f памяти

8033 memory circuit
d Speicherschaltung f
f circuit m de mémoire
r запоминающая схема f

8034 memory configuration
d Speicherkonfiguration f; Speicheraufbau m
f configuration f de mémoire
r конфигурация f памяти

8035 memory control; storage control
d Speichersteuerung f
f commande f de mémoire
r управление n памятью

**8036 memory distribution checking; store
     distribution checking**
d Speicherverteilungskontrolle f
f contrôle m de partage de mémoire
r контроль m распределения данных в
  памяти

8037 memory efficiency; storage efficiency
d Speicherausnutzung f
f efficience f de mémoire
r эффективность f [использования] памяти

8038 memory element
d Speicherelement n
f élément m de mémoire
r элемент m памяти

8039 memory equation
d Gedächtnisgleichung f

f équation *f* de mémoire
r уравнение *n* памяти

8040 memory fabrication technology
   d Speicherherstellungstechnologie *f*
   f technologie *f* de fabrication de mémoire
   r технология *f* изготовления памяти

8041 memory function; store function
   d Gedächtnisfunktion *f*
   f fonction *f* de mémoire
   r функция *f* накопления

**8042 memory inactivation; inactivation of a
         memory**
   d Stillsetzen *n* eines Speichers
   f inactivation *f* d'une mémoire
   r деактивирование *n* памяти

8043 memory location; store location
   d Speicherplatz *m*
   f position *f* de mémoire; emplacement *m* de
       mémoire; cellule *f* de mémoire
   r распределение *n* памяти

8044 memory management unit
   d Speicherverwaltungseinheit *f*
   f unité *f* de gestion de mémoire
   r устройство *n* управления памяти

   *   **memory mechanism** → 11960

8045 memory operation
   d Speicheroperation *f*
   f opération *f* de mémorisation
   r операция *f* в накопителе

8046 memory principle; store principle
   d Speicherprinzip *n*
   f principe *m* de mémoire
   r принцип *m* [работы] памяти

**8047 memory-programmed control; store-
         programmed control**
   d speicherprogrammierte Steuerung *f*
   f commande *f* programmée en mémoire
   r программное управление *n* памятью

8048 memory protection system
   d Speicherschutzvorrichtung *f*
   f système *m* de protection de mémoire
   r система *f* защиты памяти

8049 memory register
   d Speicherregister *n*
   f registre *m* mémoire
   r регистр *m* памяти; регистр накопителя

8050 memory return to zero

d Speicherrückstellung *f* auf Null
f remise *f* à zéro de la mémoire
r возвращение *n* накопителя на нуль

8051 memory system; storage system
   d Speichersystem *n*
   f système *m* de mémoire
   r система *f* памяти

8052 memory zone; memory area
   d Speicherzone *f*; Speicherbereich *m*
   f zone *f* de mémoire; domaine *m* de mémoire
   r зона *f* памяти; область *f* памяти

8053 merged technology
   d gemischte Technologie *f*; Mischtechnik *f*
   f technologie *f* mixte
   r смешанная технология *f*

8054 mesh possibility
   d Eingriffmöglichkeit *f*
   f possibilité *f* d'intervention
   r возможность *f* вмешательства

8055 message; notice
   d Mitteilung *f*; Nachricht *f*; Meldung *f*
   f message *m*
   r сообщение *n*; известие *n*

8056 message control program
   d Nachrichtenübertragungs-Steuerprogramm *n*
   f programme *m* de commande de transmission
       de messages
   r программа *f* управления сообщениями

8057 message handling
   d Nachrichtenbehandlung *f*;
       Nachrichtenverwaltung *f*
   f traitement *m* des messages
   r обработка *f* сообщений

8058 message priority
   d Nachrichtenpriorität *f*
   f priorité *f* de message
   r приоритет *m* сообщения

8059 message routing
   d Nachrichtenleitung *f*; Nachrichtenführung *f*
   f acheminement *m* de messages
   r маршрутизация *f* сообщений

8060 message source
   d Informationsquelle *f*; Nachrichtenquelle *f*
   f source *f* de message; source d'information
   r источник *m* информаций

8061 message switching
   d Nachrichtenvermittlung *f*;
       Speichervermittlung *f*

f   communication f de messages
r   коммутация f сообщений

8062   message transfer system
d   Mitteilungstransfersystem n;
Nachrichtenübertragungssystem n
f   système m de transfert de messages
r   система f передачи сообщений

8063   metering pump
d   Messpumpe f; Dosierpumpe f
f   pompe f doseuse
r   дозирующий насос m

8064   meter quality
d   Messgerätgüte f
f   coefficient m de qualité d'appareil de mesure
r   качество n измерительного прибора

8065   meter reading variations
d   Schwankungen fpl der Messgeräteanzeigen
f   variations fpl d'indications d'appareil
r   вариация f показаний измерительного
прибора

8066   method
d   Methode f; Verfahren n
f   méthode f; moyen m
r   метод m; способ m

8067   method for project planning
d   Projektplanungsmethode f
f   méthode f de planning des projets
r   метод m планирования разработки
проектов

8068   method for synthesis
d   Syntheseverfahren n
f   méthode f de synthèse
r   метод m синтеза

8069   method of charging
d   Beschickungsweise f
f   méthode f de charge; mode m de chargement;
mode d'alimentation
r   метод m загрузки

*   **method of comparison → 2695**

8070   method of correction
d   Korrekturmethode f
f   méthode f de correction
r   метод m коррекции

8071   method of decomposition
d   Dekompositionsmethode f
f   méthode f de décomposition
r   метод m декомпозиции

8072   method of determination
d   Bestimmungsmethode f
f   méthode f de détermination; méthode
d'analyse
r   метод m определения

**8073   method of determination of stability
domains**
d   Methode f zur Abgrenzung der
Stabilitätsbereiche
f   méthode f de détermination des domaines de
stabilité
r   метод m выделения областей
устойчивости

*   **method of energetic balance → 5039**

8074   method of Evans; Evans method
d   Evans-Methode f; Stabilitatsuntersuchung f
nach Evans
f   méthode f d'Evans; analyse f de stabilité
d'Evans
r   метод m Ивенса

8075   method of Huffman-Caldwell
(theory of automata)
d   Huffman-Caldwell-Verfahren n
f   méthode f d'Huffman-Caldwell
r   метод m Хафмена-Колдвелла

8076   method of Kochenburger
d   Methode f von Kochenburger;
Stabilitätsuntersuchung f nach Kochenburger
f   méthode f de Kochenburger; analyse f de
stabilité de Kochenburger; examen m de
stabilité de Kochenburger
r   метод m Кохенбургера

8077   method of least squares
d   Methode f der kleinsten Quadrate
f   méthode f des moindres carrés
r   метод m наименьших квадратов

8078   method of Nyquist; Nyquist method
d   Nyquist-Methode f; Stabilitätsuntersuchung f
nach Nyquist
f   méthode f de Nyquist; analyse f de stabilité
de Nyquist
r   метод m Найквиста

*   **method of operation → 9032**

8079   method of operation analysis
d   operationsanalytische Methode f
f   méthode f de la recherche opérationnelle
r   метод m исследования операций

8080   method of preparation
d   Vorbereitungsmethode f

f méthode *f* de préparation
r метод *m* подготовки

8081 method of proof
  d Beweismethode *f*
  f méthode *f* de démonstration
  r метод *m* доказательства

8082 method of residues
  d Residuenmethode *f*
  f méthode *f* des résidus; théorème *m* des
     résidus
  r метод *m* вычета

8083 method of small parameter
  d Methode *f* des kleinen Parameters
  f méthode *f* du petit paramètre
  r метод *m* малого параметра

8084 method of solutions sewing
  d Anstückelungsmethode *f*;
     Bereitstellungsmethode *f*
  f méthode *f* des intervalles
  r метод *m* интервалов

8085 method of steepest descent
  d Methode *f* des steilsten Abstiegs
  f méthode *f* de la plus rapide descente
  r метод *m* быстрейшего спуска

8086 method of successive approximation
  d Methode *f* der sukzessiven Approximation;
     Methode der schrittweisen Annäherung;
     Iterationsverfahren *n*
  f méthode *f* d'approximation successive
  r метод *m* последовательного приближения

  * **method of telemetering → 12256**

**8087 method of trapezoidal frequency
       responses**
  d Methode *f* trapezförmiger
     Frequenzcharakteristiken
  f méthode *f* de caractéristiques trapézoïdales
     de fréquence
  r метод *m* трапецеидальных частотных
     характеристик

8088 method of variation of constants
  d Methode *f* der Variation der Konstanten
  f méthode *f* de la variation des constantes
  r метод *m* вариации постоянных

8089 metric
  d Metrik *f*; Abstandsfunktion *f*
  f métrique *f*
  r метрика *f*

8090 metric information
  d metrische Information *f*
  f information *f* métrique
  r метрическая информация *f*

8091 metric space
  d metrischer Raum *m*
  f espace *m* métrique
  r метрическое пространство *n*

8092 metric system
  d metrisches System *n*
  f système *m* métrique
  r метрическая система *f*

8093 Michailov criterion
  d Michailovsches Kriterium *n*
  f critère *m* de Michailov
  r критерий *m* Михайлова

8094 microadjuster
  d Mikroverstellvorrichtung *f*;
     Feinstellvorrichtung *f*;
     Mikrojustiereinrichtung *f*
  f micromanipulateur *m*
  r микрометрический регулятор *m*

  * **micro-assemblage technique → 8132**

8095 microassembly
  d Mikrobaueinheit *f*
  f micro-assemblage *m*
  r микросборка *f*

8096 microblock design
  d Mikroblockbauweise *f*
  f mode *f* de fabrication de micro-éléments
  r микроблочное проектирование *n*

8097 microchecking
  d Mikrokontrolle *f*
  f microcontrôle *m*
  r микроконтроль *m*

8098 microcircuit
  d Mikroschaltung *f*
  f microcircuit *m*
  r микросхема *f*

  * **microcircuit engineering → 8099**

**8099 microcircuit technique; microcircuit
       engineering**
  d Mikroschaltungstechnik *f*
  f technique *f* des microcircuits;
     microcircuiterie *f*
  r техника *f* микросхем;
     микросхемотехника *f*

8100 micro-cleaned surface
  d superreine Oberfläche *f*
  f surface *f* superfinie
  r сверхчистая поверхность *f*

8101 microcomputer
  d Mikrorechner *m*
  f microcalculateur *m*
  r микрокомпьютер *m*

8102 microcomputer control
  d Mikrorechnersteuerung *f*
  f commande *f* de microcalculateur; commande de micro-ordinateur
  r микрокомпьютерное управление *n*

8103 microcomputer-controlled assembly head
  d mikrorechnergesteuerter Montagekopt *m*
  f tête *f* d'assemblage commandée par micro-ordinateur
  r сборочная головка *f* с управлением от микрокомпьютера

8104 microcomputer-controlled error measurement
  d mikrorechnergesteuerte Fehlermessung *f*
  f mesurage *m* d'erreur commandé par micro-ordinateur
  r измерение *n* ошибок, управляемое микрокомпьютером

8105 microcomputer-controlled fine positioning
  d mikrorechnergesteuerte Feinpositionierung *f*
  f positionnement *m* fin commandé par micro-ordinateur
  r точное позиционирование *n* с управлением от микрокомпьютера

8106 microcomputer-controlled joint[ing] movement
  d mikrorechnergesteuerte Fügebewegung *f*
  f mouvement *m* de jointage commandé par micro-ordinateur
  r движение *n* сопряжения, управляемое от микрокомпьютера

  * **microcomputer-controlled joint movement → 8106**

8107 microcomputer-controlled limitation of grip power
  d mikrorechnergesteuerte Greifkraftbegrenzung *f*
  f limitation *f* d'une force de grippion commandée par micro-ordinateur
  r ограничение *n* силы захватывания с использованием микрокомпьютера

8108 microcomputer-controlled manipulator
  d mikrorechnergesteuerter Manipulator *m*
  f manipulateur *m* commandé par micro-ordinateur
  r манипулятор *m* с управлением от микрокомпьютера

8109 microcomputer-controlled measuring device
  d mikrorechnergesteuerte Messeinrichtung *f*
  f dispositif *m* de mesurage commandé par micro-ordinateur
  r измерительное устройство n, управляемое микрокомпьютером

8110 microcomputer-controlled miniature manipulator
  d mikrorechnergesteuerter Kleinmanipulator *m*
  f manipulateur *m* miniature commandé par micro-ordinateur
  r малогабаритный манипулятор m, управляемый микрокомпьютером

8111 microcomputer-oriented automation
  d mikrorechnerorientierte Automatisierung *f*
  f automatisation *f* orientée sur le microcalculateur
  r автоматизация f, ориентированная на применение микрокомпьютера

8112 microcomputer-oriented process device
  d mikrorechnerorientierte Prozesseinrichtung *f*
  f dispositif *m* de processus orienté sur le microcalculateur
  r прибор *m* управления процессом, ориентированный на применение микрокомпьютера

8113 microcomputer software
  d Mikrorechnersoftware *f*
  f software *m* de microcalculateur
  r программное обеспечение *n* микрокомпьютера

8114 microcontrolled modem
  d mikroprozessorgesteuertes Modem *n*
  f modulateur-démodulateur *m* à commande par microprocesseur
  r модем m, управляемый микроконтроллером

8115 microcontroller
  d Mikrosteuergerät *n*
  f microcontrôleur *m*
  r устройство *n* микропрограммного управления

8116 microdeformation
  d Mikroverformung *f*

f microdéformation *f*
r микродеформация *f*

8117 microelectronic component
d mikroelektronisches Bauteil *n*
f composant *m* micro-électronique
r микроэлектронная компонента *f*

8118 microelectronics
d Mikroelektronik *f*
f microélectronique
r микроэлектроника *f*

8119 microelectronic sensor principle
d mikroelektronisches Sensorprinzip *n*
f principe *m* de senseur micro-électronique
r микроэлектронный сенсорный принцип *m*

8120 microfabrication
d Mikrofertigung *f*; Mikroherstellung *f*
f microproduction *f*
r микротехнология *f*; микрообработка *f*

8121 microinformatics
d Mikroinformatik *f*
f micro-informatique *f*
r микроинформатика *f*

8122 microinformatic tools
d mikroinformatische Werkzeuge *npl*
f outils *mpl* micro-informatiques
r микроинформационные инструменты *mpl*

8123 microinstruction
d Mikrobefehl *m*
f micro-instruction *f*
r микроинструкция *f*

8124 microinstruction memory
d Mikrobefehlsspeicher *m*
f mémoire *f* de micro-instruction
r память *f* микрокоманд

8125 micrologic
d Mikrologik *f*
f micrologique *f*
r микрологика *f*; логика *f* на микросхемах

8126 micrologic circuit
d Mikrologikschaltung *f*
f circuit *m* micrologique; montage *m* micrologique
r логическая микросхема *f*

8127 micromanipulator
d Mikromanipulator *m*
f micromanipulateur *m*
r микроманипулятор *m*

8128 microminiaturization
d Mikrominiaturisierung *f*
f microminiaturisation *f*
r микроминиатюризация *f*

8129 micromodule
d Mikromodul *m*
f micromodule *m*
r микромодуль *m*

8130 micromodule digital computer construction
d Digitalrechnerkonstruktion *f* in Mikromodulbauweise
f construction *f* de calculateurs numériques par micromodules
r микромодульная конструкция *f* вычислительных цифровых машин

8131 micromodule technique
d Mikromodultechnik *f*
f technique *f* à micromodules
r микромодульная техника *f*

8132 micro-mounting technique; micro-assemblage technique
d Mikromontagetechnik *f*
f technique *f* de micromontage; technique de microassemblage
r микромонтажная техника *f*

8133 microoperation
d Mikrooperation *f*
f micro-opération *f*
r микрооперация *f*

8134 microprocessing
d Mikroverarbeitung *f*
f microtraitement *m*
r микропрограммная обработка *f*

8135 microprocessor
d Mikroprozessor *m*
f microprocesseur *m*
r микропроцессор *m*

8136 microprocessor analyzer
d Mikroprozessor-Analysator *m*
f analyseur *m* de microprocesseur
r микропроцессорный анализатор *m*

8137 microprocessor application
d Mikroprozessoranwendung *f*
f application *f* de microprocesseurs
r применение *n* микропроцессоров

8138 microprocessor-based control
d mikroprozessorbestückte Steuerung *f*; Steuerung auf Mikroprozessorbasis

f   commande f à microprocesseur
r   управление n на основе микропроцесора

8139 microprocessor-based product
d   mikroprozessorbestücktes Produkt n
f   produit m à microprocesseur
r   изделие n на базе микропроцессора

8140 microprocessor card system
d   Mikroprozessorsteckeinheitensystem n;
    Mikroprozessorsystem n mit
    Leiterkartenmoduln
f   système m de cartes à microprocesseurs
r   система f на микропроцессорных платах

**8141 microprocessor-controlled fine
    manipulator**
d   mikroprozessorgesteuerter
    Feinmanipulator m
f   manipulateur m fin commandé par
    microprocesseur
r   точный манипулятор m с
    микропроцессорным управлением

**8142 microprocessor-controlled measuring
    technique**
d   mikroprozessorgesteuerte Messtechnik f
f   technique f de mesure commandée avec des
    microprocesseurs
r   измерительное оборудование n с
    сесорным управлением

8143 microprocessor-controlled terminal
d   mikroprozessorgesteuertes Terminal n
f   terminal m à commande par microprocesseur
r   терминал m с микропроцессорным
    управлением

8144 microprocessor control system
d   Mikroprozessorsteuersystem n
f   système m de commande à microprocesseur
r   микропроцессорная система f управления

8145 microprocessor design
d   Mikroprozessorentwurf m
f   dessin m de microprocesseur
r   проектирование n микропроцессора

8146 microprocessor development system
d   Mikroprozessor-Entwicklungssystem n
f   système m de développement de
    microprocesseur
r   система f проектирования
    микропроцессорных устройств

8147 microprocessor education system
d   Mikroprozessor-Ausbildungssystem n
f   système m d'enseignement à microprocesseur
r   микропроцессорная обучающая система f

8148 microprocessor engineering
d   Mikroprozessortechnik f
f   technique f des microprocesseurs
r   микропроцессорная техника f

8149 microprocessor function
d   Mikroprozessorfunktion f
f   fonction f de microprocesseur
r   функция f микропроцессора

8150 microprocessor hardware
d   Mikroprozessor-Hardware f
f   hardware m de microprocesseur
r   микропроцессорное оборудование n

8151 microprocessor net
d   Mikroprozessornetz n
f   réseau m de microprocesseurs
r   микропроцессорная сеть f

8152 microprocessor-piloted system
d   mikroprozessorgesteuertes System n
f   système m piloté par microprocesseur
r   система f с микропроцессорным
    управлением

8153 microprocessor simulator
d   Mikroprozessorsimulator m;
    Mikroprozessornachbildner m
f   simulateur m de microprocesseur
r   микропроцессорный имитатор m

8154 microprocessor software
d   Mikroprozessorsoftware f
f   software m de microprocesseur
r   программное обеспечение n
    микропроцессора

8155 microprocessor technology
d   Mikroprozessortechnologie f
f   technologie f de microprocesseur
r   микропроцессорная технология f

8156 microprogram control circuit
d   Mikroprogrammsteuerschaltkreis m;
    Steuerschaltkreis m für Mikroprogramm
f   circuit m de commande pour le
    microprogramme
r   схема f управления микропрограммы

**8157 microprogram-controlled automatic
    system**
d   mikroprogrammgesteuertes
    Automatiksystem n
f   système m d'automaticité commandé par
    microprogramme
r   автоматическая система f с управлением
    от микропрограммы

8158 microprogram control memory
   d Mikroprogramm-Steuerspeicher *m*
   f mémoire *f* de commande à microprogrammes
   r микропрограммная управляющая память *f*

8159 microprogram control operation
   d Mikroprogramm-Steueroperation *f*
   f opération *f* de commande d'un
      microprogramme
   r операция *f* микропрограммного
      управления

8160 microprogram control unit
   d Mikroprogrammsteuereinheit *f*
   f unité *f* de commande pour les
      microprogrammes
   r блок *m* контроля микропрограммы

8161 microprogram emulation
   d mikroprogrammierte Emulation *f*;
      Mikroprogrammemulation *f*
   f émulation *f* microprogrammée
   r микропрограммная эмуляция *f*

8162 microprogram logic
   d Mikroprogrammlogik *f*
   f logique *f* de microprogramme
   r микропрограммная логика *f*

8163 microprogrammability
   d Mikroprogrammierbarkeit *f*
   f microprogrammabilité *f*
   r микропрограммируемость *f*

8164 microprogrammed control
   d Mikroprogrammsteuerung *f*
   f commande *f* microprogrammée
   r микропрограммное управление *n*

8165 microprogramming
   d Mikroprogrammierung *f*
   f microprogrammation *f*
   r микропрограммирование *n*

8166 microprogramming techniques
   d Mikroprogrammiertechniken *fpl*
   f techniques *fpl* de microprogrammation
   r методы *mpl* микропрограммирования

8167 microprogram selection
   d Mikroprogrammauswahl *f*
   f selection *f* d'un microprogramme
   r выбор *m* микропрограммы

8168 microsecond
   d Mikrosekunde *f*
   f microseconde *f*
   r микросекунда *f*

8169 microswitch
   d Miniaturschalter *m*
   f microrupteur *m*
   r микровыключатель *m*

8170 microsynchronization
   d Mikrosynchronisation *f*
   f microsynchronisation *f*
   r микросинхронизация *f*

8171 microwave device
   d Dezimeterwellengerät *n*
   f appareil *m* à micro-ondes
   r микроволновое устройство *n*

8172 microwave integrated circuit
   d integrierte Mikrowellenschaltung *f*
   f circuit *m* intégré de microondes
   r микроволновая интегральная схема *f*

8173 microwave refractometer
   d Mikrowellenrefraktometer *n*
   f réfractomètre *m* d'ondes d'hyperfréquences
   r микроволновой рефрактометр *m*

8174 microwave spectroscopy
   d Mikrowellenspektroskopie *f*
   f spectroscopie *f* en microondes
   r микроволновая спектроскопия *f*

8175 microwave transmission
   d Mikrowellenübertragung *f*
   f transmission *f* des microondes
   r передача *f* микроволн

* **millimicrosecond** → 8518

8176 millivolt-signal group converter
   d Gruppenumformer *m* für Millivoltsignale
   f convertisseur *m* de groupe pour les signaux
      de l'ordre de millivolts
   r групповой преобразователь *m* для
      милливольтовых сигналов

8177 mimic diagram panel
   d Blinddiagrammpaneel *n*
   f panneau *m* à schéma mnémonique
   r панель *f* с мнемонической схемой

* **mingle** *v* → 1831

* **miniassemblage technique** → 8189

8178 miniature laser
   d Miniaturlaser *m*
   f minilaser *m*
   r миниатюрный лазер *m*

8179 miniature manipulator
   d Kleinmanipulator *m*
   f manipulateur *m* miniature
   r малогабаритный манипулятор *m*

8180 miniaturization
   d Miniaturisierung *f*
   f miniaturisation *f*
   r миниатюризация *f*

8181 miniaturized automation
   d miniaturisierte Automation *f*
   f automa[tisa]tion *f* miniaturisée
   r миниатюрная автоматика *f*

8182 minimal degree of freedom number
   d minimale Freiheitsgradezahl *f*
   f nombre *m* de degré de liberté minimal
   r минимальное число *n* степеней свободы

8183 minimax strategy
   d Minimaxverfahren *n*
   f procédé *m* minimax
   r минимаксная стратегия *f*

8184 minimization; minimizing
   d Minim[is]ierung *f*; Minimalisierung *f*
   f minim[al]isation *f*
   r минимизация *f*

8185 minimization of integral squared error
   d Minimisierung *f* des mittleren Fehlerquadrats
   f minimisation *f* de l'intégrale d'erreurs moyenne
   r минимизация *f* среднеквадратичной ошибки

8186 minimization of tear variables
   d Schnittzahlminimierung *f*
   f minimalisation *f* des coupes
   r минимизация *f* разрывающего множества

8187 minimize *v*
   d minimieren
   f minimiser
   r минимизировать

   * minimizing → 8184

8188 minimizing method
   d Minimisierungsmethode *f*
   f méthode *f* de minimisation
   r метод *m* минимизации

8189 minimounting technique; miniassemblage technique
   d Minimontagetechnik *f*
   f technique *f* de minimontage; technique de miniassemblage

   r минимонтажная техника *f*

8190 minimum
   d Minimum *n*
   f minimum *m*
   r минимум *m*

8191 minimum access programming
   d optimale Programmierung *f*
   f programmation *f* optimale
   r оптимальное кодирование *n*; оптимальное программирование *n*

8192 minimum access routine
   d Bestzeitprogramm *n*
   f programme *m* optimal
   r программа *f* с минимальным временем обращения

8193 minimum capacity; minimum charge
   d Mindestbelastung *f*
   f charge *f* minimum
   r минимальная нагрузка *f*

   * minimum charge → 8193

8194 minimum configuration
   d Minimalausrüstung *f*
   f configuration *f* minimale
   r минимальная конфигурация *f*

8195 minimum detectable signal
   d minimales feststellbares Signal *n*
   f signal *m* minimum détectable
   r минимальный обнаруживаемый сигнал *m*

8196 minimum deviation
   d minimale Abweichung *f*; Minimalabweichung *f*
   f écart *m* minimum
   r минимальное отклонение *n*

8197 minimum distance
   d Mindestabstand *m*
   f distance *f* minimale
   r минимальное расстояние *n*

8198 minimum efficiency
   d Mindestleistung *f*
   f puissance *f* minimum
   r минимальная мощность *f*

8199 minimum error controller
   d optimaler Regler *m*; Regler mit bestmöglicher Ausregelung
   f régulateur *m* optimal; régulateur avec erreur minimale
   r регулирующее устройство *n* с минимальной ошибкой

8200 **minimum ignition energy**
  d Mindestzündenergie *f*
  f énergie *f* minimum d'inflammation
  r минимальная энергия *f* воспламенения

8201 **minimum phase-shift system**
  d Phasen-Minimum-System *n*;
      Minimalphasenverschiebungssystem *n*;
      Kleinstphasenverschiebungsanlage *f*
  f système *m* à déphasage minimal; système à
      minimum de phase
  r система *f* с минимальным сдвигом фазы

8202 **minimum point**
  d Mindestwertpunkt *m*
  f point *m* de minimum
  r точка *f* минимума

8203 **minimum stability**
  d Mindeststabilität *f*
  f stabilité *f* minimum
  r минимальная устойчивость *f*

8204 **minimum temperature difference**
  d minimale Temperaturdifferenz *f*
  f différence *f* de température minimale
  r минимальная разность *f* температур

8205 **minimum time**
  d Mindestzeit *f*
  f temps *m* minimum
  r минимальное время *n*

8206 **minimum time to capture**
  d minimale Einfangzeit *f*
  f temps *m* de capture minimum
  r минимальное время *n* захвата

8207 **minimum voltage**
  d Mindestspannung *f*
  f tension *f* minimum
  r минимальное напряжение *n*

8208 **minority**
  d Minorität *f*
  f minorité *f*
  r подчиненность *f*

8209 **misadjustment**
  d Fehleinstellung *f*
  f ajustage *m* faux; réglage *m* incorrect
  r неправильная настройка *f*

8210 **misaligned joint**
  d schlecht ausgerichtete Verbindung *f*
  f joint *m* mal aligné
  r плохо юстированное соединение *n*

8211 **mismatch**
  d Fehlanpassung *f*
  f adaptation *f* incorrecte
  r рассогласование *n*; несовпадение *n*

8212 **mismatch loss**
  d Fehlanpassungsverlust *m*
  f perte *f* par désadaptation
  r потеря *f* рассогласования

8213 **misphasing; skew**
  d Phasenabgleichfehler *m*
  f déphasage *m*
  r расфазировка *f*

8214 **misuse failure**
  d Ausfall *m* wegen des unsachgemäßen
      Einsatzes
  f défaut *m* par inadvertance
  r отказ *m* из-за неправильного обращения

* **mix** *v* → 1831

8215 **mixed control**
  d gemischte Steuerung *f*
  f commande *f* mixte
  r комбинированное управление *n*

8216 **mixed control system**
  d kombiniertes Regelungssystem *n*
  f système *m* combiné de réglage
  r комбинированная система *f*
      регулирования

8217 **mixed-flow compressor**
  d Axial-Radial-Verdichter *m*;
      Diagonalverdichter *m*
  f compresseur *m* à écoulement mixte
  r центробежно-осевой компрессор *m*

8218 **mixed servomechanism**
  d kombinierter Servomechanismus *m*
  f système *m* de poursuite combiné
  r комбинированный сервомеханизм *m*

8219 **mixer**
  d Mischer *m*
  f mélangeur *m*
  r смеситель *m*

8220 **mixing circuit**
  d Mischkreis *m*
  f circuit *m* mélangeur
  r смесительная схема *f*

8221 **mixing of electric analog signals**
  d Mischen *n* von elektrischen Analogsignalen

f mixage *m* de signaux électriques analogiques
r смешивание *n* электрических аналоговых
   сигналов

8222 mixing stage
d Mischstufe *f*
f étage *m* mélangeur
r смесительный каскад *m*

8223 mixture analyzer
d Mischungsanalysator *m*
f analyseur *m* du mélange
r анализатор *m* смеси

8224 mnemonic
d Mnemonik *n*; mnemonische Abkürzung *f*
f mnémonique *f*
r мнемоника *f*

8225 mnemonic code
d Mnemokode *m*; mnemonischer Kode *m*;
   mnemonischer Befehlskode *m*
f code *m* mnémonique
r мнемонический код *m*

8226 mnemoscheme
d Mnemonikplan *m*
f mnémoschème *m*
r мнемосхема *f*

8227 mobile acquisition sensor
d beweglicher Erfassungssensor *m*
f senseur *m* d'acquisition mobile
r мобильный сенсор *m* для распознавания

8228 mobile device
d ortsveränderliches Gerät *n*
f appareil *m* mobile
r переносный прибор *m*

8229 mobile manipulator
d mobiler Manipulator *m*
f manipulateur *m* mobile
r подвижный манипулятор *m*

8230 mobility degree
d Beweglichkeitsgrad *m*
f degré *m* de mobilité
r степень *f* подвижности

8231 mode
d Modus *m*; Betriebsweise *f*
f mode *m*; régime *m*
r режим *m*; способ *m*

8232 mode change
d Moduswechsel *m*; Betriebsartenänderung *f*
f changement *m* de mode

r смена *f* режима

8233 model
d Modell *n*
f modèle *m*
r модель *f*

8234 model design
d Modellprojektierung *f*
f projection *f* par modèle
r модельное проектирование *n*

8235 model development
d Modellentwicklung *f*
f développement *m* de modèle
r разработка *f* модели

8236 model experiment; model test
d Modellexperiment *n*; Modellversuch *m*
f expérience *f* sur maquette; essai *m* sur
   modèle
r эксперимент *m* на модели

8237 modeling of the transfer lag
d Totzeitmodellierung *f*
f modelage *m* du retard de transfert
r моделирование *n* запаздывания передачи

8238 model of a communication process
d Modell *n* eines
   Nachrichtenübertragungsprozesses
f modèle *m* d'un procédé de la transmission de
   communications; modèle d'un procédé de la
   transmission d'informations
r модель *f* процесса коммуникации

8239 model of component parts
d Bauelementemodell *n*
f modèle *m* des composants
r модель *f* конструктивных элементов

8240 model parameter
d Modellparameter *m*
f paramètre *m* de modèle
r параметр *m* модели

8241 model recognition
d Modellerkennung *f*
f reconnaissance *f* de modèle
r распознавание *n* модели; идентификация *f*
   модели

8242 model scale
d Modellmaßstab *m*
f échelle *f* de modèle
r модельный масштаб *m*

8243 model simplification
d Modellvereinfachung *f*

f simplification f de modèle
r упрощение n модели

* **model test** → 8236

8244 **model theory**
d Modelltheorie f
f théorie f de modèle
r модельная теория f

8245 **modem; modulator-demodulator**
d Modem n; Modulator-Demodulator m
f modem m; modulateur-démodulateur m
r модем m; модулятор-демодулятор m

8246 **modem control**
d Modem-Steuerung f
f commande f de modem
r управление n модемом

8247 **mode of application**
d Anwendungsweise f
f mode f d'application
r способ m применения

8248 **mode of power supply connection**
d Art f des Kraftanschlusses
f espèce f de la connexion de force
r тип m подключаемой энергии

8249 **mode of processing**
d Verarbeitungsart f; Verarbeitungsform f
f mode m de traitement
r режим m обработки

8250 **modern information processing**
d moderne Informationsverarbeitung f
f traitement m moderne des informations
r современные методы mpl обработки
информации

8251 **modification**
d Modifikation f; Modifizierung f
f modification f; variété f
r модификация f

8252 **modification frequency**
d Änderungshäufigkeit f
f fréquence f de modification
r частота f изменений

8253 **modification of control program**
d Modifikation f des Steuerprogramms
f modification f du programme de commande
r модификация f программы управления

8254 **modified binary code**
d zyklisch-binärer Kode m

f code m binaire-cyclique
r модифицированный двоичный код m

8255 **modifier**
d Umsteuergröße f
f modificateur m
r модификатор m

8256 **modifier register**
d Modifikatorregister n
f registre m d'index
r регистр-модификатор m

8257 **modular**
d modular; bausteinartig
f modulaire
r модульный; блочный

8258 **modular assembly machine**
d modulare Montagemaschine f
f machine f modulaire d'assemblage
r сборочная машина f модульного типа

8259 **modular automaton**
d Automat m in Modulbauweise
f automate m modulaire
r модульный автомат m

8260 **modular check[ing] system**
d modulares Kontrollsystem n
f système m modulaire de contrôle
r модульная контрольная система f

* **modular check system** → 8260

8261 **modular concept**
d Modulbauweise f
f construction f modulaire
r концепция f составления устройств из
модулей

8262 **modular connecting technique**
d modulare Verbindungstechnik f
f technique f de connexion modulaire
r модульная техника f соединения

8263 **modular construction**
d modulare Bauweise f
f construction f modulaire
r модульная конструкция f

8264 **modular control system**
d modulares Steuerungssystem n
f système m de commande modulaire
r модульная система f управления

8265 **modular design**
d modularer Entwurf m

f   conception *f* modulaire
r   модульное построение *n*

8266   modular expansion
d   modulare Erweiterung *f*; Ausbau *m* nach dem Baukastenprinzip
f   élargissement *m* modulaire
r   модульное расширение *n*

8267   modular handling system
d   modulares Handhabungsgerät *n*
f   dispositif *m* de manutention modulaire
r   манипуляционная система *f* модульного типа

8268   modularity
d   Modularität *f*
f   modularité *f*
r   модульность *f*

8269   modularization
d   Modularisierung *f*
f   modularisation *f*
r   расчленение *n* на модули; модуляризация *f*

8270   modular memory system
d   modulares Speichersystem *n*
f   système *m* d'enregistrement modulaire
r   модульная запоминающая система *f*

8271   modular principle
d   Modulprinzip *n*; Baukastenprinzip *n*
f   conception *f* bloc-éléments; principe *m* modulaire
r   модульный принцип *m*

8272   modular programmable automaton
d   modularer programmierbarer Automat *m*
f   automate *m* programmable modulaire
r   модульный программируемый автомат *m*

8273   modular programmable system
d   modulares programmierbares System *n*
f   système *m* de programmation modulaire
r   модульная программируемая система *f*

8274   modular programming
d   modulare Programmierung *f*; Programmierung auf Modulbasis
f   programmation *f* modulaire
r   модульное программирование *n*

8275   modular structure
d   modulare Struktur *f*
f   structure *f* modulaire
r   модульная структура *f*

8276   modular system
d   Modularsystem *n*; modulares System *n*
f   système *m* modulaire
r   модульная система *f*

8277   modular system of automatic control
d   Modulsystem *n* automatischer Regelung
f   système *m* modulaire de réglage automatique
r   блочная система *f* автоматического регулирования

8278   modular tactile gripper system
d   modulares taktiles Greifersystem *n*
f   système *m* de grappin modulaire tactile
r   тактильная система *f* захвата модульного типа

8279   modular tactile gripper unit
d   modulare taktile Greifereinheit *f*
f   unité *f* de grappin modulaire tactile
r   тактильное захватное устройство *n* модульного типа

8280   modular tactile sensor system
d   modulares taktiles Sensorsystem *n*
f   système *m* de senseur modulaire tactile
r   тактильная сенсорная система *f* модульного типа

8281   modulate *v*
d   modulieren
f   moduler
r   модулировать

8282   modulated carrier channel
d   modulierter Trägerstromkanal *m*; modulierter Trägerfrequenzkanal *m*
f   canal *m* porteur modulé
r   канал *m* с модулированной несущей частотой

8283   modulated monochromatic light
d   moduliertes monochromatisches Licht *n*
f   lumière *f* monochromatique modulée
r   модулированный монохроматический свет *m*

8284   modulating signal
d   moduliertes Signal *n*; modulierte Größe *f*
f   signal *m* modulé; grandeur *f* modulante
r   модулированный сигнал *m*

8285   modulation
d   Modulation *f*
f   modulation *f*
r   модуляция *f*

8286   modulation capability
d   Modulierbarkeit *f*

f aptitude *f* de modulation
r модуляционная способность *f*

8287 modulation efficiency
d Modulationswirkunsgrad *m*
f efficacité *f* de modulation
r эффективность *f* модуляции

8288 modulation factor
d Modulationsgrad *m*
f taux *m* de modulation
r коэффициент *m* модуляции

8289 modulation monitor
d Modulationskontrollgerät *n*
f modulomètre *m*; contrôleur *m* de modulation
r устройство *n* для контроля модуляции

8290 modulation rate
d Tastgeschwindigkeit *f*
f rapport *m* d'impulsions; vitesse *f* de
modulation
r частота *f* модуляции

8291 modulation response
d Modulationsantwort *f*;
Modulationskennlinie *f*
f réponse *f* de modulation
r модуляционная характеристика *f*

**8292 modulation technique for data
transmission**
d Modulationsverfahren *n* für die
Datenübertragung
f méthode *f* de modulation pour la
transmission de données
r способ *m* модуляции для передачи данных

8293 modulation transfer function
d Modulationsübertragungsfunktion *f*
f fonction *f* de transfert de modulation
r функция *f* передачи модуляции

8294 modulator
d Modulator *m*
f modulateur *m*
r модулятор *m*

8295 modulator control signal
d Modulatorsteuersignal *n*
f signal-modulateur *m*; signal *m* de commande
de modulation
r сигнал m, управляющий модулятором

* modulator-demodulator → 8245

8296 module
d Baustein *m*; Modul *m*

f module *m*
r модуль *m*

8297 module map
d Modulübersicht *f*
f schéma *m* de module
r схема *f* модуля

8298 module supervision
d Modulüberwachung *f*
f monitorage *m* de module
r наблюдение *n* модуля

8299 module system technology
d Modulsystemtechnologie *f*
f technologie *f* des systèmes modulaires
r технология *f* модульной системы

8300 modulo-n check
d Modulo-n-Prüfung *f*; Modulo-n-Kontrolle *f*
f contrôle *m* modulo n; essai *m* modulo n
r контроль *m* по модулю

8301 modulo operation
d Modulo-Operation *f*
f opération *f* en modulo
r операция *f* по модулю

* modulo 2 sum gate → 5260

8302 modulus of resistance
d Festigkeitsziffer *f*
f module *m* de résistance
r модуль *m* сопротивлений

8303 moistening plant
d Befeuchtungsanlage *f*
f humidificateur *m*; installation *f* de
l'humectation
r увлажнительная установка *f*

8304 moisture content controller
d Feuchtigkeitsregler *m*
f régulateur *m* d'humidité
r регулятор *m* влажности

**8305 moisture measurement by infrared
method**
d Feuchtigkeitsmessung *f* mit der
Infrarotmethode
f mesure *f* de l'humidité par la méthode
infrarouge
r измерение *n* влажности инфракрасным
методом

8306 moisture value
d Feuchtigkeitswert *m*

f valeur *f* d'humidité
r величина *f* влажности

8307 molecular electronics
d Molek[ularelek]tronik *f*
f électronique *f* moléculaire
r молекулярная электроника *f*

8308 molecular engineering technique
d Verfahren *n* der Molekulartechnik
f procédé *m* de la technique moléculaire
r молекулярная техника *f*

8309 molecular generator
d molekularer Generator *m*
f générateur *m* moléculaire
r молекулярный генератор *m*

8310 moment
d Moment *n*
f moment *m*
r момент *m*

8311 momentary balance
d Zeitpunktbilanz *f*; Augenblicksbilanz *f*
f bilan *m* momentané
r баланс *m* к определенному моменту
времени

8312 momentary disappearance of line voltage
d vorübergehender Spannungsausfall *m*
f disparition *f* fugitive de tension
r кратковременное прекращение *n* подачи
напряжения сети

8313 momentary disturbance
d augenblickliche Störung *f*; momentane
Störung; kurzzeitige Störung
f perturbation *f* momentanée; perturbation
instantanée
r мгновенное возмущение *n*

8314 momentary phase meter
d Momentanphasenmesser *m*
f phasemètre *m* instantané
r быстрадействующий фазометр *m*

8315 momentary value
d Augenblickswert *m*
f valeur *f* instantanée
r мгновенное значение *n*

8316 momentless relay servosystem
d momentloses Relaisfolgesystem *n*;
momentfreies Relaisfolgesystem
f système *m* asservi à relais sans couple
r безмоментная релейная следящая
система *f*

8317 moment of inertia
d Trägheitsmoment *n*
f moment *m* d'inertie
r момент *m* инерции

8318 moment of load
d Belastungsmoment *n*
f moment *m* de charge; couple *m* de charge
r момент *m* нагрузки

8319 moment of motion
d Drehmoment *n*
f moment *m* cinétique
r кинетический момент *m*

8320 moment of random function
d Moment *n* der Zufallsfunktion
f moment *m* de fonction aléatoire
r момент *m* случайной функции

* **monitor → 12804**

8321 monitor *v*; verify *v*
d überwachen; prüfen
f piloter; contrôler; vérifier
r контролировать; управлять; проверять

8322 monitor checking routine
d Überwachungsprogramm *n*
f programme *m* moniteur; routine *f* de contrôle
de séquence
r контрольная программа *f*

8323 monitor control
d Monitorsteuerung *f*
f commande *f* de moniteur
r управление *n* работой при помощи
монитора

8324 monitoring; verification; supervision
d Kontrolle *f*; Überwachung *f*; Verifikation *f*;
Supervision *f*
f contrôle *m*; surveillance *f*; vérification *f*;
supervision *f*
r контроль *m*; наблюдение *n*; проверка *f*

8325 monitoring automation
d Überwachungsautomatisierung *f*
f automatisation *f* de surveillance
r автоматизация *f* контроля

* **monitoring equipment → 10192**

8326 monitoring function
d Überwachungsfunktion *f*
f fonction *f* de surveillance
r функция *f* контроля

* monitoring instrumentation → 10192

8327 monitoring machine with scanning
 d Betriebskontrolleinrichtung *f* mit
    Datenabtastung
 f dispositif *m* de contrôle multiple par
    balayage
 r контрольное устройство *n* со
    сканированием

8328 monitoring of manufacturing
 d Fertigungsüberwachung *f*
 f surveillance *f* de fabrication
 r контроль *m* за производством

8329 monitoring of sequential handling
 d Überwachung *f* von Handhabungssequenzen
 f surveillance *f* de manutentions séquentielles
 r контроль *m* отработки манипуляционных
    действий

8330 monitoring procedure
 d Monitoring-Verfahren *n*
 f procédure *f* de monitorage
 r процедура *f* диспетчерского управления

8331 monitor interface
 d Monitorinterface *n*
 f interface *f* moniteur
 r интерфейс *m* монитора

8332 monitor system
 d Überwachungssystem *n*
 f système *m* de moniteur
 r система *f* надзора

8333 monoenergetic manipulator system
 d monoenergetisches Manipulatorsystem *n*
 f système *m* de manipulateur monoénergétique
 r моноэнергетическая система *f*
    манипулятора

8334 monolithic *adj*
 d monolithisch
 f monolithique
 r монолитный

8335 monolithic circuit
 d monolithischer Schaltkreis *m*; monolithische
    Schaltung *f*
 f circuit *m* monolithique
 r монолитная схема *f*

8336 monolithic clock generator
 d monolithischer Taktgenerator *m*
 f rythmeur *m* monolithique
 r микросхема *f* тактового генератора

8337 monolithic converter
 d monolithischer Umsetzer *m*
 f convertisseur *m* monolithique
 r микросхема *f* преобразователя

8338 monolithic structure
 d monolithische Struktur *f*; monolithischer
    Aufbau *m*
 f structure *f* monolithique
 r монолитная структура *f*

8339 monophase system
 d Einphasensystem *n*
 f système *m* à phase unique
 r однофазная система *f*

8340 monoprocessor system
 d Monoprozessorsystem *n*
 f système *m* monoprocesseur
 r монопроцессорная система *f*

8341 monopulse lidar
 d Einpulslidar *m*; Monopulslidar *m*
 f lidar *m* à impulsion unique
 r моноимпульсный лазерный локатор *m*

8342 monostable circuit
 d monostabile Schaltung *f*
 f montage *m* monostable; circuit *m* monostable
 r схема *f* с одним устойчивым состоянием

**8343 monostable multivibrator; one-shot
    multivibrator**
 d monostabiler Multivibrator *m*
 f multivibrateur *m* monostable
 r мультивибратор *m* с одним устойчивым
    положением

8344 monotonic function
 d monotone Funktion *f*
 f fonction *f* monotone
 r монотонная функция *f*

8345 monotonous
 d monoton
 f monotone
 r монотонный

8346 monotonous process
 d monotoner Vorgang *m*
 f processus *m* monotone
 r монотонный процесс *m*

8347 monotonous transient response
 d monotoner Übergangsprozess *m*
 f régime *m* transitoire monotone
 r монотонная характеристика *f*
    неустановившегося режима

8348 **Monte-Carlo modelling**
  d Monte-Carlo-Simulation *f*
  f simulation *f* Monte-Carlo
  r моделирование *n* методом Монте-Карло

8349 **Monte-Carlo techniques**
  d Monte-Carlo-Verfahren *n*
  f méthode *f* de Monte-Carlo
  r метод *m* Монте-Карло

8350 **motion**
  d Bewegung *f*
  f mouvement *m*
  r движение *n*

8351 **motional impedance**
  d kinetischer Scheinwiderstand *m*
  f impédance *f* cinétique
  r кинетический импеданс *m*

8352 **motion equation**
  d Bewegungsgleichung *f*
  f équation *f* de mouvement
  r уравнение *n* движения

8353 **motor**
  d Motor *m*
  f moteur *m*
  r мотор *m*; двигатель *m*

8354 **motor board**
  d Laufwerkplatte *f*
  f plaque *f* de mécanisme
  r панель *f* привода

8355 **motor compensator with PID regulator**
  d Motorkompensator *m* mit PID-Regler
  f compensateur *m* à moteur avec régulateur PID
  r компенсатор *m* двигателя с регулятором ПИД

8356 **motor control; engine control**
  d Motorführung *f*; Kraftmaschinensteuerung *f*
  f pilotage *m* de moteur; commande *f* de moteur
  r управление *n* двигателем

8357 **motor control assembly**
  d Motorsteuerungseinheit *f*
  f ensemble *m* de commande de moteur
  r установка *f* для управления двигателями

8358 **motor element**
  d Stellantrieb *m*
  f organe *m* moteur
  r блок *m* двигателя

8359 **motor pulse control**
  d Motorimpulssteuerung *f*
  f commande *f* impulsionnelle de moteur
  r импульсное управление *n* [электро]двигателем

8360 **motor speed control**
  d Motorgeschwindigkeitssteuerung *f*; Motordrehzahlregelung *f*
  f commande *f* de vitesse du moteur
  r регулирование *n* числа оборотов двигателя

8361 **motor time constant**
  d Motorzeitkonstante *f*
  f constante *f* de temps de moteur
  r постоянная *f* времени двигателя

8362 **mounting difficulty; assembly difficulty**
  d Montageschwierigkeit *f*
  f difficulté *f* d'assemblage; difficulté de montage
  r сложность *f* сборки

  * **mounting system** → 1032

8363 **mounting technique; assemblage technique**
  d Montagetechnik *f*
  f technique *f* d'assemblage; technique de montage
  r сборочная техника *f*

8364 **mounting technology; assembly technology**
  d Montagetechnologie *f*
  f technologie *f* de montage; technologie *f* d'assemblage
  r технология *f* сборки

8365 **move**
  d Zug *m*
  f coup *m*; marche *f*
  r ход *m*

8366 **movement control**
  d Bewegungssteuerung *f*
  f commande *f* de mouvement
  r управление *n* движением

8367 **movement guiding**
  d Bewegungsführung *f*
  f guidage *m* de mouvement
  r направление *n* движения

8368 **movement instruction**
  d Bewegungsbefehl *m*
  f instruction *f* de mouvement
  r команда *f* движения

8369 **movement limitation**
 d Bewegungsbegrenzung *f*
 f limitation *f* de mouvement
 r ограничение *n* движения

8370 **movement of effectors**
 d Effektorenbewegung *f*
 f mouvement *m* d'effecteurs
 r эффекторное движение *n*

8371 **movement of element**
 d Bewegen *n* eines Bauteils
 f mouvement *m* d'élément
 r перемещение *n* элемента

8372 **movement stability**
 d Bewegungsstabilität *f*
 f stabilité *f* de mouvement
 r устойчивость *f* движения

8373 **movement strategy**
 d Bewegungsstrategie *f*
 f stratégie *f* de mouvement
 r стратегия *f* движения

8374 **moving-iron voltage regulator**
 d Dreheisenspannungsregler *m*
 f régulateur *m* de tension à fer plongeant
 r электромагнитный регулятор *m*
      напряжения

8375 **moving system**
 d bewegliches System *n*
 f système *m* mobile
 r подвижная система *f*

8376 **multiaccess**
 d Vielfachzugriff *m*
 f accès *m* multiple
 r многократный доступ *m*

8377 **multiaccess online system**
 d Mehrfachzugriff-Online-System *n*
 f système *m* online à accès multiple
 r система *f* многократного доступа в
      реальном масштабе времени

8378 **multiaddress code**
 d Mehradresskode *m*
 f code *m* à adresses multiples
 r многоадресный код *m*

8379 **multiaddress instruction**
 d Mehradressbefehl *m*
 f instruction *f* à adresses multiples
 r многоадресная инструкция *f*

8380 **multianalysis**

 d Vielfachanalyse *f*
 f analyse *f* multiple
 r многосторонный анализ *m*

8381 **multiaxis path control**
 d Mehrachsenbahnsteuerung *f*
 f commande *f* de chemin multiaxes
 r многокоординатное контурное
      управление *n*

8382 **multibeam oscilloscope**
 d Mehrstrahloszilloskop *n*
 f oscilloscope *m* à plusieurs faisceaux
 r многолучевой осциллоскоп *m*

*   **multibeam source** → 8442

8383 **multicapacity control system**
 d Mehrkapazitätsregelsystem *n*
 f système *m* de commande à capacités
      multiples
 r многоёмкостная система *f* регулирования

8384 **multicascade servomechanism**
 d mehrstufiger Servomechanismus *m*
 f servomécanisme *m* multi-cascade
 r многокаскадный сервомеханизм *m*

8385 **multichannel amplifier**
 d Mehrkanalverstärker *m*
 f amplificateur *m* à plusieurs voies
 r многоканальный усилитель *m*

8386 **multichannel communication system**
 d Mehrkanalkommunikationssystem *n*
 f système *m* de communication à canaux
      multiples
 r многоканальная система *f* доступа

8387 **multichannel controller**
 d Mehrkanalregler *m*
 f régulateur *m* à canaux multiples
 r многоканальный регулятор *m*

8388 **multichannel gamma-ray spectrometer**
 d Vielkanal-Gammaspektrometer *n*
 f spectromètre *m* gamma multicanal
 r многоканальный гамма-спектрометр *m*

8389 **multichannel measuring point amplifier**
 d Mehrkanalmessverstärker *m*
 f amplificateur *m* de mesure à canaux
      multiples
 r многоканальный измерительный
      усилитель *m*

8390 **multichannel monitoring system**
 d Vielkanal-Überwachungssystem *n*;
      Vielkanal-Überwachungsanlage *f*

f système *m* de surveillance multicanal
r многоканальная контрольно-
измерительная система *f*

8391 multichannel scattering
  d Mehrkanalstreuung *f*; Vielkanalstreuung *f*
  f diffusion *f* multivoie; diffusion sur plusieurs
  voies
  r многоканальное рассеяние *n*

8392 multicircuit control
  d Regelung *f* im vermaschten Regelkreis
  f réglage *m* à boucles multiples
  r многоконтурное управление *n*

8393 multicomputer system
  d Mehrrechnersystem *n*
  f système *m* à plusieurs calculateurs
  r мултикомпьютерная система *f*

8394 multicontroller
  d Mehrfachsteuereinheit *f*
  f unité *f* de commande multiple
  r многоконтурный контроллер *m*

8395 multicoordinate equipment
  d Mehrkoordinaten-Ausrüstung *f*
  f équipement *m* de coordonnées multiples
  r многокоординатное оборудование *n*

8396 multicoordinate manipulator
  d Mehrkoordinaten-Manipulator *m*
  f manipulateur *m* de coordonnées multiples
  r многокоординатный манипулятор *m*

8397 multidimensional
  d mehrdimensional
  f multidimensionnel
  r многомерный

*   multidimensional analysis → 8436

8398 multidimensional distribution
  d multidimensionale Verteilung *f*;
  vielfachdimensionale Verteilung
  f répartition *f* multidimensionnelle
  r многомерное распределение *n*

8399 multidimensional evaluation system
  d mehrdimensionales Bewertungssystem *n*
  f système *m* d'évaluation à plusieures
  dimensions
  r система *f* многокритериальной оценки

8400 multidimensional maximization problem
  d mehrdimensionales Optimierungsproblem *n*;
  mehrdimensionales Maximierungsproblem *n*
  f problème *m* d'optimisation

multidimensionnel
  r задача *f* многомерной оптимизации

8401 multidimensional system
  d Mehrgrößenregelungssystem *n*
  f système *m* à plusieurs variables
  r многосвязанная система *f*

8402 multielement activation analysis
  d Multielement-Aktivierungsanalyse *f*
  f analyse *f* par activation de plusieurs éléments
  r многоэлементный активационный
  анализ *m*

8403 multielement control
  d vermaschte Regelung *f*
  f régulation *f* multiple
  r мултиэлементное регулирование *n*

8404 multielement detector
  d Mehrelementdetektor *m*
  f détecteur *m* multiéléments
  r многоэлементный детектор *m*

8405 multielement separation
  d Multielementtrennung *f*; simultane
  Trennung *f*
  f séparation *f* de plusieurs éléments
  r многоэлементное разделение *n*

8406 multifunction integrated circuit
  d integrierte Schaltung *f* mit mehrfachen
  Funktionen
  f circuit *m* intégré avec fonctions multiples
  r многофункциональная интегральная
  схема *f*

8407 multifunction relay
  d Multifunktionsrelais *n*
  f relais *m* à fonction multiple
  r мултифункциональное реле *n*

8408 multifunction unit
  d Mehrzweckeinheit *f*
  f unité *f* à fonctions multiples
  r многофункциональное устройство *n*

8409 multigroup approximation
  d Mehrgruppenapproximation *f*;
  Multigruppennäherung *f*
  f approximation *f* multigroupe
  r многогрупповая аппроксимация *f*

8410 multigroup model
  d Mehrgruppenmodell *n*;
  Multigruppenmodell *n*
  f modèle *m* multigroupe
  r многогрупповая модель *f*

8411 multigroup theory
d Multigruppentheorie *f*; Mehgruppentheorie *f*
f théorie *f* à plusieurs groupes; théorie multigroupe
r многогрупповая теория *f*

8412 multiinput controller
d Regler *m* mit mehrfachem Eingang
f régulateur *m* à entrée multiple
r многовходовый регулятор *m*

8413 multilayer
d Mehrschicht-; mehrschichtig; Mehrlagen-
f multicouche
r многослойный

* multilayer coupler → 8414

8414 multilayer[ed] coupler
d Mehrschichtkoppler *m*
f coupleur *m* multicouche
r многослойный соединитель *m*

8415 multilayer[ed] structure
d Mehrschichtstruktur *f*
f structure *f* multicouche
r многослойная структура *f*

* multilayer structure → 8415

8416 multilayer interference filter
d Mehrschichteninterferenzfilter *n*
f filtre *m* interférentiel à couches multiples
r многослойный интерференционный фильтр *m*

8417 multilevel action; multistep action (US)
d Mehrpunktverhalten *n*
f action *f* à niveaux multiples
r многопозиционное [воз]действие *n*

8418 multilevel analysis
d Mehrniveauanalyse *f*
f analyse *f* multiniveaux
r многоуровневый анализ *m*

8419 multilevel communication system
d Mehrfachpegel-Fernmeldesystem *n*
f système *m* de communication à niveaux multiples
r многоуревневая система *f* связи

8420 multilevel controller; multiposition controller; multistep controller (US)
d Mehrpunktregler *m*
f régulateur *m* à action échelons multiples; régulateur à plusieurs paliers
r многопозиционный регулятор *m*

8421 multilevel interrupt
d mehrstufige Programmunterbrechung *f*
f interruption *f* de programme à plusieurs étages
r многоступенчатое прерывание *n* программы

* multilevel interrupts → 8457

8422 multilevel optimization
d Mehrebenenoptimierung *f*
f optimisation *f* à plusieurs niveaux
r многоуровневая оптимизация *f*

8423 multilevel storage
d mehrstufiger Speicher *m*
f mémoire *f* hiérarchisée
r многоуровневая память *f*

8424 multiline
d Mehrleitung-
f multiligne
r многолинейный

8425 multiline terminal
d Mehrleitungsterminal *n*
f terminal *m* multiligne
r многолинейный терминал *m*

8426 multiloop control system
d Mehrkreisregelungssystem *n*
f système *m* de réglage à plusieurs circuits
r многоконтурная система *f* регулирования

8427 multiloop digital control
d digitale Mehrgrößenregelung *f*
f réglage *m* digital à plusieurs paramètres
r многомерное цифровое регулирование *n*

8428 multiloop pulse system
d Mehrschleifenimpulssystem *n*
f système *m* impulsionnel à boucles multiples
r многоконтурная импульсная система *f*

8429 multiloop servosystem
d mehrschleifiges Folgesystem *n*
f servomécanisme *m* à boucles multiples
r многоконтурная следящая система *f*

8430 multiloop system
d Mehrschleifensystem *n*
f système *m* à boucles multiples
r многоконтурная система *f*

8431 multiloop transmission
d Übertragung *f* zu mehreren Stationen
f transmission *f* multiple
r передача *f* к нескольким станциям

* **multimeter** → 675

8432 multimicroprocessor system
d Multimikroprozessorsystem *n*;
Vielfachmikroprozessorsystem *n*
f système *m* à microprocesseurs multiples
r мултимикропроцессорная система *f*

8433 multimode behaviour
d Mehrmodenverhalten *n*
f allure *f* multimode
r многомодовый режим *m*

8434 multimode operation
d Mehrmodusbetrieb *m*; Betrieb *m* in mehreren
Arbeitsweisen
f régime *m* à mode multiple
r работа *f* в режиме мультиобработки

8435 multinomial; polynomial
d Polynom *n*
f polynôme *m*
r многочлен *m*; полином *m*

8436 multiparameter analysis;
multidimensional analysis
d Mehrparameteranalyse *f*;
Multiparameteranalyse *f*; mehrdimensionale
Analyse *f*
f analyse *f* multiparamétrique; analyse
multidimensionnelle
r многопараметрический анализ *m*;
многомерный анализ

8437 multiparameter control circuit
d Mehrparameterregelkreis *m*
f circuit *m* de réglage à plusieurs paramètres
r многопараметровый контур *m*
регулирования

8438 multipath transmission
d Mehrwegübertragung *f*
f transmission *f* par trajets multiples
r многоходовая передача *f*

8439 multiperiodic regime
d mehrperiodischer Betriebszustand *m*;
Mehrperiodenbetriebszustand *m*
f régime *m* polypériodique
r многопериодический режим *m*

8440 multiphase contactor
d Mehrphasentrenneinrichtung *f*;
Destillationsanlage *f*
f installation *f* de fractionnement
r установка *f* для фазового разделения

8441 multiple-action controller

d mehrfachwirkender Regler *m*
f régulateur *m* à action multiple
r многоточечный регулятор *m*

8442 multi[ple]beam source
d Mehrstrahlquelle *f*
f source *f* à faisceaux multiples; source
multifaisceaux
r многолучевой источник *m*

8443 multiple bound
d Mehrfachbindung *f*
f liason *f* multiple
r многократная связь *f*

8444 multiple-bus structure
d Multibusstruktur *f*; Vielfachbusstruktur *f*
f structure *f* à bus multiples
r многошинная структура *f*

8445 multiple-channel data asquisition device
d Mehrkanal-Datenerfassungsgerät *n*
f appareil *m* d'acquisition de données à canaux
multiples
r многоканальное устройство *n* сбора
данных

8446 multiple check
d Vielfachkontrolle *f*
f contrôle *m* multiple
r многократный контроль *m*

8447 multiple circuit
d Mehrfachkreis *m*
f circuit *m* multiple
r паралельная цепь *f*

8448 multiple coincidence
d Vielfachkoinzidenz *f*
f coïncidence *f* multiple
r многократное совпадение *n*

8449 multiple contact switch
d Vielfachkontaktschalter *m*
f interrupteur *m* à contacts multiples
r многоконтактный переключатель *m*

8450 multiple deflection
d mehrfache Ablenkung *f*
f déviation *f* multiple
r многократное отклонение *n*

8451 multiple-degree freedom system
d System *n* mit Mehrfachfreiheitsgraden
f système *m* à plusieurs degrés de liberté
r система *f* со многими степенями свободы

8452 multiple derivative
d mehrfache Ableitung *f*

f dérivation *f* multiple
r многократная деривация *f*

8453   multiple detector
   d Vielfachdetektor *m*
   f détecteur *m* multiple
   r многократный детектор *m*

8454   multiple effect evaporator
   d Vielkörperverdampfanlage *f*;
      Mehrfachverdampfer *m*
   f évaporateur *m* à effet multiple
   r многокорпусная выпарная установка *f*

8455   multiple excitation
   d Mehrfacherregung *f*
   f excitation *f* multiple
   r многократное возбуждение *n*

8456   multiple input selection logic
   d Mehrfacheingabeauswahllogik *f*
   f logique *f* de sélection à entrées multiples
   r многовходовая избирательная логика *f*

8457   multi[ple-]level interrupts
   d Mehrebenen-Interrupts *mpl*; Mehrebenen-
      Unterbrechungen *fpl*
   f interruptions *fpl* de niveaux multiples
   r многоуровневые прерывания *npl*

**8458   multiple-loop servomechanism; control of
      many-variable system**
   d mehrfacher Regelkreis *m*; mehrschleifiger
      Regelkreis; vermaschter Regelkreis
   f circuit *m* de réglage multiple; circuit de
      réglage à plusieurs paramètres interconnectés
   r многосвязанная система *f* регулирования;
      многоконтурная система регулирования;
      многомерная система регулирования

8459   multiple-plane interrupt control
   d Mehrebeneninterruptsteuerung *f*
   f commande *f* d'interruption à plans multiples
   r многоуровневый контроль *m* прерываний

8460   multiple precision
   d mehrfache Genauigkeit *f*
   f précision *f* multiple
   r многократно увеличенная точность *f*

8461   multi[ple-]processor operation
   d Multiprozessorarbeit *f*;
      Multiprozessorbetrieb *m*
   f régime *m* à multiprocesseurs
   r мултипроцесорная обработка *f*

8462   multiple-purpose plant
   d Mehrzweckanlage *f*

f installation *f* à plusieurs usages
r универсальная установка *f*

8463   multiple resonance
   d mehrfache Resonanz *f*; Vielfachresonanz *f*;
      Mehrfachresonanz *f*
   f résonance *f* multiple
   r многократный резонанс *m*

* **multiple scatter → 8464**

8464   multiple scatter[ing]
   d Vielfachsteuung *f*
   f diffusion *f* multiple
   r многократное рассеяние *n*

8465   multiple simultaneous optimization
   d simultane Mehrfachoptimierung *f*
   f optimisation *f* multiple simultanée
   r многократная одновременная
      оптимизация *f*

8466   multiple source
   d Mehrfachquelle *f*; Quelle *f* verschiedener
      Strahlungsarten
   f source *f* de plusieurs rayonnements
   r множественный источник *m*

8467   multiple-station machines
   d Mehrstationenmaschinen *fpl*
   f machines *fpl* à stations multiples
   r машины *fpl* с комбинированными
      технологическими позициями

8468   multiple-valued logic; many-valued logic
   d mehrwertige Logik *f*; vielwertige Logik
   f logique *f* de valeurs multiples; logique
      polyvalente
   r многозначная логика *f*

8469   multiplex
   d multiplex
   f multiplex
   r многократный; мультиплексный

8470   multiplex device
   d Multiplexeinrichtung *f*
   f dispositif *m* multiplex
   r мультиплексорное устройство *n*

* **multiplexed operation → 8475**

8471   multiplexer
   d Multiplexer *m*; Mehrfachkoppler *m*
   f multiplexeur *m*
   r мультиплексор *m*

8472   multiplexer channel control unit
   d Multiplexkanalsteuereinheit *f*

f unité *f* de commande à canal du type
  multiplex
r устройство *n* управления
  мультиплексным каналом

8473 multiplexing
  d Multiplexierung *f*; Mehrfachausnutzung *f*
  f multiplexage *m*
  r многократное использование *n*

8474 multiplexing equipment
  d Mehrkanalgerät *n*
  f équipement *m* de plusieurs canaux
  r многоканальное устройство *n*

8475 multiplex mode; multiplexed operation
  d Multiplexbetrieb *m*
  f régime *m* multiplex
  r мультиплексный режим *m*

8476 multiplex telemetering
  d Multiplexfernmessverfahren *n*
  f télémétrie *f* multiplex
  r многократная телеметрия *f*

8477 multiplication circuit
  d Multiplizierschaltung *f*
  f circuit *m* multiplicateur
  r умножающая цепь *f*

8478 multiplication process
  d Vervielfachungsprozess *m*
  f procédé *m* de multiplication
  r процесс *m* умножения

8479 multiplicative counter-current process
  d multiplikatives Gegenstromverfahren *n*
  f procédé *m* à contre-courant multiplicatif
  r мультипликативный противоточный
    способ *m*

8480 multiplicity
  d Multiplizität *f*; Vielfachheit *f*
  f multiplicité *f*
  r мултиплетность *f*

8481 multiplicity filter
  d Multiplizitätsfilter *n*
  f filtre *m* de multiplicité
  r фильтр *m* мультиплетности

8482 multiplied pulse
  d vermehrter Impuls *m*
  f impulsion *f* multipliée
  r умноженный импульс *m*

**8483 multiplier; multiplying device;
    multiplying unit**

  d Multiplikator *m*; Multiplizierer *m*;
    Multipliziereinheit *f*;
    Multipliziereinrichtung *f*
  f multiplicateur *m*; dispositif *m* multiplicateur
  r умножитель *m*; множительное
    устройство *n*

8484 multiplier gain
  d Verstärkungsfaktor *m* des Vervielfachers
  f gain *m* du multiplicateur
  r коэффициент *m* усиления электронного
    умножителя

*   **multiplying device → 8483**

*   **multiplying unit → 8483**

8485 multipoint control
  d Multipunktsteuerung *f*; Vielpunktsteuerung *f*
  f commande *f* multipoint
  r квазиконтурное управление *n*

8486 multipoint measuring instrument
  d Mehrstellenmessgerät *n*
  f mesureur *m* à points multiples de mesure
  r многоточечный измерительный прибор *m*

8487 multipoint sensor
  d Mehrpunktgeber *m*; Multipunktsensor *m*
  f capteur *m* multipoint; senseur *m* multipoint
  r многопозиционный датчик *m*

8488 multipole relay circuit connection
  d mehrpolige Schaltverbindung *f* von
    Relaiskreisen
  f connexion *f* multipôle de chaînes à relais
  r многополюсное соединение *n* релейных
    цепей

*   **multiposition controller → 8420**

8489 multiposition element
  d Mehrpunktglied *n*
  f terme *m* multiple
  r многопозиционный элемент *m*

8490 multiposition signal
  d Mehrpunktsignal *n*
  f signal *m* multiple
  r многопозиционный сигнал *m*

8491 multiposition switch
  d Mehrfachpositionsschalter *m*;
    Multipositionsschalter *m*
  f commutateur *m* à positions multiples
  r многопозиционный переключатель *m*

8492 multiprocessing
  d Mehrfachverarbeitung *f*

f traitement *m* multiple
r многократная обработка *f*

8493 multiprocessor application
d Multiprozessoranwendung *f*;
Mehrprozessoreinsatz *m*
f application *f* de multiprocesseurs
r применение *n* мультипроцессорных
систем

8494 multiprocessor control system
d Multiprozessorsteuersystem *n*;
Mehrprozessorsteuersystem *n*
f système *m* de commande à plusieurs
processeurs
r мультипроцессорная система *f* управления

* **multiprocessor operation** → 8461

8495 multiprogramming
d Mehrfachprogrammierung *f*
f programmation *f* multiple;
multiprogrammation *f*
r мультипрограммирование *n*

8496 multipulse controller
d Mehrfachimpulsregler *m*
f régulateur *m* à impulsions multiples
r многоимпульсный регулятор *m*

8497 multipurpose automatic device; universal automaton
d Mehrzweckautomat *m*; universeller
Automat *m*
f dispositif *m* automatique universel;
automate *m* universel
r универсальный автомат *m*

8498 multipurpose plant
d Mehrzweckanlage *f*
f unité *f* à plusieurs fonctions
r мобильная технологическая схема *f*

8499 multipurpose research reactor
d Mehrzweckforschungsreaktor *m*
f réacteur *m* de recherche multifonctionnel
r многоцелевой исследовательский
реактор *m*

8500 multirange instrument
d Messgerät *n* mit mehreren Messbereichen
f appareil *m* de mesure à plusieurs gammes
r многодиапазонный прибор *m*

8501 multispeed controller
d Mehrlaufregler *m*
f régulateur *m* à vitesse d'actions multiples
r многоскоростный регулятор *m*

8502 multispeed floating control
d Mehrlaufregelung *f*; astatische Regelung *f*
f réglage *m* flottant à plusieurs vitesses
r многоскоростное астатическое
регулирование *n*

8503 multistability
d Multistabilität *f*
f stabilité *f* multiple; multistabilité *f*
r мультиустойчивость *f*

8504 multistage
d mehrstufig
f à plusieurs étages
r многоступенчатый

8505 multistage allocation process
d mehrstufiger Entscheidungsprozess *m*
f procédé *m* de décision multiple; procédé de
l'allocation multiple
r многошаговый процесс *m* решения

8506 multistage process
d Vielstufenverfahren *n*
f procédé *m* à plusieurs étages
r многоступенчатый процесс *m*

8507 multistage separation
d mehrstufige Trennung *f*
f séparation *f* à plusieurs étages
r многоступенчатое разделение *n*

* **multistep action** → 8417

* **multistep controller** → 8420

8508 multistep memory system
d Vielstufenspeichersystem *n*
f système *m* de mémoire à é tages multiples
r многоступенчатая система *f* памяти

8509 multiuser spectrometer facility
d Gemeinschafts-Spektrometeranlage *f*
f installation *f* spectrométrique collective;
installation spectrométrique à utilisateurs
multiples
r коллективная спектрометрическая
установка *f*

* **multivariable control** → 6830

8510 multivariable function generator
d Funktionsumformer *m* für mehrere
Veränderliche
f générateur *m* de fonctions à variables
multiples
r функциональный преобразователь *m*
нескольких переменных

8511 multivibrator
   d Multivibrator *m*; Kippschaltung *f*
   f multivibrateur *m*
   r мультивибратор *m*

8512 mutually independent variables
   d gegenseitig unabhängige Größen *fpl*
   f valeurs *fpl* interindépendantes
   r взаимонезависимые переменные *fpl*

8513 mutually synchronized systems
   d gegenseitig synchronisierte Systeme *npl*
   f systèmes *mpl* à synchronisme mutuel
   r системы *fpl* с взаимной синхронизацией

8514 myoelectric signal
   d myoelektrisches Signal *n*
   f signal *m* myoélectrique
   r миоэлектрический сигнал *m*

N

8515 NAND-circuit
  d NAND-Schaltung *f*
  f circuit *m* NON-ET
  r [логическая] схема НЕ-И

8516 NAND-element
  d NICHT-UND-Glied *n*; NAND-Glied *n*
  f élément *m* NON-ET
  r элемент *m* НЕ-И

8517 NAND-operation
  d NICHT-UND-Operation *f*;
       NAND-Operation *f*
  f opération NON-ET
  r операция *f* НЕ-И

8518 nanosecond; millimicrosecond
  d Nanosekunde *f*
  f nanoseconde *f*
  r наносекунда *f*

8519 nanosecond impulse generator
  d Nanosekundenimpulsgenerator *m*
  f générateur *m* d'impulsions d'ordre de
       nanosecondes
  r генератор *m* наносекундных импульсов

8520 nanosecond logic
  d Logik *f* im Nanosekundenbereich
  f logique *f* nanoseconde
  r наносекундная логика *f*

8521 nanovolt chopper
  d Nanovoltzerhacker *m*
  f interrupteur *m* de tension de l'ordre de
       nanovolts
  r прерыватель *m* напряжения порядка
       нановольт

8522 narrow-angle coordinator
  d Schmalwinkelkoordinator *m*
  f coordinateur *m* à angle étroit
  r узкоугольный координатор *m*

8523 narrow-band amplifier
  d Schmalbandverstärker *m*
  f amplificateur *m* à bande étroite
  r узкополосный усилитель *m*

8524 narrow-band controller
  d Schmalbandregler *m*

  f régulateur *m* à bande étroite
  r регулятор *m* с узкой зоной регулирования

8525 narrow-band frequency range
  d Schmalbandfrequenzbereich *m*
  f gamme *f* fréquences à bande étroite
  r узкополосный диапазон *m* частот

8526 narrow-band proportional control
  d proportionale Schmalbandregelung *f*
  f réglage *m* proportionnel à bande étroite
  r пропорциональное регулирование *n* с
       узкой зоной

8527 narrow-band signal
  d Schmalbandsignal *n*
  f signal *m* à bande étroite
  r узкополосный сигнал *m*

8528 narrow gate pulse
  d schmaler Torimpuls *m*
  f impulsion *f* étroite; créneau *m* étroit
  r узкий отпирающий импульс *m*; узкий
       селекторный импульс

8529 narrow line emission
  d Schmallinienemission *f*
  f émission *f* à raie étroite
  r узкополосное излучение *n*

8530 narrow-stripe geometry
  d Geometrie *f* mit schmalem Streifen; schmale
       Streifengeometrie *f*
  f géométrie *f* en ruban étroit; géométrie en
       bande étroite
  r узкополосная геометрия *f*; геометрия с
       узкой полосой

**8531 narrow-stripe-geometry double
       heterostructure laser**
  d Doppelheterostrukturlaser *m* mit schmaler
       Streifengeometrie
  f laser *m* à double hétérostructure à géométrie
       en ruban étroit
  r лазер *m* с двойной гетероструктурой и
       узкополосной геометрией

8532 narrow-stripe semiconductor laser
  d Helbleiterlaser *m* mit schmalem Streifen
  f laser *m* semiconducteur à ruban étroit
  r узкополосный полупроводниковый
       лазер *m*

8533 narrow-wide band level indicateur
  d Schmalband-Breitband-Pegelmesser *m*
  f appareil *m* mesureur de niveau de
       transmission à bande étroite et à bande large
  r узко-широкополосный измеритель *m*
       уровня

8534 **natural**
  d natürlich
  f naturel
  r натуральный; естественный

8535 **natural attenuation; natural damping**
  d Eigenabklingen *n*
  f amortissement *m* propre
  r собственное затухание *n*

8536 **natural cooling**
  d natürliche Kühlung *f*
  f refroidissement *m* naturel
  r естественное охлаждение *n*

  * **natural damping → 8535**

8537 **natural excitation**
  d natürliche Erregung *f*
  f excitation *f* naturelle
  r естественное возбуждение *n*

8538 **natural frequency**
  d Eigenfrequenz *f*
  f fréquence *f* naturelle; fréquence propre
  r собственная частота *f*

8539 **natural frequency response of the system**
  d Eigenfrequenzkennlinie *f*
  f caractéristique *f* fréquentielle propre du
       système
  r характеристика *f* собственной частоты
       системы

8540 **natural non-linearity**
  d natürliche Nichtlinearität *f*
  f non-linéarité *f* naturelle
  r естественная нелинейность *f*

  * **natural oscillation → 1449**

8541 **natural response**
  d naturgetreue Antwort *f*; natürliches
       Ansprechen *n*
  f réponse *f* naturelle
  r естественная реакция *f*

8542 **navigation**
  d Navigation *f*
  f navigation *f*
  r навигация *f*

8543 **navigation communication**
  d Navigationskommunikation *f*
  f communication *f* de navigation
  r навигационная связь *f*

8544 **n-channel tape**

  d n-Spurenband *n*
  f bande *f* à n-canaux
  r n-канальная лента *f*

8545 **nearly single-frequency laser**
  d quasimonochromatischer Laser *m*
  f laser *m* presque monofréquence
  r квазимонохроматический лазер *m*

8546 **necessary optimality conditions**
  d notwendige Optimalitätsbedingungen *f*
  f conditions *fpl* nécessaires d'optimalité
  r необходимые условия *npl* оптимальности

8547 **negate** *v*
  d negieren
  f nier
  r отрицать

8548 **negation**
  d Negation *f*
  f négation *f*
  r отрицание *n*

8549 **negative acceleration**
  d negative Beschleunigung *f*
  f accélération *f* négative
  r отрицательное ускорение *n*

  * **negative-base number
     representation → 8550**

8550 **negative-base number representation
       [system]**
  d Zahlenschreibweise *f* mit negativer Basis;
       Zahlendarstellung *f* mit negativer Basis
  f système *m* de représentation de nombres à
       base négative
  r система *f* счисления с отрицательным
       основанием

8551 **negative booster**
  d Zusatzmaschine *f* in Gegenschaltung;
       Spannungserniedriger *m*
  f dévolteur *m*
  r отрицательный бустер *m*

8552 **negative current feedback**
  d Stromgegenkopplung *f*
  f contre-réaction *f* d'intensité; contre-réaction
       de courant
  r отрицательная обратная связь *f* по току

8553 **negative definite**
  d negativ definit
  f négativement défini
  r отрицательно определенный

* **negative feedback** → 3882

8554 negative feedback coupling resistor
d Gegenkopplungswiderstand *m*
f résistance *f* de couplage du circuit de la contre-réaction
r сопротивление *n* сочленения отрицательной обратной связи

8555 negative feedback loop
d Gegenkopplungschleife *f*
f boucle *f* de contre-réaction
r контур *m* отрицательной обратной связи

8556 negative impedance
d negativer Scheinwiderstand *m*; negative Impedanz *f*
f impédance *f* négative
r отрицательный импеданс *m*

8557 negative logic
d negative Logik *f*; negative Schaltungslogik *f*
f logique *f* négative
r отрицательная логика *f*

8558 negative-logic circuit
d Negativlogikschaltung *f*
f circuit *m* logique négatif
r схема *f* отрицательной логики

8559 negative-logic signal
d Negativlogiksignal *n*
f signal *m* logique négatif
r сигнал *m* отрицательной логики

8560 negative phase sequence relay
d negatives Phasensequenzrelais *n*
f relais *m* fonctionnant sur la composante négative de la phase
r реле *n* отрицательной последовательности фаз

8561 negative pulse
d negativer Impuls *m*
f impulsion *f* négative
r отрицательный импульс *m*

8562 negative resistance
d negativer Wirkwiderstand *m*; negativer Widerstand *m*
f résistance *f* négative
r отрицательное сопротивление *n*

8563 negative self-regulation
d negativer Selbstausgleich *m*
f autorégulation *f* négative
r отрицательное самовыравнивание *n*

8564 negative sequence power
d Gegenleistung *f*
f puissance *f* inverse
r отрицательная нагрузка *f*

8565 negative voltage feedback
d Spannungsgegenkopplung *f*
f contre-réaction *f* de tension
r обратная отрицательная связь *f* по напряжению

8566 negatoscope
d Negativschaukasten *m*
f négatoscope *m*
r негатоскоп *m*

8567 neglect *v*; disregard *v*
d vernachlässigen
f négliger; ignorer
r пренебрегать

8568 negligible error
d vernachlässigbarer Fehler *m*
f erreur *f* négligeable
r пренебрежимо малая ошибка *f*

8569 neodymium laser
d Neodym-Laser *m*
f laser *m* à néodyme
r неодимовый лазер *m*

8570 neon digital display
d Neondigitalanzeige *f*; Neondigitaldarstellung *f*
f indicateur *m* numérique à néon
r цифровой неоновый указатель *m*

8571 neper
d Neper *n*
f néper *m*
r непер *m*

8572 nephelometer
d Trübungsmesser *m*; Nephelometer *n*
f néphélomètre *m*
r нефелометр *m*

8573 nephelometric analysis
d nephelometrische Analyse *f*
f analyse *f* néphélométrique
r нефелометрический анализ *m*

8574 Nernst bridge
d Nernstbrücke *f*
f pont *m* de Nernst
r ёмкостный мост *m* Нернста

8575 net amplitude
d Gesamtamplitude *f*

 f amplitude *f* nette
 r результирующая амплитуда *f*

8576 net control
 d Netzüberwachung *f*
 f surveillance *f* de réseaux
 r контроль *m* сетевого напряжения

8577 net efficiency; overall efficiency
 d Gesamtwirkungsgrad *m*
 f rendement *m* global; efficacité *f* totale
 r общий коэффициент *m* полезного
  действия

8578 net energy
 d Nutzenergie *f*
 f énergie *f* utile
 r полезная энергия *f*; эффективная энергия *f*

**8579 net pulse rate of nuclear radiation
  detector**
 d Nettoimpulsrate *f* des
  Kernstrahlungsdetektors
 f taux *m* net d'impulsion du détecteur de
  rayonnement nucléaire
 r интенсивность *f* суммарного импульса
  детектора ядерного излучения

8580 network
 d Netzwerk *n*
 f réseau *m*
 r схема *f*; сеть *f*; цепь *f*

8581 network analysis
 d Netzwerkanalyse *f*
 f analyse *f* de réseau
 r анализ *m* цепей

8582 network analyzer
 d Netzwerkgleichungslöser *m*;
  Netzwerkanalysator *m*; Netzwerksimulator *m*
 f analyseur *m* de réseaux
 r схемный анализатор *m*; устройство *n*
  моделирования [электрических] цепей

8583 network application
 d Netzwerkanwendung *f*; Anwendung *f* in
  Netzen
 f application *f* dans des réseaux
 r применение *n* в сетевых структурах

8584 network attenuation; network damping
 d Netzdämpfung *f*
 f affaiblissement *m* du réseau
 r затухание *n* контура

8585 network communications circuit
 d Netzwerkübertragungsschaltung *f*;

  Netzwerkübertragungsschaltkreis *m*
 f circuit *m* de communication de réseau
 r схема *f* связи между элементами сети

8586 network computer
 d Netzcomputer *m*
 f ordinateur *m* en réseau
 r сетевой компьютер *m*

8587 network configuration
 d Netzstruktur *f*; Netzaufbau *m*
 f structure *f* de réseau
 r конфигурация *f* сети; структура *f* сети

8588 network constant
 d Netzkonstante *f*
 f constante *f* du réseau
 r константа *f* схемы

8589 network control system
 d Netzwerksteuerungssystem *n*
 f système *m* de commande de réseau
 r система *f* контурного управления

* **network damping** → 8584

* **network element** → 2392

8590 network model
 d Netzmodell *n*
 f modèle *m* du réseau
 r сетевая модель *f*

8591 network optimization
 d Netzoptimierung *f*
 f optimisation *f* de réseau
 r оптимизация *f* сети

8592 network phasing relay
 d Netzphasenrelais *n*
 f relais *m* de phase
 r реле *n* сдвига фаз

8593 network stability
 d Netzstabilität *f*
 f stabilité *f* de réseau
 r устойчивость *f* сети

8594 network synthesis
 d Netzwerksynthese *f*; Synthese *f* von
  Netzwerken
 f synthèse *f* de réseau
 r синтез *m* схем

8595 network technique
 d Netzplantechnik *f*
 f technique *f* de réseau
 r сетевая техника *f*

8596 **network theory**
d Netzwerktheorie *f*
f théorie *f* de réseau
r теория *f* сетей

8597 **network topology**
d Netzwerktopologie *f*
f topologie *f* de réseau
r топология *f* сети

8598 **neutral conductor**
d Nulleiter *m*
f conducteur *m* neutre
r нейтральный провод *m*

8599 **neutral connection; neutral lead**
d Nullanschluss *m*
f connexion *f* neutre
r нейтральная связь *f*

8600 **neutral-controlled plant**
d neutral gesteuertes Objekt *n*
f installation *f* réglée neutre
r нейтральная регулируемая установка *f*

8601 **neutral current; simple current; single current**
d Einfachstrom *m*
f courant *m* de conducteur neutre; courant simple
r ток *m* в нейтральном проводе; однополярный ток

8602 **neutralization**
d Neutralisieren *n*; Neutralisation *f*
f neutralisation *f*
r нейтрализация *f*

* **neutral lead → 8599**

8603 **neutral system**
d neutrales System *n*
f système *m* neutre
r нейтральная система *f*

8604 **neutral zone**
d neutrale Zone *f*
f zone *f* neutre
r нейтральная зона *f*

8605 **neutron activation analysis**
d Neutronenaktivierungsanalyse *f*
f analyse *f* d'activation par neutrons
r нейтронный активационный анализ *m*

8606 **neutron generator**
d Neutronengenerator *m*
f générateur *m* à neutrons

r нейтронный генератор *m*

8607 **neutron pulse**
d Neutronenimpuls *m*
f impulsion *f* neutronique
r нейтронный импульс *m*

8608 **Newtonian fluid mechanics**
d Newtonsche Strömungsmechanik *f*;
Mechanik *f* Newtonscher fluider Medien
f mécanique *f* des fluides newtoniens
r Ньютоновская гидромеханика *f*

8609 **Newton-Raphson method**
d Newtonsches [Iterations-] Verfahren *n*
f méthode *f* de l'itération de Newton
r метод *m* итерации Ньютона-Рафсона

8610 **Newton's motion equation**
d Newtonsche Bewegungsgleichung *f*
f équation *f* de mouvement de Newton
r уравнение *n* движения Ньютона

8611 **Nichols chart; Nichols diagram**
d Nichols-Plan *m*
f diagramme *m* de Nichols
r диаграмма *f* Николса

* **Nichols diagram → 8611**

8612 **nil report**
d Fehlanzeige *f*
f rapport *m* nul
r ложная индикация *f*

8613 **no-charge; no-load; nonchargeable**
d unbelastet
f non chargé; à vide
r ненагруженный; холостой

8614 **nodal analysis**
d Knotenpunktmethode *f*
f méthode *f* de valeurs en nœuds
r анализ *m* методом узловых точек

8615 **node; centre; junction; intersection**
d Knoten *m*; Knotenpunkt *m*
f nœud *m* d'assemblage; point *m* de jonction
r узел *m*; узловая точка *f*; транспортный узел

8616 **noise**
d Rauschen *n*; Geräusch *n*
f bruit *m*
r шум *m*

8617 **noise** *v*
d rauschen; stören

f bruiter
r шуметь

8618 noise compensation
d Rauschkompensation *f*
f compensation *f* du bruit
r компенсация *f* шума

8619 noise component
d Rauschanteil *m*
f composante *f* de bruit
r шумовая составляющая *f*; компонента *f*
шума

8620 noise dispersion
d Steuung *f* von Rauschstörungen
f dispersion *f* de bruit
r дисперсия *f* шума

8621 noise-eliminating device
d Entstörungseinrichtungen *fpl*
f dispositifs *mpl* d'élimination de bruit
r помехоподавляющие устройства *npl*

8622 noise emission
d Lärmemission *f*
f émission *f* du bruit
r эмиссия *f* шума

8623 noise equivalent circuit
d Rauschersatzschaltung *f*;
Rauschersatzschaltbild *n*
f schéma *m* équivalent de bruit
r эквивалентная схема *f* источника шума

*   **noise equivalent power → 5147**

8624 noise-free; noiseless
d rauschfrei
f sans bruit
r безшумовой; без шума

8625 noise-free optimal regulator
d rauschfreier optimaler Regler *m*
f régulateur *m* libre de bruit optimal
r оптимальный регулятор *m* без помех

8626 noise gate
d Störsperre *f*
f filtre *m* de bruit
r фильтр *m* для задерживания шума

8627 noise generator
d Rauschgenerator *m*
f générateur *m* de bruit
r генератор *m* шумов

8628 noise immunity

d Rauschunempfindlichkeit *f*; Störsicherheit *f*
f insensibilité *f* contre bruits
r помехоустойчивость *f*;
помехозащищенность *f*

8629 noise impulse
d Rauschimpuls *m*
f impulsion *f* de bruit
r шумовой импульс *m*

8630 noise interference
d Rauscheinmischung *f*
f interférence *f* de bruit
r шумовое вмешательство *n*

*   **noiseless → 8624**

8631 noise level; noise ratio
d Rauschpegel *m*; Geräuschpegel *m*;
Rauschverhältnis *n*; Störpegelabstand *m*
f niveau *m* de bruit; rapport *m* de bruit;
bande *f* du niveau de bruit
r уровень *m* шума

8632 noise-limited
d rauschbegrenzt
f limité par le bruit
r с ограничением шума

8633 noise limiter
d Rauschbegrenzer *m*; Geräuschbegrenzer *m*
f limiteur *m* du bruit
r ограничитель *m* шумов

8634 noise margin
d Störabstand *m*; Störspannungsspanne *f*
f marge *f* de bruit
r запас *m* помехоустойчивости

**8635 noise measuring instrument; noise meter;
noise test set**
d Geräuschmesser *m*; Rauschmessgerät *n*
f décibelmètre *m*; sonomètre *m*; appareil *m*
pour mesurer le bruit
r измеритель *m* шумов; шумомер *m*

*   **noise meter → 8635**

8636 noise power
d Rauschleistung *f*
f puissance *f* de bruit
r мощность *f* шумов

8637 noise properties
d Rauscheigenschaften *fpl*
f caractéristiques *fpl* de bruit
r свойства *npl* шума

8638 **noise protection**
  d Lärmschutz *m*
  f protection *f* contre les bruits
  r защита *f* от шума

* **noise pulse** → 4464

* **noise ratio** → 8631

8639 **noise reduction**
  d Rauschverminderung *f*
  f réduction *f* du bruit
  r снижение *n* уровня шума

8640 **noise simulation**
  d Geräuschsimulation *f*
  f simulation *f* de bruit
  r моделирование *n* шума

8641 **noise spectrum**
  d Rauschspektrum *n*
  f spectre *m* de bruit
  r спектр *m* шума

8642 **noise suppressor**
  d Geräuschunterdrücker *m*; Rauschfilter *n*;
      Entstörer *m*
  f dispositif *m* antiparasite; dispositif
      éliminateur de bruits
  r подавитель *m* помех

* **noise test set** → 8635

* **noise test set** → 3799

8643 **noise threshold**
  d Rauschgrenze *f*
  f seuil *m* de bruit
  r порог *m* шума

8644 **noise voltage**
  d Störspannung *f*
  f tension *f* de bruit
  r напряжение *n* шумов

8645 **noisy**
  d rauschend; mit Rauschen; verrauscht;
      rauschbehaftet
  f bruyant; avec bruits
  r шумящий; с помехами; с высоким
      уровнем шумов

8646 **noisy communication channel**
  d verrauschter
      Nachrichtenübertragungskanal *m*;
      rauschbehafteter
      Nachrichtenübertragungskanal
  f voie *f* de communication bruyante; canal *m*
      de communication bruyant
  r канал *m* связи с шумом

8647 **noisy servomechanism**
  d verrauschter Servomechanismus *m*
  f servomécanisme *m* bruyant
  r следящая система *f* с источником шума

* **no-load** → 8613

8648 **no-load characteristic; unloaded
      characteristic**
  d Leerlaufkennlinie *f*; Leerlaufcharakteristik *f*
  f caractéristique *f* à vide
  r характеристика *f* холостого хода

8649 **no-load control**
  d Leerlaufkontrolle *f*
  f contrôle *m* de marche à vide
  r контроль *m* холостого хода

8650 **no-load working**
  d Leerlaufarbeit *f*
  f marche *f* à vide
  r холостой ход *m*

8651 **nomenclature language software**
  d Fachsprachensoftware *f*
  f software *m* d'un langage de nomenclature
  r терминологическое программное
      обеспечение *n*

8652 **nominal frequency**
  d Nennfrequenz *f*
  f fréquence *f* nominale
  r номинальная частота *f*

* **nominal load** → 10635

8653 **nominal output**
  d Nennabgabe *f*; Nennleistung *f*
  f puissance *f* nominale
  r номинальная выходная мощность *f*

8654 **nominal range of use**
  d Nennbereich *m*; Nennreichweite *f*
  f domaine *m* nominal d'utilisation
  r номинальная область *f* применения

8655 **nominal steepness of wave front**
  d Nominalsteilheit *f* der Wellenfront
  f raideur *f* nominale du front d'onde
  r номинальная крутизна *f* фронта волны

8656 **nominal transformation ratio**
  d Nennumwandlungsverhältnis *n*
  f rapport *m* nominal de transformation
  r номинальный коэффициент *m*
      преобразования

8657 **nominal value; rating**
   d Nennwert *m*
   f valeur *f* nominale
   r номинальное значение *n*

8658 **nominal value comparison**
   d Sollwertvergleich *m*
   f comparaison *f* de valeur prescrite
   r сравнение *n* заданных величин

  * **nominal voltage** → 10638

8659 **nomogram**
   d Funktionsnetz *n*; Nomogramm *n*
   f nomogramme *m*; abaque *m*
   r функциональная сеть *f*; номограмма *f*

8660 **non-arithmetical operation**
   d nichtarithmetische Operation *f*
   f opération *f* non arithmétique
   r неаритметическая операция *f*

8661 **non-automated information system**
   d nichtautomatisiertes Informationssystem *n*
   f système *m* d'information non automatisé
   r неавтоматизированная информационно-поисковая система *f*

8662 **non-automatic tripping**
   d nichtautomatisches Ansprechen *n*
   f déclenchement *m* libre
   r неавтоматическое отключение *n*

8663 **nonblocking**
   d blockierungsfrei
   f non bloque; sans blocage
   r неблокированный

8664 **nonblocking configuration**
   d blockierungsfreie Anordnung *f*
   f configuration *f* sans blocage
   r конфигурация *f* без блокировки

8665 **non-chained manipulator**
   d nicht verketteter Manipulator *m*
   f manipulateur *m* non enchaîné
   r несвязанный манипулятор *m*

  * **nonchargeable** → 8613

8666 **non-closed loop control**
   d Steuerung *f* über nicht geschlossene Schleife
   f réglage *m* en boucle ouverte
   r регулирование *n* по разомкнутой цепи

  * **non-contact feeler device** → 3093

8667 **non-contacting density measurement**

   d berührungslose Dichtemessung *f*
   f densimétrie *f* sans contact [direct]; mesure *f* de densité sans contact
   r бесконтактное измерение *n* плотности

  * **non-contacting sensor** → 3093

8668 **non-contacting thickness gauging**
   d berührungslose Dickenmessung *f*
   f mesure *f* d'épaisseur sans contact
   r бесконтактное измерение *n* толщины

8669 **non-contact measurement technique**
   d berührungsfreies Messverfahren *n*
   f technique *f* de mesure sans contact
   r бесконтактный метод *m* измерений

8670 **non-contact relay element**
   d kontaktloses Relaiselement *n*
   f élément *m* de commutation sans contacts
   r бесконтактный релейный элемент *m*

8671 **non-contiguous constant**
   d unabhängige Konstante *f*
   f constante *f* indépendante
   r независимая константа *f*

8672 **non-contradictory**
   d widerspruchsfrei
   f non contradictoire
   r непротиворечивый

8673 **non-critical dimension**
   d nichtkritische Abmessung *f*
   f dimension *f* non critique
   r некритическая размерность *f*

8674 **non-critical mass**
   d nichtkritische Masse *f*
   f masse *f* non critique
   r некритическая масса *f*

8675 **non-critical point**
   d nichtsingulärer Punkt *m*
   f point *m* non singulier
   r некритическая точка *f*

8676 **non-critical reactor**
   d nichtkritischer Reaktor *m*
   f réacteur *m* non critique
   r некритический реактор *m*

8677 **non-decreasing function**
   d nicht abnehmende Funktion *f*
   f fonction *f* non décroissante
   r неубывающая функция *f*

8678 **non-degenerate energy level**
   d nicht abgewichenes Energieniveau *n*

f niveau *m* énergétique non dégénéré
r невырожденный энергетический уровень *m*

8679 non-degenerate system
d nichtentartetes System *n*
f système *m* non dégénéré
r невырожденная система *f*

8680 non-design-basis accident
d Nichtauslegungsstörfall *m*; Nichtauslegungsunfall *m*
f accident *m* hors dimensionnement
r непроектная авария *f*; нерасчётная авария

8681 non-destructive analytical technique
d zerstörungsfreies Analysenverfahren *n*
f méthode *f* non destructive d'analyse
r неразрушающий метод *m* анализа

8682 non-destructive check method
d zerstörungsfreie Kontrollmethode *f*
f méthode *f* de contrôle non destructive
r неразрушающий метод *m* испытаний

8683 non-destructive gamma-ray spectrometry
d zerstörungsfreie Gammaspektrometrie *f*
f spectrométrie *f* gamma non destructive
r неразрушающая гамма-спектроскопия *f*

8684 non-destructive materials testing
d zerstörungsfreie Werksoffprüfung *f*; zerstörungsfreie Materialprüfung *f*
f essai *m* non destructif des matériaux; examen *m* sans destruction de l'échantillon; recherche *f* de défauts des matériaux non destructive
r неразрушающий контроль *m*; испытание *n* без разрушения образца

8685 non-destructive measurement of adhesive power
d zerstörungsfreie Messung *f* der Adhäsionskraft
f mesure *f* non destructive de la force adhésive
r измерение *n* адгезионной способности

8686 non-destructive reading
d nichtzerstörendes Lesen *n*; zerstörungsfreies Auslesen *n*
f lecture *f* non destructive
r неразрушающее считывание *n*

8687 non-deterministic automaton
d nichtdeterminierter Automat *m*
f automate *m* non déterminé
r недетерминированный автомат *m*

8688 non-digital information
d nichtdigitale Information *f*
f information *f* non digitale
r нецифровая информация *f*

8689 non-dimensional
d dimensionslos
f sans dimension; non déterminé; non dénommé
r безразмерный

8690 non-dimensional coefficient
d unbenannter Koeffizient *m*
f coefficient *m* non dimensionnel
r безразмерный коэффициент *m*

8691 non-dimensional curve
d dimensionslose Kurve *f*
f courbe *f* sans dimension
r безразмерная кривая *f*

8692 non-dimensional parameter
d dimensionsloser Parameter *m*
f paramètre *m* non dimensionnel
r безразмерный параметр *m*

8693 non-dimensional response curve
d dimensionslose Kennlinie *f*
f caractéristique *f* sans dimension
r безразмерная характеристика *f* чувствительности

8694 non-dimensional time
d normierte Zeit *f*
f temps *m* réduit
r относительное время *n*

* **non-dimensional value → 4253**

* **non-dimensional variable → 4254**

8695 non-directed graph
d ungerichteter Graph *m*
f graphe *m* non dirigé
r неориентированный граф *m*

8696 non-directional current protection
d nichtgerichteter Stromschutz *m*
f disposition *f* de protection ampèremétrique non directionnelle
r ненаправленная защита *f* по току

8697 non-directional relay
d richtungsunempfindliches Relais *n*
f relais *m* non directionnel
r ненаправленное реле *n*

8698 non-dispersion infrared gas analyzer
d infraroter dispersionsloser Gasprüfer *m*

f analyseur *m* de gaz infrarouge sans
dispersion
r бездисперсионный инфракрасный
газоанализатор *m*

8699 non-dispersive spectrometry
d nichtdispersive Spektrometrie *f*
f spectrométrie *f* non dispersive
r недисперсионная спектрометрия *f*

8700 non-electric value
d nichtelektrische Größe *f*
f quantité *f* non électrique
r неэлектрическая величина *f*

8701 non-equilibrium
d Ungleichgewicht *n*; Nichtgleichgewicht *n*
f déséquilibre *m*
r неравновесие *n*

8702 non-equilibrium state
d Nichtgleichgewichtszustand *m*
f état *m* hors d'équilibre
r неравновесное состояние *n*

8703 non-equilibrium statistical mechanics
d statistische Nichtgleichgewichtsmechanik *f*
f mécanique *f* statistique hors d'équilibre
r неравновесная статистическая механика *f*

8704 non-equivalence
d Nichtäquivalenz *f*
f non-équivalence *f*
r неэквивалентность *f*; неравнозначность *f*

8705 non-equivalence element
d nichtäquivalentes Element *n*
f élément *m* de non-équivalence
r неэквивалентный элемент *m*

* **non-equivalence gate → 5260**

8706 non-failsafe mode
d nicht ausfallsichere Betriebsweise *f*
f mode *m* non tolérant aux défauts
r неотказоустойчивый режим *m*;
незащищенный режим

* **non-homogeneous magnetic field → 6613**

8707 non-indicating controller
d nichtanzeigender Regler *m*; anzeigeloser
Regler
f régulateur *m* sans indication
r бесшкальный регулятор *m*

8708 non-integrated concept
d nichtintegrierte Bauweise *f*

f conception *f* non intégrée
r неинтегральная компоновка *f*

**8709 non-interacting safety system; single-loop
safety system**
d unvermaschtes Sicherheitssystem *n*;
Sicherheitssystem in unvermaschter
Ausführungsform
f système *m* de sécurité non interconnecté;
systéme de sécurité à boucle unique
r невзаимосвязанная система *f* аварийной
защиты; одноконтурная система
аварийной защиты

8710 non-interaction
d Entkopplung *f*
f découplage *m*
r отсутствие *n* взаимозависимости;
автономность *f*

8711 non-interaction conditions
d wechselwirkungslose
Zustandsbedingungen *fpl*
f conditions *fpl* d'état sans action réciproque;
conditions d'autonomie
r условия *npl* автономности

8712 non-interaction control system
d autonomes selbsttätiges Regelungssystem *n*
f système *m* de réglage autonome
r автономная система *f* автоматического
регулирования

8713 non-isothermal measuring data
d nichtisotherme Messdaten *pl*
f données *fpl* de mesure non isothermes
r неизотермические данные *pl* измерений

8714 non-linear
d nichtlinear
f non linéaire
r нелинейный

8715 non-linear amplifier
d nichtlinearer Verstärker *m*
f amplificateur *m* non linéaire
r нелинейный усилитель *m*

* **non-linear control → 8744**

8716 non-linear control system
d nichtlineares Regelungssystem *n*
f système *m* asservi non linéaire
r нелинейная система *f* регулирования

8717 non-linear converter
d nichtlinearer Umwandler *m*

 f convertisseur *m* non linéaire
 r нелинейный преобразователь *m*

8718 non-linear coupling
 d nichtlineare Kopplung *f*
 f couplage *m* non linéaire
 r нелинейная связь *f*

8719 non-linear damping
 d nichtlineare Dämpfung *f*
 f amortissement *m* non linéaire
 r нелинейное демпфирование *n*

8720 non-linear decoupling part
 d nichtlinearer Entkopplungsanteil *m*
 f part *f* de découplage non linéaire
 r нелинейные составляющие *fpl* развязки

8721 non-linear decoupling process
 d nichlineares Entkopplungsverfahren *n*
 f procédé *m* de découplage non linéaire
 r нелинейный способ *m* развязки

8722 non-linear dependence
 d nichtlineare Abhängigkeit *f*
 f dépendance *f* non linéaire
 r нелинейная зависимость *f*

8723 non-linear differential equation
 d nichtlineare Differentialgleichung *f*
 f équation *f* différentielle non linéaire
 r нелинейное дифференциальное
  уравнение *n*

8724 non-linear distortion
 d nichtlineare Verzerrung *f*; Klirrverzerrung *f*
 f distorsion *f* non linéaire
 r нелинейное искажение *n*

8725 non-linear dynamic operation
 d nichtlineare dynamische Operation *f*
 f opération *f* dynamique non linéaire
 r нелинейная динамическая операция *f*

8726 non-linear effects in acoustical field
 d nichtlineare Erscheinungen *fpl* im
  akustischen Feld
 f effets *mpl* acoustiques non linéaires
 r нелинейные эффекты *mpl* в акустическом
  поле

 * **non-linear element** → 8733

8727 non-linear equation
 d nichtlineare Gleichung *f*
 f équation *f* non linéaire
 r нелинейное уравнение *n*

8728 non-linear filter system
 d nichtlineares Filtersystem *n*
 f système *m* non linéaire de filtrage
 r нелинейная фильтрующая система *f*

8729 non-linear function generator
 d nichtlinearer Funktionsgenerator *m*
 f générateur *m* de fonctions non linéaires
 r генератор *m* нелинейных функций

8730 non-linear interaction
 d nichtlineare Wechselwirkung *f*
 f interaction *f* non linéaire
 r нелинейное взаимодействие *n*

8731 non-linearities correction
 d Korrektur *f* von Nichtlinearitäten
 f correction *f* des non-linéarités
 r коррекция *f* нелинейностей

8732 non-linearity
 d Nichtlinearität *f*
 f non-linéarité *f*
 r нелинейность *f*

8733 non-linear link; non-linear element
 d nichtlineares Glied *n*; nichtlinearer
  Bauteil *m*; nichtlineares Element *n*
 f élément *m* non linéaire
 r нелинейное звено *n*

8734 non-linear mechanics
 d nichtlineare Mechanik *f*
 f mécanique *f* non linéaire
 r нелинейная механика *f*

8735 non-linear modulation response
 d nichtlineare Modulationskennlinie *f*
 f réponse *f* de modulation non linéaire
 r нелинейная характеристика *f* модуляции

8736 non-linear optical interaction
 d nichtlineare optische Wechselwirkung *f*
 f interaction *f* optique non linéaire
 r нелинейное оптическое взаимодействие *n*

8737 non-linear optical parametric process
 d nichtlinearer optischer parametrischer
  Prozess *m*
 f processus *m* paramétrique d'optique non
  linéaire
 r нелинейный оптический параметрический
  процесс *m*

8738 non-linear optical properties
 d nichtlineare optische Eigenschaften *fpl*
 f propriétés *fpl* optique non linéaires
 r нелинейные оптические свойства *npl*

8739 non-linear optical susceptibility
d nichtlineare optische Suszeptibilität *f*
f susceptibilité *f* optique non linéaire
r нелинейная оптическая
 восприимчивость *f*

8740 non-linear optics
d nichtlineare Optik *f*
f optique *f* non linéaire
r нелинейная оптика *f*

8741 non-linear optimalizing system
d optimales Nichtlinearsystem *n*
f système *m* non linéaire optimal
r оптимальная нелинейная система *f*

 * **non-linear part** → 8745

8742 non-linear programming
d nichtlineare Programmierung *f*
f programmation *f* non linéaire
r нелинейное программирование *n*

8743 non-linear reactance
d nichtlineare Reaktanz *f*
f réactance *f* non linéaire
r нелинейное реактивное сопротивление *n*

8744 non-linear regulation; non-linear control
d nichtlineare Regelung *f*
f régulation *f* non linéaire
r нелинейное регулирование *n*; нелинейное
 управление *n*

8745 non-linear share; non-linear part
d nichtlinearer Anteil *m*
f part *f* non linéaire
r нелинейная составляющая *f*

8746 non-linear speed controller
d nichtlinearer Geschwindigkeitsregler *m*
f régulateur *m* non linéaire de vitesse
r нелинейный регулятор *m* скорости

8747 non-linear stopper circuit
d nichtlineare Entkopplungsschaltung *f*
f circuit *m* de découplage non linéaire
r схема *f* нелинейной развязки

8748 non-linear susceptibility
d nichtlineare Suszeptibilität *f*
f susceptibilité *f* non linéaire
r нелинейная восприимчивость *f*

8749 non-linear system
d nichtlineares System *n*
f système *m* non linéaire
r нелинейная система *f*

8750 non-linear transfer circuit
d nichtlineares Übertragungsglied *n*
f élément *m* de transfert non linéaire; circuit *m*
 de transfert non linéaire
r нелинейное передаточное звено *n*

8751 non-linear transient
d nichtlineare Transiente *f*
f transitoire *m* non linéaire
r нелинейный переходный процес *m*

8752 non-moderator
d Nichtmoderator *m*
f non-modérateur *m*
r незамедлитель *m*

8753 non-normalized representation
d nichtnormalisierte Darstellung *f*
f représentation *f* non normalisée
r ненормализованное представление *n*

8754 non-numerical control
d nichtnumerische Steuerung *f*
f commande *f* non numérique
r нечисловое управление *n*

8755 non-operation region
d Ruhebereich *m*
f domaine *m* de non-fonctionnement
r область *f* покоя

 * **non-orthogonality** → 9227

8756 non-oscillating system
d nichtschwingendes System *n*
f système *m* dégénéré
r неколебательная система *f*

8757 non-perfect contour
d nichtperfekte Kontur *f*
f contour *m* non parfait
r неидеальный контур *m*

8758 non-periodical function
d nichtperiodische Funktion *f*
f fonction *f* non périodique
r непериодическая функция *f*

8759 non-polarized relay
d unpolarisiertes Relais *n*; neutrales Relais
f relais *m* non polarisé
r неполяризованное реле *n*; нейтральное
 реле

8760 non-productive operations
d organisatorische Operationen *fpl*;
 Routineoperationen *fpl*

   f opérations *fpl* accessoires
   r вспомогательные операции *fpl*

8761 non-radioactive indicator
   d nichtradioaktiver Indikator *m*
   f indicateur *m* non radioactif
   r нерадиоактивный индикатор *m*

8762 non-reciprocal parametric amplifier
   d nichtreziproker parametrischer Verstärker *m*
   f amplificateur *m* paramétrique non réciproque
   r невзаимный параметрический
      усилитель *m*

8763 non-relevant failure
   d nichtrelevanter Ausfall *m*
   f défaillance *f* à ne pas prendre en compte
   r нехарактерный отказ *m*

8764 non-required time
   d nichtgeforderte Funktionszeit *f*
   f temps *m* non exigé
   r непотребуемое время *n*

8765 non-resonance process
   d Nichtresonanzprozess *m*
   f processus *m* hors de la résonance
   r нерезонансный процесс *m*

8766 non-return flap; non-return valve
   d Rückschlagventil *n*
   f soupape *f* de retenue
   r запорный вентиль *m*; запорный клапан *m*

8767 non-return-to-zero
   d ohne Rückkehr zu Null
   f sans retour à zéro; non-retour à zéro
   r без возврата к нулю; без возвращения к
      нулю

   * **non-return valve → 8766**

8768 non-reversible control
   d nicht umkehrbare Steuerung *f*
   f commande *f* irréversible
   r нереверсируемая система *f*

8769 non-reversible counter
   d nicht umkehrbarer Zähler *m*
   f compteur *m* unidirectionnel
   r нереверсируемый счётчик *m*

8770 non-safety function
   d Nichtsicherheitsfunktion *f*
   f fonction *f* non liée à la sûreté
   r функция f, не связанная с безопасностью

8771 non-selective pneumatic detector

   d nichtselektiver pneumatischer Detektor *m*
   f palpeur *m* pneumatique non sélectif
   r неселективный пневматический
      детектор *m*

   * **non-sensitivity → 6698**

8772 non-sensitivity of element
   d Unempfindlichkeit *f* des Gliedes
   f insensibilité *f* d'élément
   r нечувствительность *f* элемента

8773 non-singular matrix
   d nichtsinguläre Matrix *f*
   f matrice *f* non singulière
   r невырожденная матрица *f*

8774 non-stationary input
   d nichtstationärer Eingang *m*
   f entrée *f* non stationnaire
   r нестационарный вход *m*

8775 non-stationary process
   d nichtstationärer Vorgang *m*; nichtstationärer
      Prozess *m*
   f processus *m* non stationnaire
   r нестационарный процесс *m*;
      неустановившийся процесс

8776 non-stationary random process
   d nichtstationärer stochastischer Prozess *m*
   f processus *m* aléatoire non stationnaire
   r нестационарный случайный процесс *m*

8777 non-stationary system
   d nichtstationäres System *n*
   f système *m* non stationnaire
   r нестационарная система *f*

8778 non-steady
   d nichtstationär
   f non stationnaire
   r неустойчивый

8779 non-steady running conditions
   d nichtstationärer Betriebszustand *m*
   f état *m* de marche non stationnaire
   r неустановившийся рабочий режим *m*

8780 non-symmetric autooscillations
   d nichtsymmetrische Selbstschwingungen *fpl*
   f auto-oscillations *fpl* non symétriques
   r несимметричные собственные
      колебания *npl*

8781 non-synchronous multiplex system
   d asynchrones Multiplexsystem *n*

f système *m* multiplex asynchrone
r несинхронная многоканальная система *f*

8782 non-trivial solution
   d nichttriviale Lösung *f*
   f solution *f* non triviale
   r нетривиальное решение *n*

8783 non-uniforme rotary movement
   d ungleichförmige Drehbewegung *f*
   f mouvement *m* de rotation non uniforme
   r неравномерное вращение *n*

8784 non-uniform laser beam
   d ungleichmäßiger Laserstrahl *m*
   f faisceau *m* non uniforme de laser
   r неоднородный лазерный луч *m*

8785 non-urgent alarm
   d nichtdringlicher Alarm *m*; nichtdringliches
      Alarmsignal *n*
   f alarme *f* non urgente
   r неспешный сигнал *m* тревоги

8786 non-zero conditions
   d Anfangsbedingungen *fpl* ungleich Null
   f conditions *fpl* initiales non nulles
   r ненулевые начальные условия *npl*

8787 NOR-circuit
   d NOR-Schaltung *f*
   f circuit *m* NOR
   r [логическая] схема *f* НЕ-ИЛИ

8788 NOR-element
   d WEDER-NOCH-Glied *n*
   f élément *m* NOR
   r элемент *m* НЕ-ИЛИ

8789 NOR-function
   d NOR-Funktion *f*
   f fonction *f* NOR
   r функция *f* НЕ-ИЛИ

8790 norm; standard
   d Norm *f*; Standard *m*
   f norme *f*; standard *m*
   r норма *f*; стандарт *m*

8791 normal
   d Normale *f*
   f normale *f*
   r нормаль *f*

8792 normal band
   d Normalband *n*; Normalmagnetband *n*
   f bande *f* normale

r нормальный диапазон *m*; нормальная
   полоса *f*

8793 normal condition; normal state
   d Normalzustand *m*
   f condition *f* normale; état *m* normal
   r нормальное состояние *n*

8794 normal contact; normally closed contact
   d Ruhekontakt *m*
   f contact *m* de repos
   r начальный контакт *m*; нормально
      замкнутый контакт

* **normal distribution** → 5875

8795 normal energy level
   d normaler Energiepegel *m*
   f niveau *m* normal énergétique
   r нормальный энергетический уровень *m*

8796 normalization
   d Normalisierung *f*
   f normalisation *f*
   r нормализация *f*

8797 normalize *v*
   d normalisieren
   f normaliser
   r нормализовать

8798 normalized data
   d normalisierte Daten *pl*
   f données *fpl* normalisées
   r нормализованные данные *pl*

8799 normalized form
   d normalisierte Form *f*
   f forme *f* normalisée
   r нормализованная форма *f*

8800 normalized power spectral density
   d normierte Spektraldichte *f*; normierte
      spektrale Leistungsdichte *f*
   f densité *f* spectrale normalisée
   r нормированная спектральная плотность *f*

8801 normalized root-mean square value
   d relative mittlere Rauschamplitude *f*
   f amplitude *f* de bruit moyenne normalisée
   r нормированная средняя амплитуда *f* шума

* **normally closed contact** → 8794

* **normally open contact** → 1540

8802 normally open gate
   d Einschaltor *n*

f porte *f* normalement ouverte; porte de travail
r нормально открытая схема *f* совпадений

8803 normal magnetization curve
  d Normalmagnetisierungskurve *f*
  f courbe *f* normale d'aimantation
  r стандартная кривая *f* намагничивания

8804 normal Markov algorithms
  d normale Markow-Algorithmen *mpl*
  f algorithmes *mpl* normaux de Markov
  r нормальные алгоритмы *mpl* Маркова

8805 normal mass effect
  d normaler Kernmasseneffekt *m*
  f effet *m* de masse normal
  r нормальный массовый эффект *m*

8806 normal operation
  d Normalbetrieb *m*; bestimmungsgemäßer
     Betrieb *m*
  f fonctionnement *m* normal; exploitation *f*
     normale
  r нормальная эксплуатация *f*; нормальное
     функционирование *n*

8807 normal permeability
  d normale Permeabilität *f*
  f perméabilité *f* normale
  r нормальная [магнитная] проницаемость *f*

8808 normal position
  d Grundstellung *f*
  f position *f* normale
  r основное положение *n*

8809 normal start-up procedure
  d normales Anfahren *n*; normaler
     Anfahrvorgang *m*; normaler Start *m*
  f démarrage *m* normal
  r нормальный пуск *m*

  * **normal state → 8793**

  * **normal voltage → 9040**

8810 norm of vector
  d Norm *f* eines Vektors
  f norme *f* d'un vecteur
  r норма *f* вектора

8811 NOR-operation
  d WEDER-NOCH-Operation *f*
  f opération *f* NOR
  r операция *f* НЕ-ИЛИ

8812 notchless control
  d stetige Regelung *f*

f réglage *m* continu
r плавное регулирование *n*

8813 NOT-circuit; NOT-gate
  d NICHT-Tor *n*; NICHT-Schaltung *f*;
     NEIN-Schaltung *f*
  f porte *f* NON; circuit *m* NON
  r схема *f* НЕТ

8814 NOT-component; NOT-element
  d NICHT-Element *n*
  f élément *m* NON
  r элемент *m* НЕТ

  * **NOT-element → 8814**

  * **NOT-gate → 8813**

  * **notice → 8055**

8815 no-voltage release; no-voltage trip
  d Nullspannungsauslöser *m*;
     Unterspannungsauslöser *m*
  f déclencheur *m* à tension nulle; interrupteur *m*
     à tension nulle
  r расцепляющее устройство *n*
     минимального напряжения

  * **no-voltage trip → 8815**

8816 noxious clearance regulation
  d Schadraumregelung *f*
  f réglage *m* d'espace nuisible
  r регулирование *n* вредного зазора

8817 nuclear energy; atomic energy
  d Kernenergie *f*
  f énergie *f* nucléaire
  r ядерная энергия *f*

8818 nuclear measuring instruments
  d Nuklearmessinstrumente *npl*
  f mesureurs *mpl* nucléaires
  r нуклеарный измерительный прибор *m*

8819 nuclear power plant
  d Kernkraftwerk *n*; Kernenergieanlage *f*
  f installation *f* énergétique nucléaire
  r ядерная энергетическая установка *f*

8820 nuclear reaction
  d Kernprozess *m*
  f réaction *f* nucléaire
  r ядерный процесс *m*

  * **nuclear reactor →1094**

8821 nuclear reactor controller
  d Kernreaktorregler *m*

f régulateur *m* de réacteur nucléaire
r регулятор *m* ядерного реактора

8822 nuclear reactor simulator
d Kernreaktorsimulator *m*
f simulateur *m* de réacteur nucléaire
r устройство *n* для моделирования ядерного реактора

8823 nuclear resonance magnetic field meter
d Kernresonanz-Magnetfeldmesser *m*
f mesureur *m* du champ magnétique à résonance nucléaire
r прибор *m* для измерения магнитного поля ядерного резонанса

* **null → 12952**

* **null balance → 1588**

8824 null-balance device
d Nullabgleichglied *n*
f dispositif *m* d'équilibrage automatique
r прибор *m* с балансировкой нуля

8825 null-balance principle; null method
d Nullmethode *f*
f méthode *f* de zéro
r нулевой метод *m*

8826 null circuit
d Kompensationskreis *m*; Abgleichkreis *m*
f circuit *m* compensateur; circuit de remise à zéro
r нулевая схема *f*

* **null cycle → 1827**

8827 null detection
d Nullpunktdetektion *f*
f dépistage *m* de zéro
r обнаружение *n* нуля

* **null device →8829**

8828 null event
d Nullereignis *n*
f événement *m* nul
r событие *n* с нулевой вероятностью

8829 null [indicating] device
d Nullanzeigegerät *n*; Nullstellungsanzeigevorrichtung *f*
f dispositif *m* indicateur de zéro; dispositif à indication de zéro
r нулевое устройство *n*; устройство индикации нуля

8830 null instrument
d Nullinstrument *n*; Instrument *n* mit Nulleinstellung
f appareil *m* à indication de zéro
r нулевой прибор *m*

8831 null matrix; zero matrix
d Nullmatrix *f*
f matrice *f* [de] zéro
r нулевая матрица *f*

* **null method → 8825**

* **null offset → 5576**

* **null point → 12977**

* **null point method of measurement → 2720**

8832 null stroke
d Nullgang *m*
f marche *f* nulle
r нулевой ход *m*

8833 null-type bridge circuit
d Brückenkreis *m* mit Nullanzeige
f circuit *m* à pont équilibré
r схема *f* с уравновешенным мостом

8834 null-type electrometer
d Nulltyp-Elektrometer *n*
f circuit *m* d'électromètre à indication de zéro
r электрометр *m* нулевого типа

8835 number of degrees of freedom
d Anzahl *f* der Freiheitsgrade
f nombre *m* de degrés de liberté
r число *n* степеней свободы

8836 number of sensors
d Sensoranzahl *f*
f nombre *m* des senseurs
r число *n* сенсоров; число датчиков

8837 number order
d Ziffernordnung *f*; Ordnung *f* der Ziffer
f ordre *m* d'un nombre
r порядок *m* числа

8838 number representation in instruction code
d Ziffernaufzeichnung *f* im Befehlskode
f représentation *f* d'un nombre en code d'instruction
r запись *f* числа в коде команды

8839 number system; positional notations
d Zahlensystem *n*

f système *m* numérique; notation *f* dans le
système de notation
r система *f* чисел; численная система

8840 numerical analysis
d numerische Analysis *f*
f analyse *f* numérique
r численный анализ *m*

8841 numerical characteristics of measurement results
d Digitaldaten *pl* von Messwerten
f caractéristiques *fpl* numériques de mesure
r числовые значения *npl* результатов измерений

8842 numerical coding; coding of numbers
d Zahlenverschlüsselung *f*; Zahlenkodierung *f*
f codage *m* des nombres
r численное кодирование *n*

8843 numerical computation schemes
d numerische Rechenschemen *npl*
f schémas *mpl* numériques à calcul
r способы *mpl* численного вычисления

8844 numerical constant
d numerische Konstante *f*
f constante *f* numérique
r численная постоянная *f*

* numerical control → 4178

8845 numerical data
d numerische Daten *pl*
f données *fpl* numériques
r цифровые данные *pl*

8846 numerical differentiation
d numerische Differentiation *f*
f différentiation *f* numérique
r численное дифференцирование *n*

8847 numerical-graphic method
d numerisch-grafische Methode *f*
f méthode *f* numérique-graphique
r численно-графический метод *m*

8848 numerical handling; numerical treatment
d numerische Behandlung *f*
f traitement *m* numérique
r числовая обработка *f*

8849 numerical impulse
d Zahlenimpuls *m*
f impulsion *f* numérique
r цифровой импульс *m*; численный импульс

8850 numerical integration
d numerische Integration *f*
f intégration *f* numérique
r численное интегрирование *n*

8851 numerical iteration
d numerische Iteration *f*
f itération *f* numérique
r числовое повторение *n*

8852 numerically controlled line wiring automaton
d numerisch gesteuerter Verdrahtungsautomat
f dispositif *m* de montage de treillis à commande numérique automatisé
r автомат *m* с цифровым управлением для электрических линий

8853 numerically controlled manipulation equipment
d numerisch gesteuerte Handhabeeinrichtung *f*
f équipement *m* de manutention commandé numérique
r манипуляционное устройство *n* с числовым управлением

8854 numerical machine tool control
d numerische Steuerung *f* von Werkzeugmaschinen
f commande *f* de machines-outils numérique
r цифровое управление *n* станками

8855 numerical position control
d numerische Positionssteuerung *f*
f commande *f* numérique de position
r числовое позиционное управление *n*

8856 numerical position indication
d numerische Positionsanzeige *f*
f indication *f* numérique de position
r числовая индикация *f* позиции

8857 numerical precision manometer
d numerisches Präzisionsmanometer *n*
f manomètre *m* numérique de précision
r цифровой прецизионный манометр *m*

8858 numerical program control
d Ziffernprogrammsteuerung *f*
f commande *f* de programme de chiffres
r числовое программное управление *n*

8859 numerical setting-up
d numerische Einstellung *f*; numerische Einrichtung *f*
f ajustage *m* numérique
r цифровая настройка *f*

8860 numerical signal
d numerisches Signal *n*
f signal *m* numérique
r цифровой сигнал *m*

8861 numerical structure
d numerische Struktur *f*
f structure *f* numérique
r цифровая структура *f*

8862 numerical techniques; digital technique
d numerische Techniken *fpl*; Digitaltechnik *f*
f techniques *fpl* numériques; technique *f* digital
r цифровая техника *f*

* **numerical treatment → 8848**

8863 numerical value
d numerischer Wert *m*; Zahlenwert *m*
f valeur *f* numérique
r численное значение *n*

8864 nutation constant
d Nutationskonstante *f*
f constante *f* de nutation
r постоянная *f* нутации

8865 Nyquist criterion
d Frequenzkriterium *n* von Nyquist;
   Nyquistsches Kriterium *n*
f critère *m* de Nyquist
r критерий *m* Найквиста

8866 Nyquist curve; Nyquist diagram
d Nyquistsches Diagramm *n*
f diagramme *m* de Nyquist
r диаграмма *f* Найквиста

* **Nyquist diagram → 8866**

* **Nyquist method → 8078**

8867 Nyquist plane
d Nyquist-Ebene *f*
f plan *m* de Nyquist
r плоскость *f* Найквиста

O

8868 object configuration
d Objektkonfiguration *f*
f configuration *f* d'objet
r объектная конфигурация *f*

8869 object description
d Objektbeschreibung *f*
f description *f* de l'objet
r описание *n* объекта

8870 object function
d Objektfunktion *f*; Zielfunktion *f*
f fonction *f* but
r целевая функция *f*

8871 object identification; pattern recognition
d Objekterkennung *f*; Gestalterkennung *f*;
Formerkennung *f*
f identification *f* d'objet
r распознавание *n* объекта

8872 object image; image of object
d Objektbild *n*
f image *f* d'objet
r изображение *n* [манипулируемого]
объекта

8873 object image analysis
d Objektbildanalyse *f*
f analyse *f* d'une image d'objet
r анализ *m* изображения объекта

8874 objective variable
d objektive Veränderliche *f*; Hilfsregelgröße *f*
f grandeur *f* objective; grandeur de réglage
auxiliaire
r объектная переменная *f*

8875 object module
d Objektmodul *m*
f module *m* objet
r объектный модуль *m*

8876 object program
d Objektprogramm *n*; übersetztes Programm *n*
f programme *m* objet
r объектная программа *f*

8877 object recognition device
d Objekterkennungseinrichtung *f*
f dispositif *m* de reconnaissance de l'objet

r устройство *n* распознавания объекта

**8878 object recognition error; object
recognition fault**
d Objekterkennungsfehler *m*
f défaut *m* de reconnaissance d'objet
r ошибка *f* распознавания объекта

* object recognition fault → 8878

**8879 object recognition system; system of object
detection**
d Objekterkennungssystem *n*; System *n* der
Objekterkennung
f système *m* de reconnaissance d'objet
r система *f* распознавания объекта

8880 objet data set
d Objektdatensatz *m*
f jeu *m* de données d'objet
r комплект *m* данных объекта

8881 observability
d Beobachtbarkeit *f*
f observabilité
r наблюдаемость *f*

8882 observe *v*
d beobachten
f observer; surveiller
r наблюдать

8883 obtainable accuracy
d erreichbare Genauigkeit *f*
f précision *f* obtenable
r достижимая точность *f*

* occurrence → 5222

8884 odd-even check; parity check
d Paritätskontrolle *f*
f essai *m* pair-impair; contrôle *m* de parité
r проверка *f* нечётности

8885 odd function
d ungerade Funktion *f*
f fonction *f* impaire
r нечётная функция *f*

8886 odd harmonic
d ungeradzahlige Harmonische *f*
f harmonique *f* impaire
r нечётная гармоника *f*

8887 odd symmetrical non-linearity
d ungerade symmetrische Nichtlinearität *f*
f non-linéarité *f* à symétrie impaire
r нечётная симметричная нелинейность *f*

8888 off; out of circuit *adj*
  d aus; off
  f hors; off
  r выключенный

8889 off-balance
  d aus dem Gleichgewicht
  f déséquilibré
  r неуровновешенный; несбалансированный

8890 off condition; off state
  d Aus-Zustand *m*
  f état *m* au repos
  r выключенное состояние *n*

  *   offering → 10372

8891 off-hook
  d ausgehängt
  f décroché; détaché
  r разъединенный

8892 offline; autonomous
  d abgetrennt; autonom; getrennt; Offline-
  f offline; autonome; hors ligne
  r офлайн; автономный

8893 offline control
  d indirekte Steuerung *f*; Offline-Steuerung *f*
  f commande *f* indirecte; réglage *m* indirect
  r автономное регулирование *n*

8894 offline converting system
  d Offline-Konvertiersystem *n*
  f système *m* convertisseur offline
  r автономная система *f* преобразования

8895 offline mode
  d getrennte Arbeitsweise *f*; Offline-Modus *m*
  f mode *m* offline; mode autonome
  r автономный режим *m*

8896 offline operation
  d indirekte Bearbeitung *f*; unabhängige Betriebsweise *f*
  f opération *f* indirecte
  r независимый режим *m* работы

8897 offline organization
  d Offline-Organisation *f*
  f organisation *f* offline
  r автономная организация *f*

8898 offline processing
  d getrennte Verarbeitung *f*; Offline-Verarbeitung *f*
  f traitement *m* offline
  r автономная обработка *f*

8899 offline programming
  d Offline-Programmierung *f*
  f programmation *f* offline
  r автономное программирование *n*

8900 off-load
  d ausgeschaltet
  f hors circuit; déchargé
  r без нагрузки

8902 offset *v*
  d abweichen; versetzen; verschieben
  f dévier; décaler
  r несовпадать; смещаться

8901 offset; steady-state error
  d bleibende Regelabweichung *f*; statischer Fehler *m*
  f écart *m* de réglage permanent; erreur *f* établie; erreur stationnaire
  r установившееся рассогласование *n*; установившаяся ошибка *f*

8903 offset characteristic
  d Regelwirkung *f* mit teilweise unterdrücktem Bereich
  f statisme *m*
  r статическая характеристика *f*

8904 offset coefficient
  d Dauerabweichungskoeffizient *m*
  f coefficient *m* d'écart permanent
  r коэффициент *m* установившегося рассогласования

8905 offset frequency
  d versetzte Frequenz *f*
  f fréquence *f* décalée
  r смешанная частота *f*

  *   offset ratio → 4046

8906 offset steady-state deviation
  d bleibende Regelabweichung *f*
  f écart *m* permanent
  r постоянное отклонение *n*

8907 off-sites
  d Nebenanlagen *fpl*
  f installations *fpl* annexes
  r вспомагательные установки *fpl*

  *   off state → 8890

8908 ohmic contact
  d ohmscher Kontakt *m*; galvanischer Kontakt
  f contact *m* ohmique
  r омический контакт *m*

8909 ohmic drop; resistance drop
  d ohmscher Spannungsabfall *m*
  f chute *f* de tension ohmique
  r омическое падение *n* напряжения

8910 ohmic heating
  d ohmsche Aufheizung *f*
  f chauffage *m* par effet Joule
  r омический нагрев *m*

8911 ohmic load; resistive load
  d ohmsche Last *f*; ohmsche Belastung *f*;
     Widerstandslast *f*
  f charge *f* ohmique
  r омическая нагрузка *f*

8912 ohmic loss
  d ohmscher Verlust *m*
  f perte *f* ohmique
  r омические потери *fpl*

8913 oil-hydraulic speed controller
  d ölhydraulischer Geschwindigkeitsregler *m*
  f régulateur *m* de vitesse oléohydraulique
  r масляно-гидравлический регулятор *m*
     скорости

8914 oil-operated control
  d hydraulische Regelung *f*
  f réglage *m* hydraulique
  r гидравлическое регулирование *n*

  * **oil-operated controller** → 6253

8915 oil-operated drive; hydraulic drive
  d hydraulischer Antrieb *m*
  f commande *f* hydraulique
  r гидравлический привод *m*

8916 oil-operated power cylinder
  d hydraulischer Servomotor *m*
  f servomoteur *m* hydraulique
  r гидравлический серводвигатель *m*

8917 oil-pneumatic
  d ölpneumatisch
  f oléopneumatique
  r масляно-пневматический

8918 oil switch
  d Ölschalter *m*
  f interrupteur *m* à huile
  r масляный выключатель *m*

8919 oil traces measuring instrument
  d Ölspurenmessgerät *n*
  f instrument *m* mesureur des traces d'huile
  r прибор *m* для определения следов масла

**8920 one-axis laser gyroscope; single-axis laser
        gyroscope**
  d Einachsenlasergyroskop *n*
  f gyroscope *m* laser à un axe
  r лазерный гироскоп *m* с одной
     измерительной осью

8921 one-dimensional
  d eindimensional
  f unidimensionnel; à une dimension
  r одномерный

8922 one-dimensional circuit
  d eindimensionale Kette *f*
  f réseau *m* à une dimension
  r линейная цепь *f*

8923 one-dimensional data field
  d eindimensionales Datenfeld *n*
  f champ *m* unidimensionnel des données
  r одномерный массив *m* данных

8924 one-dimensional evaluation
  d eindimensionale Bewertung *f*
  f évaluation *f* unidimensionnelle
  r одномерная оценка *f*

8925 one-dimensional evaluation system
  d eindimensionales Bewertungssystem *n*
  f système *m* d'évaluation à une dimension
  r одномерная система *f* оценки

8926 one-dimensional scanning
  d eindimensionale Abtastung *f*
  f exploration *f* unidimensionelle; balayage *m* à
     dimension unique
  r одноразмерное сканирование *n*

8927 one-line control
  d Steuerung *f* über eine einzelne Leitung
  f commande *f* à ligne unique
  r управление *n* поточной линией

8928 one-loop control system
  d einfaches Regelkreissystem *n*
  f système *m* asservi à une boucle
  r одноконтурная система *f* управления

8929 one-pulse delay
  d Einzelimpulsverzögerung *f*
  f retard *m* d'une impulsion
  r задержка *f* в один такт

  * **one-shot multivibrator** → 8343

8930 one-sided Laplace transformation
  d einseitige Laplace-Transformation *f*

f transformation *f* de Laplace unilatérale
r одностороннее преобразование *n* Лапласа

8931 one-sided limit
  d einseitiger Grenzwert *m*
  f limite *f* unilatérale
  r односторонний предел *m*

8932 one-stage amplifier; single-stage amplifier
  d einstufiger Verstärker *m*
  f amplificateur *m* monoétage; amplificateur à étage unique
  r однокаскадный усилитель *m*; одноступенчатый усилитель

8933 one state
  d Ein-Zustand *m*
  f état *m* un
  r единичное состояние *n*

  * **one-step operation → 11627**

8934 one-tact relay system
  d Eintaktrelaissystem *m*
  f dispositif *m* de commutation à séquence unique
  r однотактовое релейное устройство *n*

8935 one-time application
  d einmalige Anwendung *f*
  f application *f* unique
  r однократное применение *n*

8936 one-to-zero ratio
  d Verhältnis *n* Einersignal-Nullsignal
  f rapport *m* du signal un au signal zéro
  r отношение *n* единичного сигнала к нулевому сигналу

8937 one-valued function; simple function
  d eindeutige Funktion *f*; einfache Funktion
  f fonction *f* univalente; fonction simple
  r однозначная функция *f*

8938 one-way automaton
  d Einwegautomat *m*
  f automate *m* unidirectionnel
  r одноканальный автомат *m*

8939 online
  d mitlaufend; prozessgekoppelt; Online-
  f en ligne; online
  r неавтономный; зависимый

8940 online data acquisition; online data collection
  d Online-Datenerfassung *f*; prozessgekoppelte Datenerfassung *f*

f acquisition *f* données online
r неавтономный сбор *m* данных

  * **online data collection → 8940**

8941 online data processing
  d Online-Datenverarbeitung *f*; prozessgekoppelte Datenverarbeitung *f*
  f traitement *m* de données online
  r обработка *f* данных в режиме "онлайн"

8942 online diagnostics
  d Online-Diagnose *f*; mitlaufende Diagnose *f*
  f diagnostic *m* online
  r неавтономная диагностика *f*

8943 online equipment; online unit
  d angeschlossene Einheit *f*; angeschlossenes Gerät *n*
  f dispositif *m* connecté; unité *f* online
  r неавтономное оборудование *n*

8944 online system
  d Online-System *n*
  f système *m* online
  r неавтономная система *f*

8945 online test facility
  d Online-Testeinrichtung *f*; Testeinrichtung *f* für mitlaufende Prüfung
  f dispositif *m* de test online
  r средства *npl* контроля работающие в неавтономном режиме

  * **online unit → 8943**

8946 on-off
  d auf-zu; ein-aus
  f par tout ou rien; marche-arrêt
  r двухпозиционный; включенный-выключенный

8947 on-off control
  d Ein-Aus-Steuerung *f*
  f commande *f* marche-arrêt
  r двухпозиционное управление *n*

8948 on-off controller; on-off regulator; two-level controller; two-position [action] controller; two-step [action] controller
  d Ein-Aus-Regler *m*; Zweipunktregler *m*
  f régulateur *m* par tout ou rien; régulateur à deux paliers
  r релейный регулятор *m*; двухпозиционный регулятор

8949 on-off error detector
  d Zweipunktdetektor *m*

f détecteur *m* à deux paliers
r релейный детектор *m* рассогласования

8950 on-off keying
d Ein-Aus-Tastung *f*
f manipulation *f* par tout ou rien
r двухпозиционное манипулирование *n*

8951 on-off position transmitter
d Zweipunktlagegeber *m*
f transmetteur-positionneur *m* à deux paliers
r двухпозиционный датчик *m* положения

* **on-off regulator → 8948**

8952 on-off servomechanism
d Ein-Aus-Servomechanismus *m*
f système *m* asservi fonctionnant par tout ou
rien
r сервомеханизм *m* двухпозиционного типа;
прерывистый сервомеханизм

8953 on-off switch
d Ein-Aus-Schalter *m*
f commutateur *m* marche-arrêt
r двухпозиционный переключатель *m*

8954 on period
d Flußzeit *f*
f temps *m* d'ouverture
r проводящий период *m*

8955 open-channel flowmeter
d Durchflussmesser *m* für offene Gerinne
f débitmètre *m* pour canaux ouverts
r расходомер *m* для открытых каналов

8956 open circuit
d offener Stromkreis *m*
f circuit *m* ouvert
r разомкнутая цепь *f*

8957 open circuit output conductance
d Ausgangsleitwert *m* bei offenem Eingang;
Leerlaufausgangskonduktanz *f*
f conductance *f* de sortie à vide
r выходная проводимость *f* холостого хода

8958 open-circuit voltage
d Leerlaufspannung *f*
f tension *f* [de marche] à vide
r напряжение *n* холостого хода

8959 open circuit voltage transfer ratio
d Leerlaufspannungsrückwirkungsfaktor *m*
f gain *m* inverse en tension à circuit ouvert
r коэффициент *m* обратной связи по

напряжению

8960 open cycle control
d offener Wirkungskreis *m*
f boucle *f* d'action ouverte; chaîne *f* d'action
ouverte; cycle *m* d'action ouvert
r разомкнутый контур *m* регулирования

8961 opening
d Eröffnung *f*
f ouverture *f*
r открытие *n*

* **opening delay → 11505**

8962 open kinematic chain of gripper gear
d offene kinematische Kette *f* eines
Greifergetriebes
f chaîne *f* cinématique ouverte de l'engrenage
de grappin
r незамкнутая кинематическая цепь *f*
передачи захвата

8963 open loop
d offener Wirkungskreis *m*; offene Schleife *f*
f boucle *f* ouverte
r разомкнутый контур *m*

8964 open-loop control
d offene Schleifensteuerung *f*; offene
Steuerung *f*
f commande *f* en chaîne ouverte
r управление *n* по открытой цепи

* **open-loop control system → 8965**

8965 open-loop[ed] control system
d Steuerungssystem *n* mit offener Schleife
f système *m* de commande à boucle ouverte
r система *f* управления без обратной связи

8966 open-loop gain
d Verstärkung *f* des offenen Kreises
f gain *m* en boucle ouverte
r коэффициент *m* усиления разомкнутого
контура

8967 open-loop pulse system
d offenes Impulssystem *n*; aufgeschnittenes
Impulssystem
f système *m* échantilloné à impulsions à
boucle ouverte
r разомкнутая импульсная система *f*

**8968 open-loop sampled data system with
variable parameters**
d offenes Impulssystem *n* mit veränderlichen
Parametern

    f système *m* échantilloné à boucle ouverte à
       paramètres variables
    r дискретная разомкнутая система *f* с
       переменными параметрами

8969 open-loop transfer function
    d Übertragungsfunktion *f* des offenen Systems;
       Frequenzgang *m* des offenen Kreises
    f fonction *f* de transfert en boucle ouverte;
       transmittance *f* en chaîne ouverte
    r передаточная функция *f* с разомкнутым
       контуром

8970 open-phase protection
    d Leitungunterbrechungsschutz *m*
    f dispositif *m* de protection contre les coupures
       de phase
    r защита *f* от обрыва фаз

8971 open position
    d Ausschaltstellung *f*
    f position *f* d'arrêt; position d'ouverture
    r разомкнутое положение *n*

8972 open system
    d offenes System *n*
    f système *m* ouvert
    r открытая система *f*

8973 operability; operational capability
    d Betriebstüchtigkeit *f*; Betriebsbereitschaft *f*;
       Betriebsfähigkeit *f*
    f aptitude *f* opérationnelle
    r работоспособность *f*

8974 operable; workable
    d betriebsfähig
    f prêt pour le service; en ordre de
       fonctionnement
    r работоспособный; готовый к
       эксплуатации

8975 operate *v*
    d betreiben; arbeiten; bedienen
    f opérer; travailler; asservir
    r оперировать; работать

8976 operating adjustment
    d Betriebseinstellung *f*
    f mise *f* au point au cours du fonctionnement;
       mise au point au cours de l'opération
    r эксплуатационная наладка *f*

8977 operating angle
    d Arbeitswinkel *m*; Betriebswinkel *m*
    f angle *m* de fonctionnement
    r рабочий угол *m*

**8978 operating characteristic; working
      characteristic**
    d Betriebscharakteristik *f*;
       Arbeitscharakteristik *f*
    f caractéristique *f* de fonctionnement;
       caractéristique de travail
    r рабочая характеристика *f*

8979 operating condenser; process condenser
    d Betriebskondensator *m*
    f condenseur *m* de service
    r технологический конденсатор *m*

8980 operating conditions
    d Betriebsbedingungen *fpl*
    f conditions *fpl* de fonctionnement
    r условия *npl* эксплуатации

8981 operating control
    d Ablaufsteuerung *f*
    f commande *f* d'exécution; contrôle *m*
       d'exécution; commande de déroulement
    r управление *n* последовательностью
       операций

   * **operating data** → 9011

8982 operating efficiency
    d Betriebswirkungsgrad *m*
    f rendement *m* en service
    r эксплуатационный коэффициент *m*
       полезного действия

**8983 operating error; faulty operator
      intervention**
    d Fehlbedienung *f*; Bedienungsfehler *m*;
       Fehlbetätigung *f*
    f erreur *f* de manœuvre; manœuvre *f* fausse
    r неправильное управление *n*; ошибка *f*
       управления

8984 operating experience
    d Betriebserfahrungen *fpl*
    f expérience *f* acquise sur le fonctionnement;
       résultat *m* d'exploitation
    r опыт *m* эксплуатации

8985 operating frequency
    d Betriebsfrequenz *f*
    f fréquence *f* de travail; fréquence de service
    r рабочая частота *f*

8986 operating gauge
    d Betriebsmanometer *n*
    f manomètre *m* de service
    r рабочий манометр *m*

8987 operating information system
    d Betriebsinformationssystem *n*

f système m d'information d'exploitation
r информационная система f обеспечения эксплуатации

8988 operating instruction
d Betriebsanleitung f
f instruction f de service
r инструкция f по эксплуатации

8989 operating life
d Lebensdauer f
f longévité f; durée f de service
r долговечность f; эксплуатационный срок m службы

8990 operating limit
d Betriebsgrenze f
f limite f de fonctionnement
r эксплуатационный лимит m

8991 operating limiting signal; operation limiter
d Operationsabschlusssignal n
f signal m de limitation de l'opération
r сигнал m ограничения операции

8992 operating mechanism; drive gear
d Antriebsmechanismus m; Antriebsgetriebe n
f asservissement m; mécanisme m actif; engrenage m d'entraînement
r приводной механизм m

8993 operating memory
d Arbeitsspeicher m
f mémoire f d'opération
r оперативная память f

8994 operating mode switch
d Betriebsartenschalter m; Betriebsartenwahlschalter m
f commutateur m des modes; sélecteur m de fonctions
r переключатель m режимов работы

8995 operating optimization
d Betriebsoptimierung f
f optimisation f de fonctionnement
r оптимизация f функционирования

8996 operating parameter
d Betriebskennwert m
f paramètre m d'exploitation
r эксплуатационный параметр m

8997 operating point
d Betriebspunkt m
f point m de fonctionnement
r рабочая точка f

* **operating position** → 12930

8998 operating power
d Betriebsleistung f
f puissance f d'opération
r эксплуатационная мощность f

8999 operating pressure; working pressure
d Betriebsdruck m; Arbeitsdruck m
f pression f de service
r рабочее давление n

9000 operating principle
d Operationsprinzip n; Arbeitsprinzip n
f principe m d'opération
r принцип m работы

9001 operating procedure
d Arbeitsprozess m
f procédé m de travail; manœuvre f
r рабочий процесс m

9002 operating range; working range
d Operationsbereich m; Arbeitsfeld n; Arbeitsbereich m
f domaine m de fonctionnement; limites fpl d'opération; étendue f d'action
r рабочий диапазон m

9003 operating sequence
d Bediensequenz f
f séquence f opérationnelle
r последовательность f оперирования

9004 operating state; operating status
d Betriebszustand m
f état m en fonctionnement
r эксплуатационное состояние n

9005 operating state of relay circuit
d Relaiskreisbetriebszustand m
f état m de service du circuit de relais
r рабочее состояние n релейной схемы

* **operating status** → 9004

9006 operating system
d Betriebssystem n; Operationssystem n
f système m opérationnel
r операционная система f

9007 operating temperature range
d Betriebstemperaturbereich m
f champ m de température de travail
r диапазон m рабочих температур

9008 operating threshold sensibility
d Ansprechempfindlichkeit f; Arbeitsschwellwertempfindlichkeit f

f sensibilité f au seuil de fonctionnement
r рабочая пороговая чувствительность f

9009 operating time; response time
d Ansprechzeit f
f temps m de réponse; temps de démarrage;
   durée f de réponse
r время n срабатывания

9010 operating trouble
d Betriebsstörung f
f interruption f du travail; arrêt m de service
r авария f производства

9011 operating values; operating data
d Betriebsdaten pl
f données fpl de fonctionnement
r параметры mpl работы

9012 operating voltage of power-direction relay
d Ansprechspannung f des
   Leistungsrichtungsrelais
f tension f de déclenchement de relais
   directionnel de puissance
r напряжение n срабатывания реле
   направления мощности

9013 operation
d Operation f
f opération f
r операция f

9014 operational amplifier
d Operationsverstärker m
f amplificateur m opérationnel
r операционный усилитель m

9015 operational analysis
d Operationsanalyse f
f analyse f opérationnelle
r операционный анализ m

* **operational capability** → 8973

**9016 operational command; operational
      instruction**
d Operationsbefehl m
f instruction f d'opération
r операционная команда f; операционная
   инструкция f

9017 operational condition; working condition
d Betriebszustand m
f état m de marche
r режим m работы

9018 operational condition monitoring
d Betriebszustandsüberwachung f;

Überwachung f eines Betriebszustands
f surveillance f d'un état de marche
r контроль m режима работы

9019 operational diagnostics
d Betriebsdiagnostik f
f diagnostic m de fonctionnement
r эксплуатационная диагностика f

* **operational feed** → 10170

9020 operation algorithm
d Operationsalgorithmus m;
   Steuerungsalgorithmus m;
   Lösungsalgorithmus m
f algorithme m opérationnel
r алгоритм m действия

9021 operational input
d Funktionseingang m
f entrée f opérationnelle
r функциональный вход m

* **operational instruction** → 9016

9022 operational logical circuit
d logische Operationsschaltung f
f circuit m opérationnel logique
r логическая операционная схема f

9023 operational output
d Funktionsausgang m
f sortie f opérationnelle
r функциональный выход m

9024 operational programming method
d Programmierungsoperatormethode f;
   Operatormethode f der Programmierung
f méthode f opérationnelle de programmation
r операторный метод m программирования

**9025 operational reserch; enterprise
      investigation**
d Verfahrensforschung f;
   Unternehmensforschung f
f recherche f opérationnelle; investigation f
   entreprise
r исследование n операций

9026 operational shutdown
d betriebsbedingte Abschaltung f; betriebliches
   Abfahren n
f arrêt m opérationnel
r рабочий останов m

9027 operational transient
d Betriebstransiente f

f transitoire *m* de fonctionnement
r переходный технологический процесс *m*

9028 operation code
d Operationskode *m*
f code *m* des opérations
r код *m* операции

* **operation console → 1728**

9029 operation decoder
d Entschlüßler *m* der Operationen; Operationsentschlüßler *m*
f décodeur *m* d'opérations
r дешифратор *m* операции

9030 operation delay of a circuit-breaker
d Schaltverzug *m* des Leistungsschalters
f temps *m* d'intervention du disjoncteur
r выдержка *f* времени срабатывания выключателя

* **operation error → 118**

9031 operation factor
d Operationsfaktor *m*; Betriebskoeffizient *m*
f facteur *m* d'opération
r эксплуатационный коэффициент *m*

* **operation limiter → 8991**

9032 operation method; method of operation
d Operationsmethode *f*
f méthode *f* d'opération
r операционный метод *m*

9033 operation of conditional transfer of control
d bedingte Sprungoperation *f*
f opération *f* de transfertconditionnel
r операция *f* условной передачи

9034 operation of testing
d Prüfvorgang *m*
f opération *f* d'essai
r процесс *m* тестирования

9035 operation procedure
d Arbeitsablauf *m*
f succession *f* des opérations
r последовательность *f* операций

9036 operation sequence
d Betätigungsfolge *f*
f séquence *f* de manœuvres
r последовательность *f* действий

* **operation speed → 2409**

9037 operation's related defect
d betriebsbezogener Schaden *m*
f défaut *m* en exploitation; défaut relié au fonctionnement
r связанный с эксплуатацией дефект *m*

9038 operations research problem
d Operations-Research-Problem *n*; Problem *n* der Operationsforschung
f problème *m* de recherche opérationnelle
r проблема *f* исследования операций

9039 operation threshold
d Ansprechschwelle *f*
f seuil *m* d'opération
r порог *m* срабатывания

9040 operation voltage; normal voltage
d Betriebsspannung *f*
f tension *f* d'emploi
r рабочее напряжение *n*

* **operative position → 12930**

9041 operative program part; operative program section
d operativer Programmteil *m*
f part *m* de programme opératif; section *f* de programme opérative
r программная секция *f* оперативной памяти

* **operative program section → 9041**

9042 operator equation
d Operatorgleichung *f*
f équation *f* d'opérateur
r операторное уравнение *n*

9043 optical access
d optischer Zugriff *m*
f accés *m* optique
r оптический доступ *m*

9044 optical-acoustic gas analyzer
d optisch-akustischer Gasanalysator *m*
f analyseur *m* optique-acoustique de gaz
r оптико-акустический газоанализатор *m*

* **optical acquisition sensor → 9109**

9045 optical alignment
d optische Einstellung *f*; optische Ausrichtung *f*

f ajustage *m* optique
r оптическая настройка *f*

9046 optical altimeter
d optischer Höhenmesser *m*
f altimètre *m* optique
r оптический высотомер *m*

9047 optical amplification
d optische Verstärkung *f*
f amplification *f* optique
r оптическое усиление *n*

9048 optical amplifier
d optischer Verstärker *m*
f amplificateur *m* optique
r оптический усилитель *m*

9049 optical amplifier bandwidth
d Bandbreite *f* des optischen Verstärkers
f largeur *f* de bande de l'amplificateur optique
r полоса *f* пропускания лазера; ширина *f*
    полосы частот оптического усилителя

9050 optical analog computer
d optischer Analogrechner *m*
f calculateur *m* analogue optique
r оптическая аналоговая вычислительная
    машина *f*

9051 optical attenuator
d optisches Dämpfungsglied *n*
f atténuateur *m* optique
r оптический аттенюатор *m*

9052 optical balancing element
d optisches Ausgleichselement *n*
f élément *m* de balance optique
r оптический компенсатор *m*

9053 optical beam deflection
d optische Strahlenauslenkung *f*
f déviation *f* optique du faisceau
r отклонение *n* оптического луча

9054 optical beam-direction control
d optische Strahlenrichtungssteuerung *f*
f contrôle *m* optique de la direction du faisceau
r оптическое управление *n* лучом

9055 optical channel
d optischer Kanal *m*
f voie *f* optique
r оптический канал *m*

9056 optical character reader
d Klarschriftleser *m*; optischer Zeichenleser *m*

f lecteur *m* optique de caractères
r оптическое считывающее устройство *n*

* **optical check → 9057**

9057 optical check[ing]
d optische Kontrolle *f*
f contrôle *m* optique
r оптический контроль *m*

9058 optical circuit
    (of sensor)
d optischer Schaltkreis *m*
f circuit *m* optique
r оптическая схема *f*

9059 optical coherent radar
d optisches kohärentes Radar *n*
f radar *m* optique cohérent
r оптический когерентный локатор *m*

9060 optical communication device
d optisches Verbindungsgerät *n*
f dispositif *m* de communication optique
r устройство *n* оптической связи

9061 optical communication system
d optisches Kommunikationssystem *n*;
    optisches Fernmeldesystem *n*
f système *m* optique de télécommunication
r система *f* связи в оптическом диапазоне

9062 optical comparator
d optischer Vergleicher *m*; optischer
    Gleichheitsprüfer *m*
f comparateur *m* optique
r оптический компаратор *m*

9063 optical compensating filter
d optisches Kompensationsfilter *n*
f filtre *m* optique de compensation
r оптический компенсирующий фильтр *m*

9064 optical contact
d optischer Kontakt *m*
f contact *m* optique
r оптический контакт *m*

9065 optical correlation
d optische Korrelation *f*
f corrélation *f* optique
r оптическая корреляция *f*

9066 optical coupler
d optischer Koppler *m*
f coupleur *m* optique
r оптическое устройство *n* связи

9067 optical data handling
 d optische Datenverarbeitung *f*; Verarbeitung *f*
   optischer Daten
 f traitement *m* optique des données; traitement
   des données optiques
 r оптическая обработка *f* данных

9068 optical data processing system
 d optisches Datenverarbeitungssystem *n*
 f système *m* optique de traitement des données
 r оптическая система *f* обработки данных

9069 optical delay circuit
 d optische Verzögerungschaltung *f*
 f circuit *m* optique de retard
 r оптическая схема *f* задержки

9070 optical delay line
 d optische Verzögerungsleitung *f*; optisches
   Laufzeitglied *n*
 f ligne *f* à retard optique
 r оптическая линия *f* задержки

9071 optical density
 d optische Dichte *f*
 f densité *f* optique
 r оптическая плотность *f*

9072 optical detection
 d optische Detektion *f*
 f détection *f* optique
 r оптическое детектирование *n*

9073 optical detector technology
 d Infrarotstrahlendetektortechnik *f*
 f technique *f* du détecteur pour l'infrarouge
 r техническое оснащение *n* оптических
   детекторов

9074 optical direct amplification
 d optische Geradeausverstärkung *f*
 f amplification *f* optique directe
 r прямое оптическое усиление *n*

9075 optical directional coupler
 d optischer Richtkoppler *m*
 f coupleur *m* optique directif; coupleur optique
   directionnel
 r оптический направленный ответвитель *m*

**9076 optical display system; optical projection
   system**
 d optisches Anzeigesystem *n*; optische
   Anzeige *f*
 f système *m* optique d'affichage; système
   indicateur optique
 r система *f* оптического вывода; оптическая
   индикация *f*

9077 optical efficiency
 d optischer Wirkungsgrad *m*
 f rendement *m* optique
 r оптическая эффективность *f*

9078 optical-electronic coupling element
 d optoelektronisches Kopplungselement *n*
 f élément *m* de couplage optoélectronique
 r оптический электронный элемент *m* связи

9079 optical encoder
 d optischer Kodierer *m*; optischer
   Verschlüßler *m*
 f codeur *m* optique; dispositif *m* optique de
   codage
 r оптическое кодирующее устройство *n*

9080 optical energy density
 d optische Energiedichte *f*
 f densité *f* d'énergie optique
 r оптическая плотность *f* энергии

9081 optical equalization
 d optische Entzerrung *f*
 f égalisation *f* optique
 r оптическое выравнивание *n*

9082 optical feedback
 d optische Rückkopplung *f*
 f réaction *f* optique
 r оптическая обратная связь *f*

9083 optical fiber measurements
 d Lichtwellenleiter-Messtechnik *f*
 f mesures *fpl* des fibres optiques
 r световодная измерительная техника *f*

9084 optical fiber power splitter
 d optischer Leistungsteiler *m*
 f diviseur *m* de puissance à fibres optiques
 r оптический делитель *m* мощности

9085 optical fiber system
 d Lichtwellenleitersystem *n*
 f système *m* sur fibres optiques
 r световодная система *f*

9086 optical fiber transfer function
 d Lichtwellenleiter-Übertragungsfunktion *f*
 f fonction *f* de transfert de fibre optique
 r световодная передаточная функция *f*

*   optical fibres technology → 7245

9087 optical filter
 d optisches Filter *n*
 f filtre *m* optique
 r оптический фильтр *m*; светофильтр *m*

9088 optical frequency
  d optische Frequenz *f*
  f fréquence *f* optique
  r частота *f* сигнала оптического диапазона

9089 optical identification system; optical perception system
  d optisches Erkennungssystem *n*
  f système *m* de reconnaissance optique; système *m* d'identification optique
  r оптическая система *f* идентификации

9090 optical image chip
  d Bildwandlerchip *n*; Halbleiterbildwandler *m*
  f puce *f* de conversion d'image
  r микросхема *f* преобразования изображения

9091 optical imaging system
  d optisches Abbildungssystem *n*
  f système *m* d'image optique
  r оптическая система *f* отображения

9092 optical information processing system
  d optisches Informationsverarbeitungssystem *n*
  f système *m* de traitement d'information optique
  r оптическая система *f* обработки информации

9093 optical input
  d optischer Eingang *m*
  f entrée *f* optique
  r оптический входной сигнал *m*

9094 optical interference filter
  d optisches Interferenzfilter *n*
  f filtre *m* optique d'interférence
  r поляризационно-интерференционный светофильтр *m*

9095 optical laser radar
  d optisches Laserradar *n*
  f radar *m* optique à laser
  r оптический лазерный локатор *m*

9096 optical line scan equipment
  d optische Zeilenabtastungseinheit *f*
  f appareillage *m* optique de balayage de lignes
  r оптическая аппаратура *f* с линейной разверткой

9097 optical logic
  d optische Logik *f*
  f logique *f* optique
  r оптическая логика *f*

9098 optically excited
  d optisch angeregt
  f excité optiquement
  r оптически возбужденный

9099 optically pumped
  d optisch gepumpt
  f à pompage optique
  r оптически накачанный

9100 optical measuring method
  d optisches Messverfahren *n*
  f méthode *f* de mesure optique
  r оптический метод *m* измерения

9101 optical memory
  d optischer Speicher *m*
  f mémoire *f* optique
  r оптическая память *f*

9102 optical mounting
  d optische Montage *f*
  f montage *m* optique
  r сборка *f* с применением оптических средств управления

9103 optical noise autocorrelation
  d Autokorrelation *f* des optischen Rauschsignals
  f autocorrélation *f* du bruit optique
  r автокорреляция *f* помех в оптическом диапазоне

9104 optical output signal
  d optisches Ausgangssignal *n*
  f signal *m* de sortie optique
  r оптический выходной сигнал *m*

9105 [optical] parametric oscillator
  d [optischer] parametrischer Oszillator *m*
  f oscillateur *m* paramétrique [optique]; générateur *m* paramétrique [optique]
  r [оптический] параметрический генератор *m*

* optical perception system → 9089

9106 optical perception system for visual information
  d optisches Erkennungssystem *n* für visuelle informationen
  f système *m* de reconnaissance pour informations visuelles
  r оптическая система *f* распознавания визуальной информации

9107 optical phase deviation
  d optische Phasenabweichung *f*

f déviation *f* de phase optique; dérive *f* de
phase optique
r девиация *f* фазы оптического сигнала;
отклонение *n* фазы оптического сигнала

9108 optical phase-difference radar
d optisches Phasendifferenzradar *n*
f radar *m* optique à déphasage
r оптический локатор *m* со сдвигом по фазе

9109 optical pick-off; optical acquisition sensor
d optischer Geber *m*; optischer Fühler *m*;
optischer Erfassungssensor *m*
f capteur *m* [d'acquisition] optique
r оптический датчик *m*; оптический
сенсор *m*

9110 optical polarization method
d polarisationsoptisches Verfahren *n*
f méthode *f* optique de polarisation
r оптический метод *m* поляризации

9111 optical power meter
d optischer Leistungsmesser *m*
f mesureur *m* de puissance optique
r оптический измеритель *m* мощности

9112 optical processing circuit
d optischer Verarbeitungskreis *m*
f circuit *m* de traitement optique
r схема *f* преработки оптических сигналов

* **optical projection system** → 9076

9113 optical pulse code modulation
d optische Impulskodemodulation *f*
f modulation *f* optique d'impulsions
r импульсно-кодовая модуляция *f*
оптического сигнала

9114 optical radar transmitter
d optischer Radarsender *m*
f émetteur *m* du radar optique
r передатчик *m* оптического локатора

9115 optical reading
d optisches Lesen *n*
f lecture *f* optique
r оптическое считывание *n*

9116 optical resonance
d optische Resonanz *f*
f résonance *f* optique
r оптический резонанс *m*

* **optical scanner** → 9658

9117 optical scanning

d optische Abtastung *f*
f balayage *m* optique
r оптическая развёртка *f*

9118 optical scanning system
d optisches Abtastsystem *n*
f système *m* optique de balayage; système
optique d'exploration
r оптическая сканирующая система *f*

**9119 optical sensor for displacement
measurement**
d optischer Wegmesswandler *m*
f capteur *m* optique pour la mesure de
déplacements
r оптический датчик *m* перемещений

9120 optical sensorics; optical sensor technique
d optische Sensorik *f*; optische Sensortechnik *f*
f technique *f* de capteur optique; sensorique *f*
optique
r оптическая сенсорика *f*

* **optical sensor technique** → 9120

9121 optical signal; visual signal; visible signal
d optisches Signal *n*
f signal *m* optique
r оптический сигнал *m*

9122 optical signal carrier
d optischer Signalträger *m*
f porteuse *f* optique du signal
r несущая частота *f* оптического сигнала

9123 optical signal entropy
d Entropie *f* des optischen Signals
f entropie *f* du signal optique
r энтропия *f* оптического сигнала

9124 optical signal treatment
d optische Signalverarbeitung *f*
f traitement *m* de signal optique
r обработка *f* оптических сигналов

9125 optical simulation
d optische Nachbildung *f*
f simulation *f* optique
r оптическое моделирование *n*

9126 optical spectroscopy
d optische Spektroskopie *f*
f spectroscopie *f* optique
r оптическая спектроскопия *f*

9127 optical storage line
d optische Speicherschaltung *f*

    f circuit *m* optique de mémoire
    r оптическая запоминающая схема *f*

9128 optical superposition device
    d optisches Überlagerungsgerät *n*
    f dispositif *m* à superposition optique
    r устройство *n* для наложения оптического изображения

9129 optical switching circuit
    d optischer Schaltkreis *m*
    f circuit *m* de commutation optique
    r оптическая переключающая схема *f*

9130 optical transfer function
    d optische Übertragungsfunktion *f*
    f fonction *f* optique de transfert
    r оптическая передаточная функция *f*

9131 optical transmitting set
    d optischer Sender *m*; optische Sendeeinrichtung *f*
    f émetteur *m* optique
    r оптический передатчик *m*

9132 optimal action strategy
    d optimale Handlungsstrategie *f*
    f stratégie *f* d'action optimale
    r оптимальная стратегия *f* действий

9133 optimal adaptation
    d optimale Adaptation *f*; optimale Anpassung *f*
    f adaptation *f* optimale
    r оптимальное согласование *n*

9134 optimal adjustment
    d optimale Einstellung *f*
    f accord *m* optimal
    r оптимальная настройка *f*

9135 optimal assembly sequence
    d optimale Montage[reihen]folge *f*
    f séquence *f* d'assemblage optimale
    r оптимальная последовательность *f* сборочных операций

    * **optimal control → 5352**

9136 optimal control criterion
    d Kriterium *n* der optimalen Steuerung
    f critère *m* de commande optimale
    r критерий *m* оптимального регулирования

9137 optimal control system
    d optimales Regelungs- und Steuerungssystem *n*
    f système *m* optimal de réglage et commande
    r оптимальная система *f* управления

9138 optimal damping
    d optimale Dämpfung *f*
    f amortissement *m* optimal
    r оптимальное демпфирование *n*

9139 optimal digitization
    d optimale Digitalisierung *f*; optimale Digitalisation *f*
    f digitalisation *f* optimale
    r оптимальная дискретизация *f*

9140 optimal filter
    d Optimalfilter *n*
    f filtre *m* optimal
    r оптимальный фильтр *m*

9141 optimal inventory
    d optimale Lagerhaltung *f*
    f inventaire *m* optimal
    r оптимальное управление *n* запасами

9142 optimal inventory problem
    d optimales Lagerhaltungsproblem *n*
    f problème *m* de l'inventaire optimal
    r задача *f* оптимального управления запасами

9143 optimality criterion
    d Optimalitätskriterium *n*
    f critérium *m* optimum
    r критерий *m* оптимальности

9144 optimality demand
    d Optimalitätsforderung *f*
    f demande *f* d'optimalité
    r требование *n* к оптимальности

9145 optimally coded program
    d Bestzeitprogramm *n*
    f programme *m* optimum
    r оптимально кодированная программа *f*

9146 optimal parameter
    d optimale Kenngröße *f*
    f paramètre *m* optimal
    r оптимальный параметр *m*

9147 optimal path control
    d optimale Bahnsteuerung *f*
    f commande *f* de chemin optimale
    r оптимальный режим *m* контурного управления

9148 optimal point control; optimal spot control
    d optimale Punktsteuerung *f*
    f commande *f* optimale par points; commande de spot optimale
    r оптимальное позиционное управление *n*

9149 **optimal process solution**
 d optimale Prozesslösung *f*
 f dissolution *f* de processus optimale
 r оптимальное решение *n* процесса

 * **optimal spot control** → 9148

9150 **optimal strategy**
 d optimale Strategie *f*
 f stratégie *f* optimale
 r оптимальная стратегия *f*

9151 **optimal switching**
 d optimales Schalten *n*
 f couplage *m* optimal
 r оптимальное переключение *n*

9152 **optimal system; optimizing system**
 d Optimalsystem *n*
 f système *m* optimal
 r оптимальная система *f*

 * **optimal throughput** → 9179

9153 **optimization method**
 d Optimierungsmethode *f*
 f méthode *f* d'optimisation
 r метод *m* оптимизации

9154 **optimization of dynamic systems**
 d Optimierung *f* dynamischer Systeme
 f optimisation *f* des systèmes dynamiques
 r оптимизация *f* динамических систем

9155 **optimization of sustained reaction**
 d Optimierung *f* von selbst ablaufender Reaktion
 f optimisation *f* de réaction autoparcourante
 r подбор *m* оптимальных условий для незатухающих реакций

9156 **optimization problem**
 d Optimierungsproblem *n*
 f problème *m* d'optimisation
 r задача *f* оптимизации

9157 **optimization system for order processing**
 d Optimierungssystem *n* für Auftragsabwicklung
 f système *m* d'optimisation pour le déroulement des commandes
 r оптимальная система *f* для порядка обработки

9158 **optimization variable**
 d Optimierungsvariable *f*
 f variable *f* d'optimisation
 r переменная *f* оптимизация

9159 **optimize** *v*
 d optimieren
 f optimiser
 r оптимизировать

9160 **optimizer**
 d Optimisator *m*
 f optimisateur *m*
 r оптимизатор *m*

 * **optimizing control** → 5352

 * **optimizing controller** → 5349

 * **optimizing system** → 9152

9161 **optimum**
 d Optimum *n*; Extremalwert *m*
 f optimum *m*
 r оптимум *m*

9162 **optimum behaviour**
 d optimaler Betriebszustand *m*
 f fonctionnement *m* optimum
 r оптимальный режим *m*

9163 **optimum coding**
 d Bestkodierung *f*; optimale Kodierung *f*
 f codage *m* optimum
 r оптимальное кодирование *n*

9164 **optimum condition**
 d Optimalbedingung *f*
 f condition *f* optimale
 r оптимальное условие *n*

 * **optimum control** → 5352

9165 **optimum coupling**
 d optimale Kopplung *f*
 f couplage *m* optimal
 r оптимальная связь *f*

9166 **optimum detecting filter**
 d optimales Suchfilter *n*
 f filtre *m* de détection optimum
 r оптимальный фильтр *m* поиска

9167 **optimum point**
 d Optimalpunkt *m*
 f point *m* optimum
 r оптимальная точка *f*

9168 **optimum predictor**
 d optimaler Extrapolator *m*
 f extrapolateur *m* optimal; prédicteur *m* optimal
 r оптимальное прогнозирующее устройство *n*

9169 optimum process
d optimaler Prozess *m*; optimaler Verlauf *m*
f processus *m* optimal
r оптимальный процесс *m*

9170 optimum programming
d optimales Programmieren *n*
f programmation *f* optimum; programmation optimale
r оптимальное программирование *n*

9171 optimum relay servomechanism
d optimaler Relaisservomechanismus *m*
f servomécanisme *m* optimal à relais
r оптимальная релейная следящая система *f*

9172 optimum response
d Optimalantwort *f*
f réponse *f* optimale
r оптимальная чувствительность *f*

9173 optimum sampled-data system
d optimales Impulssystem *n*; optimales Datenabtast system *n*
f système *m* optimal par impulsion; système optimal d'échantillonnage de données
r оптимальная импульсная система *f*

9174 optimum seeking method
d optimale Suchmethode *f*
f méthode *f* de recherche optimale
r оптимальный метод *m* поиска

9175 optimum structure
d optimale Struktur *f*
f structure *f* optimum
r оптимальная структура *f*

9176 optimum switching function
d optimale Schaltfunktion *f*
f fonction *f* de commutation optimale
r оптимальная функция *f* переключения

9177 optimum switching line
d optimale Schaltlinie *f*
f ligne *f* de commutation optimale
r оптимальная кривая *f* переключения

9178 optimum system synthesizer
d Synthesator *m* optimaler Systeme
f synthéseur *m* de systèmes optimaux
r синтезатор *m* оптимальных систем

9179 optimum throughput; optimal throughput
d optimaler Durchsatz *m*
f débit *m* optimal; débit optimum
r оптимальная пропускная способность *f*

9180 optimum transfer function
d optimale Übertragungsfunktion *f*
f fonction *f* optimale de transfert
r оптимальная передаточная функция *f*

9181 optimum transient response
d optimaler Übergangsprozess *m*; optimale Übergangscharakteristik *f*
f réponse *f* transitoire optimale
r оптимальный переходный процесс *m*

9182 optimum value
d Optimalwert *m*
f valeur *f* optimale
r оптимальное значение *n*

9183 optional facility
d wahlweise Zusatzeinrichtung *f*; wahlweiser Zusatz *m*
f supplément *m* par option
r дополнительное средство *n*

9184 optional modification; selective modification
d wahlweise Modifikation *f*
f modification *f* au choix
r селективная модификация *f*

9185 option switch
d Wahlschalter *m*; Wähler *m*
f sélecteur *m*
r переключатель *m* выбора программы

9186 optoelectronic circuit
d optoelektronische Schaltung *f*
f circuit *m* optoélectronique
r оптоэлектронная схема *f*

9187 optoelectronic component
d optoelektronisches Bauelement *n*
f composant *m* optoélectronique
r оптоэлектронный компонент *m*

9188 optoelectronic data storage
d optoelektronische Datenspeicherung *f*
f emmagasinage *m* optoélectronique des données
r оптоэлектронное устройство *n* хранения данных

9189 optoelectronic digital logic
d optoelektronische Digitallogik *f*
f logique *f* digitale optoélectronique
r оптоэлектронная цифровая логическая схема *f*

9190 optoelectronic sensor
d optoelektronischer Sensor *m*

f senseur *m* optoélectronique
r оптоэлектронный сенсор *m*

9191 optoelectronic sensor system
d optoelektronisches Sensorsystem *n*
f système *m* de senseur optoélectronique
r оптоэлектронная сенсорная система *f*

9192 optoelectronic signal generator
d optoelektronischer Signalgeber *m*
f générateur *m* optoélectronique de signaux
r оптоэлектронный генератор *m* сигналов

9193 optoelectronic system
d optisch-elektronisches System *n*;
   optoelektronisches System
f système *m* optoélectronique
r оптоэлектронная система *f*

9194 OR-circuit
d ODER-Schaltung *f*
f circuit *m* OU
r схема *f* ИЛИ

9195 OR-component; OR-element
d ODER-Glied *n*
f élément *m* OU
r элемент *m* ИЛИ

9196 order
d Ordnung *f*
f ordre *m*
r порядок *m*

9197 order *v*
d ordnen; anordnen
f ordonner
r упорядочивать

9198 order cancel
d Befehlsaufhebung *f*
f suppression *f* de l'ordre
r отмена *f* команды; аннулирование *n*
   команды

9199 order code
d Befehlskode *m*
f code *m* d'instructions
r код *m* команды

9200 order device
d Ordnungseinrichtung *f*
f dispositif *m* d'ordre
r устройство *n* ориентирования

9201 ordered structure
d geordnete Struktur *f*
f structure *f* ordonnée

r упорядоченная структура *f*

9202 order element
d Befehlselement *n*
f élément *m* d'instruction
r элемент *m* команды

9203 order from outside
d Außenbefehl *m*
f commande *f* extérieure
r внешняя команда *f*

9204 order of connexion
d Kopplungsreihenfolge *f*;
   Kopplungsanordnung *f*
f ordre *m* d'accouplement
r порядок *m* соединения

9205 order of controlled system
d Ordnung *f* der Regelstrecke
f ordre *m* du système réglé
r порядок *m* действия регулируемой
   системы

9206 order of logic function
d Ordnung *f* der logischen Funktion
f ordre *m* de fonction logique
r порядок *m* логической функции

9207 order of magnitude
d Größenordnung *f*
f ordre *m* de grandeur
r порядок *m* величины

9208 order of priority
d Prioritätsordnung *f*
f ordre *m* de priorité
r порядок *m* приоритета

9209 order of switching
d Schaltreihenfolge *f*; Schaltanordnung *f*
f ordre *m* de commutation
r порядок *m* включения

9210 order parameter
d Ordnungsparameter *m*
f paramètre *m* d'ordre
r порядковый параметр *m*

9211 order register
d Befehlsregister *n*
f registre *m* d'instructions
r регистр *m* команд

9212 order structure
d Befehlsanordnung *f*; Befehlsaufbau *m*;
   Befehlsstruktur *f*

f structure *f* de l'instruction
r форма *f* инструкции

9213 order transmission
d Befehlsübertragung *f*
f transmission *f* d'ordres
r передача *f* команды

9214 ordinary point
d gewöhnlicher Punkt *m*
f point *m* ordinaire
r регулярная точка *f*

* **OR-element → 9195**

9215 organization of industrial management
d Organisation *f* der Betriebsführung und Betriebswirtschaft
f organisation *f* de gestion industrielle
r организация *f* управления производством

9216 orginal state
d Ausgangszustand *m*; ursprünglicher Zustand *m*
f état *m* original
r исходное состояние *n*

9217 orientation of handling objects
d Orientierung *f* von Handhabungsobjekten
f orientation *f* d'objets de manutention
r ориентация *f* манипулируемых объектов

9218 orientation system
d Orientierungssystem *n*
f système *m* d'orientation
r система *f* ориентации

9219 oriented
d orientiert; gerichtet
f orienté
r ориентированный

9220 oriented automaton
d orientierter Automat *m*
f automate *m* orienté
r ориентированный автомат *m*

9221 oriented graph
d orientierter Graph *m*
f graphe *m* orienté
r ориентированный граф *m*

* **orifice plate → 7951**

9222 original data
d Ausgangsdaten *pl*; Primärdaten *pl*; Originaldaten *pl*
f données *fpl* originales; données primaires

r исходные данные *pl*

9223 original function
d Originalfunktion *f*
f fonction *f* orginale
r функция-оргинал *m*

9224 OR-operation
d ODER-Operation *f*
f opération *f* OU
r операция *f* ИЛИ

9225 orthogonal data sequence
d orthogonale Datenfolge *f*
f séquence *f* orthogonale des données
r ортогональная цепочка *f* данных

9226 orthogonal filter; rectangular filter
d orthogonales Filter *n*
f filtre *m* orthogonal
r ортогональный фильтр *m*

9227 orthogonality error; non-orthogonality (of an oscilloscope)
d Orthogonalitätsfehler *m*
f non-orthogonalité *f*
r погрешность *f* ортогональности

9228 orthogonal pulse
d Rechteckimpuls *m*
f impulsion *f* rectangulaire
r ортогональный импульс *m*

9229 orthogonal representation
d orthogonale Darstellung *f*
f représentation *f* orthogonale
r ортогональное представление *n*

9230 orthogonal transformation
d orthogonale Transformation *f*
f transformation *f* orthogonale
r ортогональное преобразование *n*

9231 oscillating circuit; resonator circuit
d Schwingkreis *m*
f circuit *m* oscillant; circuit oscillatoire
r колебательный контур *m*

9232 oscillating contact
d Schwingkontakt *m*
f contact *m* oscillant
r колебательный контакт *m*

9233 oscillating controller; oscillating regulator; vibrating controller
d Schwingregler *m*; Vibrationsregler *m*
f régulateur *m* oscillatoire; régulateur vibratoire
r вибрационный регулятор *m*

9234 oscillating control servomechanism
 d oszillierender Regelkreis *m*; schwingender Regelkreis
 f circuit *m* de réglage oscillant
 r колеблющаяся система *f* регулирования

9235 oscillating parallel circuit
 d Parallelschwingkreis *m*
 f circuit *m* antirésonant
 r параллельный колебательный контур *m*

9236 oscillating phase
 d Schwingungsphase *f*
 f phase *f* d'oscillation
 r фаза *f* колебания

9237 oscillating process
 d Schwingungsvorgang *m*
 f processus *m* oscillatoire
 r колебательный процесс *m*

9238 oscillating quantity
 d Schwinggröße *f*
 f grandeur *f* oscillante
 r колебательная величина *f*

9239 oscillating regime
 d Schwingungszustand *m*
 f régime *m* oscillatoire
 r колебательный режим *m*

 * **oscillating regulator** → 9233

9240 oscillating relay
 d Schwingrelais *n*
 f relais *m* vibratoire
 r вибрационное реле *n*

9241 oscillation
 d Schwingung *f*; Oszillation *f*
 f oscillation *f*
 r колебание *n*

9242 oscillation capability; property to oscillate
 d Schwingfähigkeit *f*
 f propriété *f* oscillatrice
 r колебательность *f*

9243 oscillation element
 d Schwingungsglied *n*
 f élément *m* d'oscillation
 r колебательное звено *n*

9244 oscillation excitation
 d Schwingungserregung *f*
 f excitation *f* d'oscillations; amorçage *f* d'oscillations
 r возбуждение *n* колебаний

9245 oscillation frequency
 d Schwingungsfrequenz *f*
 f fréquence *f* oscillatrice
 r частота *f* колебаний

9246 oscillation function
 d Schwingungsfunktion *f*
 f fonction *f* oscillatrice
 r колебательная функция *f*

9247 oscillation synchronization
 d Schwingungssynchronisation *f*
 f synchronisation *f* d'oscillations
 r синхронизация *f* колебаний

9248 oscillation system property
 d Schwingfähigkeit *f* des Systems
 f propriété *f* oscillatoire du système
 r колебательное свойство *n* системы

9249 oscillator
 d [harmonischer] Oszillator *m*
 f oscillateur *m* [harmonique]
 r осциллятор *m*

9250 oscillator circuit
 d Oszillatorschaltung *f*
 f circuit *m* oscillateur
 r схема *f* генератора

9251 oscillatory gripper system
 d schwingendes Greifersystem *n*
 f système *m* de grappin oscillatoire
 r колеблющееся захватное устройство *n*

9252 oscillatory induction transmitter
 d Oszillatorinduktionsgeber *m*
 f détecteur *m* inductif à oscillateur
 r колебательный индукционный датчик *m*

9253 oscillatory laser state
 d Laserschwingzustand *m*
 f régime *m* oscillatoire de laser
 r колебательный режим *m* работы лазера

9254 oscillatory motion
 d oszillatorische Bewegung *f*; Schwingung *f*
 f mouvement *m* oscillatoire
 r колебательное движение *n*

9255 oscillogram time-marks
 d Oszillogrammzeitmarken *fpl*
 f repères *mpl* de temps d'oscillogramme
 r осциллограмма *f* с метками времени

9256 oscillographic presentation of processes
 d oszillografische Darstellung *f* von Vorgängen

f représentation *f* oscillographique de
   processus
r осциллографическое изображение *n*
   процессов

9257 oscilloscope
  d Oszilloskop *n*
  f oscilloscope *m*
  r осциллоскоп *m*

9258 oscillotitrator
  d Oszillotitrator *m*
  f oscillateur *m* titrateur
  r осциллотитратор *m*

9259 outage time
  d Ausfalldauer *f*; Außerbetriebsdauer *f*
  f durée *f* de panne; durée de coupure
  r время *n* перебоя

9260 outer feedback
  d äußere Rückkopplung *f*
  f réaction *f* extérieure
  r внешняя отрицательная обратная связь *f*

9261 outgoing laser beam
  d Ausgangslaserstrahl *m*
  f faisceau *m* de sortie du laser
  r выходящий лазерный луч *m*

9262 outlet automatics
  d Ablaßautomatik *f*
  f automaticité *f* de décharge
  r автоматика *f* выпусков

9263 outlet pressure
  d Austrittsdruck *m*
  f pression *f* de sortie
  r выходное давление *n*

  * **outline** → 3170

  * **out of circuit** *adj*→ 8888

9264 output action
  d Austrittsverfahren *n*; Ausgabeverfahren *n*
  f action *f* de sortie
  r выходное воздействие *n*

9265 output amplifier
  d Ausgangsverstärker *m*
  f amplificateur *m* de sortie
  r выходной усилитель *m*

9266 output amplitude
  d Ausgangsamplitude *f*; Amplitude *f* am
    Ausgang
  f ampltude *f* de sortie

r амплитуда *f* выходного сигнала

9267 output axis
  d Ausgangsachse *f*; Ausgangsfolgeachse *f*
  f axe *m* de sortie
  r выходная ось *f*

9268 output capacitance
  d Ausgangskapazität *f*
  f capacité *f* de sortie
  r ёмкость *f* на выходе

9269 output cascade; output state
  d Ausgangsstufe *f*; Ausgangszustand *m*
  f étage *m* final; étage de sortie
  r выходной каскад *m*

9270 output control
  d Ausgabesteuerung *f*
  f commande *f* de sortie
  r управление *n* выводом

9271 output conversion
  d Ausgabekonvertierung *f*
  f conversion *f* de sortie
  r преобразование *n* на выходе

9272 output element
  d Ausgangselement *n*
  f élément *m* de sortie
  r выходной элемент *m*

9273 output information signals
  d Ausgabeinformationssignale *npl*
  f signaux *mpl* de sortie des informations
  r выходные информационные сигналы *mpl*

9274 output level
  d Ausgangspegel *m*
  f niveau *m* de sortie
  r уровень *m* выходного сигнала

9275 output line
  d Ausgabeleitung *f*; Ausgangsleitung *f*
  f ligne *f* de sortie
  r линия *f* вывода

9276 output logic
  d Ausgabelogik *f*
  f logique *f* de sortie; logique d'extraction
  r логика *f* вывода

9277 output of control information
  d Ausgabe *f* von Steuerinformationen
  f sortie *f* des informations de commande
  r вывод *m* управляющей информации

9278 output of measured value
  d Messwertausgabe *f*

f sortie *f* des valeurs de mesure
r вывод *m* измеряемых значений

9279 output of sensor information
d Ausgabe *f* der Sensorinformation
f sortie *f* de l'information de senseur
r вывод *m* сенсорной информации

9280 output pulse
d Ausgangsimpuls *m*; Ausgabeimpuls *m*
f impulsion *f* de sortie
r выходной импульс *m*

9281 output-request signal
d Ausgabe-Anforderungssignal *n*
f signal *m* de demande de sortie
r сигнал *m* запроса на вывод

9282 output signal
d Ausgangssignal *n*
f signal *m* de sortie
r выходной сигнал *m*

9283 output speed
d Ausgabegeschwindigkeit *f*
f vitesse *f* de sortie
r скорость *f* вывода

* **output state → 9269**

9284 output strobe signal
d Ausgabemarkiersignal *n*; Ausgabetor-Aktivierungssignal *n*
f signal *m* de marquage de sortie
r стробирующий сигнал *m* вывода

9285 output time constant
d Ausgangszeitkonstante *f*
f constante *f* de temps de sortie
r постоянная *f* времени выходной цепи

* **output transfer function → 2481**

9286 output variable
d Ausgangsvariable *f*
f quantité *f* variable de sortie
r выходная переменная *f*

9287 output winding
d Ausgangswicklung *f*
f enroulement *m* de sortie
r выходная обмотка *f*

9288 overall accuracy
d Gesamtgenauigkeit *f*
f précision *f* totale
r общая надёжность *f*

9289 overall cavity gain
d Totalgewinn *m* des Hohlraumes
f gain *m* total de la cavité
r общий коэффициент *m* усиления резонатора

9290 overall design problems
(of a computer)
d gesamte Entwurfsprobleme *npl*
f problèmes *mpl* de projet complets
r общие проблемы *fpl* проектирования

9291 overall dimension
d Gesamtabmessung *f*; äußere Abmessung *f*
f dimension *f* totale; dimension globale
r предельный размер *m*

* **overall efficiency → 8577**

9292 overall loss
d Gesamtverlust *m*
f perte *f* totale
r общие потери *fpl*

9293 overall machine characteristic
d gesamte Maschinencharakteristik *f*
f caractéristique *f* de machine complète
r общая характеристика *f* машины

9294 overall plant; complete plant
d Gesamtanlage *f*
f installation *f* totale
r полная установка *f*

9295 overall process theory
d allgemeine Prozesstheorie *f*
f théorie *f* générale des processus
r общая теория *f* процессов

9296 overall starting-time relay
d Anlaufzeitbegrenzerrelais *n*
f relais *m* limiteur de temps de démarrage
r предельное стартовое реле *n* времени

9297 overall steady-flow coefficient
d Gesamtbeiwert *m* der stationären Strömung
f coefficient *m* total du courant stationnaire
r суммарный коэффициент *m* расхода установившегося потока

* **overall system → 2837**

9298 overall system performance
d Gesamtsystemleistung *f*
f performance *f* du système entier
r общая производительность *f* системы

9299 overall system speed
d Arbeitsgeschwindigkeit *f* eines Systems

f   vitesse f d'opération globale d'un système
r   общее быстродействие n системы

9300 overcurrent protection
d   Überstromschutz m
f   dispositif m de protection à maximum de
    courant
r   защита f от сверхтока

9301 overdamping
d   Überdämpfung f
f   suramortissement m
r   сильное затухание n

9302 overdriven amplifier
d   überbeanspruchter Verstärker m; überlasteter
    Verstärker
f   amplficateur m surchargé; amplificateur
    ecréteur
r   искажающий усилитель m

9303 overflow
d   Überlauf m
f   dépassement m
r   переполнение n

9304 overflow alarm
d   Überlaufanzeige f
f   avertissement m du dépassement
r   сигнализация f переполнения

9305 overflow register
d   Überlaufregister n
f   registre m de dépassement
r   регистр m переполнения

9306 overfrequency protection
d   Überfrequenzschutz m
f   protection f contre le maximum de fréquence
r   защита f от превышения частоты

9307 overheat control
d   Überhitzungsschutz m
f   protection f contre la surchauffe
r   предохранитель m от перегрева

9308 overheating degree
d   Überhitzungsgrad m
f   degré m de surchauffe
r   степень f перегрева

9309 overlap
d   Überlappen n
f   recouverment m; fourchette f
r   перекрытие n

9310 overlap action
d   Überdeckungswirkung f;

Überdeckungsregelung f
f   action f de recouvrement; régulation f de
    recouvrement
r   двухпозиционное регулирование n с
    двумя значениями интервалов

9311 overlapped access
d   überlappter Zugriff m; überlappender Zugriff
f   accès m chevauché
r   совмещенный доступ m

9312 overlapping operations
d   Überlappungsoperationen fpl;
    Überdeckungsoperationen fpl
f   recouvrement m d'opérations
r   совмещенные операции fpl

9313 overlapping pulses
d   Überdeckungsimpulse mpl
f   impulsions fpl de recouvrement
r   импульсы mpl с перекрытием

9314 overlay
d   Überlagerung f
f   overlay m
r   оверлей m; наложение n

9315 overlay area
d   Überlagerungsbereich m
f   zone f de recouvrement
r   область f перекрытия

9316 overlay structure
d   Überlagerungsstruktur f
f   structure f d'overlay
r   оверлейная структура f

*   **overload → 9320**

9317 overload capacity
d   Überlastbarkeit f
f   capacité f de surcharge
r   перегрузочная способность f

9318 overload controller
d   Überlastungsregler m
f   régulateur m de surcharge
r   регулятор m перегрузки

9319 overload detector
d   Überlastungsdetektor m;
    Überlastungsmelder m
f   détecteur m de surcharge
r   детектор m перегрузки

9320 overload[ing]
d   Überlast[ung] f; Übersteuerung f
f   surcharge f
r   перегрузка f; перенагрузка f

9321 overload protection
d Überlastungsschutz m
f protection f contre les surcharges
r защита f от перегрузки

9322 overload warning system
d Übersteuerungsanzeige f
f indication f de surcharge; système m de surcharge
r сигнализация f предела напряжения

9323 overpower protection
d Leistungsbegrenzungsschutz m
f dispositif m de protection à maximum du puissance
r защита f от максимальной мощности

9324 overrun
(of data)
d Datenverlust m
f perte f de données
r потеря f данных

9325 overrun control signal
d Überlaufkontrollsignal n
f signal m de contrôle de dépassement
r сигнал m контроля переполнения

9326 oversampling
d Überabtastung f
f suréchantillonnage m
r сверхразвёртка f

* overshoot → 9328

9327 overshoot impulse
d Ausschlagimpuls m
f impulsion f de rebondissement
r импульс m отклонения

9328 overshoot[ing]
d Überregelung f
f surréglage m
r перерегулирование n

9329 overshoot period
d Überschwingzeit f
f temps m de rebondissement
r период m перерегулирования

9330 overshoot ratio
d Überregelungsfaktor m
f coefficient m de dépassement
r коэффициент m перерегулирования

9331 overspeed limiter
d Übergeschwindigkeitsbegrenzer m
f limiteur m de survitesse

r ограничитель m скорости

9332 oversynchronization
d Übersynchronisierung f
f sursynchronisation f
r пересинхронизация f

9333 overtone; higher harmonic
d Oberwelle f; höhere Harmonische f
f onde f harmonique; harmonique f supérieure
r высшая гармоника f

9334 overvoltage
d Überspannung f
f surtension f; survoltage m
r перенапряжение n

9335 overvoltage device; voltage limiter
d Überspannungsgerät n; Spannungsbegrenzer m
f appareil m à maximum de tension; limiteur m de tension
r ограничитель m напряжения

9336 overvoltage protection
d Überspannungsschutz m
f protection f contre surtensions
r защита f от перенапряжения

9337 overvoltage tripping
d Überspannungsabschalten n
f déclenchement m à maximum de tension
r выключение n высокого напряжения

P

9338 pacemaker
d Herztaktgeber *m*; Schrittmacher *m*
f pacemaker *m*; stimulateur *m* cardiaque
r тактовый датчик *m*

9339 packaged circuit
d gepackte Schaltung *f*
f circuit *m* groupé
r компактная схема *f*

9340 packaged control unit
d Reglerbaueinheit *f*
f bloc *m* de régulateur
r составная единица *f* цепи управления

9341 packing density
d Informationsdichte *f*; Packungsdichte *f*
f densité *f* d'information
r плотность *f* информации

9342 packing scheme
d Packungsschema *n*
f schéma *m* de compression
r схема *f* прокладки

9343 Pade approximation
d Pade-Approximation *f*
f approximation *f* de Pade
r аппроксимация *f* Падэ

9344 pairing measuring device
d Paarungsmesseinrichtung *f*
f appareil *m* mesureur d'appairage; instrument *m* mesureur d'appairement
r устройство *n* для измерения спаривания

9345 Paley-Wiener criterion
d Paley-Wiener-Kriterium *n*
f critère *m* de Paley-Wiener
r критерий *m* Винера

9346 panel; switch board
d Schalttafel *f*
f panneau *m* de distribution
r распределительный коммутационный щит *m*

9347 parabola
d Parabel *f*
f parabole *f*
r парабола *f*

9348 parabolic characteristic
d parabolische Charakteristik *f*; parabolförmige Charakteristik
f caractéristique *f* parabolique
r параболическая характеристика *f*

9349 parabolic charge
d parabolische Belastung *f*
f charge *f* parabolique
r параболическая нагрузка *f*

9350 parabolic function
d Parabelfunktion *f*
f fonction *f* parabolique
r параболическая функция *f*

9351 parabolic interpolation
d Parabelinterpolation *f*
f interpolation *f* parabolique
r параболическая интерполяция *f*

9352 parabolic partial differential equation
d parabolische partielle Differentialgleichung *f*
f équation *f* différentielle partielle parabolique
r параболическое дифференциальное уравнение *n* в частных производных

9353 parabolic reflector
d Parabolreflektor *m*
f réflecteur *m* parabolique
r параболический рефлектор *m*

9354 parabolic velocity
d parabolische Geschwindigkeit *f*
f vitesse *f* parabolique
r параболическая скорость *f*

9355 parallax adjusting
d Parallaxeneinstellung *f*
f réglage *m* parallactique
r корректировка *f* параллакса

9356 parallax error
d Parallaxefehler *m*
f erreur *f* parallactique
r параллактическая погрешность *f*

9357 parallax-free reading
d parallaxfreies Ablesen *n*
f lecture *f* sans parallaxe
r считывание *n* без учета параллактического смещения

9358 parallax in altitude
d Höhenparallaxe *f*
f parallaxe *f* de hauteur
r параллакс *m* по высоте

9359 parallel access
d paralleler Zugriff *m*
f accès *m* parallèle
r параллельная выборка *f*

9360 parallel arithmetic unit
d Parallelrechenwerk *n*
f unité *f* de calcul parallèle
r параллельное арифметическое
    устройство *n*

9361 parallel cascade action
d Parallelkaskadenverhalten *n*
f action *f* parallèle en cascade
r параллельное каскадное регулирование *n*

*   **parallel circuit → 9365**

9362 parallel combination
d Parallelschaltung *f*
f couplage *m* en parallèle; connexion *f* en
    parallèle
r параллельное включение *n*

**9363 parallel combination of control loop
       elements**
d Parallelschaltung *f* von Regelkreisgliedern
f combinaison *f* parallèle des éléments du
    système asservi
r параллельное соединение *n* звеньев цепи
    регулирования

9364 parallel-connected
d nebeneinandergeschaltet; parallelgeschaltet
f couplé en parallèle
r параллельно включенный

9365 parallel connection; parallel circuit
d parallele Verbindung *f*; Parallelschaltung *f*
f connexion *f* en parallèle; montage *m* en
    parallèle
r параллельное соединение *n*

9366 parallel control
d Parallelsteuerung *f*
f commande *f* parallèle
r параллельное управление *n*

9367 parallel control loop
d nebengeschalteter Regelkreis *m*
f boucle *f* parallèle de réglage
r параллельная петля *f* регулирования

9368 parallel correcting element
d parallel geschaltetes Korrekturglied *n*
f élément *m* parallèle de correction
r параллельное корректирующее
    устройство *n*

9369 parallel curves
d Parallelkurven *fpl*
f courbes *fpl* parallèles
r параллельные кривые *fpl*

9370 parallel data channel
d Paralleldatenkanal *m*
f canal *m* des données parallèles
r параллельный информационный канал *m*

9371 parallel data controller
d Paralleldatensteuergerät *n*
f dispositif *m* de commande pour les données
    parallèles
r устройство *n* управления для
    параллельной передачи данных

9372 parallel data converter
d Paralleldatenumwandler *m*
f convertisseur *m* pour les données parallèles
r параллельный преобразователь *m* данных

9373 parallel digital computer
d Paralleldigitalrechner *m*
f calculatrice *f* digitale parallèle
r параллельная вычислительная машина *f*

9374 parallel feedback operational amplifier
d parallelrückgekoppelter
    Funktionsverstärker *m*
f amplificateur-computeur *m* à réaction
    parallèle
r решающий усилитель *m* с параллельной
    обратной связью

9375 parallel input
d Paralleleingabe *f*; Paralleleingang *m*
f entrée *f* parallèle
r параллельный вход *m*

9376 parallel interface
d Parallelinterface *n*; Parallelschnittstelle *f*
f interface *f* parallèle
r параллельный интерфейс *m*

9377 parallel manipulator
d Parallelmanipulator *m*
f manipulateur *m* en parallèle
r синхронный манипулятор *m*

9378 parallel operation
d Parallelbetrieb *m*
f opération *f* en parallèle; fonctionnement *m* en
    parallèle
r параллельная операция *f*

9379 parallel-operation recognition
d Parallelaufauswahl *f*

f reconnaissance f d'opération parallèle
r распознавание n параллельного действия

9380 parallel organization
(of computers)
d parallele Organisation f
f organisation f parallèle
r параллельная организация f

* **parallel phase resonance** → 907

9381 parallel rectifier circuit
d Parallelschaltung f des Gleichrichters;
Gleichrichterparallelschaltung f
f montage m en parallèle de redresseur
r параллельная выпрямительная схема f

9382 parallel register
d parallelwirkendes Register n
f registre m parallèle
r регистр m параллельного действия

9383 parallel representation
d Paralleldarstellung f
f représentation f parallèle
r параллельное представление n

9384 parallel run controller
d Parallelregler m
f régulateur m de marche parallèle
r регулятор m параллельного хода

9385 parallel search
d Parallelsuche f; Parallelabfrage f
f recherche f parallèle
r параллельный поиск m

9386 parallel sensors
d parallele Sensoren mpl
f capteurs mpl parallèles
r параллельные датчики mpl

9387 parallel-serial structure
d Parallel-Serien-Struktur f
f structure f parallèle-série
r параллельно-последовательная
структура f

9388 parallel stabilization
d parallele Stabilisation f; parallele
Stabilisierung f
f stabilisation f parallèle
r параллельная стабилизация f

9389 parallel switching circuit
d Parallelschaltkreis m
f circuit m parallèle de commutation; circuit de
commutation parallèle

r параллельная коммутирующая схема f

9390 parallel system
d Parallelsystem n
f système m en parallèle
r параллельная система f

9391 parallel-to-serial conversion
d Parallel-Seriell-Umsetzung f
f conversion f parallè en série
r параллельно-последовательное
преобразование n

9392 parallel transfer
d Parallelübertragung f
f transfert m parallèle
r параллельная передача f; параллельный
перенос m

9393 parallel transmission of information
d Parallelübertragung f der Information
f transmission f parallèle de l'information
r параллельная передача f информации

9394 paramagnetic system
d paramagnetisches System n
f système m paramagnétique
r парамагнитная система f

9395 parameter
d Parameter m
f paramètre m
r параметр m

9396 parameter adjustment control
d Bedienungselemente npl zur
Prametereinstellung
f éléments mpl de réglage des paramètres
r устройства npl для установки параметров

9397 parameter delimiter
d Parameterbegrenzer m;
Begrenzungszeichen n für Parameter
f délimiteur m de paramètre
r ограничитель m параметра

9398 parameter determination
d Parameterbestimmung f
f détermination f de paramètres
r определение n параметров

9399 parameter hierarchy
d Parameterrangfolge f
f suite f hiérarchique des paramètres
r последовательность f параметров по их
значимости

9400 parameter identification
d Parametererkennung f

f identification *f* des paramètres
r идентификация *f* параметров

9401 parameter image
d Parameterabbildung *f*
f image *f* de paramètre
r параметрическое изображение *n*

9402 parameter limit
d Parametergrenze *f*
f limite *f* de paramètre
r предельное значение *n* параметра

9403 parameter region; parametric domain
d Parametergebiet *n*; Parameterbereich *m*
f domaine *m* paramétrique; région *f* paramétrique
r область *f* параметра

9404 parameter value
d Wert *m* des Parameters; Parameterwert *m*
f valeur *f* de paramètre
r значение *n* параметра

9405 parametric amplifier bandwidth
d Bandbreite *f* des parametrischen Verstärkers
f largeur *f* de bande de l'amplificateur paramétrique
r полоса *f* пропускания параметрического усилителя

9406 parametric damping
d parametrische Dämpfung *f*
f amortissement *m* paramétrique
r параметрическое демпфирование *n*

* parametric domain → 9403

9407 parametric electronic component
d parametrisches elektronisches Bauelement *n*
f composant *m* électronique paramétrique
r параметрический электронный элемент *m*

9408 parametric equation
d Parametergleichung *f*
f équation *f* paramétrique
r параметрическое уравнение *n*

9409 parametric excitation
d parametrische Erregung *f*
f excitation *f* paramétrique
r параметрическое возбуждение *n*

9410 parametric frequency conversion
d parametrische Frequenzkonversion *f*; parametrische Frequenzumsetzung *f*
f conversion *f* paramétrique de fréquence; changement *m* paramétrique de fréquence

r параметрическое преобразование *n* частоты

9411 parametric gain
d parametrischer Gewinn *m*
f gain *m* paramétrique
r параметрическое усиление *n*

9412 parametric generality
d parametrische Universalität *f*
f généralité *f* paramétrique
r параметрическая универсальность *f*

9413 parametric interaction
d parametrische Wechselwirkung *f*
f interaction *f* paramétrique
r параметрическое взаимодействие *n*

9414 parametric multiplier
d parametrischer Vervielfacher *m*
f multiplicateur *m* paramétrique
r параметрический умножитель *m*

9415 parametric optimization
d parametrische Optimierung *f*
f optimisation *f* paramétrique
r параметрическая оптимизация *f*

* parametric oscillator → 9105

9416 parametric programming
d parametrische Programmierung *f*
f programmation *f* paramétrique
r параметрическое программирование *n*

9417 parametric pumping energy
d parametrische Pumpenenergie *f*
f énergie *f* paramétrique de pompage
r энергия *f* параметрической накачки

9418 parametric resonance
d parametrische Resonanz *f*
f résonance *f* paramétrique
r параметрический резонанс *m*

9419 parametric space
d Parameterraum *m*
f espace *m* paramétrique
r пространство *n* параметров

9420 parametric variation
d Parameteränderung *f*
f variation *f* paramétrique
r изменение *n* параметров

9421 parametrized operation
d parametrisierte Operation *f*

f opération *f* paramétrisée
r параметрическая операция *f*

9422   paraphase amplifier
d Phasenumkehrverstärker *m*
f amplificateur *m* inverseur de phase
r парафазный усилитель *m*

9423   parasitic autooscillations
d parasitische Selbstschwingungen *fpl*
f auto-oscillations *fpl* parasites
r паразитные собственные колебания *npl*

9424   parasitic capacitance
d parasitäre Kapazität *f*
f capacité *f* parasitaire
r паразитная ёмкость *f*

9425   parasitic connection
d Streukopplung *f*
f liaison *f* parasite
r паразитная связь *f*

9426   parasitic echo
d Echo *n* durch Gerätefehler
f écho *m* interne
r паразитное эхо *n*

9427   parasitic frequency
d Störfrequenz *f*
f fréquence *f* parasitaire
r частота *f* возмущающего воздействия

*   **parasitic oscillations** → 6229

9428   parasitics
d Störeffekte *mpl*
f parasites *mpl*
r паразитные помехи *fpl*

9429   parity bit
d Prüfbit *n*; Paritätsbit *n*
f bit *m* de parité; bit de contrôle
r бит *m* для проверки на чётность или
    нечётность

*   **parity check** → 8884

*   **parity function** → 5218

9430   Parseval theorem
d Parsevalsches Theorem *n*
f théorème *m* de Parseval
r теорема *f* Паресеваля

*   **part** → 9431

9431   part[ial]

d partiell
f partiel; en partie
r частичный; парциальный; неполный

9432   partial automatization
d Teilautomatisierung *f*
f automatisation *f* partielle
r частичная автоматизация *f*

9433   partial automaton
d Teilautomat *m*
f automate *m* partiel
r частичный автомат *m*

9434   partial-capacity operation
d Teillastbetrieb *m*
f opération *f* à charge partielle
r работа *f* с частичной
    производительностью

9435   partial differential equation
d partielle Differentialgleichung *f*
f équation *f* aux dérivées partielles; équation
    différentielle aux dérivées partielles
r дифференциальное уравнение *n* в частных
    производных

9436   partial information
d Teilinformation *f*
f information *f* partielle
r частичная информация *f*

9437   partially automated machining process
d teilautomatischer Bearbeitungsvorgang *m*
f processus *m* de l'usinage automatique partiel
r частично-автоматизированный процесс *m*
    обработки

9438   partially determined function
d teilweise bestimmte Funktion *f*
f fonction *f* déterminée partielle
r частично определенная функция *f*

9439   partially symmetrical function
d partiell symmetrische Funktion *f*
f fonction *f* symétrique partielle
r частично симметричная функция *f*

9440   partial operation
d Teiloperation *f*
f opération *f* partielle
r отдельная операция *f*

9441   partial pressure
d Partialdruck *m*
f pression *f* partielle
r парциальное давление *n*

9442 **partial-read pulse**
   d Teilleseimpuls *m*
   f impulsion *f* de lecture partielle
   r импульс *m* частичной выборки

9443 **partial-select output**
   d teilweise selektive Ausgabe *f*
   f sortie *f* partiellement sélective
   r частичный селекторный выход *m*

9444 **partial system model**
   d Partialsystemmodell *n*
   f modèle *m* d'un système partiel
   r модель *f* подсистемы

9445 **partial system synchronization**
   d Teilsystemsynchronisation *f*
   f synchronisation *f* du système partiel
   r синхронизация *f* подсистемы

9446 **partial volume**
   d Partialvolumen *n*
   f volume *m* partiel
   r парциальный объём *m*

9447 **partial-write pulse**
   d Teilschreibeimpuls *m*
   f impulsion *f* d'enregistrement partielle
   r импульс *m* частичного ввода

9448 **particular conductivity**
   d partikuläre Leitfähigkeit *f*
   f conductibilité *f* particulière
   r собственная проводимость *f*

9449 **particular integral**
   d partikuläres Integral *n*
   f intégrale *f* particulière
   r частный интеграл *m*

9450 **particular solution**
   d partikuläre Lösung *f*
   f solution *f* particulière
   r частное решение *n*

9451 **particular system**
   d spezielles System *n*; besonderes System
   f système *m* particulier
   r особая система *f*

9452 **particular value**
   d Einzelwert *m*; Teilwert *m*
   f valeur *f* particulière
   r частное значение *n*

  * **partition** → 9455

9453 **partition** *v*

   d aufteilen; unterteilen
   f partager; segmenter
   r разделять; секционировать

9454 **partition coefficient; distribution coefficient**
   d Verteilungskoeffizient *m*; Aufteilungskoeffizient *m*; Teilungsverhältnis *n*
   f coefficient *m* de distribution
   r коэффициент *m* распределения

9455 **partition[ing]**
   d Partition *f*; Zerlegung *f*; Aufteilung *f*
   f partage *m*; découpage *m*
   r разчленение *n*; разложение *n*; разбиение *n*

9456 **passing band**
   d Durchlassband *n*
   f bande *f* passante
   r полоса *f* пропускания

9457 **passive circuit**
   d passiver Kreis *m*; passive Schaltung *f*
   f circuit *m* passif
   r пассивная цепь *f*

9458 **passive compliance device**
   d passive Ausgleichseinrichtung *f*
   f dispositif *m* de compensation passif
   r пассивное уравнивающее устройство *n*

  * **passive component** → 9459

9459 **passive element; passive component**
   d passives Glied *n*; passives Bauelement *n*
   f organe *m* passif; élément *m* passif
   r пассивный элемент *m*; пассивный компонент *m*

9460 **passive infrared rangefinder**
   d passiver Ultrarotentfernungsmesser *m*
   f télémètre *m* passif à rayons infrarouges
   r пассивный инфракрасный дальномер *m*

9461 **passive infrared system**
   d passives Infrarotsystem *n*
   f système *m* infrarouge passif
   r пассивная инфракрасная система *f*

9462 **passive optical component**
   d passives optisches Element *n*; passive optische Komponente *f*
   f composant *m* optique passif; élément *m* optique passif
   r пассивный оптический компонент *m*; пассивный оптический элемент *m*

9463 **path control**
    d Bahnsteuerung *f*
    f commande *f* de voie
    r управление *n* маршрутом

9464 **path-controlled production**
    d bahngesteuerte Fertigung *f*
    f production *f* commandée par voie
    r производство *n* с контурным управлением

9465 **path measuring system**
    d Wegmesssystem *n*
    f système *m* de mesure de chemin
    r система *f* измерения перемещений

9466 **pattern classification**
    d Musterklassifizierung *f*
    f classification *f* d'échantillons
    r классификация *f* образцов

  * **pattern recognition** → 8871

9467 **pause signal control**
    d Pausenzeichen-Steuerung *f*
    f contrôle *m* de signal de pause
    r управление *n* сигналом паузы

  * **P-control** → 10352

  * **P-controller** → 10348

9468 **peak amplitude**
    d Spitzenamplitude *f*
    f amplitude *f* de pointe
    r максимальная амплитуда *f*

9469 **peak current**
    d Spitzenstrom *m*
    f courant *m* de crête
    r пиковой ток *m*

9470 **peak data transfer rate**
    d Spitzen-
       Datenübertragungsgeschwindigkeit *f*;
       Spitzenrate *f* der Datenübertragung
    f vitesse *f* de transmission de données
       maximale
    r максимальная скорость *f* передачи данных

9471 **peak energy**
    d Spitzenenergie *f*
    f énergie *f* de crête
    r пиковая энергия *f*; энергия в импульсе

  * **peak factor** → 3516

9472 **peak flux density**
    d Spitzenflussdichte *f*

    f densité *f* de flux de crête
    r пиковое значение *n* плотности магнитного
       потока

9473 **peak frequency**
    d Spitzenfrequenz *f*
    f fréquence *f* de pointe
    r максимальная частота *f*

  * **peakholding control** → 5352

  * **peakholding controller** → 5349

9474 **peak-holding optimizing control**
    d Extremwertregelung *f*;
       Höchstwertoptimalregelung *f*
    f réglage *m* à valeur optimale de crête
    r позиционная система *f* экстремального
       регулирования

9475 **peak intensity**
    d höchste Intensität *f*
    f intensité *f* de crête
    r максимальная интенсивность *f*

  * **peak inverse voltage** → 9485

9476 **peak limiter**
    d Amplitudenbegrenzer *m*
    f limiteur *m* d'amplitude; écrêteur *m*
    r ограничитель *m* пика

9477 **peak load**
    d Spitzenbelastung *f*
    f charge *f* maximum
    r пиковая нагрузка *f*

9478 **peak magnetizing force**
    d Spitzenwert *m* der magnetischen Erregung
    f force *f* d'aimantation de crête
    r пиковое значение *n* намагничивающей
       силы

9479 **peak making current**
    d Einschaltstromspitze *f*;
       Stoßeinschaltstrom *m*; Scheitelwert *m* des
       Einschaltstromes
    f valeur *f* de crête du courant de fermeture
    r удар *m* тока при включении

9480 **peak optical power**
    d optische Spitzenleistung *f*
    f puissance *f* optique de crête
    r оптическая пиковая мощность *f*

9481 **peak output power**
    d Ausgangsspitzenleistung *f*

f puissance *f* de pointe de sortie
r максимальная выходная мощность *f*

9482 peak point
d Scheitelpunkt *m*
f point *m* de crête
r пиковая точка *f*

9483 peak pressure meter
d Spitzendruckmesser *m*
f indicateur *m* de pression de pointe
r индикатор *m* максимального давления

9484 peak pulse power
d Impulsspitzenleistung *f*
f puissance *f* de pointe d'impulsion
r максимальная мощность *f* импульса

9485 peak reverse voltage; peak inverse voltage
d Spitzensperrspannung *f*
f tension *f* inverse de crête
r пиковое запорное напряжение *n*; пиковое обратное напряжение

9486 peak signal
d Spitzensignal *n*
f signal *m* de pointe
r пиковый сигнал *m*

* peak spectral sensitivity → 9487

9487 peak spectral [threshpld] sensitivity
d maximale spektrale Empfindlichkeit *f*
f sensibilité *f* spectrale de crête
r максимум *m* спектральной [пороговой] чувствительности

9488 peak time
d Spitzenzeit *f*
f temps *m* de crête
r время *n* пика

* peak-to-peak amplitude → 4477

9489 peak-to-zero voltage
d Spitze-Null-Spannung *f*
f tension *f* de crête à zéro
r разность *f* между максимальным и нулевым потенциалом

* peak value → 3517

9490 peak voltage
d Spitzenspannung *f*; Scheitelspannung *f*
f tension *f* de crête
r пиковое напряжение *n*

9491 pecking motor; stepping motor; repeat

motor
d Schrittmotor *m*; Fortschaltmotor *m*
f moteur *m* pas à pas
r шаговый двигатель *m*

9492 pedipulator system
d Pedipulatorsystem *n*
f système *m* pédipulateur
r педипуляторная система *f*

9493 peer *adj*
d seinesgleichen; gleichberechtigt
f égal
r равноправный

9494 Peirce function
d Peirce-Funktion *f*
f fonction *f* de Peirce
r функция *f* Пирса

9495 pending
d wartend
f pendant; suspendu
r ожидающий; отсроченный; отложенный

9496 pending interrupt
d anstehender Interrupt *m*; anstehende Unterbrechungsanforderung *f*
f interruption *f* pendante
r отложенное прерывание *n*

9497 pending interruption
d schwebende Unterbrechung *f*
f interruption *f* suspendue
r ждущее прерывание *n*

9498 pendulate *v*; swing *v*
d schwingen
f pendouiller
r колебаться; качаться

* penetrating → 9500

9499 penetrating power; penetration power
d Durchdringungsvermögen *n*
f pouvoir *m* de pénétration
r способность *f* проникновения

9500 penetration; penetrating
d Eindringen *n*; Eindringtiefe *f*; Durchdringung *f*
f pénétration *f*
r проникновение *n*

9501 penetration energy
d Eindringenenergie *f*
f énergie *f* de pénétration
r проникающая энергия *f*

*   **penetration power** → 9499

9502   percentage by volume
  d   Volum[en]prozent *n*
  f   pour-cent *m* volumique
  r   объёмный процент *m*

9503   percentage differential protection
  d   Prozentvergleichsschutz *m*
  f   protection *f* différentielle à pourcentage
  r   дифференциальная защита *f* с
        торможением

9504   percent ripple voltage
  d   Brummspannungsverhältnis *n*
  f   taux *m* d'ondulation résiduelle
  r   коэффициент *m* пульсации напряжения

9505   perception
  d   Wahrnehmung *f*; Empfindung *f*
  f   perception *f*
  r   восприятие *n*

9506   perception sensor; identification sensor
  d   Erkennungssensor *m*
  f   capteur *m* de vision; senseur *m*
        d'identification
  r   сенсор *m* распознавания

9507   perceptive motoric
  d   Wahrnehmungsmotorik *f*
  f   motorice *f* d'aperception
  r   моторные функции *fpl* наблюдения

9508   perfect
  d   perfekt; vollkommen
  f   parfait; idéal
  r   совершенный; полный

9509   perfect information
  d   vollständige Information *f*
  f   information *f* complète; information partfaite
  r   полная информация *f*

9510   perfectly mixed reactor
  d   ideal durchmischter Reaktor *m*; Reaktor mit
        idealer Vermischung
  f   réacteur *m* parfaitement mélangé
  r   реактор *m* идеального перемешивания

9511   performance
  d   Leistung *f*
  f   performance *f*
  r   производительность *f*

9512   performance characteristics
  d   Leistungscharakteristik *f*;
        Leistungskennwerte *mpl*
  f   caractéristiques *mpl* de performance

  r   рабочие характеристики *fpl*

9513   performance conditions
  d   Gütebedingungen *fpl*
  f   conditions *fpl* de qualité
  r   условия *npl* работы

9514   performance evaluation
  d   Leistungsauswertung *f*
  f   évaluation *f* de performance
  r   оценка *f* производительности

9515   performance index
  d   Gütekriterium *n* für die Regelung;
        Kriterium *n* für das Übergangsverhalten;
        verallgemeinerte Regelfläche *f*
  f   chiffre *m* de qualité; indice *m* de performance
  r   критерий *m* качества переходной
        характеристики

9516   performance limitation
  d   Übergangsbegrenzung *f*; Begrenzung *f* des
        Übergangsverhaltens
  f   limitation *f* transitoire
  r   ограничение *n* переходной
        характеристики

9517   performance reference
  d   Leistungsreferenz *f*
  f   référence *f* de performance
  r   эталонная производительность *f*

9518   performance requirement
  d   Leistungsanforderung *f*
  f   exigence *f* de performance
  r   требование *n* к техническим
        характеристикам

9519   performing operation speed
  d   Operationsgeschwindigkeit *f*;
        Geschwindigkeit *f* der
        Operationsdurchführung
  f   vitesse *f* d'exécution des opérations
  r   скорость *f* выполнения операций

9520   period
  d   Periode *f*; Zeitraum *m*
  f   période *f*
  r   период *m*

9521   periodic coefficient
  d   periodischer Koeffizient *m*
  f   coefficient *m* périodique
  r   периодический коэффициент *m*

**9522   periodic controller; sampled-data
         controller**
  d   Impulsregler *m*

    f régulateur *m* impulsionnel
    r импульсный регулятор *m*

9523 periodic duty
    d Dauerbetrieb *m* mit periodisch veränderlicher Belastung
    f fonctionnement *m* périodique
    r периодический режим *m* работы

9524 periodic element
    d periodisches Element *n*
    f élément *m* périodique
    r периодический элемент *m*

9525 periodic frequency modulation
    d periodische Frequenzmodulation *f*
    f modulation *f* périodique de fréquence
    r периодическая частотная модуляция *f*

9526 periodic function
    d periodische Funktion *f*
    f fonction *f* périodique
    r периодическая функция *f*

9527 periodic intensity distribution
    d periodische Intensitätsverteilung *f*
    f distribution *f* d'intensité
    r периодическое распределение *n* интенсивности

9528 periodic quantity
    d periodische Größe *f*
    f grandeur *f* périodique
    r периодическая величина *f*

9529 periodic quantity phase
    d Phase *f* periodischer Große
    f phase *f* de grandeur périodique
    r фаза *f* периодической величины

9530 periodic rating
    d Nennleistung *f* für Aussetzbetrieb
    f puissance *f* nominale pour régime intermittent
    r номинальное значение *n* периодической нагрузки

9531 periodic solution
    d periodische Lösung *f*; Lösung *f* im eingeschwungenen Zustand
    f solution *f* périodique
    r периодическое решение *n*

9532 periodic solution stability
    d Stabilität *f* der periodischen Lösung
    f stabilité *f* de solution périodique
    r устойчивость *f* периодического решения

9533 periodic test signal
    d periodisches Testsignal *n*
    f signal *m* de test périodique
    r периодический эталонный сигнал *m*

9534 periodic velocity fluctuation
    d periodische Geschwindigkeitsschwankung *f*
    f fluctuation *f* de vitesse périodique
    r периодические колебания *npl* скорости

9535 period of use
    d Benutzungsdauer *f*; Nutzungsdauer *f*
    f durée *f* d'utilisation
    r продолжительность *f* использования

9536 peripheral
    d peripher
    f périphérique
    r периферийный

9537 peripheral control
    d periphere Steuerung *f*; Peripheriesteuerung *f*
    f commande *f* périphérique
    r управление *n* периферийным оборудованием

9538 peripheral controller
    d Peripheriesteuerungsbaustein *m*
    f module *m* commande périphérique
    r контроллер *m* периферийного оборудования

9539 peripheral device allocation
    d Peripheriegerätezuordnung *f*
    f attribution *f* d'appareils périphériques
    r распределение *n* внешних устройств

9540 peripheral interface adapter
    d Peripherieschnittstellen-Anpassungsbaustein *m*
    f adapteur *m* d'interface périphérique
    r адаптер *m* периферийного оборудования

9541 peripheral processor
    d Peripherieprozessor *m*
    f processeur *m* périphérique
    r периферийный процессор *m*

9542 peripheral signal
    d Signal *n* aus der Peripherie
    f signal *m* périphérique
    r периферийный сигнал *m*

9543 peripheral subsystem
    d Peripherie-Untersystem *n*
    f sub-système *m* de périphérie
    r периферийная подсистема *f*

9544 **periphery**
d Peripherie *f*
f périphérie *f*
r периферия *f*

9545 **periphery of hybrid computer**
d Hybridrechnerperipherie *f*; Peripherie *f* eines Hybridrechners
f périphérie *f* d'un calculateur hybride
r периферийные устройства *npl* гибридного компьютера

9546 **permanent action**
d Dauerbetrieb *m*; Dauerwirkung *f*
f action *f* permanente
r непрерывное [воз]действие *n*

9547 **permanent data**
d gleichbleibende Daten *pl*; permanente Daten
f données *fpl* permanentes
r постоянные данные *pl*

* **permanent delay** → 3856

9548 **permanent droop**
d dauernde Ungleichförmigkeit *f*; bleibende Ungleichförmigkeit
f non-uniformité *f* permanente; irrégularité *f* permanente
r остаточная неравномерность *f*

9549 **permanent load**
d Dauerbelastung *f*
f charge *f* permanente
r постоянная нагрузка *f*

9550 **permanent main control program**
d bleibendes Hauptsteuerprogramm *n*
f programme *m* de commande principal permanent
r резидентная управляющая программа *f*

* **permanent memory** → 5534

9551 **permeability**
d Permeabilität *f*
f perméabilité
r проницаемость *f*

9552 **permissible noise level**
d zugelassener Rauschpegel *m*
f niveau *m* admissible de bruit
r допустимый уровень *m* помех

9553 **permissible peak inverse voltage**
d zulässiger Spitzenwert *m* der Sperrspannung
f tension *f* inverse de crête admissible
r допустимая пиковая величина *f* обратного напряжения

* **permissible solution** → 532

9554 **permissible supply deviation**
d zulässige Abweichung *f* der Energieversorgung
f déviation *f* d'alimentation permise
r допустимое отклонение *n* напряжения питания

9555 **persistency checking**
d Stetigkeitsprüfung *f*
f vérification *f* de persistance
r проверка *f* устойчивости

9556 **personal error**
d Beobachterfehler *m*
f erreur *f* d'observateur
r субъективная ошибка *f* наблюдения

9557 **perturbated system**
d gesstörtes System *n*
f système *m* perturbé
r возмущенная система *f*

9558 **perturbation**
d Störung *f*
f perturbation *f*; défaut *m*; défaillance
r возмущение *n*

9559 **perturbation coefficient**
d Störungskoeffizient *m*
f coefficient *m* de perturbation
r коэффициент *m* возмущений

9560 **perturbation method**
d Methode *f* der Störungen
f méthode *f* de perturbations
r метод *m* возмущений

9561 **perturbation theory**
d Störungstheorie
f théorie *f* de perturbations
r теория *f* помех

9562 **phantom circuit**
d Phantomschaltkreis *m*
f circuit *m* fantôme
r фантомная схема *f*

* **phantom load** → 5464

9563 **phase**
d Phase *f*
f phase *f*
r фаза *f*

9564 **phase advance; phase lead**
d Phasenvoreilung *f*; Voreilen *n* der Phase
f avance *f* de phase
r опережение *n* фазы

9565 **phase advancer**
d Phasenschieber *m*
f avanceur *m* de phase
r фазокомпенсатор *m*; компенсатор *m* фаз

9566 **phase advancing; phase shift**
d Phasenschiebung *f*
f décalage *m* de phase
r фазовый сдвиг *m*

9567 **phase angle**
d Phasenwinkel *m*
f angle *m* de déphasage; angle de phase
r фазовый угол *m*

9568 **phase angle locus**
d Phasenwinkelort *m*
f lieu *m* de l'angle de phase; lieu de l'angle de déphasage
r годограф *m* фазового угла

9569 **phase-asynchronous interface**
d phasenasynchrone Schnittstelle *f*
f interface *f* asynchrone de phase
r фазо-асинхронный интерфейс *m*

9570 **phase averaging**
d Phasenmitteilung *f*
f prise *f* en moyenne de phase
r усреднение *n* фазы

9571 **phase characteristic; phase response**
d Phasenkennlinie *f*; Phasengang *m*
f caractéristique *f* de déphasage; réponse *f* en phase
r фазовая характеристика *f*

9572 **phase coincidence**
d Phasenübereinstimmung *f*
f coïncidence *f* des phases
r синфазность *f*; совпадение *n* по фазе

9573 **phase compensator; phase equalizer**
d Phasenentzerrer *m*
f compensateur *m* de phase; égaliseur *m* de phase
r фазовыравнитель *m*

9574 **phase constant**
d Phasenkonstante *f*
f constante *f* de phase
r фазовая постоянная *f*

9575 **phase contour**
d Linie *f* konstanter Phase
f courbe *f* de déphasage
r фазовый контур *m*

9576 **phase control**
d Phasenregelung *f*
f réglage *m* de phase
r регулирование *n* фазы

9577 **phase control circuit**
d Phasenregelungsschema *n*
f circuit *m* de réglage de phase
r схема *f* регулирования фазы

9578 **phase counter**
d Phasenzähler *m*
f compteur *m* de phase
r счётчик *m* фаз

9579 **phase cross-over frequency**
d Phasenschnittfrequenz *f*
f fréquence *f* de coupure de phase
r частота *f* разделения по фазе

9580 **phase curve**
d Phasenkurve *f*
f courbe *f* de phase
r фазовая кривая *f*

9581 **phase delay**
d Phasenlaufzeit *f*; Phasenverzögerung *f*
f décalage *m* de phase
r задержка *f* по фазе

9582 **phase detector; phase discriminator**
d Phasendetektor *m*; Phasendiskriminator *m*
f détecteur *m* de différence de phases; discriminateur *m* de phase
r фазовый детектор *m*; фазовый дискриминатор *m*

9583 **phase diagram**
d Phasendiagramm *n*
f diagramme *m* de phase
r фазовая диаграмма *f*

* **phase discriminator → 9582**

9584 **phase distortion; delay distortion**
d Phasenverzerrung *f*; Laufzeitverzerrung *f*
f distorsion *f* de phase
r фазовое искажение *n*

* **phase equalizer → 9573**

* **phase error → 9587**

9585 phase index
d Phasenbrechzahl *f*
f indice *m* de phase
r фазовый коэффициент *m* преломления

9586 phase indicator
d Phasenwinkelanzeiger *m*
f indicateur *m* de phase
r указатель *m* сдвига фаз

9587 phase[ing] error
d Phasenfehler *m*
f erreur *f* de phase
r фазовая погрешность *f*

9588 phase inverter
d Phasenumkehrer *m*; Phasenwender *m*
f inverseur *m* de phase
r фазоинвертор *m*

9589 phase inverter circuit
d Phasenumkehrschaltung *f*
f circuit *m* d'inversion de phase
r фазоинверсная схема *f*

9590 phase lag
d Phasennachlauf *m*
f retard *m* de phase
r отставание *n* по фазе

* **phase lead** → 9564

9591 phase-lead circuit
d Phasenvoreilungsstromkreis *m*;
Phasenvoreilungsschaltung *f*
f circuit *m* d'avance de phase
r фазоопережающий контур *m*

9592 phase-lead network
d Phasenvorhaltglied *n*
f circuit *m* à avance de phase
r фазоопережающее звено *n*

9593 phase lock
d Phasenkopplung *f*; Phasensynchronisierung *f*
f asservissement *m* de phase
r синхронизация *f* фазы

9594 phase-locked
d phasenstarr [synchronisiert]; mit
phasenstarrer Kopplung
f à verrouillage de phase; à accrochage de
phase
r с постоянной фазой; синхронизованный
по фазе

9595 phase lock loop
d Phasensynchronisierungsschleife *f*

f boucle *f* de synchronisation de phase
r замкнутая схема *f* фазовой синхронизации

9596 phase locus
d Phasenkennlinie *f*
f lieu *m* de phase
r фазовый годограф *m*

9597 phase manipulation
d Phasenumtastung *f*
f modulation *f* par déplacement de phase
r фазовая манипуляция *f*

9598 phase margin
d Phasenrand *m*; Phasenreserve *f*
f marge *f* de phase
r запас *m* по фазе

9599 phase matching
d Phasenanpassung *f*
f adaptation *f* de phase
r фазовое согласование *n*

9600 phase-matching relation
d Phasenanpassungsbeziehung *f*
f relation *f* d'adaptation de phase
r отношение *n* фазового согласования

9601 phase-modulated carrier
d phasenmodulierter Träger *m*;
phasenmodulierte Trägerwelle *f*;
phasenmodulierte Trägerfrequenz *f*
f porteuse *f* modulée en phase
r фазо-модулированная несущая *f*

9602 phase-modulated optical signal
d phasenmoduliertes optisches Signal *n*
f signal *m* optique à modulation de phase
r фазомодулированный оптический
сигнал *m*

9603 phase modulation
d Phasenmodulation *f*
f modulation *f* de phase
r фазовая модуляция *f*

9604 phase modulator
d Phasenmodulator *m*
f modulateur *m* de phase
r фазовый модулятор *m*

9605 phase noise
d Phasenrauschen *n*
f bruit *m* de phase
r фазовый шум *m*

9606 phase path
d Phasenbahn *f*; Phasentrajektorie *f*

 f trajectoire *f* de phase
 r фазовая траектория *f*

9607 phase plane method
 d Methode *f* der Phasenebene
 f méthode *f* du plan de phase
 r метод *m* фазовой плоскости

 * **phase-pulse modulation** → 4389

9608 phase relationship
 d Phasenverhältnis *n*
 f rapport *m* de phase
 r соотношение *n* фаз

 * **phase response** → 9571

9609 phase response measurement
 d Messung *f* des Phasengangs
 f mesure *f* de la réponse en phase; mesure du déphasage
 r измерение *n* фазовой характеристики

 * **phase reversal** → 11042

9610 phase reversal relay
 d Phasenumkehrrelais *n*
 f relais *m* à inversion de phase
 r реле *n* обратного вращения фазы

9611 phase-sensitive amplifier
 d Phasendiskriminator *m*
 f amplificateur *m* sensible à la phase
 r фазочувствительный усилитель *m*

9612 phase-sensitive detector
 d phasenempfindliches Nachweisgerät *n*; phasenempfindlicher Detektor *m*
 f détecteur *m* sensible au changement de phase
 r фазочувствительный детектор *m*

9613 phase-sensitive null indicator
 d phasenempfindlicher Nullanzeiger *m*
 f indicateur *m* de zéro sensible à la phase
 r фазочувствительный нульиндикатор *m* фазы

9614 phase-sensitive rectifier
 d phasenempfindlicher Gleichrichter *m*
 f redresseur *m* sensible à la phase; redresseur sensible aux variations de phase
 r фазочувствительный выпрямитель *m*

9615 phase separation
 d Phasentrennung *f*
 f séparation *f* de phase
 r разделение *n* фаз

9616 phase-sequence indicator
 d Drehfeldrichtungsanzeiger *m*; Phasenfolgeanzeiger *m*
 f indicateur *m* d'ordre de phases
 r индикатор *m* последовательности фаз

 * **phase shift** → 9566

9617 phase-shift circuit
 d Phasenverschiebungskreis *m*; Phasenverschiebungskette *f*
 f circuit *m* de déphasage
 r фазосдвигающий контур *m*

9618 phase-shifting transformaer
 d Synchronphasenschieber *m*
 f synchro-déphaseur *m*
 r синхронный фазокомпенсатор *m*

9619 phase space
 d Phasenraum *m*; Zustandsraum *m*
 f espace *m* de phase
 r фазовое пространство *n*

9620 phase spectrum
 d Phasenspektrum *n*
 f spectre *m* de phase
 r фазовый спектр *m*

9621 phase splitter
 d Phasenteiler *m*
 f diviseur *m* de phase
 r схема *f* расщепления фазы

9622 phase stability
 d Phasenstabilität *f*
 f stabilité *f* de phase
 r стабильность *f* фазы

9623 phase synchronization
 d Phasensynchronisierung *f*
 f synchronisation *f* de phase
 r фазовая синхронизация *f*

9624 phase timing signal
 d Phasentaktsignal *n*
 f signal *m* de rythme de phase
 r фазирующий тактовый сигнал *m*

9625 phase undervoltage relay
 d Phasenunterspannungsrelais *n*
 f relais *m* à minimum de tension de phase
 r реле *n* пониженного фазового напряжения

9626 phase velocity
 d Phasengeschwindigkeit *f*
 f vitesse *f* de phase
 r фазовая скорость *f*

9627 phasing
d Phaseneinstellung *f*; Phasenabgleich *m*
f mise *f* en phase; calage *m*
r фазировка *f*; фазирование *n*

9628 phasing adjustment
d Phaseneinstellungsregulierung *f*
f réglage *m* de déphasage
r регулировка *f* фазирования

9629 phasing capacitor
d Phasenabgleichkondensator *m*
f condensateur *m* d'ajustement de phase
r фазовыравнивающий конденсатор *m*

9630 phonemic analysis
d phonemische Analyse *f*
f analyse *f* phonémique
r фонемический анализ *m*

*   photocell → 9640

9631 photocell amplifier
d Fotozellenverstärker *m*
f amplificateur *m* de cellule photoélectrique
r усилитель *m* фотоэлемента

9632 photocell pick-up; photoelectric sensor
d Fotozellenfühler *m*;
   Fotozellentonabnehmer *m*; fotoelektrischer
   Sensor *m*
f capteur *m* photoélectrique
r фотоэлектрический датчик *m*

9633 photoconductive
d lichtelektrisch leitend
f photoconductible
r фотопроводящий

9634 photoconductive cell; photoresistance
d Fotowiderstand *m*; Fotowiderstandszelle *f*
f cellule *f* photorésistante
r фотосопротивление *n*

9635 photoconductive effect
d innerer Fotoeffekt *m*; innerer
   lichtelektrischer Effekt *m*
f effet *m* photoélectrique interne; effet de
   photoconduction
r внутренный фотоэффект *m*

9636 photoconductivity
d Fotoleitfähigkeit *f*
f conductibilité *f* photoélectrique
r фотопроводимость *f*

9637 photoelectric absorption
d Absorption *f* durch Fotoeffekt

f absorption *f* photoélectrique
r фотоэлектрическое поглощение *n*

9638 photoelectric analog divider
d fotoelektrischer Analogteiler *m*
f diviseur *m* analogique photoélectrique
r фотоэлектрическое аналоговое
   делительное устройство *n*

9639 photoelectric building block element
d fotoelektrisches Bauelement *n*
f élément *m* modulaire photoélectrique
r фотоэлектрический конструкционный
   элемент *m*

9640 photoelectric cell; photocell; photoelement
d Lichtelement *n*; lichtelektrische Zelle *f*;
   Fotozelle *f*; Fotoelement *n*
f cellule *f* photoélectrique; photocellule *f*;
   photo-élément *m*
r фотоэлемент *m*

9641 photoelectric circuit
d fotoelektrischer Stromkreis *m*
f circuit *m* photoélectrique
r схема *f* с фотоэлементом

9642 photoelectric colorimeter
d fotoelektrischer Farbmesser *m*
f colorimètre *m* photoélectrique
r фотоэлектрический колориметр *m*

9643 photoelectric comparator
d fotoelektrischer Komparator *m*
f comparateur *m* photoélectrique
r фотоэлектрический компаратор *m*

9644 photoelectric constant
d fotoelektrische Konstante *f*
f constante *f* photoélectrique
r фотоэлектрическая постоянная *f*

9645 photoelectric control
d fotoelektrische Steuerung *f*
f commande *f* photoélectrique
r фотоэлектрическое управление *n*

9646 photoelectric control equipments
d lichtelektrische Steueranlagen *fpl*
f installations *fpl* de commande
   photoélectriques
r фотоэлектрическое контрольное
   оборудование *n*

9647 photoelectric densitometer
d fotoelektrischer Schwärzungsmesser *m*
f densitomètre *m* photoélectrique
r фотоэлектрический денситометр *m*

9648 **photoelectric displacement transmitter**
  d fotoelektrischer Verschiebungsgeber *m*;
  lichtelektrischer Verschiebungsgeber
  f capteur *m* photoélectrique du déplacement
  r фотоэлектрический датчик *m*
  перемещений

9649 **photoelectric emission; photoemission**
  d lichtelektrische Emission *f*; Fotoemission *f*
  f émission *f* photoélectronique;
  photo-émission *f*
  r фотоэлектронная эмиссия *f*

9650 **photoelectric function generator**
  d fotoelektrischer Funktionsgenerator *m*
  f générateur *m* photoélectrique de fonctions
  r фотоэлектрический функциональный
  преобразователь *m*

9651 **photoelectric inspection of quality**
  d fotoelektrische Qualitätskontrolle *f*
  f contrôle *m* photo-électrique de la qualité
  r контроль *m* качесва фотоэлектрическим
  методом

9652 **photoelectric interaction**
  d fotoelektrische Wechselwirkung *f*
  f interaction *f* photoélectrique
  r фотоэлектрическое взаимодействие *n*

9653 **photoelectric measurement by null
  method**
  d fotoelektrische Messung *f* mittels
  Nullmethode
  f mesure *f* photoélectrique par méthode de
  zéro
  r фотоэлектрическое измерение *n* нулевым
  методом

9654 **photoelectric method**
  d fotoelektrisches Verfahren *n*
  f méthode *m* photo-électrique
  r фотоэлектрический метод *m*

9655 **photoelectric polarimeter**
  d lichtelektrisches Polarimeter *n*
  f polarimètre *m* photoélectrique
  r фотэлектрический поляриметр *m*

9656 **photoelectric position controller**
  d fotoelektrischer Stellungsregler *m*
  f régulateur *m* photoélectrique de position
  r фотоэлектрический регулятор *m*
  положения

9657 **photoelectric pulse maker**
  d lichtelektrischer Impulsgeber *m*
  f transmetteur *m* d'impulsions photoélectrique

  r передатчик *m* фотоэлектрических
  импульсов

9658 **photoelectric reader; optical scanner**
  d fotoelektrischer Leser *m*; optischer
  Abtaser *m*
  f balayeur *m* optique; lecteur *m*
  photoélectrique
  r фотосчитывающее устройство *n*

9659 **photoelectric receiver**
  d fotoelektrischer Empfänger *m*
  f récepteur *m* photoélectrique
  r фотоэлектрический приёмник *m*

9660 **photoelectric scanner**
  d lichtelektrischer Abtaster *m*
  f analyseur *m* photoélectrique
  r фотоэлектрическое сканирующее
  устройство *n*

9661 **photoelectric scanning**
  d fotoelektrische Abtastung *f*
  f exploration *f* photoélectrique
  r фотоэлектрическое сканирование *n*

* **photoelectric sensor** → 9632

9662 **photoelectric spectrophotometer**
  d fotoelektrisches Spektralfotometer *n*
  f spectrophotomètre *m* photoélectrique
  r фотоэлектрический спектрофотометр *m*

9663 **photoelectronic installation**
  d lichtelektronische Anlage *f*
  f installation *f* photoélectronique
  r фотоэлектронное устройство *n*

* **photoelement** → 9640

* **photoemission** → 9649

9664 **photoemission pick-off**
  d Fotoemissionswandler *m*
  f capteur *m* photoémissif
  r фотоэмиссионный чувствительный
  элемент *m*

9665 **photoemissive detector**
  d Fotoemissionsdetektor *m*
  f détecteur *m* photo-émissif
  r фотоэмиссионный детектор *m*

9666 **photogrammetric measuring method**
  d fotogrammetrische Messmethode *f*
  f méthode *f* photogrammétrique de mesure
  r фотограмметрический метод *m* измерения

9667   **photogrammetric technology**
  d   fotogrammetrische Technik *f*
  f   technique *f* photogrammétrique
  r   фотограмметрическая техника *f*

9668   **photogrammetry; picture measuring;**
       image measuring
  d   Bildvermessung *f*
  f   photogrammétrie *f*
  r   фотограмметрия *f*; измерение *n*
       изображения

9669   **photoinduced**
  d   fotoinduziert
  f   photoinduit
  r   фотоиндуцированный

9670   **photoluminescence**
  d   Fotolumineszenz *f*
  f   photoluminescence *f*
  r   фотолюминесценция *f*

9671   **photomagnetic effect**
  d   fotomagnetischer Effekt *m*
  f   effet *m* photomagnétique
  r   фотомагнитный эффект *m*

9672   **photometer**
  d   Fotometer *n*
  f   photomètre *m*
  r   фотометр *m*

9673   **photometric computer**
  d   Fotometerrechner *m*;
       Lichtmessungsrechner *m*
  f   calculatrice *f* photométrique
  r   фотометрический компьютер *m*

*   **photoresistance** → 9634

9674   **phototransistor circuit**
  d   Fototransistorschaltung *f*
  f   circuit *m* du phototransistor
  r   схема *f* на фототранзисторах

9675   **phototube circuit**
  d   Fotozellenkreis *m*
  f   circuit *m* de la photocellule
  r   схема *f* включения фотоэлемента

9676   **physical address**
  d   physikalische Adresse *f*
  f   adresse *f* physique
  r   физический адрес *m*

9677   **physical analog; physical model**
  d   physikalisches Modell *n*
  f   modèle *m* physique

  r   физическая модель *f*; физический
       аналог *m*

9678   **physical assembly object data**
  d   physikalische Montageobjektdaten *pl*
  f   données *fpl* physiques d'un objet
       d'assemblage
  r   физические параметры *mpl* сборочных
       объектов

9679   **physical channel**
  d   physischer Kanal *m*
  f   canal *m* physique
  r   физический канал *m*

9680   **physical connection**
  d   physische Verbindung *f*
  f   connexion *f* physique
  r   физическая связь *f*

9681   **physical input-output control system**
  d   physisches Eingabe-Ausgabe-Steuersystem *n*
  f   système *m* de commande d'entrée-sortie
       physique
  r   система *f* управления физическим
       устройством

9682   **physical manipulator size**
  d   physikalische Manipulatorgröße *f*
  f   grandeur *f* de manipulateur physique
  r   физический параметр *m* манипулятора

*   **physical model** → 9677

9683   **physical photometer**
  d   physikalisches Fotometer *n*; objektives
       Fotometer
  f   photomètre *m* physique; photomètre objectif
  r   физический фотометр *m*

9684   **physical properties**
  d   physikalische Eigenheiten *fpl*
  f   propriétés *fpl* physiques
  r   физические свойства *npl*

9685   **physical simulation**
  d   physikalische Simulation *f*
  f   simulation *f* physique
  r   физическое моделирование *n*

9686   **physical unit block**
  d   physischer Einheitsblock *m*
  f   bloc *m* d'unité physique
  r   физический блок *m*

9687   **physical value**
  d   physikalische Größe *f*
  f   grandeur *f* physique
  r   физическая величина *f*

* **PI-control** → 10369

* **PI-controller** → 10362

9688 picoprocessor
d Picoprozessor *m*
f picoprocesseur *m*
r пикопроцессор *m*

9689 picoprogramming
d Picoprogrammierung *f*
f picoprogrammation *f*
r пикопрограммирование *n*

9690 picosecond
d Picosekunde *f*
f picoseconde *f*
r пикосекунда *f*

9691 picture control
d Bildsteuerung *f*
f commande *f* vidéo
r управление *n* изображений

9692 picture evaluation system
d Bildauswertesystem *n*
f système *m* d'évaluation d'un vidéo
r система *f* обработки изображения

* **picture measuring** → 9668

9693 picture signal
d Bildsignal *n*
f signal *m* d'image
r сигнал *m* изображения

9694 picture signal amplitude
d Bildsignalamplitude *f*
f amplitude *f* du signal d'image
r амплитуда *f* сигнала изображения

* **PID-control** → 10363

* **PID-controller** → 10364

* **piece adjusting analysis** → 9695

9695 piece fit analysis; piece adjusting analysis
d Teilepassungsanalyse *f*; Analyse *f* der
Teilepassungen
f analyse *f* d'ajustement de pièce
r анализ *m* юстировки деталей

9696 piecewise approximation
d stückweise Approximation *f*
f approximation *f* partielle
r кусочно-линейная апроксимация *f*

9697 piecewise continuous function
d stückweise stetige Funktion *f*
f fonction *f* continue par sections
r непрерывная кусочно-линейная функция *f*

9698 piecewise linear characteristic
d stückweise lineare Kennlinie *f*
f charactéristique *f* linéaire par sections
r кусочно-линейная характеристика *f*

9699 PI-element
d PI-Glied *n*; Isodromglied *n*
f élément *m* PI
r изодром *m*

9700 piercing voltage
d Durchbruchspannung *f*
f tension *f* de perçage
r напряжение *n* пробоя

9701 piezoelectric device
d piezoelektrisches Bauteil *n*
f module *m* piézo-électrique
r пьезоэлектрический элемент *m*

9702 piezoelectric effect
d piezoelektrischer Effekt *m*
f effet *m* piézo-électrique
r пьезоэлектрический эффект *m*

9703 piezoelectric laser modulator
d piezoelektrischer Lasermodulator *m*
f modulateur *m* piézo-électrique du laser
r пьезоэлектрический модулятор *m* лазера

9704 piezoelectric measuring instrument
d piezoelektrisches Messgerät *n*
f instrument *m* de mesure piézo-électrique
r пьезоэлектрический измерительный
прибор *m*

9705 piezoelectric pressure gauge
d piezoelektrischer Druckmesser *m*;
piezoelektrisches Manometer *n*
f manomètre *m* piézo-électrique
r пьезоэлектрический манометр *m*;
пьезоманометр *m*

9706 piezoelectric sensor
d piezoelektrischer Sensor *m*
f senseur *m* piézo-électrique
r пьезоэлектрический сенсор *m*

9707 piezoelectric strain gauge
d piezoelektrischer Dehnungsmessstreifen *m*
f jauge *f* de contrainte piézo-électrique
r пьезотензометр *m*

9708 piezoelectric transducer
d piezoelektrischer Wandler *m*
f transducteur *m* piézo-électrique
r пьезоэлектрический преобразователь *m*

9709 piezoelectric vibration
d piezoelektrische Schwingung *f*
f vibration *f* piézo-électrique
r пьезоэлектрическая вибрация *f*

9710 piezo-optic
d piezooptisch
f piézo-optique
r пьезооптический

9711 piezoresistance effect measuring method
d Piezowiderstandseffekt-Messmethode *f*
f méthode *f* de mesure par effet de résistance piézo-électrique
r метод *m* измерения на основе эффекта пьезосопротивления

9712 pilot *v*
d vorsteuern
f piloter
r управлять; стабилизировать

9713 pilot cell
d Steuerelement *n*; Leitelement *n*
f élément *m* de manœuvre; élément de contrôle
r контрольный элемент *m*

9714 pilot circuit
d Leitstromkreis *m*; Steuerstromkreis *m*; Pilotstromkreis *m*; Kontrollstromkreis *m*
f circuit-pilote *m*; circuit *m* de contrôle
r контрольная цепь *f*

9715 pilot frequency generator
d Pilotfrequenzgenerator *m*
f générateur *m* de fréquence pilote
r генератор *m* контрольной частоты

9716 pilot-operated controller
d hilfsgesteuerter Regler *m*; indirekt wirkender Regler
f régulateur *m* à signal pilote
r регулятор *m* с вспомогательным источником энергии

9717 pilot operation
d Vorsteuerung *f*
f opération *f* de pilotage
r предварительное регулирование *n*

9718 pilot plant; experimental unit
d Pilotanlage *f*; Modellanlage *f*; Versuchsstand *m*; Teststand *m*; Versuchsanlage *f*
f installation *f* d'essai; installation pilote; banc *m* d'épreuve
r опытная установка *f*; модельная установка; пилотная установка; экспериментальный стенд *m*; испытательный стенд

9719 pilot protection with direct comparison
d Streckenschutz *m* mit direktem Vergleich
f protection *f* par pilote à comparaison directe
r контрольная защита *f* линии с непосредственным сравнением

9720 pilot protection with indirect comparison
d Streckenschutz *m* mit indirektem Vergleich
f protection *f* par pilote à comparaison indirecte
r контрольная защита *f* [линии] с косвенным сравнением

9721 pilot run
d Piloteinsatz *m*; Ersteinsatz *m*
f application *f* pilote
r опытное применение *n*

9722 pilot selector
d Zeitselektor *m*
f sélecteur-pilote *m*
r контрольный искатель *m*

9723 pilot system
d Pilotsystem *n*; Erkundungssystem *n*
f système *m* pilote
r система *f* сбора экспериментальных данных

*   **pilot tape → 3347**

9724 pilot wire regulator
d Zähladerregler *m*; Messdrahtregler *m*
f régulateur *m* à fil pilote
r авторегулятор *m* усиления

9725 pipelined architecture
d zeitverschachtelt arbeitende Struktur *f*; Pipeline-Architektur *f*
f structure *f* à travail à la chaîne; structure pipeline
r конвейерная архитектура *f*

9726 pipeline flowmeter
d Rohrleitungsdurchflussmesser *m*
f débitmètre *m* pour conduites
r расходомер *m* для трубопроводов

9727 pipeline system
d Rohrleitungssystem *n*

   f système *m* de tuyautage
   r система *f* трубопроводов

9728 pipelining
   d Zeitverschachtelung *f* der
      Befehlsabwicklung; Pipeline-Arbeitsweise *f*
   f principe *m* à travail à la chaîne; principe
      pipeline
   r конвейерный принцип *m*

9729 Pirani gauge
   d Pirani-Messgerät *n*; Heizdrahtmanometer *n*
   f jauge *f* de Pirani; manomètre *m* à fil chaud
   r манометр *m* Пирани; тепловой манометр

9730 pivoted manipulator
   d Schwenkmanipulator *m*
   f manipulateur *m* pivotant
   r поворотный манипулятор *m*

9731 pivoting sensor
   d schwenkbarer Sensor *m*
   f senseur *m* pivotant
   r поворотный сенсор *m*

9732 plane graphic structure
   d ebene grafische Struktur *f*
   f structure *f* graphique plane
   r двухмерная графическая структура *f*

   * **planning** → 10332

9733 planning documents
   d Planungsgrundlage *f*; Planungsunterlagen *fpl*
   f documents *mpl* de planification
   r документация *f* планирования

9734 planning method
   d Planungsmethode *f*
   f méthode *f* de planification
   r метод *m* планирования

9735 planning models
   d Planungsmodelle *npl*
   f modèles *mpl* de planification
   r модели *fpl* планирования

9736 planning of process information system
   d Prozessinformationssystemplanung *f*
   f planning *m* pour systèmes d'information de
      processus
   r проектирование *n* информационно-
      управляющих систем

9737 plan of operation sequence
   d Arbeitsablaufplan *m*
   f planification *f* de la suite des opérations
   r операционный график *m*

9738 plan subject control
   d überwachungspflichtige Anlage *f*
   f installation *f* soumise à la surveillance
   r установка f, требующая постоянного
      контроля

9739 plant (US)
   d Regelstrecke *f*; Automatisierungsobjekt *n*
   f système *m* à régler; installation *f* réglée
   r объект *m* управления

9740 plant attenuation
   d Regelstreckendämpfung *f*
   f affaiblissement *m* global du système de
      réglage
   r затухание *n* колебаний в регулируемом
      объекте

9741 plant characteristic
   d Regelstreckencharakteristik *f*;
      Regelstreckenkennlinie *f*
   f caractéristique *f* du système réglé
   r собственная характеристика *f*
      регулируемого объекта

9742 plant computer system
   d Betriebsrechnersystem *n*
   f système *m* de calculateurs de production
   r вычислительная система *f* для управления
      производством

9743 plant control
   d Anlagensteuerung *f*; Anlagenüberwachung *f*
   f commande *f* de l'unité
   r управление *m* установкой

9744 plant design
   d Anlagenauslegung *f*; anlagentechnischer
      Entwurf *m*
   f projet *m* de l'installation
   r проект *m* технической установки

9745 plant identification
   d Regelstreckenanalyse *f*
   f identification *f* d'objets à asservir
   r идентификация *f* объекта

9746 plant operation
   d Betrieb *m* von Anlagen
   f exploitation *f* d'installation
   r эксплуатация *f* установки

9747 plant preparation
   d Anlagenvorbereitung *f*
   f préparation *f* d'installation
   r подготовка *f* установки

9748 plant safety
   d Anlagensicherheit *f*

f sécurité *f* de l'installation
r безопасность *f* установки

9749 plant structure
d Anlagenstruktur *f*
f structure *f* de l'installation
r структура *f* установки

9750 plasma
d Plasma *n*
f plasma *m*
r плазма *f*

9751 plasma generator
d Plasmagenerator *m*
f générateur *m* de plasma
r генератор *m* плазмы

9752 plasma phase shifter
d Plasmaphasenschieber *m*
f déphaseur *m* à plasma
r плазменный фазорегулятор *m*

9753 plastic potentiometer
d plastisches Potentiometer *n*
f potentiomètre *m* plastique
r пластический потенциометр *m*

9754 plastometer
d Plastizitätsmesser *m*
f plasticimètre *m*
r пластометр *m*; измеритель *m*
   пластичности

*   **platinotron** → 750

9755 plot of the function
d Kurvenbild *n*
f diagramme *m* de fonction
r график *m* функции

9756 plotter
d Plotter *m*; Kurvenschreiber *m*
f traceur *m* graphique
r графопостроитель *m*

9757 plotter configuration
d Plotterkonfiguration *f*
f configuration *f* de traceur
r конфигурация *f* графопостроителя

9758 plug
d Stecker *m*; Steckkontakt *m*
f connecteur *m*; fiche *f*
r штепсель *m*; штекер *m*

9759 plug connection
d Steckverbindung *f*; Steckvorrichtung *f*

f connexion *f* à fiches
r штепсельное соединение *n*

9760 pluggable; plug-in; plug-type; removable
d austauschbar; steckbar
f enfichable; changeable; remplaçable; à fiches
r сменный; вставной

9761 pluggable telephone channel selector
d Kanalwähler *m*
f sélecteur *m* de canaux
r переключатель *m* каналов

9762 plugging chart
d Schalttafeldiagramm *n*; Steckplan *m*
f schèma *m* des fiches
r схема *f* коммутаций; коммутационная
   диаграмма *f*

*   **plug-in** → 9760

9763 plug-in amplifier
d Einsteckverstärker *m*; Einschubverstärker *m*
f amplificateur *m* enfichable
r сменный усилительный блок *m*

9764 plug-in circuit
d Einsteckkreis *m*; Einschubkreis *m*
f circuit *m* enfichable
r блочная схема *f*

9765 plug-in module
d Modulsteckbaugruppe *f*
f unité *f* enfichablebloc *m* interchangeable
r модуль *m* с разъемом

9766 plug-in relay
d Steckrelais *n*
f relais *m* à fiches
r реле *n* штепсельной конструкции

9767 plug-in system
d Einschubsystem *n*
f système *m* enfichable; système d'unités
   interchangeable à fiches
r система f, содержащая штепсельные
   соединения

*   **plug-type** → 9760

*   **pneumatic actuator** → 600

9768 pneumatically delayed control
d pneumatisch verzögerte Steuerung *f*
f commande *f* retardée pneumatiquement
r пневматическая система *f* управления с
   запаздыванием

* **pneumatically operated regulator** → 597

* **pneumatic amplifier** → 595

9769 pneumatic analog computing system
 d pneumatisches Analogierechensystem *n*
 f système *m* de calcul analogique pneumatique
 r пневматическая аналоговая вычислительная система *f*

9770 pneumatic analog model
 d pneumatisches Analogmodell *n*
 f modèle *m* analogique pneumatique
 r пневматическая аналоговая модель *f*

9771 pneumatic analogy
 d pneumatische Analogie *f*
 f analogie *f* pneumatique
 r пневматическая аналогия *f*

9772 pneumatic approximation sensor
 d pneumatischer Annäherungssensor *m*
 f senseur *m* pneumatique de proximité
 r пневматический сенсор *m* приближения

* **pneumatic automation installations** → 1340

9773 pneumatic bridge circuit
 d pneumatische Brückenschaltung *f*
 f couplage *m* en pont pneumatique
 r пневматическая мостовая схема *f*

9774 pneumatic channel
 d pneumatischer Kanal *m*
 f canal *m* pneumatique
 r пневматический канал *m*

9775 pneumatic circuit
 d pneumatischer Kreis *m*; pneumatische Schaltung *f*
 f circuit *m* pneumatique
 r пневматическая цепь *f*

* **pneumatic control** → 596

* **pneumatic controller** → 597

* **pneumatic control system** → 598

9776 pneumatic control technique
 d pneumatische Steuerungstechnik *f*
 f technique *f* de la commande pneumatique
 r техника *f* пневматического управления

9777 pneumatic diaphragm servomotor
 d pneumatischer Membranservomechanismus *m*

 f servomoteur *m* pneumatique à membrane
 r пневматический мембранный сервомеханизм *m*

9778 pneumatic dimensions tranducer
 d pneumatischer Maßwandler *m*
 f capteur *m* pneumatique de dimensions
 r пневматический датчик *m* размеров

9779 pneumatic divider
 d pneumatischer Teiler *m*
 f appareil *m* pneumatique à diviser; diviseur *m* pneumatique
 r пневматический делитель *m*

9780 pneumatic drive feedback
 d Rückführung *f* des pneumatischen Antriebes
 f asservissement *m* de commande pneumatique
 r обратная связь *f* пневматического привода

9781 pneumatic drive of rotation angle
 d pneumatischer Drehwinkelantrieb *m*
 f entraînement *m* pneumatique de l'angle de rotation
 r пневматический сельсин-привод *m*

9782 pneumatic drive system
 d pneumatisches Antriebssystem *n*
 f système *m* d'entraînement pneumatique
 r система *f* пневмопривода

9783 pneumatic element
 d pneumatisches Element *n*
 f élément *m* pneumatique
 r пневматический элемент *m*

9784 pneumatic equipment
 d pneumatisches Gerät *n*
 f équipement *m* pneumatique
 r пневматическая аппаратура *f*

9785 pneumatic high-pressure control
 d pneumatische Hochdruckregelung *f*
 f réglage *m* pneumatique à haute pression
 r пневматическое регулирование *n* высокого давления

9786 pneumatic-hydraulic controller
 d pneumatisch-hydraulischer Regler *m*
 f régulateur *m* pneumatique-hydraulique
 r пневмо-гидравлический регулятор *m*

9787 pneumatic-hydraulic control system
 d pneumatisch-hydraulische Steuerung *f*
 f système *m* de réglage pneumatique-hydraulique
 r пневмо-гидравлическая система *f* управления

9788 pneumatic-hydraulic drive
d pneumatisch-hydraulischer Antrieb *m*
f mécanisme *m* de commande pneumatique-hydraulique
r пневмо-гидравлический привод *m*

9789 pneumatic indicator
d pneumatisches Anzeigegerät *n*
f indicateur *m* pneumatique
r пневматический индикатор *m*

9790 pneumatic integrator
d pneumatischer Integrator *m*
f intégrateur *m* pneumatique
r пневматический интегратор *m*

9791 pneumatic level control
d pneumatische Pegelregelung *f*; pneumatische Niveauregelung *f*
f commande *f* pneumatique de niveau; réglage *m* pneumatique de niveau
r пневматическая регуляция *f* уровня

9792 pneumatic logical elements
d pneumatische Logikelemente *npl*
f éléments *mpl* logiques pneumatiques
r пневматические логические элементы *mpl*

9793 pneumatic logical installations
d pneumatische Logikanlagen *fpl*
f installations *fpl* logiques pneumatiques
r пневматические логические устройства *npl*

9794 pneumatic low-pressure control
d pneumatische Niederdruckregelung *f*
f réglage *m* pneumatique à basse pression
r пневматическая система *f* регулирования низкого давления

9795 pneumatic manipulation assembly unit
d pneumatischer Handhabebaukasten *m*
f boîte *f* de construction de manipulation pneumatique
r пневматическая система *f* манипулирования агрегатного типа

9796 pneumatic manipulator drive
d pneumatischer Manipulatorantrieb *m*
f entraînement *m* de manipulateur pneumatique
r пневмопривод *m* манипулятора

9797 pneumatic measuring transducer
d pneumatischer Messumformer *m*
f convertisseur *m* mesureur pneumatique
r пневматический измерительный

преобразователь *m*

9798 pneumatic microswitch
d pneumatischer Mikroschalter *m*
f micro-interrupteur *m* pneumatique
r пневматический микроключ *m*

9799 pneumatic model
d pneumatisches Modell *n*
f modèle *m* pneumatique
r пневматическая модель *f*

9800 pneumatic multifunction gripper
d pneumatischer Mehrfunktionsgreifer *m*
f grappin *m* pneumatique à fonctions multiples
r пневматический многофункциональный захват *m*

9801 pneumatic operational amplifier
d pneumatischer Rechenverstärker *m*
f amplificateur *m* opérationnel pneumatique
r пневматический операционный усилитель *m*

9802 pneumatic pressure guard
d pneumatischer Druckwächter *m*
f dispositif *m* pneumatique de protection de pression
r пневматический ограничитель *m* давления

9803 pneumatic program timer
d pneumatischer Programmgeber *m*
f donneur *m* de programme pneumatique
r пневматический программный датчик *m*

9804 pneumatic pulse-controlled program control
d pneumatisch impulsgesteuerte Programmsteuerung *f*
f commande *f* pneumatique à impulsions programmée
r пневматическая система *f* импульсного программного управления

9805 pneumatic recorder
d pneumatischer Schreiber *m*
f enregistreur *m* pneumatique
r пневматическое регистрирующее устройство *n*

9806 pneumatic remote control
d pneumatische Fernsteuerung *f*
f commande *f* pneumatique à distance
r пневматическое дистанционное управление *n*

9807 pneumatic remote measuring technique
d pneumatische Fernmesstechnik *f*

f technique *f* pneumatique de mesure à
distance

r пневматическая аппаратура *f* для
дистанционных измерений

9808 pneumatic remote transmission
d pneumatische Fernübertragung *f*
f transmission *f* pneumatique à distance
r пневматическая дистанционная передача *f*

9809 pneumatic resistance
d pneumatischer Widerstand *m*
f résistance *f* pneumatique
r пневматическое сопротивление *n*

* pneumatic rotation air motor → 9810

9810 pneumatic rotation air [pneumatic] motor
d pneumatischer Rotationsdruckluftmotor *m*
f moteur *m* pneumatique sur rotative à air
comprimé
r вращательный пневмодвигатель *m*

9811 pneumatic rotation angle motor
d pneumatischer Drehwinkelmotor *m*
f moteur *m* d'angle de rotation pneumatique
r поворотный пневмодвигатель *m*

9812 pneumatic sensor
d pneumatischer Geber *m*
f capteur *m* pneumatique
r пневматический сенсор *m*

9813 pneumatic setting drive
d pneumatischer Stellantrieb *m*
f commande *f* pneumatique de réglage
r регулировочный пневмопривод *m*

9814 pneumatic setting vane
d pneumatisches Stellglied *n*
f organe *m* de réglage pneumatique
r звено *n* пневматической регулировки

9815 pneumatic signal
d pneumatisches Signal *n*
f signal *m* pneumatique
r пневматический сигнал *m*

9816 pneumatic signal converter
d pneumatischer Signalumformer *m*
f convertisseur *m* de signaux pneumatique
r преобразователь *m* пневматических
сигналов

9817 pneumatic simulator
d pneumatischer Simulator *m*
f simulateur *m* pneumatique
r пневматическое моделирующее

устройство *n*

9818 pneumatic single-purpose controller
d pneumatischer Einzweckregler *m*
f régulateur *m* pneumatique à application
spéciale
r пневматический регулятор *m*
специального назначения

9819 pneumatic switchboard
d pneumatischer Schalttisch *m*
f pupitre *m* de commande pneumatique
r пневматический коммутатор *m*

9820 pneumatic temperature controller
d pneumatischer Temperaturregler *m*
f régulateur *m* pneumatique de température
r пневматический регулятор *m* температуры

9821 pneumatic throttle
d pneumatische Drossel *f*
f étrangleur *m* pneumatique
r пневматический дроссель *m*

9822 pneumatic time constant
d pneumatische Zeitkonstante
f constante *f* de temps pneumatique
r постоянная *f* времени пневматических
устройств

9823 pneumatic time delay relay
d Druckluftverzögerungsrelais *n*
f relais *m* temporisé à air comprimé
r пневматическое реле *n* с выдержкой
времени

9824 pneumatic transmission line
d pneumatische Übertragungsleitung *f*
f ligne *f* de transmission pneumatique
r пневматическая линия *f* передачи

9825 pneumatic vibrating drive
d pneumatischer Vibrationsantrieb *m*
f commande *f* pneumatique vibratoire
r вибрационный пневмопривод *m*

9826 pneumatic Wheatstone bridge
d pneumatische Wheatstonesche Brücke *f*
f pont *m* de Wheatstone pneumatique
r пневматический мостик *m* Уитстона

9827 pneumoelectric
d pneumoelektrisch
f pneumo-électrique
r пневмоэлектрический

9828 pneumohydraulic
d pneumohydraulisch

f pneumohydraulique
r пневмогидравлический

9829 pneumonic building block elements
d pneumonische Bauteile *mpl*
f éléments *mpl* de construction pneumoniques
r пневмонические конструктивные элементы *mpl*

9830 pneumonics
d Pneumonik *f*
f pneumonique *f*
r пневмоника *f*

9831 pneumonic system
d Pneumoniksystem *n*
f système *m* pneumonique
r пневмоническая система *f*

9832 pneutronic level control
d elektropneumonische Pegelregelung *f*
f réglage *m* électropneumonique de niveau
r электронно-пневматический регулятор *m* уровня

9833 point approximation
d punktweise Annäherung *f*
f approximation *f* par points
r точечное приближение *n*

9834 point contact
d Spitzenkontakt *m*
f contact *m* de pointe
r точечный контакт *m*

9835 pointer register
d Zeigerregister *n*; Hinweisregister *n*
f registre *m* de pointeur
r регистр-указатель *m*

9836 point mapping method
d Methode *f* der Punkttransformation
f méthode *f* de transformation ponctuelle
r метод *m* точечного преобразования

* point of connection → 3025

9837 point of control
d Stellort *m*
f emplacement *m* d'action du réglage
r место *n* приложения регулирующего воздействия

9838 point of discontinuity
d Unstetigkeitsstelle *f*
f point *m* de discontinuité
r точка *f* разрыва

9839 point of inflection
d Wendepunkt *m*
f point *m* d'inflexion
r точка *f* перегиба

9840 point of invocation
d Aufrufstelle *f*
f point *m* d'invocation
r точка *f* вызова

9841 point of measurement
d Messstelle *f*
f lieu *m* de mesure; endroit *m* de mesure
r точка *f* измерения

9842 point of reversal
d Umkehrpunkt *m*
f point *m* de retour
r точка *f* возврата

9843 point of support
d Stützpunkt *m*
f point *m* de support
r опорная точка *f*

9844 point-to-point control
d Punkt-zu-Punkt-Steuerung *f*
f commande *f* point par point
r управление *n* по точкам

9845 point-to-point mapping graph
d Diagramm *n* mit gepunkteten Werten; Punkt-für-Punkt-Diagrammaufzeichnung *f*
f diagramme *m* ponctuel
r точечное составление *n* диаграммы

9846 point-to-point positioning control
d Einzelpunktsteuerung *f*
f commande *f* de position par point; mise *f* en position point par point
r автоматическое позиционирование *n* по точкам

9847 point-to-point scanning system
d Punktabtastsystem *n*
f système *m* de surveillance point par point
r обегающая система *f* развёртки

9848 point transformation
d Punkttransformation *f*
f transformation *f* ponctuelle; transformation par points
r точечное преобразование *n*

9849 Poisson's distribution
d Poissonsche Verteilung *f*
f répartition *f* de Poisson
r распределение *n* Пуанссона

9850 Poisson's ratio
d Querdehnungszahl *f*
f coefficient *m* de Poisson
r коэффициент *m* Пуассона

9851 polar angle
d Polarwinkel *m*
f angle *m* polaire
r полярный угол *m*

9852 polar coordinates
d Polarkoordinaten *fpl*
f coordonnées *fpl* polaires
r полярные координаты *fpl*

9853 polar diagram
d Polardiagramm *n*
f diagramme *m* polaire
r полярная диаграмма *f*

9854 polarity detector
d Polaritätsdetektor *m*; Polungsweiser *m*
f détecteur *m* de polarité
r детектор *m* полярности

9855 polarizability
d Polarisierbarkeit *f*
f polarisabilité *f*
r способность *f* поляризации

9856 polarization
d Polarisation *f*
f polarisation *f*
r поляризация *f*

9857 polarization analyzer
d Polarisationsanalysator *m*
f analyseur *m* de polarisation
r анализатор *m* поляризации

9858 polarization behaviour
d Polarisationsverhalten *n*
f caracteristiques *fpl* de polarisation
r поляризационная характеристика *f*

9859 polarization dispersion
d Polarisationsdispersion *f*
f dispersion *f* de la polarisation
r дисперсия *f* поляризации

9860 polarization effects
d Polarisationseffekte *mpl*
f effets *mpl* de polarisation
r поляризационные эффекты *mpl*

9861 polarization-independent
d polarisationsunabhängig
f indépendant de la polarisation

r независимый от поляризации

9862 polarization optical method
d optische Polarisationsmethode *f*
f méthode optique *f* de polarisation
r оптический поляризационный метод *m*

9863 polarization stability
d Polarisationsstabilität *f*
f stabilité *f* de polarisation
r стабильность *f* поляризации

9864 polarization state
d Polarisationszustand *m*
f état *m* de polarisation
r поляризованное состояние *n*

9865 polarize *v*
d polarisieren
f polariser
r поляризовать

9866 polarized radiation
d polarisierte Strahlung *f*
f rayonnement *m* polarisé
r поляризированное излучение *n*

9867 polarizer
d Polarisator *m*
f polariseur *m*
r поляризатор *m*

9868 polarizing component
d polarisierendes Bauelement *n*
f composant *m* polarisant
r поляризирующий компонент *m*

9869 polarizing filter
d Polarisationsfilter *n*
f filtre *m* polarisateur
r поляризующий светофильтр *m*

9870 policy space
d Entscheidungsraum *m*
f espace *m* de décision
r пространство *n* решений

9871 polling interrupt system
d Abruf-Interruptsystem *n*
f système *m* d'interruption d'interrogation
r система *f* определения источника
   прерывания методом опроса

9872 polling phase
d Abrufphase *f*
f phase *f* d'interrogation
r фаза *f* опроса

9873 **polling signal**
d Abrufsignal n
f signal m d'interrogation
r опрашивающий сигнал m

9874 **polling technique**
d Abruftechnik f
f technique f d'interrogation
r способ m выполнения процедуры опроса

* **polynomial** → 8435

9875 **polynomial approximation**
d Polynom-Approximation f
f approximation f de polynômes
r аппроксимация f функции при помощи полиномов

9876 **polynomial input**
d Eingangsgröße f mit Polynomcharakteristik
f grandeur f d'entrée de polynôme
r входная величина f в виде многочлена

9877 **polyoptimization**
d Polyoptimierung f; mehrkriterielle Optimierung f
f optimisation f multiple; polyoptimisation f
r полиоптимизация f; многокритериальная оптимизация f

9878 **polyphase current**
d Mehrphasenstrom m
f courant m polyphasé
r многофазный ток m

* **Pontrjagin maximum principle** → 7913

9879 **pop**
d Auskellern n; Entkellern n
f dépilage m; dépilement m
r извлечение n; выталкивание n

9880 **pop up** v
d hervorholen
f relever; surgir
r доставать; извлекать

9881 **portable data collection terminal**
d tragbares Datenerfassungsterminal n
f terminal m de collection de données portable
r переносное устройство n сбора данных

9882 **portable terminal**
d tragbares Terminal n
f terminal m portable
r переносный терминал m

9883 **portable use**

d Trageinsatz m
f usage m sous forme portable
r применение портативного устройства n

9884 **port controller**
d Torsteuerung f
f commande f de porte
r контроллер m порта

9885 **positional checking**
d Stellungsmelder m
f contrôle m de position
r позиционный контроль m

9886 **positional notation**
d Stellenschreibweise f
f notation f en position
r позиционное представление n

* **positional notations** → 8839

* **positional servosystem** → 7056

9887 **position code**
d Positionskode m
f code m à position
r позиционный код m

9888 **position control**
d Lageregelung f
f commande f en position
r регулирование n по положению

9889 **position control accuracy**
d Positionsregelungsgenauigkeit f
f précision f de réglage de la position
r точность f регулирования позиции

* **position controlled manipulator** → 7812

* **position controller** → 9907

9890 **position control servomechanism**
d Stellservomechanismus m; Positionierungsservomechanismus m
f servomécanisme m positionneur
r позиционная следящая система f

9891 **position control system**
d Positionssteuerungssystem n
f système m de contrôle de position
r система f позиционного управления

9892 **position data**
d Positionsdaten pl
f données fpl de position
r параметры mpl позиции

9893 position data of manipulator
  d Manipulatorpositionsdaten *pl*;
      Positionsdaten *pl* eines Manipulators
  f données *fpl* de position d'un manipulateur
  r параметры *mpl* положения манипулятора

9894 position-dependent code
  d positionsabhängiger Kode *m*;
      stellungsabhängiger Kode
  f code *m* dépendant de position
  r позиционно-зависимый код *m*

9895 position deviation
  d Lageabweichung *f*
  f déviation *f* de position
  r отклонение *n* от позиции

9896 position encoder
  d Stellungskodierer *m*
  f codeur *m* de position
  r кодирующее устройство *n* положения

9897 positioner; position mechanism
  d Stellwerk *n*; Stellknebel *m*;
      Positionsmechanismus *m*
  f positionneur *m*; mécanisme *m* de position
  r позиционер *m*; позиционирующее
      устройство *n*

  * **position error → 9910**

9898 position feedback
  d Stellungsrückkopplung *f*
  f réaction *f* de mise en position
  r обратная связь *f* по положению

9899 position finder
  d Abstandsmesser *m*; Lagebestimmungsgerät *n*
  f dispositif *m* de relèvement du gisement
  r прибор *m* для определения
      местоположения

  * **position fixage → 9900**

9900 position fixing; position fixage
  d Lagefixierung *f*
  f fixage *m* de position
  r фиксация *f* положения

9901 position function
  d Ortsfunktion *f*
  f fonction *f* de position
  r позиционная функция *f*

9902 position-independent code
  d positionsunabhängiger Kode *m*;
      stellungsunabhängiger Kode
  f code *m* indépendant de position

  r позиционно-независимый код *m*

9903 position indicator
  d Stellungsanzeiger *m*
  f indicateur *m* de position
  r указатель *m* положения

9904 position information
  d Lageinformation *f*
  f information *f* de position
  r информация *f* о позиции

9905 positioning
  d Steuerung *f* nach Lage
  f commande *f* en position
  r регулировка *f* положения

  * **positioning accuracy → 9911**

9906 positioning control accuracy
  d Positionierregelungsgenauigkeit *f*;
      Regelungsgenauigkeit *f* der Positionierung
  f précision *f* de réglage de positionnement
  r точность *f* регулирования
      позиционирования

9907 position[ing] controller
  d Lageregler *m*
  f régulateur *m* de position
  r регулятор *m* положения

9908 positioning deviation compensation
  d Positionierabweichungsausgleich *m*;
      Ausgleich *m* von Positionierabweichungen
  f compensation *f* de déviations de
      positionnement
  r компенсация *f* отклонения
      позиционирования

9909 positioning drive
  d Positionierantrieb *m*
  f entraînement *m* de positionnement
  r привод *m* позиционирования

9910 position[ing] error
  d statischer Fehler *m*; Positionsfehler *m*
  f erreur *f* statique; erreur de position[nement]
  r позиционная погрешность *f*; ошибка *f*
      положения

**9911 positioning exactitude; positioning
      accuracy**
  d Positioniergenauigkeit *f*
  f exactitude *f* de positionnement
  r точность *f* позиционирования

9912 positioning of handling object
  d Positionierung *f* eines Handhabungsobjekts

f positionnement *m* d'un objet de manutention
r позиционирование *n* объекта манипулирования

9913 positioning of stop
d Anschlageinstellung *f*
f réglage *m* en position de butée
r установка *f* упора

9914 positioning sensor
d Positionierungssensor *m*
f senseur *m* de positionnement
r сенсор *m* позиционирования

9915 positioning technique
d Positionierungstechnik *f*
f technique *f* de positionnement
r техника *f* позиционирования

9916 position measuring instrument
d Stellungsmessgerät *n*
f appareil *m* de mesure de position
r прибор *m* для измерения положения

9917 position measuring system
d Positionsmesssystem *n*
f système *m* de mesure de position
r измерительная система *f* положения

* position mechanism → 9897

9918 position sensor system
d Lagesensorsystem *n*
f système *m* de senseur de position
r система *f* локационных сенсоров

9919 position servocontrol
d Stellungsservosteuerung *f*
f servocommande *f* de positionnement
r следящая система *f* управления положением

9920 positive coupling
d positive Kopplung *f*
f couplage *m* positif
r положительная связь *f*

9921 positive definite
d positiv definit
f positivement défini
r положительно определенный

9922 positive difference
d positive Differenz *f*
f différence *f* positive
r положительная разность *f*

9923 positive feedback

d positive Rückkopplung *f*
f réaction *f* positive
r положительная обратная связь *f*

9924 positive level
d positives Niveau *n*
f niveau *m* positif
r положительный уровень *m*

9925 positive logic
d positive Logik *f*; positive Schaltungslogik *f*
f logique *f* positive
r положительная логика *f*

9926 positive-logic signal
d Positivlogiksignal *n*
f signal *m* à logique positive
r сигнал *m* положительной логики

9927 positive-negative three-level action
d Dreipunktverhalten *n* mit Nullwert
f action *f* à trois paliers
r положительно-отрицательное трехпозиционное действие *n*

9928 positive-phase sequence relay
d Mitphasensystemrelais *n*
f relais *m* fonctionnant sur la composante positive de la phase
r реле *n* положительной последовательности фаз

9929 positive potential
d positives Potential *n*
f potentiel *m* positif
r положительный потенциал *m*

9930 positive pulse
d positiver Impuls *m*
f impulsion *f* positive
r положительный импульс *m*

9931 positive quadratic form
d positive quadratische Form *f*
f forme *f* quadratique positive
r положительная квадратичная форма *f*

9932 positive self-regulation
d positiver Selbstausgleich *m*
f autorégulation *f* positive
r положительное саморегулирование *n*

9933 positive signal
d positives Signal *n*
f signal *m* positif
r положительный сигнал *m*

9934 possible event
d mögliches Ereignis *n*

f événement *m* possible
r возможное событие *n*

9935 post-detection gain
d Gewinn *n* nach der Demodulation
f gain *m* en aval du détection
r усиление *n* после детектирования

* postulate → 1532

9936 potential analogy
d Potentialanalogie *f*
f analogie *f* potentielle
r аналогия *f* потенциала

9937 potential barrier
d Potentialschwelle *f*; Potentialwall *m*
f seuil *m* de potentiel; barrière *f* de potentiel
r потенциальный барьер *m*

9938 potential correction
d Potentialkorrektur *f*
f correction *f* de potentiel
r коррекция *f* потенциала

9939 potential diagram
d Potentialbild *n*; Potentialdiagramm *n*
f diagramme *m* de potentiel
r потенциальная диаграмма *f*

9940 potential distribution
d Potentialverteilung *f*; Potentialverlauf *m*
f distribution *f* de potentiel
r распределение *n* потенциала

9941 potential distribution control
d Potential[verteilungs]steuerung *f*
f contrôle *m* de distribution du potentiel
r регулирование *n* [распределения] потенциала

* potential divider → 12857

9942 potential energy
d potentielle Energie *f*
f énergie *f* potentielle
r потенциальная энергия *f*

9943 potential-energy curve
d Potentialkurve *f*
f courbe *f* d'énergie potentielle
r кривая *f* потенциальной энергии

9944 potential function
d Potentialfunktion *f*
f fonction *f* de potentiel
r потенциальная функция *f*

9945 potential jump
d Potentialsprung *m*
f saut *m* de potentiel
r потенциальный скачок *m*

9946 potential of vector field
d Potential *m* des Vektorfeldes
f potentiel *m* du champ vectoriel
r потенциал *m* векторного поля

9947 potential theory
d Potentialtheorie *f*
f théorie *f* du potentiel
r теория *f* потенциалов

9948 potentiometer
d Potentiometer *n*
f potentiomètre *m*
r потенциометр *m*

9949 potentiometer controller
d Potentiometerregler *m*
f régulateur *m* potentiométrique
r потенциометрический регулятор *m*

9950 potentiometer method
d Potentiometermethode *f*
f méthode *f* potentiométrique
r потенциометрический метод *m*

9951 potentiometer pick-off
d Potentiometerabgriff *m*; Widerstandsabgriff *m*
f capteur *m* potentiométrique
r потенциометрический датчик *m*

9952 potentiometer-setting system
d Einstellsystem *n* füt Potentiometer
f système *m* de réglage pour des potentiomètres; système d'ajustage pour des potentiomètres
r система *f* настройки потенциометров

9953 potentiometric error measuring system
d potentiometrisches Fehlermesssystem *n*
f système *m* potentiométrique de mesure d'erreurs
r потенциометрическая система *f* измерения ошибок

9954 potentiostat
d Potentiostat *m*
f potentiostat *m*
r потенциостат *m*; стабилизатор *m* напряжения

9955 power amplification
d Leistungsverstärkung *f*

f amplification *f* de puissance
r усиление *n* мощности

9956 power amplifier
d Leistungsverstärker *m*
f amplificateur *m* de puissance
r усилитель *m* мощности

* **power balance method → 5039**

9957 power circuit
d Leistungskreis *m*
f circuit *m* de puissance
r силовая цепь *f*

9958 power consumption; power demand
d Kraftbedarf *m*; Kraftverbrauch *m*;
Energiebedarf *m*
f force *f* nécessaire; demande *f* d'énergie
r потребность *f* в энергии; силовая
потребность

9959 power contactor
d Leistungsschütz *n*
f contacteur *m* de puissance
r силовой контактор *m*

9960 power conversion
d Energieumformung *f*
f transformation *f* d'énergie
r преобразование *n* энергии

9961 power current
d Kraftstrom *m*
f force *f* motrice; courant *m* force
r силовой ток *m*

9962 power cylinder; servomotor
d Servomotor *m*; Stellmotor *m*
f servomoteur *m*
r серводвигатель *m*; сервомотор *m*;
сервопривод *m*

* **power demand → 9958**

9963 power density
d Leistungsdichte *f*
f densité *f* de puissance
r плотность *f* мощности

9964 power detector
d Leistungsdetektor *m*
f détecteur *m* de puissance
r мощный детектор *m*

9965 power dissipation
d Verlustleistung *f*
f puissance *f* dissipée

r рассеяние *n* мощности

9966 power distribution
d Energieverteilung *f*
f distribution *f* d'énergie
r распределение *n* энергии

9967 power divider
d Leistungsteiler *m*
f diviseur *m* de puissance
r делитель *m* мощности

9968 power-down
d Ausschalten *n* der Stromversorgung
f mise *f* au repos de l'alimentation
r отключение *n* электропитания

9969 power driver
d Leistungstreiber *m*
f basculeur *m* de puissance
r силовой формирователь *m*

* **power exponent → 5299**

* **power fail → 9970**

9970 power fail[ure]
d Netzausfall *m*; Stromausfall *m*
f manque *m* d'alimentation
r отказ *m* [источника] питания

9971 power failure interrupt
d Stromausfallinterrupt *m*
f interruption *f* de manque de courant
r прерывание *n* при отказе источника
питания

9972 power formula
d Leistungsformel *f*
f formule *f* de puissance
r формула *f* мощности

9973 power frequency
d Netzfrequenz *f*
f fréquence *f* du secteur
r частота *f* сети электропитания

9974 power-frequency flashover voltage
d Überschlagspannung *f* bei
Kraftstromfrequenz
f tension *f* d'amorçage à fréquence industrielle
r напряжение *n* пробоя промышленной
частоты

9975 power loss
d Leistungsverlust *m*
f perte *f* d'énergie; perte de puissance
r потери *fpl* мощности

9976 **power of performance curve**
 d Leistungskennlinie *f*
 f courbe *f* de puissance
 r кривая *f* характеристики мощности

9977 **power-operated holding tool**
 d kraftbetätigte Spannvorrichtung *f*;
   kraftbetätigtes Spannmittel *n*
 f mécanisme *m* d'élasticité
 r зажимное устройство *n* с механическим
   приводом

9978 **power operation**
 d Betätigung *f* mit Kraftantrieb; Betätigung
   durch Stellmotor
 f commande *f* par servomoteur
 r управление *n* с помощью сервопривода

9979 **power stabilization**
 d Leistungsstabilisierung *f*
 f stabilisation *f* de la puissance
 r стабилизация *f* мощности

9980 **power supply system**
 d Energieversorgungssystem *n*
 f système *m* d'alimentation en énergie
 r система *f* энергообеспечения

9981 **power supply unit**
 d Netzgerät *n*
 f bloc *m* d'alimentation; appareil *m*
   d'alimentation
 r устройство *n* питания

9982 **power switch**
 d Leistungsschalter *m*
 f disjoncteur *m*
 r силовой выключатель *m*

9983 **power transformer**
 d Leistungstransformator *m*
 f transformateur *m* de puissance
 r силовой трансформатор *m*

9984 **power transient**
 d Leistungstransiente *f*
 f transitoire *m* de puissance
 r переходный процесс *m* мощности

 * **power transmission → 5068**

9985 **power transmission line**
 d Energietransportleitung *f*;
   Energiefernleitung *f*
 f ligne *f* de transport d'énergie
 r линия *f* энергопередачи

9986 **power winding**

 d Arbeitswicklung *f*
 f enroulement *m* de puissance
 r силовая обмотка

9987 **practical steady state**
 d praktischer Beharrungszustand *m*
 f régime *m* permanent en pratique
 r рабочий стационарный режим *m*

9988 **preaccelerator**
 d Vorbeschleuniger *m*
 f préaccélérateur *m*
 r предускоритель *m*

 * **preamplifier → 6091**

 * **precision → 199**

9989 **precision control**
 d Präzisionssteuerung *f*
 f commande *f* de précision
 r прецизионное управление *n*

9990 **precision device engineering**
 d Präzisionsgerätetechnik *f*
 f technique *f* des appareils de précision
 r прецизионная приборная техника *f*

9991 **precision engineering**
 d Präzisionstechnik *f*
 f technique *f* de précision
 r прецизионная техника *f*

9992 **precision mounting of components**
 d Baugruppenpräzisionsmontage *f*
 f montage *m* de précision des éléments
 r прецизионна сборка *f* узлов

9993 **precision robotics**
 d Präzisionsrobotertechnik *f*
 f robotique *f* de précision
 r прецизионная робототехника *f*

 * **preconditioning → 10022**

9994 **precooling**
 d Vorkühlung *f*
 f refroidissement *m* primaire; refroidissement
   préliminaire
 r предохлаждение *n*; предварительное
   охлаждение *n*

9995 **precriticality**
 d vorkritischer Zustand *m*
 f période *f* avant la première divergence
 r предкритичность *f*; предкритическое
   состояние *n*

9996 predetermined pulse counter
d voreingestellter Impulszähler *m*
f compteur *m* d'impulsion préréglé
r счётчик *m* импульсов с предварительной
    установкой

9997 predetermined value
d voreingestellter Wert *m*
f valeur *f* prédéterminée
r заранее заданная величина *f*

9998 predicate logic
d Prädikatenlogik *f*
f logique *f* des prédicats
r предикатная логика *f*

9999 predicting equation
d Voraussagegleichung *f*
f équation *f* de prédiction
r уравнение *n* предсказания

10000 predicting filter
d prädiktives Filter *n*
f filtre *m* prédicteur
r упреждающий фильтр *m*

10001 prediction; prognosis; prognostication
d Prognosearbeit *f*
f prévision *f*; prédiction *f*
r прогнозирование *n*

10002 prediction relay control system
d prädiktives Relaissystem *n*
f système *m* de relais à prédiction
r релейная система *f* регулирования с
    предсказанием

10003 prediction theory
d Vorhersagetheorie *f*
f théorie *f* de prédiction
r теория *f* предсказывания

10004 prediction transfer function
d Vorhaltübertragungsfunktion *f*
f transmittance *f* de prédiction
r функция *f* преобразования с
    предсказанием

10005 predictive control
d vorausschauende Steuerung *f*
f commande *f* préventive
r управление *n* с предсказанием

10006 predictor network
d Prediktornetzwerk *n*; predizierendes
    Netzwerk *n*; Vorhersagenetzwerk *n*
f réseau *m* de prédicteurs

r схема *f* предсказывания

10007 pre-emphasis
d Preemphase *f*
f préemphase *f*; préaccentuation *f*;
    précorrection *f*
r предварительная коррекция *f*

10008 pre-equilibrium process
d Präcompoundprozess *m*;
    Vorgleichgewichtsprozess *m*
f processus *m* du noyau pré-composé
r предравновесный процесс *m*

10009 prefabricate *v*
d vorfertigen
f préfabriquer
r предварительно изготовлять

10010 prefabrication
d Vorfertigung *f*
f préfabrication *f*
r предварительное изготовление *n*

10011 preferred direction
d Vorzugsrichtung *f*
f direction *f* préférentielle
r предпочтительное направление *n*

10012 prefetch buffer
d Puffer *m* für vorausgeholte Informationen
f tampon *m* pour informations cherchées
    d'avance
r буфер *m* предварительной выборки

10013 prefiltration; preliminary filtration
d Vorfilterung *f*; Grobfilterung *f*
f préfiltration *f*; filtration *f* préliminaire
r предварительная фильтрация *f*

*   **preionization** → 1132

10014 preknock impulse
d Vorimpuls *m*
f préimpulsion *f*
r предимпульс *m*

10015 preliminary design
d Vorprojektierung *f*
f projection *f* de départ
r эскизное проектирование *n*

*   **preliminary filtration** → 10013

10016 preliminary investigation
d Voruntersuchung *f*
f examen *m* préliminaire
r предварительное исследование *n*

10017 **preliminary logic**
   d vorläufige Logik *f*; vorläufige
      Logikschaltung *f*
   f logique *f* préliminaire
   r логические схемы *fpl* предварительной
      обработки

10018 **preliminary test**
   d Vorprüfung *f*; Vorversuch *m*
   f essai *m* préliminaire
   r предварительное испытание *n*

10019 **preloaded**
   d voraus eingerichtet
   f préchargé
   r предварительно загруженный

10020 **premodifications**
   d Vorwegänderungen *fpl*
   f prémodifications *fpl*
   r предварительные изменения *npl*

10021 **pre-operational [system] test**
   d Vorbetriebsprüfung *f*; vorbetriebliche
      Prüfung *f*
   f essai *m* préopérationnel
   r предэксплуатационное испытание *n*

   * **pre-operational test** → 10021

10022 **preparation; pretreatment; dressing;**
      preconditioning
   d Vorbereitung *f*; Vorbehandlung *f*
   f préparation *f*; travaux *mpl* préparatoires;
      traitement *m* préalable
   r подготовка *f*; предварительная
      обработка *f*; первичная обработка *f*

10023 **preparatory function**
   d Wegbedingung *f*
   f fonction *f* préparatoire
   r подготвительная операция *f*

10024 **preparatory signal**
   d Vorbereitungszeichen *n*
   f signal *m* de préparation
   r подготвительный сигнал *m*

10025 **prepressurization; initial internal pressure**
   d Vorinnendruck *m*
   f pression *f* interne initiale
   r начальное внутреннее давление *n*

10026 **preprogramming**
   d Vorprogrammierung *f*
   f préprogrammation *f*
   r предварительное программирование *n*

10027 **prepulse**
   d Vorimpuls *m*
   f impulsion *f* préalable
   r предварительный импульс *m*

10028 **preregulator**
   d Grobregler *m*
   f régulateur *m* grossier
   r регулятор *m* для грубого регулирования

10029 **prerotation controller; prerotation**
      regulator
   d Vordrallregler *m*
   f régulateur *m* de prérotation
   r предвключенный регулятор *m*

   * **prerotation regulator** → 10029

10030 **prescribed value**
   d Aufgabenwert *m*
   f valeur *f* assignée
   r заданное значение *n* регулируемой
      величины

10031 **preselector**
   d Vorwähler *m*
   f présélecteur *m*
   r преселектор *m*

10032 **presence**
   d Präsenz *f*; Anwesenheit *f*
   f présence *f*
   r присуствие *n*; наличие *n*

10033 **presence signal**
   d Anwesenheitszeichen *n*
   f signal *m* de présence
   r сигнал *m* присуствия

10034 **presentation device**
   d Darstellungseinrichtung *f*
   f dispositif *m* de reproduction
   r устройство *n* воспроизведения

10035 **preset** *v*
   d voreinstellen
   f ajuster d'avance
   r предварительно установлять

10036 **preset adjustment**
   d Voreinstellung *f*; vorgegebene Einstellung *f*
   f mise *f* au point préalable; préréglage *m*;
      prépositionnement *m*
   r предварительная настройка *f*

10037 **preset counter**
   d Vorwahlzähler *m*

f compteur *m* à présélection
r счётчик *m* с предварительной установкой

10038 preset guidance
d Programmlenkung *f*
f guidage *m* programmé
r программное наведение *n*

10039 preset parameter
d vorgegebener Parameter *m*
f paramètre *m* préfixé
r заданный параметр *m*

10040 preset timer
d Zeitvorwähler *m*
f présélecteur *m* de temps
r задатчик *m* времени

10041 presetting
d Voreinstellen *n*; Festlegen *n* von
    Anfangsbedingungen
f fixation *f* d'une condition initiale
r предварительная установка *f*

10042 pressure[-actuated] switch
d Druckschalter *m*
f interrupteur *m* manométrique; interrupteur à
    pression
r переключатель m, приводимый в действие
    давлением

*   **pressure adapter** → **10059**

10043 pressure adjustment
d Druckeinstellung *f*
f mise *f* au point de la pression
r регулировка *f* давления

*   **pressure balance** → **10059**

10044 pressure balance method
d Druckausgleichmethode *f*
f méthode *f* de balance des pressions
r метод *m* баланса давлений

10045 pressure balancing; pressure equalization
d Druckausgleich *m*
f compensation *f* de pression; égalisation *f* de
    pression
r компенсация *f* давления; выравнивание *n*
    давления

10046 pressure buildup
d Druckaufbau *m*
f établissement *m* de pression; accumulation *f*
    de pression
r накопление *n* давления

10047 pressure-compensated flowmeter

f débitmètre *m* à compensation de pression
r расходомер *m* с компенсацией давления

10048 pressure connection
d Druckentnahme *f*
f prise *f* de pression
r подсоединение *n* давления

*   **pressure contact** → **2002**

10049 pressure controlled
d druckgesteuert
f contrôlé par la pression
r управляемый давлением

*   **pressure difference indicator** → **4134**

10050 pressure difference transmitter
d Druckdifferenzgeber *m*
f capteur *m* de différence de pression
r датчик *m* дифференциального давления

10051 pressure-dividing diagram
d Druckteiler-Schaltung *f*
f circuit *m* répartiteur de pression
r схема *f* делителя давления

10052 pressure drop
d Druckabfall *m*
f chute *f* de pression
r перепад *m* давления

10053 pressure drop rate
d Druckverlustwert *m*
f valeur *f* de la perte de charge
r величина *f* потерь давления

*   **pressure equalization** → **10045**

10054 pressure feedback
d Druckrückführung *f*
f réaction *f* de pression
r обратная связь *f* по давлению

10055 pressure filtration
d Druckfiltration *f*
f filtration *f* sous pression
r фильтрация *f* под давлением

10056 pressure gradient
d Druckgradient *m*
f gradient *m* de pression
r градиент *m* давления

10057 pressure guard
d Druckwächter *m*
f dispositif *m* de surveillance de pression
r ограничитель *m* давления

d druckkompensierter Durchflussmesser *m*

10058 pressure reduction
d Druckreduzierung f
f réduction f de pression
r редуцирование n давления

10059 pressure regulator; pressure adapter; pressure balance
d Druckregler m
f régulateur m de pression; manostat m
r регулятор m давления

10060 pressure relief technology
d Druckentlastungstechnik f
f technique f de détention
r техника f разгрузки давления

10061 pressure sensitive element; pressure sensor
d Druckfühler m; Drucksensor m
f élément m sensible à pression; capteur m de pression; senseur m de pression
r элемент m, чувствительный к давлению; датчик m давления; сенсор m давления

* **pressure sensor → 10061**

* **pressure switch → 10042**

10062 pressure ventilation
d Überdrucklüftung f
f ventilation f à surpression
r нагнетательная вентиляция f

10063 pressure welded junction
d druckgeschweißte Verbindung f
f jonction f soudée à pression
r перход m, полученный сваркой под давлением

10064 pressurization
d Druckbelüftung f; Druckbeaufschlagung f
f pressurisation f; mise f sous pression
r герметизация f

10065 prestage
d Vorstufe f
f étage m préliminaire
r предварительный каскад m

10066 prestored information
d vorgespeicherte Information f
f information f préenregistrée
r заранее накопленная информация f

10067 pretreatement of sensor informations
d Vorverarbeitung f von Sensorinformationen
f prétraitement m d'informations de capteur
r предварительная обработка f сенсорной информации

* **pretreatment → 10022**

10068 prevailing value
d übereiegender Wert m
f valeur f prédominante
r преобладающая величина f

10069 prevent v
d vorbeugen
f prévenir
r предупреждать; предохранять; предотвращать

10070 preventive maintenance
d vorbeugende Wartung f
f maintenance f prophylactique
r профилактическое [техническое] обслуживание n

10071 previous mode
d vorhergehender Modus m; vorherige Betriebsart f
f mode m précédent
r предшествующий режим m

10072 previous status
d vorhergehender Status m; vorheriger Status
f état m précédent
r предыдущее состояние n

* **primary algorithm → 4976**

10073 primary cell
d Primärelement n
f pile f primaire
r первичный элемент m

10074 primary control element
d primäres Regelelement n
f élément m primaire de réglage
r первичный регулирующий элемент m

10075 primary data
d Primärdaten pl
f données f pl primaires
r первичные данные pl

10076 primary detector
d primärer Fühler m
f capteur m primaire
r первичный датчик m

10077 primary feedback signal
d äußeres Rückkopplungssignal n
f signal m de rétroaction externe
r сигнал m основной обратой связи

10078 primary memory; primary store
d Primärspeicher m
f mémoire f primaire
r первичное запоминающее устройство n

10079 primary parameter
d Primärparameter m
f paramètre m primaire
r первичный параметр m

10080 primary process
d Primärvorgang m; Primärprozess m
f procédé m primaire
r первичный процесс m

10081 primary regulation
d Primärregelung f
f régulation f primaire
r первичное регулирование n

* **primary store** → 10078

10082 primary voltage
d Urspannung f
f tension f initiale
r эталонное напряжение n

* **principle drawing** → 10091

10083 principle of adaptive time constant
d Prinzip n der adaptiven Zeitkonstante
f principe m de la constante de temps adaptive
r принцип m адаптивной постоянной времени

10084 principle of assembly unit
d Baukastenprinzip n
f principe m de boîte de construction
r агрегатный принцип m

10085 principle of causality
d Kausalitätsprinzip n
f principe m de la causalité
r принцип m причинности

10086 principle of construction
d Aufbauprinzip n
f principe m de construction
r принцип m построения

10087 principle of finite induction
d Methode f der vollständigen Induktion
f méthode f de l'induction complète
r метод m полной индукции

10088 principle of impulse reflection
d Impulsreflexionsprinzip n
f principe m de la réflexion des impulsions
r метод m отражательных импульсов

10089 principle of optimality
d Optimalitätsprinzip n
f principe m optimal
r принцип m оптимальности

10090 principle of simultaneous operations; concept of simultaneous operations
d Prinzip n gleichzeitig ablaufender Operationen
f principe m de fonctionnements de simultanéité; notion f de fonctionnements de simultanéité
r принцип m одновременной работы

10091 principle scheme; principle drawing
d Prinzipschema n; Prinzipzeichnung f
f schéma m de principe
r принципиальная схема f

10092 print control
d Drucksteuerung f
f commande f d'impression
r управление n выводом на печать

10093 printer
d Drucker m; Printer n
f imprimante f; imprimeur m
r печатающее устройство n; принтер m

10094 printer control logic
d Druckersteuerlogik f
f logique f de commande d'imprimeuse
r логический блок m управления печатающим устройством

10095 printing technique
d Drucktechnik f
f technique f d'impression
r техника f печати

10096 print unit
d Schreibwerk n; Druckwerk n
f unité f d'impression
r блок m печати

10097 priority check
d Prioritätskontrolle f
f contrôle m de priorité
r проверка f приоритета; контроль m приоритета

10098 priority circuit
d Prioritätsstromkreis m
f circuit m de priorité
r предпочтительная цепь f

10099 priority control
   d Vorrangsteuerung *f*; Prioritätssteuerung *f*
   f commande *f* de priorité
   r приоритетное управление *n*

10100 priority dispatching
   d Prioritätszuteilung *f*
   f attribution *f* de priorité
   r приоритетная диспетчеризация *f*

10101 priority feature
   d Vorrangmerkmal *n*; Vorrangmeldung *f*
   f caractéristique *f* de priorité
   r признак *m* приоритета

10102 priority function
   d Vorrangfunktion *f*
   f fonction *f* prioritaire
   r приоритетная функция *f*

10103 priority grading
   d Prioritätsgrad *m*; Prioritätsstufe *f*
   f degré *m* de priorité
   r степень *f* приоритета

10104 priority group
   d Prioritätengruppe *f*
   f groupe *m* de priorité
   r приоритетная группа *f*

10105 priority indicateur
   d Prioritätsanzeiger *m*; Vorranganzeiger *m*
   f indicateur *m* de priorité
   r индикатор *m* приоритета

10106 priority interrupt control
   d Prioritätsinterruptsteuerung *f*;
     Vorrangunterbrechungssteuerung *f*
   f commande *f* d'interruption de priorité
   r управление *n* приоритетными
     прерываниями

10107 priority interrupt system
   d vorranggestuftes Unterbrechungssystem *n*;
     hierarchisches Unterbrechungssystem;
     Prioritätsinterruptsystem *n*
   f système *m* d'interruption de priorité
   r система *f* приоритетных прерываний

10108 priority processing
   d Prioritätsverarbeitung *f*; vorranggerechte
     Verarbeitung *f*
   f traitement *m* prioritaire
   r обработка *f* по приоритетам

10109 priority program
   d Vorrangprogramm *n*; Prioritätsprogramm *n*
   f programme *m* de priorité
   r программа *f* с приоритетом

10110 priority relation
   d Vorrangverhältnis *n*
   f rapport *m* de priorité
   r соотношение *n* приоритетов

10111 priority structure
   d Prioritätsstruktur *f*; Vorrangstruktur *f*
   f structure *f* de priorités
   r приоритетная структура *f*

10112 probabilistic
   d probabilistisch
   f probabilistique
   r вероятностный

10113 probabilistic automaton
   d probabilistischer Automat *m*
   f automate *m* probabilistique
   r вероятностый автомат *m*

10114 probabilistic logics
   d Wahrscheinlichkeitslogik *f*; probabilistische
     Logik *f*
   f logique *f* de probabilité
   r вероятностная логика *f*

10115 probabilistic machine
   d Wahrscheinlichkeitsmaschine *f*
   f machine *f* de probabilité
   r вероятностная машина *f*

10116 probability
   d Wahrscheinlichkeit *f*
   f probabilité *f*
   r вероятность *f*

10117 probability calculation
   d Wahrscheinlichkeitsrechnung *f*
   f calcul *m* de probabilité
   r расчёт *m* вероятностей

10118 probability density
   d Wahrscheinlichkeitsdichte *f*
   f densité *f* de probabilité
   r плотность *f* вероятности

10119 probability density function
   d Wahrscheinlichkeitsdichtefunktion *f*
   f fonction *f* de densité de probabilité
   r функция *f* плотности вероятностей

10120 probability distribution
   d Wahrscheinlichkeitsverteilung *f*
   f distribution *f* de probabilité
   r распределение *n* вероятностей

10121 probability function
   d Wahrscheinlichkeitsfunktion *f*
   f fonction *f* de probabilité
   r функция *f* вероятности

10122 probability integral
   d Wahrscheinlichkeitsintegral *n*
   f intégrale *f* de probabilité
   r интеграл *m* вероятности

   * **probability low** → 7585

10123 probability method
   d Wahrscheinlichkeitsmethode *f*
   f méthode *f* de probabilité
   r вероятностный метод *m*

10124 probability of occurrence
   d Eintrittswahrscheinlichkeit *f*;
      Auftretenswahrscheinlichkeit *f*
   f probabilité *f* d'occurrence; probabilité
      d'apparition
   r вероятность *f* наступления

10125 probability of stability
   d Stabilitätswahrscheinlichkeit *f*
   f probabilité *f* de stabilité
   r вероятность *f* устойчивости

10126 probability theory
   d Wahrscheinlichkeitstheorie *f*
   f théorie *f* des probabilités
   r теория *f* вероятности

10127 probable error
   d wahrscheinlicher Fehler *m*
   f erreur *f* probable
   r вероятная ошибка *f*

10128 probable structure
   d wahrscheinliche Struktur *f*
   f structure *f* probable
   r вероятная структура *f*

10129 problem algorithm
   d Problemalgorithmus *m*
   f algorithme *m* de problème
   r проблемный алгоритм *m*

10130 problem definition
   d Problemdefinition *f*; Problembestimmung *f*
   f définition *f* de problème
   r постановка *f* задачи

10131 problem description
   d Problembeschreibung *f*
   f description *f* de problème
   r описание *n* проблемы

10132 problem formulation check
   d Kontrolle *f* des Aufgabenkomplexes
   f contrôle *m* de la composition du problème
   r контроль *m* постановки задачи

10133 problem-oriented control
   d problemorientierte Steuerung *f*
   f commande *f* orientée sur problème
   r проблемно-ориентированное
      управление *n*

10134 problem-oriented language
   d problemorientierte Programmiersprache *f*
   f langage *m* orienté problèmes
   r проблемно-ориентированный язык *m*

10135 problem-oriented software package
   d problemorientiertes Softwarepaket *n*;
      problemorientiertes Systemunterlagenpaket *n*
   f paquet *m* de logiciel orienté problème
   r проблемно-ориентированный пакет *m*
      системного программного обеспечения

10136 problem-oriented work piece description
   d problemorientierte Werkstückbeschreibung *f*
   f description *f* de pièce à usiner orienté sur
      problème
   r проблемно-ориентированное описание *n*
      [обрабатываемых] деталей

10137 problem preparation process
   d Problembearbeitungsprozess *m*;
      Problemlösungsprozess *m*
   f processus *m* de traitement de problèmes; voie
      f de solution
   r алгоритм *m* решения проблем

10138 procedure
   d Prozedur *f*
   f procédure *f*
   r процедура *f*

10139 procedure declaration
   d Prozedurvereinbarung *f*
   f déclaration *f* de procédure
   r описание *n* процедуры

10140 procedure-oriented language
   d prozedurorientierte Programmiersprache *f*;
      verfahrensorientierte Programmiersprache
   f langage *m* orienté procédé
   r процедурно-ориентированный язык *m*

10141 proceeding control
   d Prozesssteuerung *f*
   f commande *f* des procédés
   r управление *n* техническими процессами

10142 **process**
  d Prozess *m*
  f processus *m*
  r процесс *m*

10143 **process algorithm**
  d Prozessalgorithmus *m*
  f algorithme *m* de processus
  r алгоритм *m* процесса

10144 **process algorithmization**
  d Prozessalgorithmisation *f*
  f algorithmisation *f* du procédé
  r алгоритмизация *f* процесса

10145 **process analysis**
  d Prozessanalyse *f*
  f analyse *f* de procédé
  r анализ *m* процесса

10146 **process automation**
  d Prozessautomatisierung *f*
  f automatisation *f* d'un processus
  r автоматизация *f* процесса

10147 **process card**
  d Fertigungskarte *f*
  f carte *f* d'opération
  r технологическая карта *f*

10148 **process characteristic**
  d Prozesskennlinie *f*; Arbeitsverfahren-
      Kennlinie *f*
  f caractéristique *f* du procédé
  r характеристика *f* процесса

10149 **process chart**
  d Arbeitsablaufdiagramm *n*;
      Verfahrensdiagramm *n*
  f diagramme *m* des opérations successives
  r график *m* процесса

10150 **process communication**
  d Prozesskommunikation *f*
  f communication *f* de processus
  r связь *f* с управляемым процессом

10151 **process communication system**
  d Prozessdatenübertragungseinrichtung *f*
  f système *m* de communication des données de
      procédé
  r система *f* передачи данных

  * **process condenser → 8979**

10152 **process conditions**
  d Verfahrensbedingungen *fpl*
  f conditions *fpl* de fonctionnement; conditions

de service
  r режим *m* процесса

10153 **process control**
  d Prozessregelung *f*; Prozesssteuerung *f*
  f régulation *f* d'un processus; commande *f*
      [directe] d'un processus
  r управление *n* промышленным процессом

10154 **process control analysis**
  d Prozesssteuerungsanalyse *f*
  f analyse *f* de la commande de processus
  r анализ *m* управления процессом

10155 **process control at manipulation of
      mechanical parts**
  d Prozesskontrolle *f* bei Manipulation
      mechanischer Teile
  f contrôle *m* de processus de manipulation de
      pièces mécaniques
  r контроль *m* процесса манипулирования
      механическими деталями

10156 **process-controlled manipulator workplace**
  d prozessgesteuerter Manipulatorarbeitsplatz *m*
  f emplacement *m* de travail de manipulateur
      commandé par processus
  r рабочее место *n* манипулятора с
      управлением процессом

10157 **process control simulation**
  d Nachbildung *f* der Verfahrenssteuerung;
      Modellierung *f* der Verfahrenssteuerung
  f simulation *f* de la commande de processus
  r моделирование *n* управления процесса

10158 **process control system**
  d Prozesssteuersystem *n*;
      Verfahrenssteuersystem *n*
  f système *m* de commande de processus
  r технологическая контрольная система *f*

10159 **process convergence**
  d Prozesskonvergenz *f*
  f convergence *f* de processus
  r сходимость *f* процесса

10160 **process cycle controller**
  d Prozesszyklusregler *m*
  f régulateur *m* du cycle d'un processus
  r регулятор *m* производственного цикла

10161 **process data**
  d Prozessdaten *pl*; Prozessinformationen *fpl*
  f données *fpl* de processus
  r данные *pl* о процессе

10162 **process data control unit**
  d Prozessdaten-Steuereinheit *f*

f organe *m* de commande pour données de
  procédé
r устройство *n* управления
  технологическими данными

10163 process data processing system
d Prozessdatenverarbeitungssystem *n*
f système *m* de traitement des données de
  procédé
r система *f* обработки технологических
  данных

10164 process data treatment
d Prozessdatenverarbeitung *f*
f traitement *m* des données de processus
r обработка *f* данных технологического
  процесса

10165 process development
d Prozessentwicklung *f*;
  Verfahrensentwicklung *f*
f développement *m* de procédé;
  développement technologique
r разработка *f* процесса; разработка
  технологии

10166 process diagram; flow sheet
d [schematisches] Fließbild *n*; Fließschema *n*;
  Verfahrensfließbild *n*
f flow-sheet *m* [schématique]; flow-sheet
  technologique
r схематическая диаграмма *f* процесса

10167 process disturbance
d Prozessstörung *f*
f perturbation *f* du processus
r возмущение *n* процесса

10168 process equipment design
d Ausrüstungsentwurf *f*;
  Ausrüstungsauslegung *f*
f projet *m* des installations; projet des
  équipements technologiques
r проектирование *n* оборудования

10169 process evaluation
d Verfahrenseinschätzung *f*;
  Verfahrensbewertung *f*
f évaluation *f* de procédé
r оценка *f* процесса

10170 process feed; operational feed
d Betriebseinspeisung *f*
f alimentation *f* de service
r рабочее питание *n*

10171 process hardware
d Prozesshardware *f*; Prozesseinrichtung *f*

f dispositif *m* de processus
r аппаратные средства *npl* управления
  процессом

10172 process identification
d Verfahrenserkennung *f*;
  Prozessidentifikation *f*
f identification *f* d'un processus
r распознавание *n* процессов

10173 process information system
d Prozessinformationssystem *n*
f système *m* d'information de processus
r производственная информационная
  система *f*

10174 processing
d Verarbeitung *f*
f traitement *m*
r обработка *f*

10175 processing capability
d Verarbeitungsfähigkeit *f*
f faculté *f* de traitement
r возможность *f* обработки

10176 processing capacity
d Verarbeitungskapazität *f*
f capacité *f* de traitement
r производительность *f* обработки

10177 processing cycle
d Verarbeitungszyklus *m*
f cycle *m* de traitement
r цикл *m* обработки

10178 processing equipment
d Fertigungsausrüstung *f*
f équipement *m* de traitement
r производственное оборудование *n*

10179 processing information
d Verarbeitungsinformation *f*
f information *f* de traitement
r обрабатываемая информация *f*

10180 processing method
d Behandlungsmethode *f*
f méthode *f* de traitement
r метод *m* обработки

10181 processing module
d Verarbeitungsmodul *m*
f module *m* de traitement
r модуль *m* обработки

10182 processing monitor
d Verarbeitungsmonitor *m*

    f moniteur *m* de traitement
    r процессорный монитор *m*

10183 processing monitoring
    d Prozessüberwachung *f*
    f surveillance *f* de traitement
    r контроль *m* обработки

10184 processing of geometric information
    d Verarbeitung *f* geometrischer Informationen
    f traitement *m* des informations géométriques
    r обработка *f* геометрической информации

10185 processing of measured data
    d Messwertverarbeitung *f*
    f traitement *m* des valeurs mesurées;
    traitement de données
    r обработка *f* данных измерений

10186 processing operation
    d Verarbeitungsoperation *f*
    f opération *f* de traitement
    r операция *f* обработки

10187 processing period
    d Bearbeitungsperiode *f*
    f cycle *m* de traitement
    r период *m* обработки

10188 processing speed
    d Verarbeitungsgeschwindigkeit *f*
    f vitesse *f* de traitement
    r скорость *f* обработки

10189 processing step
    d Prozessschritt *m*; Bearbeitungsschritt *m*;
    Verarbeitungsschritt *m*
    f pas *m* de traitement
    r шаг *m* обработки

10190 processing system for sensor signals
    d Verarbeitungssystem *n* für Sensorsignale
    f système *m* de traitement pour signals de
    capteur
    r система *f* обработки сенсорных сигналов

    * **processing unit → 10197**

10191 process input
    d Prozesseingabe *f*
    f entrée *f* de processus
    r ввод *m* данных о процессе

**10192 process instrumentation; monitoring
instrumentation; monitoring equipment**
    d Kontroll- und Überwachungsgeräte *npl*;
    Betriebsüberwachungsgeräte *npl*
    f appareils *mpl* de mesure et surveillance;

    équipement *m* de surveillance;
    instrumentation *f* de surveillance
    r контрольно-измерительные приборы *mpl*

10193 process intensification
    d Prozessintensivierung *f*
    f intensification *f* de processus
    r интенсификация *f* процесса

10194 process model control
    d Prozessmodellsteuerung *f*
    f commande *f* par modèle de processus
    r управление *n* моделью технологического
    процесса

10195 process node
    d Prozessknoten *m*
    f nœud *m* fonctionnel
    r функциональный узел *m*

10196 process of superheating
    d Überhitzungsprozess *m*
    f processus *m* de surchauffage
    r процесс *m* перегрева

10197 processor; processing unit
    d Prozessor *m*; Verarbeitungseinheit *f*
    f processeur *m*
    r процессор *m*

10198 processor clock
    d Prozessortakt *m*
    f rythme *m* de processeur
    r синхронизация *f* процессора

10199 processor-controlled exchange
    d prozessorgesteuerte Vermittlungsstelle *f*
    f point *m* de communication à commande par
    processeur
    r коммутатор *m* с процессорным
    управлением

10200 processor-controlled logic analysis
    d prozessorgesteuerte Logikanalyse *f*
    f analyse *f* logique commandée par processeur
    r логический анализ *m* с процессорным
    управлением

10201 processor cycle
    d Prozessorzyklus *m*
    f cycle *m* de processeur
    r цикл *m* процессора

10202 processor for pretreatment
    d Prozessor *m* für Vorverarbeitung
    f processeur *m* pour prétraitement
    r процессор *m* предварительной обработки

10203 process-oriented industry
d verfahrensorientierte Industrie f
f industrie f orientée sur le processus
r технологически-ориентированная
   промышленность f

10204 processor interface
d Prozessorschnittstelle f
f interface f de processeur
r интерфейс m процессора

10205 processor interrupt
d Prozessorunterbrechung f
f interruption f de processeur
r прерывание n процессора

10206 processor module
d Prozessormodul m
f module m de processeur
r процессорный модуль m

10207 processor power
d Prozessorleistung f
f performance f de processeur
r производительность f процессора

10208 processor register
d Prozessorregister n
f registre m de processeur
r регистр m процессора

10209 process output
d Prozessausgabe f
f sortie f de processus
r вывод m данных о процессе

10210 process remote control
d Prozessfernsteuerung f
f télécommande f de processus industriels;
   commande f à distance de processus
   industriels
r дистанционное управление n процессами

10211 process sensor
d Prozesssensor m
f senseur m de processus
r сенсор m параметра технологического
   процесса

10212 process stabilization
d Prozessstabilisierung f
f stabilisation f du processus
r стабилизация f процесса

10213 process stream
d Betriebsstrom m
f courant m de processus
r технологический поток m

10214 process supervision
d Prozessüberwachung f
f supervision f de processus
r контроль m процесса

10215 process timer
d Zeitplangeber m;
   Arbeitsvorgangszeitmesser m
f compte-pose m
r реле n времени процесса

10216 process unit
d Prozesseinheit f
f unité f de procédé
r элемент m процесса

10217 process utilization factor
d Prozessausnutzungsfaktor m
f facteur m d'utilisation du procédé
r коэффициент m использования процесса

10218 producer process
d produzierender Prozess m; Quellenprozess m
f processus m producteur
r процесс-источник m

10219 production automation
d Produktionsautomatisierung f
f automation f de production
r автоматизация f производства

* **production chain → 10226**

10220 production control
d Betriebsüberwachung f
f surveillance f d'exploitation
r контроль m производством

10221 production control automation
d Automatisierung f der Fertigungssteuerung
f automa[tisa]tion f de la gestion industrielle
r автоматизация f управления
   производством

10222 production control scheme
d Produktionssteuerungsplan m
f schéma m de commandde de production
r схема f управления производством

10223 production control system
d Produktionskontrollsystem n
f système m de contrôle de la production
r система f управления производством

10224 production equipment
d Fertigungsanlage f
f installation f de fabrication
r производственная установка f

10225 production in series; serial production; series manufacture
  d Serienproduktion *f*
  f production *f* en série
  r серийное производство *n*

10226 production line; production chain
  d Fertigungsstraße *f*; Fertigungskette *f*
  f chaîne *f* de fabrication; chaîne de production
  r поточная линия *f*

10227 production measuring technique
  d Betriebsmesstechnik *f*
  f technique *f* de mesure industrielle
  r техника *f* промышленных измерений

10228 production of binary image
  d Binärbilderzeugung *f*; Erzeugen *n* eines Binärbildes
  f production *f* d'une image binaire
  r выработка *f* двоичного изображения

10229 production rate
  d Produktionsrate *f*; Fertigungsquote *f*
  f cote *f* de fabrication
  r скорость *f* изготовления

10230 production supervision
  d Produktionsüberwachung *f*
  f surveillance *f* de la production
  r надзор *m* над производством

10231 production system
  d Produktionssystem *n*
  f système *m* de production
  r система *f* производства

10232 production technology
  d Herstellungstechnologie *f*
  f technoligie *f* de fabrication
  r технология *f* производства

** * production test → 10233**

10233 production test[ing]
  d Fertigungsprüfung *f*
  f test *m* de fabrication
  r производственное испытание *n*

10234 productive sampling test
  d Produktionsstichprobe *f*
  f test *m* d'échantillon de production
  r выборочный производственный контроль *m*

10235 product line
  d Produktlinie *f*
  f ligne *f* de produit

  r производственная линия *f*

10236 product scheduling
  d Produktionsplanung *f*
  f planning *m* de production
  r планирование *n* производства

10237 profile dispersion parameter
  d Profoldispersionskoeffizient *m*
  f coefficient *m* de dispersion de profil
  r коэффициент *m* профильной дисперсии

** * prognosis → 10001**

** * prognostication → 10001**

10238 program
  d Programm *n*
  f programme *m*
  r программа *f*

10239 program *v*
  d programmieren
  f programmer
  r программировать

10240 program algorithm
  d Programmalgorithmus *m*
  f algorithme *m* de programme
  r алгоритм *m* программы

10241 program analyzer
  d Programmanalysator *m*
  f analyseur *m* de programme
  r программный анализатор *m*

10242 program carrier
  d Programmträger *m*
  f support *m* de programme
  r программный носитель *m*

10243 program circle
  d Programmkreis *m*
  f circuit *m* de programme
  r контур *m* программного управления

10244 program compatibility
  d Programmkompatibilität *f*; Programmverträglichkeit *f*
  f compatibilité *f* de programmes
  r программная совместимость *f*

10245 program concept
  d Programmkonzept *n*
  f concept *m* de programme
  r концепция *f* программы

10246 program control; time-cycle control
  d Programmregelung *f*; Programmsteuerung *f*

f réglage *m* à programme
r программное регулирование *n*;
     программное управление *n*

**10247 program control device with
         coordinatographs**
d Programmsteuerung *f* mit Koordinatografen
f commande *f* à programme à
     coordinatographes
r устройство *n* для программного
     управления с координатографами

10248 program control gear
d Programmsteuergerät *n*
f dispositif *m* de commande à programme
r программно-управляющий механизм *m*

10249 program controling element
d Programmgeber *m*
f programmateur *m*
r программный элемент *m*

10250 program control instruction
d Programmsteuerbefehl *m*
f instruction *f* de commande de programme
r команда *f* программного управления

10251 program-controlled
d programmgesteuert
f commandé par programme
r управляемый программой

10252 program-controlled automaton
d programmgesteuerter Automat *m*
f automate *m* commandé par [le] programme
r автомат *m* с программным управлением

**10253 program-controlled data processing
         system**
d programmgesteuerte
     Datenverarbeitungsanlage *f*;
     programmgesteuertes
     Datenverarbeitungssystem *n*
f système *m* de traitement des données
     commandé par programme
r система *f* обработки данных с
     программным управлением

10254 program-controlled device
d programmgesteuertes Gerät *n*
f appareil *m* à commande programmée
r программно-управляемое устройство *n*

10255 program-controlled macroprocessor
d programmgesteuerter Makroprozessor *m*
f macroprocesseur *m* commandé par
     programme
r макропроцессор *m* с программным

     управление

**10256 program-controlled priority interrupt
         system**
d programmgesteuertes
     Vorrangunterbrechungssystem *n*
f système *m* d'interruption de priorités à
     commande programmée
r программно-управляемая система *f*
     приоритетных прерываний

10257 program-controlled switching system
d programmgesteuertes Schaltsystem *n*
f système *m* de commutation commandé par
     programme
r система *f* коммутации с программным
     управлением

**10258 program controller; time-cycle controller;
         time-schedule controller** (US)
d Programmregler *m*; Zeitplanregler *m*
f régulateur *m* à programme
r программный регулятор *m*

**10259 program control of technological
         processes**
d Programmsteuerung *f* von technologischen
     Prozessen
f commande *f* à programme des procédés
     technologiques
r программное управление *n*
     технологическими процессами

10260 program control system
d Zeitplanregelungssystem *n*;
     Programmregelungssystem *n*
f système *m* de commande à programme
r система *f* программного регулирования

10261 program control unit
d Programmsteuereinheit *f*
f unité *f* de commande de programme
r устройство *n* управления прохождением
     программы

10262 program-dependent
d programmabhängig
f dépendant de programme
r программно-зависимый

10263 program-dependent operating
d programmabhängiger Betrieb *m*
f exploitation *f* dépendante de programme
r режим *m* программного управления

10264 program development system
d Programmentwicklungssystem *n*
f système *m* de développement de programmes
r система *f* разработки программ

10265 program error; program fault
d Programmfehler *m*
f erreur *f* de programme; défaut *m* de programme
r ошибка *f* программы

10266 program error detection
d Programmfehlerermittlung *f*
f détection *f* d'erreurs de programme
r обнаружение *n* ошибок программы

* **program fault** → 10265

* **program flow** → 2932

10267 program flow chart
d Programmablaufplan *m*
f diagramme *m* de cheminement d'un programme
r графическая блок-схема *f* программы

10268 program flow control
d Programmablaufsteuerung *f*
f commande *f* de déroulement de programme
r управление *n* выполнением программы

10269 program function
d Programmfunktion *f*
f fonction *f* de programme
r функция *f* программы

10270 program generation
d Programmgenerierung *f*
f génération *f* de programme
r выработка *f* программы

10271 program generator
d Programmgenerator *m*
f générateur *m* de programme
r генератор *m* программ

10272 program-hardware interaction
d Wechselwirkung *f* zwischen Programm und technischer Anlage
f interaction *f* entre le programme et la machine
r программно-аппаратное взаймодействие *n*

10273 program identification
d Programmidentifikation *f*
f identification *f* de programme
r идентификация *f* программы

10274 program identifier
d Programmkennzeichner *m*; Programmidentifizierer *m*
f identificateur *m* de programme

r идентификатор *m* программы

10275 program interrupt
d Programmunterbrechung *f*
f interruption *f* de programme
r прерывание *n* программы

10276 program library
d Programmbibliothek *f*
f bibliothèque *f* de programmes
r библиотека *f* программ

10277 programmable communication iterface
d programmierbares Kommunikationsinterface *n*
f interface *m* de communication programmable
r программируемый интерфейс *m* системы связи

10278 programmable communication line
d programmierbare Kommunikationsleitung *f*
f ligne *f* de communication programmable
r программируемая линия *f* связи

10279 programmable control
d programmierbare Steuerung *f*
f commande *f* programmable
r программируемое управление *n*

10280 programmable conveyor device
d programmierbare Fördereinrichtung *f*
f convoyeur *m* programmable
r программируемое транспортное устройство *n*

10281 programmable display
d programmierbare Anzeige *f*
f indication *f* programmable
r программируемая индикация *f*

10282 programmable fluidic control
d programmierbare fluidische Steuerung *f*
f commande *f* fluidique programmable
r программируемое струйное управление *n*

10283 programmable industrial automaton
d programmierbarer Industrieautomat *m*
f automate *m* programmable industriel
r программируемый промышленный автомат *m*

10284 programmable interface adapter
d programmierbarer Interface-Adapter *m*; programmierbare Schnittstellenanpassung *f*
f adapteur *m* d'interface programmable
r программируемый интерфейсный адаптер *m*

10285 programmable interrupt controller
  d programmierbare Interruptsteuereinheit *f*;
     programmierbare
     Unterbrechungssteuereinheit *f*
  f unité *f* de contrôle d'interruptions
     programmable
  r программируемый контроллер *m*
     прерываний

10286 programmable logic
  d programmierbare Logik *f*; programmierbare
     Schaltung *f*
  f logique *f* programmable
  r программируемая логика *f*

**10287 programmable manipulation devices
       without logic functions**
  d programmierbare Handhabungsgeräte *npl*
     ohne Logikfunktionen
  f appareils *mpl* d'opération programmable sans
     fonctions logiques
  r программируемые манипуляционные
     устройства *npl* без логических функций

10288 programmable manipulator
  d programmierbarer Manipulator *m*
  f télémanipulateur *m* programmable
  r программируемый манипулятор *m*

10289 programmable mini-automaton
  d programmierbarer Miniautomat *m*
  f mini-automate *m* programmable
  r программируемый мини-автомат *m*

10290 programmable multiplexer
  d programmierbarer Multiplexer *m*
  f multiplexeur *m* programmable
  r программируемый мультиплексор *m*

10291 programmable peripheral interface
  d programmierbarer
     Peripherieschnittstellenbaustein *m*
  f interface *f* périphérique programmable
  r программируемый интерфейс *m*
     периферийных устройств

10292 programmable regulation of position
  d programmierbare Lageregelung *f*
  f régulation *f* de position programmable
  r программируемое регулирование *n*
     положения

10293 programmable sensor control
  d programmierbare Sensorsteuerung *f*
  f commande *f* programmable pour capteurs
  r программируемое сенсорное управление *n*

10294 programmable sorting device
  d programmierbare Sortiereinrichtung *f*
  f dispositif *m* à triage programmable
  r программируемое сортировочное
     устройство *n*

10295 programmed action safety assembly
  d programmierte Sicherheitsgruppe *f*
  f ensemble *m* de sécurité programmé
  r программированный узел *m* безопасности

10296 programmed data change
  d programmierter Datenaustausch *m*
  f changement *m* programmé des données
  r запрограммированный обмен *m* данными

10297 programmed error indication
  d programmierte Fehleranzeige *f*
  f indication *f* programmée d'erreurs
  r программируемая индикация *f* ошибок

10298 programmed graphic processor
  d programmierter Grafikprozessor *m*
  f processeur *m* graphique programmé
  r программированный графический
     процессор *m*

10299 programmed interlock
  d programmierte Verriegelung *f*;
     programmierte Sperre *f*
  f verrouillage *m* programmé
  r программная блокировка *f*

10300 programmed learning
  d programmiertes Lernen *n*
  f enseignement *m* programmé
  r программированное обучение *n*

10301 programmed machine tool system
  d programmiertes
     Werkzeugmaschinensystem *n*
  f système *m* de programmation de machine-
     outil
  r система *f* программного управления
     производством

10302 programmed operation mode
  d programmierte Betriebsart *f*
  f mode *m* de service programmé
  r запрограммированный рабочий режим *m*

  * **programmed telemetry control** → 10942

  * **programmed tolerance check** → 10303

10303 programmed tolerance check[ing]
  d programmierte Toleranzprüfung *f*
  f essai *m* de tolérance programmé
  r программированный допусковый
     контроль *m*

10304 program memory; program storage
  d Programmspeicher *m*
  f mémoire *f* de programme
  r программная память *f*

10305 programmer
  d Programmgeber *m*
  f programmateur *m*
  r программатор *m*

10306 programming
  d Programmierung *f*
  f programmation *f*
  r программирование *n*

10307 programming code
  d Programmierungskode *m*
  f code *m* de programmation
  r программный код *m*

10308 programming in natural language
  d Programmieren *n* in natürlicher Sprache
  f programmation *f* en langage naturel
  r программирование *n* на естественном языке

10309 programming instruction
  d Programmieranleitung *f*
  f instruction *f* de programmation
  r инструкция *f* по программированию

10310 programming language
  d Programmsprache *f*
  f langage *m* de programmation
  r программный язык *m*

10311 programming language family
  d Programmiersprachenfamilie *f*
  f famille *f* de langage de programmation
  r семейство *n* языков программирования

10312 programming of graphic devices
  d Programmierung *f* grafischer Geräte
  f programmation *f* des dispositifs graphiques
  r программирование *n* графических устройств

10313 programming problem
  d Programmier[ungs]problem *n*
  f problème *m* de programmation
  r проблема *f* программирования

10314 programming strategy
  d Programmierstrategie *f*
  f stratégie *f* de programmation
  r методика *f* программирования

10315 programming system

10316 programming technique
  d Programmiertechnik *f*
  f technique *f* de programmation
  r техника *f* программирования

10317 programming technology
  d Programmiertechnologie *f*
  f technologie *f* de programmation
  r технология *f* программирования

10318 program pulse
  d Programmimpuls *m*
  f impulsion *f* de programme
  r программный импульс *m*

10319 program request
  d Programmanforderung *f*
  f exigence *f* de programme
  r запрос *m* программы

10320 program sequence
  d Programmfolge *f*
  f séquence *f* de programmes
  r последовательность *f* программ

10321 program simulation
  d Programmsimulation *f*;
    Programmablaufnachbildung *f*
  f simulation *f* de programme
  r моделирование *n* программы

  * **program storage → 10304**

10322 program structure
  d Programmaufbau *m*; Programmstruktur *f*
  f structure *f* de programme
  r структура *f* программы

10323 program system
  d Programmsystem *n*
  f système *m* de programmes
  r система *f* программ

10324 program table for coding
  d Programmtafel *f* zur Verschlüsselung
  f tableau *m* de programmation pour la codification
  r программная панель *f* для кодирования

10325 program-technical solution
  d programmtechnische Lösung *f*
  f solution *f* programme-technique
  r программно-техническое решение *n*

d Programmiersystem *n*
f système *m* de programmation
r система *f* программирования

10326 program testing
d Programmprüfung *f*
f essai *m* de programme; contrôle *m* de programme
r отладка *f* программы

10327 prograssive motion servomotor
d Servomotor *m* mit fortschreitender Bewegung
f servomoteur *m* à mouvement progressif
r серводвигатель *m* с поступательным движением

10328 progressive action
d progressive Wirkung *f*
f action *f* progressive
r нарастающее воздействие *n*

10329 progressive failure
d Änderungsausfall *m*; Driftausfall *m*
f défaillance *f* progressive
r постепенный отказ *m*

10330 progressive speed control
d stufenlose Geschwindigkeitsregelung *f*
f réglage *m* progressif de la vitesse
r непрерывное регулирование *n* скорости

10331 project documents
d Projektunterlagen *fpl*
f documents *mpl* de projet
r проектная документация *f*

10332 projecting; planning; designing
d Projektierung *f*
f projection *f*
r проектирование *n*

10333 projecting model
d Projektierungsmodell *n*
f modèle *m* de projet
r проектная модель *f*

10334 projection display
d Projektionsanzeige *f*
f affichage *m* de projection
r проекционный дисплей *m*

10335 prompt criticality; prompt critical state
d prompte Kritikalität *f*; promptkritischer Zustand *m*; prompte Kritizität *f*
f criticité *f* instantanée; état *m* critique instantané; état immédiatement critique
r мгновенная критичность *f*; мгновенно-критическое состояние *n*

* **prompt critical state → 10335**

* **prompt jump → 10336**

10336 prompt [reactivity] jump
d prompter Sprung *m*; prompter Reaktivitätssprung *m*
f saut *m* instantané; saut de réactivité instantané
r мгновенный скачок *m* [реактивности]; внезапное увеличение *n* реактивности

10337 propagating beam analysis
d Strahlenausbreitungsanalyse *f*
f analyse *f* de la propagation de faisceaux
r анализ *m* лучевого распространения

10338 propagation
d Ausbreitung *f*; Verbreitung *f*; Fortpflanzung *f*
f propagation *f*
r распространение *n*

10339 propagation delay
d Ausbreitungsverzögerung *f*
f délai *m* de propagation; retard *m* de propagation
r задержка *f* разпространения

10340 propagation equation
d Ausbreitungsgleichung *f*
f équation *f* de propagation
r уравнение *n* распространения

10341 propagation loss
d Ausbreitungsdämpfung *f*; Ausbreitungsverlust *m*
f affaiblissement *m* de la propagation
r затухание *n* при распространении

10342 propagation mode
d Ausbreitungsmodus *m*
f mode *m* de propagation
r способ *m* распространения

10343 propagation model
d Ausbreitungsmodell *n*
f modèle *m* de propagation
r модель *f* распространения

10344 propagation ratio
d Fortpflanzungsverhältnis *n*; Ausbreitungsverhältnis *n*
f rapport *m* de propagation
r коэффициент *m* распространения

10345 propagation theory
d Ausbreitungstheorie *f*
f théorie *f* de la propagation
r теория *f* распространения

* property to oscillate → 9242

* proportion → 10645

10346 **proportional**
  d proportional; proportionell
  f proportionnel
  r пропорциональный

10347 **proportional action; proportional input**
  d proportionales Verhalten *n*;
     Proportionaleinwirkung *f*
  f action *f* proportionnelle
  r пропорциональное воздействие *n*

10348 **proportional [action] controller; P-controller; throttling controller**
  d proportional wirkender Regler *m*;
     P-Regler *m*
  f régulateur *m* proportionnel; régulateur P
  r пропорциональный регулятор *m*;
     П-регулятор *m*

10349 **proportional band; proportional control zone**
  d Proportionalbereich *m*; P-Bereich *m*
  f bande *f* [de réglage] proportionnelle
  r зона *f* пропорциональности
     [регулирования]

10350 **proportional behaviour**
  d proportionales Verhalten *n*
  f comportement *m* proportionnel
  r пропорциональный режим *m* [работы];
     линейное поведение *n*

10351 **proportional component**
  d Proportionalkomponente *f*
  f composante *f* proportionnelle
  r пропорциональная составляющая *f*

10352 **proportional control; P-control**
  d Proportionalregelung *f*; P-Regelung *f*
  f commande *f* progressive; régulation *f*
     proportionnelle; réglage *m* rigide
  r пропорциональное регулирование *n*;
     П-регулирование *n*

* proportional controller → 10348

10353 **proportional controller with disturbance-variable compensation**
  d PZ-Regler *m*; P-Regler *m* mit
     Störgrößenausschaltung *f*
  f régilateur *m* proportionel avec introduction
     d'une grandeur compensatrice
  r пропорциональный регулятор *m* с
     компенсацией [переменных] возмущений

10354 **proportional control limits**
  d Proportionalregelungsgrenzen *fpl*
  f limites *fpl* du réglage proportionnel
  r пределы *mpl* пропорционального
     регулирования

* proportional control zone → 10349

10355 **proportional correction factor**
  d proportionaler Berichtigungsfaktor *m*
  f facteur *m* de correction proportionnel
  r пропорциональный поправочный
     коэффициент *m*

10356 **proportional counter**
  d proportionaler Zähler *m*
  f compteur *m* proportionnel
  r пропорциональный счётчик *m*

10357 **proportional counter spectrometer**
  d Proportionalzählrohrspektrometer *n*;
     Spektrometer *n* mit Proportionalzählrohr
  f spectromètre *m* à tube compteur
     proportionnel
  r спектрометр *m* с пропорциональным
     счётчиком

10358 **proportional coupling**
  d starre Kopplung *f*
  f couplage *m* proportionnel
  r пропорциональная связь *f*

10359 **proportional divider**
  d Proportionalteiler *m*
  f diviseur *m* proportionnel
  r пропорциональный делитель *m*

10360 **proportional feedback; rigid feedback**
  d proportionale Rückkopplung *f*; starre
     Rückkopplung
  f réaction *f* proportionnelle; réaction rigide
  r жёсткая обратная связь *f*;
     пропорциональная обратная связь

* proportional-floating controller → 10362

* proportional-floating-derivative control → 10363

* proportional-floating-derivative controller → 10364

10361 **proportional gain**
  d proportionale Verstärkung *f*
  f gain *m* proportionnel
  r линейное усиление *n*

* proportional input → 10347

**10362 proportional-integral controller;
PI-controller; proportional-floating
controller**
d Proportional-Integral-Regler *m*; PI-Regler *m*
f régulateur *m* proportionnel et intégral
r пропорционально-интегральный
    регулятор *m*; ПИ-регулятор *m*

**10363 proportional-integral-derivative control;
proportional-floating-derivative control;
PID-control**
d Proportional-Integral-Derivativ-Regelung *f*;
    PID-Regelung *f*
f réglage *m* à action proportionnelle, intégrale
    et dérivée
r пропорционально-интегрально-
    дифференциальное регулирование *n*;
    ПИД-регулирование *n*

**10364 proportional-integral-derivative
controller; PID-controller;
proportional-floating-derivative
controller**
d Proportional-Integral-Derivativ-Regler *m*;
    PID-Regler *m*
f régulateur *m* à action proportionnelle,
    intégrale et dérivée
r пропорционально-интегрально-
    дифференциальный регулятор *m*;
    ПИД-регулятор *m*

10365 proportionality
d Proportionalität *f*
f proportionnalité *f*
r пропорциональность *f*

10366 proportionality factor
d Verhältniszahl *f*
f nombre *m* proportionnel
r фактор *m* пропорциональности

10367 proportional navigation
d Proportionalitätsnavigation *f*
f navigation *f* proportionnelle
r пропорциональное наведение *n*

10368 proportional-plus-derivative controller
d Proportional-Differential-Regler *m*;
    PD-Regler *m*
f régulateur *m* à action proportionnelle et
    dérivée; régulateur PD
r [автоматический] статический
    регулятор *m* с воздействием по
    производной

**10369 proportional-plus-integral control;
proportional-plus-reset control; PI-control**
d Proportional-Integral-Regelung *f*;

PI-Regelung *f*
f réglage *m* à action proportionnelle et
    intégrale
r пропорционально-интегральное
    регулирование *n*

*   **proportional-plus-reset control** → 10369

10370 proportional-rate action
d proportional-derivative Einwirkung *f*
f action *f* proportionnelle et dérivée
r пропорциональное воздействие *n* по
    скорости

10371 proportional speed floating regulator
d ausschlagabhängiger astatischer Regler *m*
f régulateur *m* astatique à vitesse dépendante
r астатический регулятор *m* с зависимой
    скоростью

10372 proposition; offering
d Aussage *f*
f proposition *f*
r предложение *n*; высказывание *n*

10373 protection check
d Schutzprüfung *f*
f vérification *f* de la protection
r контроль *m* защиты

10374 protection facility
d Schutzeinrichtung *f*
f dispositif *m* de protection
r устройство *n* защиты

10375 protection in remote control system
d Fernsteuerungsschutz *m*
f protection *f* en télécommande
r защита *f* в системе дистанционного
    управления

**10376 protection of alternating current supply
lines**
d Wechselstromleitungsschutz *m*
f protection *f* des lignes d'énergie à courant
    alternatif
r защита *f* электросети переменного тока

10377 protection of direct current supply lines
d Schutz *m* von Gleichstromfernleitungen
f protection *f* des lignes d'énergie à courrant
    continu
r защита *f* электросети постоянного тока

10378 protection system; protective system
d Schutzsystem *n*
f système *m* de protection
r система *f* защиты

10379 **protective action**
  d Schutzhandlung *f*
  f action *f* de protection
  r защитное действие *n*

10380 **protective device to prevent switching errors**
  d Schaltfehlerschutz *m*
  f dispositif *m* de protection contre les erreurs de commutation
  r защита *f* от неправильного включения

10381 **protective equipment**
  d Schutzeinrichtung *f*; Schutzvorrichtung *f*
  f protecteur *m*
  r защитное приспособление *n*

10382 **protective function**
  d Schutzfunktion *f*
  f fonction *f* de protection
  r защитная функция *f*

10383 **protective gas contactor**
  d Schutzgaskontaktschütz *n*
  f contacteur *m* à gaz protectif
  r защитный газовый контактор *m*

10384 **protective resistance**
  d Schutzwiderstand *m*
  f résistance *f* de protection; résistance de sécurité
  r предохранительное сопротивление *n*

  * **protective system** → 10378

10385 **prototype**
  d Prototyp *m*
  f prototype *m*
  r прототип *m*

10386 **prototyping system**
  d Prototypsystem *n*
  f système *m* prototype
  r резидентная отладочная система *f*

10387 **proximity detector**
  d Näherungsdetektor *m*
  f détecteur *m* de proximité
  r дистанционный датчик *m*

10388 **proximity effect**
  d Näherungseffekt *m*
  f effet *m* de proximité
  r эффект *m* близости; влияние близости обратного провода

10389 **proximity sensor**
  d Nahgeber *m*; nahangeordneter Sensor *m*

  f senseur *m* proximité
  r датчик *m* близости

  * **pseudo-code** → 107

10390 **pseudo-code system**
  d Pseudokodesystem *n*
  f système *m* pseudo-code
  r система *f* псевдокода

10391 **pseudoharmonic oscillation**
  d pseudoharmonische Schwingung *f*
  f oscillation *f* pseudoharmonique
  r псевдогармоническое колебание *n*

10392 **pseudoinstruction**
  d Pseudobefehl *m*
  f pseudo-instruction *f*
  r псевдокоманда *f*

10393 **pseudolinear**
  d pseudolinear
  f pseudolinéaire
  r псевдолинейный

10394 **pseudolinear system**
  d pseudolineares System *n*
  f système *m* pseudolinéaire
  r псевдолинейная система *f*

10395 **pseudometric space**
  d pseudometrischer Raum *m*
  f espace *m* pseudométrique
  r псевдометрическое пространство *n*

10396 **pseudooperation**
  d Pseudooperation *f*
  f pseudo-opération *f*
  r псевдооперация *f*

10397 **pseudoprogram**
  d Pseudoprogramm *n*
  f pseudoprogramme *m*
  r псевдопрограмма *f*

10398 **pseudorandom number method**
  d Methode *f* der Pseudozufallszahlen
  f méthode *f* des nombres pseudo-aléatoires
  r метод *m* псевдослучайных чисел

10399 **pseudorandom numbers**
  d Pseudozufallszahlen *fpl*
  f nombres *mpl* pseudo-aléatoires
  r псевдослучайные числа *npl*

10400 **pseudorandom sequence**
  d Pseudozufallsfolge *f*
  f séquence *f* pseudo-aléatoire
  r псевдослучайная последовательность *f*

10401 pseudoscalar coupling
   d pseudoskalare Kopplung f
   f couplage m pseudoscalaire
   r псевдоскалярная связь f

10402 pseudoscalar quantity
   d Pseudoskalar m; pseudoskalare Große f
   f grandeur f pseudo-scalaire
   r псевдоскалярная величина f

10403 pseudovector coupling
   d Pseudovektorkopplung f; pseudovektorielle
      Bindung f
   f couplage m pseudovectoriel
   r псевдовекторная связь f

10404 pull-in frequency
   d Mitnahmefrequenz f; Mitziehfrequenz f
   f fréquence f d'accrochage
   r частота f захвата

   * pull-in voltage → 2491

10405 pull-up torque
   d Ansprechmoment n
   f couple m au démarrage
   r минимальный пусковой момент m

10406 pulsating voltage
   d pulsierende Spannung f
   f tension f pulsée
   r пульсирующее напряжение n

   * pulsation → 1717

10407 pulsation coefficient
   d Pulsationskoeffizient m; Welligkeit f
   f coefficient m de pulsation
   r коэффициент m пульсации

10408 pulsation instability
   d Pulsationsinstabilität f
   f instabilité f due à la pulsation
   r пульсационная неустойчивость f

   * pulse → 6373

10409 pulse amplifier
   d Impulsverstärker m
   f amplificateur m d'impulsions
   r импульсный усилитель m

10410 pulse-amplitude analyzer
   d Impulsamplitudenprüfer m
   f analyseur m d'amplitude d'impulsion
   r анализатор m амплитуд импульсов

   * pulse amplitude analyzing → 753

10411 pulse-amplitude-modulated carrier
   d impulsamplitudenmodulierter Träger m
   f porteuse f modulée en amplitude
      d'impulsions
   r амплитудно-импульсно модулированная
      несущая f

10412 pulse attenuator
   d Impulsdämpfungsglied n;
      Impulsabschwächer m
   f ligne f d'affaiblissement d'impulsions
   r импульсный аттенюатор m

10413 pulse bandwidth
   d Impulsbandbreite f
   f largeur f de bande d'impulsion
   r частотная полоса f импульса

10414 pulse base
   d Impulsbasis f
   f base f d'impulsion
   r основание n импульса

   * pulse build-up time → 10448

10415 pulse burst
   d Impulsbündel n
   f faisceau m d'impulsions
   r пучок m импульса

10416 pulse carrier
   d Impulsträger m
   f support m d'impulsion
   r несущая f [частота] импульса

10417 pulse circuits theory
   d Theorie f der Impulskreise
   f théorie f des circuits impulsionnels
   r теория f импульсных схем

10418 pulse clipping
   d Impulsbegrenzung f
   f limitation f d'impulsion
   r ограничение n импульсов

10419 pulse-code-modulation transmission
      system
   d PCM-Übertragungssystem n;
      Pulsekodemodulationsübertragungssystem n
   f système m MIC de transmission; système de
      transmission à modulation par impulsions
      codées
   r система f передачи с кодово-импульсной
      модуляцией

10420 pulse-code telemetering system
   d Impulskode-Fernmesssystem n

f système *m* de mesure à distance à code d'impulsion
r телеизмерительная система *f* с импульсным кодом

10421 pulse-controlled
d impulsgesteuert
f à commande par impulsions
r с импульсным управлением

10422 pulse correction
d Impulskorrektion *f*; Impulskorrektur *f*
f correction *f* par impulsion
r импульсная коррекция *f*

10423 pulse curve
d Impulskurve *f*
f courbe *f* d'impulsion
r импульсная кривая *f*

10424 pulse diagram
d Impulsdiagramm *n*
f diagramme *m* des impulsions
r импульсная диаграмма *f*

10425 pulsed injection laser
d Injektionsimpulslaser *m*
f laser *m* impulsionnel à injection
r импульсный инжекционный лазер *m*

10426 pulse discrimination circuit
d Impulsdiskriminatorkreis *m*
f circuit *m* de discrimination d'impulsions
r схема *f* дискриминатора импульсов

10427 pulse distortion
d Impulsverzerrung *f*
f déformation *f* d'impulsions
r искажение *n* импульса

10428 pulse disturbance
d Impulsstörung *f*
f perturbation *f* impulsionnelle
r импульсное возмущение *n*; импульсная помеха *f*

10429 pulse dividing circuit
d Impulsuntersetzerschaltung *f*
f démultiplicateur *m* d'impulsions
r схема *f* [раз]деления импульсов

10430 pulsed laser welding
d Impulslaserschweißung *f*; Schweißung *f* mit pulsierendem Laser
f soudage *m* à laser en régime impulsionnel
r импульсная лазерная сварка *f*

10431 pulsed lidar

d impulsbetriebener Lidar *m*; Impulslidar *m*
f lidar *m* impulsionnel
r импульсный лазерный локатор *m*

10432 pulsed operation
d Impulsbetrieb *m*
f fonctionnement *m* pulsatoire; marche *f* en impulsion
r импульсный режим *m*

10433 pulsed supply
d schubweise Versorgung *f*
f alimentation *f* pulsée
r пульсирующее питание *n*

10434 pulse emission
d Impulsemission *f*
f émission *f* impulsionnelle
r передача *f* импульсов

10435 pulse enable
d Impulsfreigabe *f*
f validation *f* d'impulsion
r разрешение *n* импульса

10436 pulse equalizer
d Impulsentzerrer *m*
f égaliseur *m* d'impulsion
r импульсный выравниватель *m*

10437 pulse-frequency modulation
d Pulsfrequenzmodulation *f*
f modulation *f* d'impulsions en fréquence
r частотно-импульсная модуляция *f*

* **pulse function** → 6386

10438 pulse generator
d Impulsgenerator *m*
f générateur *m* d'impulsion
r генератор *m* импульсов

* **pulse height analizing assembly** → 755

10439 pulse interleaving
d Impulsverschachtelung *f*
f imbriquage *m* d'impulsions
r чередование *n* импульсов

10440 pulse interval
d Impulsabstand *m*
f intervalle *m* d'impulsions
r интервал *m* между импульсами

10441 pulse line
d Impulsleitung *f*
f ligne *f* à impulsions
r импульсная линия *f*

10442 pulse loading
d Impulsbelastung *f*
f charge *f* impulsionnelle
r импульсная нагрузка *f*

*   **pulse-recurrence frequence → 6383**

10443 pulse recurrence rate
d Impulsfolgefrequenz *f*
f fréquence *f* de répétition des impulsions
r частота *f* последовательности импульсов

10444 pulse refining
d Impulsaufbereitung *f*
f préparation *f* d'impulsions
r первичная обработка *f* импульсов

10445 pulse regeneration circuit
d Impulsregenerationsschaltung *f*
f circuit *m* de régénération d'impulsions
r схема *f* регенерации импульсов

10446 pulse repeater; transponder
d Impulswiederholer *m*
f répétiteur *m* d'impulsions
r импульсный повторитель *m*

*   **pulse response → 6391**

10447 pulse restoration
d Impulserneuerung *f*
f rétablissement *m* de forme d'impulsions
r востановление *n* импульсов

10448 pulse rise time; pulse build-up time
d Impulsanstiegszeit *f*
f temps *m* de montée d'impulsion
r время *n* нарастания импульса

10449 pulse sampling
d Impulsabtastung *f*
f échantillonnage *m* d'impulsions
r отбор *m* импульсов

10450 pulse shaping
d Impulsformung *f*
f formage *m* d'impulsion
r формирование *n* импульса

*   **pulse signal → 6392**

10451 pulse spectrograph
d Impulsspektrograf *m*
f spectrographe *m* à impulsions
r импульсный спектрограф *m*

10452 pulse spectrometry
d Impulsspektrometrie *f*

f spectrométrie *f* à impulsions
r импульсная спектрометрия *f*

10453 pulse stabilization
d Impulsstabilisation *f*; Impulsstabilisierung *f*
f stabilisation *f* impulsionnelle
r импульсная стабилизация *f*

10454 pulse step function
d Impulsübergangsfunktion *f*
f réponse *f* impulsionnelle
r переходная импульсная функция *f*

10455 pulse system differential analyzer
d Impulsdifferentialanalysator *m*
f analyseur *m* différentiel impulsionnel
r дифференциальный анализатор *m* импульсных систем

10456 pulse system simulation
d Modellierung *f* von Impulssystemen
f simulation *f* des systèmes à impulsions
r моделирование *n* импульсных систем

10457 pulse system with delay; pulse system with lag time; pulse system with retardation
d Impulssystem *n* mit Verzögerung
f système *m* impulsionnel à retard
r импульсная система *f* с запаздыванием [передачи]

10458 pulse system with extrapolators
d Impulssystem *n* mit Extrapolatoren
f système *m* impulsionnel à extrapolateurs
r импульсная система *f* с экстраполяторами

*   **pulse system with lag time → 10457**

*   **pulse system with retardation → 10457**

10459 pulse telemetering method
d Impulsmessverfahren *n*
f méthode *f* de télémesure par impulsions
r импульсный метод *m* телеметрии

10460 pulse threshold energy
d Impulsschwellenenergie *f*
f énergie *f* de seuil de l'impulsion
r пороговая импульсная энергия *f*

10461 pulse-time modulation
d Impulszeitmodulation *f*
f modulation *f* de temps d'impulsion
r время-импульсная модуляция *f*

10462 pulse-type telemeter
d Impulsfernmessgerät *n*

    f télémètre *m* impulsionnel
    r импульсное телеметрическое
       устройство *n*

10463 pulse-width control
    d Impulsbreitenregelung *f*
    f réglage *m* en durée d'impulsion
    r регулирование *n* по длительности
       импульса

    * **pulse width modulation** → 6389

    * **pulsing** → 1717

    * **pump** → 10464

10464 pump[ing]
    d Pumpen *n*
    f pompage *m*
    r накачка *f*

10465 pumping plant
    d Pumpenanlage *f*
    f installation *f* de pompes
    r насосная установка *f*

10466 pump pulse
    d Pumpimpuls *m*
    f impulsion *f* de pompage
    r импульс *m* накачки

10467 pure time delay; real dead time
    d reine Laufzeit *f*; reine Verzögerung *f*; echte
       Totzeit *f*
    f retard *m* pur
    r чистое запаздывание *n*

10468 push-button pulse
    d Drucktastenimpuls *m*
    f impulsion *f* déclenchée par bouton-poussoir
    r импульс *m* кнопки

10469 push-button station
    d Druckknopfsteuerungsstation *f*
    f pupitre *m* de boutons-poussoirs
    r кнопочный пункт *m* управления

10470 push-down automaton
    d Rückstellautomat *m*
    f automate *m* push-down
    r автоматическое устройство *n* возврата

10471 push-pull modulation
    d Zweitaktmodulation *f*; Gegenmodulation *f*
    f modulation *f* en push-pull
    r двухтактная модуляция *f*

10472 push-pull stage

    d Gegentaktstufe *f*
    f étage *m* symétrique; étage push-pull
    r двухтактных каскад *m*

10473 putting into operation
    d Inbetriebnahme *f*
    f mise *f* en exploitation
    r ввод *m* в эксплуатацию

Q

10474 Q-factor; quality factor; figure of merit
 d Gütefaktor *m*; Gütegrad *m*; Faktor *m* Q
 f coefficient *m* de qualité; facteur Q
 r коэффициент *m* добротности;
      добротность *f*

10475 Q-meter
 d Q-Messer *m*
 f Q-mètre *m*
 r куметр *m*; измеритель *m* добротности

10476 quadratic
 d quadratisch
 f quadratique
 r квадратичный

10477 quadratic criterion
 d quadratisches Kriterium *n*
 f critère *m* quadratique
 r квадратичный критерий *m*

10478 quadratic error area
 d quadratische Fehlerfläche *f*
 f surface *f* quadratique d'erreur
 r область *f* квадратичных отклонений

10479 quadrature-axis synchronous impedance
 d synchrone Querimpedanz *f*
 f impédance *f* synchrone transversale
 r синхронный импеданс *m* шунта

10480 quadrature modulation
 d quadratische Modulation *f*
 f modulation *f* quadratique
 r квадратурная модуляция *f*

10481 quadrature oscillator
 d Quadraturoszillator *m*
 f oscillateur *m* à quadrature
 r квадратурный генератор *m* [колебаний]

10482 qualification testing
 d Qualifikationsprüfung *f*
 f examen *m* de qualification
 r квалификационное испытание *n*

10483 qualitative analysis
 d qualitative Analyse *f*
 f analyse *f* qualitative
 r качесственный анализ *m*

10484 qualitative methods
 d Verfahren *npl* mit Güteparametern
 f méthodes *fpl* qualitatives
 r качественные методы *mpl*

10485 quality checking
 d Qualitätskontrolle *f*
 f contrôle *m* de qualité
 r контроль *m* качества

10486 quality direction
 d Qualitätslenkung *f*
 f direction *f* de qualité
 r управление *n* качеством

10487 quality engineering
 d Qualitätstechnik *f*
 f technique *f* de qualité
 r техника *f* проверки качества

 * **quality factor** → 10474

10488 quality factor circuit
 d Kreisgüte *f*
 f surtension *f* du circuit
 r добротность *f* схемы

 * **quality index** → 6453

10489 quality of prediction
 d Vorhaltsgüte *f*
 f qualité *f* de prédiction
 r качество *n* упреждения

10490 quantification
 d Quantisierung *f*; Quantelung *f*
 f quantification *f*; échantillonnage *m*
 r квантификация *f*; дискретизация *f*

10491 quantitative analysis
 d quantitative Analyse *f*
 f analyse *f* quantitative; dosage *m*
 r количественный анализ *m*

10492 quantitative measurement of gas pressure
 d quantitative Gasdruckmessung *f*
 f mesure *f* quantitative de pression du gaz
 r количественное измерение *n* давления
      газа

10493 quantitative model
 d quantitatives Modell *n*
 f modèle *m* quantitatif
 r количественная модель *f*

10494 quantity control
 d Mengenregelung *f*

f réglage *m* de quantité
r регулирование *n* количества

10495 quantity controller
d Mengenregler *m*
f régulateur *m* de quantité
r регулятор *m* количества

10496 quantity flow-sheet
d Mengenfließbild *n*
f flow-sheet *m* quantitatif
r схема *f* количественных потоков

* **quantization distortion** → **10503**

* **quantization noise** → **10503**

10497 quantization problem
d Problem *n* der Quantisierung
f problème *m* de la quantification
r проблема *f* квантования

10498 quantization step
d Quantisierungsschritt *m*
f pas *m* de quantification; pas de découplage
r шаг *m* квантования

10499 quantize *v*
d quantisieren; quanteln
f quantifier
r квантовать

10500 quantized feedback equalization
d Entzerrung *f* durch quantisierte Rückkopplung
f égalisation *f* à rétroaction quantifiée
r компенсация *f* при квантованной обратной связи

10501 quantized signal
d quantisiertes Signal *n*
f signal *m* quantifié
r дискретизированный сигнал *m*; квантованный сигнал

10502 quantizing coder
d Quantisierungskodierer *m*
f codeur *m* quantificateur
r преобразователь *m* непрерывной величины в код

10503 quantizing distortion; quantizing noise; quantization distortion; quantization noise
d Quantisierungsverzerrung *f*; Quantisierungsgeräusch *n*; Quantisierungsrauschen *n*
f distorsion *f* de quantification; bruit *m* de quantification
r искажение *n* квантования

10504 quantizing error
d Quantisierungsfehler *m*
f erreur *f* de découpage; erreur de quantification
r ошибка *f* квантования

* **quantizing noise** → **10503**

10505 quantum amplifier
d Quantenverstärker *m*
f amplificateur *m* quantique
r квантовый усилитель *m*

10506 quantum condition
d Quantenbedingung *f*
f condition *f* quantique
r квантовое условие *n*

10507 quantum detector
d Quantendetektor *m*
f détecteur *m* quantique
r квантовый детектор *m*

10508 quantum efficiency
d Quantenwirkungsgrad *m*
f rendement *m* quantique
r квантовая эффективность *f*

10509 quantum electronics
d Quantenelektronik *f*
f électronique *f* quantique
r квантовая электроника *f*

10510 quantum frequency conversion
d Quantenfrequenzumsetzung *f*
f conversion *f* quantique de fréquence
r квантовое преобразование *n* частоты

10511 quantum level
d Quantenniveau *n*
f niveau *m* quantique
r квантовый уровень *m*

10512 quantum limit
d Quantengrenze *f*
f limite *f* quantique
r квантовый предел *m*

10513 quantum noise
d Quantenrauschen *n*
f bruit *m* quantique
r квантовый шум *m*

10514 quantum optical generator
d Quantenoptikgenerator *m*
f générateur *m* à optique quantique
r квантовый оптический генератор *m*

10515   quantum system
d Quantensystem *n*
f système *m* quantique
r квантовая система *f*

10516   quantum theory
d Quantentheorie *f*
f théorie *f* quantique
r квантовая теория *f*

10517   quasi-associated
d quasiassoziiert
f quasi-associé
r квази-связанный

10518   quasi-associated mode
d quasiassoziierter Modus *m*
f mode *m* quasi-associé
r квази-связанный режим *m*

10519   quasi-automaton
d Quasiautomat *m*
f quasi-automate *m*
r квазиавтомат *m*

10520   quasi-balanced bridge
d quasi-abgeglichene Brücke *f*
f pont *m* quasi-équilibre
r квазиуравновешенный мост *m*

10521   quasi-critical damping
d quasikritische Dämpfung *f*
f amortissement *m* quasi-critique
r квазикритическое демпфирование *n*

10522   quasi-harmonic oscillation
d quasiharmonische Schwingung *f*
f oscillation *f* quasi-harmonique
r квазигармоническое колебание *n*

10523   quasi-harmonic system
d quasiharmonisches System *n*
f système *m* quasi-harmonique
r квазигармоническая система *f*

10524   quasi-linearization
d Quasilinearisierung *f*
f quasi-linéarisation *f*
r квазилинеаризация *f*

10525   quasi-linear system
d quasilineares System *n*
f système *m* quasi-linéaire
r квазилинейная система *f*

10526   quasi-machine structure
d Quasimaschinenstruktur *f*
f structure *f* quasi-machine

r квазимашинная структура *f*

10527   quasi-parallel program execution
d quasiparallele Programmausführung *f*
f exécution *f* de programme quasi-parallèle
r квазипараллельное выполнение *n*
   программы

*   **quasi-periodic behaviour → 678**

10528   quasi-stable
d quasistabil
f métastable
r квазиустойчивый

10529   quasi-static
d quasistatisch
f quasi statique
r квазистатический

10530   quasi-stationary
d quasistationär
f quasi-stationnaire
r квазистационарный

10531   quasi-stationary behaviour
d qasistationäres Verhalten *n*
f comportement *m* quasi-stationnaire
r квазистационарное поведение *n*

*   **quenching circuit → 1828**

10532   quenching resistance
d Tilgungswiderstand *m*; Löschwiderstand *m*
f résistance *f* d'étouffement
r [искро]гасящее сопротивление *n*

**10533   quick-acting regulator; high-speed action
         controller**
d Schnellregler *m*; schnellwirkender Regler *m*
f régulateur *m* agissant instantanément;
   régulateur à action rapide
r быстродействующий регулятор *m*

**10534   quick-action switch; snap-action switch;
         snap-switch**
d Schnappschalter *m*; Sprungschalter *m*
f commutateur *m* instantané; interrupteur à
   grande vitesse
r мгновенный выключатель *m*

10535   quick-break feeder fuse
d unverzögerte Streckensicherung *f*
f coupe-circuit *m* rapide de ligne
   d'alimentation
r быстродействующий предохранитель *m*

10536   quick-make
d Schnelleinschaltung *f*

    f fermeture *f* rapide
    r быстрое замыкание *n*

10537 quick positioning system
    d schnelles Positionierungssystem *n*;
       Schnellpositioniersystem *n*
    f système *m* de positionnement rapide
    r система *f* быстрого позиционирования

10538 quick release
    d schnelle Auslösung *f*
    f déclenchement *m* rapide
    r немедленное разъединение *n*

10539 quiescent-carrier modulation
    d Ruheträgermodulation *f*; Modulation *f* mit
       Trägerwellenunterdrückung
    f modulation *f* à suppression de l'onde porteuse
    r модуляция *f* с подавлением несущей

10540 quiescent current
    d Ruhestrom *m*
    f courant *m* de repos
    r ток *m* покоя

10541 quiescent point
    d statischer Arbeitspunkt *m*
    f point *m* de repos
    r точка *f* покоя

10542 quiescent state
    d Ruhezustand *m*
    f état *m* de repos
    r состояние *n* покоя

10543 quiescent value
    d Ruhewert *m*; nichtgestörter Wert *m*
    f valeur *f* de repos
    r невозмущенное значение *n*; величина *f*
       покоя

10544 quiet operation
    d geräuscharmer Lauf *m*; geräuscharmer
       Funktionsablauf *m*
    f fonctionnement *m* silencieux
    r малошумная операция *f*

R

10545 radar
d Radar *n*
f radar *m*
r радар *m*; радиолокатор *m*

10546 radar command post
d Radarbefehlsstelle *f*
f poste *m* de commande radar
r радиолокационный пункт *m* управления

10547 radar data
d Radardaten *pl*
f données *fpl* du radar
r радиолокационные данные *pl*

10548 radar echo
d Radarecho *n*
f écho *m* radar
r отражeнный радиолокационный сигнал *m*

10549 radar frequency
d Radarfrequenz *f*
f bande *f* de fréquence d'un radar
r радиолокационная частота *f*

10550 radar reflector
d Radarreflektor *m*
f réflecteur *m* de radar
r радиолокационный отражатель *m*

10551 radiant flux density
d Strahlungsflussdichte *f*
f densité *f* de flux rayonnant
r плотность *f* [потока] излучения

* **radiation detector** → 10553

10552 radiation dosimeter
d Strahlendosimeter *n*; Strahlungsdosimeter *n*
f dosimètre *m* d'irradiation
r дозиметр *m* излучения

10553 radiation indicator; radiation detector
d Strahlungsindikator *m*; Strahlungsdetektor *m*
f indicateur *m* de rayonnement; détecteur *m* de rayonnement
r индикатор *m* излучения

10554 radiation measuring detector
d Strahlungsmessdetektor *m*
f détecteur *m* mesureur de rayonnement

r детектор *m* для измерения излучения

10555 radiation monitor
d Strahlenmonitor *m*
f moniteur *m* de rayonnement
r радиационный монитор *m*; контрольный монитор

10556 radiation pattern
d Strahlungsdiagramm *n*
f diagramme *m* de rayonnement
r диаграмма *f* [направленности] излучения

10557 radiation pyrometer
d Strahlungspyrometer *n*; Ardometer *n*
f pyromètre *m* à radiation
r радиационный пирометр *m*

10558 radiation-resistant
d strahlungsfest
f résistant au radiation
r нечувствительный к излучению

10559 radioactive warning device
d radioaktives Warngerät *n*
f indicateur *m* avertisseur radioactif
r радиоактивное устройство *n* [предупредительной сигнализации]

* **radioautogram** → 1459

* **radioautograph** → 1459

10560 radio control
d Funkleitung *f*; Funksteuerung *f*; drahtlose Steuerung *f*
f radioguidage *m*; guidage *m* hertzien; radioconduite *f*; radiotélécommande *f*
r радиоуправление *n*

10561 radio-control system
d System *n* der Funkfernsteuerung
f système *m* opérant à distance par radio; système de radiotélécommande
r радиоуправляемая система *f*

10562 radio-direction finder
d Radioortungsgerät *n*
f radiogoniomètre
r радиогониометр *m*

10563 radioelectric sensor
d radioelektrischer Sensor *m*
f capteur *m* radioélectrique
r радиоэлектрический сенсор *m*

10564 radioelectronics
d Radioelektronik *f*

f radio-électronique *f*
r радиоэлектроника *f*

10565 radio frequency; high frequency
d Hochfrequenz *f*
f haute fréquence *f*; radio-fréquence *f*
r высокая частота *f*

* **radio-frequency alternator** → 6150

10566 radioisotopic measuring method
d Radioisotopenmessmethode *f*
f méthode *f* de mesure radio-isotopique
r радиоизотопный метод *m* измерения

10567 radio-link protection
d Streckenschutz *m* mit Funkverbindung
f protection *f* de section par
   radiocommunication
r защита *f* радиосвязи

10568 radiometric analyzer
d radiometrischer Analysator *m*
f analyseur *m* radiométrique
r радиометрический анализатор *m*

10569 radiometric method of density measuring
d radiometrisches Dichtemessverfahren *n*
f méthode *f* radiométrique de mesurer de la
   densité
r радиометрический метод *m* измерения
   плотности

10570 radio remote control
d Funkfernlenkung *f*
f télécommande *f* par radio
r радиотелеуправление *n*

10571 radiospectroscopy
d Radiospektroskopie *f*
f radiospectroscopie *f*
r радиоспектроскопия *f*

10572 radiotelemetering
d Funkfernmessung *f*
f radiotélémesure *f*
r радиотелеизмерение *n*

10573 radius of convergence
d Konvergenzradius *m*
f rayon *m* de convergence
r радиус *m* сходимости

10574 Raman spectrometry
d Ramansche Spektrometrie *f*
f spectrométrie *f* Raman
r спектрометрия *f* Рамана

10575 ramp-forced response; ramp function response
d Antwort *f* auf Rampenfunktion als Eingang;
   Anstiegsantwort *f*
f réponse *f* à la fonction échelon de vitesse;
   réponse à la fonction rampe
r ступенчатая характеристика *f*

10576 ramp function
d Anstiegsfunktion *f*; linear wachsende
   Funktion *f*
f fonction *f* augmentante; fonction rampe
r возрастающая функция *f*; пилообразная
   функция

* **ramp function response** → 10575

10577 ramp signal
d rampenförmiges Signal *n*
f signal *m* échelon de vitesse
r ступенчатый сигнал *m*

10578 random
d stochastisch; regellos; zufällig
f aléatoire; irrégulier
r случайный; стохастический

10579 random access device; random access storage
d Einrichtung *f* mit wahlfreiem Zugriff;
   Speicherung *f* mit wahlfreiem Zugriff
f dispositif *m* à accès aléatoire; mise *f* en
   mémoire à accès libre
r устройство *n* с произвольной выборкой;
   запоминающее устройство с
   произвольным доступом

* **random access storage** → 10579

10580 random action
d stochastische Einwirkung *f*
f action *f* aléatoire
r случайное воздействие *n*

10581 random bend loss
d Dämpfung *f* durch zufallsverteilte
   Krümmungen
f affaiblissement *m* par courbures aléatoires
r затухание *n* при случайно
   распределенных изгибах

10582 random dependence
d Zufallsabhängigkeit *f*
f dépendance *f* stochastique
r случайная зависимость *f*

10583 random distribution
d stochastische Verteilung *f*; willkürliche
   Verteilung *f*

f distribution *f* aléatoire; répartition *f* aléatoire
r случайное распределение *n*

10584 random disturbance
  d zufällige Einwirkung *f*; Zufallsstörung *f*
  f perturbation *f* aléatoire
  r случайное возмущение *n*

*   **random error → 189**

10585 random event
  d Zufallsereignis *n*; zufälliges Ereignis *n*
  f événement *m* aléatoire; événement accidentel
  r случайное событие *n*

10586 random failure
  d Zufallsausfall *m*
  f défaillance *f* accidentelle
  r случайный отказ *m*

10587 random filling
  d regellose Füllung *f*
  f garnissage *m* désordonné
  r случайная загрузка *f*

10588 random fluctuation
  d Zufallsschwankung *f*
  f fluctuation *f* aléatoire
  r случайное колебание *n*

10589 random function
  d zufällige Funktion *f*; regellose Funktion;
    stochastische Fonktion
  f fonction *f* aléatoire
  r случайная функция *f*

10590 random input sampled-data system
  d Abtastsystem *n* mit stochastischen Eingaben
  f système *m* d'échantillonnage à entrées
    aléatoires
  r импульсная система *f* со случайными
    воздействиями

10591 randomization
  d zufällige Anordnung *f*
  f disposition *f* par hasard
  r рандомизация *f*

10592 randomize *v*
  d zufällig zuordnen
  f disposer au hasard
  r рандомизировать

10593 random logic
  d frei gestaltete Logik *f*; willkürlich
    angeordnete Logik
  f logique *f* de structure libre
  r логические схемы *fpl* с нерегулярной

структурой

10594 random noise
  d Zufallsstörung *f*
  f bruit *m* aléatoire
  r случайный шум *m*

10595 random number
  d Zufallszahl *f*
  f nombre *m* aléatoire
  r случайное число *n*

10596 random numbers transducer
  d Zufallsgrößengeber *m*
  f transmetteur *m* de nombres aléatoires
  r датчик *m* случайных чисел

10597 random optimalizer
  d Zufallsoptimisator *m*
  f optimisateur *m* aléatoire
  r выборочный оптимализатор *m*

10598 random parameter variation
  d beliebige Parametervariation *f*
  f variation *f* de paramètre aléatoire
  r случайный разброс *m* параметров

10599 random process
  d Zufallsprozess *m*
  f processus *m* aléatoire
  r случайный процесс *m*

10600 random sample
  d Stichprobe *f*
  f échantillon *m* pris au hasard
  r выборка *f*

10601 random step function
  d Zufallsstufenfunktion *f*
  f fonction *f* aléatoire par échelon
  r случайная ступенчатая функция *f*

**10602 random value variance; variance of
          random value**
  d Dispersion *f* der Zufallsgröße;
    Zufallsgrößendispersion *f*
  f dispersion *f* de grandeu aléatoire
  r дисперсия *f* случайной величины

10603 random variable
  d Zufallsvariable *f*
  f grandeur *f* aléatoire; variable *f* aléatoire
  r случайная переменная *f*

10604 random vector
  d Zufallsvektor *m*; stochastischer Vektor *m*
  f vecteur *m* aléatoire; grandeur *f* vectorielle
    aléatoire
  r случайный вектор *m*

10605 random vibration
d statistische Schwingung *f*
f vibration *f* aléatoire
r неупорядоченные колебания *npl*

10606 range
d Bereich *m*
f étendue *f*; domaine *m*
r диапазон *m*; область *f*

10607 range accuracy
d Genauigkeit *f* der Entfernungsmessung
f précision *f* de distance
r точность *f* определения дальности

* **range adjustment → 1629**

10608 range changing; range switching
d Bereichsumschaltung *f*
f commutation *f* de domaine
r переключение *n* диапазонов

10609 range circuit
d Entfernungsmesskreis *m*
f circuit *m* télémétrique
r дальномерная схема *f*

10610 range correction
d Entfernungsrichtigstellung *f*;
    Abstandsberichtigung *f*
f correction *f* de distance
r поправка *f* по дальности

10611 range discriminator
d Abstandsdiskriminator *m*
f discriminateur *m* de distance
r дискриминатор *m* диапазона

10612 range extension
d Bereitserweiterung *f*
f extension *f* de gamme
r расширение *n* диапазона

10613 range marker generator
d Entfernungsmarkierergenerator *m*
f générateur *m* de marques d'étalonnage
r генератор *m* масштабных импульсов

10614 range-measurement infrared system
d infrarotes Abstandsmesssystem *n*; infrarotes
    Entfernungsmesssystem *n*
f système *m* mesureur de distance à rayons
    infrarouges
r инфракрасная система *f* измерения
    дальности

10615 range of disturbance
d Störbereich *m*

f domaine *m* de perturbation
r область *f* возмущений

* **range of indication → 6462**

* **range of linearity → 12997**

10616 range of manipulateur
d Manipulatorbereich *m*
f gamme *f* de manipulateur
r диапазон *m* манипулятора

10617 range of rated voltage
d Nennspannungsbereich *m*
f gamme *f* des tension nominales
r диапазон *m* номинального напряжения

10618 range of sensitivity; sensitivity region
d Empfindlichkeitsbereich *m*
f gamme *f* de sensibilité; région *f* sensible
r диапазон *m* чувствительности; область *f*
    чувствительности

10619 range of set value
d Sollwertbereich *m*; Einstellbereich *m*
f domaine *m* de consigne
r область *f* заданных значений

10620 range of time-lag settings
d Zeitverzögerungsbereich *m*
f plage *f* des réglages de temporisation
r диапазон *m* настройки запаздывания во
    времени

10621 range of values
d Wertebereich *m*
f domaine *m* de valeurs; champ *m* de valeurs
r диапазон *m* значений

* **range switching → 10608**

10622 ranging laser
d Laserentfernungsmesser *m*
f laser *m* télémétrique
r лазерный дальномер *m*

10623 rapid-action control
d schnellwirkende Steuerung *f*
f commande *f* à action rapide
r быстродействующее управление *n*

10624 rapid deceleration
d schnelle Verzögerung *f*
f décélération *f* rapide
r быстрое замедление *n*

10625 rapid filter
d Schnellfilter *n*

f filtre *m* rapide
r быстродействующий фильтр *m*

10626 rapid inversion system
d schnelles Inversionssystem *n*
f système *m* rapide réversible; système rapide d'inversion
r система *f* с быстрой инверсией

10627 rapid-scan spectrometer
d Spektrometer *n* mit schneller Abtastung; Spektrometer mit großer Analysiergeschwindigkeit
f spectromètre *m* à balayage rapide
r быстродействующий спектрометр *m*

* **rapid stop** → 5398

10628 rate action
d differential wirkende Regelung *f*
f réglage *m* à action dérivée
r регулирование *n* по [первой] производной

10629 rate action controller; differential controller; D-controller
d Regler *m* mit Vorhalt; differentieller Regler *m*; D-Regler *m*
f régulateur *m* à action dérivée; régulateur *m* différentiel; régulateur *m* D
r дифференциальный регулятор *m*

10630 rated breaking capacity
d Nennausschaltvermögen *n*
f pouvoir *m* nominal de coupure
r номинальная мощность *f* выключения

10631 rated capacity
d Nennlast *f*; Betriebslast *f*
f charge *f* de régime
r номинальная ёмкость *f*

10632 rated current
d Nennstrom *m*; Sollstrom *m*
f courant *m* nominal
r номинальный ток *m*

10633 rated life
d Nennlebensdauer *f*
f durée *f* de vie nominale
r номинальный срок *m* службы

10634 rated linear speed
d lineare Nenngeschwindigkeit *f*
f vitesse *f* nominale linéaire
r номинальная линейная скорость *f*

10635 rated load; nominal load
d Nennbelastung *f*

f charge *f* nominale
r номинальная нагрузка *f*

10636 rated motor torque
d nominales Motoranzugsmoment *n*
f couple *m* nominal de démarrage du moteur
r номинальный крутящий момент *m* двигателя

10637 rated quantity
d Nenngröße *f*
f grandeur *f* nominale
r номинальная величина *f*

10638 rated voltage; nominal voltage
d Nennspannung *f*
f tension *f* nominale
r номинальное напряжение *n*

10639 rate feedback; velocity feedback
d Geschwindigkeitsrückführung *f*
f réaction *f* tachymétrique
r обратная связь *f* по скорости

10640 rate of convergence
d Konvergenzgeschwindigkeit *f*
f vitesse *f* de convergence
r скорость *f* сходимости

10641 rate of decrease
d Abfallwert *m*
f valeur *f* de mise au repos
r коэффициент *m* убывания

10642 rate of inherent regulation; rate of self-regulation
d Geschwindigkeit *f* des Selbstausgleiches
f vitesse *f* d'autorégulation
r параметр *m* автоматического регулирования

* **rate of self-regulation** → 10642

10643 rate test
d Bewertungstest *m*
f test *m* d'évaluation
r оценочное испытание *n*

* **rating** → 8657

10644 rating chart
d Leistungsdiagramm *n*
f diagramme *m* de charge
r нагрузочная способность *f*

10645 ratio; proportion
d Verhältnis *n*; Proportion *f*
f proportion *f*; rapport *m*
r пропорция *f*; соотношение *n*

10646 ratio analyzer
 d Verhältnisanalysator *m*
 f analyseur *m* de rapport
 r анализатор *m* относительного содержания

10647 ratio control
 d Verhältnisregelung *f*
 f régulation de rapport; réglage *m* proportion
 r регулирование *n* по соотношению

10648 ratio controller
 d Verhältnisregler *m*
 f régulateur *m* de rapport
 r регулятор *m* соотношений

10649 ratio flow controller
 d Durchflussverhältnisregler *m*
 f régulateur *m* du rapport des courants
 r регулятор *m* соотношения потоков

10650 ratio indicator
 d Verhältnisanzeiger *m*
 f indicateur *m* de proportion
 r индикатор *m* соотношения

10651 ratio measuring instrument
 d Verhältnismessgerät *n*
 f appareil *m* de mesure de rapport
 r прибор *m* для измерения соотношения

10652 ratiometer
 d Quotientenmesser *m*
 f quotientmètre *m*; logomètre *m*
 r логометр *m*; измеритель *m* отношения

10653 rational combination
 d rationelle Kombination *f*
 f combinaison *f* rationnelle
 r рациональная комбинация *f*

10654 rational integral function
 d rationale ganzzahlige Funktion *f*
 f fonction *f* intégrale rationnelle
 r рациональная целевая функция *f*

10655 rationalization effect
 d Rationalisierungseffekt *m*
 f effet *m* de rationalisation
 r эффект *m* от рационализации

10656 rationalization investment
 d Rationalisierungsinvestition *f*
 f investissement *m* de rationalisation
 r капиталовложение *n* на проведение
    рационализации

10657 rational numerical system
 d rationales numerisches System *n*

 f système *m* numérique rationnel
 r рациональная числовая система *f*

10658 Rayleigh-Ritz method
 d Ritz-Verfahren *n*; Ritzsche Methode *f*
 f procédé *m* de Rayleigh-Ritz
 r метод *m* Рица

10659 reactance
 d Blindwiderstand *m*; Reaktanz *f*
 f réactance *f*
 r реактивное сопротивление *n*

10660 reactance drop
 d Reaktanzspannungsabfall *m*
 f chaute *f* de tension sur réactance
 r реактивное падение *n* напряжения

10661 reaction force
    (on automatic assembly)
 d Reaktionskraft *f*
 f force *f* de réaction
 r противодействующая сила *f*

10662 reaction moment
    (on automatic assembly)
 d Reaktionsmoment *m*
 f moment *m* de réaction
 r противодействующий момент *m*

10663 reaction time
 d Reaktionszeit *f*
 f temps *m* de réaction
 r время *n* реакции

10664 reactive-energy meter; var-hour meter
 d Blindstromverbrauchsmesser *m*
 f compteur *m* d'énergie réactive;
    varheuremètre *m*
 r счётчик *m* реактивной энергии

10665 reactive power measurement
 d Blindleistungsmessung *f*
 f mesure *f* de puissance réactive
 r измерение *n* реактивной мощности

10666 reactor control
 d Reaktorregelung *f*
 f commande *f* du réacteur
 r управление *n* реактором

10667 reactor dynamics
 d Reaktordynamik *f*; Dynamik *f* von Reaktoren
 f dynamique *f* de réacteur
 r динамика *f* реактора

10668 reactor modelling
 d Reaktormodellierung *f*

    f modélisation *f* de réacteur
    r моделирование *n* реактора

10669 read data
    d Lesedaten *pl*
    f données *fpl* de lecture
    r считываемые данные *pl*

   * **reader** → **10670**

10670 reading device; reader
    d Leser *m*; Lesegerät *n*
    f lecteur *m*; dispositif *m* de lecture
    r считывающее устройство *n*

10671 reading function
    d Ablesefunktion *f*
    f fonction *f* de lecture
    r функция *f* считывания

10672 reading of data byte
    d Datenbytelesen *n*
    f lecture *f* de byte de données
    r считывание *n* байта данных

10673 read[ing] operation
    d Leseoperation *f*; Lesebetrieb *m*
    f opération *f* de lecture
    r операция *f* считывания

10674 reading pulse
    d Abtastimpuls *m*; Leseimpuls *m*
    f impulsion *f* de lecture
    r импульс *m* считывания

   * **readjustability** → **3253**

10675 readjustment
    d Nachregelung *f*
    f [r]ajustage *m*
    r дополнительная регулировка *f*

   * **read operation** → **10673**

   * **read-out time** → **187**

10676 read-write current source
    d Lesestrom-Schreibstrom-Quelle *f*; Quelle *f* für Lese-Schreibstrom
    f source *f* de courant lecture-écriture
    r блок *m* питания схем считывания и записи

10677 read-write cycle
    d Lese-Schreib-Zyklus *m*
    f cycle *m* de lecture-écriture
    r цикл *m* чтения-записи

10678 ready line

    d Fertigmeldeleitung *f*
    f ligne *f* de libération
    r линия *f* готовности

10679 real actual workplace position
    d reale aktuelle Arbeitsplatzposition *f*
    f position *f* d'emplacement de travail réelle actuelle
    r текущая позиция *f* рабочего места

10680 real automaton
    d Realautomat *m*
    f automate *m* réel
    r реальный автомат *m*

   * **real dead time** → **10467**

10681 real diagram
    d Realdiagramm *n*
    f diagramme *m* réel
    r вещественная диаграмма *f*

10682 real element
    d reelles Element *n*
    f élément *m* réel
    r вещественная элемент *m*

10683 real frequency characteristic
    d reelle Frequenzcharakteristik *f*
    f réponse *f* de fréquence réelle
    r вещественная частотная характеристика *f*

   * **real gas** → **333**

10684 realistic reproduction
    d realistische Wiedergabe *f*
    f reproduction *f* réelle
    r реальная воспроизводимость *f*

10685 realized experimental arrangement
    d realisierter Versuchsaufbau *m*
    f dispositif *m* d'essai réalisé
    r реализованное опытное устройство *n*

   * **real power** → **306**

10686 real structure
    d Realstruktur *f*
    f structure *f* réelle
    r реальная структура *f*

10687 real-time application
    d Echtzeitanwendung *f*; Echtzeiteinsatz *m*
    f application *f* en temps réel
    r применение *n* в системах реального времени

10688 real-time base
    d Echtzeitbasis *f*

532

f base *f* en temps réel
r база *f* в истинном масштабе времени

10689 real-time clock
  d Realzeituhr *f*
  f horloge *f* en temps réel
  r датчик *m* реального времени

10690 real-time control program
  d Echtzeitsteuerprogramm *n*
  f programme *m* de commande en temps réel
  r программа *f* управления в реальном масштабе времени

10691 real-time data processing
  d Echtzeitdatenverarbeitung *f*
  f traitement *m* de données en temps réel
  r обработка *f* данных в реальном масштабе времени

10692 real-time data transmission control
  d Real-Time-Datenübertragungssteuerung *f*
  f commande *f* de transmission de données en temps réel
  r управление *n* передачей данных в реальном масштабе времени

10693 real-time demand
  d Echtzeitanforderung *f*
  f demande *f* de temps réel
  r требование *n* в реальном масштабе времени

10694 real-time image acquisition
  d Echtzeitbilderfassung *f*
  f acquisition *f* d'image en temps réel
  r сбор *m* данных отображения в реальном масштабе времени

10695 real-time input
  d Echtzeiteingabe *f*; Echtzeitdateneingang *m*
  f entrée *f* en temps réel
  r ввод *m* [данных] в реальном времени

10696 real-time operating system
  d Echtzeitbetriebssystem *n*
  f système *m* opérationnel en temps réel
  r операционная система *f* в реальном масштабе времени

10697 real-time output
  d Echtzeitausgabe *f*; Echtzeitdatenausgang *m*
  f sortie *f* en temps réel
  r вывод *m* [данных] в реальном времени

10698 real-time processing
  d Echtzeitverarbeitung *f*; Sofortverarbeitung *f*
  f traitement *m* en temps réel

r обработка *f* в реальном времени

10699 real-time simulator
  d Echtzeitsimulator *m*
  f simulateur *m* en temps réel
  r моделирующее устройство *n* с реальным масштабом времени

10700 real-time system
  d Echtzeitverarbeitungssystem *n*
  f système *m* en temps réel
  r система *f* реального времени

10701 real-time telemetry
  d Echtzeittelemetrie *f*
  f télémesure *f* en temps réel
  r телеметрия *f* в реальном маштабе времени

* **real value → 346**

10702 real variable
  d reelle Veränderliche *f*
  f variable *f* réelle
  r вещественная переменная *f*

10703 rearranging of work pieces
  d Umlagern *n* von Werkstücken
  f relogement *m* de pièces
  r перегрупировка *f* обрабатываемых деталей

10704 receive channel
  d Empfangskanal *m*
  f canal *m* de réception
  r приёмный канал *m*

10705 receive clock
  d Empfängertakt *m*
  f rythme *m* de récepteur
  r синхронизация *f* приема

10706 receive data
  d Empfangsdaten *pl*
  f données *fpl* de réception
  r принимаемые данные *pl*

10707 receiver
  d Empfänger *m*
  f récepteur *m*
  r приёмник *m*

10708 receiver of force
  d Kraftaufnehmer *m*
  f capteur *m* de force
  r датчик *m* усилия

10709 receiver program
  d Empfängerprogramm *n*

f programme *m* de récepteur
r приёмная программа *f*

* **receiving device** → 10712

* **receiving gear** → 10712

10710 receiving plant
d Empfangsanlage *f*
f poste *m* de récepteur
r приёмная установка *f*

10711 receiving terminal
d Empfangsterminal *n*
f terminal *m* récepteur
r принимающий терминал *m*

**10712 reception device; receiving device;
receiving gear**
d Aufnahmeeinrichtung *f*;
Empfangseinrichtung *f*
f dispositif *m* de réception; équipement *m* de
réception
r приёмное устройство *n*

10713 reciprocal variable
d Umkehrgröße *f* einer Variablen
f variable *f* réciproque
r обратное значение *n* переменной

10714 reclaim *v*; regenerate *v*
d zurückgewinnen; regenerieren;
wiederherstellen
f récupérer; régénérer
r регенерировать; востанавливать

10715 recloser
d Wiedereinschalter *m*; Leistungsschalter *m*
mit Schnellwiedereinschaltung
f réenclencheur *m*; disjoncteur *m* à
réenclenchement
r автоматичсский включатель *m*

10716 recognition
d Erkennung *f*
f reconnaissance *f*
r распознавание *n*

10717 recognition algorithm
d Erkennungsalgorithmus *m*
f algorithme *m* de reconnaissance
r алгоритм *m* распознавания

10718 recognition automaton
d Erkennungsautomat *m*
f automate *m* de reconnaissance
r распознающий автомат *m*

10719 recognition data
d Erkennungsdaten *pl*
f données *fpl* de reconnaissance
r данные *pl* процесса распознавания

10720 recognition equipment
d Erkennungsinrichtung *f*
f équipement *m* de reconnaissance
r устройство *n* распознавания

10721 recognition function
d Erkennungsfunktion *f*
f fonction *f* d'identification
r опознавательная функция *f*

10722 recognition information
d Erkennungsinformation *f*
f information *f* d'identification, information de
reconnaissance
r информация f, необходимая для
распознавания

10723 recognition logic
d Erkennungslogik *f*
f logique *f* de reconnaissance
r логика *f* распознавания

10724 recognition method
d Erkennungsmethode *f*
f méthode *f* de reconnaissance
r метод *m* распознавания

10725 recognition system of mobile objects
d Erkennungssystem *n* mobiler Objekte
f système *m* de reconnaissance d'objets
mobiles
r система *f* распознавания подвижных
объектов

10726 recognition task
d Erkennungsaufgabe *f*
f tâche *f* de reconnaissance
r задача *f* распознавания

10727 recognition time
d Erkennungszeit *f*
f temps *m* de reconnaissance
r время *n* распознавания

10728 recognizing machine
d Identifizierungsmaschine *f*
f machine *f* à identifier
r распознающая машина *f*

10729 recombination
d Rekombination *f*
f recombinaison *f*
r рекомбинация *f*

10730 recombination coefficient
 d Rekombinationskoeffizient m
 f coefficient m de recombinasion
 r коэффициент m рекомбинации

10731 recombination mechanism
 d Rekombinationsmechanismus m
 f mécanisme m de recombinaison
 r механизм m рекомбинации

10732 recombination model
 d Rekombinationsmodell n
 f modèle m de recombinaison
 r модель f рекомбинации

10733 recombination spectrum
 d Rekombinationsspektrum n
 f spectre m de recombinaison
 r спектр m рекомбинации

10734 recorded value
 d registrierter Wert m
 f valeur f enregistrée
 r записанная величина f

10735 recorder adjustment
 d Einstellen n des Aufzeichnungsgerätes
 f mise f au point de l'appareil enregistreur
 r настройка f регистрирующего устройства

10736 recorder controller
 d Registrierregler m
 f régulateur m enregistreur
 r регистратор-регулятор m

10737 recorder driver amplifier
 d Verstärker m des Aufzeichnungsantriebes
 f amplificateur m de l'attaque de l'enregistreur
 r усилитель m регистрирующего устройства

10738 recorder with linear recording
 d Linienschreiber m
 f enregistreur m à ligne continue
 r прибор m с линейной записью

10739 recording acceloremeter
 d registrierender Beschleunigungsmesser m
 f acceleromètre m enregisstreur
 r регистрирующий акцелерометр m

10740 recording densitometer
 d registrierender Dichtemesser m;
   registrierendes Densitometer n
 f densitomètre m enregistreur
 r записывающий денситометр m

10741 recording element
 d Registrierelement n

 f élément m d'enregistrement
 r регистрирующий элемент m

10742 recording frequency meter
 d Registrierfrequenzmesser m
 f fréquencemètre m enregisstreur
 r регистрирующий частотомер m

10743 recording gas analyzer
 d registrierender Gasanalysator m
 f analyseur m enregistreur de gaz
 r регистрирующий газоанализатор m

10744 recording infrared tracking instrument
 d Infrarotstrahlennachlaufregistriergerät n
 f appareil m enregistreur de poursuite à rayons infrarouges
 r устройство n для сопровождения и регистрации траектории с использованием инфракрасного излучения

10745 recording level gauge
 d registrierender Pegelanzeiger m
 f jauge f enregistreurse de niveau
 r регистрирующий уровнемер m

* recording meter → 1361

10746 recording microdensitometer
 d registrierendes Mikrodensitometer n;
   registrierender Schwärzungsmesser m
 f microdensitomètre-enregistreur m;
   microdensigraphe m
 r микроденситограф m

10747 recording of digital results
 d Aufzeichnung f digitaler Messergebnisse
 f enregistrement m de résultats digitaux
 r запись f цифровых результатов

10748 recording potentiometer
 d Registrierpotentiometer n
 f potentiomètre m enregistreur
 r регистрирующий потенциометр m

10749 recording spectrophotometer
 d registrierendes Spektralfotometer n
 f spectrophotomètre m enregistreur
 r регистрирующий спектрофотометр m

10750 recording unit
 d Registriersatz m
 f bloc m enregistreur
 r регистрирующий блок m

10751 recovery management
 d Organisation f der Wiederherstellung

f organisation *f* de restauration
r управление *n* востановлением

10752 recovery procedure
   d Rückstellvorgang *m*
   f procédure *f* de restauration
   r процедура *f* востановления

10753 recovery time
   d Erholungszeit *f*; Regenerierungszeit *f*
   f temps *m* d'adaptation; durée *f* de
      rétablissement
   r время *n* восстановления

10754 [rectangular] Cartesian components;
       [rectangular] Cartesian coordinates
   d [rechtwinklige] Kartesische
      Komponenten *fpl*; [rechtwinklige]
      Kartesische Koordinaten *fpl*
   f coordonnées *fpl* [rectangulaires];
      composantes *fpl* [rectangulaires]
   r [прямоугольные] декартовые
      координаты *fpl*

10755 rectangular distribution
   d rechtwinklige Verteilung *f*
   f répartition *f* rectangulaire
   r прямоугольное распределение *n*

   * **rectangular filter** → 9226

10756 rectangular hysteresis loop
   d rechteckige Hystereseschleife *f*
   f boucle *f* d'hystérésis rectangulaire
   r прямоугольная кривая *f* гистерезиса

10757 rectangular matrix
   d Rechteckmatrix *f*
   f matrice *f* rectangulaire
   r прямоугольная матрица *f*

10758 rectangular modulated
   d rechteckmoduliert; rechteckwellenmoduliert
   f modulé rectangulaire; modulé de l'onde
      rectangulaire
   r модулированный в виде прямоугольника

10759 rectangular pulse
   d Rechteckimpuls *m*
   f top *m* rectangulaire; créneau *m*
   r прямоугольный импульс *m*

10760 rectangular waveform
   d Rechteckwellenform *f*
   f forme *f* d'onde rectangulaire
   r прямоугольная форма *f* колебания

10761 rectification

   d Gleichrichtung *f*
   f redressage *m*
   r выпрямление *n*

10762 rectification efficiency
   d Gleichrichtungswirkungsgrad *m*
   f rendement *m* de redressement
   r эффективность *f* действия выпрямления

10763 rectified alternating current
   d gleichgerichteter Wechselstrom *m*
   f courant *m* alternatif redressé
   r выпрямленный переменный ток *m*

10764 rectified signal
   d gleichgerichtetes Signal *n*
   f signal *m* redressé
   r выпрямленный сигнал *m*

10765 rectifier converter
   d Gleichrichterwandler *m*
   f convertisseur *m* à redresseur
   r выпрямленный преобразователь *m*

10766 rectifier instrument
   d Gleichrichtermessgerät *n*
   f appareil *m* mesureur à redresseur incorporé
   r выпрямительное устройство *n*;
      детекторный прибор *m*

10767 rectifier photoelectric cell
   d Gleichrichterfotozelle *f*
   f élément *m* photoélectrique redresseur
   r вентильный фотоэлемент *m*

10768 recurrence relation
   d Rekursionsrelation *f*
   f relation *f* de récurrence
   r рекуррентное соотношение *n*

10769 recursion
   d Rekursion *f*; Wiederkehr *f*
   f récursion *f*
   r рекурсия *f*

10770 recursion formula
   d Rekursionsformel *f*
   f formule *f* de récursion
   r рекурсивная формула *f*

10771 recursive
   d rekursiv
   f récursif
   r рекурсивный

10772 recursive filter
   d rekursives Filter *n*

    f filtre *m* récursif
    r рекурсивный фильтр *m*

10773 recursive function
    d Rekursionsfunktion *f*
    f fonction *f* de récurrence
    r рекурсивная функция *f*

10774 recursive procedure
    d rekursives Verfahren *n*;
       Rekursionsverfahren *n*
    f procédure *f* récursive
    r рекурсивная процедура *f*

10775 recursive self-call
    d rekursiver Selbstaufruf *m*
    f auto-appel *m* récursif
    r рекурсивный автоматический вызов *m*

10776 recursive structure
    d rekursive Struktur *f*
    f structure *f* récursive
    r рекурсивная структура *f*

10777 reduced error
    d reduzierter Fehler *m*
    f erreur *f* réduite
    r приведенная погрешность *f*

10778 reduced frequency
    d reduzierte Frequenz *f*
    f fréquence *f* réduite
    r редуцированная частота *f*

    * reduced variable → 4254

10779 reducible
    d reduzibel
    f réductible
    r приводимый

10780 reduction
    d Reduktion *f*
    f réduction *f*
    r приведение *n*; сведение *n*

10781 redundancy
    d Redundanz *f*
    f redondance *f*
    r избыточность *f*

10782 redundancy check
    d Redundanzprüfung *f*
    f contrôle *m* de redondance
    r контроль *m* с введением избыточности

10783 redundant circuit
    d redundante Schaltung *f*

    f circuit *m* à redondance
    r избыточная схема *f*

10784 redundant module
    d redundanter Modul *m*; Redundanzbaustein *m*
    f module *m* à redondance
    r избыточный модуль *m*

10785 redundant system
    d redundantes System *n*
    f système *m* à redondance
    r избыточная система *f*

10786 reference clock
    d Bezugstaktgeber *m*
    f horloge *f* de référence; horloge maîtresse
    r опорный тактовый датчик *m*

10787 reference data
    d Referenzdaten *pl*; Bezugsdaten *pl*;
       Bezugsangaben *fpl*
    f données *fpl* repères; données de référence
    r справочные данные *pl*

10788 reference element
    d Bezugselement *n*
    f élément *m* de référence
    r элемент *m* опорного напряжения

10789 reference feedback
    d Bezugsrückkopplung *f*
    f réaction *f* à repère; réaction de référence
    r исходная обратная связь *f*

10790 reference instrument
    d Vergleichsapparat *m*; Bezugsgerät *n*
    f appareil *m* de référence
    r образцовый прибор *m*

10791 reference level
    d Bezugspegel *m*
    f niveau *m* de référence
    r опорный уровень *m*

10792 reference point
    d Bezugspunkt *m*
    f point *m* repère; point de référence
    r исходная точка *f*

10793 reference quantity
    d Bezugsgröße *f*; Bezugsmenge *f*
    f grandeur *f* de référence
    r исходная величина *f*

10794 reference signal
    d Bezugssignal *n*
    f signal *m* de référence
    r опорный сигнал *m*

10795   reference source
    d Vergleichsquelle *f*
    f source *f* de référence
    r источник *m* сравнения

10796   reference system
    d Bezugssystem *n*
    f système *m* de référence
    r система *f* отсчёта

10797   reference value
    d Referenzwert *m*; Referenzgröße *f*;
         Bezugsgröße *f*
    f valeur *f* de consigne; valeur de référence
    r величина *f* отсчёта

10798   reference voltage
    d Bezugsspannung *f*; Referenzspannung *f*
    f tension *f* d'étalonnage; tension de référence
    r эталон *m* напряжения

10799   reference voltage stabilizer
    d Bezugsspannungsstabilisator *m*
    f stabilisateur *m* de la tension de référence
    r стабилизатор *m* опорного напряжения

10800   reference winding
    d Bezugswicklung *f*
    f enroulement *m* de référence
    r основная обмотка *f*

10801   reflex amplifier
    d Reflexverstärker *m*
    f amplificateur *m* réflexe
    r рефлексный усилитель *m*

    *   reflex circuit → 4476

10802   refractometric analysis
    d refraktometrische Analyse *f*
    f analyse *f* réfractométrique
    r рефрактометрический анализ *m*

10803   refresh circuit
    d Refresh-Schaltung *f*;
         Auffrischungschaltung *f*;
         Datenregenerierungsschaltung *f*
    f circuit *m* de rafraîchissement
    r схема *f* регенерации

10804   refresh control
    d Refresh-Steuerung *f*;
         Auffrischungssteuerung *f*;
         Datenregenerierungssteuerung *f*
    f commande *f* de rafraîchissement
    r управление *n* регенерацией

10805   refresh cycle

    d Refresh-Zyklus *m*; Auffrischungszyklus *m*;
         Datenregenerierungszyklus *m*
    f cycle *m* de rafraîchissement
    r цикл *m* регенерации

10806   refresh period
    d Refresh-Periode *f*; Auffrischungsperiode *f*;
         Datenregenerierungsperiode *f*
    f période *f* de rafraîchissement
    r период *m* регенерации

    *   regenerate *v* → 10714

10807   regeneration
    d Regenerierung *f*; Wiederherstellung *f*
    f régénération *f*
    r регенерация *f*; востановление *n*

10808   regenerative
    d rückkoppelnd
    f régénératif
    r регенеративный

10809   regenerative amplifier
    d Rückkopplungsverstärker *m*
    f amplificateur *m* à réaction
    r регенеративный усилитель *m*

10810   regenerative detector
    d Rückkopplungsdetektor *m*
    f détecteur *m* à réaction
    r регенеративный детектор *m*

10811   regenerative pulse generator
    d regenerativer Impulsgenerator *m*;
         rückgekoppelter Impulsgenerator
    f générateur *m* régénératif d'impulsion
    r импульсный генератор *m* с обратной
         связью

10812   regenerative repeater; intermediate
        repeater; regenerator
    d Zwischenregenerator *m*; Regenerator *m*
    f répéteur *m* régénérateur; répéteur
         intermédiaire; régénérateur
    r промежуточный повторитель *m*;
         регенератор *m*

    *   regenerator → 10812

10813   region control task
    d Bereichssteuerungstask *f*
    f tâche *f* régulatrice en zone
    r региональная задача *f* управления

10814   region of admissible deviations
    d Bereich *m* für die zulässigen Abweichungen

f domaine *m* d'écarts admissibles
r область *f* допустимых отклонений

10815 register
   d Register *n*
   f registre *m*
   r регистр *m*

10816 register direct addressing
   d Registerdirektadressierung *f*
   f adressage *m* direct de registre
   r прямая регистровая адресация *f*

10817 register indirect addressing
   d Registerindirektadressierung *f*
   f adressage *m* indirect à registre
   r косвенная регистровая адресация *f*

10818 registering mechanism
   d Registriermechanismus *m*
   f mécanisme *m* enregistreur
   r регистрирующий механизм *m*

10819 register instruction
   d Registerbefehl *m*;
      Registeroperationsbefehl *m*
   f instruction *f* de registre
   r инструкция *f* регистра

10820 register memory; register store
   d Registerspeicher *m*
   f mémoire *f* de registre
   r регистровая память *f*

   * **register store** → **10820**

10821 register structure
   d Registerstruktur *f*
   f structure *f* de registres
   r регистровая структура *f*

10822 regression
   d Regression *f*
   f régression *f*
   r регресия *f*

10823 regression analysis
   d Regressionsanalyse *f*
   f analyse *f* de recours
   r регресивный анализ *m*

10824 regression coefficient
   d Regressionskoeffizient *m*
   f coefficient *m* de régression
   r коэффициент *m* регресии

10825 regular
   d regulär; analytisch

f régulier
r регулярный; аналитический

10826 regular cell structure
   d reguläre Zellenstruktur *f*; regelmäßige
      Zellenstruktur
   f structure *f* de cellule régulière
   r регулярная клеточная структура *f*

10827 regular element
   d reguläres Element *n*
   f élément *m* régulier
   r регулярный элемент *m*

10828 regular logic
   d reguläre Logik *f*; regelmäßige
      Logikschaltungsstruktur *f*
   f logique *f* régulière
   r регулярная логика *f*

10829 regular part of the function
   d regulärer Teil *m* der Funktion
   f partie *f* régulière de la fonction
   r регулярная часть *f* функции

10830 regulated rectifier
   d geregelter Gleichrichter *m*; stabilisierter
      Gleichrichter
   f redresseur *m* réglé; redresseur stabilisé
   r стабилизированный выпрямитель *m*

10831 regulated supply; stabilized power supply
   d geregelte Stromversorgung *f*; stabilisierte
      Stromversorgung
   f alimentation *f* réglée; alimentation stabilisée
   r стабилизированный источник *m* питания

10832 regulating circuit
   d Regelschaltung *f*
   f circuit *m* de réglage
   r схема *f* регулирования

   * **regulating element** → **351**

10833 regulating inductor
   d Regeldrossel *f*
   f inducteur *m* de réglage
   r регулирующий дроссель *m*

10834 regulating quantity
   d regelbare Größe *f*; einstellbare Größe
   f grandeur *f* réglable; quantité *f* réglable
   r регулирующая величина *f*

10835 regulating resistor
   d Regelwiderstand *m*
   f résistance *f* de réglage
   r переменное сопротивление *n*

10836   regulating transformer
  d Reguliertransformator *m*
  f transformateur *m* de réglage
  r регулирующий трансформатор *m*

  *   **regulating unit → 351**

10837   regulating unit position
  d Steuergliedstellung *f*; Stellung *f* des
       Steuergliedes
  f position *f* de l'organe de réglage
  r положение *n* регулирующего органа

  *   **regulating valve → 615**

  *   **regulation band → 3190**

10838   regulation by visual supervision
  d Regelung *f* durch visuelle Überwachung
  f réglage *m* par inspection visuelle
  r супервизорный контроль *m*

10839   regulation of grip force
  d Greifkraftregelung *f*
  f régulation *f* de force de grippion
  r регулирование *n* захватного усилия

  *   **regulation precision → 205**

10840   regulation process
  d Regelungsprozess *m*; Regelungsvorgang *m*
  f processus *m* de réglage; procédé *m* de
       réglage
  r процесс *m* регулирования

10841   regulation scale; tuning scale; tuning dial
  d Einstellskale *f*; Abstimmskale *f*
  f échelle *f* de réglage; cadran *m* d'accord
  r шкала *f* настройки

  *   **regulation speed → 518**

10842   regulation structure
  d Regelungsstruktur *f*
  f structure *f* de réglage
  r структура *f* регулирования

  *   **regulation with fixed set point → 5539**

10843   regulator distance
  d Reglerstrecke *f*
  f distance *f* de régulateur
  r линия *f* регулятора

10844   regulator supply
  d Reglerspeisung *f*; Speisung *f* des Reglers
  f alimentation *f* du régulateur
  r источник *m* питания регулятора

  *   **relation → 10846**

10845   relational operator
  d Vergleichsoperator *m*
  f opérateur *f* de relation
  r оператор *m* отношения

10846   relation[ship]
  d Beziehung *f*; Relation *f*
  f relation *f*
  r [со]отношение *n*

10847   relative address
  d relative Adresse *f*; Relativadresse *f*
  f adresse *f* relative
  r относительный адрес *m*

10848   relative amplitude
  d relative Amplitude *f*
  f amplitude *f* relative
  r относительная амплитуда *f*

10849   relative angle acceleration
  d relative Winkelbeschleunigung *f*
  f accélération *f* d'angle relative
  r относительное угловое ускорение *n*

10850   relative angle speed
  d relative Winkelgeschwindigkeit *f*
  f vitesse *f* d'angle relative
  r относительная угловая скорость *f*

10851   relative code
  d relativer Kode *m*
  f code *m* relatif
  r относительный код *m*

10852   relative control range
  d bezogener Regelbereich *m*
  f étendue *f* relative de réglage
  r диапазон *m* непрямого регулирования

10853   relative damping
  d relative Dämpfung *f*
  f amortissement *m* relatif
  r относительное затухание *n*;
       относительное демпфирование *n*

10854   relative deviation of controlled variable
  d relative Regelabweichung *f*
  f écart *m* de réglage relatif
  r относительное отклонение *n*
       регулируемой величины

10855   relative deviation of manipulated variable
  d relative Abweichung *f* der Stellgröße
  f écart *m* relatif de la grandeur réglante
  r относительное отклонение *n*
       регулирующей величины

10856 relative duration
  d relative Breite
  f durée *f* relative
  r относительная продолжительность *f*

10857 relative error
  d relativer Fehler *m*
  f erreur *f* relative
  r относительная ошибка *f*

10858 relative frequency
  d relative Häufigkeit *f*
  f fréquence *f* relative
  r относительная частота *f*

10859 relative harmonic cotent
  d relativer harmonischer Anteil *m*
  f teneur *f* relative en harmoniques; résidu *m* relatif harmonique
  r относительное содержание *n* гармоник

10860 relative maximum
  d relatives Maximum *n*
  f maximum *m* relatif
  r относительный максимум *m*

10861 relative minimum
  d relatives Minimum *n*
  f minimum *m* relatif
  r относительный минимум *m*

10862 relative parameter
  d relativer Parameter *m*
  f paramètre *m* relatif
  r относительный параметр *m*

10863 relative permittivity
  d relative Dielektrizitätskonstante *f*
  f facteur *m* de permittivité relatif
  r относительная диэлектрическая постоянная *f*

10864 relative programming
  d relatives Programmieren *n*
  f programmation *f* relative
  r относительное программирование *n*

10865 relative proportional band
  d relativer Proportionalitätsbereich *m*; relativer Regelbereich *m*
  f bande *f* relative proportionnelle; étendue *f* relative de réglage
  r относительный диапазон *m* пропорциональности

10866 relative scattering function
  d relative Streufunktion *f*
  f finction *f* relative de diffusion
  r относительная функция *f* рассеяния

10867 relative scatter intensity
  d relative Streuung *f*
  f intensité *f* relative de diffusion
  r относительная интенсивность *f* рассеяния

10868 relative sensibility
  d relative Empfindlichkeit *f*
  f sensibilité *f* relative
  r относительная чувствительность *f*

10869 relative speed drop
  d Relativgeschwindigkeitsabfall *m*
  f chute *f* relative de vitesse
  r относительное падение *n* скорости

10870 relative stability
  d relative Stabilität *f*
  f stabilité *f* relative
  r относительная устойчивость *f*

10871 relative system
  d relatives System *n*
  f système *m* relatif
  r относительная система *f*

10872 relaxation circuit
  d Kippkreis *m*; Relaxationskreis *m*
  f circuit *m* basculeur; circuit à relaxation
  r релаксационный контур *m*

* relaxation generator → 10874

10873 relaxation oscillations
  d Relaxationsschwingungen *fpl*; Kippschwingungen *fpl*
  f oscillations *fpl* de la relaxation
  r релаксационные колебания *npl*

* relaxation oscillator → 10874

10874 relaxation [pulse] generator; relaxation [pulse] oscillator
  d Relaxationsgenerator *m*; Kippschwinger *m*; Kippgenerator *m*
  f générateur *m* à relaxation; oscillateur *m* à relaxation
  r релаксационный генератор *m*

* relaxation pulse oscillator → 10874

10875 relaxation spectrum
  d Relaxationsspektrum *n*
  f spectre *m* de relaxation
  r релаксационный спектр *m*

10876 relaxation time
  d Relaxationszeit *f*

f temps *m* de relaxation
r время *n* релаксации

10877 relay action
d Relaiswirkung *f*
f action *f* de relais
r релейное действие *n*

10878 relay chain system
d Relaiskettensystem *n*
f système *m* à chaînes de relais
r каскадная релейная система *f*

10879 relay characteristic
d Relaiskennlinie *f*; Relaischarakteristik *f*
f caractéristique *f* de relais
r релейная характеристика *f*

10880 relay characteristic with dead zone
d Relaiskennlinie *f* mit Unempfindlichkeitszone
f caractéristique *f* de relais à temps mort
r релейная характеристика *f* с мертвой зоной

10881 relay circuit forbidden condition
d verbotener Relaiskreiszustand *m*
f état *m* interdit de circuit relais
r запрещенное состояние *n* релейной цепи

10882 relay circuit structural formula
d Relaiskettenstrukturformel *f*
f formule *f* de structure du circuit de relais
r структурная формула *f* релейного контура

10883 relay circuit theory
d Theorie *f* der Relaisschaltungen; Relaisschaltungstheorie *f*
f théorie *f* des circuits à relais
r теория *f* релейных схем

10884 relay compensation
d Relaiskompensation *f*
f compensation *f* de relais
r релейная компенсация *f*

10885 relay contact
d Relaiskontakt *m*
f contact *m* de relais
r контакт-реле *n*

10886 relay control
d Relaissteuerung *f*; Relaisregelung *f*
f réglage *m* à relais
r релейное регулирование *n*

10887 relay controller
d Relaisregler *m*

f régulateur *m* à relais
r релейный регулятор *m*

10888 relay device operation cycle
d Arbeitstakt *m* des Relaisgerätes
f période *f* de fonctionnement d'un dispositif à relais
r такт *m* работы релейного устройства

10889 relay device structure
d Relaisgerätanordnung *f*
f structure *f* du dispositif à relais
r структура *f* релейного устройства

10890 relay fitted contactor
d Schütz *n* mit Relais; Relaisschütz *n*
f contacteur *m* à relais
r аппроксимирующий контактор-реле *n*

10891 relay group
d Relaisgruppe *f*; Relaissatz *m*
f groupe *m* de relais
r релейный искатель *m*

10892 relay interrupter
d Relaisunterbrecher *m*
f interrupteur *m* à relais
r пульс-реле *n*

10893 relay matrix
d Relaismatrize *f*
f matrice *f* de relais
r релейная матрица *f*

10894 relay network
d Relaisschaltung *f*
f réseau *m* à relais; montage *m* à relais
r релейная схема *f*

10895 relay non-linearity
d Relaisnichtlinearität *f*
f non-linéarité *f* de relais
r релейная нелинейность *f*

10896 relay-operated controller
d Regler *m* mit Hilfsenergie; indirekter Regler
f régulateur *m* indirect
r регулятор *m* непрямого действия

10897 relay protection channel
d Relaisschutzkanal *m*
f canal *m* de protection à relais
r канал *m* релейной защиты

10898 relay pyramid; contact pyramid
d Kontaktpyramide *f*
f pyramide *f* de contact
r контактная пирамида *f*

10899 relay reset coefficient
 d Relaisrückgangsfaktor *m*
 f coefficient *m* de retour de relais
 r коэффициент *m* возврата реле

10900 relay selecting circuit
 d Relaiswählkreis *m*
 f circuit *m* sélectif à relais
 r релейная избирательная схема *f*

10901 relay servomechanism
 d Relaisregelkreis *m*; Relaisregelung *f*
 f servomécanisme *m* [à relais]
 r релейная система *f* регулирования

10902 relay system
 d Relaissystem *n*
 f système *m* de relais
 r релейная система *f*

10903 relay system structure
 d Relaissystemanordnung *f*
 f structure *f* du système à relais
 r структура *f* релейной системы

10904 relay telemetering system
 d Relaisfernmesssystem *n*
 f système *m* de télémesure à relais
 r релейная система *f* телеизмерения

 * **release pulse** → 6634

10905 releasing current
 d Auslösestrom *m*
 f courant *m* de déclenchement
 r ток *m* размыкания

 * **relevant digit** → 12742

 * **reliability** → 3954

10906 reliability analysis
 d Zuverlässigkeitsanalyse *f*
 f analyse *f* de fiabilité
 r анализ *m* надёжности

10907 reliability control
 d Zuverlässigkeitskontrolle *f*
 f contrôle *m* d'admissibilité
 r проверка *f* надёжности

10908 reliability curve
 d Zuverlässigkeitskurve *f*
 f courbe *f* de fiabilité; courbe de sûreté
 r кривая *f* надёжности

10909 reliability data
 d Zuverlässigkeitsdaten *pl*

 f données *fpl* de fiabilité
 r данные *pl* надёжности

10910 reliability of static system
 d Zuverlässigkeit *f* des statischen Systems
 f fiabilité *f* du système statique
 r надёжность *f* статической системы

10911 reliability parameter
 d Zuverlässigkeitskenngröße *f*
 f caractéristique *m* de fiabilité
 r параметр *m* надёжности

10912 reliability testing
 d Zuverlässigkeitsprüfung *f*
 f test *m* de fiabilité
 r испытание *n* на надёжность

10913 reliable
 d zuverlässig
 f fiable
 r надёжный

10914 reliable sensor
 d zuverlässiger Geber *m*
 f capteur *m* reliable
 r надёжный датчик *m*

10915 reliable sensorics; reliable sensor technique
 d zuverlässige Sensorik *f*; zuverlässige Sensortechnik *f*
 f sensorique *f* reliable; technique *f* de senseur reliable
 r надёжная сенсорика *f*; надёжная сенсорная техника *f*

 * **reliable sensor technique** → 10915

10916 relocatable module
 d verschiebbarer Modul *m*
 f module *m* décalable; module relocatable
 r перемещаемый модуль *m*

10917 remanence
 d Remanenz *f*
 f rémanence *f*
 r остаточная намагниченность *f*

10918 remote access
 d Fernzugriff *m*
 f accès *m* à distance
 r дистанционный доступ *m*

10919 remote action system
 d Fernwirksystem *n*
 f système *m* télémécanique
 r телемеханическая система *f*

10920 remote communications
d Fernkommunikation f;
  Nachrichtenfernverkehr m;
  Nachrichtenfernverbindung f
f communication f à distance;
  télécomunications fpl
r дистанционная связь f

* remote control → 4405

10921 remote control channel
d Fernsteuerungskanal m
f canal m de télécommande
r канал m телеуправления

10922 remote control coding
d Verschlüsselung f in der Fernsteuerung
j codage m de telecommande
r кодирование n в дистанционном
  управлении

10923 remote control distributor
d Verteiler m in Fernwirkanlagen
f distributeur m de dispositifs de
  télémécanique
r распределительное устройство n
  механизмов телеуправления

10924 remote control equipment
d Fernsteuereinrichtung f
f dispositif m de commande à distance
r оборудование n телеуправления

10925 remote control installation
d Fernsteuerungsanlage f
f installation f de commande à distance
r установка f телеуправления

10926 remote controller
d Fernregler m
f régulateur m à distance
r дистанционный регулятор m

10927 remote control modulation
d Fernwirktechnikmodulation f
f modulation f de télécommande
r модуляция f с использованием
  телеуправления

10928 remote diagnostic
d Ferndiagnose f; Fernfehlersuche f
f diagnostic m à distance
r дистанционная диагностика f

10929 remote display
d Fernanzeige f
f affichage m à distance
r удаленный дисплей m

10930 remote feed control
d Fernvorschubsteuerung f
f avance f télécommandée
r дистанционное регулирование n подачи

10931 remote indication
d Fernanzeige f
f indication f à distance
r вынесенная индикация f

10932 remote indicator
d Fernanzeiger m
f téléindicateur m
r дистанционный указатель m

10933 remote input
d Ferneingabe f
f introduction f de données à distance
r дистанционный ввод m

10934 remote inquiry
d Fernabfrage f
f interrogation f à distance
r дистанционный запрос m

10935 remote inquiry unit adapter
d Adapter m für Fernabfrageeinheit
f adapteur m pour unité d'interrogation à
  distance
r адаптер m устройства дистанционного
  опроса

* remote level indicator → 7234

10936 remotely controlled
d ferngesteuert
f commandé à distance
r дистанционно-управляемый

10937 remote maintenance
d Fernwartung f
f maintenance f à distance
r дистанционное обслуживание n

**10938 remote measurement; distance
measurement**
d Fernmessung f
f télémesure f; mesure f à distance
r дистанционное измерение n

10939 remote measurement feedback converter
d Fernmesswandler m; Fernmessumsetzer m
f convertisseur m de télémesure
r телеметрический преобразователь m

10940 remote monitoring; remote supervision
d Fernüberwachung f
f surveillance f à distance
r дистанционный контроль m

10941 remote pressure controller
d Ferndruckregler *m*
f régulateur *m* de pression à distance
r дистанционный регулятор *m* давления

10942 remote program control; programmed telemetry control
d Programmfernsteuerung *f*; programmierte Fernsteuerung *f*
f télécommande *f* suivant un programme; commande *f* programmée à distance
r дистанционное управление *n* программированием; программированное телеуправление *n*

10943 remote selsyn transmission
d Drehmelderfernübertragung *f*
f transmission *f* à distance par selsyn
r дистанционная сельсинная передача *f*

10944 remote signalling
d Fernmeldung *f*
f signalisation *f* à distance
r дистанционная сигнализация *f*

* **remote station** → 10946

* **remote supervision** → 10940

10945 remote temperature control
d Ferntemperaturregelung *f*
f régulation *f* de température à distance
r дистанционное регулирование *n* температуры

10946 remote terminal; remote station
d Fernterminal *n*; entfernt aufgestellte Datenendstation *f*
f terminal *m* à distance; station *f* à distance
r удаленный терминал *m*; удаленная станция *f*

* **remote transfer** → 12266

* **remote transmission** → 12266

* **removable** → 9760

10947 removable singularity
d behebbare Singularität *f*
f singularité *f* fausse
r устранимая особенность *f*

10948 rendering inert
d Inertisierung *f*
f inertisation *f*
r инертизация *f*

10949 renewal theory
d Erneuerungstheorie *f*
f théorie *f* de renouvellement
r теория *f* востановления

* **repeatability** → 10965

10950 repeat cycle
d Zykluswiederholung *f*
f reprise *f* du cycle
r цикл *m* повторения

10951 repeater station
d Relaisstelle *f*; Zwischenstation *f*
f station *f* de relais
r ретранслятор *m*

10952 repeating functional unit
d Wiederholfunktionseinheit *f*
f unité *f* de fonction de répétition
r функциональное устройство *n* воспроизведения

* **repeat motor** → 9491

10953 repeat scanning
d nochmalige Abtastung *f*
f balayage *m* répété
r повторное сканирование *n*

10954 repetition accuracy of prototype
d Wiederholgenauigkeit *f* des Prototyps
f précision *f* de répétition de prototype
r точность *f* повторения прототипа

10955 repetition rate
d Folgefrequenz *f*
f fréquence *f* de répétition; fréquence de récurrence
r частота *f* посылок [импульсов]

10956 replacement instruments
d austauschbare Geräte *npl*
f instruments *mpl* interchangeables
r взаимозаменяемые приборы *mpl*

10957 replacement model
d Ersatzmodell *n*
f modèle *m* équivalent
r модель *f* замены

10958 replacement theory
d Ersatztheorie *f*
f théorie *f* de replacement
r теория *f* замещений

10959 replacing discrete logic
d Ersatz *m* diskreter Logikbausteine

 f remplacement *m* de modules logiques
  discrets
 r замена *f* дискретной логики

10960 replay circuit
 d Antwortschaltkreis *m*
 f circuit *m* de réponse
 r схема *f* ответа

10961 replay pulse
 d Antwortimpuls *m*
 f impulsion *f* de réponse
 r ответный импульс *m*

10962 representation oriented acquisition
 d darstellungsorientierte Erfassung *f*
 f acquisition *f* orientée sur la représentation
 r представительно-ориентированный
  сбор *m*

10963 representative parameter
 d repräsentativer Parameter *m*
 f paramètre *m* représentatif
 r характерный параметр *m*

10964 reprocessing
 d Aufarbeitung *f*
 f remaniement *m*
 r переработка *f*

10965 reproducibility; repeatability
 d Reproduzierbarkeit *f*; Repetierbarkeit *f*
 f reproductibilité *f*; répétabilité *f*
 r воспроизводимость *f*

10966 reproducibility degree
 d Reproduzierbarkeitsgrad *m*
 f degré *m* de reproductibilité
 r степень *f* воспроизводимости

10967 reproducible
 d reproduzierbar
 f reproductible
 r воспроизводимый

10968 reproduction voltage
 d Wiedergabespannung *f*
 f tension *f* de reproduction
 r напряжение *n* воспроизведения

10969 request *v*
 d anfordern; anfragen
 f demander; exiger
 r запрашивать; требовать

10970 request repeat system
 d Übertragungssystem *n* mit Wiederholung;
  Abfragesystem *n* mit Wiederholung

 f système *m* de requête à répétition
 r система *f* циклического опроса

10971 request-repeat system
 d System *n* mit automatischer
  Wiederholungsanforderung
 f système *m* à demande de répétition
 r система *f* с автоматическим переспросом

10972 request signal
 d Anforderungssignal *n*
 f signal *m* de demande
 r сигнал *m* запроса

10973 request to send
 d Sendeanforderung *f*
 f demande *f* d'émission
 r запрос *m* передачи

10974 required object position
 d erforderliche Objektlage *f*
 f position *f* d'objet requise
 r требуемое положение *n* объекта

10975 required value
 d Sollwert *m*
 f valeur *f* désirée
 r искомая величина *f*

10976 requirement description
 d Bedarfsdeskription *f*
 f description *f* des nécessités
 r описание *n* требований; техническое
  задание *n*

10977 rerun
 d Wiederholungslauf *m*
 f marche *f* de répétition
 r повторный пуск *m*

10978 rerun *v*
 d wiederablaufen
 f répéter la marche
 r запускать повторно

10979 rerun routine; rollback routine
 d Wiederholungsprogramm *n*
 f routine *f* de répétition
 r программа *f* повторения;
  востанавливающая программа

10980 research
 d Forschung *f*; Suche *f*
 f recherche *f*
 r исследование *n*; поиск *m*

10981 research centre
 d Forschungszentrum *n*

f centre *m* de recherches
r исследовательский центр *m*

10982 research method
d Forschungsmethode *f*
f méthode *f* de recherche
r метод *m* исследования

10983 research-oriented processing
d forschungsorientierte Verarbeitung *f*;
wissenschaftlich-technische
Datenverarbeitung *f*
f traitement *m* orienté recherche
r обработка *f* результатов научных
исследований

10984 research work
d Forschungsarbeit *f*
f travail *m* de recherche
r исследовательская работа *f*

* **reserve protection** → 1578

10985 reservoir capacitor
d Speicherkondensator *m*
f condensateur *m* réservoir
r накопительный конденсатор *m*

* **reset** → 10989

10986 reset circuit
d Rückstellkreis *m*
f circuit *m* de remise
r цепь *f* повторного включения

10987 reset component
d Integralgröße *f*; Nachstellglied *n*
f composante *f* intégrale
r интегральная составляющая *f*

10988 reset condition
d Nullstellungszustand *m*
f état *m* de zéro
r предпусковой режим *m*

* **reset controller** → 1052

10989 reset[ing]
d Neueinstellung *f*; Nulleinstellung *f*;
Rückstellung *f*; Auslösung *f*
f remise *f* à zéro; réenclenchement *m*
r возврат *m* в исходное положение;
установка *f* нуля

* **reset of hardware system** → 1572

* **reset of software system** → 1573

10990 reset pulse
d Rückstellimpuls *m*; Löschimpuls *m*;
Nullimpuls *m*
f impulsion *f* de remise [à zéro]
r импульс *m* сброса

10991 resetting value
d Rückgangswert *m*
f valeur *f* de retour
r параметр *m* возврата

10992 resident software
d residente Systemunterlagen *fpl*
f logiciels *mpl* résidents
r резидентное программное обеспечение *n*

10993 resistance balance system
d Widerstandsgleichsystem *n*
f système *m* d'équilibrage à résistances
r резистивная уравновешивающая система *f*

10994 resistance controller
d Widerstandsregler *m*
f régulateur *m* à résistances
r реостатный регулятор *m*

* **resistance drop** → 8909

10995 resistance strain gauge
d Widerstandsdehnungsmessstreifen *m*
f jauge *f* de contrainte à résistance
r тензометр *m* сопротивления

10996 resistance teletransmitter
d Widerstandsferngeber *m*
f télétransmetteur *m* à résistance
r резистивный телепередатчик *m*

10997 resistance temperature detector
d Widerstandstemperaturwandler *m*;
Widerstandstemperaturdetektor *m*
f palpeur *m* résitif de température
r резистивный термочувствительный
элемент *m*

10998 resistance transmitters adapter
d Adapter *m* für Widerstandsgeber
f adapteur *m* pour transmetteurs à résistance
r адаптер *m* для датчиков сопротивления

10999 resistive component
d ohmsche Komponente *f*
f composante *f* ohmique
r резистивный компонент *m*

11000 resistive feedback
d Widerstandsrückkopplung *f*
f réaction *f* à résistance
r резистивная обратная связь *f*

* resistive load → 8911

11001 resistor coupling
   d Widerstandskopplung *f*
   f couplage *m* par résistances
   r реостатная связь *f*

* **resolution capability** → 11005

11002 resolution function
   d Auflüsungsfunktion *f*
   f fonction *f* de résolution
   r функция *f* разрешения

* **resolution power** → 11005

11003 resolver
   d Resolver *m*
   f résolveur *m*
   r решающее устройство *n*

11004 resolver potentiometer
   d Resolverpotentiometer *n*; Potentiometer *n* für
      Koordinatenwandlung
   f résolveur *m* potentiométrique
   r потенциометр *m* преобразования
      координат

**11005 resolving power; resolution power;
         resolution capability**
   d Auflösungsvermögen *n*
   f pouvoir *m* résolvant; pouvoir de résolution
   r разрешающая способность *f*

11006 resonance amplifier; tuned amplifier
   d Resonanzverstärker *m*
   f amplificateur *m* à résonance
   r резонансный усилитель *m*

11007 resonance circuit
   d Resonanzschaltung *f*
   f circuit *m* à résonance
   r резонансный контур *m*

11008 resonance curve
   d Resonanzkurve *f*
   f courbe *f* de résonance
   r резонансная кривая *f*

11009 resonance frequency meter
   d Resonanzfrequenzmesser *m*
   f fréquencemètre *m* à résonance
   r резонансный частотомер *m*

11010 resonance measuring method
   d Resonanzmessverfahren *n*
   f méthode *f* de mesure à résonance
   r резонансный метод *m* измерения

11011 resonant frequency
   d Resonanzfrequenz *f*
   f fréquence *f* de résonance
   r резонансная частота *f*

11012 resonant shunt
   d Resonanznebenschluß *m*
   f shunt *m* résonnant
   r резонансный шунт *m*

* **resonator circuit** → 9231

11013 resource
   d Ressource *f*
   f ressource *f*
   r ресурс *m*

11014 resource allocation processor
   d Ressourcen-Zuordnungsprozessor *m*
   f processeur *m* d'attribution de ressources
   r процессор *m* распределения ресурсов

11015 resource control
   d Ressourcensteuerung *f*
   f commande *f* des ressources
   r управление *n* ресурсами

11016 responder
   d Antwortsender *m*; Antwortbake *f*
   f répondeur *m*
   r ответчик *m*

11017 responding value
   d Ansprechwert *m*
   f valeur *f* minimale; valeur-seuil *f*
   r величина *f* на выходе

11018 response
   d Antwort *f*
   f réponse *f*
   r ответ *m*

* **response capacity** → 5

11019 response limit
   d Ansprechgrenze *f*
   f limite *f* de réponse
   r граница *f* чувствительности

11020 response mode
   d Antwortbetriebsart *f*
   f mode *m* de réponse
   r асинхронный ответ *m*

11021 response threshold
   d Ansprechschwelle *f*
   f seuil *m* de réponse
   r порог *m* чувствительности

* **response time** → 9009

11022 response time error
d Ansprechzeitfehler *m*
f erreur *f* de temps de réponse
r ошибка *f* времени реакции

11023 response to unit impulse
d Einheitsimpulsreaktion *f*
f réponse *f* à impulsion unitaire
r реакция *f* на единичный импульс

11024 response voltage
d Ansprechspannung *f*
f tension *f* de réponse
r напряжение *n* срабатывания

11025 responsive time constant
d Ansprechzeitkonstante *f*
f constante *f* de temps de réponse
r постоянная *f* времени срабатывания

11026 rest-current release
d Ruhestromauslöser *m*
f déclencheur *m* à courant de repos
r выключатель *m* тока покоя

11027 rest duration
d Aussetzdauer *f*
f durée *f* de repos
r промеждуток *m* времени между двумя
  импульсами

* **resting contact** → 1924

11028 restoring torque
d Rückstellmoment *n*; Richtmoment *n*
f couple *m* antagoniste; couple de rappel
r противодействующий момент *m*

11029 rest position
d Ruhestellung *f*
f position *f* de repos
r положение *n* покоя

* **restrictor** → 2444

11030 restrictor valve
d Drosselventil *n*
f valve *f* d'étranglement
r ограничивающий клапан *m*

11031 restriking voltage
d Wiederzündspannung *f*
f tension *f* de rallumage
r потенциал *m* повторного зажигания

11032 result of research

d Forschungsergebnis *n*
f résultat *m* de recherche
r результат *m* исследования

11033 result recording
d Aufzeichnung *f* von Ergebnissen
f enregistrement *m* graphique des résultats
r [графическая] запись *f* результатов

* **retardation** → 3903

11034 retarded argument
d retardiertes Argument *n*
f argument *m* retardé
r запаздывающий аргумент *m*

11035 retarded control
d verzögerte Regelung *f*
f réglage *m* à retard
r регулирование *n* с запаздыванием

11036 retrieval
d Wiederherstellung *f* verzerrter Information
f rétablissement *m* d'information déformée
r воспроизведение *n* искаженной
  информации

11037 return
d Rückkehr *f*; Rücksprung *m*
f retour *m*; saut *m* de retour
r возврат *m*

11038 return *v*
d zurückkehren; zurückführen
f retourner
r возвращать[ся]

11039 return laser beam
d Laserrückstrahl *m*
f faisceau *m* de retour du laser
r отраженный лазерный луч *m*

* **return trace** → 1575

* **return transfer function** → 6980

11040 reversal control
d Umkehrsteuerung *f*
f commande *f* réversible
r управление *n* обратным ходом

11041 reversal method
d Inversionsmethode *f*
f méthode *f* d'inversions
r метод *m* реверсирования

11042 reversal of phase; phase reversal
d Phasenumkehrung *f*

f inversion *f* de phase
r обращение *n* фазы

11043 reverse-acting control element
d Umkehrsteuerelement *n*
f élément *m* de commande à action inverse
r регулирующий элемент *m* обратного
      действия

11044 reverse conduction current
d Übergangsrückstrom *m*;
      Übergangsgegenstrom *m*
f courant *m* de conduction inverse
r ток *m* обратной проводимисти

11045 reverse conduction voltage
d Übergangsgegenspannung *f*
f tension *f* de conduction inverse
r напряжение *n* обратной проводимисти

11046 reverse current
d Rückstrom *m*
f courant *m* inverse
r обратный ток *m*

11047 reverse direction
d Rücklaufrichtung *f*
f direction *f* retour
r обратное направление *n*

*   **reversed motion → 1575**

11048 reverse phase
d Gegenphase *f*
f phase *f* inverse
r противоположная фаза *f*

11049 reverse-power relay
d Rückleistungsrelais *n*
f relais *m* à retour de puissance
r реле *n* обратной мощности

*   **reverse run → 1575**

11050 reversibility
d Reversibilität *f*
f réversibilité
r обратимость *f*

11051 reversibility degree
d Reversibilitätsgrad *m*
f degré *m* de réversibilité
r степень *f* обратимости

11052 reversible
d umkehrbar
f reversible
r реверсируемый; обратимый

11053 reversible booster
d Umkehrspannungserhöher *m*
f survolteur *m* réversible
r обратимый бустер *m*

11054 reversible control
d Reversiersteuerung *f*
f commande *f* réversible
r реверсируемое управление *n*

11055 reversible process
d umkehrbarer Vorgang *m*
f processus *m* reversible
r обратимый процесс *m*

11056 revolution indicator
d Drehzahlmesser *m*
f compte-tours *m*
r указатель *m* числа оборотов

11057 rewriting device
d Überschreibungseinrichtung *f*
f enregistreur-récepteur *m*
r перезаписывающее устройство *n*

11058 rhythm of production
d Produktionsrhythmus *m*
f rythme *m* de production
r ритм *m* производства

11059 Riemann integrable
d Riemann-integrierbar
f intégrable en sens de Riemann
r интегрируемый по Риману

11060 Riemann integral
d Riemannsches Integral *n*
f intégrale *f* de Riemann
r интеграл *m* Римана

*   **rigid feedback → 10360**

11061 rigid feedback controller
d Regler *m* mit starrer Rückführung
f régulateur *m* à contreréaction rigide
r регулятор *m* с жёсткой обратной связью

11062 rigid limitation
d starre Begrenzung *f*
f limitation *f* rigide
r жёсткое ограничение *n*

11063 rigid linkage
d starre Verkettung *f*
f enchaînement *m* rigide
r жёсткая связь *f*

11064 ring-balance differential manometer
d Differentialringmanometer *n*

f manomètre *m* différentiel à tore pendulaire
r кольцевой дифференциальный
манометр *m*

* **ringing relay → 2056**

11065 ripple potential difference
d Brummspannungsdifferenz *f*; welliger
Gleichspannungsunterschied *m*
f différence *f* de tension d'ondulation
r разность *f* потенциалов пульсаций

11066 ripple voltage
d Brummspannung *f*; wellige Spannung *f*
f tension *f* d'ondulation
r [слабо] пульсирующее напряжение *n*

11067 rise speed
d Anstieggeschwindigkeit *f*
f vitesse *f* de montée
r скорость *f* нарастания

* **rise time → 1967**

11068 rise time et maximal amplitude
d Anstiegszeit *f* bei maximaler Amplitude
f temps *m* de montée lors de l'amplitude
maximum
r время *n* нарастания при максимальной
амплитуде

* **robot → 6517**

11069 robot acceptance checking
d Roboterabnahmeprüfung *f*
f contrôle *m* de réception des robots
r приёмочные испытания *npl* роботов

11070 robot action system
d Roboteraktionssystem *n*
f système *m* d'action de robot
r система *f* воздействия робота

11071 robot-aided automatic assembly
d robotergestützte automatische Montage *f*
f assemblage *m* à l'aide du robot
r автоматическая сборка *f* с использованием
робота

11072 robot application characteristic
d Robotereinsatzcharakteristik *f*
f caractéristique *f* d'emploi de robots
r характеристика *f* применения роботов

11073 robot application condition
d Robotereinsatzbedingung *f*
f condition *f* d'application de robot
r условия *npl* применения промышленных

роботов

11074 robot assembly system
d Robotermontagesystem *n*; Montagesystem *n*
mit Robotern
f système *m* d'assemblage de robot
r сборочная система *f* с использованием
роботов

11075 robot automation
d Roboterautisierung *f*; Automatisierung *f*
mittels Industrierobotern
f automatisation *f* de robot; robotisation *f*
r автоматизация *f* с помощью
[промышленных] роботов

11076 robot building-block system
d Roboterbaukastensystem *n*
f système *m* de construction par blocs pour
robots
r модульная система *f* робота

11077 robot checking control system
d Roboterkontrollsteuerungssystem *n*;
Kontroll-Steuerungssystem *n* von Robotern
f système *m* de contrôle-commande de robots
r система *f* контроля управления роботом

11078 robot computer control
d Roboterrechnersteuerung *f*
f commande *f* d'un calculateur de robot
r система *f* управления роботом с помощью
компьютера

11079 robot control algorithm
d Robotersteueralgorithmus *m*
f algorithme *m* de commande de robot
r алгоритм *m* управления роботом

**11080 robot control by means of programmable
automatons**
d Robotersteuerung *f* mittels programmierbarer
Automaten
f commande *f* de robots par automates
programmables
r управление *n* роботами от
программируемых автоматов

11081 robot-controlled process
d robotergesteuerter Prozess *m*
f procédé *m* commandé par robot
r [технологический] процесс m,
управляемый роботом

**11082 robot control unit; control unit of
industrial robot**
d Robotersteuereinheit *f*; Robotersteuergerät *n*;
Steuereinheit *f* eines Industrieroboters

   f unité *f* de commande d'un robot [industriel]
   r блок *m* управления [промышленным] роботом

    * **robot drive** → 6518

11083 robot drive efficiency
   d Roboterantriebswirkungsgrad *m*
   f efficience *f* d'entraînement de robot
   r коэффициент *m* полезного действия привода робота

11084 robot for nuclear problems
   d Roboter *m* für nukleare Probleme
   f robot *m* pour problèmes nucléaires
   r робот *m* для ядерной техники

11085 robot gear plan
   d Robotergetriebeplan *m*
   f plan *m* d'engrenage de robot
   r кинематическая схема *f* робота

    * **robot hardware solution** → 6519

11086 robotic problems
   d Robotikprobleme *npl*
   f problèmes *mpl* de robotique
   r проблемы *fpl* робототехники

11087 robotics; industrial robot[s] technique
   d Robotertechnik *f*; Robotik *f*; Industrierobotertechnik *f*
   f robotique *f* [industrielle]
   r робототехника *f*; робботика *f*

11088 robotic system
   d Robotiksystem *n*
   f système *m* robotique
   r система *f* робототехники

11089 robot information electronics
   d Roboterinformationselektronik *f*; Informationselektronik *f* eines Roboters
   f électronique *f* d'information d'un robot
   r информационная электроника *f* робота

11090 robot information processing
   d Roboterinformationsverarbeitung *f*; Informationsverarbeitung *f* eines Industrieroboters
   f traitement *m* d'information de robot [industriel]
   r обработка *f* информации [промышленного] робота

11091 robot-integrated assembly system
   d roboterintegriertes Montagesystem *n*
   f système *m* d'assemblage intégré sur le robot

   r интегральная сборочная система *f* робота

11092 robotization
   d Robotisierung *f*; Roboterisierung *f*
   f robotisation *f*
   r роботизация *f*

11093 robotized assemblage system
   d robotisiertes Montagesystem *n*
   f système *m* robotisé d'assemblage
   r роботизированная сборочная система *f*

11094 robotized checking system
   d robotisiertes Kontrollsystem *n*
   f système *m* de contrôle robotisé
   r роботизированная система *f* контроля

11095 robotized handling
   d robotisierte Handhabung *f*
   f manutention *f* robotisée
   r роботизированное манипулирование *n*

11096 robot memory; robot store
   d Roboterspeicher *m*; Speicher *m* eines Industrieroboters
   f mémoire *f* de robot
   r запоминающее устройство *n* робота

11097 robot operating possibility
   d Roboterfunktionsmöglichkeit *f*
   f possibilité *f* de fonctionnement de robot
   r операционные возможности *fpl* робота

    * **robot operation** → 936

11098 robot operation performance
   d Roboterbetriebsverhalten *n*
   f comportement *m* d'opération d'un robot
   r рабочая характеристика *f* робота

11099 robot process computer
   d Roboterprozessrechner *m*
   f calculateur *m* de procédé de robot
   r управляющий компьютер *m* робота

11100 robot process information
   d Roboterprozessinformation *f*
   f information *f* de procédé de robot
   r управляющая информация *f* робота

11101 robot program control system
   d Roboterprogrammsteuerungssystem *n*
   f système *m* de commande pour programme de robot
   r система *f* программного управления роботом

    * **robot regulating system** → 6520

   * **robot regulator** → 6521

11102 robot regulating technology
  d Roboterregelungstechnologie *f*
  f technologie *f* de régulation de robot
  r технология *f* регулирования робота

   * **robot store** → 11096

11103 robot study
  d Roboterstudie *f*
  f étude *f* de robot
  r анализ *m* робота

11104 robot supervision system
  d Roboterüberwachungssystem *n*
  f système *m* de surveillance de robot
  r супервизорная система *f* робота

11105 robot test equipment
  d Roboterprüfeinrichtung *f*
  f équipement *m* de test d'un robot
  r устройство *n* для испытания робота

11106 robot with artificial intelligence
  d Roboter *m* mit künstlicher Intelligenz
  f robot *m* à intelligence artificielle
  r робот *m* с искусственным интеллектом

11107 robot with network control system
  d Roboter *m* mit Netzwerksteuerungssystem
  f robot *m* à système de commande de réseau
  r робот *m* с системой контурного управления

11108 robot with numerical structure
  d Roboter *m* mit numerischer Struktur
  f robot *m* à structure numérique
  r робот *m* с цифровой структурой

11109 rocking-contact speed regulator
  d Schwenkkontaktgeschwindigkeitsregler *m*
  f régulateur *m* de vitesse à contact basculant
  r регулятор *m* скорости с качающимися контактами

   * **rollback routine** → 10979

11110 root
  d Wurzel *f*
  f racine *f*
  r корень *m*

   * **root exponent** → 5314

11111 root-locus method
  d Wurzelortverfahren *n*
  f méthode *f* du lieu des pôles; méthode du lieu des racines
  r метод *m* корневого годографа

11112 root-mean square
  d quadratischer Mittelwert *m*
  f valeur *f* moyenne quadratique
  r среднеквадратичное значение *n*

11113 rotary actuator
  d rotierender Effektor *m*
  f élément *m* rotatif de commande
  r вращательный привод *m*

11114 rotary amplifier control of electric drives
  d Steuerung *f* eines Elektroantriebes mittels Drehverstärker
  f commande *f* d'entraînement électromécanique par amplificateur rotatif
  r электромашинное управление *n* электроприводом

11115 rotary impulse
  d Drehimpuls *m*
  f quantité *f* de mouvement angulaire
  r импульс *m* вращения

11116 rotary switch
  d Drehschalter *m*; Umlaufschalter *m*
  f commutateur *m* rotatif
  r вращающий переключатель *m*; пакетный выключатель *m*

11117 rotary table
  (for feeding of work pieces)
  d Drehteller *m*
  f plateau *m* tournant
  r ротационный питатель *m*

11118 rotating converter
  d Drehumformer *m*
  f convertisseur *m* rotatif
  r вращающийся преобразователь *m*

11119 rotation[al] speed
  d Umlaufgeschwindigkeit *f*
  f vitesse *f* de rotation
  r скорость *f* вращения

11120 rotational viscosimeter
  d Rotationsviskosimeter *n*
  f viscosimètre *m* rotatif
  r ротационный вискозиметр *m*

   * **rotation speed** → 11119

11121 rotation unit
  d Rotationseinheit *f*; Dreheinheit *f*

 f unité *f* de rotation
 r блок *m* вращения

11122 rotatoric manipulator drive
 d rotatorischer Manipulatorantrieb *m*
 f entraînement *m* rotatoire de manipulateur
 r вращательный привод *m* манипулятора

11123 rough approximation
 d grobe Annäherung *f*
 f approximation *f* rude
 r грубое приближение *n*

11124 rounding error
 d Rundungsfehler *m*
 f erreur *f* d'arrondissement
 r ошибка *f* округления

11125 roundness measuring instrument
 d Rundungsmessgerät *n*
 f instrument *m* pour mesurer la courbure
 r прибор *m* для измерения округлости

11126 Routh criterion
 d Routhsches Kriterium *n*
 f critère *m* de Routh
 r критерий *m* Payca

11127 Routh inequality
 d Routhsche Ungleichung *f*
 f inégalité *f* de Routh
 r неравенство *n* Руса

11128 routine operation
 d Routineoperation *f*
 f opération *f* de routine
 r рутинная операция *f*

11129 routing indicator
 d Leitweganzeiger *m*
 f indicateur *m* d'acheminement
 r индикатор *m* маршрута

**11130 run-off control; series control; sequential
    control**
 d Ablaufsteuerung *f*; Folgesteuerung *f*;
   sequentielle Steuerung *f*
 f commande *f* séquentielle
 r последовательное управление *n*

S

11131 safe reactor control
d sichere Reaktorregelung *f*
f commande *f* sûre du réacteur
r безопасное управление *n* реактором

11132 safety
d Sicherheit *f*
f sécurité *f*; sûreté *f*
r безопасность *f*

11133 safety belt with automatic suspension
d Sicherheitsgurt *m* mit automatischer Aufhängung
f ceinture *f* de sûreté à suspension automatique
r предохранительный пояс *m* с автоматической подвесной системой

11134 safety circuit
d Sicherheitskreis *m*
f circuit *m* de sûreté
r схема *f* предохранения

11135 safety factor
d Sicherheitsfaktor *m*
f coefficient *m* de sécurité
r коэффициент *m* безопасности

11136 safety factor for drop-out
d Abfallsicherheitsfaktor *m*
f facteur *m* de sécurité pour la mise au repos
r коэффициент *m* запаса выпадения

11137 safety factor for holding
d Haltesicherheitsfaktor *m*
f facteur *m* de sécurité au maintien
r коэффициент *m* запаса для удержания

11138 safety factor for pick-up
d Ansprechsicherheitsfaktor *m*
f facteur *m* de sécurité pour la mise au travail
r коэффициент *m* надёжности чувствительного элемента

11139 safety function
d Sicherheitsfunktion *f*
f fonction *f* de sûreté
r функция *f* надёжности

11140 safety fuse
d Sicherung *f*; Schmeizsicherung *f*
f fusible *m*; fusible protecteur

r [плавкий] предохранитель *m*

11141 safety in data transmission
d Sicherheit *f* der Datenübertragung
f sécurité *f* de transmission des données
r надёжность *f* передачи данных

11142 safety interlock
d Sicherheitssperre *f*; Sicherheitsverriegelung *f*
f verrouillage *m* de sécurité
r предохранительная блокировка *f*

* **safety of operation** → 5133

11143 safety regulator
d Sicherheitsregler *m*
f régulateur *m* de sûreté
r предохранительный регулятор *m*

11144 safety relay
d Sicherheitsrelais *n*
f relais *m* de sûreté
r реле *n* защиты

11145 safety sensor
d Sicherheitssensor *m*
f senseur *m* de sûreté
r сенсор *m* безопасности

11146 safety-technical device
d sicherheitstechnische Einrichtung *f*
f dispositif *m* de sûreté technique
r предохранительное приспособление *n*

11147 safety technology
d Sicherheitstechnik *f*
f technique *f* de sécurité
r техника *f* безопасности

11148 safety valve
d Sicherheitsventil *n*
f soupape *f* de sûreté
r предохранительный вентиль *m*

11149 sag adjustment
d Durchhangseinstellung *f*
f réglage *m* de flèche
r установка *f* провеса

11150 sample
d Muster *n*; Probe *f*
f échantillon *m*
r замер *m*; проба *f*

11151 sampled data
d Abtastwerte *mpl*
f données *fpl* échantillonnées
r данные *pl* развёртки

11152 sampled-data control
   d Abtastregelung *f*
   f réglage *m* par impulsions échantillonnées
   r импульсное регулирование *n*

* **sampled-data controller → 9522**

11153 sampled-data control system
   d Abtastregelsystem *n*
   f système *m* de réglage par échantillons
   r импульсная система *f* регулирования

11154 sampled-data filter
   d Filter *n* mit Abtastung
   f filtre *m* de données d'exploration
   r фильтр *m* с квантованием

11155 sampled-data system
   d Tastsystem *n*; System *n* mit
       Zeitquantisierung
   f système *m* d'exploration; système de données
       d'exploration
   r система *f* прерывистого действия

11156 sampler
   d Abtaster *m*; Abtasteinrichtung *f*;
       Abtastglied *n*
   f lecteur *m*; dispositif *m* d'exploration;
       explorateur *m*; palpeur *m*
   r квантизатор *m*

11157 sample space
   d Stichprobenraum *m*; Probenraum *m*
   f espace *m* d'échantillonnage
   r выборочное пространство *n*

11158 samples per second
   d Stichproben *fpl* pro Sekunde
   f échantillons *fpl* par seconde
   r выборки *fpl* в секунде

* **sampling → 11159**

11159 sampling [action]
   d Abtastung *f*; Signalprobenabnahme *f*
   f échantillonnage *m*; prise *f* d'impulsions
       échantillons
   r отбор *m* проб; отбор образцов

11160 sampling analysis
   d Stichprobenanalyse *f*; Prüfpunktanalyse *f*
   f analyse *f* par échantillonage
   r выборочный метод *m* анализа; анализ *m*
       отобранных образцов

11161 sampling circuit
   d Abtastkreis *m*; Samplingkreis *m*
   f circuit *m* d'échantillonnage

11162 sampling interval
   d Abtastintervall *n*
   f intervalle *m* d'échantillonage
   r интервал *m* измерений

11163 sampling oscillograph
   d Samplingoszillograf *m*; Abtastoszillograf *m*
   f oscillographe *m* à échantillonnage
   r стробирующий осциллограф *m*

11164 sampling period
   d Abtastperiode *f*
   f période *f* d'échantillonnage
   r цикл *m* отбора проб

11165 sampling rate
   d Abtastgeschwindigkeit *f*
   f vitesse *f* d'exploration
   r скорость *f* квантования

11166 sampling servosystem
   d Abtastservosystem *n*
   f système *m* asservi échantillonneur
   r импульсная следящая система *f*

11167 sampling synchronization
   d Samplingsynchronisation *f*
   f synchronisation *f* sampling
   r синхронизация *f* развёртки

11168 sampling test
   d Stichprobenprüfung *f*
   f contrôle *m* [effectué] par prélèvement
   r выборочное испытание *n*

11169 sampling theorem
   d [Shannonsches] Abtasttheorem *n*; Theorem *n*
       von Shannon
   f théorème *m* de balayage [de Shannon];
       théorème d'exploration [de Shannon];
       théorème de Shannon
   r теорема *f* квантования

11170 sampling time; scanning time
   d Abtastzeit *f*
   f temps *m* d'exploration
   r время *n* развёртки

11171 sanity
   d Intaktheit *f*
   f état *m* sain; état de marche
   r исправность *f*

11172 satisfaction
   d Erfüllung *f*

f satisfaction *f*
r выполнение *n*; удовлетворение *n*

11173 satisfiability
d Erfüllbarkeit *f*
f satisfaisabilité *f*
r выполнимость *f*

11174 satisfy *v*
d erfüllen; befriedigen
f satisfaire
r удовлетворять

11175 saturable magnetometer
d Magnetometer *n* mit sättigungsfähigem Kern
f magnétomètre *m* à noyau saturable
r магнитометр *m* с насыщением

11176 saturated; saturating
d saturiert; gesättigt
f saturé
r насыщенный

* **saturating → 11176**

11177 saturation
d Sättigung *f*
f saturation *f*
r насыщение *m*

11178 saturation level
d Sättigungspegel *m*
f palier *m* de saturation
r уровень *m* насыщения

11179 saturation non-linearity
d Sättigungsinchtlinearität *f*
f non-linéarité *f* de saturation
r нелинейность *f* насыщения

11180 saturation process
d Sättigungsprozess *m*
f processus *m* de saturation
r процесс *m* насыщения

11181 saturation reactor
d Sättigungsdrossel *f*
f bobine *f* d'inductance saturable
r дроссель *m* с насыщением

11182 saturation state
d Sättigungszustand *m*
f régime *m* de saturation
r режим *m* насыщения

* **saturation zone → 12998**

11183 saw-tooth converter
d Sägezahnumwandler *m*
f convertisseur *m* de tension en dents de scie
r преобразователь *m* пилообразного сигнала

11184 saw-tooth current generator
d Sägezahnstromgenerator *m*
f générateur *m* à courant en dents de scie
r генератор *m* пилообразного тока

11185 saw-tooth wave form generator
d Generator *m* für sägezahnförmige Schwingungen
f générateur *m* en dents de scie
r генератор *m* пилообразных сигналов

11186 scalar axis
d skalare Achse *f*
f axe *m* scalaire
r скалярная ось *f*

11187 scalar fielad theory
d skalare Feldtheorie *f*
f théorie *f* du champ scalaire
r теория *f* скалярного поля

11188 scalar function
d Skalarfunktion *f*
f fonction *f* scalaire
r скалярная функция *f*

11189 scalar quantity
d skalare Größe *f*; Skalar *m*
f grandeur *f* scalaire
r скалярная величина *f*

11190 scalar variational problem
d skalares Variationsproblem *n*
f problème *m* de variation scalaire
r скалярная вариационная задача *f*

11191 scale
d Skale *f*; Maßstab *m*
f échelle *f*
r шкала *f*; масштаб *m*

11192 scale division
d Skalenteilung *f*
f graduation *f*
r деление *n* шкалы

11193 scale interval
d Skalenintervall *n*
f intervalle *m* d'échelle
r интервал *m* шкалы

11194 scale positioning
 d Skaleneinstellung *f*
 f mise *f* au point d'échelle graduée
 r юстировка *f* шкалы

11195 scale time
 d Skalendurchlaufzeit *f*
 f durée *f* de parcours d'échelle
 r время *n* прохождения

11196 scale unit
 d Skaleneinheit *f*
 f unité *f* d'échelle
 r единица *f* шкалы

11197 scale zero
 d Nullpunkt *m* der Skale
 f zéro *m* d'échelle
 r нуль *m* шкалы

11198 scaling
 d Maßstabänderung *f*;
     Zahlenbereichsänderung *f*
 f introduction *f* de l'échelle
 r пересчёт *m*

11199 scanner; scanning unit
 d Abtastblock *m*; Abtaster *m*; Zerleger *m*;
     Scanner *m*
 f bloc *m* de balayage; bloc *m* d'exploration;
     balayeur *m*; scanner *m*
 r развёртывающее устройство *n*; сканер *m*;
     сканирующее устройство

 * **scanning** → 1627

11200 scanning action
 d Abtastvorgang *m*
 f action *f* de balayage
 r процесс *m* сканирования

11201 scanning circuit
 d Abtaststrkrcis *m*
 f circuit *m* de balayage; circuit d'exploration
 r схема *f* развёртки

11202 scanning control
 d Abtastregelung *f*
 f commande *f* à échantillonnage
 r сканирующее управление *n*

11203 scanning cycle
 d Abtastzyklus *m*
 f cycle *m* de balayage
 r цикл *m* сканирования

11204 scanning laser radar
 d Laserabtastradar *n*; abtastendes

     Laserradargerät *n*
 f radar *m* explorateur à laser
 r сканирующий лазерный локатор *m*

11205 scanning linearization
 d Linearisierung *f* der Abtastung
 f linéarisation *f* du balayage
 r линеаризация *f* развёртки

11206 scanning signal input
 d Abtastsignaleingang *m*
 f entrée *f* du signal de balayage
 r входной сигнал *m* сканирования

11207 scanning spectrometer
 d Abtastspektrometer *n*
 f spectromètre *m* à balayage
 r сканирующий спектрометр *m*

11208 scanning speed
 d Abtastgeschwindigkeit *f*
 f vitesse *f* de balayage; vitesse d'exploration
 r скорость *f* сканирования

11209 scanning speed by sensors
 d Abtastgeschwindigkeit *f* durch Sensoren
 f vitesse *f* d'exploration par capteurs
 r скорость *f* сканирования с помощью
     датчиков

11210 scanning system
 d Abtastsystem *n*
 f système *m* de balayage
 r система *f* развёртки

11211 scanning technique
 d Abtastverfahren *n*
 f méthode *f* de balayage
 r метод *m* сканирования

 * **scanning time** → 11170

 * **scanning unit** → 11199

11212 scanning voltage
 d Abtastspannung *f*
 f tension *f* de balayage
 r напряжение *n* развёртки

11213 scanning X-ray microanalyzer
 d Abtaströntgenstrahlmikroanalysator *m*
 f microanalyseur *m* capteur à rayons X
 r ренгеновский сканирующий
     микроанализатор *m*

11214 scatter *v*
 d streuen; zerstreuen
 f disséminer; diffuser; disperser
 r разбрасывать; рассеивать

11215 scattering frequency
d Streufrequenz *f*
f fréquence *f* de dispersion
r частота *f* рассеяния

11216 scheduling model; sequencing model
d Ablaufplanungsmodell *n*;
Reihenfolgemodell *n*
f modèle *m* de déroulement
r модель *f* последовательностей

11217 schematic circuit; schematic diagram
d Grundschaltung *f*; Prinzipschaltung *f*;
Prinzipschaltbild *n*
f schéma *m* général; schéma de principe
r принипиальная схема *f*

* **schematic diagram** → 11217

* **schematic diagram** → 1838

11218 schematic model
d schematisches Modell *n*
f modèle *m* schématique
r схематическая модель *f*

11219 Schmidt orthonormalization
d Schmidtsches
Orthonormalisierungsverfahren *n*
f méthode *f* d'orthonormalisation de Schmidt
r метод *m* ортонормирования Шмидта

11220 scientific and technic calculations
d wissenschaftliche und technische
Berechnungen *fpl*
f calculs *mpl* techniques et scientifiques
r научные и технические расчёты *mpl*

11221 scientific instrumentation
d wissenschaftliche Gerätetechnik *f*;
wissenschaftliche Instrumentierung *f*
f instrumentation *f* scientifique
r приборы *mpl* для научных исследований

11222 scientific instrument manufacture
d wissenschaftlicher Gerätebau *m*
f fabrication *f* d'appareils scientifiques
r производство *n* приборов для научных
исследований

11223 scientific microcomputer
d wissenschaftlicher Mikrorechner *m*
f microcalculateur *m* scientifique
r микрокомпьютер *m* для научных
исследований

11224 screen device
d Bildschirmgerät *n*

f appareil *m* à écran
r устройство *n* визуального отображения

11225 screen factor
d Schirmfaktor *m*
f facteur *m* d'écran
r коэффициент *m* экранирования

11226 screen monitoring
d Bildschirmüberwachung *f*
f surveillance *f* d'écran
r мониторный контроль *m*

11227 screen-oriented data structure
d bildschirmorientierte Datenstruktur *f*
f structure *f* de données orientée sur l'écran de
vision
r структура *f* данных, ориентированная на
вывод на экран

11228 screen terminal
d Bildschirmterminal *n*
f terminal *m* à écran
r экранный терминал *m*

11229 search check; seek check
d Suchprüfung *f*
f test *m* de recherche
r контроль *m* поиска

11230 search circuit; finding circuit
d Suchschaltung *f*; Suchkreis *m*
f circuit *m* de recherche
r поисковая схема *f*

11231 search device
d Suchgerät *n*
f dispositif *m* de recherche
r поисковое устройство *n*

11232 searching operating; seek operation
d Suchvorgang *m*; Suchoperation *f*
f opération *f* de recherche
r операция *f* поиска

11233 searching technique
d Suchtechnik *f*
f technique *f* de recherche
r техника *f* поиска

11234 search loss
d Suchverlust *m*
f perte *f* de recherche
r потери *fpl* на поиск

11235 search memory
d Suchspeicher *m*

f mémoire *f* de recherche
r ассоциативная память *f*

11236 search process
d Suchprozess *m*
f procédé *m* de recherche
r процесс *m* поиска

11237 search strategy
d Suchstrategie *f*
f stratégie *f* de recherche
r методика *f* поиска

11238 search theory
d Suchtheorie *f*
f théorie *f* de recherche
r теория *f* поиска

11239 search time
d Suchzeit *f*
f temps *m* recherche
r время *n* поиска

11240 secants method
d Sekantenmethode *f*
f méthode *f* de sécantes
r метод *m* секущих

11241 secondary function
d Nebenfunktion *f*
f fonction *f* secondaire
r вторичная функция *f*

11242 secondary method
d Sekundärverfahren *n*
f procédé *m* secondaire
r вторичный метод *m*

11243 secondary radar
d Sekundärradar *n*
f radar *m* secondaire
r вторичный радиолокатор *m*

11244 secondary regulation
d Sekundärregelung *f*
f régulation *f* secondaire
r вторичное регулирование *n*

11245 secondary storage system
d Sekundärspeichersystem *n*
f système *m* secondaire de mémoire
r система *f* вторичной памяти

11246 secondary trip
d Sekundärauslöser *m*
f déclencheur *m* secondaire
r вторичный разъединитель *m*

11247 second derivative control
d Regelung *f* gemäß zweiter Ableitung
f réglage *m* par seconde dérivée
r регулирование *n* по ускорению

11248 second-order servo
d Servomechanismus *m* zweiter Ordnung
f servomécanisme *m* de seconde ordre
r следящая система *f* второго порядка

11249 sectional automation
d unterteilte Automatisierung *f*
f automation *f* à sections
r система *f* автоматизации [линий] с
   разбивкой на секции

11250 section control
d Streckensteuerung *f*
f commande *f* par sections
r секционное регулирование *n*

**11251 sector-alignment indicator; threshold
         value indicator**
d Schwellwertgeber *m*
f indicateur *m* de valeur de seuil
r индикатор *m* порогового значения

* **seek check** → 11229

11252 seek error diagnostic procedure
d Diagnoseverfahren *n* für Suchfehler
f procédé *m* de diagnostic pour erreur de
   recherche
r диагностическая процедура *f* выявления
   ошибок поиска

* **seek operation** → 11232

11253 segment *v*
d segmentieren; unterteilen
f segmenter
r сегментировать

11254 segmentation
d Segmentierung *f*; Unterteilung *f*
f segmentation *f*
r сегментирование *n*; сегментация *f*

11255 segmented encoding law
d segmentierte Kodierungskennlinie *f*
f loi *f* de quantification à segments
r линейно-ломанная характеристика *f*
   кодирования

11256 segment management
d Segmentverwaltung *f*
f gestion *f* de segment
r управление *n* сегментами

11257 **selection circuit**
  d Auswahlschaltung *f*; Ansteuerschaltung *f*
  f circuit *m* de sélection
  r схема *f* селекции

11258 **selection method**
  d Selektionsmethode *f*
  f méthode *f* de sélection
  r селективный метод *m*

11259 **selection of assembly documentation**
  d Auswahl *f* von Montageunterlagen
  f sélection *f* de documentation d'assemblage
  r выбор *m* сборочной документации

11260 **selection principle**
  d Auswahlprinzip *n*
  f principe *m* de sélection
  r принцип *m* селекции

11261 **selection ratio**
  d Auswahlverhältnis *n*
  f taux *m* de sélection
  r отношение *n* выбора

11262 **selection rule**
  d Auswahlregel *f*
  f règle *f* de sélection
  r правило *n* селекции

11263 **selective adjustment**
  d Teilregelung *f*; Selektivregelung *f*
  f réglage *m* sélectif
  r избирательное регулирование *n*

11264 **selective amplifier**
  d Selektivverstärker *m*
  f amplificateur *m* sélectif
  r избирательный усилитель *m*

11265 **selective control**
  d Selektivsteuerung *f*; Auswahlsteuerung *f*
  f réglage *m* sélectrif
  r избирательное управление *n*

  * **selective modification** → 9184

11266 **selective protection**
  d wahlweiser Schutz *m*
  f protection *f* sélective
  r селективная защита *f*

11267 **selectivity characteristic**
  d Selektivitätskurve *f*
  f caractéristique *f* de sélectivité
  r характеристика *f* избирательности

11268 **selector**
  d Selektor *m*; Auswahlschalter *m*
  f sélecteur *m*; commutateur *m* de sélection
  r селектор *m*

11269 **selector channel**
  d Selektorkanal *m*
  f canal *m* sélecteur
  r селекторный канал *m*

  * **selector system** → 5871

  * **self-acting** → 1165

11270 **self-acting shutter**
  d automatisches Schützentor *n*
  f hausse *f* automatique
  r автоматический затвор *m*

11271 **self-adaptable manipulation equipment**
  d selbstanpassungsfühige Handhabeeinrichtung *f*
  f équipement *m* de manutention auto-adaptable
  r адаптивное манипуляционное устройство *n*

11272 **self-adapted grip organ**
  d selbstadaptiertes Greiforgan *n*
  f organe *m* de grippion adapté automatiquement
  r самоприспосабливающийся захватный орган *m*

11273 **self-adaptive control**
  d selbstanpassende Regelung *f*
  f commande *f* auto-adaptative
  r адаптивное управление *n*

11274 **self-adjoint system**
  d selbstadjungiertes System *n*
  f système *m* d'auto-adjonction
  r самосопряженная система *f*

11275 **self-adjusting**
  d selbsteinstellend
  f à auto-ajustement
  r самонастраивающийся

11276 **self-adjusting model**
  d selbsteinstellendes Modell *n*; selbstabstimmendes Modell
  f simulateur *m* auto-régleur
  r самоустанавливающаяся модель *f*

  * **self-adjusting system** → 381

11277 **self-adjustment**
  d Selbsteinstellung *f*
  f réglage *m* automatique
  r самоустановка *f*

11278 **self-admittance**
d Eigenadmittanz *f*; Eigenscheinleitwert *m*
f admittance *f* propre; self-admittance *f*
r собственная проводимость *f*

11279 **self-aligning**
d selbstausrichtend
f auto-alignant
r самовыравнивающийся

11280 **self-balancing**
d selbstabgleichend
f auto-équilibré
r самобалансирующийся

11281 **self-balancing magnetic amplifier**
d selbstabgleichender magnetischer
   Verstärker *m*
f amplificateur *m* magnétique à compensation
   automatique
r самобалансирующийся магнитный
   сервоусилитель *m*

11282 **self-balancing potentiometer**
d selbstabgleichendes Potentiometer *n*
f potentiomètre *m* automatique
r потенциометр *m* с автоматической
   балансировкой

11283 **self-binary function**
d selbstduale Funktion *f*
f fonction *f* autobinaire
r автодуальная функция *f*

11284 **self-check; self-test**
d Selbstprüfung *f*; Selbstkontrolle *f*
f autocontrôle *m*
r самоконтроль *m*

11285 **self-checking; self-testing; auto-checking;**
       auto-testing *adj*
d selbstprüfend
f autocontrôlé
r самоконтролирующийся

11286 **self-clocking**
d Selbsttaktierung *f*
f rythme *m* propre; autosynchronisation *f*
r автосинхронизация *f*;
   самосинхронизация *f*

11287 **self-clocking system**
d Selbsttaktierungssystem *n*
f système *m* autohorloge
r самосинхронизирующаяся система *f*

11288 **self-contained instrument**
d unabhängiges Messinstrument *n*

f appareil *m* autonome; appareil indépendant
r автономный прибор *m*

11289 **self-contained supply**
d autonome Speisung *f*
f alimentation *f* autonome
r автономное питание *n*

11290 **self-correcting memory**
d selbstkorrigierender Speicher *m*
f mémoire *f* autocorrectrice
r самокорректирующая память *f*

11291 **self-correcting system**
d selbstkorrigierendes System *n*
f système *m* autocorrecteur
r самокорректирующаяся система *f*

* **self-correlation function** → 1123

11292 **self-diagnostic ability**
d Selbstdiagnosefähigkeit *f*
f possibilité *f* d'autodiagnostic
r возможность *f* самодиагностики

11293 **self-diagnostic method**
d Selbstdiagnosemethode *f*
f méthode *f* de propre diagnostic
r метод *m* самодиагностики

* **self-excitation** → 1128

11294 **self-excited oscillation**
d selbsterregte Schwingung *f*
f auto-oscillation *f*
r самовозбужденное колебание *n*

11295 **self-focusing**
d Selbstfokussierung *f*
f autofocalisation *f*
r самофокусирование *n*

11296 **self-focusing effect**
d Selbstfokussicrungseffkt *m*
f effet *m* d'autofocalisation
r эффект *m* самофокусирования

11297 **self-gating**
d Selbststeuerung *f*
f autocommande *f*
r самоуправление *n*

11298 **self-holding circuit**
d Selbsthaltestromkreis *m*
f circuit *m* d'automaintien
r схема *f* автоблокировки

11299 **self-homing device**
d Zielanfluggerät *n*; Zielsuchgerät *n*

f dispositif *m* d'autoguidage
r система *f* самонаведения

11300 self-inductance
   d Selbstinduktivität *f*
   f auto-inductance *f*
   r самоиндукция *f*

11301 self-interrupting circuit
   d Selbstunterbrechungsschaltung *f*
   f circuit *m* auto-interrupteur
   r цепь *f* с автоматическим прерыванием

11302 self-learning
   d selbstlernend
   f autodidacteur
   r самообучающийся

11303 self-learning system
   d Lernsystem *n*; selbstlernendes System *n*
   f système *m* auto-éducateur
   r самообучающаяся система *f*

11304 self-modulation
   d eigene Modulation *f*
   f automodulation *f*
   r самомодуляция *f*; автомодуляция *f*

11305 self nulling
   d automatische Rückstellung *f* auf Null
   f remise *f* à zéro automatique
   r самоустановка *f* на нуль

11306 self-operated measuring unit
   d selbsttätige Messeinheit *f*
   f unité *f* de mesure automatique
   r автоматическое измерительное устройство *n*

11307 self-optimizing control
   d selbstoptimierende Regelung *f*
   f commande *f* auto-optimisante
   r экстремальное управление *n*

11308 self-organizing automaton
   d selbstorganisierender Automat *m*
   f automate *m* auto-organisé
   r самоорганизирующийся автомат *m*

  * **self-oscillation** → 1449

11309 self-positioning
   d automatische Positionierung *f*
   f autopositionnement *m*
   r самоустановление *n*

11310 self-powered device
   d selbstversorgtes Gerät *n*; Gerät mit
netzunabhängiger Stromversorgung
   f appareil *m* à alimentation autonome
   r устройство *n* с автономным питанием

  * **self-programming** → 1352

11311 self-recovery
   d Selbstausgleich *m*; Ausgleich *m*
   f auto-équilibrage *m*
   r самовостановление *n*

11312 self-regulating controlled system
   d Selbstregelstrecke *f*
   f système *m* autoréglé
   r система [управления] с
самовыравниванием

  * **self-regulation** → 1221

11313 self-regulation
   d Selbstregelung *f*; Selbstausgleich *m*
   f autoréglage *m*; autorégulation *f*
   r саморегулирование *n*

11314 self-reproducting system
   d selbstreproduzierendes System *n*
   f système *m* de reproduction automatique
   r самовоспроизводящаяся система *f*

11315 self-reproduction
   d Selbstreproduktion *f*
   f reproduction *f* automatique
   r самовозпроизведение *n*

11316 self-resetting
   d automatische Rückstellung *f*
   f réenclenchement *m* automatique
   r самовозврат *m*

11317 self-saturated magnetic amplifier
   d magnetischer Verstärker *m* mit
Selbstsättigung
   f amplificateur *m* magnétique à autosaturation
   r магнитный усилитель *m* с
самонасыщением

11318 self-stabilization
   d Eigenstabilisierung *f*
   f autostabilisation *f*; stabilisation *f* propre
   r самостабилизация *f*

11319 self-starter
   d Selbstanlasser *m*
   f autodémarreur *m*
   r автоматический стартер *m*

11320 self-starting hysteresis motor
   d Hysteresemotor *m* mit Selbstanlauf

f moteur *m* autodémarreur hystérétique
r гистерезисный электродвигатель *m* с автоматическим разбегом

11321 self-starting synchronous motor
d selbststartender Synchronmotor *m*
f moteur *m* synchrone à autodémarrage
r синхронный двигатель *m* с самопуском

*   **self-sustained oscillation system → 1452**

11322 self-sustained pulsation
d selbsterhaltende Schwingung *f*
f oscillation *f* auto-entretenue
r самосохраняющееся колебание *n*

*   **self-teaching system of automatic optimization → 1181**

*   **self-test → 11284**

11323 self-test electronics
d Selbsttestelektronik *f*; Selbstprüfungselektronik *f*
f électronique *f* à test propre
r самотестируемая электроника *f*

*   **self-testing → 11285**

11324 self-test method
d Selbstprüfmethode *f*
f méthode *f* de test autonome
r метод *m* самопроверки; метод самоконтроля

*   **self-vibration → 1449**

11325 selsyn; synchro
d Selsyn *m*; Synchro *m*; Drehmelder *m*
f selsyn *m*
r сельсин *m*

11326 selsyn control
d Selsynsteuerung *f*; Drehmeldersteuerung *f*
f commande *f* à selsyn; commande par synchro
r сельсинное управление *n*

11327 selsyn-type synchronous system
d Selsynsynchronsystem *n*
f système *m* synchrone à selsyns
r сельсинная синхронная система *f*

11328 semi-automatic
d halbautomatisch
f semi-automatique
r полуавтоматический

11329 semi-automatic adjusting; semi-automatic

regulation
d halbautomatische Einstellung *f*
f réglage *m* semi-automatique
r полуавтоматическое регулирование *n*

11330 semi-automatic controller
d halbautomatischer Regler *m*
f régulateur *m* semi-automatique
r полуавтоматический регулятор *m*

11331 semi-automatic data acquisition system
d halbautomatisches Datenerfassungssystem *n*
f système *m* d'acquisition semi-automatique des données
r полуавтоматическая система *f* сбора данных

11332 semi-automatic machine
d halbautomatische Maschine *f*
f machine *f* semi-automatique
r машина-полуавтомат *m*

11333 semi-automatic operation
d halbautomatischer Betrieb *m*
f marche *f* semi-automatique
r полуавтоматическая работа *f*

11334 semi-automatic procedure
d halbautomatisches Verfahren *n*
f procédé *m* semi-automatique
r полуавтоматическая процедура *f*

*   **semi-automatic regulation → 11329**

11335 semi-automatic tester
d halbautomatisches Prüfgerät *n*
f essayeur *m* semi-automatique
r полуавтоматическое контрольное устройство *n*

11336 semicircular deviation
d halbkreisartige Abweichung *f*
f déviation *f* semi-circulaire
r полукруговая девиация *f*; полукруговое отклонение *n*

11337 semiconductor
d Halbleiter *m*
f semi-conducteur *m*
r полупроводник *m*

11338 semiconductor amplifier
d Halbleiterverstärker *m*
f amplificateur *m* à semi-conducteurs
r полупроводниковый усилитель *m*

11339 semiconductor building block element
d Halbleiterbauelement *n*

f composant *m* semi-conducteur
r полупроводниковый конструкционный
элемент *m*

11340 semiconductor film laser
  d Dünnschicht-Halbleiterlaser *m*; Halbleiter-
    Dünnschichtlaser *m*
  f laser *m* semiconducteur à couche mince;
    laser à couches minces semiconductrices
  r тонкослойный полупроводниковый
    лазер *m*

11341 semiconductor instruments parameters
  d Halbleitergeräteparameter *mpl*
  f paramètres *mpl* d'appareils à semi-
    conducteur
  r параметры *mpl* полупроводниковых
    приборов

11342 semiconductor pressure sensing device
  d Druckfühler *m* auf Halbleiterbasis
  f palpeur *m* de pression à semi-conducteur
  r полупроводниковый датчик *m* давления

11343 semiconductor technology
  d Halbleitertechnologie *f*
  f technologie *f* des semiconducteurs
  r полупроводниковая технология *f*

11344 semicontinuous function
  d halbstetige Funktion *f*
  f fonction *f* semi-continue
  r полунепрерывная функция *f*

  *   **semicycle → 6017**

11345 semidefiniteness
  d Semidefinitheit *f*
  f semi-définité *f*
  r полуопределенность *f*

11346 semiduplex optical fiber transmission
  d Semiduplex-Lichtwellenleiterübertragung *f*
  f transmission *f* à l'alternat sur fibre optique
  r полудуплексная световодная передача *f*

11347 semigraphical method
  d grafisch-analytische Methode *f*
  f méthode *f* grapho-analytique
  r графо-аналитический метод *m*

11348 semi-logarithmic
  d halblogarithmisch; einfachlogarithmisch
  f semi-logarithmique
  r полулогарифмический

11349 semi-partial processing unit
  d teilparallele Verarbeitungseinheit *f*

f unité *f* de traitement semi-partielle
r блок *m* полупаралельной обработки

11350 semipermanent
  d semipermanent
  f semi-permanent
  r полупостоянный; полуустойчивый

11351 semipermanent data
  d semipermanente Daten *pl*
  f données *fpl* semi-permanentes
  r полупостоянные данные *pl*

11352 sensibility reciprocal
  d Empfindlichkeitskehrwert *m*
  f valeur *f* réciproque de sensibilité
  r обратная чувствительность *f*

11353 sensing component member
  d Fühlglied *n*
  f organe *m* sensible
  r воспринимающее устройство *n*

  *   **sensing element → 4017**

  *   **sensing unit → 6691**

11354 sensitive
  d empfindlich
  f sensible
  r чувствительный

11355 sensitive force sensor
  d empfindlicher Kraftsensor *m*
  f senseur *m* de force sensible
  r чувствительный датчик *m* усилия

11356 sensitivity analysis
  d Empfindlichkeitsanalyse *f*
  f analyse *f* de sensibilité
  r анализ *m* чуствительности

11357 sensitivity criterion
  d Empfindlichkeitskriterium *n*
  f critère *m* de sensibilité
  r критерий *m* чувствительности

11358 sensitivity curve
  d Empfindlichkeitskurve *f*
  f courbe *f* de sensibilité
  r кривая *f* чувствительности

11359 sensitivity function
  d Empfindlichkeitsfunktion *f*
  f fonction *f* de sensibilité
  r функция *f* чувствительности

11360 sensitivity loss
  d Empfindlichkeitsverlust *m*

f perte *f* de sensibilité
r потеря *f* чувствительности

* **sensitivity region** → **10618**

* **sensor** → **4017**

11361 sensor accuracy
d Sensorgenauigkeit *f*
f exactitude *f* de senseur
r точность *f* сенсора

11362 sensor component; sensor part
d Sensorbauelement *n*
f composant *m* de senseur; élément *m*
    [de construction] de capteur
r сенсорный элемент *m*

11363 sensor-connected oscillatory motion
d sensorgeschaltete Schwingbewegung *f*
f mouvement *m* oscillant connecté par senseur
r колебательное движение *n* управляемое
    сенсором

11364 sensor control
d Sensorsteuerung *f*
f commande *f* de senseur
r сенсорное управление *n*

11365 sensor-controlled assembly gripper
d sensorgesteuerter Montagegreifer *m*
f grappin *m* d'assemblage commandé par
    senseur
r сборочное захватное устройство *n* с
    сенсорным управлением

11366 sensor-controlled axis
d sensorgesteuerte Achse *f*
f axe *m* commandé par senseur
r ось *f* с сенсорным управлением

11367 sensor-controlled change gripper
d sensorgesteuerter Wechselgreifer *m*
f grappin *m* de changement commandé par
    senseur
r сменный захват *m* с сенсорным
    управлением

11368 sensor-controlled element
d sensorgesteuertes Element *n*
f élément *m* commandé par senseur
r элемент *m* с сенсорным управлением

11369 sensor-controlled fine motion
d sensorgesteuerte Feinbewegung *f*
f mouvement *m* fin commandé par senseur
r ориентирующее движение *n* с сенсорным
    управлением

11370 sensor-controlled industrial robot
d sensorgesteuerter Industrieroboter *m*
f robot *m* industriel commandé par senseur
r промышленный робот *m* с сенсорным
    управлением

* **sensor-controlled joint** → **11371**

11371 sensor-controlled joint[ing]
d sensorgesteuertes Fügen *n*
f jointage *m* commandé par senseur
r сопряжение *n* с управлением от сенсора

11372 sensor data acquisition
d Sensordatenerfassung *f*
f acquisition *f* de données à senseur
r сбор *m* сенсорных данных

11373 sensor data evaluation
d Sensordatenauswertung *f*
f évaluation *f* des données de senseur
r обработка *f* сенсорных данных

* **sensor design** → **4031**

11374 sensor equipment
d Sensorausrüstung *f*
f équipement *m* de senseur
r сенсорное устройство *n*

11375 sensor function
d Sensorfunktion *f*
f fonction *f* de capteur
r сенсорная функция *f*

11376 sensor industry
d Sensorindustrie *f*
f industrie *f* des capteurs
r сенсорная промышленность *f*

11377 sensor information
d Sensorinformation *f*
f information *f* de capteur
r сенсорная информация *f*

11378 sensor measurement
d Sensormessung *f*
f mesure *f* au moyen du senseur
r измерение *n* с использованием сенсоров

11379 sensor models
d Sensormodelle *npl*
f modèles *mpl* de capteurs
r сенсорные модели *fpl*

11380 sensor object identification
d Sensorobjektidentifizierung *f*;
    Identifizieren *n* von Objekten durch Sensoren

    f identification *f* d'objets par senseurs
    r распознавание *n* объектов сенсорами

*   **sensor part** → **11362**

11381 sensor position information
    d Sensorlageinformation *f*
    f information *f* de position avec senseur
    r сенсорная информация *f* о позиции

11382 sensor processor
    d Sensorprozessor *m*
    f processeur *m* de capteur
    r сенсорный процессор *m*

11383 sensor pulse
    d Sensorimpuls *m*
    f impulsion *f* de senseur
    r сенсорный импульс *m*

11384 sensor scanning device
    d Sensorabtastgerät *n*
    f dispositif *m* de balayage de capteur
    r сенсорное устройство *n* сканирования

11385 sensor structure
    d Sensoraufbau *m*
    f structure *f* de senseur
    r структура *f* сенсора

11386 sensor survey
    d Sensorüberwachung *f*
    f surveillance *f* de senseur
    r сенсорный контроль *m*

11387 sensor system
    d Sensorsystem *n*; sensorisches System *n*
    f système *m* de senseur
    r сенсорная система *f*

11388 sensor with great resolving power
    d Sensor *m* mit hohem Auflösevermögen
    f capteur *m* à grand pouvoir de résolution
    r сенсор *m* с большой разрешающей
       способностью

11389 sensor with luminescence diode and
       phototransistor
    d Sensor *m* mit Lumineszenzdiode und
       Fototransistor
    f capteur *m* à diode luminescente et
       phototransistor
    r сенсор *m* с люминесцентным диодом и
       фототранзистором

11390 separate component
    d Einzelbauelement *n*
    f composant *m* solitaire

    r отдельный компонент *m*

11391 separated data
    d separierte Daten *pl*; getrennte Daten
    f données *fpl* séparées
    r отдельные данные *pl*

11392 separated manipulation cycle
    d getrennter Handhabezyklus *m*
    f cycle *m* de manutention séparé
    r отдельный цикл *m* манипулирования

11393 separated unit
    d getrennte Einheit *f*
    f unité *f* séparée
    r отдельное устройство *n*

11394 separate module
    d separates Modul *n*
    f module *m* séparé
    r автономный модуль *m*

*   **separating** → **11395**

11395 separation; separating
    d Auszug *m*; Trennung *f*; Separation *f*
    f séparation *f*
    r разделение *n*; отделение *n*; сепарация *f*

11396 separation process
    d Trennungsvorgang *m*
    f processus *m* de séparation
    r процесс *m* разделения

11397 sequence
    d Reihenfolge *f*; Folge *f*
    f séquence *f*; succession *f*
    r последовательность *f*

11398 sequence alternator
    d Folgealternator *m*
    f alternateur *m* de séquence
    r устройство *n* для измерения
       последовательности команд

11399 sequence automatics
    d Folgeautomatik *f*
    f automatique *f* séquentielle
    r автоматика *f* порядка следования

11400 sequence checking
    d Ablauffolgeprüfung *f*
    f vérification *f* de séquence
    r контроль *m* последовательности

11401 sequence control
    d Ablauffolgesteuerung *f*

f commande *f* de séquence
r управление *n* последовательностью

11402 sequence control element
d Programmschaltelement *n*
f élément *m* de commutation séquentielle
r блок *m* последовательного контроля

11403 sequence controller
d Folgeschalter *m*; Programmschalter *m*
f combinateur *m* séquentiel
r последовательный регулятор *m*

11404 sequence-control register
d Befehlsfolgeregister *n*
f registre *m* de contrôle de séquence;
   compteur *m* ordinal
r регистр *m* последовательного управления

11405 sequence of nominal value
d Sollwertfolge *f*
f séquence *f* de la valeur prescrite
r последовательность *f* заданных величин

11406 sequence of operation
d Bearbeitungsfolge *f*
f suite *f* d'opération
r порядок *m* обработки

11407 sequence of processes
d Prozessfolge *f*
f suite *f* de procédés
r последовательность *f* процессов

11408 sequence of switches
d Schaltfolge *f*
f séquence *f* de commutations
r последовательность *f* переключений

11409 sequence programming
d Folgeprogrammierung *f*
f programmation *f* de séquence
r последовательное программирование *n*

11410 sequence selector switch
d Folgewahlschalter *m*
f sélecteur *m* séquentiel asservi
r последовательный селективный
   коммутатор *m*

* **sequencing model** → 11216

11411 sequential; serial
d sequentiell; seriell; nacheinander
f séquentiel; sériel
r последовательный

11412 sequential analysis

d Sequenzanalyse *f*
f analyse *f* séquentielle
r последовательный анализ *m*

11413 sequential automation
d Folgeautomatik *f*
f automatique *f* séquentielle
r последовательная автоматизация *f*

11414 sequential automaton
d sequentieller Automat *m*
f automate *m* [à fonctionnement] séquentiel
r автомат *m* последовательного действия

11415 sequential calculation
d sequentielle Berechnung *f*
f calcul *m* séquentiel
r последовательное вычисление *n*

11416 sequential circuit with memories
d Folgeschaltung *f* mit Speicherkreisen
f circuit *m* séquentiel à mémoires
r следящая система *f* с запоминающими
   элементами

* **sequential control** → 11130

11417 sequential correcting element
d reihengeschaltetes Korrekturglied *n*;
   sequentielles Korrekturglied
f élément *m* correcteur en série
r последовательное корректирующее
   устройство *n*

11418 sequential decision process
d sequentieller Entscheidungsprozess *m*
f procédé *m* de décision séquentiel
r последовательный процесс *m* решения

11419 sequential digital servomechanism
d sequentieller digitaler Servomechanismus *m*
f servomécanisme *m* séquentiel digital
r цифровая следящая система *f*

11420 sequential handling
d Handhabungssequenz *f*
f manutention *f* séquentielle
r отработка *f* манипуляционных действий

11421 sequential logic
d sequentielle Logik *f*
f logique *f* séquentielle
r последовательная логика *f*

11422 sequential operation
d sequentieller Betrieb *m*
f régime *m* séquentiel
r последовательный режим *m* работы

11423 **sequential processing**
d sequentielle Verarbeitung *f*
f traitement *m* séquentiel
r последовательная обработка *f*

11424 **sequential switching circuit**
d Folgeschaltkreis *m*
f circuit *m* de commutation séquentielle
r схема *f* последовательного действия

* **serial** → 11411

11425 **serial access**
d serienweiser Zugriff *m*; serieller Zugriff *m*
f accès *m* en série
r последовательное обращение *n*

11426 **serial adder**
d Serienadder *m*; Serienaddierer *m*
f addeur *m* à fonctionnement séquentiel
r сумматор *m* последовательного действия

11427 **serial arithmetic unit**
d Serienrechenwerk *n*
f unité *f* arithmétique en série
r устройство *n* последовательного
вычисления

11428 **serial data channel**
d serieller Datenkanal *m*
f canal *m* de données sériel
r последовательный канал *m* обмена
данными

11429 **serial data input**
d serieller Dateneingang *m*
f entrée *f* de données sérielle
r последовательный вход *m* данных

11430 **serial data line**
d serielle Datenleitung *f*
f ligne *f* de données sérielle
r линия *f* последовательного обмена
данными

11431 **serial-parallel arithmetic**
d Serien-Parallel-Arithmetik *f*
f arithmétique *f* parallèle en série
r последовательно-паралельная
арифметика *f*

11432 **serial principle**
d Serienprinzip *n*
f principe *m* de fonctionnement séquentiel
r принцип *m* последовательного действия

* **serial production** → 10225

11433 **serial transfer; serial transmission**
d serielle Übertragung *f*; Serienübertragung *f*
f transfert *m* en série
r последовательная передача *f*

* **serial transmission** → 11433

11434 **series capacitor**
d Serienkondensator *m*;
Vorschaltkondensator *m*
f condensateur *m* série
r добавочная ёмкость *f*

11435 **series cascade action**
d Reihenkaskadenwirkung *f*
f action *f* en cascade série
r последовательное каскадное действие *n*

* **series connection** → 2148

* **series control** → 11130

* **series manufacture** → 10225

11436 **series-parallel circuit**
d Reihenparallelschaltung *f*
f circuit *m* série-parallèle
r последовательно-параллельная цепь *f*

11437 **series-parallel control**
d Reihenparallelregelung *f*
f réglage *m* série-parallèle
r последовательно-параллельное
регулирование *n*

11438 **series-parallel system**
d serienparalleles System *n*;
Reihenparallelsystem *n*
f système *m* série-parallèle
r последовательно-параллельная система *f*

11439 **series resonance**
d Serienresonanz *f*; Reihenresonanz *f*
f résonance *f* série
r последовательный резонанс *m*

11440 **series stabilization**
d Serienstabilisierung *f*
f stabilisation *f* série
r последовательная стабилизация *f*

11441 **series transmissions of informations**
d serienweise Informationsübertragung *f*
f transmission *f* en série des informations
r последовательная передача *f* информации

11442 **service failure**
d Betriebsstörung *f*

f panne *f* de service
r повреждение *n* при эксплуатации

11443 service function
d Servicefunktion *f*
f fonction *f* de service
r функция *f* обслуживания

11444 service program
d Bedienprogramm *n*
f programme *m* de service
r программа *f* обслуживания

* **servicing equipment → 7759**

11445 servoanalyzer
d Servoanalysator *m*
f analyseur *m* à servomécanisme
r анализатор *m* следящий систем

11446 servocontrol
d Servosteuerung *f*
f servocommande *f*
r сервоуправление *n*

11447 servo-driven
d mit Servomotor; servobetrieben
f à commande servomécanique
r сервоуправляемый; с сервоприводом

* **servofollower → 5629**

11448 servo-hydraulic manipulator drive
d servohydraulischer Manipulatorantrieb *m*
f entraînement *m* de manipulateur
   servohydraulique
r гидравлический сервопривод *m*
   манипулятора

11449 servoloop
d Servoschleife *f*
f boucle *f* de servomécanisme
r контур *m* следящей системы

11450 servomechanism characteristic constants
d charakteristische
   Servomechanismuskonstanten *fpl*
f constantes *fpl* caractéristiques de
   servomécanisme
r константы *fpl* характеристик следящей
   системы

11451 servomechanism for continuous operation
d Servomechanismus *m* für Dauerbetrieb
f servomécanisme *m* pour action continue
r сервомеханизм *m* длительной работы

* **servomotor → 9962**

11452 servomultiplier
d Servomultiplizierer *m*
f servomultiplicateur *m*
r сервоумножитель *m*

11453 servo-operated control
d Regelung *f* mit Reglerverstärkung
f réglage *m* à servomécanisme
r регулирующее устройство *n* с
   сервоприводом

11454 servo-operated inductance bridge circuit
d Induktanzmessbrücke *f* mit Servoregelung
f pont *m* à inductances à servomécanisme
r схема *f* индукционного моста с
   сервоприводом

11455 servo-output signal
d Servogerätausgangssignal *n*
f signal *m* de sortie du système asservi
r выходной сервосигнал *m*

11456 servopotentiometer
d Servopotentiometer *n*
f servopotentiomètre *m*
r сервопотенциометр *m*

* **servoregulator → 5629**

11457 servostability
d Servosystemstabilität *f*
f stabilité *f* du système asservi
r устойчивость *f* сервосистемы

* **servosystem → 5630**

* **session mode → 3388**

11458 set *v*
d einstellen
f consigner
r устанавливать

11459 set of curves
d Kurvenschar *f*
f famille *f* de courbes; faisceau *m* de courbes
r семейство *n* кривых

11460 set point; set value
d Einstellwert *m*
f valeur *f* de consigne; grandeur *f* de consigne
r заданная величина *f*

11461 set-point adjuster
d Sollwerteinsteller *m*; Sollwertgeber *m*
f dispositif *m* de changement de la valeur de
   consigne
r регулировка *f* заданной величины

11462 set-point adjustment
  d Sollwerteinstellung f
  f ajustage m de la valeur de consigne
  r установка f контрольной точки

11463 set-point mechanism
  d Führungsglied n
  f source f de référence
  r управляющее звено n

11464 set pulse; setting pulse
  d Einstellimpuls m
  f impulsion f d'ajustage
  r установочный импульс m

11465 setting
  d Abstimmung f; Einstellung f; Einstellwert m
  f ajustage m; consigne f
  r установка f

11466 setting accuracy
  d Einstellgenauigkeit f
  f précision f d'ajustage
  r точность f установки

  * setting device → 511

11467 setting in operation
  d Inbetriebnahme f
  f mise f en service; mise en mouvement
  r пуск m в ход

  * setting pulse → 11464

11468 setting time
  d Einstellzeit f
  f temps m d'ajustement
  r время n установки

11469 set unit
  d Gebereinheit f; Geberglied n
  f unité f de capteur
  r задающий блок m

  * setup v → 6705

11470 set-up of problem
  d Problemeinstellung f
  f composition f du problème
  r набор m задачи

  * set value → 11460

11471 shape control
  d Formsteuerung f
  f commande f de la forme
  r регулирование n формы

11472 shape-forming device
  d Formierungseinrichtung f
  f dispositif m de formation
  r формирователь m

11473 shaping circuit; shaping network
  d Formierkreis m; Formierungsglied n
  f circuit m conformateur
  r формирующая цепь f; формирующая схема f

11474 shaping filter
  d formgebendes Filter n
  f filtre m conformateur
  r формирующий фильтр m

  * shaping network → 11473

11475 shareable
  d gemeinsam benutzbar
  f partageable
  r совместно используемый

11476 sharp beam
  d scharf gebündelter Strahl m
  f faisceau m aigu
  r узкий луч m

11477 sharpness of resonance
  d Resonanzbreite f
  f acuité f de résonance
  r острота f резонанса

11478 sharp pulse
  d scharfer Impuls m
  f impulsion f pointue; impulsion tranchante
  r острый импульс m

11479 Sheffer's function
  d Sheffer-Funktion f
  f fonction f de Sheffer
  r функция f Шеффера

11480 shielded line
  d abgeschirmte Leitung f
  f ligne f blindée
  r экранированная линия f

  * shift → 11483

11481 shift v
  d schieben; verschieben
  f décaler
  r сдвигать

11482 shift circuit
  d Verschiebungskreis m
  f circuit m de décalage
  r фазосдвигающая схема f

11483 shift[ing]
  d Schieben *n*; Verschiebung *f*
  f décalage *m*
  r сдвиг *m*

11484 shifting instruction
  d Verschiebebefehl *m*
  f instruction *f* de décalage
  r команда *f* сдвига

11485 shift register
  d Schieberegister *n*
  f registre *m* de décalage
  r регистр *m* сдвига

11486 shift signal
  d Verschiebesignal *n*
  f signal *m* de décalage
  r сигнал *m* сдвига

11487 short
  d kurz; verkürzt; gekürzt
  f court; écourté; bref
  r короткий; укороченный

11488 short-circuit current peak value
  d Kurzschlußstromspitzenwert *m*
  f valeur *f* maximum de courant de court-circuit
  r пиковое значение *n* тока короткого замыкания

11489 short-circuit detector
  d Kurzschlußsucher *m*
  f détecteur *m* de courts-circuits
  r прибор *m* для нахождения короткого замыкания

11490 short-circuited line
  d kurzgeschlossene Leitung *f*
  f ligne *f* court-circuitée
  r короткозамкнутая линия *f*

11491 short-circuit protection
  d Kurzschlußschutz *m*
  f protection *f* contre les courts-circuits
  r защита *f* от короткого замыкания

11492 short-distance system
  d Kurtzstreckensystem *n*
  f système *m* [à] courte distance
  r система *f* коротких дистанций

11493 shorted out
  d überbrückt
  f court-circuité
  r шунтированный; закороченный

11494 short infrared

  d kurzwellige Infrarotstrahlung *f*
  f rayonnement *m* infrarouge à ondes courtes
  r коротковолновая область *f* [спектра] инфракрасного излучения

11495 short optical pulse
  d kurzer optischer Impuls *m*
  f impulsion *f* optique de courte durée
  r короткий оптический импульс *m*

11496 short-pulse laser
  d Laser *m* für kurze Impulse
  f laser *m* à impulsions de courte durée
  r лазер *m* коротких импульсов

11497 short-range Doppler
  d Kurzstrecken-Dopplerverfahren *n*
  f système *m* de trajectographie courte portée par effet Doppler
  r допплеровская система *f* траекторных измерений ближнего действия

11498 short recovery time
  d kurze Erholzeit *f*
  f temps *m* de repos court
  r короткое время *n* восстановления

11499 short scanning of measuring signals
  d kurzzeitige Abtastung *f* von Messsignalen
  f exploration *f* à courte durée de signaux à mesurer
  r кратковременное сканирование *n* сигналов измерительных устройств

11500 short-term phase stability
  d Kurzzeitphasenstabilität *f*
  f stabilité *f* à court terme de la phase
  r кратковременная фазовая стабильность *f*

11501 short-time duty
  d Kurzzeitbetrieb *m*
  f régime *m* à temps court
  r режим *m* кратковременной работы

11502 short-time measuring apparatus
  d Kurzzeitmessgerät *n*
  f appareil *m* de mesure pour temps courts
  r прибор *m* для измерения малых интервалов времени

11503 short-time memory
  d Kurzzeitspeicher *m*
  f mémoire *f* temps court
  r запоминающее устройство *n* кратковременного действия

  *   shunt *v* → 2003

11504 shunt tripping
   d Spannungsauslösung *f*
   f déclenchement *m* par bobine en dérivation
   r отключение *n* напряжением

11505 shut-down delay; opening delay
   d Ausschaltverzung *m*; Abschaltverzögerung *f*
   f retard *m* à l'ouverture; retard d'arrêt
   r запаздывание *n* выключения

11506 shut-off contact
   d Ausschaltkontakt *m*
   f contact *m* interrupteur; contact de coupure
   r контакт *m* остановки

11507 side-band transmission
   d Seitenbandübertragung *f*
   f transmission *f* sur bande latérale
   r передача *f* на боковой полосе [частот]

11508 side control
   d Seitensteuerung *f*
   f commande *f* latérale
   r боковое управление *n* машиной

   * **sign → 2278**

11509 signal acquisition
   d Signalerfassung *f*
   f acquisition *f* de signal
   r регистрация *f* сигнала

11510 signal bandwidth
   d Signalbandbreite *f*
   f largeur *f* de bande du signal
   r спектр *m* частот сигнала

11511 signal block
   d Signalblock *m*
   f bloc *m* des signaux
   r сигнальный блок *m*

11512 signal button
   d Signaltaste *f*
   f bouton-poussoir *m* d'appel
   r сигнальная кнопка *f*

11513 signal character
   d Signalcharakter *m*
   f caractère *m* de signal
   r характер *m* сигнала

11514 signal code
   d Signalkode *m*
   f code *m* de signal
   r сигнальный код *m*

11515 signal conditioning
   d Signalformung *f*; Signalaufbereitung *f*
   f formation *f* de signaux
   r приведение *n* сигнала к заданным нормам

11516 signal conversion equipment
   d Signalwandlungseinrichtung *f*
   f équipement *m* à conversion de signaux; dispositif *m* à conversion de signaux
   r устройство *n* преобразования сигнала

11517 signal converter
   d Signalkonverter *m*; Signalumsetzer *m*; Signalwandler *m*
   f convertisseur *m* de signal
   r преобразователь *m* сигналов

11518 signal correlation
   d Signalkorrelation *f*
   f corrélation *f* du signal
   r корреляция *f* сигнала

11519 signal data
   d Signaldaten *pl*
   f données *fpl* de signal
   r параметры *mpl* сигналов

11520 signal decoding
   d Signalentschlüsselung *f*
   f décodage *m* de signal
   r декодирование *n* сигнала

11521 signal delay
   d Signalverzögerung *f*
   f retardement *m* de signal
   r задержка *f* сигнала

11522 signal delay time
   d Signalverzögerungszeit *f*
   f temps *m* de délai du signal
   r время *n* запаздывания сигнала

11523 signal distributor
   d Signalverteiler *m*
   f distributeur *m* de signaux
   r распределитель *m* сигналов

11524 signal duration
   d Signaldauer *f*
   f durée *f* de signal
   r длительность *f* сигнала

11525 signal evaluation
   d Signalauswertung *f*
   f évaluation *f* de signal
   r обработка *f* сигнала

11526 signal generator
   d Signalgeber *m*; Prüfsender *m*

f générateur *m* de signaux
r генератор *m* сигналов

* **signal grid → 3240**

11527 signal handling; signal processing
d Signalbehandlung *f*; Signalverarbeitung *f*
f manipulation *f* de signaux; traitement *m* de signaux
r обработка *f* сигналов

11528 signal level
d Signalpegel *m*
f niveau *m* de signal
r уровень *m* сигнала

11529 signal line
d Signalleitung *f*
f ligne *f* de signaux
r линия *f* передачи сигналов

11530 signalling
d Signalisierung *f*; Signalgebung *f*
f signalisation *f*
r сигнализация *f*

11531 signal main line
d Signalhauptleitung *f*
f ligne *f* principale de signaux
r сигнальная магистраль *f*

11532 signal of automatic blocking
d Signal *n* automatischer Blockierung
f signal *m* de blocage automatique
r сигнал *m* автоматической блокировки

* **signal processing → 11527**

11533 signal recognition
d Signalerkennung *f*
f reconnaissance *f* de signaux
r распознавание *n* сигналов

11534 signal representation
d Signaldarstellung *f*
f représentation *f* de signaux
r представление *n* сигналов

11535 signal reproduction
d Wiederherstellung *f* des Signals
f reconstitution *f* du signal
r воспроизведение *n* сигнала

11536 signal scanning
d Signalabtastung *f*
f balayage *m* de signaux
r сканирование *n* сигналов

11537 signal selection
d Signalauswahl *f*
f sélection *f* de signal
r селекция *f* сигналов

11538 signal selector
d Signalselektor *m*; Signalauswähler *m*
f sélecteur *m* des signaux
r селектор *m* сигналов

11539 signal storage system
d Signalspeichersystem *n*
f système *m* d'emmagasinage de signaux
r система *f* запоминания сигналов

11540 signal strength adjustment
d Signalstärkeregelung *f*
f réglage *m* de niveau du signal
r юстировка *f* уровня сигнала

11541 signal threshold
d Signalschwelle *f*
f seuil *m* du signal
r порог *m* различимости сигнала

11542 signal-to-noise ratio
d Signal-Rausch-Verhältnis *n*; Signal-Rausch-Abstand *m*
f rapport *m* signal-bruit
r отношение *n* сигнал-шум

11543 signal tracer
d Signalverfolger *m*; Signalnachspürer *m*
f tráceur *m* de signal; ondoscope *m* dépanneur
r прибор *m* для проверки прохождения сигнала

11544 signal transmission level
d Signalübertragungsstärke *f*; Zeichenübertragungspegel *m*
f niveau *m* de transmission des signaux
r уровень *m* передачи сигнала

11545 signal transmission speed
d Signalübertragungsgeschwindigkeit *f*
f rapidité *f* de la transmission du signal
r скорость *f* передачи сигнала

11546 significance criterion
d Signifikanzkriterium *n*
f critère *m* de signification
r критерий *m* значимости

11547 significance study
d Signifikanzuntersuchung *f*
f étude *f* de signification
r исследование *n* значимости

*   similarity → 3005

11548 similarity conditions; similitude conditions
 d Ähnlichkeitsbedingungen *fpl*
 f conditions *fpl* de similitude
 r условия *npl* подобия

11549 similarity transformation
 d ähnliche Transformation *f*
 f transformation *f* de similitude
 r подобное преобразование *n*

*   similitude → 3005

*   similitude conditions → 11548

11550 simple assembly automaton
 d einfacher Montageautomat *m*
 f automate *m* de montage simple
 r простой сборочный автомат *m*

11551 simple assembly operation
 d einfache Montageoperation *f*
 f opération *f* d'assemblage simple
 r элементарная монтажная операция *f*

11552 simple assembly system
 d einfaches Montagesystem *n*
 f système *m* d'assemblage simple
 r простая сборочная система *f*

11553 simple check element
 d einfaches Kontrollelement *n*
 f élément *m* de contrôle simple
 r простой элемент *m* контроля

11554 simple controller
 d einfacher Regler *m*
 f régulateur *m* simple
 r простой регулятор *m*

*   simple current → 8601

11555 simple decomposition
 d einfache Dekomposition *f*
 f décomposition *f* simple
 r простая декомпозиция *f*

11556 simple event
 d elementares Ereignis *n*
 f événement *m* simple
 r элементарное событие *n*

*   simple function → 8937

11557 simple handling technique
 d einfache Handhabetechnik *f*
 f technique *f* de manutention simple

 r простая манипуляционная техника *f*

11558 simple lag network
 d einfaches Verzögerungsnetzwerk *n*
 f réseau *m* de retard simple
 r простая цепь *f* задерживания

11559 simple register
 d Einfachregister *n*
 f registre *m* simple
 r простой регистр *m*

*   simple sensor → 4988

11560 simple sensor structure
 d einfacher Sensoraufbau *m*
 f structure *f* de senseur simple
 r простая сенсорная структура *f*

11561 simple visual sensor
 d einfacher Sichtsensor *m*
 f capteur *m* visuel simple
 r простой визуальный сенсор *m*

11562 simple work piece assembly
 d einfache Werkstückmontage *f*
 f assemblage *m* de pièce à travailler simple
 r простой процесс *m* сборки деталей

11563 simplex algorithm
 d Simplexalgorithmus *m*
 f algorithme *m* simplex
 r симплексный алгоритм *m*

11564 simplex channel
 d Simplexkanal *m*
 f canal *m* simplex
 r симплексный канал *m*

11565 simplex criterion
 d Simplexkriterium *n*
 f critère *m* simplex
 r симплексный критерий *m*

11566 simplex method; simplex technique
 d Simplexverfahren *n*; Simplexmethode *f*
 f méthode *f* simplex; technique *f* simplex
 r симплекс-метод *m*

11567 simplex operation
 d Simplexbetrieb *m*; Einfachbetrieb *m*
 f régime *m* simplex
 r симплексный режим *m*

*   simplex technique → 11566

11568 simplified automatic multiplication
 d vereinfachte automatische Multiplikation *f*

f multiplication *f* automatique simplifiée
r упрощенное автоматическое умножение *n*

11569 simplified model
d vereinfachtes Modell *n*
f modèle *m* simplifié
r упрощенная модель *f*

11570 simplified regulation algorithm
d vereinfachter Regelalgorithmus *m*
f algorithme *m* de régulation simplifié
r упрощенный алгоритм *m* регулирования

11571 simply connected
d einfach zusammenhängend
f simplement connecté
r односвязаный

11572 Simpson's rule
d Simpsonsche Regel *f*; Simpsonsche Formel *f*
f formule *f* de Simpson
r формула *f* Симпсона

11573 simulate *v*
d nachbilden; simulieren
f simuler
r моделировать

11574 simulated design
d simulierter Entwurf *m*
f conception *f* simulée
r разработка *f* с применением
    моделирования

11575 simulated program
d simuliertes Programm *n*
f programme *m* simulé
r моделирующая программа *f*

11576 simulation
d Simulation *f*; Modellierung *f*
f simulation *f*
r моделирование *n*

11577 simulation computation
d Simulationsrechnung *f*
f calcul *m* de simulation
r расчёт *m* математической модели

11578 simulation language
d Simulationssprache *f*
f langage *m* de simulation
r язык *m* моделирования

**11579 simulation of continuous multiloop control
          system**
d Modellierung *f* kontinuierlicher
    Mehrfachsysteme

f simulation *f* de systèmes asservis continue à
    plusieurs boucles
r моделирование *n* непрерывных
    многоконтурных систем регулирования

11580 simulation of cybernetic system
d Simulation *f* eines kybernetischen Systems;
    Nachbildung *f* eines kybernetischen Systems
f simulation *f* d'un système cybernétique
r моделирование *n* кибернетической
    системы

11581 simulation of logical operations
d Modellierung *f* logischer Operationen
f simulation *f* des opérations logiques
r моделирование *n* логических операций

11582 simulation of movement
d Bewegungssimulation *f*; Laufsimulation *f*
f simulation *f* de la marche
r моделирование *n* движения

11583 simulation of non-linear equation
d Simulation *f* nichtlinearer Gleichungen
f simulation *f* d'équations non linéaires
r моделирование *n* нелинейных уравнений

11584 simulation of physical phenomenon
d Simulation *f* eines physikalischen Vorgangs
f simulation *f* d'un phénomène physique
r моделирование *n* физического процесса

11585 simulation procedure
d Simulationsverfahren *n*
f procédé *m* de simulation
r метод *m* симулирования

11586 simulation result
d Simulationsergebnis *n*
f résultat *m* de simulation
r результат *m* моделирования

11587 simulation system
d Simulationssystem *n*
f système *m* de simulation
r система *f* моделирования

11588 simulation techniques
d Simulationstechnik *f*
f technique *f* de simulation
r техника *f* моделирования

11589 simulation testing
d Simulationsprüfung *f*
f test *m* de simulation
r проверка *f* методом моделирования

11590 simulator
d Simulator *m*

    f simulateur *m*
    r моделирующее устройство *n*; имитатор *m*

11591 simulator control
    d Simulatorsteuerung *f*
    f commande *f* de simulateur
    r управление *n* моделирующей системой

11592 simulator processor
    d Simulationsprozessor *m*
    f processeur *m* de simulation
    r процессор *m* моделирующей системы

11593 simultaneous
    d simultan
    f simultané
    r совместный; одновременный

11594 simultaneous control in distribution networks
    d Simultansteuerung *f* in Verteilungsnetzen
    f commande *f* simultanée dans réseaux de distribution
    r совместное управление *n* в распределительных сетях

11595 simultaneous executing
    d simultane Ausführung *f*
    f exécution *f* simultanée
    r совместное исполнение *n*

11596 simultaneous movements
    d Simultanbewegungen *fpl*
    f mouvements *mpl* simultanés
    r синхронные движения *npl*

11597 simultaneous operation
    d gleichzeitiger Betrieb *m*
    f opération *f* simultanée; fonctionnement *m* simultané
    r синхроннопротекающая операция *f*

11598 simultaneous processing
    d Simultanverarbeitung *f*
    f traitement *m* simultané
    r совместная обработка *f*

11599 sine-cosine potentiometer
    d Sinus-Kosinus-Potentiometer *n*
    f potentiomètre *m* sinus-cosinus
    r синусно-косинусный потенциометр *m*

    * **single accuracy** → **11615**

    * **single-axis laser gyroscope** → **8920**

11600 single-board microprocessor
    d Einzelsteckeinheit-Mikroprozessor *m*
    f microcalculateur *m* sur circuit imprimé

    unique
    r одноплатная микропроцессорная система *f*

11601 single-bus structure
    d Einzelbusstruktur *f*
    f structure *f* à bus unique
    r одношинная структура *f*

11602 single-channel analyzer
    d Einkanalanalysator *m*
    f analyseur *m* à voie unique
    r одноканальный анализатор *m*

11603 single-circuit
    d einschleifig
    f à circuit unique
    r одноконтурный; односхемный

    * **single current** → **8601**

11604 single-cycle equivalent
    d Eintaktzyklusäquivalent *n*
    f équivalent *m* à séquence unique
    r однотактный эквивалент *m*

11605 single-duty controller
    d Einzweckregler *m*
    f régulateur *m* à application spéciale
    r специальный регулятор *m*

11606 single instrumentation
    d Einzelinstrumentierung *f*
    f instrumentation *f* locale
    r индивидуальная контрольно-измерительная аппаратура *f*

11607 single-layer board
    d Einzelfehlerkorrektur *f*
    f correction *f* d'erreur unique
    r исправление *n* одиночных ошибок

11608 single-level interrupt system
    d Einebenen-Interruptsystem *n*
    f système *m* d'interruption à niveau unique
    r одноуровневая система *f* прерываний

11609 single-line controller
    d Einzelleitungssteuereinheit *f*
    f unité *f* de commande de ligne unique
    r одноконтурный контроллер *m*

    * **single-loop safety system** → **8709**

11610 single-mode operation
    d Monomodebetrieb *m*; Einmodenbetrieb *m*
    f fonctionnement *m* sur un mode unique; fonctionnement monomode
    r одномодовый режим *m*

11611 single-mode optical fiber sensor
d Monomode-Lichtwellenleitersensor *m*
f capteur *m* à fibre optique unimodale
r одномодовый световодный датчик *m*

11612 single-mode optical pulse
d einmodiger optischer Impuls *m*
f impulsion *f* optique en mode unique
r одномодовый оптический импульс *m*

11613 single-phase clock
d Einphasentakt *m*
f rythme *m* à phase unique
r однотактная синхронизация *f*

11614 single-phase earth-fault protection set
d Schutz *m* bei Einphasenerdschluß
f protection *f* contre les défauts monophasés à la terre
r защита *f* при однофазных замыканиях на землю

11615 single precision; single accuracy
d einfache Genauigkeit *f*
f précision *f* simple
r одинарная точность *f*

11616 single-program mode
d Einzelprogrammbetriebsart *f*; Einprogrammbetrieb *m*
f mode *m* à programm unique
r однопрограммный режим *m*

11617 single-purpose automatic machine; single-purpose automaton
d Einzweckautomat *m*
f automaton *m* à un seul but; machine *f* automatique à un seul but
r специализированный автомат *m*

* **single-purpose automaton → 11617**

11618 single-purpose manipulator
d Einzweckmanipulator *m*
f manipulateur *m* à un seul but
r целевой манипулятор *m*

11619 single-shot blocking oscillator
d monostabiler Sperrschwinger *m*
f oscillateur *m* bloqué monostable
r моностабильный блокинг-генератор *m*

11620 single-signal receiver
d Einzeichensignalempfänger *m*; hochselektiver Empfänger *m*
f récepteur *m* à haute selectivité
r узкополосный приёмник *m*

11621 single-speed floating action
d gleitendes Verhalten *n* mit konstanter Geschwindigkeit
f action *f* flottante à vitesse unique
r астатическое действие *n* с постоянной скоростью

11622 single-speed floating control
d Integrationsregelung *f*; astatische Regelung *f* mit konstanter Geschwindigkeit
f réglage *m* flottant à vitesse constante
r астатическое регулирование *n* с постоянной скоростью

* **single-stage amplifier → 8932**

11623 single-stage process
d Einstufenprozess *m*
f procédé *m* à simple effet
r одноступенчатый процесс *m*

11624 single-stand unit
d Einstranganlage *f*
f installation *f* en une ligne
r единичная установка *f*

11625 single-stationary detector system
d System *n* mit einem ruhenden Detektor
f système *m* à détecteur stable unique
r система *f* с одним неподвижным детектором

11626 single-step method
d Einzelschrittverfahren *n*
f méthode *f* pas à pas
r одношаговый метод *m*

11627 single-step mode; single-step operation; one-step operation
d Einzelschrittarbeitsweise *f*; Schrittbetrieb *m*; schrittweise Operation *f*
f mode *m* de travail pas à pas; operation *f* pas à pas
r работа *f* в пошаговом режиме

* **single-step operation → 11627**

11628 single-switching test
d Einzelschaltkontrolle *f*
f contrôle *m* en commutation simple
r контроль *m* при одиночном переключении

11629 single-variable control system
d Einfachregelkreis *m*
f système *m* à une variable
r однопараметровая система *f* регулирования

11630 **singular automaton**
  d singulärer Automat *m*
  f automate *m* singulier
  r сингулярный автомат *m*

11631 **singular matrix**
  d singuläre Matrix *f*
  f matrice *f* singulaire
  r вырожденная матрица *f*

11632 **singular point**
  d singulärer Punkt *m*
  f point *m* singulier; singularité *f*
  r особая точка *f*

11633 **singular trajectory**
  d singuläre Trajektorie *f*
  f trajectoire *f* singulière
  r особая траектория *f*

11634 **sinusoidal disturbance**
  d Sinusstörung *f*
  f perturbation *f* sinusoïdale
  r синусоидальная помеха *f*

11635 **sinusoidal input**
  d sinusförmiges Eingangssignal *n*
  f signal *m* sinusoïdal d'entrée
  r синусоидальный входной сигнал *m*

11636 **sinusoidal modulation**
  d sinusförmige Modulation *f*
  f modulation *f* sinusoïdale
  r синусоидальная модуляция *f*

11637 **sinusoidal signal generator**
  d Sinussignalgenerator *m*; Generator *m*
     sinusförmiger Signale
  f générateur *m* d'ondes sinusoïdales
  r генератор *m* синусоидальных сигналов

11638 **situation display**
  d Situationsanzeige *f*
  f affichage *m* de situation
  r индикация *f* ситуации

11639 **size control**
  d Größensteuerung *f*
  f commande *f* de la dimension
  r контроль *m* размеров

  *   **skew → 8213**

  *   **slave processor → 12047**

11640 **slide movement**
  d Schlittenbewegung *f*
  f mouvement *m* par le chariot

  r скольжение *n*

11641 **slide reference point**
  d Schlittenbezugspunkt *m*
  f point *m* de référence d'outil
  r скользящая опорная точка *f*

11642 **sliding regime**
  d Gleitzustand *m*
  f régime *m* pulsant
  r скользящий режим *m*

11643 **slip control with logical control element**
  d Schlupfsteuerung *f* mit logischem
     Schaltelement
  f contrôle *f* du glissement à l'aide d'élément
     logique de branchement
  r проверка *f* скольжения при помощи
     логического регулирующего элемента

11644 **slip regulator**
  d Schlupfregler *m*
  f régulateur *m* de glissement
  r регулятор *m* скольжения

11645 **slope**
  d Steilheit *f*
  f pente *f*
  r крутизна *f*

11646 **slope of characteristics**
  d Steilheit *f* von Kennlinien
  f pente *f* des caractéristiques
  r наклон *m* характеристик

11647 **slow-changing function method**
  d Methode *f* der langsam veränderlichen
     Funktionen
  f méthode *f* de fonctions à variations lentes
  r метод *m* медленно изменяющихся
     функций

11648 **slowing-down circuit**
  d Verzögerungsschaltung *f*
  f circuit *m* retardateur
  r схема *f* замедления

  *   **slugging → 1925**

11649 **small automatization**
  d kleine Automatisierung *f*
  f petite automatisation *f*
  r малая автоматизация *f*

11650 **small deflection method linearization**
  d Linearisierung *f* durch kleine Abweichungen
  f linéarisation *f* par petites déviations
  r линеаризация *f* методом малых
     отклонений

579

11651 **small electropneumatical regulateur**
d elektropneumatischer Kleinregler *m*
f petit régulateur *m* électropneumatique
r малый электропневматический
    регулятор *m*

11652 **small parameter**
d kleiner Parameter *m*
f petit paramètre *m*
r малый параметр *m*

11653 **small parameter method**
d Methode *f* des kleinen Parameters
f méthode *f* du petit paramètre
r метод *m* малого параметра

11654 **small perturbation method**
d Kleinstörungsmethode *f*
f méthode *f* des petites perturbations
r метод *m* малых возмущений

11655 **small signal**
d Kleinsignal *n*
f signal *m* faible
r слабый сигнал *m*

11656 **small signal analysis**
d Kleinsignalanalyse *f*
f analyse *f* en régime de signal faible
r анализ *m* слабых сигналов

11657 **small signal capacitance**
d Kleinsignalkapazität *f*
f capacité *f* aux signaux faibles
r ёмкость *f* слабого сигнала

11658 **small signal current gain**
d Kleinsignalstromverstärkung *f*
f amplification *f* de courant de petits signaux
r усиление *n* по току в режиме малого
    сигнала

11659 **smooth** *v*
d glätten
f glisser
r сглаживать

11660 **smoothed non-linearity**
d glatte Nichtlinearität *f*
f non-linéarité *f* nivelée
r плавная нелинейность *f*

11661 **smoothing**
d Glättung *f*
f amortissement *m*; aplatissement *m*
r сглаживание *n*

11662 **smoothing circuit**

d Glättungskreis *m*; Glättungsschaltung *f*
f circuit *m* épurateur; circuit de filtrage;
    montage *m* à aplatissement
r сглаживающий контур *m*

11663 **smoothing coefficient**
d Glättungskoeffizient *m*
f coefficient *m* d'aplatissement
r коэффициент *m* сглаживания

11664 **smoothing filter**
d Glättungsfilter *n*; glättendes Filter *n*
f filtre *m* épurateur
r сглаживающий фильтр *m*

11665 **smoothing reactor**
d Glättungsdrossel *f*; Ausgleichdrosselspule *f*
f bobine *f* de filtrage
r сглаживающий дроссель *m*

11666 **snap action**
d Schnappwirkung *f*; Schnellwirkung *f*
f action *f* brusque
r мгновенное действие *n*

11667 **snap-action contacts**
d Schnappkontakte *mpl*
f contacts *mpl* à déclic
r контакты *mpl* мгновенного действия

* **snap-action switch** → 10534

11668 **snap actuation**
d Schnappbetätigung *f*
f action *f* brusque
r мгновенное включение *n*

11669 **snap closing**
d Momenteinschaltung *f*;
    Schnappeinschaltung *f*
f enclenchement *m* brusque
r мгновенное замыкание *n*

* **snap-switch** → 10534

11670 **soft oscillations**
d weiche Schwingungen *fpl*
f oscillations *fpl* douces
r мягкие колебания *npl*

11671 **soft self-excitation**
d weiche Selbsterregung *f*
f accrochage *m* doux
r мягкое самовозбуждение *n*

11672 **software**
d Software *f*; Programmierhilfen *fpl*

f software *m*; logiciel *m*
r программное обеспечение *n*;
программные системы *fpl*

11673 softwareaided identification system
   d softwaregestütztes Erkennungssystem *n*
   f système *m* d'identification à l'aide de
     software
   r программное обеспечение *n* системы
     идентификации

11674 software application
   d Softwareanwendung *f*; Softwareapplikation *f*
   f application *f* de software
   r применение *n* программного обеспечения

11675 software architecture
   d Software-Struktur *f*;
     Programmsystemstruktur *f*
   f structure *f* logicielle
   r структура *f* программного обеспечения

11676 software-controlled
   d software-gesteuert; programmgesteuert
   f à commande logicielle
   r программно управляемый

11677 software development system
   d Software-Entwicklungssystem *n*
   f système *m* de développement logiciel
   r система *f* разработки программного
     обеспечения

11678 software diagnostic
   d Software-Diagnose *f*;
     Software-Fehlersuche *f*; Diagnose *f* mittels
     Programms
   f diagnostic *m* logiciel
   r программная диагностика *f*

11679 software project
   d Softwareprojekt *n*
   f projet *m* de software
   r проект *m* программного обеспечения

11680 software research
   d Software-Forschung *f*;
     Systemunterlagenforschung *f*
   f recherche-software *f*
   r исследование *n* средств
     программирования

11681 software technique
   d Softwaretechnik *f*
   f technique *f* de software
   r техника *f* программного обеспечения

11682 solenoid actuator

   d Solenoidantrieb *m*
   f commande *f* par solénoïde
   r соленоидный привод *m*

11683 solenoid servomechanism
   d Solenoidservomechanismus *m*
   f servomécanisme *m* à solénoïde
   r соленоидный сервомеханизм *m*

11684 solenoid valve
   d Solenoidventil *n*; Magnetventil *n*
   f vanne *f* à solénoïde
   r соленоидный вентиль *m*

11685 solid-state circuits
   d Festkörperschaltkreise *mpl*
   f circuits *mpl* à état solide
   r твердотельная схема *f*

11686 solution error
   d Lösungsfehler *m*
   f erreur *f* de solution
   r погрешность *f* решения

11687 solution of initial value problems
   d Lösung *f* von Anfangswertproblemen
   f résolution *f* des problèmes de conditions
     initiales
   r решение *n* задач с начальными условиями

11688 solution satisfying stability conditions
   d Stabilitätsbedingungen erfüllende Lösung *f*
   f solution *f* assurant stabilité
   r решение n, удовлетворяющее условиям
     устойчивости

11689 solvability
   d Lösbarkeit *f*; Auflösbarkeit *f*
   f résolubilité *f*
   r разрешимость *f*

* **sonic delay line** → 229

11690 sonic vibrations
   d akustische Vibrationen *fpl*
   f vibrations *fpl* acoustiques
   r звуковые колебания *npl*

11691 sort *v*
   d sortieren
   f assortir; trier
   r сортировать

11692 sorting process
   d Sortiervorgang *m*
   f processus *m* de tri
   r процесс *m* сортировки

11693 **sound field measurement of ultrasonic instruments**
  d Ausmessung *f* von Schallfeldern von Ultraschallgeräten
  f mesurage *m* des champs sonores d'instruments ultrasoniques
  r измерение *n* звуковы полей ультразвуковых приборов

11694 **sound information**
  d Schallinformation *f*
  f information *f* de son
  r звуковая информация *f*

11695 **sound pressure method**
  d Schalldruckverfahren *n*
  f procédé *m* de pression sonore
  r метод *m* звукового давления

11696 **sound sensor**
  d Schallsensor *m*; Geräuschsensor *m*
  f capteur *m* sonique
  r акустический сенсор *m*

  * **sound signal** → 1112

11697 **source configuration**
  d Quellenkonfiguration *f*
  f configuration *f* d'origine
  r исходная конфигурация *f*

11698 **source impedance**
  d Quellwiderstand *m*
  f impédance *f* de sources
  r внутреннее сопротивление *n* [источника]

11699 **source module**
  d Quellenmodul *m*
  f module *m* d'origine
  r исходный модуль *m*

11700 **source of control pulses**
  d Steuerimpulsquelle *f*
  f source *f* d'impulsions de commande
  r источник *m* управляющих импульсов

11701 **source of defects; source of error; error source**
  d Fehlerquelle *f*
  f source *f* de défauts; source d'erreur[s]
  r источник *m* ошибок

11702 **source of energy**
  d Energiequelle *f*
  f source énergétique
  r источник *m* энергии

  * **source of error** → 11701

11703 **source program**
  d Quellprogramm *n*
  f programme *m* source
  r исходная программа *f*

11704 **space of control actions**
  d Raum *m* der Steuerwirkungen
  f espace *m* des actions de commande
  r пространство *n* управляющих воздействий

11705 **space-time yeld**
  d Raum-Zeit-Ausbeute *f*
  f rendement *m* espace-temps
  r пространственно-временной выход *m*

11706 **spare channel**
  d Reservekanal *m*
  f canal *m* de réserve
  r запасной канал *m*

11707 **spark-type induction heating generator**
  d Funkenoszillator *m* für induktive Erwärmung
  f générateur *m* à étincelles pour chauffage par induction
  r искровой генератор *m* для индукционного нагрева

11708 **spatial coordinate system**
  d räumliches Koordinatensystem *n*
  f système *m* de coordonnées spatial
  r пространственная координатная система *f*

11709 **special application**
  d Spezialanwendung *f*
  f application *f* spéciale
  r специальное применение *n*

11710 **special circuitry**
  d Spezialschaltung *f*
  f circuit *m* spécial
  r специальные схемы *fpl*

  * **special dialogue system** → 12232

11711 **special equipment**
  d Sonderausrüstung *f*
  f équipement *m* spécial
  r специальное оборудование *n*

11712 **special function**
  d Sonderfunktion *f*
  f fonction *f* spéciale
  r специальная функция *f*

11713 **special instruction**
  d Spezialbefehl *m*; Sonderbefehl *m*
  f instruction *f* spéciale
  r специальная инструкция *f*

11714 specialized device
d Spezialbaustein *m*; spezialisierter Baustein *m*
f module *m* spécialisé
r специализированное устройство *n*

11715 specialized manipulator use
d spezialisierter Manipulatoreinsatz *m*
f utilisation *f* de manipulateur spécialisée
r специализированное применение *n*
    манипулятора

11716 special machine
d Spezialmaschine *f*
f machine *f* spéciale
r машина *f* специального назначения

11717 special programming languages
d spezielle Programmiersprachen *fpl*
f langages *mpl* de programmation spécialisés
r специальные языки *mpl*
    программирования

11718 special programming strategy
d spezielle Programmierstrategie *f*
f stratégie *f* de pragrammation spéciale
r специальная стратегия *f*
    программирования

11719 special-purpose computer
d Spezial[zweck]rechner *m*; spezialisierter
    Rechner *m*
f ordinateur *m* spécial
r специализированный компьютер *m*

11720 special-purpose logic chip
d Spezialzweck-Logikschaltkreis *m*
f circuit *m* logique spécial
r специализированная логическая
    микросхема *f*

11721 special sensor processor
d spezieller Sensorprozessor *m*
f processeur *m* de senseur spécial
r специализированный сенсорный
    процессор *m*

11722 special states
d spezielle Zustände *mpl*
f états *mpl* spéciaux
r специальные состояния *npl*

11723 special test program; special test routine
d Spezialtestprogramm *n*; spezielles
    Prüfprogramm *n*
f programme *m* de test spécial; routine *f* d'essai
    spéciale
r специальная тестпрограмма *f*

* **special test routine** → 11723

11724 specification
d Spezifikationssprache *f*
f spécification
r спецификация *f*

11725 specific impulse
d spezifischer Impuls *m*
f impulsion *m* spécifique
r удельный импульс *m*

11726 specific shape of work piece
d spezifische Form *f* eines Werkstücks
f forme *f* spécifique de pièce d'œvre
r специфичная форма *f* обрабатываемой
    детали

11727 specific use
d spezifische Verwendung *f*
f usage *m* spécifique
r специфичное применение *n*

11728 specified data byte
d spezifiziertes Datenbyte *n*
f byte *m* spécifié des données
r специфицированный байт *m* данных

11729 spectral absorptance
d spektraler Absorptionsgrad *m*
f facteur *m* d'absorption spectral
r спектральный коэффициент *m*
    поглощения

11730 spectral analysis in hydraulic systems
d Spektralanalyse *f* in hydraulischen Systemen
f analyse *f* spectrale dans les systèmes
    hydrauliques
r спекттральный анализ *m* в
    гидравлических системах

11731 spectral attenuation
d Dämpfungsspektrum *n*
f affaiblissement *m* spectral
r спектр *m* затухания

11732 spectral curve
d Spektralkurve *f*
f courbe *f* spectrale
r спектральная кривая *f*

11733 spectral emissivity
d spektraler Emissionsgrad *m*; spektrale
    Emissionsfähigkeit *f*
f émissivité *f* spectrale
r спектральный коэффициент *m* излучения

11734 spectral error density
d spektrale Fehlerdichte *f*

f densité *f* spectrale d'erreurs
r спектральная плотность *f* ошибки

11735 spectral function
d Spektralfunktion *f*
f fonction *f* spectrale
r спектральная функция *f*

11736 spectral index
d Spektralindex *m*
f index *m* spectral
r спектральный показатель *m*

11737 spectral information
d Spektralangaben *fpl*
f données *fpl* spectrales
r спектральная информация *f*

11738 spectral loss curve
d spektrale Dämpfungskurve *f*
f courbe *f* d'aaénuation spectrale
r спектральная характеристика *f* затухания

* **spectral response → 11740**

11739 spectral-response characteristic; spectral-response curve
d Spektralcharakteristik *f*; spektrale Verteilungscharakteristik *f*
f caractéristique *f* spectrale; courbe *f* de réponse spectrale
r спектральная характеристика *f*

* **spectral-response curve → 11739**

11740 spectral responsivity; spectral response
d Spektralempfindlichkeit *f*
f sensibilité *f* spectrale
r спектральная чувствительность *f*

11741 spectral selectivity
d spektrale Selektivität *f*; Spektralselektivität *f*
f sélectivité *f* spectrale
r спектральная избирательность *f*

11742 spectrobolometer
d Spektralbolometer *n*
f bolomètre *m* spectral
r спектроболометр *m*

11743 spectrochemical measurement with digital counter
d spektrochemische Messung *f* mit Digitalzähler
f mesure *f* spectrochimicale à compteur digital
r спектрохимическое измерение *n* с применением цифрового счётчика

11744 spectrometer automatism
d Spektrometerautomatik *f*
f système *m* automatique de spectromètre
r автоматика *m* спектрометра

11745 spectrum analysis
d Spektralanalyse *f*
f analyse *f* spectrale
r спектральный анализ *m*

11746 spectrum analyzer
d Spektralanalysator *m*
f analyseur *m* spectral
r спектроанализатор *m*

11747 speed control
d Geschwindigkeitsregelung *f*; Geschwindigkeitssteuerung *f*
f régulation *f* de vitesse; réglage *m* de vitesse
r регулирование *n* скорости

11748 speed difference measuring device
d Geschwindigkeitsdifferenzmessgerät *n*
f appareil *m* mesureur de la différence de vitesses
r прибор *m* для измерения разности скоростей

11749 speeding-up of operations
d Operationsbeschleunigung *f*
f accélération *f* des opérations
r ускорение *n* операций

11750 speed modulation
d Geschwindigkeitsmodulation *f*
f modulation *f* par vitesse
r модуляция *f* по скорости

11751 speedometer
d Geschwindigkeitsmesser *m*
f indicateur *m* de vitesse; tachymètre *m*
r измеритель *m* скорости; тахометр *m*

11752 speed range
d Geschwindigkeitsbereich *m*
f champ *m* de vitesse
r диапазон *m* скоростей

11753 split control section
d getrennter Steuerteil *m*
f section *f* de commande splittée
r секционный блок *m* управления

11754 split load
d aufgeteilte Belastung *f*
f charge *f* partagée
r разделенная нагрузка *f*

11755 **spot analysis; drop analysis**
    d Tüpfelanalyse *f*
    f analyse *f* à la touche; analyse par goutte
    r капельный анализ *m*

11756 **spurious modulation**
    d parasitische Modulation *f*;
       Nebenmodulation *f*
    f modulation *f* parasite
    r паразитная модуляция *f*

11757 **spurious pulse**
    d Fehlimpuls *m*
    f impulsion *f* fautive
    r паразитный импульс *m*

11758 **spurious signal**
    d Störsignal *n*
    f signal *m* parasite
    r паразитный сигнал *m*

11759 **square hysteresis loop**
    d rechteckige Hystereseschleife *f*
    f boucle *f* d'hystérésis rectangulaire
    r прямоугольная петля *f* гистерезиса

11760 **square matrix**
    d quadratische Matrix *f*
    f matrice *f* carrée
    r квадратичная матрица *f*

11761 **square-wave modulation**
    d rechteckige Impulsmodulation *f*
    f modulation *f* par impulsions rectangulaires
    r модуляция *f* прямоугольными импульсами

    * **SRAM** → 11862

    * **stability analysis** → 845

11762 **stability behaviour**
    d Stabilitätsverhalten *n*
    f comportement *m* de stabilité
    r поведение *n* [процесса] в устойчивом
       состоянии

11763 **stability behaviour of motion**
    d stabiles Bewegungsverhalten *n*
    f comportement *m* de mouvement stable
    r устойчивые характеристики *fpl* движения

11764 **stability behaviour of two-loop control
       systems**
    d Stabilitätsverhalten *n* von
       Zweifachregelkreisen
    f comportement *m* de stabilité de systèmes
       asservis à deux boucles
    r устойчивость *f* работы двух контуров
       регулирования

    * **stability boundary** → 3540

11765 **stability check device**
    d Stabilitätsprüfeinrichtung *f*
    f contrôleur *m* d'étalonnage
    r устройство *n* для проверки устойчивости

11766 **stability conditions**
    d Stabilitätsbedingungen *fpl*
    f conditions *fpl* de stabilité
    r условия *npl* устойчивости

11767 **stability criterion**
    d Stabilitätskriterium *n*
    f critère *m* de stabilité
    r критерий *m* устойчивости

    * **stability degree** → 3897

11768 **stability domain determination**
    d Stabilitätsbereichabgrenzung *f*
    f détermination *f* du domaine de stabilité
    r выделение *n* областей устойчивости

11769 **stability error**
    d Stabilitätsfehler *m*
    f erreur *f* de stabilité
    r погрешность *f* устойчивости

    * **stability estimation** → 5206

11770 **stability in the large; global stability**
    d Stabilität *f* im Großen; globale Stabilität
    f stabilité *f* globale
    r устойчивость *f* в целом

11771 **stability in the small; local stability**
    d Stabilität *f* im Kleinen; lokale Stabilität
    f stabilité *f* locale
    r устойчивость *f* в малом

    * **stability limit** → 3540

    * **stability line** → 7408

11772 **stability margin**
    d Stabilitätsreserve *f*; Stabilitätsrand *m*
    f marge *f* de stabilité
    r запас *m* устойчивости

11773 **stability of perturbed motion**
    d Störbewegungsstabilität *f*
    f stabilité *f* de mouvement perturbé
    r устойчивость *f* возмущенного движения

11774 **stabilization**
    d Stabilisierung *f*

f stabilisation f
r стабилизация f

11775 stabilization system
d Stabilisierungssystem n
f système m de stabilisation
r система f стабилизации

11776 stabilization time
d Stabilisierungszeit f; Stabilisierungsdauer f
f temps m de stabilisation; durée f de
   stabilisation
r время n стабилизации

11777 stabilized current supply
d stabilisierte Stromversorgung f
f alimentation f stabilisée en courant
r питание n стабилизированным током

* **stabilized power supply → 10831**

11778 stabilized superconductor
d stabilisierter Supraleiter m
f supraconducteur m stabilisé
r стабилизированный сверхпроводник m

11779 stabilizer
d Stabilisator m
f stabiliseur m
r стабилизатор m

11780 stabilizing circuit
d Stabilisierungsstromkreis m
f chaîne f stabilsante; circuit m stabilisateur
r стабилизирующая цепь f

11781 stabilizing feedback network
d stabilisierendes Rückkopplungsnetzwerk n
f réseau m de réaction; réseau à rétrocouplage
r стабилизирующая сеть f обратной связи

11782 stable component; stable element
d stabiles Bauelement n; stabiles Element n
f composant m stable; élément m stable
r статическое звено n; устойчивое звено n

11783 stable control
d stabile Regelung f
f réglage m stable
r устойчивый процесс m регулирования

11784 stable data
d stabile Daten pl; eingeschwungene
   Datensignale npl
f données fpl stables
r установившиеся данные pl

* **stable element → 11782**

11785 stable equilibrium point
d Punkt m stabilen Gleichgewichtes
f point m d'équilibre stable
r точка f устойчивого равновесия

11786 stable nodal point
d stabiler Knotenpunkt m
f nœud m stable
r устойчивая узловая точка f

11787 stable node point quantity
d stabile Knotenpunktmenge f
f quantité f de point nodal stable
r количество n устойчивых узловых точек

11788 stable state
d stabiler Zustand m
f état m stable
r устойчивое состояние n

11789 stable system
d stabiles System n
f système m stable
r стабильная система f

11790 stage
d Stufe f; Stadium n
f étage m; degré m
r каскад m; стадия f; ступень f

11791 staggered circuits
d versetzte Stromkreise mpl
f circuits mpl décalés
r взаимно расстроенные контуры mpl

11792 stagnation pressure
d Staudruck m; Ruhedruck m
f pression f d'arrêt; pression au point de repos
r давление n [полного] торможения

* **standard → 8790**

11793 standard block; standard unit
d Standardblock m; Standardeinheit f
f bloc m standard; unité f standard
r стандартный блок m

11794 standard component
d Baueinheit f
f élément m de construction
r элемент m стандартного нормирования

11795 standard control circuit
d Standardregelkreis m
f circuit m de réglage standard; boucle f
   d'asservissement standard
r контур m стандартного регулирования

11796 standard design
 d Standardauslegung *f*
 f conception *f* standard
 r стандартный расчёт *m*

* **standard deviation** → 7925

11797 standard equilibrium potential
 d Standardgleichgewichtspotential *n*
 f potentiel *m* d'équilibre standard
 r стандартный равновесный потенциал *m*

11798 standard equipment
 d Standardausrüstung *f*; Normalausrüstung *f*
 f équipement *m* standard
 r стандартное оборудование *n*

11799 standard instrument
 d Standardgerät *n*; Normalgerät *n*
 f appareil *m* standard; appareil étalon
 r стандартный прибор *m*

11800 standard interface
 d Standardinterface *n*; Standardschnittstelle *f*
 f interface *f* standard
 r стандартный интерфейс *m*

11801 standardization; uniformalization
 d Vereinheitlichung *f*; Unifizierung *f*
 f standardisation *f*; unification *f*
 r стандартизация *f*; унификация *f*

11802 standardization principle
 d Normierungsprinzip *n*
 f principe *m* de normalisation
 r принцип *m* нормализации

11803 standardized application program
 d standardisiertes Anwendungsprogramm *n*
 f programme *m* d'application standardisé
 r стандартизованная прикладная программа *f*

11804 standardized device
 d standardisierter Baustein *m*
 f module *m* standardisé
 r стандартизированный модуль *m*

11805 standardized equipment
 d standardisierte Ausrüstung *f*
 f équipement *m* standardisé
 r стандартизованное оборудование *n*

11806 standardized modular system
 d standardisiertes Bausteinsystem *n*
 f système *m* modulaire standardisé
 r система *f* нормализованных модулей

11807 standardized structure
 d standardisierte Struktur *f*; standardisierter Aufbau *m*
 f structure *f* standardisée
 r стандартизированная структура *f*

11808 standard method
 d Standardmethode *f*; Einheitsmethode *f*
 f méthode *f* normale
 r стандартный метод *m*

11809 standard response spectrum
 d Standardantwortspektrum *n*; Standardresponsespektrum *n*
 f spectre *m* de réponse type; spectre de réponse étalon
 r стандартный спектр *m* чувствительности

11810 standard signal
 d Normsignal *n*
 f signal *m* étalon
 r эталонный сигнал *m*

11811 standard software
 d Standardprogrammausstattung *f*
 f software *m* standard
 r стандартное программное обеспечение *n*

11812 standard system
 d Einheitssystem *n*
 f système *m* standardisé
 r стандартная система *f*

11813 standard type
 d Standardmodell *n*
 f modèle *m* standard
 r стандартная модель *f*

* **standard unit** → 11793

11814 standby condition; standby status
 d Bereitschaftszustand *m*
 f état *m* de disponibilité
 r состояние *n* резерва; резервное состояние

11815 standby connection; back-up connection
 d Reservezuschaltung *f*
 f mise *f* en circuit de réserve; mise en circuit de secours
 r резервное подключение *n*

11816 standby consumption
 d Leistungsverbrauch *m* im Bereitschaftszustand
 f puissance *f* consommée en état d'attente
 r энергопотребление *n* в режиме резерва

11817 standby current
 d Nortstrom *m*; Bereitschaftsbetriebsstrom *m*

  f courant *m* de régime d'attente
  r ток *m* в режиме резерва

11818 standby emergency cooling system
  d Reserve-Notkühlsystem *n*;
      Reservekühlsystem *n*
  f système *m* de refroidissement de réserve
  r резервная система *f* аварийного
      охлаждения

11819 standby mode
  d Bereitschaftsbetrieb *m*; Reservebetrieb *m*
  f mode *m* de réserve
  r режим *m* резерва

  * **standby status** → 11814

  * **standby system** → 1579

11820 standby unit
  d Reserveeinheit *f*
  f unité *f* de réserve
  r резервное устройство *n*

11821 star connection
  d Sternschaltung *f*
  f couplage *m* en étoile; connexion *f* en étoile
  r схема *f* соединения звездой

11822 starting and breaking curve
  d Anlauf- und Bremskurve *f*
  f courbe *f* de démarrage et freinage
  r характеристика *f* пуска и торможения

  * **starting conditions** → 6617

11823 starting device
  d Inbetriebnahmegerät *n*
  f dispositif *m* mise en marche
  r пусковое устройство *n*

11824 starting of regulating circuits
  d Anfahren *n* von Regelkreisen
  f amorçage *m* de circuits de réglage
  r запуск *m* регулирующих схем

11825 starting-pulse action
  d Startimpulsverhalten *n*; Startimpulswirkung *f*
  f action *f* par impulsion initiale
  r пусковое воздействие *n*

11826 start point
  d Startpunkt *m*
  f point *m* start
  r начальная точка *f*

11827 start routine
  d Startprogramm *n*

  f programme *m* de démarrage
  r стартовая программа *f*

11828 start time
  d Startzeit *f*
  f durée *f* de start
  r время *n* старта

11829 start-up behaviour
  d Anfahrverhalten *n*
  f comportement *m* à la mise en marche;
      comportement au démarrage
  r поведение *n* в период пуска

11830 start-up control
  d Anlaufkontrolle *f*
  f contrôle *m* de démarrage
  r контроль *m* перед пуском

11831 start-up filter
  d Anfahrfilter *n*
  f filtre *m* de démarrage
  r пусковой фильтр *m*

11832 start-up instrumentation
  d Anfahrinstrumentierung *f*
  f instrumentation *f* de démarage
  r пусковая измерительная аппаратура *f*

11833 start-up operation
  d Anfahrbetrieb *m*
  f régime *m* de démarrage
  r пусковой режим *m*

11834 start-up ramp
  d Anfahrrampe *f*; Startrampe *f*
  f rampe *f* de démarrage
  r пусковое скачкообразное изменение *n*

11835 start-up test
  d Anfahrprüfung *f*
  f essai *m* de démarrage
  r пусковое испытание *n*

11836 start-up transient
  d Anfahrtransiente *f*
  f transitoire *m* de démarrage
  r пусковой переходный процесс *m*

  * **start value** → 6630

11837 state; status
  d Zustand *m*; Status *m*
  f état *m*; statut *m*
  r состояние *n*

11838 state analysis
  d Zustandsanalyse *f*

    f analyse *f* d'état
    r анализ *m* состояния

    *   **state change → 11898**

11839 state coding; coding of states
    d Zustandskodierung *f*
    f codification *f* d'état
    r кодирование *n* состояний

11840 state density
    d Zustandsdichte *f*
    f densité *f* d'états
    r плотность *f* состояний

11841 state function; function of state
    d Zustandsfunktion *f*
    f fonction *f* d'état
    r функция *f* состояния

11842 state space equation
    d Zustands[raum]gleichung *f*
    f équation *f* [de l'espace] d'état
    r уравнение *n* в пространстве состояний

11843 state transition
    d Zustandsübergang *m*
    f transition *f* d'état
    r переход *m* между состояниями

11844 static
    d statisch
    f statique
    r статический

11845 static accuracy
    d statische Genauigkeit *f*
    f précision *f* statique
    r статическая точность *f*

11846 static allocation
    d statische Zuordnung *f*
    f allocation *f* statique
    r статическое распределение *n*

11847 static analog device
    d statisches Analogongerät *n*
    f appareil *m* analogue statique
    r статическое моделирующее устройство *n*

11848 static balance
    d statisches Gleichgewicht *n*
    f équilibre *m* statique
    r статическое равновесие *n*

    *   **static calibrating plot → 11849**

**11849 static calibration curve; static calibrating
      plot**

    d statische Eichkurve *f*
    f courbe *f* de calibration statique; courbe
       d'étalonnage statique
    r статическая тарировочная
       характеристика *f*; статическая
       калибровочная кривая *f*

11850 static circuitry
    d statische Schaltung *f*
    f circuit *m* statique
    r статические схемы *fpl*

11851 static controller; static regulator
    d statischer Regler *m*
    f régulateur *m* statique
    r статический регулятор *m*

11852 static design
    d statische Berechnung *f*
    f calcul *m* statique
    r статический расчёт *m*

11853 static display
    d statische Sichtanzeige *f*
    f affichage *m* statique
    r статическая визуальная индикация *f*

11854 static efficiency
    d statischer Wirkungsgrad *m*
    f rendement *m* statique
    r коэффициент *m* полезного действия по
       статическим параметрам

11855 static instability
    d statische Instabilität *f*
    f instabilité *f* statique
    r статическая неустойчивость *f*

    *   **static load → 11856**

11856 static load[ing]
    d statische Belastung *f*; Ruhebelastung *f*
    f charge *f* statique
    r статическая нагрузка *f*

11857 static luminous sensitivity
    d statische Lichtempfindlichkeit *f*
    f sensibilité *f* lumineuse statique
    r статическая световая чувствительность *f*

11858 static magnetic logic element
    d statisches Magnetlogikelement *n*
    f élément *m* logique statique magnétique
    r статический логический магнитный
       элемент *m*

11859 static magnetic storage
    d statische Magnetspeicherung *f*

f emmagasinage *m* magnétique statique
r запоминание *n* на статических магнитных
   элементах

11860 static optimization
d statische Optimierung *f*
f optimisation *f* statique
r статическая оптимизация *f*

11861 static performance
d statisches Verhalten *n*
f comportement *m* statique
r статическое свойство *n*

11862 static random access memory; SRAM
d statischer Speicher *m* mit wahlfreiem
   Zugriff; SRAM
f mémoire *f* statique à accès libre
r статическое оперативное запоминающее
   устройство *n*

11863 static reactivity; global reactivity
d statische Reaktivität *f*; globale Reaktivität
f réactivité *f* statique; réactivité globale
r статическая реактивность *f*; глобальная
   реактивность

11864 static regime
d statischer Zustand *m*
f régime *m* statique
r статический режим *m*

* **static regulator** → 11851

11865 statics analysis
d statische Analyse *f*
f analyse *f* statique
r статический анализ *m*

11866 static sensitivity
d statische Empfindlichkeit *f*
f sensibilité *f* statique
r статическая чувствительность *f*

11867 static setting dimension
d statisches Einrichtmaß *n*
f dimension *f* de l'ajustage statique
r диапазон *m* статической настройки

11868 static solution
d statische Lösung *f*
f solution *f* statique
r статическое решение *n*

11869 static stability
d Standsicherheit *f*; Stabilität *f*
f stabilité *f*
r стабильность *f*; устойчивость *f*

11870 static switching unit
d statische Umschalteinheit *f*
f unité *f* de commutation statique
r статическое переключающее устройство *n*

11871 static system
d statisches System *n*
f système *m* statique
r статическая система *f*

11872 static temperature
d statische Temperatur *f*; Temperatur der
   Anströmung
f température *f* statique
r статическая температура *f*

11873 stationary
d stationär
f stationnaire
r стационарный

11874 stationary behaviour
d stationäres Verhalten *n*
f comportement *m* stationnaire; régime *m*
   permanent
r стационарное поведение *n*

11875 stationary function converter
d stationärer Funktionsumformer *m*
f convertisseur *m* stationnaire de fonction
r стационарный преобразователь *m*
   функции

11876 stationary iterative process
d stationärer Iterationsprozess *m*
f procédé *m* itératif stationnaire
r стационарный итерационный процесс *m*

11877 stationary linear system
d lineares stationäres System *n*
f système *m* stationnaire linéaire
r стационарная линейная система *f*

11878 stationary process
d stationärer Prozess *m*
f processus *m* stationnaire
r стационарный процесс *m*

11879 stationary random action
d stationäre stochastische Einwirkung *f*
f action *f* aléatoire stationnaire
r стационарное случайное воздействие *n*

11880 stationary random function
d stationäre stochastische Funktion *f*
f fonction *f* aléatoire stationnaire
r стационарная случайная функция *f*

11881 **stationary reactor**
   d stationärer Reaktor *m*
   f réacteur *m* stationnaire
   r стационарный реактор *m*

11882 **station control block**
   d Stationssteuerblock *m*
   f bloc *m* de commande de station
   r блок *m* управления станцией

11883 **station identifikation**
   d Stationsbestimmung *f*
   f identification *f* de la station
   r идентификация *f* станции

11884 **statistical compensation**
   d statistische Kompensation *f*
   f compensation *f* statistique
   r статистическая компенсация *f*

11885 **statistical estimations**
   d statistische Bewertungen *fpl*
   f évaluations *fpl* statistiques
   r статистические оценки *fpl*

11886 **statistical factor**
   d statistischer Faktor *m*
   f facteur *m* statistique
   r статистический коэффициент *m*

11887 **statistical linearization**
   d statistische Linearisierung *f*
   f linéarisation *f* statistique
   r статистическая линеаризация *f*

11888 **statistical model of system**
   d statistisches Modell *n* des Systems
   f modèle *m* statistique du système
   r статистическая модель *f* системы

   * **statistical processing** → **11892**

11889 **statistical program**
   d statistisches Programm *n*;
      Statistikprogramm *n*
   f programme *m* statistique
   r программа *f* обработки стстистических
      данных

11890 **statistical property**
   d statistische Eigenschaft *f*
   f propriété *f* statistique
   r статистическое свойство *n*

11891 **statistical test**
   d statistischer Test *m*
   f test *m* statistique
   r статистическая проверка *f*

11892 **statistical treatment; statistical processing**
   d statistische Verarbeitung *f*; statistische
      Auswertung *f*
   f traitement *m* statistique; exploitation *f*
      statistique
   r статистическая обработка *f*

11893 **statistical variable**
   d statistische Variable *f*
   f variable *f* statistique
   r статистическая переменная *f*

11894 **statistic analysis**
   d statistische Analyse *f*
   f analyse *f* statistique
   r статистический анализ *m*

11895 **statistics**
   d Statistik *f*
   f statistique *f*
   r статистика *f*

   * **status** → **11837**

11896 **status bit**
   d Statusbit *n*
   f bit *m* d'état
   r бит *m* состояния

11897 **status byte**
   d Zustandsbyte *n*; Statusbyte *n*
   f byte *m* d'état
   r байт *m* состояния

11898 **status change; state change**
   d Statuswechsel *m*; Zustandswechsel *m*
   f changement *m* d'état
   r изменение *n* состояния

11899 **status data**
   d Zustandsdaten *pl*
   f données *fpl* d'état
   r данные *pl* о состоянии

11900 **status-saving hardware**
   d Statusrettungsschaltung *f*; technische
      Lösung *f* zur Statusrettung
   f montage *m* de sauvegarde d'état
   r аппаратные средства *npl* для сохранения
      состояния

11901 **steady**
   d beharrend
   f établi
   r установившийся

11902 **steady state; steady-state regime**
   d stationärer Zustand *m*; stabiler Zustand;
      Beharrungszustand *m*; Dauerzustand *m*

f état *m* stable; état stationnaire; régime *m*
stationnaire
r стационарное состояние *n*;
установившийся режим *m*

11903 steady state characteristic
d stationärer Kennwert *m*
f caractéristique *f* stationnaire
r характеристика *f* стационарного режима

11904 steady-state conditions
d stationäre Zustandsbedingungen *fpl*
f conditions *fpl* d'état stationnaire
r условия *npl* установившегося процесса

* **steady-state error → 8901**

11905 steady-state oscillation
d stationäre Schwingung *f*
f oscillation *f* stationnaire
r установившиеся колебания *npl*

* **steady-state regime → 11902**

11906 steam apparatus
d Dampfapparat *m*
f appareil *m* à vapeur
r пароиспользующий аппарат *m*

11907 steam regulator
d Dampfregler *m*
f régulateur *m* de vapeur
r регулятор *m* притока пара

* **steering circuit → 3202**

11908 steering function
d Steuerungsfunktion *f*; Leitfunktion *f*
f fonction *f* de guidage
r функция *f* управления

11909 steering gate
d Steuergatter *n*
f porte *f* de commande
r управляющая вентильная схема *f*

11910 steering program
d Steuerprogramm *n*; organisatorisches
Programm *n*
f programme *m* directeur; routine *f* d'exécution
r программа *f* контроля

11911 step
d Schritt *m*
f pas *m*
r шаг *m*; ступень *f*

11912 step action

d Schrittwirkung *f*
f action *f* par échelons
r ступенчатое воздействие *n*

11913 step-by-step control; stepping control
d Schrittregelung *f*; Stufenregelung *f*; Schritt-
für-Schritt-Steuerung *f*
f commande *f* pas à pas; rélage *m* par
échelons; commande par paliers
r шаговое регулирование *n*; ступенчатое
управление *n*

11914 step-by-step operation
d Einzelschrittbetrieb *m*
f opération *f* pas à pas
r шаговый режим *m* работы

11915 step-by-step system
d Schritt-für-Schritt-System *n*
f système *m* pas à pas
r шаговая система *f*

11916 step-by-step tuning
d stufenweise Abstimmung *f*
f accord *m* pas à pas
r ступенчатая настройка *f*

* **step controlller → 11920**

* **step function → 7039**

11917 step function transformation
d Bildübertragung *f* der Stufenfunktion
f transformation *f* d'image de fonction étagée
r преобразование *n* ступенчатой функции

11918 step input
d Stufeneinwirkung *f*;
Stufeneingangswirkung *f*
f action *f* échelonnée; action d'entrée par
échelon
r входное ступенчатое воздействие *n*

11919 stepped curve distance-time protection
d Distanzschutz *m* mit Stufenkennlinie
f dispositif *m* de protection de distance à
caracteristique discontinue
r дистанционная защита *f* выдержки
времени со ступенчатой характеристикой

* **stepping control → 11913**

11920 step[ping] controller
d Schrittregler *m*
f régulateur *m* à action pas à pas
r шаговый регулятор *m*

11921 stepping distributor
d Schrittschaltwerk *n*

f distributeur *m* pas à pas
r шаговый распределитель *m*

11922 stepping extremal system
   d Extremalschrittsystem *n*;
      Extremalschrittschaltsystem *n*
   f système *m* extrémal pas à pas
   r экстремальная система *f* шагового типа

  * **stepping motor** → 9491

11923 stepping switch operation
   d Schrittschaltoperation *f*
   f opération *f* de commutation pas à pas
   r операция *f* шагового переключателя

11924 step response
   d Sprungantwort *f*
   f réponse *f* étagée
   r реакция *f* на ступенчатое возмущение

11925 step selectors for automatic operations
   d Schrittwähler *mpl* für selbsttätige
      Operationen
   f sélecteurs *mpl* à pas pour opérations
      automatiques
   r шаговые искатели *mpl* для
      автоматических операций

11926 step unit disturbance
   d Störung *f* durch Einheitssprung
   f perturbation *f* par échelon unitaire
   r возмущение *n* в виде единичного скачка

11927 step velocity input
   d sprungartige Geschwindigkeitsänderung *f*
   f variation *f* à échelon de vitesse
   r скачкообразное изменение *n* скорости
      входного сигнала

11928 stepwise correlation calculation
   d stufenweise Korrelationsberechnung *f*
   f méthode *f* successive de calcul de corrélation
   r расчёт *m* методом постадийной
      корреляции

11929 stepwise integration
   d schrittweise Integration *f*
   f intégration *f* graduelle
   r ступенчатая интеграция *f*

11930 sticking voltage
   d Haftspannung *f*; Klebespannung *f*
   f tension *f* limite; tension de collage
   r напряжение *n* прилипания

11931 stiffness coefficient
   d Steifigkeitskoeffizient *m*

f coefficient *m* de rigidité
r коэффициент *m* жёсткости

11932 stimulated transition frequency
   d Frequenz *f* des angeregten Überganges
   f fréquence *f* de transition stimulée
   r частота *f* стимулированного перехода

11933 stochastically disturbed system
   d stochastisch gestörtes System *n*
   f système *m* à perturbation aléatoire
   r система *f* со стохастическими помехами

11934 stochastic approximation
   d stochastische Approximation *f*
   f approximation *f* stochastique
   r стохастическая аппроксимация *f*

11935 stochastic automaton
   d stochastischer Automat *m*
   f automate *m* stochastique
   r стохастический автомат *m*

11936 stochastic control
   d stochastische Steuerung *f*
   f réglage *m* aléatoire; commande *f* stochastique
   r стохастическое управление *n*

11937 stochastic control process
   d stochastischer Regelungsprozess *m*
   f procédé *m* de réglage stochastique
   r стохастический процесс *m* регулирования

11938 stochastic elements
   d stochastische Elemente *npl*
   f éléments *mpl* stochastiques
   r стохастические элементы *mpl*

11939 stochastic hypothesis
   d stochastische Hypothese *f*
   f hypothèse *f* stochastique
   r стохастическая гипотеза *f*

11940 stochastic learning model
   d stochastisches Lernmodell *n*
   f modèle *m* à apprentissage stochastique
   r стохастическая модель *f* обучения

11941 stochastic optimizing method
   d stochastisches Optimierungsverfahren *n*
   f méthode *f* stochastique d'optimisation
   r стохастический метод *m* оптимизации

11942 stochastic simulation
   d stochastische Simulation *f*
   f simulation *f* stochastique
   r стохастическое моделирование *n*

11943 **stochastic system**
    d stochastisches System *n*
    f système *m* aléatoire
    r стохастическая система *f*

11944 **stop at end of sequence**
    d Anhalten *n* bei Ende des Satzes; Anhalten am Ende der Reihe
    f arrêt *m* en fin de séquence
    r ограничитель *m* на конце последовательности

11945 **stop band**
    d Sperrbereich *m*
    f bande *f* de coupure
    r полоса *f* запирания

11946 **stop cycle**
    d Stoppzyklus *m*
    f cycle *m* arrêt
    r цикл *m* остановки

11947 **stopping time**
    d Stoppzeit *f*
    f temps *m* d'arrêt
    r время *n* останова

11948 **stop-start control**
    d Ein-Aus-Schaltung *f*
    f commande *f* par tout ou rien; commande par bouton-poussoir
    r кнопочное управление *n*

11949 **storage assignment control**
    d steuerbare Speicherzuweisung *f*
    f assignation *f* réglable de la mémoire; répartition *f* réglable de la mémoire
    r управляемое распределение *n* памяти

11950 **storage automation**
    d Automatisierung *f* der Lagerung
    f automatisation *f* du magasinage
    r автоматизация *f* хранения

   * **storage control** → 8035

   * **storage efficiency** → 8037

11951 **storage function**
    d Speicherfunktion *f*
    f fonction *f* mémoire
    r функция *f* памяти

11952 **storage oscillograph**
    d Speicheroszillograf *m*
    f oscillographe *m* à mémoire
    r осциллограф *m* с накопителем

11953 **storage regeneration**
    d Speichererneuerung *f*; Speicherwiederherstellung *f*
    f régénération *f* de mémoire
    r обновление *n* памяти

11954 **storage speed**
    d Speichergeschwindigkeit *f*
    f vitesse *f* de mémoire; vitesse d'emmagasinage
    r быстродействие *n* запоминающего устройства

   * **storage system** → 8051

11955 **store** *v*
    d speichern; lagern; stapeln; aufspeichern
    f emmagasiner; mettre en mémoire; accumuler
    r накоплять; запасать

   * **store assembly** → 8030

11956 **store contens**
    d Speicherinhalt *m*
    f contenu *m* de mémoire
    r содержание *n* памяти

11957 **store data**
    d Speicherdaten *pl*
    f données *fpl* de mémoire
    r данные *pl* памяти

   * **stored error** → 193

11958 **store display unit**
    d Speicherbildgerät *n*
    f écran *m* à mémoire
    r блок *m* индикации содержимого памяти

   * **store distribution checking** → 8036

   * **store function** → 8041

11959 **store integrator**
    d Speicherintegrator *m*
    f intégrateur *m* de mémoire
    r интегратор *m* с памятью

   * **store location** → 8043

11960 **store mechanism; memory mechanism**
    d Speicherwerk *n*
    f mécanisme *m* de mémoire
    r механизм *m* памяти

   * **store principle** → 8046

   * **store-programmed control** → 8047

*   **store system arrangement** → 995

11961 stream factor; utilization factor
  d Auslastungsfaktor *m*
  f facteur *m* d'utilisation
  r коэффициент *m* использования

11962 streamline regime
  d Laminarbetriebszustand *m*
  f régime *m* laminaire
  r ламинарный режим *m*

11963 stream of handling objects
  d Strom *m* von Handhabeobjekten
  f flux *m* des objets de manutention
  r поток *m* объектов манипулирования

11964 stress deviator
  d Spannungsdeviator *m*
  f déviateur *m* de tension
  r девиатор *m* напряжения

11965 stress optical coefficient
  d optischer Spannungskoeffizient *m*
  f coefficient *m* photo-élastique
  r оптический коэффициент *m* напряжения

11966 stringed transducer; vibrating wire gauge
  d Saitengeber *m*
  f capteur *m* à corde
  r струнный датчик *m*

11967 strip-width controller
  d Bandbreitenregler *m*
  f régulateur *m* de largeur de bande de tôle
  r регулятор *m* ширины полосы

11968 strobe-pulse generator; strobe-pulse oscillator
  d Strobimpulsgenerator *m*
  f générateur *m* d'impulsions de fixation
  r генератор *m* стробимпульсов

*   **strobe-pulse oscillator** → 11968

11969 strobing
  d Strobierung *f*; Storben *n*
  f déclenchement *m* périodique
  r стробирование *n*

11970 strobometry
  d Strobometrie *f*
  f strobométrie *f*
  r стробометрия *f*

11971 stroboscope
  d Stroboskop *n*
  f stroboscope *m*

  r стробоскоп *m*

11972 stroboscopic method
  d stroboskopisches Verfahren *n*
  f méthode *f* stroboscopique
  r стробоскопический метод *m*

11973 strongly monotonic
  d eigentlich monoton; streng monoton
  f exactement monotone; strictement monotone
  r строго монотонный

11974 structural
  d strukturell
  f structurel
  r структурный

11975 structural change
  d Strukturwandlung *f*
  f changement *m* de structure
  r изменение *n* структуры

11976 structural formula
  d Strukturformel *f*
  f formule *f* structurale
  r структурная формула *f*

11977 structural instability
  d strukturelle Instabilität *f*
  f instabilité *f* structurelle
  r структурная неустойчивость *f*

11978 structural-kinetic unit
  d strukturkinetische Einheit *f*
  f unité *f* structurale-cinétique
  r структурно-кинетическая единица *f*

11979 structurally stable system
  d strukturell stabiles System *n*
  f système *m* à stabilité structurelle
  r структурно устойчивая система *f*

11980 structurally unstable system
  d strukturell unstabiles System *n*
  f système *m* à structure instable
  r структурно неустойчивая система *f*

*   **structural member** → 2815

11981 structural model
  d Strukturmodell *n*
  f modèle *m* structural
  r модель *f* структуры

11982 structural parameter
  d Strukturparameter *m*
  f paramètre *m* de structure
  r параметр *m* структуры

11983 structural property
  d strukturelle Eigenschaft *f*
  f propriété *f* structurale
  r структурное свойство *n*

11984 structural reliability
  d strukturelle Zuverlässigkeit *f*
  f fiabilité *f* structurelle
  r конструкционная надёжность *f*

11985 structural theory of automata
  d strukturelle Automatentheorie *f*
  f théorie *f* structurale d'automates
  r структурная теория *f* автоматов

11986 structure
  d Struktur *f*
  f structure *f*
  r структура *f*

11987 structure analysis
  d Strukturanalyse *f*
  f analyse *f* de structure
  r структурный анализ *m*

11988 structured circuit
  d strukturierte Schaltung *f*
  f circuit *m* structuré
  r структурированная схема *f*

11989 structured programming
  d strukturierte Programmierung *f*
  f programmation *f* structurée
  r структурное программирование *n*

11990 structure element
  d Strukturelement *n*
  f élément *m* de structure
  r структурный элемент *m*

11991 structure input
  d Struktureingabe *f*
  f entrée *f* de structure
  r ввод *m* структуры

11992 structure model
  d Strukturmodell *n*
  f modèle *m* de structure
  r структурная модель *f*

11993 structure modelling; structure simulation
  d Strukturmodellierung *f*
  f modélisation *f* structurale
  r моделирование *n* структуры

11994 structure of computer system; computer system structure
  d Struktur *f* des Rechnersystems;
     Rechnersystemstruktur *f*

  f structure *f* du système de calculateur
  r структура *f* компьютерной системы

11995 structure of digits
  d Ziffernstruktur *f*; Struktur *f* der Ziffern
  f structure *f* des chiffres
  r структура *f* цифр

11996 structure optimization
  d Strukturoptimierung *f*
  f optimisation *f* structurale
  r оптимизация *f* структуры

11997 structure output
  d Strukturausgabe *f*
  f sortie *f* de structure
  r вывод *m* структуры

11998 structure parameter method
  d Strukturparametermethode *f*
  f méthode *f* des paramètres structurals
  r метод *m* структурных параметров

* structure simulation → 11993

11999 structure simulator
  d Struktursimulator *m*
  f simulateur *m* de structure
  r устройство *n* для моделирования структуры

12000 structure stability
  d Strukturstabilität *f*
  f stabilité *f* de structure
  r структурная устойчивость *f*

12001 structure synthesis
  d strukturelle Synthese *f*
  f synthèse *f* structurelle
  r структурный синтез *m*

12002 subcarrier frequency
  d Hilfsträgerfrequenz *f*
  f fréquence *f* de sous-porteuse
  r поднесущая частота *f*

12003 subcircuit
  d Teilschaltung *f*; Nebenstromkreis *m*
  f sous-circuit *m*
  r подсхема *f*; часть *f* схемы

12004 subharmonic oscillation
  d subharmonische Schwingung *f*
  f oscillation *f* sous-harmonique
  r субгармоническое колебание *n*

12005 subharmonic phase-locked oscillation
  d subharmonischer phasengekoppelter Oszillator *m*

f oscillateur *m* à accrochage de phase sous-harmonique
r субгармонический генератор *m* фазовой синхронизации

12006 subharmonic resonance
d subharmonische Resonanz *f*
f résonance *f* sous-harmonique
r субгармонический резонанс *m*

12007 submodular phase
d Submodularphase *f*
f phase *f* sousmodulaire
r подмодульная фаза *f*

12008 suboptimization
d Suboptimierung *f*
f sous-optimisation *f*
r субоптимизация *f*

12009 subordered structure
d untergeordnete Struktur *f*
f structure *f* subordonnée
r подчиненная структура *f*

12010 subpermanent magnetism
d subpermanenter Magnetismus *m*
f magnétisme *m* rémanent
r устойчивый остаточный магнетизм *m*

12011 subprocess
d Teilprozess *m*
f procédé *m* partiel
r субпроцесс *m*; частичный процесс

* **subsequent treatment** → 422

12012 subsidiary feedback
d Hilfsrückführung *f*
f réaction *f* secondaire
r вспомагательная обратная связь *f*

12013 substitute mode
d Substitutionsmodus *m*
f mode *m* de substitution
r режим *m* подстановки

12014 substructure
d Teilstruktur *f*
f structure *f* partielle
r субструктура *f*

12015 subsynchronous rectifier cascade
d untersynchrone Stromrichterkaskade *f*
f cascade *f* de redresseur sous-synchrone
r подсинхронный каскадный преобразователь *m*

12016 subsystem
d Untersystem *n*
f sous-système *m*
r подсистема *f*

12017 subtract pulse
d Subtraktionsimpuls *m*
f impulsion *f* de soustraction
r импульс *m* вычитания

12018 successive stages
d aufeinanderfolgende Stufen *fpl*
f étages *mpl* successifs
r последовательные стадии *fpl*

12019 suitability
d Eignung *f*
f qualification *f*
r пригодность *f*

12020 summation action
d Gesamtverhalten *n*
f action *f* totalisatrice
r суммарное управляющее воздействие *n*

* **summation circuit** → 405

12021 summation element; summing element
d Summationsglied *n*
f sommateur *m*; additionneur *m*
r суммирующий блок *m*

* **summator** → 404

* **summing element** → 12021

12022 superconductivity
d Supraleitfähigkeit *f*
f supraconductivité *f*
r сверхпроводимость *f*

12023 supercritical pressure
d überkritischer Druck *m*
f pression *f* hypercritique
r сверхкритическое давление *n*

12024 supercritical temperature
d überkritische Temperatur *f*
f température *f* hypercritique
r сверхкритическая температура *f*

12025 superheating plant
d Überhitzeranlage *f*
f installation *f* de surchauffe
r перегревательная установка *f*

12026 superimposed coding method
d überlagernde Kodierungsmethode *f*; überlagernde Verschlüsselungsmethode *f*

f méthode f de codification superposée;
   méthode de codage superposée
r суперпозиционный метод m кодирования

12027 superimposed interference
d überlagerte Störung f
f bruit m superposé
r наложенная помеха f

12028 superimposing principle; superposition principle
d Überlagerungsprinzip n;
   Superpositionsprinzip n
f principe m de superposition
r принцип m наложения; принцип
   суперпозиции

* **superphantom circuit → 4481**

12029 superposition
d Superposition f
f superposition f
r суперпозиция f

* **superposition principle → 12028**

12030 superpressure
d Überdruck m
f surpression f
r избыточное давление n

12031 supersensitive communication system
d hochempfindliches Nachrichtensystem n
f système m ultrasensible de
   télécommunications
r сверхчувствительная система f связи

12032 supersonic delay line; ultrasonic delay line
d Ultraschallverzögerungsleitung f;
   Ultraschallaufzeitglied n;
   Ultraschalllaufzeitkette f
f ligne f à retard ultrasonore
r ультразвуковая линия f задержки

12033 supersonic detector
d Ultraschalldetektor m
f détecteur m ultrasonore
r ультразвуковой детектор m

* **supervision → 8324**

12034 supervision of automatons
d Automatenüberwachung f
f inspection f d'automates
r контроль m автоматов

12035 supervision of production facilities
d Betriebsmittelüberwachung f

f supervision f des moyens de production
r управление n средствами производства

12036 supervisory control
d Überwachungskontrolle f; Fernkontrolle f
f contrôle m à distance; appareillage m de
   surveillance
r телеконтроль m

* **supervisory control system → 4383**

12037 supplay frequency
d Speisefrequenz f
f fréquence f d'alimentation
r частота f питания

12038 supplementary controlled system constans
d Ersatzregelstreckenkonstanton fpl
f constantes fpl de systèmes réglés
   supplémentaires
r параметры mpl дополнительных
   регулируемых систем

12039 supplementary controlled systems
d Ersatzregelstrecken fpl
f systèmes mpl réglés supplémentaires
r запасные регулируемые системы fpl

12040 supplementary half-step method
d Methode f des zusätzlichen Halbschrittes
f méthode f de demi-pas supplémentaire
r метод m добавочного полушага

12041 supply block
d Speiseeinheit f
f bloc m d'alimentation
r блок m питания

12042 supply current
d Netzstrom m; Speisestrom m
f courant m d'alimentation
r ток m питания

12043 supply current measurement
d Versorgungsstrommessung f
f mesure f du courant électrique d'alimentation
r измерение n тока источника питания

12044 supply frequency
d Speisefrequenz f
f fréquence f d'alimentation
r частота f источника питания

12045 supply pressure
d Speisedruck m
f pression f d'alimentation
r подводимое давление n

* **supply winding** → 6680

12046 supporting function
d Stützfunktion *f*
f fonction *f* de support
r опорная функция *f*

12047 support processor; slave processor
d Nebenprozessor *m*; Zusatzprozessor *m*;
   Slave-Prozessor *m*
f processeur *m* complémentaire
r дополнительный процессор *m*

12048 support system
d Unterstützungsprogrammsystem *n*
f système *m* d'assistance
r система *f* поддержки

12049 suppose *v*; assume *v*
d voraussetzen; vermuten
f supposer; admettre
r допускать; предполагать

12050 suppression
d Unterdrückung *f*; Austastung *f*
f suppression *f*; annulation *f*
r подавление *n*; гашение *n*

12051 suppression of data
d Datenunterdrückung *f*
f suppression *f* des données
r блокировка *f* данных

12052 suppression of self-oscillations
d Unterdrückung *f* der Selbstschwingungen
f suppression *f* des auto-oscillations;
   élimination *f* des auto-oscillations
r подавление *n* автоколебаний

12053 survey instrument
d Überwachungsgerät *n*
f appareil *m* de surveillance
r контрольный прибор *n*

* **susceptibility of failure** → 5408

12054 suspended body flowmeter
d Schwebekörperdurchflussmesser *m*
f débitmètre *m* à corps flottant
r плавающий расходомер *m*

12055 sustained deviation
d bleibende Regelabweichung *f*
f écart *m* permanent
r устойчивое отклонение *n*

12056 sweep balance
d dynamischer Ausgleich *m*; dynamische
   Kompensation *f*
f compensation *f* dynamique
r балансировка *f* развёртки

12057 sweep circuit
d Kippkreis *m*; Zeitablenkungskreis *m*;
   Wobbelkreis *m*
f circuit *m* vobulateur; circuit de base de temps
r цепь *f* развёртки

12058 sweep phase
d Abtastphase *f*; Kippphase *f*
f phase *m* de balayage
r фаза *f* сигналов развёртки

* **swing *v*** → 9498

12059 switchable
d umschaltbar
f commutable
r переключаемый; коммутируемый

* **switch board** → 9346

12060 switchgear; switching equipment
d Schaltgerät *n*; Schalteinrichtung *f*
f mécanisme *m* de couplage; dispositif *m* de
   commutation
r коммутационный аппарат *m*;
   распределительное устройство *n*

12061 switching algebra
d Schaltalgebra *f*
f algèbre *f* de commutation
r алгебра *f* переключательных схем

12062 switching characteristic
d Schaltcharakteristik *f*
f caractéristique *f* de commutation
r характеристика *f* переключения

12063 switching check with simultaneous timing
d Umschaltkontrolle *f* mit gleichzeitiger
   Zeitmessung
f contrôle *m* de commutateur avec
   chronométrage simultané
r контроль *m* переключения с
   одновременным хронированием

12064 switching circuit
d Schaltkreis *m*; Umschaltkreis *m*
f circuit *m* commutateur
r переключающая схема *f*

12065 switching circuit failure
d Schaltkreisfehler *m*
f défaut *m* du circuit de commutation
r отказ *m* переключающей схемы

12066 **switching coefficient**
   d Schaltkoeffizient *m*
   f coefficient *m* de commutation
   r коэффициент *m* коммутации

12067 **switching cycle**
   d Schaltzyklus *m*
   f cycle *m* d'opération
   r цикл *m* переключения

12068 **switching devices synthesis**
   d Synthese *f* von Relaisanlagen
   f synthèse *f* des dispositifs de commutation
   r синтез *m* коммутирующих устройств

   *   **switching element → 2392**

12069 **switching element of control system**
   d Schaltelement *n* eines Steuerungssystems
   f élément *m* de circuit d'un système de
        commande
   r переключающий элемент *m* управляющей
        системы

   *   **switching equipment → 12060**

12070 **switching frequency**
   d Schaltfrequenz *f*
   f fréquence *f* de commutation
   r частота *f* переключения

12071 **switching information**
   d Schaltinformation *f*
   f information *f* de commutation
   r информация *f* для коммутации

12072 **switching logic**
   d Schaltlogik *f*
   f logique *f* de commutation
   r логика *f* коммутации

   *   **switching network → 2676**

12073 **switching off; trip out; tripping**
   d Abschalten *n*; Ausschalten *n*; Abschaltung *f*;
        Ausschaltung *f*
   f déclenchement *m*; interruption *f*
   r выключение *n*; отключение *n*

12074 **switching-on**
   d Einschaltung *f*; Einschalten *n*
   f mise *f* en circuit; mise sous tension
   r включение *n*

   *   **switching period → 12082**

12075 **switching point**
   d Schaltpunkt *m*

   f point *m* de commutation
   r точка *f* переключения

12076 **switching pulse**
   d Schaltimpuls *m*
   f impulsion *f* de commutation
   r переключающий импульс *m*

12077 **switching signal**
   d Schaltsignal *n*; Umschaltsignal *n*
   f signal *m* de commutation
   r сигнал *m* коммутации

   *   **switching speed → 2409**

12078 **switching speed of transistors**
   d Schaltgeschwindigkeit *f* von Transistoren
   f vitesse *f* de commutation de transistors
   r скорость *f* переключения транзисторов

12079 **switching technique**
   d Schalttechnik *f*
   f technique *f* de commutation
   r техника *f* коммутации

12080 **switching theory; theory of relay systems**
   d Theorie *f* der Relaiseinrichtungen;
        Schalttheorie *f*
   f théorie *f* des dispositifs de commutation
   r теория *f* релейных устройств

12081 **switching threshold**
   d Schaltschwelle *f*
   f seuil *m* de commutation
   r порог *m* переключения

12082 **switching time; switching period;
       switch-over time**
   d Umschaltzeit *f*
   f temps *m* de commutation
   r время *n* переключения

12083 **switching transistor**
   d Schalttransistor *m*
   f transistor *m* de commutation
   r коммутационный транзистор *m*

12084 **switch-on position**
   d Einschaltstellung *f*
   f position *f* de fermeture
   r положение *n* включения [тока]

   *   **switch-over time → 12082**

12085 **switch technology**
        (for computer devices)
   d Schalttechnologie *f*
   f technologie *f* de commutation
   r технология *f* переключательных схем

12086 **Sylvester's interpolation formula**
    d Sylvester's Interpolationsformel *f*
    f formule *f* d'interpolation de Sylvester
    r интерполяционная формула *f* Сильвестра

12087 **symbolic analysis**
    d symbolische Analyse *f*
    f analyse *f* symbolique
    r символический анализ *m*

12088 **symbolic circuit; functional scheme**
    d Funktionsschema *n*
    f circuit *m* symbolique; schéma *m* fonctionnel
    r символическая схема *f*; функциональная
        схема

    * **symbolic device → 12093**

12089 **symbolic instruction**
    d symbolischer Befehl *m*
    f instruction *f* symbolique
    r символическая команда *f*

12090 **symbolic logic**
    d Symbollogik *f*
    f logique *f* symbolique
    r символическая логика *f*

12091 **symbolic operation**
    d symbolische Operation *f*
    f opération *f* symbolique
    r символическая операция *f*

12092 **symbolic programming system**
    d symbolisches Programmier[ungs]system *n*
    f système *m* de programmation symbolique
    r система *f* символического
        программирования

12093 **symbolic unit; symbolic device**
    d symbolisches Gerät *n*
    f dispositif *m* symbolique
    r символическое устройство *n*

12094 **symmetrical binomial distribution**
    d symmetrische Binomialverteilung *f*
    f distribution *f* binomiale symétrique
    r симметрическое биноминальное
        распределение *n*

    * **symmetrical circuit → 1591**

12095 **symmetrical equation system**
    d symmetrisches Gleichungssystem *n*
    f système *m* d'équations symétrique
    r симметрическая система *f* уравнений

12096 **symmetrical function**

12097 **symmetrically cyclically magnetized
        condition**
    d symmetrisch-zyklisch magnetisierter
        Zustand *m*
    f état *m* d'aimantation cyclisymétrique
    r состояние *n* симметрично-цикличного
        намагничивания

12098 **symmetric code**
    d symmetrischer Kode *m*
    f code *m* symétrique
    r симметричный код *m*

12099 **symmetric logic function**
    d symmetrische logische Funktion *f*
    f fonction *f* logique symétrique
    r симметричная логическая функция *f*

12100 **symmetric non-linearity**
    d symmetrische Nichtlinearität *f*
    f non-linéarité *f* symétrique
    r симметричная нелинейность *f*

12101 **symmetric oscillations**
    d symmetrische Schwingungen *fpl*
    f oscillations *fpl* symétriques
    r симметричные колебания *npl*

    * **synchro → 11325**

12102 **synchro-angle transmission**
    d Gleichaufwinkelübertragung *f*
    f transmission *f* d'angle synchrone
    r синхронная передача *f* угла

12103 **synchro-control differential transmitter**
    d Steuerdrehmelder-Differentialgeber *m*;
        Synchrodifferentialsender *m*;
        Synchroausgleichsübertrager *m*
    f synchro-transmetteur *m* différentiel
    r дифференциальный сельсин-датчик *m*

12104 **synchro-control transmitter**
    d Steuerdrehmeldergeber *m*
    f synchro-transmetteur *m* de commande
    r управляющий сельсин-датчик *m*

12105 **synchro dephaser**
    d Synchrophasenverschieber *m*
    f synchro-déphaseur *m*
    r синхронный трансформатор *m* фаз

12106 **synchronism check**
    d Gleichlaufprüfung *f*

f essai *m* synchrone; vérification *f* de synchronisme
r контроль *m* синхронизма

12107 synchronization
d Synchronisation *f*; Synchronisierung *f*; Gleichlaufsteuerung *f*
f mise *f* en phase; synchronisation *f*
r синхронизация *f*

* **synchronization pulse** → 3763

12108 synchronization signal
d Synchronisationssignal *n*
f signal *m* de synchronisation
r сигнал *m* синхронизации

12109 synchronization unit
d Synchronisierungssatz *m*; Synchronisierungsgruppe *f*
f bloc *m* de synchronisation; ensemble *m* de synchronisation
r блок *m* синхронизации

12110 synchronize *v*
d synchronisieren
f synchroniser
r синхронизировать

12111 synchronized total operation
d synchronisierte Gesamtoperation *f*
f opération *f* totale synchronisée
r синхронная [общая] операция *f*

12112 synchronizing circuit
d Synchronisierschaltung *f*
f circuit *m* de synchronisation
r схема *f* синхронизации

12113 synchronizing frequency
d Synchronisierungsfrequenz *f*
f fréquence *f* de synchronisation
r синхронизирующая частота *f*

12114 synchronous adapter
d Synchronadapter *m*
f adaptateur *m* synchrone
r синхронный адаптер *m*

12115 synchronous communication inductive system
d Induktivsystem *n* der Synchronverbindung
f système *m* de liaison synchrone inductive
r индуктивная система *f* синхронной связи

12116 synchronous communication interface
d synchrone Übertragungsschnittstelle *f*
f interface *f* de communication synchrone

r синхронный связной интерфейс *m*

12117 synchronous data link control
d synchrone Datenleitungssteuerung *f*
f commande *f* de liaison de données synchrone
r синхронное управление *n* каналом передачи

* **synchronous data transfer** → 12118

12118 synchronous data transmission; synchronous data transfer
d synchrone Datenübertragung *f*
f transmission *f* synchrone de données
r синхронная передача *f* данных

12119 synchronous detector
d Synchrondetektor *m*
f détecteur *m* synchrone
r синхронный детектор *m*

12120 synchronous input
d Synchroneingabe *f*
f entrée *f* synchrone
r синхронный ввод *m*

12121 synchronous logic
d synchrone Logik *f*; getaktete Logik
f logique *f* synchrone
r синхронная логика *f*

12122 synchronous mode
d Synchronmodus *m*; synchrone Betriebsart *f*
f mode *m* synchrone
r синхронный режим *m*

12123 synchronous motor drive
d Synchronmotorantrieb *m*; Synchronantrieb *m*
f commande *f* par moteur synchrone
r привод *m* синхронного двигателя

12124 synchronous operation
d Taktbetrieb *m*; Zeitgeberbetrieb *m*; Synchronbetrieb *m*
f opération *f* synchrone
r синхронная работа *f*

12125 synchronous optical network
d synchrones optisches Netzwerk *n*
f réseau *m* optique synchrone
r синхронная оптическая сеть *f*; синхронная сеть оптической связи

12126 synchronous output
d Synchronausgabe *f*
f sortie *f* synchrone
r синхронный вывод *m*

12127 synchronous serial system
d synchrones Folgesystem *n*; synchrones Sequenzsystem *n*
f système *m* synchrone séquentiel
r синхронная последовательная система *f*

12128 synchronous storage method
d Synchronspeicherungsverfahren *n*
f méthode *f* d'accumulation synchrone
r метод *m* накопления

12129 synchronous tact
d Synchrontakt *m*
f rythme *m* synchrone
r синхронный такт *m*

12130 synchronous working
d synchrone Betriebsweise *f*
f manière *f* d'opérer synchrone
r синхронный режим *m* работы

12131 synchro-torque receiver
d Synchroempfänger *m*
f synchro-récepteur *m*
r моментный сельсин-приёмник *m*

12132 synchro-trigonometer
d Synchrotrigonometer *n*
f synchro-trigonomètre *m*; synchro-analyseur *m*
r синхро-тригонометр *m*; синхро-анализатор *m*

12133 syntactic algorithm
d syntaktischer Algorithmus *m*
f algorithme *m* syntactique
r синтактический алгоритм *m*

12134 synthesis of control systems with process computers
d Synthese *f* von Regelkreisen mit Prozessrechnern
f synthèse *f* de systèmes asservis au moyen de calculateurs de processus
r синтез *m* систем управления при помощи вычислительных устройств

12135 synthesis of linear single-loop control systems
d Synthese *f* linearer einschleifiger Regelungssysteme
f synthèse *f* de systèmes linéaires à une boucle
r синтез *m* линейных одноконтурных систем регулирования

12136 synthesizer
d Synthesator *m*
f synthéseur *m*; appareil *m* de synthèse

r синтезатор *m*; синтезирующее устройство *n*

12137 system
d System *n*
f système *m*
r система *f*

12138 system analysis
d Systemanalyse *f*
f analyse *f* du système
r анализ *m* системы

12139 system approach
d Systemlösung *f*
f solution *f* de système
r системное решение *n*

12140 system architecture
d Systemarchitektur *f*; Systemaufbau *m*
f architecture *f* de système
r архитектура *f* системы

12141 systematic classification
d systematische Klassifizierung *f*
f classification *f* systématique
r систематическая классификация *f*

12142 systematic data checking
d systematische Datenprüfung *f*
f essai *m* systématique des données
r систематический контроль *m* данных

* systematic error → 1735

12143 systematics of gripper
d Greifersystematik *f*
f systématique *f* de grappin
r систематика *f* захватов

12144 systematic software technique
d systematische Softwaretechnik *f*
f technique *f* de software systématique
r систематическая техника *f* программного обеспечения

12145 systematic solution
d systematische Lösung *f*
f solution *f* systématique
r систематическое решение *n*

12146 system behaviour
d Verhalten *n* des Systems
f comportement *m* du système
r поведение *n* системы

12147 system boundary
d Systemgrenze *f*

f limite *f* de système
r граница *f* системы

12148 system calculaion
d Systemberechnung *f*
f calcul *m* de système
r расчёт *m* системы

12149 system capacity
d Anlagenleistung *f*
f capacité *f* de système
r производительность *f* системы

* **system check → 12150**

12150 system check[ing]; system test[ing]
d Systemprüfung *f*; Systemtestung *f*
f essai *m* de système; test *m* de système
r проверка *f* системы; испытание *n* системы

12151 system clock
d Systemtakt *m*
f rythme *m* de système
r такт *m* системы

12152 system compatibility
d Systemkompatibilität *f*
f compatibilité *f* de système
r системная совместимость *f*

12153 system-compatible device
d systemkompatibles Gerät *n*
f dispositif *m* système-compatible
r устройство n, совместимое с системой

12154 system component
d Systemkomponente *f*
f composant *m* de système
r системный компонент *m*

12155 system conception
d Systemkonzeption *f*
f conception *f* du système
r концепция *f* системы

12156 system configuration
d Systemkonfiguration *f*; Systemgestaltung *f*
f configuration *f* de système
r конфигурация *f* системы

12157 system control
d Systemsteuerung *f*
f commande *f* de système
r системное управление *n*

12158 system controller
d Systemsteuerbaustein *m*
f appareil *m* de commande de système

r системный контроллер *m*

* **system crash → 8**

12159 system description
d Systembeschreibung *f*
f description *f* du système
r описание *n* системы

12160 system design
d Systementwurf *m*
f projet *m* de système; conception *f* de système
r разработка *f* системы

12161 system designer
d Systementwerfer *m*
f concepteur *m* de système
r разработчик *m* системы

12162 system development
d Systementwicklung *f*
f développement *m* de système
r развитие *n* системы

12163 system diagnostic
d Systemdiagnose *f*
f diagnostic *m* de système
r системная диагностика *f*

* **system efficiency → 4660**

12164 system-element-relations
d System-Element-Relationen *fpl*
f relations *fpl* système-élément
r система-элемент-отношения *npl*

12165 system engineering; system technology
d Systemplanung *f*; System[entwurfs]technik *f*
f ingénierie *f* de système; technique *f* du système
r системотехника *f*

12166 system environment
d Systemumgebung *f*
f environnement *m* de système
r окружение *n* системы

12167 system frequency response
d Systemfrequenzgang *m*
f caractéreistique *f* fréquentielle du système
r частотная характеристика *f* системы

12168 system function
d Systemfunktion *f*
f fonction *f* de système
r функция *f* системы

12169 system generation
d Systemgenerierung *f*

f génération f de système
r генерация f системы

12170 system hierarchy
d Systemhierarchie f
f hiérarchie f de système
r иерархия f системы

12171 system interlock
d Systemblockierung f
f blocage m de système
r блокировка f системы

12172 system level
d Systemebene f
f niveau m de système
r системный уровень m

12173 system logical device; system logical unit
d systemlogisches Gerät n
f unité f logique adaptée sur le système
r системное логическое устройство n

*   **system logical unit → 12173**

12174 system lybrary
d Systembibliothek f
f bibliothèque f de système
r системная библиотека f

12175 system memory chip
d systemspezifischer Speicherschaltkreis m
f circuit m de mémoire spécifique de système
r микросхема f системной памяти

12176 system model
d Systemmodell n
f modèle m de système
r системная модель f

12177 system of contactless switching
d System n kontaktloser Schaltung
f système m de commutation sans contacts
r система f бесконтактного переключения

12178 system of logical equations
d System n logischer Gleichungen
f système m d'équations logiques
r система f логических уравнений

*   **system of object detection → 8879**

12179 system of units
d Einheitensystem n
f système m d'unités
r система f единиц

*   **system of variable coefficients → 12757**

12180 system order
d Systemordnung f
f ordre m de système
r порядок m системы

12181 system parameter
d Systemparameter m
f paramètre m du système
r параметр m системы

12182 system planning
d Systemplanung f
f planification f de système
r системное планирование n

12183 system programming
d Systemprogrammierung f
f programmation f générale
r системное программирование n

12184 system reliability
d Systemzuverlässigkeit f
f fiabilité f de système
r надёжность f системы

12185 system residence
d Systemresidenz f
f résidence f de système
r размещение n системы

12186 system resource
d Systemmittel n
f ressource f de système
r системный ресурс m

12187 system restart
d Systemwiederanlauf m
f redémarrage m de système
r рестарт m системы

12188 system security
d Systemsicherheit f
f sûreté f de système
r безопасность f системы

12189 system simulation
d Systemmodellierung f
f simulation f de système
r моделирование n системы

12190 system software
d System-Software f; Systemprogrammpaket n
f logiciels mpl de système
r системное программное обеспечение n

12191 system stability analysis
d Systemstabilitätsanalyse f
f analyse f de stabilité des systèmes
r анализ m устойчивости системы

12192 system state
 d Systemzustand *m*
 f état *m* du système
 r состояние *n* системы

12193 system statistical analysis
 d statistische Analyse *f* des Systems
 f analyse *f* statistique du système
 r статистический анализ *m* системы

12194 system structure
 d Systemstruktur *f*
 f structure *f* de système
 r структура *f* системы

12195 system surroundings
 d Systemumgebung *f*
 f environnement *m* de système
 r среда f, окружающая систему

 * **system technology** → 12165

 * **system test** → 12150

 * **system testing** → 12150

12196 system theory
 d Systemtheorie *f*
 f théorie *f* des systèmes
 r теория *f* о системах

12197 system transfer function
 d Systemübertragungsfunktion *f*
 f transmittance *f* du système
 r переходная функция *f* системы

12198 system variable
 d Systemvariable *f*
 f variable *f* de système
 r переменная *f* системы

12199 system with additional pressure feedback
 d System *n* mit zusätzlicher Druckrückführung
 f système *m* avec retour complémentaire sous pression
 r система *f* с дополнительной обратной связью по давлению

12200 system with a simple position feedback
 d System *n* mit einfacher Stellenrückführung
 f système *m* avec retour simple en position [initiale]
 r система *f* с обратной связью по положению

12201 system with damping
 d System *n* mit Dämpfung
 f système *m* à amortissement
 r система *f* с демпфированием

12202 system with one degree of freedom
 d System *n* mit einem Freiheitsgrad
 f système *m* à un degré de liberté
 r система *f* с одной степенью свободы

12203 system without clock
 d ungetaktetes System *n*
 f système *m* sans cadence
 r нетактированная система *f*

 * **system with power amplification** → 6474

12204 system with several degrees of freedom
 d System *n* mit mehreren Freiheitsgraden
 f système *m* à plusieurs degrés de liberté
 r система *f* со многими степенями свободы

T

12205 tactile sensor survey
d taktile Sensorüberwachung f; Überwachung f mit taktilem Sensor
f surveillance f de senseur tactile
r контроль m с помощью тактильного сенсора

12206 tact period
d Taktperiode f
f période f de récurrence
r тактовый период n

12207 tag line in
d Eingangskennwortleitung f; Eingangsanzeigeleitung f
f ligne f d'identification d'entrée
r входная шина f идентификации

12208 tag line out
d Ausgangskennwortleitung f; Ausgangsanzeigeleitung f
f ligne f d'identification de sortie
r выходная шина f идентификации

12209 tail plane adjustment
d Höhenflossenverstellung f
f réglage m d'incidence du stabilisateur
r установка f стабилизатора

12210 tandem compensation network
d Serienkompensationsnetzwerk n
f réseau m de compensation tandem
r цепь f последовательной компенсации

12211 tandem network
d zwei gekoppelte Netzwerke npl [mit Rückführung]
f réseau m tandem
r каскадное соединение n двух контуров

12212 tangent
d Tangente f
f tangente f
r касательная f

12213 tangent condition
d Tangentenbedingung f
f condition f de tangentes
r условие n касательной

12214 tangents method

d Tangentenmethode f
f méthode f de tangentes
r метод m касательных

12215 tank gauge
d Pegelmesser m
f indicateur m de niveau
r уровнемер m для резервуаров

12216 tape
d Band n; Streifen m
f bande f; ruban m
r лента f

12217 tape controlled; tape operated
d streifengesteuert; bandgesteuert
f commandé par bande
r управляемый лентой

* **tape device → 7716**

12218 tape drive
d Bandlaufwerk n; Bandantrieb m; Bandvorschubeinrichtung f
f mécanisme m d'entraînement de bande; transporteur m de bande
r лентопротяжный механизм m

* **tape mode → 7717**

* **tape operated → 12217**

* **tape unit → 7716**

12219 tapping switch
d Stufenschalter m
f commutateur m de branchements
r переключатель m ответвлений

12220 target function
d Zielfunktion f
f fonction f de but
r целевая функция f

12221 target system
d Zielsystem n
f système m de destination
r целевая система f

12222 task abnormal end
d Taskabnormalhalt m
f arrêt m anormal de tâche
r аварийное завершание n задачи

12223 task control block
d Aufgabensteuerblock m; Task-Steuerblock m
f bloc m pour la commande de tâches
r блок m управления задачами

12224 task coordination
d Aufgabenzuordnung *f*
f coordination *f* de tâche
r координация *f* задач

12225 task management
d Aufgabenverwaltung *f*; Task-Management *n*
f management *m* de tâches
r управление *n* задачами

12226 task selection mechanism
d Task-Auswahlmechanismus *m*
f mécanisme *m* pour la sélection de tâches
r механизм *m* выбора задач

12227 task supervision
d Aufgabenüberwachung *f*
f supervision *f* de tâches
r контроль *m* задач

* **task switch → 12228**

12228 task switch[ing]
d Aufgabenwechsel *m*; Aufgabenumschaltung *f*
f commutation *f* de tâches
r переключение *n* задач

12229 Taylor's series
d Taylor-Reihe *f*
f série *f* de Taylor
r ряд *m* Тейлора

12230 teaching automaton; teaching machine
d lehrender Automat *m*
f automate *m* enseignant; machine *f* enseignante
r обучающий автомат *m*

* **teaching machine → 12230**

12231 technical control system
d technisches Regelsystem *n*
f système *m* de commande technologique
r техническая система *f* управления

12232 technical dialogue system; special dialogue system
d Fachdialogsystem *n*
f système *m* de dialogue technique; système de dialogue spécialisé
r специализированная диалоговая система *f*

12233 technical drawing; mechanical drawing
d technische Zeichnung *f*
f dessin *m* technique
r технический чертеж *m*

12234 technical evaluation
d techn[olog]ische Bewertung *f*
f évaluation *f* technique; estimation *f* technologique
r техническая оценка *f*; оценка по технологическим критериям

12235 technical requisition
d technische Anforderung *f*
f demande *f* technique
r техническое требование *n*

12236 technological auxiliary equipment
d technologische Hilfsausrüstung *f*
f équipement *m* auxiliaire technologique
r вспомогательное технологическое оборудование *n*

12237 technological criteria
d technologische Kriterien *npl*
f critères *mpl* technologiques
r технологические критерии *mpl*

12238 technological equipment
d technologische Ausrüstung *f*
f équipement *m* technologique
r технологическая оснастка *f*

12239 technological multicoordinate equipment
d technologische Mehrkoordinaten-Ausrüstung *f*
f équipement *m* de coordonnées multiples technologiques
r технологическое многокоординатное оборудование *n*

12240 technological property
d technologische Eigenschaft *f*
f propriété *f* technologique
r технологическое свойство *n*

12241 technological structure
d technologische Struktur *f*
f structure *f* technologique
r технологическая структура *f*

12242 technological unit
d technologische Einheit *f*
f unité *f* technologique
r технологическая единица *f*

12243 technology
d Technologie *f*
f technologie *f*
r технология *f*

12244 technology description
d Technologiebeschreibung *f*

f description *f* de technologie
r описание *n* технологии

12245 technology of absorption
d Absorptionstechnik *f*
f technique *f* de l'absorption
r абсорбционная техника *f*

12246 teleautomation; telecontrol engineering
d Fernwirktechnik *f*
f technique *f* d'opérations à distance; technique de télécommande
r телеавтоматизация *f*; техника *f* телеуправления

12247 telecharge *v*
d fernladen
f télécharger
r загружать дистанционно

12248 telecommunication access method
d Fernübertragungs-Zugriffsmethode *f*
f méthode *f* d'accès de télécommunication
r телекоммуникационный метод *m* доступа

12249 telecommunication control unit
d Datenfernverarbeitungssteuereinheit *f*
f unité *f* de commande pour le télécommunication
r блок *m* управления дистанционной обработкой данных

*   **telecontrol engineering** → 12246

12250 telecontrol system
d Fernsteuersystem *n*
f système *m* de télécommande
r система *f* телеуправления

12251 telemanipulator
d Telemanipulator *m*
f télémanipulateur *m*
r дистанционно-управляемый манипулятор *m*

12252 telemechanic contactor
d telemechanisches Schütz *n*
f contacteur *m* télémécanique
r телемеханический контактор *m*

12253 telemechanics
d Telemechanik *f*
f télémécanique *f*
r телемеханика *f*

12254 telemetering; telemetry
d Fernmessung *f*; Fernmessen *n*; Telemetrie *f*

f télémesure *f*; mesure *f* à distance
r телеметрия *f*; дистанционные измерения *npl*

12255 telemetering device
d Fernmesseinrichtung *f*
f dispositif *m* de télémesure
r дистанционное измерительное устройство *n*

12256 telemetering method; method of telemetering
d Fernmessmethode *f*
f méthode *f* de mesure à distance
r метод *m* телеизмерения

12257 telemetering pulse-position system
d Zeit-Impuls-Fernmesssystem *n*
f système *m* de télémesure par déplacement d'impulsions
r время-импульсная телеизмерительная система *f*

12258 telemetering system
d Fernmesssystem *n*
f système *m* de mesure à distance
r дистанционная измерительная система *f*

12259 telemetering transducer
d Fernmessgeber *m*
f transmetteur *m* de télémesure
r телеметр *m*

12260 telemetric circuit
d Fernmessstromkreis *m*
f circuit *m* de télémesure
r телеметрическая схема *f*

*   **telemetry** → 12254

12261 teleprinter; teletype [writer]; teletype printer
d Fernschreiber *m*; Ferndrucker *m*; Fernschreibgerät *n*
f téléimprimeur *m*; téléscripteur *m*; télétype *m*
r телетайп *m*; дистанционное печатающее устройство *n*

12262 teleprocessing system
d Fernverarbeitungssystem *n*
f système *m* de traitement à distance
r система *f* телеобработки

12263 telestatic equipment
d telestatische Steuerung *f*
f équipement *m* téléstatique
r телестатическая аппаратура *f*

12264   **telestatic long range transmission**
        equipment
        d   telestatische Fernübertragungsausrüstung *f*
        f   équipement *m* télestatique de
             télétransmission
        r   телестатическая установка *f* для
             дистанционной передачи

12265   **teleswitching**
        d   Fernschaltung *f*
        f   télécommande *f* d'interrupteurs
        r   дистанционное выключение *n*

12266   **teletransmission; remote transfer; remote**
        transmission
        d   Fernübertragung *f*
        f   télétransmission *f*
        r   телепередача *f*; дистанционная передача *f*

12267   **teletransmitter**
        d   Ferngeber *m*
        f   transmetteur *m* à distance
        r   дистанционный передатчик *m*

*   **teletype** → **12261**

*   **teletype printer** → **12261**

*   **teletype writer** → **12261**

12268   **temperature anomaly**
        d   Temperaturanomalie *f*
        f   anomalie *f* de température
        r   температурная аномалия *f*

12269   **temperature coefficient**
        d   Temperaturkoeffizient *m*
        f   coefficient *m* de température
        r   температурный коэффициент *m*

12270   **temperature-compensated**
        d   temperaturkompensiert
        f   compensé en température
        r   температурно-компенсированный

12271   **temperature compensation limits**
        d   Temperaturkompensationsgrenzen *fpl*
        f   limites *fpl* de compensation de la température
        r   пределы *mpl* температурной компенсации

12272   **temperature conditions**
        d   Temperaturzustand *m*;
             Temperaturbedingungen *fpl*
        f   conditions *fpl* de température
        r   температурный режим *m*

12273   **temperature contrast**
        d   Temperaturkontrast *m*;

        Temperaturunterschied *m*
        f   contraste *m* de température
        r   температурный контраст *m*

12274   **temperature control equipment**
        d   Temperaturregeleinrichtung *f*
        f   équipement *m* de régulation de température
        r   оборудование *n* для регулирования
             температуры

12275   **temperature controller**
        d   Temperaturregler *m*
        f   régulateur *m* de température
        r   регулятор *m* температуры

12276   **temperature control of induction heating**
        d   Temperaturkontrolle *f* der
             Induktionserwärmung
        f   contrôle *m* des températures de chauffage par
             induction
        r   контроль *m* температуры при
             индукционном нагреве

12277   **temperature correction factor**
        d   Temperaturfehler *m*
        f   erreur *m* de température
        r   температурная поправка *f*

12278   **temperature dependence**
        d   Temperaturabhängigkeit *f*
        f   variation *f* avec la température
        r   температурная зависимость *f*

12279   **temperature detecting device**
        d   Temperaturfühleinrichtung *f*
        f   appareil *m* controlleur de température
        r   термочувствительное устройство *n*

12280   **temperature detector**
        d   Temperaturdetektor *m*;
             Temperaturwandler *m*
        f   palpeur *m* de température
        r   термочувствительный элемент *m*

12281   **temperature equilibrium**
        d   Temperaturgleichgewicht *n*
        f   équilibre *m* de température
        r   температурное равновесие *n*

12282   **temperature error**
        d   Temperaturfehler *m*
        f   erreur *f* de température
        r   температурная погрешность *f*

12283   **temperature gradient**
        d   Temperaturgradient *m*; Temperaturgefälle *n*
        f   gradient *m* thermique; chute *f* de température
        r   температурный градиент *m*

12284 temperature profile
 d Temperaturprofil *n*
 f profil *m* de température
 r температурный профиль *m*

12285 temperature range
 d Temperaturbereich *m*
 f gamme *f* de température
 r температурная зона *f*

12286 temperature stabilization
 d Temperaturstabilisierung *f*
 f stabilisation *f* de température
 r температурная стабилизация *f*

12287 tensile load-test
 d Zugfestigkeitsprüfung *f*; Zugtest *m*
 f essai *m* de résistance à la traction
 r испытание *n* на прочность при
   растяжении

 * **tension theory** → **12336**

12288 tensometric apparatus
 d tensometrischer Apparat *m*
 f appareil *m* tensométrique
 r тензометрическая аппаратура *f*

12289 terminal
 d Endstelle *f*; Klemme *f*; Anschlussklemme *f*;
   Terminal *n*
 f borne *f*; terminal *m*
 r зажим *m*; клемма *f*

12290 terminal control
 d Endwertregelung *f*
 f régulation *f* de valeur finale; réglage *m* de
   valeur finale
 r регулирование *n* по конечному состоянию

12291 terminal control unit
 d Steuereinheit *f* für die Außenstation
 f unité *f* de commande de la station terminale
 r устройство *n* управления внешней
   станции

12292 terminal-oriented system
 d terminalorientiertes System *n*
 f système *m* orienté sur le terminal
 r терминально-ориентированная система *f*

12293 terminal station
 d Außenstation *f*
 f station *f* terminale; poste *m* terminal
 r внешняя станция *f*

12294 terminate *v*
 d beenden

 f terminer
 r завершить

12295 terminating reaction
 d Abbruchreaktion *f*
 f réaction *f* terminale
 r реакция *f* обрыва цепи

12296 test and control console
 d Test- und Kontrollpult *n*
 f pupitre *m* d'analyse et de contrôle
 r пульт *m* контроля и управления

 * **test assembly** → **12316**

 * **test bit** → **2323**

 * **test board** → **6753**

12297 test chamber
 d Testkammer *f*; Versuchsraum *m*
 f chambre *f* d'essai
 r испытательная камера *f*

12298 test condition
 d Prüfbedingung *f*
 f condition *f* de test
 r условие *n* испытания

12299 test data
 d Testdaten *pl*
 f données *fpl* de test
 r тестовые данные *pl*

12300 testing instrument; check instrument
 d Prüfgerät *n*; Kontrollgerät *n*; Prüfer *m*
 f instrument *m* de contrôle
 r контрольно-измерительный прибор *m*

12301 testing service monitor
 d Testhilfsmonitor *m*
 f moniteur *m* de service de test
 r монитор *m* испытательных средств

12302 testing service program
 d Testhilfsprogramm *n*
 f programme *m* de service de test
 r программа *f* испытательных средств

 * **testing technique** → **2339**

12303 test[ing] tools
 d Testhilfsmittel *npl*; Testwerkzeuge *npl*;
   Prüfwerkzeuge *npl*
 f moyens *mpl* de test
 r испытательные средства *npl*

12304 test instruction
 d Testbefehl *m*

f instruction *f* de test
r команда *f* теста

12305 test latch
d Prüfselbsthalteschaltung *f*
f bascule *f* à verrouillage de test
r контрольная схема *f* с самоблокировкой

12306 test machine
d Prüfmaschine *f*
f machine *f* d'essai
r машина *f* для испытаний

12307 test method
d Prüfmethode *f*
f méthode *f* d'essai
r метод *m* испытания

12308 test monitor
d Testmonitor *m*; Prüfablaufüberwacher *m*
f moniteur *m* de test
r тест-монитор *m*

12309 test of process input
d Test *m* der Prozesseingabe
f test *m* de l'entrée de processus
r контроль *m* входных данных процесса

12310 test of process output
d Test *m* der Prozessausgabe
f test *m* de la sortie de processus
r контроль *m* выходных данных процесса

12311 test operation
d Prüfvorgang *m*; Testoperation *f*
f opération *f* de test
r операция *f* тестирования

12312 test panel computer
d Prüffeldrechner *m*
f ordinateur *m* du banc d'essai
r компьютер *m* испытательного стенда

12313 test pattern
d Testmuster *n*
f échantillon *m* de test
r тестовая комбинация *f*

12314 test pattern generation
d Testmustererzeugung *f*;
   Prüffolgenerzeugung *f*
f génération *f* d'échantillons de test
r генерация *f* тестовых комбинаций

12315 test period
d Testzeitraum *m*; Prüfperiode *f*; Prüfzeit *f*
f période *f* d'essai
r отрезок *m* времени проверки

12316 test place; test assembly
d Prüfplatz *m*
f poste *m* d'essai
r контрольно-испытательная станция *f*

12317 test plan
d Versuchsplan *m*
f plan *m* de recherche; plan d'expérience
r испытательный план *m*

12318 test point
d Testpunkt *m*
f point *m* de test
r контрольная точка *f*

* **test procedure** → 2339

* **test program beginning** → 2351

* **test program end** → 2352

12319 test program error; check program error
d Prüfprogrammfehler *m*
f erreur *f* de programme de test
r ошибка *f* в тестовой программе

* **test program time** → 2353

12320 test reading; control sensing
d Kontrollabtastung *f*
f lecture *m* d'essai; lecture de contrôle
r контрольное считывание *n*

12321 test requirements
d Prüfbedingungen *fpl*
f conditions *fpl* de test
r требования *npl* контроля

12322 test result; experimental result
d Versuchsergebnis *n*
f résultat *m* expérimental; résultat de test
r результат *m* опыта

12323 test run; trial run
d Probebetrieb *m*; Versuchsbetrieb *m*
f fonctionnement *m* d'essai
r контрольный запуск *m*; пробное
   испытание *n*

12324 test selector
d Prüfwähler *m*; Messwähler *m*
f sélecteur *m* d'essai; connecteur *m* de test
r пробный искатель *m*

* **test sequence** → 2342

12325 test signal
d Testsignal *n*; Kontrollsignal *n*; Prüfsignal *n*

f signal *m* d'essai; signal d'épreuve; signal de contrôle
r испытательный сигнал *m*

* test stand → 6753

12326 test system
d Testsystem *n*
f système *m* de test; système d'essai
r система *f* проверки

* test table → 6753

12327 test tension; test voltage
d Prüfspannung *f*
f tension *f* d'essai
r испытательное напряжение *n*

* test tools → 12303

* test voltage → 12327

12328 theoretical model
d theoretisches Modell *n*
f modèle *m* théorique
r теоретическая модель *f*

12329 theory of automatic control
d Theorie *f* der automatischen Steuerung; Theorie der selbsttätigen Regelung
f théorie *f* de commande automatique; théorie du réglage automatique
r теория *f* автоматического управления

12330 theory of dynamics
d Theorie *f* dynamischer Systeme
f théorie *f* des systèmes dynamiques
r теория *f* динамических систем

* theory of error → 5201

12331 theory of game; game theory
d Theorie *f* der Spiele; Spieltheorie *f*
f théorie *f* des jeux
r теория *f* игр

12332 theory of mapping
d Abbildungstheorie *f*; Theorie *f* der Abbildungen
f théorie *f* des transformations des représentations conformes
r теория *f* отображения

12333 theory of numerical treatment
d Theorie *f* der numerischen Behandlung
f théorie *f* du traitement numérique
r теория *f* числовой обработки

* theory of relay systems → 12080

12334 theory of reliability
d Zuverlässigkeitstheorie *f*
f théorie *f* de fiabilité; théorie de sécurité
r теория *f* надёжности

12335 theory of servomechanism
d Regelungstheorie *f*
f théorie *f* de réglage; théorie de régulation automatique
r теория *f* регулирования

12336 theory of strain; tension theory
d Spannungstheorie *f*
f théorie *f* de tension
r теория *f* напряжения

12337 theory of technical stability
d technische Stabilitätstheorie *f*
f théorie *f* de stabilité technique
r теория *f* технической стабильности

12338 thermal admittance
d thermische Leitfähigkeit *f*; thermischer Leitwert *m*
f admittance *f* thermique
r теримическая проводимость *f*

12339 thermal compensation
d thermische Kompensation *f*; Wärmekompensation *f*
f compensation *f* thermique
r компенсация *f* теплового воздействия

12340 thermal computing element
d thermisches Rechenelement *n*
f élément *m* de computation thermique
r тепловой решающий элемент *m*

12341 thermal conductivity
d thermische Leitfähigkeit *f*; Wärmeleitfähigkeit *f*
f conductibilité *f* thermique
r удельная теплопроводность *f*

12342 thermal conductivity measurement
d Wärmeleitfähigkeitsfernmessung *f*
f télémesure *f* de transfert de chaleur
r измерение *n* теплопроводности

12343 thermal control
d Wärmekontrolle *f*
f contrôle *m* thermique
r термический контроль *m*

12344 thermal cut-out
d thermische Sicherung *f*

   f coupe-circuit *m* thermique
   r тепловой предохранитель *m*

12345 thermal feedback
   d thermische Rückführung *f*
   f réaction *f* thermique; rétroaction *f* thermique
   r тепловая обратная связь *f*

12346 thermal flowmeter
   d Wärmedurchflussmesser *m*
   f débitmètre *m* calorifique; débitmètre thermique
   r тепловой расходомер *m*

12347 thermal inertia
   d Wärmeträgheit *f*
   f inertie *f* thermale
   r тепловая инерция *f*

12348 thermal interaction
   d Wärmewechselwirkung *f*
   f interaction *f* thermique
   r тепловое взаимодействие *n*

12349 thermal noise
   d thermisches Rauschen *n*
   f bruit *m* thermique
   r термический шум *m*

12350 thermal overload capacity
   d thermische Überlastbarkeit *f*
   f capacité *f* de surcharge thermique
   r тепловая перегрузочная способность *f*

12351 thermal power
   d Wärmekraft *f*
   f puissance *f* thermique
   r тепловая мощность *f*

12352 thermal printer; electrothermic printer; thermoprinter
   d Thermodrucker *m*; elektrothermischer Drucker *m*
   f imprimante *f* thermique; imprimante thermographique
   r термографический принтер *m*; устройство *n* термопечати

12353 thermal process
   d thermischer Prozess *m*; Wärmevorgang *m*; Wärmeprozess *m*
   f procédé *m* thermique
   r тепловой процесс *m*

12354 thermal time constant of the thermal converter
   d thermische Zeitkonstante *f* des Thermoumformers

   f constante *f* de temps thermique du convertisseur thermique
   r тепловая постоянная *f* времени термопреобразователя

12355 thermistor
   d Heißleiter *m*
   f thermistor *m*
   r термистор *m*

12356 thermochemical gas analyzer
   d thermochemischer Gasanalysator *m*
   f analyseur *m* thermochimique de gaz
   r термохимический газоанализатор *m*

12357 thermocontact
   d Thermokontakt *m*
   f thermocontact *m*
   r термоэлектрический контакт *m*

12358 thermocouple
   d Thermoelement *n*
   f thermocouple *m*
   r термоэлемент *m*

12359 thermocouple ammeter
   d Thermopaarstrommesser *m*; Thermokreuzstrommesser *m*
   f ampèremètre *m* à thermocouple
   r термопарный амперметр *m*

12360 thermocouples adapter
   d Adapter *m* für Thermoelemente
   f adapteur *m* pour thermocouples
   r адаптер *m* для термоэлементов

12361 thermodynamic analysis
   d thermodynamische Analyse *f*
   f analyse *f* thermodynamique
   r термодинамический анализ *m*

12362 thermodynamic coordinate
   d thermodynamische Koordinate *f*
   f coordonnée *f* thermodynamique
   r термодинамический параметр *m*

12363 thermodynamic equilibrium
   d thermodynamisches Gleichgewicht *n*
   f équilibre *m* thermodynamique
   r термодинамическое равновесие *n*

12364 thermodynamic functions
   d thermodynamische Funktionen *fpl*
   f fonctions *fpl* thermodynamiques
   r термодинамические функции *fpl*

12365 thermodynamic potential
   d thermodynamisches Potential *n*

f potentiel *m* thermodynamique
r термодинамический потенциал *m*

12366 thermodynamic probability
  d thermodynamische Wahrscheinlichkeit *f*
  f probabilité *f* thermodynamique
  r термодинамическая вероятность *f*

12367 thermodynamic process
  d thermodynamischer Prozess *m*
  f processus *m* thermodynamique
  r термодинамический процесс *m*

12368 thermodynamic properties
  d thermodynamische Eigenschaften *fpl*
  f propriétés *fpl* thermodynamiques
  r термодинамические свойства *npl*

12369 thermodynamic relationship
  d thermodynamische Beziehung *f*
  f relation *f* thermodynamique
  r термодинамическое соотношение *n*

12370 thermodynamics
  d Thermodynamik *f*
  f thermodynamique *f*
  r термодинамика *f*

12371 thermoelectric comparator
  d thermoelektrischer Komparator *m*
  f comparateur *m* thermoélectrique
  r термоэлектрический компаратор *m*

12372 thermomagnetic analysis
  d thermomagnetische Analyse *f*
  f analyse *f* thermomagnétique
  r термомагнитный анализ *m*

  * **thermoprinter** → **12352**

  * **thermoregulator** → **6094**

12373 thermostatic
  d thermostatisch
  f thermostatique
  r термостатический

12374 thermostatic control
  d thermostatische Regelung *f*
  f réglage *m* thermostatique
  r термостатическое регулирование *n*

12375 thermostatic controller
  d thermostatischer Regler *m*
  f régulateur *m* thermostatique
  r терморегулятор *m*

12376 thermostatic gas analyzer
  d thermostatischer Gasanalysator *m*
  f analysateur *m* de gaz thermostatique

  r термостатический газоанализатор *m*

12377 thermostatic liquid level control
  d thermostatischer Flüssigkeitsregler *m*
  f régulateur *m* thermostatique de niveau de liquide
  r регулятор *m* уровня жидкости с использованием терморегулирующего вентиля

12378 three-dimensional heat flow
  d dreidimensionale Wärmeströmung *f*
  f transport *m* de chaleur tridimensionnel
  r трехмерный процесс *m* теплопередачи

12379 three-dimensional motor control
  d dreidimensionale Motorführung *f*
  f pilotage *m* de moteur tridimensionnel
  r трехмерное управление *n* двигателем

12380 three-dimensional optical sensor
  d dreidimensionaler optischer Sensor *m*
  f senseur *m* optique tridimensionnel
  r трехмерный оптический сенсор *m*

12381 three-level system
  d Dreipegelsystem *n*
  f système *m* à trois niveaux
  r трехуровневая система *f*

12382 three-mode control
  d Dreiwegsteuerung *f*
  f réglage *m* à trois termes
  r регулирование *n* по трем параметрам

12383 three-phase supply
  d Dreiphasenspeisung *f*
  f alimentation *f* triphasée
  r трехфазная сеть *f*

12384 three-phase system
  d Dreiphasensystem *n*
  f système *m* de trois phases
  r трехфазная система *f*

12385 three-point action; three-step action
  d Dreipunktregelung *f*
  f action *f* à trois points
  r трехпозиционное воздействие *n*

12386 three-position control
  d Dreipunktregelung *f*
  f réglage *m* à trois positions
  r трехпозиционное регулирование *n*

12387 three-stage management information system
  d dreistufiges Management-Informationssystem *n*

f système *m* d'information de la direction à
  trois degrés
r трехкаскадная информационная система *f*
  управления

12388 three-state device; tristate device
  d Tri-State-Bauelement *n*; Bauelement *n* mit
    Tri-State-Ausgang
  f élément *m* à sortie à trois états
  r устройство *n* с тремя [устойчивыми]
    состояниями

* **three-step action → 12385**

12389 three-step control
  d Dreistufensteuerung *f*
  f réglage *m* à trois échelons
  r трехступенчатое управление *n*

12390 three-valued functions
  d dreiwertige Funktionen *fpl*
  f fonctions *fpl* ternaires
  r трехзначные функции *fpl*

12391 threshold
  d Schwellenwert *m*
  f valeur *f* de seuil
  r порог *m*; пороговое значение *n*

12392 threshold circuit
  d Schwellwertschaltung *f*
  f circuit *m* de seuil
  r пороговая схема *f*

12393 threshold detector
  d Schwellendetektor *m*
  f détecteur *m* de seuil
  r пороговый детектор *m*

12394 threshold effect
  d Einsatzeffekt *m*; Schwelleneffekt *m*;
    Grenzeffekt *m*
  f effet *m* de seuil
  r пороговый эффект *m*

12395 threshold field
  d Schwellenfeld *n*
  f champ *m* de seuil
  r поле *n* пороговых значений

12396 threshold frequency
  d Schwellenfrequenz *f*; Frequenz *f* der
    Ansprechschwelle
  f fréquence *f* du seuil
  r пороговая частота *f*

12397 threshold function
  d Schwellenfunktion *f*

f fonction *f* de seuil
r пороговая функция *f*

12398 threshold logic
  d Schwellwertlogik *f*
  f logique *f* de seuil
  r пороговая логика *f*

12399 threshold network
  d Schwellennetzwerk *n*
  f réseau *m* de seuil
  r пороговая сеть *f*

12400 threshold of operation
  d Operationsschwelle *f*; Funktionsschwelle *f*
  f seuil *m* de fonction; seuil de réponse
  r порог *m* функционирования

12401 threshold sensitivity
  d Schwellenempfindlichkeit *f*
  f sensibilité *f* de seuil
  r пороговая чувствительность *f*

12402 threshold signal
  d Schwellensignal *n*
  f signal *m* de seuil
  r пороговый сигнал *m*

* **threshold theorem → 1901**

* **threshold value indicator → 11251**

12403 threshold voltage
  d Schwellspannung *f*; Ansprechspannung *f*
  f tension *f* de seuil
  r пороговое напряжение *n*

12404 throttling action
  d drosselnde Wirkung *f*; Drosselwirkung *f*
  f réglage *m* par papillon
  r дросселирующее действие *n*

12405 throttling control
  d Drosselregelung *f*; Proportionalregelung *f*
  f réglage *m* de l'étranglement
  r регулирование *n* дросселированием

* **throttling controller → 10348**

* **throttling index → 2364**

12406 throttling process
  d Drosselprozess *m*
  f processus *m* d'étranglement
  r процесс *m* дросселирования

12407 throttling zone
  d proportionales Band *n*; proportionaler
    Bereich *m*

f zone *f* d'action proportionnelle
r пропорциональная зона *f*

12408 throughput
d Durchsatz *m*; Durchlauf *m*
f débit *m*
r пропускная способность *f*

12409 time-averaged
d zeitgemittelt; über die Zeit gemittelt
f moyenné dans le temps
r усредненный по времени

12410 time base
d Zeitbasis *f*
f base *f* de temps
r базис *m* времени

12411 time-base unit
d Kippeinheit *f*; Zeitbasiseinheit *f*
f unité *f* de base de temps
r блок *m* развёртки

12412 time check
d Zeitkontrolle *f*
f contrôle *m* de temps
r контроль *m* по времени

12413 time constraints
d Zeitbeschränkung *f*; Zeiteinschränkungen *fpl*
f contrainte *f* de temps
r временные ограничения *npl*

12414 time-consuming operation
d zeitaufwendige Operation *f*
f opération *f* nécessitant beaucoup de temps
r операция *f* с большими затратами времени

* **time-cycle control** → 10246

* **time-cycle controller** → 10258

12415 timed acceleration
d zeitgesteuerte Beschleunigung *f*; programmierte Beschleunigung
f accélération *f* temporisée
r регулируемое по времени ускорение *n*

12416 time-decoding device
d Zeitdekodiergerät *n*
f décodeur *m* de temps
r устройство n, декодирующее время

* **time delay circuit** → 6609

12417 time-delayed
d zeitverzögert

f retardé
r с запаздыванием; с временной задержкой

12418 time-delay simulation
d Zeitverzögerungsnachbildung *f*
f simulation *f* de retard de temps
r моделирование *n* временного запаздывания

12419 time-dependent
d zeitabhängig
f dépendant du temps
r времязависимый

12420 time-dependent behaviour
d zeitabhängiges Verhalten *n*
f comportement *m* dépendant du temps
r нестационарное поведение *n*

12421 time-dependent coupled power theory
d Theorie *f* der zeitabhängigen Leistungskopplung
f théorie *f* du couplage de puissance dépendant du temps
r теория *f* связи мощностей с временной зависимостью

* **time division** → 12453

12422 time-domain backscatter[ing] measuring set
d Rückstreumesseinrichtung *f* im Zeitbereich
f appareil *m* de mesure par rétrodiffusion dans le domaine temporel
r устройство *n* измерения обратного рассеяния во временной области

* **time-domain backscatter measuring set** → 12422

12423 time duration
d Zeitdauer *f*
f durée *f*
r длительность *f*; продолжительность *f*

12424 time element
d Zeitglied *n*
f organe *m*; élément *m* de temps; membre *m* de temps
r звено *n* времени

12425 time function; function of time
d Zeitfunktion *f*
f fonction *f* temporelle
r временная функция *f*

* **time-independent** → 6448

12426 time interval
 d Zeitabschnitt *m*; Zeitintervall *n*; Zeitraum *m*;
   Zeitspanne *f*
 f intervalle *m* de temps
 r интервал *m* времени

12427 time-invariant filter
 d stationäres Filter *n*
 f filtre *m* stationnaire
 r стационарный [по времени] фильтр *m*

12428 time-invariant system
 d zeitinvariantes System *n*
 f système *m* invariant de temps
 r инвариантная по времени система *f*

12429 time lag
 d zeitliche Nacheilung *f*; Zeitverzögerung *f*
 f temporisation *f*
 r запаздывание *n* по времени

12430 time-lag action
 d verzögerte Wirkung *f*
 f action *f* retardée
 r действие *n* с выдержкой времени

12431 time-lag system
 d System *n* mit Totzeit; Totzeitsystem *n*
 f système *m* avec temps mort; système avec
   temps de retard
 r система *f* с запаздыванием

12432 time limit
 d Zeitgrenze *f*
 f limite *f* de temps
 r предел *m* по времени

12433 time-limit protection
 d verzögerter Schutz *m*
 f protection *f* à action différée
 r защита *f* с выдержкой времени

12434 time of response
 d Einschwingzeit *f*; Reaktionszeit *f*
 f temps *m* de réaction; temps de réponse
 r время *n* установления

12435 time-optimal algorithm
 d zeitoptimaler Algorithmus *m*
 f algorithme *m* optimal de temps
 r оптимальный временной алгоритм *m*

**12436 time-optimal control of sampled-data
   systems**
 d zeitoptimale Steuerung *f* in linearen
   Abtastregelkreisen
 f systèmes *mpl* échantillonnés linéaires
   optimaux en temps

 r оптимальное по быстродействию
   управление *n* в дискретных или
   импульсных системах

12437 time-optimal handling
 d zeitoptimale Handhabung *f*
 f manipulation *f* optimale de temps
 r оптимальное временное
   манипулирование *n*

12438 timeout
 d Zeitsperre *f*; Auszeit *f*; Timeout *n*
 f temps *m* de suspension; temporisation *f*
 r блокировка *f* по времени; тайм-аут *m*

12439 timeout interval
 d Auszeitintervall *n*
 f intervalle *m* de suspension
 r длительность *f* тайм-аута

12440 timeout mode
 d Timeout-Betrieb *m*
 f mode *m* de limitation de temps
 r режим *m* ограничения по времени

12441 time-pulse converter
 d Zeit-Impuls-Umwandler *m*
 f convertisseur *m* d'impulsions à déplacement
   dans le temps
 r время-импульсный преобразоавтель *m*

12442 time-pulse distributor
 d Zeitimpulsverteiler *m*
 f distributeur *m* d'impulsions de temps
 r распределитель *m* тактовых импульсов

**12443 time-pulse generator; timing [pulse]
   generator**
 d Zeit[takt]geber *m*; Zeitglied *n*;
   Zeitsignalgenerator *m*
 f générateur *m* d'impulsion d'horloge;
   rythmeur *m*
 r генератор *m* хронирующих импульсов;
   генератор синхросигналов; таймер *m*

12444 timer
 d Zeitrelais *n*; Zeiteinstellgerät *n*;
   Zeitmessinstrument *n*
 f temporisateur *m*; dispositif *m* de
   synchronisation; chronométreur *m*
 r реле *n* времени; хронизатор *m*

12445 time-recording equipment
 d Zeitregistriergerät *n*
 f appareil *m* d'enregistrement de temps
 r аппаратура *f* регистрации времени

12446 time regulator
 d Zeitregler *m*

   f chronorégulateur *m*
   r хронорегулятор *m*

12447 time requirement
   d Zeiterfordernis *n*; Zeitbedarf *m*
   f exigence *f* de temps
   r потребное время *n*

12448 time response
   d Zeitverhalten *n*; Zeitverlauf *m*
   f réponse *f* de temps
   r временная характеристика *f*

12449 timer event control block
   d Zeitgebersteuerblock *m*
   f bloc *m* de contrôle pour temporisateur
   r управляющий блок *m* тактового
   генератора

12450 timesaving
   d zeitsparend
   f à économie de temps
   r не требующий больших затрат времени

   * **time-schedule controller → 10258**

12451 time-schedule controller
   d Zeitablaufplan-Steuereinrichtung *f*
   f dispositif *m* de commande de déroulement
   r контроллер *m* распределения времени

12452 time-shared operation
   d Paralleloperation *f*
   f opération *f* à répartition temporelle
   r операция *f* с разделением времени

12453 timesharing; time division
   d Zeit[auf]teilung *f*; Time-Sharing *n*
   f répartition *f* temporelle; partage *m* de temps
   r разделение *n* времени

12454 timesharing system
   d Zeitteilungssystem *n*
   f système *m* de répartition temporelle
   r система *f* распределения времени

12455 timesharing terminal
   d Time-Sharing-Terminal *n*;
   Zeitteilungsterminal *n*
   f terminal *m* de répartition temporelle
   r терминал *m* в системе с разделением
   времени

12456 time-variable control
   d zeitveränderliche Regelung *f*
   f réglage *m* variable en temps
   r регулирование *n* по времени

12457 time-varying gradient
   d zeitlich veränderlicher Gradient *m*
   f gradient *m* variable dans le temps
   r градиент m, изменяющийся по времени

12458 time-varying system
   d zeitvariantes System *n*; zeitvariables System
   f système *m* variable de temps
   r переменная во времени система *f*

12459 timing; clocking
   d Takt[ier]ung *f*; Zeitsteuerung *f*
   f commande *f* de temps; minutage *m*
   r тактирование *n*; синхронизация *f*

12460 timing characteristics
   d Taktierungskennwerte *mpl*
   f caractéristiques *fpl* de rythme
   r характеристики *fpl* синхронизации

12461 timing comparator circuit
   d Zeitvergleichsstromkreis *m*
   f circuit *m* de comparaison de temps
   r схема *f* сравнения времени

12462 timing control
   d Taktregelung *f*
   f contrôle *m* de minutage; commande *f* de
   synchronisation
   r управление *n* синхронизацией

12463 timing error; clocking error
   d Taktierungsfehler *m*; Zeitsteuerungsfehler *m*
   f défaut *m* de rythme
   r ошибка *f* синхронизации

12464 timing estimations
   d Zeitabschätzungen *fpl*; Zeitbewertungen *fpl*
   f estimations *fpl* du temps
   r оценки *fpl* времени

12465 timing generation
   d Takterzeugung *f*
   f génération *f* d'impulsions d'horloge
   r генерирование *n* синхроимпульсов

   * **timing generator → 12443**

12466 timing logic element
   d logisches Zeitelement *n*
   f élément *m* logique de temporisation
   r временной логический элемент *m*

   * **timing pulse generator → 12443**

12467 timing verification
   d Zeitablaufverifizierung *f*

f vérification f de relations temporelles
r верификация f временных соотношений

12468 timing wave
d Zeitmesswelle f
f base f de mesure de temps
r хронирующий сигнал m

12469 toggle v
d kippen
f basculer
r генерировать релаксационные колебания

12470 tolerance
d Toleranz f; zulässiger Fehler m;
Fehlergrenze f
f tolérance f; erreur f admissible; limite f
d'erreur
r допуск m; допустимое отклонение n

12471 tolerance factor
d Toleranzfaktor m
f facteur m de tolérance
r фактор m стабильности

12472 tolerance on rated capacitance
d zulässige Abweichung f von der
Nennkapazität
f tolérance f sur la capacité nominale
r допустимое отклонение n от номинальной
ёмкости

12473 tolerance system
d Toleranzsystem n
f système m des tolérances
r система f допусков

12474 tool
d Werkzeug n; Hilfsmittel n; Instrument n
f outil m; instrument m
r средство n; инструмент m

12475 tool coding
d Werkzeugkodierung f
f codage m d'outil
r кодирование n инструмента

**12476 tool fracture checking; tool rupture
checking**
d Werkzeugbruchkontrolle f
f contrôle m de la rupture d'outil
r контроль m поломки инструмента

12477 tool handling
d Werkzeughandhabung f
f manutention f d'outil
r манипулирование n инструментами

12478 tool manipulator assembly
d Werkzeugmanipulatoraufbau m
f montage m d'un manipulateur d'outil
r компоновка f инструментального
манипулятора

12479 tool providing system
d Werkzeugversorgungssystem n
f système m d'alimentation d'outil
r система f подвода инструмента

* **tool rupture checking** → 12476

12480 top technology
d Spitzentechnologie f
f technologie f de crête
r передовая технология f

12481 torque amplifier
d Drehmomentverstärker m
f amplificateur m de couple
r усилитель m вращающего момента

12482 torque characteristics
d Momentkennlinie f
f caractéristique f de moment
r характеристика f моментов

12483 torque motor
d Drehmomentantrieb m
f moteur m à couple constant
r поворотный двигатель m

12484 torsional vibration damper
d Torsionsschwingungsdämpfer m
f amortisseur m de vibration à torsion
r успокоитель m крутильных колебаний

12485 total closing time
d Gesamteinschaltzeit f
f durée f totale de fermeture
r общее время n замыкания

* **total error** → 6004

12486 total structure of hybrid system
d Gesamtstruktur f eines Hybridsystems
f structure f totale d'un système hybride
r общая структура f гибридной системы

12487 total system
d Gesamtsystem n
f système m total; système complet
r полная система f

12488 touchless revolution counter
d berührungsfreier Drehzahlmesser m
f compteur m de révolutions sans contact
r бесконтактный тахометр m оборотов

12489 traced fault
d aufgespürter Fehler *m*
f défaut *m* détecté
r обнаруженная неисправность *f*

12490 tracer contact
d Fühlerkontakt *m*
f contact *m* du palpeur
r следящий контакт *m*

12491 tracking infrared system
d Infrarotstrahlennachlaufsystem *n*
f système *m* de poursuite à rayons infrarouges
r следящая инфракрасная система *f*

12492 tracking lag
d Nachlaufverzögerung *f*
f retard *m* de poursuite
r запаздывание *n* сопровождения

12493 transceiver code
d Transceiver-Kode *m*
f transceiver-code *m*
r передаточный код *m*

12494 transcendental equation
d transzendente Gleichung *f*
f équation *f* transcendante
r трансцендентное уравнение *n*

* transcoder → 2513

* transducer → 3402

12495 transducer system
d Wandlersystem *n*
f système *m* de transducteur
r система *f* преобразователя

* transductor → 3402

12496 transductor element
d Verstärkerdrossel *f*
f organe *m* transducteur
r преобразовательный элемент *m*

* transfer → 2133

12497 transfer *v*
d übertragen
f transmettre
r передавать; пересылать

12498 transfer capacitance
d Durchgangskapazität *f*
f capacité *f* de transfert; capacité de passage
r проходная ёмкость *f*

* transfer channel → 12543

12499 transfer characteristic
d Übertragungskennlinie *f*;
   Übertragungscharakteristik *f*
f caractéristique *f* de transfert
r передаточная характеристика *f*

12500 transfer check
d Übertragungskontrolle *f*
f vérification *f* de transfert
r контроль *m* передачи

12501 transfer circuit
d Übertragungsglied *n*
f circuit *m* de transmission
r передаточное звено *n*

* transfer contact → 2251

12502 transfer device
d Übertragungsgerät *n*
f dispositif *m* de transfert
r передающее устройство *n*

12503 transfer element
d Übertragungselement *n*
f élément *m* de transfert
r передаточный элемент *m*

12504 transfer error; transmission error
d Übertragungsfehler *m*
f erreur *f* de transmission
r ошибка *f* передачи

12505 transfer function
d Übertragungsfunktion *f*
f fonction *f* de transfert
r передаточная функция *f*

12506 transfer line; transmission line
d Übertragungsleitung *f*
f ligne *f* de transfert
r линия *f* передачи

* transfer number → 5881

* transfer of control → 3358

12507 transfer operation
d Übertragungsoperation *f*
f opération *f* de transfert
r операция *f* передачи

12508 transfer path
d Übertragungsweg *m*
f voie *f* de transfert
r тракт *m* передачи

12509 transfer rate; transfer speed
d Übertragungsrate *f*;
Übertragungsgeschwindigkeit *f*
f vitesse *f* de transfert
r скорость *f* передачи

* **transfer ratio** → 5881

* **transfer speed** → 12509

12510 transfer technique
d Übergabetechnik *f*
f technique *f* du transfert
r техника *f* передачи

12511 transfer time
d Übertragungszeit *f*; Transferzeit *f*
f temps *m* de transfert
r время *n* передачи

12512 transform *v*
d transformieren
f transformer
r преобразовать

* **transformation** → 3389

12513 transformation section
d Transformationsstück *n*
f section *f* de transformation
r трансформирующий участок *m*

12514 transformer
d Transformator *m*; Umformer *m*
f transformateur *m*
r трансформатор *m*

* **transforming** → 3389

12515 transform matrix
d Transformationsmatrix *f*
f matrice *f* de transformation
r матрица *f* преобразования

* **transient** → 12525

12516 transient *adj*
d vorübergehend
f transitoire
r переходный

12517 transient analyzer
d Schwingungsmodell *n*
f analyseur *m* des procédés transitoires
r анализатор *m* неустанововшихся
процессов

12518 transient characteristic; transient curve
d Übergangsprozesskurve *f*

f courbe *f* de réponse transitoire
r кривая *f* переходного процесса

12519 transient condition
d Übergangsbedingung *f*
f condition *f* transitoire
r условие *n* перехода

12520 transient current
d Ausgleichsstrom *m*; Übergangsstrom *m*
f courant *m* transitoire
r ток *m* переходного процесса

* **transient curve** → 12518

12521 transient deviation
d vorübergehende Abweichung *f*;
vorübergehende Regelabweichung *f*
f écart *m* de consigne transitoire
r промеждуточное отклонение *n*

12522 transient effect
d Einschwingeffekt *m*
f effet *m* transitoire
r влияние *n* переходного процесса

12523 transient overshoot
d Überschwingweite *f*
f dépassement *m* transitoire
r переходное перерегулирование *n*

12524 transient performance; transient response
d Übergangserscheinung *f*;
Übergangscharakteristik *f*
f performance *f* transitoire; régime *m*
transitoire
r характеристика *f* переходного процеса

* **transient period** → 1967

12525 transient [phenomenon]
d Übergangsvorgang *m*; Übergangsprozess *m*;
Einschwingvorgang *m*;
Einschwingzustand *m*
f phénomène *m* transitoire; régime *m*
transitoire; transitoire *m*
r переходный процесс *m*;
неустановившийся режим *m*

12526 transient pulse
d Einschwingimpuls *m*
f impulsion *f* transitoire
r импульс *m* переходного процесса

* **transient response** → 12524

12527 transient response analysis
d Analyse *f* der Übergangsvorgänge; Analyse
instationärer Vorgänge

f analyse *f* des régimes transitoires
r анализ *m* переходных процессов

12528 transient signal
d Übergangssignal *n*; Einschwingsignal *n*
f signal *m* transitoire
r неустановившийся сигнал *m*

12529 transient state
d Einschwingzustand *m*; Übergangszustand *m*
f état *m* transitoire
r переходное состояние *n*

12530 transient temperature gradient
d vergänglicher Temperaturgradient *m*
f gradient *m* transitoire de température
r температурный градиент *m* в переходном процессе

12531 transient time
d Übergangsprozessdauer *f*
f durée *f* du régime transitoire; période *f* transitoire
r длительность *f* переходного процесса

12532 transistor
d Transistor *m*; Halbleitertriode *f*
f transistor *m*; triode *f* à cristal
r транзистор *m*

12533 transistorized building-block switching units
d transistorisierte Schaltbaukasteneinheiten *fpl*
f unités *fpl* logiques transistorisées
r коммутационные блочные элементы *mpl* на тразисторах

12534 transistor-transistor logic; TTL
d Transistor-Transistor-Logik *f*; TTL
f logique *f* transistor-transistor; TTL
r транзисторно-транзисторная логика *f*; ТТЛ

12535 transition
d Transition *f*; Übergang *m*
f transition *f*
r переход *m*

12536 transition element
d Übergangselement *n*
f élément *m* de transition
r переходный элемент *m*

12537 transition graph
d Übergangsgraf *m*
f graphe *m* de transition
r граф *m* перехода

12538 transition probability
d Übergangswahrscheinlichkeit *f*
f probabilité *f* transitoire; probabilité de transition
r вероятность *f* перехода

12539 transition regime
d Übergangszustand *m*
f régime *m* de transition
r переходный режим *m*

12540 transition state theory
d Theorie *f* des Übergangszustandes
f théorie *f* de l'état de transition
r теория *f* переходного состояния

12541 transitivity
d Transitivität *f*
f transitivité *f*
r транзитивность *f*

12542 translation
d Translation *f*
f translation *f*
r трансляция *f*

12543 transmission channel; transfer channel
d Übertragungskanal *m*
f canal *m* de transfert; canal de transmission
r канал *m* передачи

12544 transmission characteristic
d Durchlasskennlinie *f*
f caractéristique *f* de passage; caractéristique de passe-bande
r характеристика *f* пропускания

*   **transmission direction → 4307**

*   **transmission error → 12504**

12545 transmission interface
d Übertragungsschnittstelle *f*
f interface *f* de transmission
r интерфейс *m* с системой передачи

*   **transmission line → 12506**

12546 transmission loss
d Übertragungsdämpfung *f*
f affaiblissement *m* de transmission
r затухание *n* передачи

12547 transmission monitor
d Endkontrollgerät *n*
f moniteur *m* final
r оконечное устройство *n* контроля

12548 **transmission properties**
  d Übertragungseigenschaften *fpl*
  f propriétés *fpl* de transmission
  r свойства *npl* передачи

12549 **transmit data**
  d Sendedaten *pl*
  f données *fpl* de transmission
  r передаваемые данные *pl*

12550 **transmittance**
  d Transmissionsgrad *m*
  f facteur *m* de transmission
  r коэффициент *m* пропускания

12551 **transmitter**
  d Transmitter *m*; Übertrager *m*; Wandler *m*
  f transmetteur *m*
  r трансмиттер *m*

12552 **transmitter module**
  d Sende[r]modul *m*
  f module *m* émetteur
  r модуль *m* передатчика

12553 **transmitting distortion**
  d Sendeverzerrung *f*
  f distorsion *f* à l'émission
  r искажение *n* при передаче

12554 **transparency**
  d Transparenz *f*; Durchsichtigkeit *f*
  f transparence *f*
  r прозрачность *f*

  * **transponder** → 10446

12555 **transportation algorithm**
  d Transportalgorithmus *m*
  f algorithme *m* de transport
  r транспортный алгоритм *m*

12556 **transport automation**
  d Transportautomatisierung *f*
  f automatisation *f* de transport
  r автоматизация *f* транспорта

12557 **transport optimization**
  d Transportoptimierung *f*
  f optimisation *f* de transport
  r транспортная оптимизация *f*

12558 **transport technology**
  d Transporttechnologie *f*
  f technologie *f* de transport
  r транспортная технология *f*

12559 **treatment automation**

  d Bearbeitungsautomatisierung *f*
  f automatisation *f* de traitement
  r автоматизация *f* операций обработки

12560 **treatment of handling objects**
  d Bearbeiten *n* von Handhabungsobjekten
  f traitement *m* d'objets de manutention
  r обработка *f* объектов манипулирования

12561 **treatment system**
  d Bearbeitungssystem *n*
  f système *m* de traitement
  r система *f* обработки

12562 **tree structure**
  d Baumstruktur *f*
  f structure *f* d'arbre
  r древовидная структура *f*

12563 **trial and error**
  d Versuch *m* und Irrtum *m*
  f essai *m* et erreur *f*
  r проба *f* и ошибка *f*

12564 **trial equipment**
  d Versuchsanlage *f*
  f équipement *m* d'essai
  r экспериментальное оборудование *n*

12565 **trial production**
  d Versuchsproduktion *f*; Probeproduktion *f*
  f production *f* d'essai
  r опытная продукция *f*

  * **trial run** → 12323

12566 **trigger criterion**
  d Schaltkriterium *n*
  f critérium *m* de commutation
  r критерий *m* пуска

12567 **trigger pair circuit**
  d bistabile Kippschaltung *f*
  f multivibrateur *m* bistable; circuit *m* flip-flop;
     débrayeur *m*; embrayeur *m*
  r пусковая схема *f*

  * **trim** *v* → 479

12568 **triple device control**
  d Dreiersteuerung *f*
  f contrôle *m* triple du dispositif
  r тройной контроль *m* устройства

  * **trip out** → 12073

  * **tripping** → 12073

12569 **tripping of power supply**
  d Energiezufuhrabschaltung *f*
  f coupure *f* d'alimentation énergétique
  r выключение *n* источника питания

  * **tristate device** → 12388

12570 **trivial problem**
  d triviales Problem *n*
  f problème *m* trivial
  r тривиальная проблема *f*

12571 **trouble indication**
  d Störungsanzeige *f*
  f indication *f* de pannes; signalisation *f* de pannes
  r индикация *f* неисправностей

12572 **trouble location**
  d Störungslokalisierung *f*
  f localisation *f* de panne
  r локализация *f* неисправностей

12573 **trouble-location problem**
  d Störuchproblem *n*
  f problème *m* de dépannage
  r задача *f* определения местоположения неисправностей

  * **trouble shooting** → 5367

12574 **true**
  d Wahrheit *f*
  f vrai *m*
  r истина *f*; логическая единица *f*

12575 **true control signal**
  d wahres Steuersignal *n*
  f signal *m* de commande vrai
  r действительный управляющий сигнал *m*

  * **true disabling signal** → 12576

12576 **true inhibiting signal; true disabling signal**
  d Wahr-Sperrsignal *n*; wahres Sperrsignal *n*
  f signal *m* d'inhibition vrai
  r истинный запирающий сигнал *m*

12577 **true input signal**
  d wahres Eingangssignal *n*
  f signal *m* d'entrée vrai
  r истинный входной сигнал *m*

12578 **true output signal**
  d wahres Ausgangssignal *n*
  f signal *m* de sortie vrai
  r истинный выходной сигнал *m*

12579 **true state**
  d wahrer Zustand *m*
  f état *m* vrai
  r истинное состояние *n*

12580 **true value**
  d wahrer Wert *m*
  f valeur *f* vraie
  r истинное значение *n*

12581 **truncation condition**
  d Abbruchbedingung *f*
  f condition *f* de tronquage
  r условие *n* прерывания

12582 **truth function**
  d Wahrheitsfunktion *f*
  f fonction *f* de vérité
  r функция *f* истинности

  * **TTL** → 12534

  * **tune** *v* → 479

  * **tuned amplifier** → 11006

12583 **tuned detector**
  d abgestimmter Detektor *m*
  f détecteur *m* accordé
  r синхронизированный детектор *m*

12584 **tuning board**
  d Abstimmtafel *f*
  f pupitre *m* d'accord
  r пульт *m* настройки

12585 **tuning control**
  d Abstimmregelung *f*
  f réglage *m* d'accord
  r настройка *f*

  * **tuning dial** → 10841

  * **tuning fork control** → 5652

  * **tuning range** → 523

  * **tuning scale** → 10841

12586 **turbulent energy transport**
  d turbolenter Energietransport *m*
  f transport *m* turbulent d'énergie
  r турбулентный перенос *m* энергии

12587 **turnkey system**
  d schlüsselfertiges System *n*
  f système *m* clés en main
  r система f, готовая к непосредственному использованию

12588 turn-off delay
  d Ausschaltverzögerung *f*
  f retardement *m* de mise au repos
  r задержка *f* выключения

  * **turn-off time** → 1931

12589 turn-on delay
  d Einschaltverzögerung *f*
  f retardement *m* de mise en marche
  r задержка *f* включения

12590 turn-on time
  d Einschaltzeit *f*
  f temps *m* de mise en marche
  r время *n* включения

12591 two-channel switching
  d Zweikanalschaltung *f*
  f commutation *f* pour deux pistes
  r переключение *n* по двум каналам

12592 two-element regulator
  d Zwei-Element-Regler *m*
  f régulateur *m* à deux éléments
  r регулятор *m* для двух цепей

12593 two-frequency laser
  d Zweifrequenzlaser *m*
  f laser *m* bifréquence
  r двухчастотный лазер *m*

  * **two-level controller** → 8948

  * **two-loop servomechanism** → 4480

12594 two-parameter control
  d Zweiparameterregelung *f*
  f réglage *m* à deux paramètres
  r двухпараметровое регулирование *n*

12595 two-phase clock
  d Zweiphasentakt *m*
  f rythmeur *m* biphasé
  r двухфазная синхронизация *f*

12596 two-phase system
  d Zweiphasensystem *n*
  f système *m* de deux phases
  r двухфазная система *f*

  * **two-position action controller** → 8948

  * **two-position controller** → 8948

12597 two-position differential gap control
  d Zweipunktregelung *f* mit Totband
  f réglage *m* à deux paliers séparés

  r двухпозиционное регулирование *n* с
       нейтральной зоной

  * **two-sided Laplace transformation** → 1745

12598 two-speed controller
  d Zweilaufregler *m*
  f régulateur *m* à deux vitesses d'action
  r двухскоростной регулятор *m*

12599 two-stage amplifier
  d zweistufiger Verstärker *m*
  f amplificateur *m* à deux étages
  r двухкаскадный усилитель *m*

12600 two-stage circuit
  d zweistufiger Kreislauf *m*
  f circuit *m* à deux étages
  r двухступенчатая схема *f*

  * **two-stage control** → 1641

12601 two-stage cycle
  d zweistufiger Prozess *m*
  f cycle *m* à deux étages; procédé *m* à deux
       étappes
  r двухступенчатый цикл *m*

12602 two-stage operation
  d zweistufige Arbeitsweise *f*
  f mode *m* de fonctionnement à deux étages
  r двухступенчатая операция *f*

12603 two-stage servomechanism
  d zweistufiger Servomechanismus *m*
  f servomécanisme *m* à deux étages
  r двухкаскадный сервомеханизм *m*

  * **two-step action controller** → 8948

12604 two-step action with overlap
  d Zweistellenwirkung *f* mit Überlappung
  f réglage *m* à deux paliers à recouvrement
  r двухпозиционное регулирование *n* с
       перекрытием

  * **two-step control** → 1641

  * **two-step controller** → 8948

12605 two-step distance protection set
  d zweistufiger Distanzschutz *m*
  f protection *f* de distance à deux échelons
  r двухступенчатая дистанционная защита *f*

12606 two-valued decision element
  d zweiwertiges Entscheidungselement *n*
  f élément *m* de décision à deux valeurs
  r двухзначный решающий элемент *m*

12607 two-valued logic
 d zweiwertige Logik *f*
 f logique *f* [à valeur] binaire
 r двухзначная логика *f*

 * **two-way contact → 4487**

12608 type model
 d Typemodell *n*
 f modèle *m* type
 r типовая модель *f*

12609 type of digital control
 d Digitalsteuerungstyp *m*
 f type *m* de commande numérique
 r тип *m* цифрового управления

12610 type of program control
 d Programmsteuerungsart *f*
 f sorte *f* de commande programmée
 r тип *m* программного управления

12611 typical forward processing
 d typische Vorwärtsverarbeitung *f*
 f traitement *m* typique en avant
 r типичная приоритетная обработка *f*

12612 typical non-linearity simulation
 d Modellierung *f* typischer Nichtlinearitäten;
      Simulierung *f* typischer Nichtlinearitäten
 f simulation *f* de non-linéarités typiques
 r моделирование *n* типичных
      нелинейностей

U

12613 ultimate pressure
d Enddruck *m*; Endvakuumdruck *m*
f pression *f* limite; vide *m* limite
r предельное давление *n*; предельный вакуум *m*

12614 ultrahigh-frequency filter
d UHF-Filter *n*
f filtre *m* à hyperfréquence
r фильтр *m* ультравысоких частот

12615 ultrasonic absorption measurement
d Ultraschallabsorptionsmessung *f*
f mesure *f* d'absorption d'ultrasons
r измерение *n* поглощения ультразвука

* **ultrasonic delay line** → 12032

12616 ultrasonic material testing apparatus
d Ultraschallmaterialprüfgerät *n*
f appareil *m* de contrôle des matériaux par ultrasons
r аппарат *m* для ультразвукового испытания материала

12617 ultrasonic measurement
d Ultraschallmessverfahren *n*
f méthode *f* de mesure par ultrasons
r ультразвуковой измерительный метод *m*

12618 ultrasonic pulse generator
d Ultraschallimpulsgenerator *m*
f générateur *m* d'impulsions ultrasonores
r ультразвуковой генератор *m* импульсов

12619 ultrasonics
d Ultraschall *m*
f ultrasons *m*
r ультразвук *m*

12620 ultrasonic scanner
d Ultraschallabtastgerät *n*
f dispositif *m* de balayage à ultrasons
r ультразвуковое сканирующее устройство *n*

12621 ultrasonic sensor
d Ultraschallsensor *m*
f capteur *m* à ultrasons
r ультразвуковой сенсор *m*

12622 ultrasonic technology
d Ultraschalltechnologie *f*
f technologie *f* ultrasonore
r ультразвуковая технология *f*

12623 ultrastability
d Ultrastabilität *f*; Selbststabilisierung *f*
f ultrastabilité *f*
r сверхустойчивость *f*

12624 unbalance
d Unausgeglichenheit *f*; Unwucht *f*
f désalignement *m*; erreur *f* d'alignement
r рассогласование *n*

12625 unbiased
d nichtvorgespannt
f sans polarisation
r несмещенный

* **unblock** *v* → 12696

12626 unbounded; boundless
d unbeschränkt; unbegrenzt
f absolu; illimité
r неограниченный

12627 uncertainty function
d Unbestimmtheitsfunktion *f*
f fonction *f* d'incertitude
r функция *f* неопределенности

12628 unclocked
d ungetaktet
f non rythmé
r несинхронизированный

12629 unconditional
d unbedingt
f inconditionnel
r безусловный

12630 unconditional jump
d unbedlingter Sprung *m*
f saut *m* inconditionnel; transfert *m* inconditionnel
r безусловный переход *m*

12631 uncontrolled
d ungeregelt
f non réglé
r нерегулирумый

12632 uncontrolled join[ing] mechanism
d ungesteuerter Fügemechanismus *m*
f mécanisme *m* de jointage non commandé
r неуправляемый механизм *m* сопряжения

* uncontrolled join mechanism → 12632

* uncorrectable error → 12702

12633 uncorrected delay
  d unkorrigierte Laufzeit f; unkorrigierte
    Verzögerung f
  f retard m non corrigé
  r некорректированное запаздывание n

12634 undamped control
  d ungedämpfte Regelung f
  f réglage m non amorti
  r расходящийся процесс m регулирования

12635 undercurrent tripping
  d Unterstromausschalten n;
    Unterstromauslösen n
  f déclenchement m à minimum de courant;
    fonctionnement m à baisse d'inensité
  r отключение n минимального тока

12636 underdamping
  d periodische Dämpfung f
  f amortissement m souscritique
  r периодическое демпфирование n

12637 underdetermined
  d unterdeterminiert; unterbestimmt
  f incomplètement déterminé
  r недоопределенный

12638 undervoltage device
  d Unterspannungsgerät n
  f appareil m à minimum de tension
  r прибор m с пониженным напряжением

12639 undisturbed one signal
  d ungestörtes Einersignal n
  f signal m un non perturbé
  r неискаженный единичный сигнал m

12640 undisturbed-zero output
  d ungestörtes Nullsignal n
  f signal m zéro non perturbé
  r неискаженный нулевой выходной
    сигнал m

12641 undo v
  d rückgängig machen; rückgängig erstellen;
    den vorherigen Zustand wiederherstellen
  f annuler; défaire
  r отменять последнее действие

12642 undoing actions
  d rückgängige Wirkungen fpl
  f actions fpl d'annulation
  r отменяющие действия npl; возвратные
    действия

12643 unequal impulse
  d Ungleichimpuls m
  f impulsion f d'inégalité
  r неравный импульс m

12644 unidirectional circuit
  d Kreis m mit einseitiger Richtwirkung;
    gerichtete Kette f
  f chaîne f unidirectionelle
  r однонаправленная цепь f

12645 uniform
  d gleichförmig; uniform
  f uniforme
  r равномерный; однородный

* uniformalization → 11801

12646 uniform convergence
  d gleichmäßige Konvergenz f
  f convergence f uniforme
  r равномерная сходимость f

12647 uniformity
  d Gleichmäßigkeit f
  f uniformité f
  r равномерность f

12648 uniformity factor
  d Gleichmäßigkeitsfaktor m
  f facteur m d'uniformité
  r фактор m однородности

12649 uniformly bounded
  d gleichmäßig beschränkt
  f limité uniformément
  r равномерно ограниченный

* uniform scale → 5220

12650 uniform steady state
  d Gleichgewichtszustand m;
    Gleichgewichtslage f
  f état m d'équilibre; point m d'équilibre;
    position f d'équilibre
  r положение n равновесия

12651 unilateral hybrid system
  d einseitiges Hybridsystem n
  f système m hybride unilatéral
  r односторонная гибридная система f

12652 unimodal distribution
  d Einmodenverteilung f
  f distribution f monomode
  r унимодальное распределение n

12653 unimodal laser
  d Einmodenlaser m

f laser *m* à mode unique
r унимодальный лазер *m*

12654 unimplemented instruction
d nichtrealisierter Befehl *m*
f instruction *f* non réalisée
r нереализованная инструкция *f*

12655 uninterruptable mode
d nichtunterbrechbare Betriebsweise *f*
f mode *m* non interruptible
r режим *m* с запрещенными прерываниями

12656 union
d Vereinigung *f*
f union *f*
r объединение *n*

12657 unique
d eindeutig; einmalig
f unique
r уникальный; единственный

12658 uniquely determined
d eindeutig bestimmt
f déterminé [non ambigu]
r однозначно определенный

12659 uniqueness of solution
d Eindeutigkeit *f* einer Lösung
f unicité *f* d'une solution
r единственность *f* решения

12660 uniqueness theorem
d Eindeutigkeitssatz *m*
f théorème *m* de l'unicité
r теорема *f* единственности

12661 unitary transformation
d unitäre Transformation *f*
f transformation *f* unitaire
r унитарное преобразование *n*

12662 unit control block
d Gerätesteuerblock *m*
f bloc *m* pour la commande de dispositifs
r блок *m* управления устройством

12663 unit impulse
d Einheitsimpuls *m*
f impulsion *f* unitaire
r единичный импульс *m*

* **unit impulse function → 3937**

12664 unit interval
d Einheitsintervall *n*; Elementarintervall *n*
f intervalle *m* unitaire

r единичный интервал *m*

12665 unit operation
d Grundoperation *f*
f opération *f* unitaire
r основная операция *f*

12666 unit process
d Grundverfahren *n*
f procédé *m* de base
r основной процесс *m*

* **unit status → 4061**

12667 unit step
d Einheitssprung *m*
f échelon *m* unitaire; échelon d'unité
r единичный скачок *m*

* **unit test → 4062**

12668 universal aggregate system
d universales Baukastensystem *n*
f système *m* d'agrégat universal
r универсальная агрегатная система *f*

12669 universal algorithm
d universeller Algorithmus *m*
f algorithme *m* universel
r универсальный алгоритм *m*

* **universal automaton → 8497**

12670 universal auxiliary relay
d Universalhilfsrelais *n*
f relais *m* auxiliaire universel
r промежуточное универсальное реле *n*

12671 universal checking machine
d Universalkontrollgerät *n*
f dispositif *m* universel de contrôle
r универсальная испытательная машина *f*

12672 universal computer-oriented language
d universelle rechner-orientierte Sprache *f*
f langage *m* universel adaptable à toutes
calculatrices; langage universel orienté sur
les calculatrices
r универсальный
машинно-ориентированный язык *m*

12673 universal computer program
d Programm *n* eines Universalrechners
f programme *m* d'un calculateur universel;
programme d'une calculatrice universelle
r универсальная компьютерная программа *f*

12674 universal control
d universelle Steuerung *f*

f commande *f* universelle
r универсальное управление *n*

12675 universal control automaton with free program selection
d Universalsteuerautomat *m* mit freier Programmauswahl
f automate *m* de commande universel à sélection libre du programme
r универсальный управляющий автомат *m* со свободным выбором программы

12676 universal development system
d universelles Entwicklungssystem *n*
f système *m* universel de développement
r универсальная система *f* проектирования

12677 universal element
d Universalelement *n*
f élément *m* universel
r универсальный элемент *m*

12678 universal function converter
d Universalfunktionswandler *m*
f convertisseur *m* de fonction universel
r преобразователь *m* универсальных функций

12679 universal impulse model
(of controlled system)
d Universalimpulsmodell *n*
f modèle *m* universel d'impulsion
r универсальная импульсная система *f*

12680 universal instruction
d universeller Befehl *m*
f instruction *f* universelle
r универсальная инструкция *f*

12681 universal interface; general-purpose interface
d universelles Interface *n*; Universalschnittstelle *f*
f interface *f* universelle
r универсальный интерфейс *m*

12682 universal machine
d Universalmaschine *f*
f machine *f* universelle
r универсальная машина *f*

12683 universal manipulator
d Universalmanipulator *m*
f manipulateur *m* universel
r универсальный манипулятор *m*

12684 universal manipulator application
d universeller Manipulatoreinsatz *m*;

universelle Manipulatoranwendung *f*
f application *f* de manipulateur universelle
r применение *n* универсального манипулятора

12685 universal measuring laboratory-type automaton
d universaler Labormessautomat *m*; Universallabormessautomat *m*
f dispositif *m* automatique universel de mesure de laboratoire
r универсальный измерительный автомат *m* лабораторного типа

12686 universal microscope with automatic exposure
d Universalmikroskop *n* mit automatischer Belichtungsregelung
f microscope *m* universel à réglage automatique d'exposition
r универсальный микроскоп *m* с автоматической экспозиций

12687 universal peripheral controller; UPC
d universeller Peripheriesteuerungs-Schaltkreis *m*
f circuit *m* de commande périphérique universel
r универсальный периферийный контроллер *n*

12688 universal peripheral interface
d universelle Peripherieschnittstelle *f*; Universalinterface *n* für Peripherieanschluss
f interface *f* périphérique universelle
r универсальный периферийный интерфейс *m*

12689 universal processor
d universeller Prozessor *m*
f processeur *m* universel
r универсальный процессор *m*

12690 universal program transmitter
d Universalprogrammgeber *m*
f émetteur *m* universel de programme
r универсальный программный датчик *m*

12691 universal robot command
d universelle Robotersteuerung *f*
f commande *f* de robot universel
r универсальное управление *n* роботом

12692 universal semi-automatic tester
d universelles halbautomatisches Prüfgerät *n*
f essayeur *m* semi-automatique universel
r универсальное полуавтоматическое контрольное устройство *n*

12693 universal sensor type
d   universeller Sensortyp *m*
f   type *m* de senseur universel
r   универсальный сенсор *m*

12694 unknown state
d   unbekannter Zustand *m*
f   état *m* inconnu
r   неизвестное состояние *n*

12695 unlabelled
d   nichtetikettiert; nichtmarkiert
f   sans label; non étiquetté; non marqué
r   непомеченный

*   **unloaded characteristic → 8648**

12696 unlock *v*; unblock *v*; deblock *v*
d   freigeben; befreien; entsperren; entblocken
f   déverrouiller; débloquer; libérer; ouvrir
r   разблокировать; деблокировать

12697 unmanned production with robots
d   unbemannte Fertigung *f* mit Robotern
f   production *f* non habitée avec robots
r   автоматизированное производство *n* с применением роботов

12698 unmatched
d   ungleich; unpaarig
f   inégal; impair
r   несогласованный

12699 unmodulated carrier
d   nichtmodulierte Trägerwelle *f*; unmodulierte Trägerwelle
f   porteuse *f* non modulée
r   немодулированная несущая *f*

12700 unmonitored control system
d   rückführungsloses Steuersystem *n*
f   système *m* de commande à boucle ouverte
r   система *f* управления с открытым циклом

12701 unobservable system
d   nichtbeobachtbares System *n*
f   système *m* inobservable
r   ненаблюдаемая система *f*

*   **unrecoverable → 6989**

12702 unrecoverable error; uncorrectable error
d   nichtbehebbarer Fehler *m*; unkorrigierbarer Fehler
f   erreur *f* incorrigible
r   неисправимая ошибка *f*

12703 unregulated continuous voltage
d   ungeregelte Gleichspannung *f*
f   tension *f* continue non stabilisée
r   нестабилизированное постоянное напряжение *n*

12704 unsettled condition
d   unbeständiger Zustand *m*; unstetiger Zustand
f   régime *m* non établi; régime non stationnaire
r   неустановившееся состояние *n*; неустановившееся условие *n*

12705 unsigned integer
d   vorzeichenlose ganze Zahl *f*
f   entier *m* sans signe
r   целое *n* [число] без знака

12706 unstable control operation
d   unstabiler Regelvorgang *m*
f   procès *m* instable de réglage
r   неустойчивый процесс *m* регулирования

12707 unstable equilibrium point
d   instabiler Gleichgewichtspunkt *m*
f   point *m* d'équilibre instable
r   неустойчивая точка *f* равновесия

12708 unstable focus
d   unstabiler Brennpunkt *m*
f   foyer *m* instable
r   неустойчивый фокус *m*

12709 unstable limit cycle
d   unstabiler Grenzzyklus *m*
f   cycle *m* limite instable
r   неустойчивый предельный цикл *m*

12710 unstable nodal point
d   instabiler Knotenpunkt *m*
f   point *m* nodal instable
r   неустойчивая узловая точка *f*

12711 unstable node
d   unstabiler Knoten *m*
f   nœud *m* instable
r   неустойчивый узел *m*

*   **unstable state → 6703**

12712 unstable system
d   instabiles System *n*
f   système *m* instable
r   неустойчивая система *f*

12713 unsteadiness
d   Unstetigkeit *f*
f   discontinuité *f*
r   нерегулярность *f*; разрывность *f*

12714 unsteady motion
d nichtstationäre Bewegung *f*; unstabile
  Bewegung
f mouvement *m* non stationnaire
r неустановившееся движение *n*

* **UPC → 12687**

12715 upgrade
d Steigerung *f*
f montée *f*
r надстройка *f*

12716 upgrade *v*
d ausbauen; aufstocken; seigern
f augmenter
r поднимать; надстраивать

12717 upper bound
d obere Schranke *f*
f borne *f* supérieure
r верхний предел *m*

12718 upper limit
d oberer Grenzwert *m*
f valeur *f* limite supérieure
r верхнее предельное значение *n*

12719 upper limit of integration
d obere Integrationsgrenze *f*
f limite *f* supérieure d'intégration
r верхний предел *m* интегрирования

12720 upward compatibility
d Aufwärskompatibilität *f*
f compatibilité *f* vers le haut
r совместимость *f* снизу вверх

12721 usable range
d brauchbarer Bereich *m*
f étendue *f* utile
r эффективный диапазон *m*

* **use → 923**

* **useful component → 3998**

12722 useful performance; effective power
d Nutzleistung *f*
f effet *m* utile; puissance *f* utile
r полезная мощность *f*

12723 useful signal; desired signal
d Nutzsignal *n*
f signal *m* utile
r полезный сигнал *m*

12724 use of control
d Benutzung *f* der Steuerung
f utilisation *f* de commande
r использование *n* управления

* **use of measuring system → 930**

* **use of robots → 936**

12725 user-oriented design
d nutzerorientierter Entwurf *m*;
  anwendungsorientierter Entwurf
f conception *f* orientée usager
r разработка, *f* ориентированная на
  пользователя

12726 utility program
d Nutzprogramm *n*
f programme *m* d'utilisation
r вспомагательная программа *f*

* **utilization factor → 11961**

12727 utilize quantity
  (for computer)
d nutzbare Große *f*
f grandeur *f* utilisable
r воспринимаемая величина *f*

V

12728 vacuum apparatus
  d Vakuumapparat *m*
  f appareil *m* à vide
  r вакуум-аппарат *m*

12729 vacuum chamber
  d Vakuumkammer *f*
  f chambre *f* à vide
  r вакуумная камера *f*

**12730 vacuum controller; vacuum governor;
      vacuum regulator**
  d Vakuumregler *m*
  f régulateur *m* du vide
  r стабилизатор *m* вакуума; регулятор *m*
      вакуума

  * **vacuum electric discharge gauge** → 4712

12731 vacuum fluorescent display
  d Vakuumfluoreszenzanzeige *f*
  f affichage *m* fluorescent au vide
  r вакуумный флюоресцентный
      индикатор *m*

12732 vacuum gauge control circuit
  d Regelkreis *m* eines Vakuummanometres
  f circuit *m* de réglage de vacuomètre
  r схема *f* управления вакуумметра

  * **vacuum governor** → 12730

12733 vacuum indicator
  d Vakuumanzeiger *m*
  f indicateur *m* du vide
  r индикатор *m* вакуума

12734 vacuum measuring instrument
  d Vakuummessgerät *n*
  f vacuomètre *m*
  r вакуумметр *m*

12735 vacuum photocell
  d Vakuumfotozelle *f*; Fotozelle *f* mit äußerem
      lichtelektrischem Effekt
  f tube *m* photo-électronique à vide; cellule *f*
      photo-électrique à vide
  r вакуумный фотоэлемент *m*

12736 vacuum plant
  d Vakuumanlage *f*

  f installation *f* sous vide
  r вакуумная установка *f*

12737 vacuum process
  d Vakuumverfahren *n*
  f procédé *m* au vide
  r процесс *m* под вакуумом

12738 vacuum radiometer gauge
  d Strahlungsvakuummeter *n*;
      Radiometermanometer *n*
  f vacuomètre *m* radiométrique
  r радиометрический вакуумметр *m*

  * **vacuum regulator** → 12730

12739 vacuum technique
  d Vakuumtechnik *f*
  f technique *f* de vide
  r вакуумная техника *f*

12740 validation
  d Gültigkeitsbestätigung *f*
  f confirmation *f* de validité
  r подтверждение *n* достоверности

12741 valid data
  d gültige Daten *pl*
  f données *fpl* valables
  r истинные данные *pl*

12742 valid digit; relevant digit
  d gältige Ziffer *f*
  f chiffre *m* valable
  r значащая цифра *f*

12743 validity
  d Gültigkeit *f*; Richtigkeit *f*
  f validité
  r истинность *f*; достоверность *f*

  * **validity check** → 12744

12744 validity check[ing]
  d Gültigkeitskontrolle *f*
  f contrôle *m* de validité
  r проверка *f* на значимость

12745 value
  d Wert *m*
  f valeur *f*
  r величина *f*

12746 value of input signal
  d Eingangssignalwert *m*
  f valeur *f* de signal d'entrée
  r значение *n* входного сигнала

12747   value of self-inductance
   d   Selbstinduktionswert *m*
   f   valeur *f* d'auto-induction
   r   величина *f* самоиндукции

12748   value of the check; check value
   d   Kontrollwert *m*; Wert *m* der Kontrolle
   f   valeur *f* de contrôle
   r   контрольное значение *n*

12749   value sampling
   d   Wertabfrage *f*
   f   exploration *f* de valeur
   r   развёртка *f* величин

12750   values inquest order
   d   Abfrageordnung *f* der Größen
   f   ordre *m* d'examen des grandeurs
   r   порядок *m* исследования величин

12751   valve
   d   Ventil *n*
   f   vanne *f*; soupape *f*
   r   клапан *m*; вентиль *m*

12752   valve positioner
   d   Stellrelais *n*; Regler *m* zur Ventilstellung;
       Positioner *m*
   f   positionneur *m* [de vanne]
   r   клапанный позиционер *m*

   *   var-hour meter → 10664

12753   variable
   d   Variable *f*; Veränderliche *f*
   f   variable *f*
   r   переменная *f*

12754   variable autotransformer
   d   regelbarer Autotransformator *m*
   f   autotransformateur *m* variable
   r   регулируемый автотрансформатор *m*

12755   variable block length
   d   variable Blocklänge *f*
   f   longueur *f* de bloc variable
   r   переменная длина *f* блока

12756   variable capacitor
   d   regelbarer Kondensator *m*
   f   condensateur *m* variable
   r   переменный конденсатор *m*

12757   variable coefficient system; [linear] system
        of variable coefficients
   d   [lineares] System *n* mit variablen
       Koeffizienten
   f   système *m* linéaire aux coefficients variables
   r   [линейная] система *f* с переменными
       коэффициентами

12758   variable coefficient unit; variable scale
        factor unit
   d   variabler Faktorgeber *m*
   f   unité *f* avec coefficient variable
   r   блок *m* переменных коэффициентов

12759   variable command control
   d   Führungsregelung *f*
   f   asservissement *m*; régulation *f* de
       correspondance
   r   следящее управление *n*

12760   variable component
   d   variable Komponente *f*
   f   composante *f* variable
   r   переменная составляющая *f*

12761   variable coordinate
   d   veränderliche Koordinaten *fpl*
   f   coordonnées *fpl* variables
   r   переменные координаты *fpl*

12762   variable cycle duration
   d   veränderliche Zyklusdauer *f*
   f   durée *f* variable de cycle
   r   длительность *f* переменного цикла

12763   variable damping system
   d   System *n* mit veränderlicher Dämpfung
   f   système *m* à amortissement variable
   r   система *f* с переменным демпфированием

12764   variable data
   d   variable Daten *pl*
   f   données *fpl* variables
   r   переменные данные *pl*

12765   variable delay
   d   veränderliche Verzögerung *f*
   f   retard *m* variable
   r   переменное запаздывание *n*

12766   variable density
   d   veränderliche Dichte *f*
   f   densité *f* variable
   r   переменная плотность *f*

   *   variable feedback → 4670

   *   variable feedback controller → 4671

12767   variable field length system
   d   System *n* variabler Feldlänge
   f   système *m* de longueur de champ variable
   r   система *f* с переменной длиной полей

12768 variable flow control
   d veränderbare Durchflussregelung *f*
   f réglage *m* variable du débit
   r регулирование *n* переменного расхода

12769 variable frequency chopper; variable speed chopper
   d Zerhacker *m* mit veränderbarer Frequenz
   f interrupteur *m* à fréquence variable
   r прерыватель *m* потока, вращающийся с переменной скоростью

12770 variable frequency generator
   d Generator *m* mit veränderlicher Frequenz
   f générateur *m* à fréquence variable
   r генератор *m* регулируемой частоты

12771 variable handling requirement
   d variable Handhabungsanforderung *f*
   f demande *f* de manipulation variable
   r меняющееся требование *n* к манипулированию

12772 variable inductor
   d regelbare Drosselspule *f*; Induktionsspule *f*
   f inducteur *m* variable
   r катушка *f* с плавнорегулируемой индуктивностью

12773 variable instruction
   d veränderlicher Befehl *m*
   f instruction *f* variable
   r переменная команда *f*

12774 variable master clock
   d veränderlicher Haupttakt *m*
   f rythme *m* principal variable
   r переменная тактовая частота *f*

12775 variable modulation
   d Wechselmodulation *f*
   f modulation *f* variable
   r переменная модуляция *f*

   * **variable recorder** → 3420

12776 variable resistance transducer
   d Geber *m* mit veränderlichem Widerstand
   f transmetteur *m* à résistance variable
   r датчик *m* с переменным сопротивлением

   * **variable scale factor unit** → 12758

12777 variables of automatic control
   d Veränderliche *fpl* automatischer Regelung
   f grandeurs *fpl* variables du réglage automatique
   r переменные автоматического регулирования

   * **variable speed chopper** → 12769

12778 variable speed device
   d Gerät *n* mit veränderbarer Geschwindigkeit
   f dispositif *m* à vitesse variable
   r устройство *n* переменной скорости

12779 variable time-delay block; variable time-lag unit
   d variable Verzögerungsblock *m*
   f bloc *m* de retard variable
   r блок *m* переменных задержек

   * **variable time-lag unit** → 12779

12780 variable tuning
   d veränderliche Abstimmung *f*
   f accord *m* variable
   r переменная настройка *f*

12781 variance
   d Varianz *f*; Streuung *f*; Dispersion *f*
   f variance *f*; dispersion *f*
   r дисперсия *f*; рассеяние *n*

   * **variance of random value** → 10602

12782 variation
   d Variation *f*
   f variation *f*
   r вариация *f*

12783 variational equation
   d Variationsgleichung *f*
   f équation *f* de variation
   r вариационное уравнение *n*

12784 variation of constants
   d Variation *f* der Konstanten
   f modification *f* des constantes; variation *f* des constantes
   r вариация *f* постоянных

12785 variation principle
   d Variationsprinzip *n*
   f principe *m* de variation
   r вариационный принцип *m*

12786 variation range
   d Variationsbereich *m*
   f domaine *m* de variation
   r диапазон *m* вариации

12787 varieties of program
   d Abarten *fpl* eines Programms
   f catégories *fpl* d'un programme
   r варианты *mpl* программы

12788   variety
  d   Vielfalt f; Mannigfaltigkeit f; Varietät f
  f   variété f
  r   разнообразие n

  *   variolosser → 480

  *   variometer → 491

12789   various units
  d   verschiedene Einheiten fpl
  f   unités fpl diverses
  r   различные устройства npl

12790   varistor
  d   Varistor m; spannungsabhängiger
      Widerstand m
  f   varistor m; varistance f
  r   варистор m; регулируемое
      сопротивление n

12791   varmeter
  d   Varmeter n; Blindleistungsmesser m
  f   varmètre m; varheuremètre m
  r   вармерт m; реактивный ваттметр m

12792   vector
  d   Vektor m
  f   vecteur m; grandeur f vectorielle
  r   вектор m

12793   vector diagram
  d   Vektordiagramm n
  f   diagramme m vectoriel
  r   векторная диаграмма f

12794   vector function
  d   Vektorfunktion f
  f   fonction f vectorielle
  r   векторная функция f

12795   vectorial theorie
  d   Vektortheorie f
  f   théorie f vectorielle
  r   векторная теория f

12796   vectorial wave analysis
  d   Vektorwellenanalyse f
  f   analyse f en ondes vectorielles
  r   анализ m векторных волн

12797   vector optimization
  d   Vektoroptimierung f
  f   optimisation f vectorielle
  r   векторная оптимизация f

12798   vector parameter
  d   Vektorparameter m

  f   paramètre m vectoriel
  r   векторный параметр m

12799   vector representation
  d   vektorielle Darstellung f; Vektordarstellung f
  f   représentation f vectorielle
  r   представление n в векторной форме

12800   velocity
  d   Geschwindigkeitskoordinate f
  f   coordonnée f de vitesse
  r   скорость f; быстрота f

  *   velocity feedback → 10639

12801   velocity of propagation
  d   Fortpflanzungsgeschwindigkeit f
  f   vitesse de propagation
  r   скорость f распространения

12802   velocity rating
  d   Sollgeschwindigkeit f
  f   vitesse f nominale
  r   номинальная скорость f

12803   velocity servomechanism
  d   Geschwindigkeitsregelung f
  f   servomécanisme m de vitesse
  r   регулятор m скорости

  *   verification → 8324

  *   verify v → 8321

12804   verifying device; monitor
  d   Kontrolleinrichtung f;
      Überwachungseinrichtung f; Monitor m
  f   dispositif m de contrôle; moniteur m
  r   контрольное устройство n; монитор m

12805   vertical check
  d   Querprüfung f; vertikale Prüfung f
  f   vérification f verticale
  r   вертикальный контроль m

12806   vertical microprogramming
  d   vertikale Mikroprogrammierung f
  f   microprogrammation f verticale
  r   вертикальное микропрограммирование n

12807   vertical redundancy check
  d   Querredundanzprüfung f; Querprüfung f über
      Sicherungszeichen
  f   contrôle m vertical de redondance
  r   вертикальный контроль m избыточным
      кодом

  *   vibrating controller → 9233

*   vibrating wire gauge → 11966

12808 vibration
  d Vibration *f*
  f vibration *f*
  r вибрация *f*

12809 vibrational transition
  d Schwingungsübergang *m*
  f transition *f* vibronique
  r переход *m* вибраций

12810 vibration analyzer
  d Schwingungsanalysator *m*
  f analyseur *m* de vibrations
  r виброанализатор *m*

12811 vibration damper
  d Schwingungsdämpfer *m*
  f amortisseur *m* de vibration
  r виброгаситель *m*

12812 vibration-detecting laser apparatus
  d Laserschwingungsdetektor *m*
  f détecteur *m* de vibration à laser
  r лазерный прибор *m* для обнаружения вибраций; лазерный вибродатчик *m*

12813 vibration energy
  d Schwingungsenergie *f*
  f énergie *f* oscillatoire
  r энергия *f* колебания

12814 vibration measuring equipment
  d Vibrationsmessvorrichtung *f*
  f vibromètre *m*; mesureur *m* de vibrations
  r виброизмерительное оборудование *n*

12815 vibration research
  d Schwingungsforschung *f*
  f recherche *f* de vibration
  r исследование *n* вибраций

12816 vibration resistance
  d Vibrationsfestigkeit *f*
  f résistance *f* aux vibrations
  r вибропрочность *f*; вибростойкость *f*

12817 vibration spectrum analyzer
  d Schwingungsspektrumanalysator *m*
  f analyseur *m* du spectre des vibrations
  r анализатор *m* спектра вибраций

12818 vibration test
  d Vibrationsprobe *f*
  f essai *m* aux vibrations
  r испытание *n* на вибростойкость

12819 vibrator
  d Vibrator *m*
  f vibrateur *m*
  r вибратор *m*

12820 vibrator conveyor
  d Vibrationsspeiser *m*
  f transporteur *m* par vibration; alimentateur *m* vibrant
  r вибрационный питатель *m*

12821 vibrograph
  d Vibrograf *m* schreibender Schwingungsmesser *m*
  f vibrographe *m*; enregistreur *m* de vibrations
  r виброграф *m*

12822 vibrotron
  d Vibrotron *n*
  f triode *f* à anode mobile
  r вибротрон *m*

12823 video amplifier
  d Videoverstärker *m*
  f amplificateur *m* vidéo
  r видеоусилитель *m*

12824 video amplitude
  d Videosignalamplitude *f*; Bilddsignalamplitude *f*
  f amplitude *f* du signal vidéo
  r амплитуда *f* видеосигнала

12825 video data terminal
  d Datensichtstation *f*
  f station *f* de visualisation
  r видеотерминал *m*

12826 video detector
  d Bildgleichrichter *m*
  f détecteur *m* vidéo
  r видеодетектор *m*

12827 video-signal processing
  d Videosignalverarbeitung *f*
  f traitement *m* du signal vidéo
  r обработка *f* видеосигнала

12828 view orientation
  d Sichtorientierung *f*
  f orientation *f* visuelle
  r визуальная ориентация *f*

12829 virtual machine
  d virtuelle Maschine *f*
  f machine *f* virtuelle
  r виртуальная машина *f*

12830 virtual memory; virtual storage
d virtuelle Speicher *m*
f mémoire *f* virtuelle
r виртуальная память *f*

12831 virtual memory concept
d virtuelles Speicherkonzept *n*;
    Virtuellspeicherkonzept *n*
f concept *m* de mémoire virtuelle
r концепция *f* виртуальной памяти

* **virtual storage → 12830**

12832 virtual storage technique
d virtuelle Speichertechnik *f*
f technique *f* de mémoire virtuelle
r методы *mpl* виртуальной памяти

* **virtual value → 4643**

* **viscometer → 12833**

12833 visco[si]meter
d Viskosimeter *n*
f viscosimètre *m*
r вискозиметр *m*

12834 viscosity control
d Zähigkeitsregelung *f*
f réglage *m* de viscosité
r контроль *m* вязкости

12835 viscosity measurement
d Zähigkeitsmessung *f*
f mesure *f* de viscosité
r вискозиметрия *f*

12836 viscous damping
d Zähflüssigkeitsdämpfung *f*; flüssige
    Dämpfung *f*
f amortissement *m* par liquide visqueux;
    amortissement visqueux
r вязкостное затухание *n*; вязкое
    демпфирование *n*

12837 viscous friction
d zähflüssige Reibung *f*
f frottement *m* visqueux
r вязкое трение *n*

* **visible signal → 9121**

12838 visual alarm system
d visuelles Alarmsystem *n*
f système *m* d'alarme visuel
r визуальная система *f* оповещения
    [аварийной сигнализации]

* **visual check → 12839**

12839 visual check[ing]
d Sichtkontrolle *f*
f contrôle *m* visuel
r визуальный контроль *m*

12840 visual communication
d visuelle Kommunikation *f*
f communication *f* visuelle
r визуальная связь *f*

12841 visual differential refractometer
d visuelles Differentialrefraktometer *n*
f réfractomètre *m* différentiel visuel
r визуальный дифференциальный
    рефрактометр *m*

12842 visual display
d visuelle Anzeige *f*; visuelle Darstellung *f*;
    Sichtanzeige *f*
f représentation *f* visuelle
r визуальная индикация *f*

12843 visual identification data
d visuelle Erkennungsdaten *pl*
f données *fpl* d'identification visuelles
r данные *pl* визуального распознавания

12844 visual image processor
d optisches Bildverarbeitungsgerät *n*
f dispositif *m* visuel de traitement d'images
r устройство *n* обработки изображений

12845 visual perception system
d optisches Wahrnehmungssystem *n*; visuelles
    Erkennungssystem *n*
f système *m* de perception visuelle
r система *f* визуального восприятия

12846 visual sensorics; visual sensor technique
d visuelle Sensorik *f*; visuelle Sensortechnik *f*
f technique *f* de senseur visuelle; sensorique *f*
    visuelle
r визуальная сенсорная техника *f*

* **visual sensor technique → 12846**

* **visual signal → 9121**

12847 visual system generation
d Sichtsystemgeneration *f*
f génération *f* d'un système visuel
r поколение *n* визуальной системы

12848 visual system structure
d Sichtsystemstruktur *f*

    f structure *f* d'un système visuel
    r структура *f* визуальной системы

12849 visual tuning
    d visuelle Abstimmung *f*
    f accord *m* visuel
    r визуальная настройка *f*

12850 voltage
    d [elektrische] Spannung *f*
    f tension *f*
    r напряжение *n*

12851 voltage analog
    d Spannungsanalogon *n*
    f analogue *m* de tension
    r аналог *m* напряжения

12852 voltage calibrator
    d Spannungseichgerät *n*
    f étalon *m* de tension; générateur *m* étalonné de tension
    r калибратор *m* напряжения

12853 voltage comparator
    d Spannungskomparator *m*; Spannungsvergleicher *m*
    f comparateur *m* de tension
    r компаратор *m* напряжения

12854 voltage control
    d Spannungsregelung *f*
    f réglage *m* de tension
    r регулирование *n* напряжения

12855 voltage-controlled oscillator
    d spannungsgesteuerter Oszillator *m*
    f oscillateur *m* commandé en tension
    r осциллятор m, управляемый напряжением

12856 voltage direction
    d Spannungsrichtung *f*
    f sens *m* de tension
    r направление *n* напряжения

12857 voltage divider; potential divider
    d Spannungsteiler *m*
    f diviseur *m* de tension
    r делитель *m* напряжения

12858 voltage-doubling circuit
    d Spannungsverdopplungsschaltung *f*
    f montage *m* doubleur de tension
    r схема *f* удвоения напряжения

12859 voltage-driven device
    d spannungsgesteuertes Bauelement *n*
    f composant *m* commandé par tension

    r [электронный] прибор m, управляемый напряжением

12860 voltage drop
    d Spannungsabfall *m*
    f chute *f* de tension
    r падение *n* напряжения

12861 voltage function
    d Spannungsfunktion *f*
    f fonction *f* de tension électrique
    r функция *f* напряжения

12862 voltage gain
    d Spannungsverstärkung *f*
    f gain *m* de tension
    r усиление *n* напряжения

12863 voltage inquiry group
    d Spannungsabfragegruppe *f*
    f groupe *m* d'interrogation de tension
    r блок *m* опроса напряжения

12864 voltage jump
    d Spannungssprung *m*
    f saut *m* de tension
    r скачок *m* напряжения

12865 voltage level
    d Spannungspegel *m*
    f niveau *m* de tension
    r уровень *m* напряжения

    * **voltage limiter** → 9335

12866 voltage range
    d Spannungsbereich *m*
    f gamme *f* de tension
    r диапазон *m* напряжений

12867 voltage regulating system
    d Spannungsstabilisierungssystem *n*
    f système *m* stabiliseur de tension
    r система *f* регулирования напряжения

12868 voltage regulator
    d Spannungsregler *m*; Netzregler *m*
    f régulateur *m* de tension électrique
    r регулятор *m* напряжения

    * **voltage stabilizer** → 3064

12869 voltage-to-current characteristic
    d Strom-Spannungs-Kennlinie *f*
    f caractéristique *f* courant-tension
    r вольт-амперная характеристика *f*

12870 voltage-type telemetering (US)
    d Spannungsfernmessung *f*

    f télémesure *f* à couplage par tension
    r потенциальная система *f* телеизмерений

12871 volume governor
    d Volumenregler *m*
    f régulateur *m* de volume
    r регулятор *m* объёма

12872 volume potential
    d Volumenpotential *n*
    f potentiel *m* de volume
    r объёмный потентиал *m*

12873 volumetric gas analyzer
    d volumetrisch-manometrisches
        Gasanalysengerät *n*
    f analyseur *m* volumétrique de gaz
    r объёмно-манометрический
        газоанализатор *m*

r привод *m* по схеме Леонарда

W

12874 wait condition; wait state; wait[ing] status
d Wartebedingung *f*; Wartezustand *m*
f condition *f* d'attente; état *m* d'attente
r состояние *n* ожидания

12875 wait counter
d Wartezähler *m*
f compteur *m* d'attente
r счётчик *m* ожидания

12876 waiting cycle; waiting loop
d Wartezyklus *m*
f cycle *m* d'attente
r цикл *m* ожидания

* **waiting loop** → 12876

12877 waiting mode
d Wartemodus *m*
f mode *m* d'attente
r режим *m* ожидания

* **waiting status** → 12874

12878 waiting theory
d Wartetheorie *f*
f théorie *f* d'attente
r теория *f* очередей

12879 waiting time equation
d Wartezeitgleichung *f*
f équation *f* de temps d'attente
r уравнение *n* времени ожидания

12880 waiting time factor
d Wartezeitfaktor *m*
f facteur *m* de temps d'attente
r коэффициент *m* времени простоя

* **wait state** → 12874

* **wait status** → 12874

12881 wanted system
d erwünschtes System *n*; gesuchtes System *n*
f système *m* désiré
r желаемая система *f*

12882 Ward-Leonard drive
d Ward-Leonard-Antrieb *m*
f groupe *m* Ward-Leonard

12883 warning device
d Warngerät *n*; Frühanzeigegerät *n*
f appareil *m* avertisseur; appareil d'alarme
r устройство *n* предупредительной сигнализации

12884 warning diagnostic
d Warnungsdiagnostik *f*
f diagnostic *m* d'avertissement
r профилактическая диагностика *f*

12885 warp *v*
d verbiegen
f gondoler; gaucher
r искажать; деформировать

12886 waste-free process
d abproduktfreier Prozess *m*
f procédé *m* sans déchets
r безотходный процесс *m*

12887 waste-free technology
d abproduktfrcie Technologie *f*
f technologie *f* sans déchets
r безотходная технология *f*

12888 water flow control
d Wassermengenregelung *f*
f régulation *f* du débit d'eau
r регулирование *n* расхода воды

12889 water power system
d Wasserkraftsystem *n*
f système *m* hydroélectrique
r гидроэнергетическая система *f*

12890 water supply control
d Speisewasserregelung *f*
f réglage *m* d'alimentation en eau
r регулирование *n* питания водой

12891 wave action
d Wellenwirkung *f*
f action *f* de l'onde
r волновое воздействие *n*

12892 wave analysis
d Wellenanalyse *f*; Wellenformuntersuchung *f*; Signalanalyse *f*
f analyse *f* d'onde; analyse de forme de l'onde
r анализ *m* формы сигналов

12893 wave analyzer
d Wellenanalysator *m*
f analyseur *m* d'ondes
r анализатор *m* формы сигнала

* **wave detector** → 3655

12894 wave equation
d Wellengleichung *f*
f équation *f* d'onde
r волновое уравнение *n*

12895 wave filter
d Wellensieb *n*
f filtre *m* d'ondes
r [электрический] волновой фильтр *m*

* **waveform distortion** → 760

12896 waveguide laser
d Wellenleiterlaser *m*
f laser *m* à guide d'ondes
r волноводной лазер *m*

12897 waveguide structure
d Wellenleiterstruktur *f*
f structure *f* en guide d'onde
r структура *f* волновода

* **wave impedance** → 2287

12898 wavelength-division multiplexer
d Wellenlängenmultiplexer *m*
f multiplexeur *m* à répartition en longueur d'onde
r устройство *n* для спектрального уплотнения

12899 wave level gauge
d Füllstandsmessgerät *n*
f jauge *f* de niveau à ondes; indicateur *m* de niveau à ondes
r волновой уровнемер *m*

12900 wave mechanics
d Wellenmechanik *f*
f mécanique *f* ondulatoire
r волновая механика *f*

12901 wave-shape monitor
d Wellenformmonitor *m*
f contrôleur *m* de forme d'onde
r устройство *n* для контроля формы волны

12902 wave train
d Wellenzug *m*
f train *m* d'ondes
r группа *f* волн

* **way of behaviour** → 1720

* **weak coupling** → 7566

12903 wear-testing gauge
d Verschleißprüfer *m*
f machine *f* à vérifier l'usure
r аппаратура *f* для испытаний на износ

12904 weighted code
d stellenbewerteter Kode *m*
f code *m* pondéré
r взвешанный код *m*

12905 weight factor
d Gewichtsfaktor *m*
f facteur *m* de poids
r весовой множитель *m*

12906 weighting function
d Gewichtsfunktion *f*
f fonction *f* pondérée
r весовая функция *f*

12907 welding part reception
d Schweißteilaufnahme *f*
f réception *f* de part soudée; prise *f* de pièce de soudage
r приёмное устройство *n* свариваемых деталей

12908 Weston cell
d Westonelement *n*
f élément *m* de Weston
r элемент *m* Вестона

12909 white noise
d weißes Rauschen *n*
f bruit *m* blanc
r белый шум *m*

12910 white noise limiting circuit
d Begrenzerschaltung *f* für weißes Rauschen
f circuit *m* limiteur du bruit blanc
r схема *f* ограничения белого шума

12911 wide-angle coordinator
d Breitwinkelkoordinator *m*
f coordinateur *m* à angle large
r широкоугольный координатор *m*

12912 wideband communication system
d breitbandiges Fernmeldesystem *n*
f système *m* de télécommunication à large bande
r широкополосная система *f* связи

12913 wideband controller
d Breitbandregler *m*; Breitbereichregler *m*
f régulateur *m* à large bande de réglage
r регулятор *m* с широкой зоной регулирования

12914 wideband oscillograph
  d Breitbandoszillograf *m*
  f oscillographe *m* à large bande
  r широкополосный осциллограф *m*

12915 wideband pulse amplifier
  d Breitbandimpulsverstärker *m*
  f amplificateur *m* d'impulsions à bande large
  r импульсный широкополосный
      усилитель *m*

12916 wide-passband infrared system
  d breitbandiges Infrarotsystem *n*
  f système *m* infrarouge à large bande passante
  r широкополосная инфракрасная система *f*

12917 wide range temperature controller
  d Breitbandtemperaturregler *m*
  f régulateur *m* de température à large bande
  r широкодиапазонный регулятор *m*
      температуры

12918 width adjustment
  d Breiteneinstellung *f*
  f ajustage *m* de largeur
  r регулировка *f* ширины [сигнала]

12919 Wiener-Khintchine relation
  d Wiener-Khintschine-Beziehung *f*; Wiener-
      Chintchin-Beziehung *f*
  f relation *f* de Wiener-Khintchine
  r соотношение *n* Винера-Хинчина

12920 Wiener's filtering problem
  d Wienersches-Filterproblem *n*
  f problème *m* de filtrage de Wiener; problème
      de triage de Wiener
  r задача *f* Винера о фильтрации

12921 wire link telemetry
  d drahtgebundene Fernmesstechnik *f*
  f télémesure *f* à liaison par fil
  r телеметрия *f* по проводам

12922 wire-wound potentiometer
  d Drahtpotentiometer *n*
  f potentiomètre *m* bobine
  r проволочный потенциометр *m*

12923 wobble frequency
  d Wobbelfrequenz *f*
  f fréquence *f* de balayage
  r частота *f* качания

12924 wobbler; wobbulator
  d Wobbler *m*
  f wobbulateur *m*; vobulateur *m*
  r вобулятор *m*

* **wobbulator** → 12924

* **workable** → 8974

* **work check** → 12925

12925 work check[ing]
  d Arbeitskontrolle *f*
  f contrôle *m* du travail
  r проверка *f* работы

12926 work conditions of switching device
  d Arbeitsbedingungen *fpl* einer
      Schalteinrichtung
  f régime *m* de dispositif de commutation
  r условия *npl* работы коммутирующего
      устройства

* **work cycle** → 4554

12927 work elements
  d Arbeitselemente *npl*
  f éléments *mpl* de travail
  r рабочие элементы *mpl*

12928 working capacity
  d Arbeitsvermögen *n*
  f pouvoir *m* de travail
  r рабочая мощность *f*

* **working characteristic** → 8978

* **working condition** → 9017

12929 working intensity
  d Arbeitsintensität *f*
  f intensité *f* de travail
  r интенсивность *f* работы

**12930 working position; operative position;
        operating position**
  d Arbeitsstellung *f*; Betriebsstellung *f*
  f position *f* de fonctionnement; position de
      travail
  r рабочее положение *n*

* **working pressure** → 8999

* **working range** → 9002

12931 working temperature
  d Arbeitstemperatur *f*
  f température *f* d'opération; température de
      fonctionnement
  r рабочая температура *f*

12932 work instruction
  d Arbeitsbefehl *m*

f instruction *f* de travail
r инструкция *f* обработки

12933 work piece parts geometry
d Werkstückteilegeometrie *f*
f géométrie *f* de pièces des produits manufacturés
r геометрическая характеристика *f* обрабатываемых деталей

12934 work piece positioning
d Werkstückpositionierung *f*
f positionnement *m* de pièce à travailler
r позиционирование *n* обрабатываемых деталей

12935 work piece quality checking
d Werkstückqualitätskontrolle *f*
f contrôle *m* de qualité des produits manufacturés
r контроль *m* качества обрабатываемых деталей

12936 work piece recognition
d Werkstückerkennung *f*
f récognition *f* d'une pièce à travailler
r распознавание *n* обрабатываемых деталей

12937 workplace analysis
d Arbeitsplatzanalyse *f*
f analyse *f* de position d'opérateur
r анализ *m* рабочего места

12938 work station
d Arbeitsstation *f*
f station *f* de travail
r рабочая станция *f*

12939 worst-case circuit analysis
d Schaltungsanalyse *f* unter Grenzbedingungen
f analyse *f* de circuit sous conditions limites
r анализ *m* схемы в граничных условиях

12940 writable control store
d beschreibbarer Steuerspeicher *m*; umschreibbarer Steuerspeicher
f mémoire *f* de commande à écrire
r управляющая память *f* с возможностью перезаписи

12941 write control
d Schreibsteuerung *f*
f commande *f* d'écriture
r управление *n* записью

12942 write enable signal
d Schreib-Freigabesignal *n*
f signal *m* d'autorisation d'écriture
r сигнал *m* разрешения записи

12943 write operation
d Schreiboperation *f*
f opération *f* d'écriture
r операция *f* записи

12944 write pulse
d Schreibimpuls *m*; Schreibtaktimpuls *m*
f impulsion *f* d'écriture
r импульс *m* записи

X

* X-axis → 12

* XOR → 5261

12945 X-ray control
d Röntgenstrahlkontrolle *f*
f contrôle *m* à rayons X
r ренгеновский контроль *m*

12946 X-ray diffraction phase analysis
d Röntgenbeugungs-Phasenanalyse *f*
f analyse *f* de phase par diffraction de
   rayons X
r ренгеноструктурный анализ *m* фаз

12947 X-ray fluorescence spectrometer
d Röntgenstrahlenfluoreszenzspektrometer *n*
f spectromètre *m* à fluorescence à rayons X
r рентгеновский флуоресцентный
   спектрометр *m*

12948 X-Y-coupling capacitor
d X-Y-Koppler *m*
f coupleur *m* en X et en Y
r X-Y-элемент *m* связи

12949 X-Y-recorder
d Schreiber *m*; X-Y-Schreiber *m*
f X-Y-enregistreur *m*
r двухкоординатное регистрирующее
   устройство *n*

Y

* **Y-axis → 1535**

12950 yield-time diagram
 d Weg-Zeit-Diagramm *n*
 f courbe *f* d'affaiblissement en fonction de temps
 r диаграмма *f* зависимости оседания от времени

Z

12951 Zener breakdown; Zener effect
  d Zenereffekt *m*; Zenerdurchbruch *m*
  f rupture *f* Zener
  r эффект *m* Зенера; зинеровский пробой *m*

  *   **Zener effect → 12951**

12952 zero; null
  d Nullstelle *f*; Null *f*
  f zéro *m*; point *m* zéro
  r нуль *m*

**12953 zero adjuster; zero adjusting device; zero
        resetting device; zero set[ting] control**
  d Nullpunkteinstellvorrichtung *f*; Nullsteller *m*;
       Nullpunkteinstellung *f*
  f dispositif *m* de mise à zéro; ajustage *m* à zéro
  r устройство *n* для установки па нуль

  *   **zero adjusting device → 12953**

12954 zero adjustment; zero resetting
  d Nulleinstellung *f*
  f mise *f* au point de zéro; remise *f* à zéro
  r настройка *f* на нуль

  *   **zero balance → 1588**

12955 zero-balance amplifier
  d Nullindikatorverstärker *m*
  f amplificateur *m* à détection d'équilibre
  r усилитель *m* с уравновешиванием

12956 zero beat
  d Schwebungslücke *f*; Schwebungsnull *f*
  f battement *m* nul
  r нулевое биение *n*

12957 zero beat indicator wavemeter
  d Schwebungsnullmesser *m*;
       Zeigerwellenmesser *m*
  f ondemètre *m* hétérodyne
  r волномер *m* с индикатором нулевого
       биения

12958 zero bias
  d Nullvorspannung *f*
  f polarisation *f* nulle
  r нулевое смещение *n*

12959 zero byte; byte of zero
  d Nullbyte *n*
  f byte *m* nul
  r нулевой байт *m*

12960 zero charge
  d Nullast *f*
  f charge *f* nulle
  r нулевая нагрузка *f*

12961 zero data length
  d Nulldatenlänge *f*
  f longueur *f* de données nulle
  r длина *f* нулевых данных

12962 zero defects program
  d Nullfehler-Programm *n*
  f programme *m* à défaut nul
  r программа *f* нулевых ошибок

12963 zero dimension
  d Nulldimension *f*
  f dimension *f* zéro
  r нулевая размерность *f*; безразмерность *f*

12964 zero direction
  d Nullpunktrichtung *f*
  f direction *f* zéro
  r нулевое направление *n*

12965 zero dispersion
  d Dispersionsnullstelle *f*; Nulldispersion *f*
  f dispersion *f* nulle
  r нулевая дисперсия *f*

12966 zero drift
  d Nullpunktwanderung *f*
  f dérive *f* de zéro
  r нулевой дрейф *m*

12967 zero error
  d Nullpunktabweichung *f*
  f erreur *f* de zéro
  r нулевая погрешность *f*

12968 zero error position system
  d Nullfehlerstellungssystem *n*
  f système *m* de position à déviation zéro
  r позиционная система *f* с нулевой
       погрешностью

12969 zero form
  d Nullform *f*
  f forme *f* de zéro
  r нулевая форма *f*

12970 zero frequency
  d Nullfrequenz *f*

    f fréquence *f* zéro
    r нулевая частота *f*

12971 zero-level sensitivity
    d Nullpunktempfindlichkeit *f*
    f sensibilité *f* rapportée au niveau zéro
    r чувствительность *f* по отношению к нулевому уровню

    * zero matrix → 8831

12972 zero method
    d Nullmethode *f*
    f méthode *f* de zéro
    r нулевой метод *m*

12973 zero offset control
    d Nullvorlaufsteuerung *f*
    f réglage *m* astatique
    r управление *n* для нулевой установки

12974 zero output
    d Nullausgangssignal *n*
    f signal *m* zéro de sortie
    r нулевой сигнал *m* выхода

12975 zero-phase-sequence coordinate system
    d Nullkoordinatensystem *n*
    f système *m* de coordonnées homopolaire
    r гомополярная система *f* координат

12976 zero-phase-sequence protection
    d Nullsystemschutz *m*
    f protection *f* homopolaire
    r гомополярная защита *f*; защита нулевой последовательности

12977 zero point; null point
    d Nulldurchgang *m*; Nullpunkt *m*
    f point *m* de passage par zéro
    r нулевая точка *f*

12978 zero point energy
    d Nullpunktenergie *f*
    f énergie *f* au zéro
    r энергия *f* в нулевой точке

12979 zero position
    d Nullstellung *f*
    f position *f* de zéro
    r нулевое положение *n*

12980 zero potential
    d Nullpotential *n*
    f potentiel *m* zéro
    r нулевой потенциал *m*

12981 zero power

    d Leistung *f* Null
    f puissance *f* zéro
    r нулевая мощность *f*

12982 zero probability
    d Nullwahrscheinlichkeit *f*
    f probabilité *f* nulle
    r нулевая вероятность *f*

12983 zero radiation level
    d Strahlungsnullpegel *m*
    f niveau *m* zéro de radiation
    r нулевой уровень *m* излучения

    * zero resetting → 12954

    * zero resetting device → 12953

12984 zero-resistance ammeter
    d widerstandsloser Strommesser *m*
    f ampèremètre *m* à résistance nulle
    r амперметр *m* с нулевым сопротивлением

12985 zero series
    d Nullfolge *f*
    f suite *f* de zéros
    r нулевая последовательность *f*

    * zero set control → 12953

    * zero setting control → 12953

12986 zero setting of selsyns
    d Nulleinstellen *n* von Selsynen
    f mise *f* à zéro de selsyns
    r установка *f* сельсинов на нуль

12987 zero stability
    d Nullpunktstabilität *f*
    f stabilité *f* du zéro
    r устойчивость *f* нуля

12988 zero state
    d Nullzustand *m*
    f état *m* zéro
    r нулевое состояние *n*

12989 zero suppression
    d Nullenunterdrückung *f*
    f suppression *f* de zéros
    r подавление *n* нуля

12990 zero value
    d Nullwert *m*
    f valeur *f* nulle
    r нулевое значение *n*

12991 zero variation
    d Nullpunktabweichung *f*

  f déviation *f* du zéro; déviation résiduelle
  r нулевое изменение *n*

12992 zero velocity
  d Nullgeschwindigkeit *f*
  f vitesse *f* nulle
  r нулевая скорость *f*

12993 zero voltage
  d Nullspannung *f*
  f tension *f* zéro
  r нулевое напряжение *n*

12994 zone
  (of a computer)
  d Sonderspeicher *m*
  f mémoire *f* spéciale
  r зона *f*; область *f*

12995 zone of action
  d Regelband *n*
  f bande *f* d'action; étendue *f* d'action
  r зона *f* регулирования

12996 zone of ambiguity
  d Unbestimmtheitsbereich *m*
  f zone *f* d'ambiguïté
  r зона *f* неоднозначности

12997 zone of linearity; range of linearity
  d Linearitätsbereich *m*
  f zone *f* linéaire; domaine *m* de linéarité;
      gamme *f* de linéarité
  r зона *f* линейности; диапазон *m*
      линейности

12998 zone of saturation; saturation zone
  d Sättigungszone *f*; Sättigungsbereich *m*
  f zone *f* de saturation
  r зона *f* насыщения

12999 zone selector
  d Zonenwähler *m*
  f sélecteur *m* de zone
  r искатель *m* зоны

13000 z-transform
  d z-Transformation *f*; diskrete Laplace-
      Transformation *f*
  f transformation *f* z; transformation discrète de
      Laplace
  r z-преобразование *n*

Deutsch

Effektorgeometrie 4651
Effektorkoordinaten 4645
Effektorkraftmessung 4649
Effektororientierungsdaten 4654
Effektorort in Basiskoordinaten
    4655
Effektorposition 4656
Effektorrechner 4644
Effektorregelung 4658
Effektorzustand 4659
effizientes Programmiersystem
    4663
Effizienztheorem 4661
Eichen 2040
eichen 5872
Eichfrequenz 2046
Eichgenauigkeit eines
    Steuerungssystems 2042
Eichgerät 2047
Eichimpuls 2048
Eichkreis 2043
Eichkurve 2045
Eichmessteilung 2050
Eichplatz 6753
Eichpotentiometer 2038
Eichsignal 2039
Eichskale 2050
Eichspannungsteiler 2038
Eichstrom 2044
Eichtemperatur 2051
Eichtransformation 5873
Eichung 2040
Eichung des Regelstabes 3326
Eichungsgenauigkeit 2041
Eichung von Messgeräten 7957
Eichwiderstand 2049
Eigenabklingen 8535
Eigenadmittanz 11278
Eigenbedarfsanlage 1477
Eigenbedarfsgenerator 1484
eigene Modulation 11304
Eigenenergie 2284
Eigenfrequenz 8538
Eigenfrequenzkennlinie 8539
Eigenfunktion 4665
Eigenkinetik 6962
Eigenlösung 4665
eigenmächtige Funktion 966
Eigenrauschen 1682
eigenrelative Adresse 4104
Eigenscheinleitwert 11278
Eigenschwingung 1449
Eigenschwingungen in Servo-
    systemen 1451
Eigenstabilisierung 11318
Eigenstabilität 6606
eigentlich monoton 11973
Eigenvektor 4669

Eigenwertaufgabe 4668
Eigenwertgleichung 4667
Eigenwertproblem 2301, 4668
Eigenzustand 2297
Eignung 12019
Einachsenlasergyroskop 8920
ein-aus 8946
Ein-Aus-Regler 8948
Ein-Aus-Schalter 8953
Ein-Aus-Schaltung 11948
Ein-Aus-Servomechanismus 8952
Ein-Aus-Steuerung 8947
Ein-Aus-Tastung 8950
einbauen 6705
eindeutig 12657
eindeutig bestimmt 12658
eindeutige Funktion 8937
Eindeutigkeit einer Lösung 12659
Eindeutigkeitssatz 12660
eindimensional 8921
eindimensionale Abtastung 8926
eindimensionale Bewertung 8924
eindimensionale Kette 8922
eindimensionales Bewertungs-
    system 8925
eindimensionales Datenfeld
    8923
Eindringen 9500
Eindringenenergie 9501
Eindringtiefe 9500
Einebenen-Interruptsystem
    11608
Einerrücklauf 5031
Einfachbetrieb 11567
einfache Dekomposition 11555
einfache Funktion 8937
einfache Genauigkeit 11615
einfache Handhabetechnik 11557
einfache Montageoperation 11551
einfacher Montageautomat 11550
einfacher Regler 11554
einfacher Sensoraufbau 11560
einfacher Sichtsensor 11561
einfaches Kontrollelement 11553
einfaches Montagesystem 11552
einfaches Regelkreissystem 8928
einfaches Verzögerungsnetzwerk
    11558
einfache Werkstückmontage
    11562
einfachlogarithmisch 11348
Einfachregelkreis 11629
Einfachregister 11559
Einfachstrom 8601
einfach zusammenhängend 11571
Einfahrverhalten 944
Einflußgröße 359, 6535, 6538
Einflußkorrektur 260

Eingabe analoger Informationen
    814
Eingabe-Ausgabe-Einheit 6672
Eingabe-Ausgabe-Gerät 6672
Eingabe-Ausgabe-Interface 6673
Eingabe-Ausgabe-Modell 6675
Eingabe-Ausgabe-Pufferspeicher
    6668
Eingabe-Ausgabe-Steuersystem
    6670
Eingabe-Ausgabe-Uberwachung
    6676
Eingabe-Ausgabe-Zyklus 6671
Eingabebereich 6646
Eingabe-Bustyp 6647
Eingabedaten 6652
Eingabefluss 6657
Eingabe-Freigabesignal 6654
Eingabefunktion 6658
Eingabe geometrischer Strukturen
    5924
Eingabegerät 6655, 6691
Eingabekanal 6648
Eingabemaschine 6665
Eingabemechanismus 6665
Eingabespeicher 6688
Eingabesteuereinheit 6649
Eingabestrom 6657
Eingabetor-Aktivierungssignal
    6654
Eingabeverfahren 6643
Eingabe von Steuerinformationen
    6666
Eingangsamplitudenbereich 6645
Eingangsanzeigeleitung 12207
Eingangsdaten 6652
Eingangsempfindlichkeit 6685
Eingangsfilter 6656
Eingangsgitterkapazität 6659
Eingangsgleichspannung 4278
Eingangsgröße 6683
Eingangsgröße des Systems 6684
Eingangsgröße mit Polynom-
    charakteristik 9876
Eingangs-Highpegelspannung
    6660
Eingangsimpuls 6682
Eingangsinformation 6662
Eingangsinformationsdauer 6663
Eingangsinfrarotstrahl 6590
Eingangskennwortleitung 12207
Eingangskontrolle 6421
Eingangskoordinate 6650
Eingangsleistung 6679
Eingangs-Lowpegelspannung
    6664
Eingangsscheinwiderstand 6661
Eingangssignal 6686

positive Differenz 9922
positive Kopplung 9920
positive Logik 9925
positive quadratische Form 9931
positiver Impuls 9930
positiver Selbstausgleich 9932
positive Rückkopplung 9923
positive Schaltungslogik 9925
positives Niveau 9924
positives Potential 9929
positives Signal 9933
Positivlogiksignal 9926
Postenzähler 7002
Postulat 1532
Potentialanalogie 9936
Potentialbild 9939
Potential des Vektorfeldes 9946
Potentialdiagramm 9939
Potentialfunktion 9944
Potentialkorrektur 9938
Potentialkurve 9943
Potentialschwelle 9937
Potentialsprung 9945
Potentialsteuerung 9941
Potentialtheorie 9947
Potentialverlauf 9940
Potentialverteilung 9940
Potentialverteilungssteuerung
    9941
Potentialwall 9937
potentielle Energie 9942
Potentiometer 9948
Potentiometerabgriff 9951
Potentiometer für Koordinaten-
    wandlung 11004
Potentiometermethode 9950
Potentiometer mit Kräfteausgleich
    5634
Potentiometerregler 9949
potentiometrisches Fehlermess-
    system 9953
Potentiostat 9954
Präcompoundprozess 10008
Prädikatenlogik 9998
prädiktives Filter 10000
prädiktives Relaissystem 10002
Präionisation 1132
praktischer Beharrungszustand
    9987
Präsenz 10032
Präzision 199
Präzisionsgerätetechnik 9990
Präzisionsschweißautomat 1345
Präzisionssteuerung 9989
Präzisionstechnik 9991
Prediktornetzwerk 10006
predizierendes Netzwerk 10006
Preemphase 10007

P-Regelung 10352
P-Regler 10348
P-Regler mit Störgrößen-
    ausschaltung 10353
Primärdaten 10075
Primärdaten 9222
Primärelement 10073
primärer Fühler 10076
primäres Regelelement 10074
Primärparameter 10079
Primärprozess 10080
Primärregelung 10081
Primärserienauslöser 4321
Primärspeicher 10078
Primärvorgang 10080
Printer 10093
Prinzip der adaptiven Zeit-
    konstante 10083
Prinzip der Gruppenrangordnung
    3242
Prinzip des Argumentes 979
Prinzip gleichzeitig ablaufender
    Operationen 10090
Prinzipschaltbild 11217
Prinzipschaltung 11217
Prinzipschema 10091
Prinzipzeichnung 10091
Prioritätengruppe 10104
Prioritätsanzeiger 10105
Prioritätsgrad 10103
Prioritätsinterruptsteuerung 10106
Prioritätsinterruptsystem 10107
Prioritätskontrolle 10097
Prioritätsordnung 9208
Prioritätsprogramm 10109
Prioritätssteuerung 10099
Prioritätsstromkreis 10098
Prioritätsstruktur 10111
Prioritätsstufe 10103
Prioritätsverarbeitung 10108
Prioritätszuteilung 10100
probabilistisch 10112
probabilistische Logik 10114
probabilistischer Automat 10113
Probe 11150
Probebetrieb 12323
Probenraum 11157
Probeproduktion 12565
Problemalgorithmus 10129
Problembearbeitungsprozess
    10137
Problembeschreibung 10131
Problembestimmung 10130
Problemdefinition 10130
Problem der Operationsforschung
    9038
Problem der Quantisierung 10497
Problemeinstellung 11470

Problemlösungsprozess 10137
problemorientierte Programmier-
    sprache 10134
problemorientiertes Software-
    paket 10135
problemorientiertes Systemunter-
    lagenpaket 10135
problemorientierte Steuerung
    10133
problemorientierte Werkstück-
    beschreibung 10136
Processsteuersystem 10158
Produktionsautomatisierung
    10219
Produktionskontrollsystem 10223
Produktionsplanung 10236
Produktionsrate 10229
Produktionsrhythmus 11058
Produktionssteuerungsplan 10222
Produktionsstichprobe 10234
Produktionssystem 10231
Produktionsüberwachung 10230
Produktivitätssteigerung 6430
Produktlinie 10235
produzierender Prozess 10218
Profoldispersionskoeffizient
    10237
Prognosearbeit 10001
Programm 10238
programmabhängig 10262
programmabhängiger Betrieb
    10263
Programmablauf 2932
Programmablaufänderung 4601
Programmablaufnachbildung
    10321
Programmablaufplan 10267
Programmablaufsteuerung 10268
Programmalgorithmus 10240
Programmanalysator 10241
Programmanforderung 10319
Programmaufbau 10322
Programmbibliothek 10276
Programm des bedingten
    Übergangs 1919
Programm eines Universal-
    rechners 12673
Programmentwicklungssystem
    10264
Programmfehler 10265
Programmfehlerermittlung 10266
Programmfernsteuerung 10942
Programmfolge 10320
Programmfunktion 10269
Programmgeber 10249, 10305
Programmgenerator 10271
Programmgenerierung 10270
programmgesteuert 10251, 11676

Français

amplificateur symétrique push-pull 1605
amplificateur tampon 1957
amplificateur vidéo 12823
amplification à croissance exponentielle 5311
amplification de courant de petits signaux 11658
amplification de l'onde porteuse 2130
amplification de puissance 9955
amplification optique 9047
amplification optique directe 9074
amplifier 746, 1881
amplistat 749
amplitude 751
amplitude complexe 2766
amplitude crête à crête 4477
amplitude de bruit moyenne normalisée 8801
amplitude de déviation 4041
amplitude de la densité efficace du courant d'excitation 4634
amplitude de pointe 9468
amplitude d'une grandeur alternative 779
amplitude du signal d'image 9694
amplitude du signal vidéo 12824
amplitude nette 8575
amplitude porteuse 2117
amplitude relative 10848
ampltude de sortie 9266
analogie 837
analogie électrique 4676
analogie électrodynamique 4757
analogie hydraulique 6251
analogie pneumatique 9771
analogie potentielle 9936
analogique 835
analogue 834, 835
analogue de tension 12851
analuseur logique 7540
analysateur à circuits fluidiques 5617
analysateur calorimétrique de traces de gaz 2062
analysateur de fréquences 5719
analysateur de gaz thermostatique 12376
analysateur de liquides infrarouge 7434
analyse 840
analyse à black-box 1824
analyse à la touche 11755
analyse au spectrographe de masse 7851
analyse chromatographique en phase gazeuse 5836

analyse cinématique 7050
analyse d'activation par neutrons 8605
analyse d'activité 318
analyse d'ajustement de pièce 9695
analyse d'amplitudes des impulsions 753
analyse de calculateur 2863
analyse de circuit 2380
analyse de circuit sous conditions limites 12939
analyse de comportement 841
analyse de comportement d'un automate 1721
analyse de concept 3978
analyse de contours 3171
analyse de défaut de machine 7626
analyse de défaut de manipulateur 7794
analyse de fiabilité 10906
analyse de forme de l'onde 12892
analyse de la commande de processus 10154
analyse de langage 7087
analyse de la propagation de faisceaux 10337
analyse de l'immiscibilité 6349
analyse de matrice 7885
analyse de phase par diffraction de rayons X 12946
analyse de position d'opérateur 12937
analyse de procédé 10145
analyse de processus de production 846
analyse de projet 3978
analyse de recours 10823
analyse de réseau 8581
analyse des dangers 6090
analyse des dérangements 4449
analyse de sensibilité 11356
analyse des fréquences 5685
analyse des impuretés 6403
analyse des niveaux d'énergie 5055
analyse des objets géométriques 843
analyse des régimes transitoires 12527
analyse des structures géométriques 5923
analyse des systèmes de réglage et de commande 842
analyse de stabilité 845
analyse de stabilité de Bode 1865

analyse de stabilité de Kochenburger 8076
analyse de stabilité de Nyquist 8078
analyse de stabilité des systèmes 12191
analyse de stabilité d'Evans 8074
analyse de structure 11987
analyse d'état 11838
analyse d'image 6331
analyse dimensionelle 4249
analyse d'onde 12892
analyse d'oscillation 844
analyse du gaz 5833
analyse d'une image d'objet 8873
analyse d'un système digital par calculateur 2850
analyse du système 12138
analyse dynamique 4561
analyse électrique 4752
analyse électrique de Fourier 4685
analyse élémentaire 4977
analyse en amplitude 753
analyse en ondes vectorielles 12796
analyse en régime de signal faible 11656
analyse en régime signal fort 7094
analyse entièrement automatique 5736
analyse fonctionnelle 5759
analyse graphique 5944
analyse harmonique 5685, 6075
analyse logique 7495
analyse logique commandée par processeur 10200
analyse multidimensionnelle 8436
analyse multiniveaux 8418
analyse multiparamétrique 8436
analyse multiple 8380
analyse néphélométrique 8573
analyse numérique 8840
analyse opérationnelle 9015
analyse par absorption 70
analyse par activation 275
analyse par activation de plusieurs éléments 8402
analyse par échantillonage 11160
analyse par émission 5011
analyse par goutte 11755
analyse par rétrodiffusion 1567
analyse phonémique 9630
analyse qualitative 10483
analyse quantitative 10491
analyser 838
analyse réfractométrique 10802
analyse séquentielle 11412

analyse spectrale 11745

analyse spectrale dans les systèmes hydrauliques 11730

analyse spectrale d'émission 5018

analyse statique 11865

analyse statistique 11894

analyse statistique du système 12193

analyse symbolique 12087

analyse thermique différentielle 4150

analyse thermodynamique 12361

analyse thermomagnétique 12372

analyseur à servomécanisme 11445

analyseur automatique des réseaux 1326

analyseur à voie unique 11602

analyseur chromatographique 2375

analyseur continu 3126

analyseur d'amplitude 754

analyseur d'amplitude d'impulsion 10410

analyseur de circuit 2382

analyseur de courbes 3605

analyseur de distorsions 4413

analyseur de fonctions de distribution 5785

analyseur de gaz 5834

analyseur de gaz à fonctionnement automatique 1178

analyseur de gaz infrarouge sans dispersion 8698

analyseur de microprocesseur 8136

analyseur d'énergie 5041

analyseur de polarisation 9857

analyseur de programme 10241

analyseur de rapport 10646

analyseur de réseaux 8582

analyseur des gaz d'échappment 5613

analyseur de spectres optiques intégré 6799

analyseur des procédés transitoires 12517

analyseur détecteur 4015

analyseur de vibrations 12810

analyseur différentiel 4106

analyseur différentiel digital 4196

analyseur différentiel électro-mécanique 4818

analyseur différentiel hydraulique 6254

analyseur différentiel impulsionnel 10455

analyseur différentiel numérique 4196

analyseur digital de processus transitoires 4235

analyseur d'ondes 12893

analyseur du champ lointain 5385

analyseur du mélange 8223

analyseur du spectre des vibrations 12817

analyseur du système de réglage 3342

analyseur électrique 4677

analyseur électronique à canaux multiples 4910

analyseur enregistreur de gaz 10743

analyseur harmonique 5665

analyseur infrarouge de gaz 6582

analyseur infrarouge des gaz 6587

analyseur magnétique 7673

analyseur magnétique de gaz 7695

analyseur magnétomécanique de gaz 7729

analyseur mécanique 7984

analyseur optique-acoustique de gaz 9044

analyseur photoélectrique 9660

analyseur pour circuits intégrés 6783

analyseur radiométrique 10568

analyseur spectral 11746

analyseur thermochimique de gaz 12356

analyseur volumétrique de gaz 12873

analytique 848

anémomètre 862

anémostat 863

angle affiché 6456

angle critique d'erreur 3530

angle d'acceptance 151

angle d'admission 151

angle d'avance 542

angle de déphasage 9567

angle de divergence 875

angle de fonctionnement 8977

angle de phase 9567

angle de retard 7075

angle polaire 9851

annulation 12050

annuler 2069, 12641

anomalie de température 12268

anormal 892

antifading à déclenchement périodique 5867

antilogarithme 903

antiparallèle 905

antirésonance 907

apériodique 908

aplatissement 11661

appareil 7024

appareil à alimentation autonome 11310

appareil à commande programmée 10254

appareil à copier 3433

appareil à dessiner à commande automatique 1176

appareil à écran 11224

appareil à écran magnétique 6755

appareil à indication de zéro 8830

appareil à laser 7107

appareil à lecture directe 4317

appareil alimentateur 5421

appareil à maximum de tension 9335

appareil à mesurer le brouillage 6868

appareil à micro-ondes 8171

appareil à minimum de tension 12638

appareil analogue statique 11847

appareil apériodique 915

appareil astatique 1053

appareil autonome 11288

appareil à vapeur 11906

appareil avertisseur 12883

appareil à vide 12728

appareil contre-courant 3489

appareil controlleur de température 12279

appareil cryogénique de traitement des données 3561

appareil d'acquisition de données à canaux multiples 8445

appareil d'alarme 12883

appareil d'alimentation 9981

appareil d'analyse 839

appareil de commande de disque 4375

appareil de commande de système 12158

appareil de commande d'interruption 6939

appareil de commande électrique 4706

appareil de commande magnétique 7679

appareil de comptage 3494

appareil de contact 2166

appareil de contrôle de la fente lumineuse 7250

appareil de contrôle des matériaux par ultrasons 12616

appareil de décomposition 3821

élément sélectif de fréquence 5720
élément sensible à pression 10061
élément sensible de réglage 3328
éléments logiques pneumatiques 9792
élément sommateur 404
éléments stochastiques 11938
élément stable 11782
élément supplémentaire de commutation 420
élément unité 6313
élément universel 12677
élévateur de charge électrique 4727
élimination automatique du défaut 1271
élimination d'amortissement 3672
élimination des auto-oscillations 12052
élimination non synchrone 1082
élimineur automatique du bruit 1328
élongation de dépassement 7910
embranchement 1913, 1921
embrayeur 12567
émetteur à modulation en amplitude 776
émetteur de télémesure différentielle 4148
émetteur du radar optique 9114
émetteur optique 9131
émetteur-récepteur de données 3754
émetteur universel de programme 12690
émission à raie étroite 8529
émission du bruit 8622
émission du laser 7134
émission électronique 4847
émission impulsionnelle 10434
émission photoélectronique 9649
émissivité spectrale 11733
emmagasinage 4456
emmagasinage d'information électrique 4721
emmagasinage du report 2143
emmagasinage électrique des données 4709
emmagasinage magnétique statique 11859
emmagasinage optoélectronique des données 9188
emmagasiner 11955
emplacement d'action du réglage 9837
emplacement de mémoire 8043

emplacement de travail de manipulateur commandé par processus 10156
emploi 2644
émulateur interne au circuit 6412
émulation interne au circuit 6411
émulation microprogrammée 8161
émulseur 5027
émulsionneuse 5027
encapsulation à étanchéité 7211
encastrement fonctionnel 5771
enchaînement de machineoutil 7653
enchaînement flexible 5567
enchaînement rigide 11063
enchaîner 7423
enclenchement automatique d'installation de réserve 1365
enclenchement brusque 11669
encodeur angulaire 864
endroit de mesure 9841
endurance 3954
énergie 5040
énergie au zéro 12978
énergie au zéro absolu 5060
énergie cinétique du mouvement thermique 7057
énergie d'activation 277
énergie de crête 9471
énergie de décomposition 4370
énergie de désintégration 4370
énergie de laser 7135
énergie de liaison 1789
énergie de pénétration 9501
énergie d'équilibre 5120
énergie de seuil de l'impulsion 10460
énergie d'excitation moyenne 1505
énergie interne 6908
énergie mécanique d'un manipulateur 7999
énergie minimum d'inflammation 8200
énergie nucléaire 8817
énergie oscillatoire 12813
énergie paramétrique de pompage 9417
énergie potentielle 9942
énergie propre 2284
énergie utile 8578
energistrement binaire 1772
enfichable 9760
engrenage de différentiel 4125
engrenage de grappin 5997

engrenage d'entraînement 8992
enregistrement actif des données 290
enregistrement à pointe par étrier mobile 2369
enregistrement automatique des données 1245
enregistrement automatique des résultats 1307
enregistrement continu 3160
enregistrement de perturbations 4454
enregistrement de résultats digitaux 10747
enregistrement digital 4222
enregistrement d'information automatique 1320
enregistrement électronique des valeurs mesurées 4903
enregistrement graphique des résultats 11033
enregistreur 1130
enregistreur à compensation 2723
enregistreur à ligne continue 10738
enregistreur automatique 7493
enregistreur de convergence 3382
enregistreur de coordonnées 3420
enregistreur de débit du gaz 5850
enregistreur de données 3717
enregistreur de données analogiques 804
enregistreur de maximum 7915
enregistreur de pression différentielle 4136
enregistreur de pression sous-marine 3960
enregistreur de vibrations 12821
enregistreur digital 4221
enregistreur d'impulsions 6390
enregistreur électrique 4695
enregistreur graphique 5971
enregistreur pneumatique 9805
enregistreur-récepteur 11057
enregistreur traceur de gaz 5864
enregistreuse couplée 3019
enrichissement laser 7136
enroulement amplificateur 745
enroulement d'alimentation 6680
enroulement de commande 3363
enroulement de compensation 2733
enroulement d'entrée 6680
enroulement de puissance 9986
enroulement de référence 10800
enroulement de sortie 9287
enroulement d'excitation 5251
enseignement programmé 10300

indication 6466
indication à distance 10931
indication automatique 1294
indication de code 7067
indication de déconnexion 4330
indication de pannes 12571
indication de surcharge 9322
indication en retour 1559
indication numérique de position 8856
indication programmable 10281
indication programmée d'erreurs 10297
indice de performance 9515
indice de phase 9585
indice de ramification 1915
indice d'oscillation 6452
indice intégral de performance 6775
inductance 6478
inductance distribuée 4428
inductance variable 3148
inducteur 6494
inducteur de chauffage 6486
inducteur de réglage 10833
inducteur variable 12772
induction complète 5518
induction critique 3533
induction magnétique 7697
industrie de calculateur 2915
industrie des capteurs 11376
industrie électronique 4927
industrie orientée sur le processus 10203
inégal 12698
inégalité de Cauchy-Schwarz 2185
inégalité de Routh 11127
inégalité de Tchebychev 2320
inertie 6527
inertie thermale 12347
inertisation 10948
infini 6530
influence d'un grandeur perturbatrice 6534
information 6539
information additionnelle 416
information alphabétique 684
information binaire 1763
information certaine 2235
information complète 9509
information complexe 2786
information conservée 3034
information considérable 3035
information de caméra 2065
information de capteur 11377
information de commande 3243

information de commutation 12071
information de diagnostic 4068
information de gestion 7783
information de grande capacité 1978
information d'entrée 6662
information de position 9904
information de position avec senseur 11381
information de procédé de robot 11100
information de reconnaissance 10722
information de son 11694
information d'essai 2336
information de traitement 10179
information détruite 4000
information d'identification 10722
information élémentaire 4981
information engendrée par l'ordinateur 2904
information fictive 4539
information géométrique 5919
information géométrique codée alphanumériquement 687
information graphique 5959
information métrique 8090
information mutilée 5831
information non digitale 8688
information orientée sur décision 3811
information parfaite 9509
information partielle 9436
information préenregistrée 10066
informations de température d'ambiance 723
informatique à distance 3743
infrarouge moyen 8020
ingénierie de connaissance 7065
ingénierie de système 12165
inhiber 6608
inhibiteur cathodique 2180
initialisation 6621
initialisation de calcul 2025
initiateurs de positions finales 5501
insensibilité 6698
insensibilité aux parasites électromagnétiques 6350
insensibilité contre bruits 8628
insensibilité d'élément 8772
insensibilité en boucle ouverte 7563
insensible 6696
insertion de machine spéciale 932
insolubilité algorithmique 657

inspection 6693
inspection d'automates 12034
inspection de commande directe 4271
inspection de commande indirecte 6473
inspection finale 5500
inspection visuelle directe 4326
inspection visuelle indirecte 6476
instabilité 6229, 6428, 6701
instabilité due à la pulsation 10408
instabilité statique 11855
instabilité structurelle 11977
installation 5125, 6706
installation à plusieurs usages 8462
installation automatique pour le taillage 1237
installation centrale d'introduction de données 2209
installation d'absorption 64
installation d'aération 618
installation d'automatisation 1420
installation de calculateur 2916
installation de caméra 2066
installation de chauffage 6102
installation d'éclairage 7254
installation de commande à distance 10925
installation de conditionnement d'air 591
installation de dépoussiérage 4553
installation de déshydratation 3899
installation de déshydrogénation 3901
installation de distribution 4447
installation de fabrication 10224
installation de fabrication complètement intégrée 5746
installation de filtrage 5494
installation de fractionnement 8440
installation de laboratoire 1729
installation de l'humectation 8303
installation de liquéfaction de gaz 5855
installation de manipulateur 6704
installation de manipulateur industriel 6708
installation de mesurage de force 5647
installation de mesure 7955
installation de montage 5154
installation de pompes 10465

ordre de système 12180
ordre d'examen des grandeurs
    12750
ordre d'un nombre 8837
ordre du système réglé 9205
organe 12424
organe actif 292, 304
organe binaire de mémoire 1779
organe central 2210
organe commandé 3264
organe comparateur 2688
organe coordonnateur 3431
organe correcteur 351
organe d'action intégrale 6805
organe d'ajustement 511
organe de boucle 7560
organe de calcul 981
organe de calcul d'adresse 438
organe de commande 5268
organe de commande d'accélé-
    ration 128
organe de commande pour
    données de procédé 10162
organe de grippion adapté
    automatiquement 11272
organe de maintien 6197
organe de mémoire 8028
organe d'entrée 6691
organe de positionnement 351
organe de réglage 351
organe de réglage direct 5660
organe de réglage final 5497,
    5504
organe de réglage pneumatique
    9814
organe de retard 7078
organe d'estimation 5207
organe de verrouillage 7484
organe d'exécution 5268
organe différentiateur 3968
organe directeur 7854
organe électrique de réglage
    4673
organe exécutif 5497
organe linéaire 7391
organe moteur 360, 8358
organe non linéaire asymétrique
    1063
organe passif 9459
organes de commande 4720
organe sensible 11353
organe sensible capacitif 2105
organe sensible d'accélération 129
organe sensible du système
    d'autoguidage 6203
organe tampon 1963
organe transducteur 12496
organigramme 1838

organisation avancée des données
    549
organisation de gestion
    industrielle 9215
organisation de montage 1026
organisation de restauration
    10751
organisation fonctionnelle 5777
organisation offline 8897
organisation parallèle 9380
orientation d'effecteur absolue 29
orientation définie de coor-
    données 3847
orientation d'objets de manuten-
    tion 9217
orientation visuelle 12828
orienté 9219
orienté application 933
orienté machine 7642
original construit par ordinateur
    2903
origine de défauts 2189
origine de panne 5372
origine d'un système de coor-
    données 3426
oscillateur 9249
oscillateur à accrochage de phase
    sous-harmonique 12005
oscillateur à magnétostriction
    7735
oscillateur à quadrature 10481
oscillateur à réaction retardée
    3922
oscillateur à relaxation 10874
oscillateur bloqué monostable
    11619
oscillateur commandé en tension
    12855
oscillateur de blocage 1847
oscillateur électromécanique
    basse fréquence 4823
oscillateur harmonique 6086,
    9249
oscillateur laser 7161
oscillateur paramétrique 9105
oscillateur paramétrique optique
    9105
oscillateur résistance-capacité
    2078
oscillateur surcouplé 1847
oscillateur titrateur 9258
oscillation 9241
oscillation auto-entretenue 11322
oscillation croissante 6431
oscillation d'une fonction 5791
oscillation fondamentale 1685
oscillation naturelle 1449
oscillation propre 1449

oscillation pseudoharmonique
    10391
oscillation quasi-harmonique
    10522
oscillations amorties 3385
oscillations continues 3151
oscillations convergentes 3385
oscillations de la relaxation 10873
oscillations divergentes 4469
oscillations douces 11670
oscillations électromagnétiques
    4804
oscillations électromagnétiques
    cohérentes 2553
oscillations forcées 5642
oscillations modulées en
    amplitude 772
oscillation sous-harmonique
    12004
oscillations parasitaires 6229
oscillations symétriques 12101
oscillation stationnaire 11905
oscillographe à échantillonnage
    11163
oscillographe à faisceau
    électronique 4834
oscillographe à large bande
    12914
oscillographe à mémoire 11952
oscillographe à rayons catho-
    diques 2175
oscillographe cathodique 2175
oscilloscope 9257
oscilloscope à plusieurs faisceaux
    8382
oscilloscope à rayons cathodiques
    2176
oscilloscope compact 2685
outil 12474
outillage de montage 1017
outils de synchronisation basés
    1690
outils micro-informatiques 8122
outils programmés auto-
    matiquement 1179
ouverture 8961
ouverture du laser 7105
ouvrir 12696
overlay 9314

pacemaker 9338
paire de bornes 1905
palier de saturation 11178
palpation électrostatique 4972
palpeur 3082, 11156
palpeur à rayons infrarouges 6597
palpeur capacitif 2083
palpeur d'approximation 952

structure de senseur simple 11560
structure de système 12194
structure de terme 7233
structure différentiable 4102
structure digitalisée 4242
structure d'instruction 6745
structure d'interruption en chaîne 3657
structure d'inversion 6978
structure d'overlay 9316
structure du dispositif à relais 10889
structure d'unité linéaire 7395
structure d'un système visuel 12848
structure du système à relais 10903
structure du système de calculateur 11994
structure électrique du bus 4742
structure en guide d'onde 12897
structure fixe 5540
structure fonctionnelle 5781
structure géométrique 5922
structure graphique 5969
structure graphique de données 5955
structure graphique plane 9732
structure grossière 2501
structure hiérarchique pour le management des données 6132
structure intégrée 6801
structurel 11974
structure linéaire 7410
structure locale 7476
structure logicielle 11675
structure mathématique 7882
structure mécanique 7988
structure modulaire 8275
structure monolithique 8338
structure multicouche 8415
structure numérique 8861
structure optimum 9175
structure ordonnée 9201
structure organisée bus 1994
structure orientée sur les blocs 1860
structure parallèle dynamique du programme 4590
structure parallèle-série 9387
structure partielle 12014
structure pipeline 9725
structure plate du circuit de relais 5550
structure probable 10128
structure quasi-machine 10526
structure récursive 10776

structure réelle 10686
structure standardisée 11807
structure subordonnée 12009
structure technologique 12241
structure totale d'un système hybride 12486
substitution automatique d'adresse 1171
substitution d'adresse 459
substitution des variables 2250
sub-système de périphérie 9543
succession 11397
succession des opérations 9035
suite automatique en processus de manipulation 1332
suite de manipulateur alternative 711
suite de manutention 6049
suite de procédés 11407
suite de zéros 12985
suite d'opération 11406
suite hiérarchique des paramètres 9399
suiveur 5625
superposition 12029
supervision 8324
supervision de processus 10214
supervision des moyens de production 12035
supervision de tâches 12227
supplément par option 9183
support central 2219
support de programme 10242
support de représentation de base 1680
support d'impulsion 10416
supposer 12049
supposition 6277
suppression 12050
suppression de l'ordre 9198
suppression des auto-oscillations 12052
suppression des données 12051
suppression de synchronisme 2653
suppression de zéros 12989
supraconducteur stabilisé 11778
supraconductivité 12022
suraccéléromètre 7021
suralimentation 7778
suramortissement 9301
surcharge 9320
suréchantillonnage 9326
sûreté 3954, 11132
sûreté automatique CN de données 1329
sûreté de service 5133
sûreté de système 12188

surface absorbante 95
surface active 286, 313
surface apparente de contact 921
surface de réglage 3187
surface des écarts 4042
surface effective 4633
surface quadratique d'erreur 10478
surface superfinie 8100
surgir 9880
surpression 12030
surréglage 9328
sursynchronisation 9332
surtension 9334
surtension de concentration 2963
surtension du circuit 10488
surveillance 8324
surveillance à distance 10940
surveillance à laser 7185
surveillance automatique 1323
surveillance automatique de la suspension 1292
surveillance automatique des données 1247
surveillance centrale 2200
surveillance d'écran 11226
surveillance de fabrication 8328
surveillance de fabrication automatisée 1154
surveillance de force de grippion 5990
surveillance de la production 10230
surveillance d'élévateurs à chaînes 2241
surveillance de manutentions séquentielles 8329
surveillance d'entrée-sortie 6676
surveillance de réseaux 8576
surveillance de senseur 11386
surveillance de senseur tactile 12205
surveillance de traitement 10183
surveillance d'exploitation 10220
surveillance d'un état de marche 9018
surveillance fonctionnelle 5775
surveillance permanente 3149
surveiller 8882
survoltage 9334
survolteur différentiel 4107
survolteur réversible 11053
susceptibilité aux perturbations 5408
susceptibilité initiale 6627
susceptibilité non linéaire 8748
susceptibilité optique non linéaire 8739
suspendu 9495

Русский

аппроксимация во временной
     области 956
аппроксимация Падэ 9343
аппроксимация функции при
     помощи полиномов 9875
аппроксимация функций
     времени 959
аппроксимация Чебышева 2319
аппроксимация экспонен-
     циальных функций 958
аппроксимирующий контактор-
     реле 10890
аргоновый лазер 976
аргумент 977
аргумент функции 978
арифметикологическое устрой-
     ство 988
арифметическая задача 2955
арифметическая запись 990
арифметическая команда
     присвоения 985
арифметическая операция 983
арифметическая проверка 980
арифметическая функция 982
арифметические данные 986
арифметический процессор 991
арифметический сдвиг 984
арифметический элемент 981
арифметическое звено 993
арифметическое и логическое
     процессорное устройство
     987
арифметическое правило 2954
арифметическое среднее 989
арретирующее устройство 999
архитектура, базирующаяся на
     регистрах общего назна-
     чения 5899
архитектура системы 12140
асимметричная модуляция 1061
асимметричная нелинейность
     1058
асимметричное изделие 1064
асимптота 1065
асимптотическая устойчивость
     1072
асимптотический 1066
асимптотический метод 1070
асимптотический оптимум 1067
асимптотический поток 1069
асимптотический процесс 944
асимптотическое значение 1073
асимптотическое отношение
     1071
асимптотическое разложение
     1068
асинхронная вычислительная
     машина 1076

асинхронная логическая схема
     1081
асинхронная релейная система
     1083
асинхронная связь 1077
асинхронная система передачи
     1087
асинхронная следящая схема
     1085
асинхронное возбуждение 1080
асинхронное гашение 1082
асинхронное управление 1078
асинхронный адаптер коммута-
     ции 1074
асинхронный ответ 11020
асинхронный разъедини-
     тельный режим 1079
асинхронный режим работы
     1088
асинхронный сервомотор 1086
ассоциативная память 11235
ассоциативная связь 1047
ассоциативная структура 1042
ассоциативное программиро-
     вание 1049
ассоциативный 1044
ассоциативный закон 1046
ассоциативный индексный
     метод 1045
ассоциативный принцип 1048
ассоциация 1043
ассоциировать 1040
астатическая система 1054
астатическая система управле-
     ния 1051
астатический прибор 1053
астатический регулятор 1052
астатический регулятор с зави-
     симой скоростью 10371
астатический регулятор с
     постоянной скоростью 3056
астатический элемент 5575
астатическое действие 5574
астатическое действие с посто-
     янной скоростью 11621
астатическое регулирование
     5576
астатическое регулирование с
     постоянной скоростью
     11622
атмосферная оптика 1090
атмосферное торможение 1089
атмосферные помехи 1091
аттоматическая замкнутая
     сервосистема 1209
атомная теплоёмкость 1093
атомный реактор 1094
аттенюатор 1106

аудиовизуальное представление
     1114
аудиовизуальный 1113
ацидометр 215
аэродинамические свойства 557
аэродинамические характе-
     ристики 557

бесконтактное управление 3086
база в истинном масштабе
     времени 10688
база данных 3688
базис 1646
базис времени 12410
базис кода 1658
базис-мерного пространства
     состояний 1696
базисная полоса 1649
базисная функция 1697
базисное решение 1698
базисные синхронные
     инструменты 1690
базисный блок передачи 1694
базисный вектор 1699
базисный последовательный
     метод доступа 1687
базисный регистр 1660
базисный телекоммуника-
     ционный метод доступа
     1692
базис реализации 1659
базовая конфигурация мани-
     пулятора 1667
базовая координата 1653
базовая машина 1679
базовая модель 1681
базовая операция 1684
базовая переменная 1654
базовая поддержка отобра-
     жения 1680
базовая система 1691
базовая система эффектора
     4657
базовая функция 1674
базовое соотношение 1655
базовый адрес 1647
базовый алгоритм 1648
базовый метод 1661
базовый формат 1673
байт 2006
байт данных 2009
байт состояния 11897
балансировка 1610
балансировка нуля 1588
балансировка развёртки 12056
баланс к определенному
     моменту времени 8311
балансная схема 1617

быстрота 12800
быстрый останов 5398

вакуум-аппарат 12728
вакуумметр 12734
вакуумная камера 12729
вакуумная техника 12739
вакуумная установка 12736
вакуумный флюоресцентный
    индикатор 12731
вакуумный фотоэлемент 12735
вариант размещения 998
варианты программы 12787
вариационное исчисление 2034
вариационное уравнение 12783
вариационный принцип 12785
вариация 12782
вариация показаний измери-
    тельного прибора 8065
вариация постоянных 12784
вариометр 491
варистор 12790
вармметр 12791
введение интегрирующего
    звена 6802
ввод аналогового сигнала 814
ввод в реальном времени 10695
ввод в эксплуатацию 10473
ввод геометрических структур
    5924
ввод данных 3711, 6699
ввод данных в аналоговую
    вычислительную машину
    3712
ввод данных в реальном
    времени 10695
ввод данных в режиме диалога
    3387
ввод данных в цифровую
    вычислительную машину
    3713
ввод данных о процессе 10191
ввод измеряемых величин 6667
вводить в действие 1719
вводное устройство 6653
ввод структуры 11991
ввод управляющей информации
    6666
ведущее устройство 7855
вектор 12792
вектор нагрузки 7459
вектор намагниченности 7724
векторная диаграмма 12793
векторная оптимизация 12797
векторная теория 12795
векторная функция 12794
векторный параметр 12798
величина 12745

величина в графической форме
    5966
величина влажности 8306
величина воздействия 355
величина для сопоставления
    7869
величина емкости 2107
величина затухания 1105
величина изменения энергети-
    ческого уровня 5056
величина на выходе 11017
величина отклонения 4047
величина отсчёта 10797
величина оценки 5216
величина покоя 10543
величина потерь давления
    10053
величина расхода 5609
величина самоиндукции 12747
величина ускорения 136
величина фактора 5358
величина частоты 5729
вентиль 12751
вентильный фотоэлемент 10767
вентиль прямого действия 4310
вентиляционная установка 618
верификация временных
    соотношений 12467
верность 199
вероятная ошибка 10127
вероятная структура 10128
вероятностная логика 10114
вероятностная машина 10115
вероятностное решение 2242
вероятностный 10112
вероятностный метод 10123
вероятостый автомат 10113
вероятность 10116
вероятность испускания 5015
вероятность наступления 10124
вероятность отказа 5368
вероятность перехода 12538
вероятность поглощения 96
вероятность устойчивости
    10125
вертикальное микропро-
    граммирование 12806
вертикальный контроль 12805
вертикальный контроль избы-
    точным кодом 12807
верхнее предельное значение
    12718
верхний предел 12717
верхний предел интегрирования
    12719
весовая функция 12906
весовой множитель 12905
вести 4499

ветвящийся процесс 1918
вещественная диаграмма 10681
вещественная переменная
    10702
вещественная частотная харак-
    теристика 10683
вещественная элемент 10682
взаимная блокировка 1840
взаимная корреляция 3548
взаимная модуляция 6904
взаимная спектральная
    плотность 3559
взаимное запирание 1840
взаимно компенсирующиеся
    ошибки 2714
взаимно расстроенные контуры
    11791
взаимный импеданс 6363
взаимодействие между
    процессами 6927
взаимодействовать 6829
взаимозаменяемые приборы
    10956
взаимозаменяемый 6839
взаимозаменяемый механизм
    для промеждуточного про-
    цесса 6894
взаимонезависимые пере-
    менные 8512
взаимосвязанная система авто-
    матического регулирования
    6831
взаимосвязанное автомати-
    ческое регулирование 6830
взвешанный код 12904
вибратор 12819
вибрационное реле 9240
вибрационный питатель 12820
вибрационный пневмопривод
    9825
вибрационный регулятор 9233
вибрация 5621, 12808
виброанализатор 12810
виброгаситель 12811
виброграф 12812
виброизмерительное оборудо-
    вание 12814
вибропрочность 12816
вибростойкость 12816
вибротрон 12822
видеодетектор 12826
видеосигнал базисной полосы
    1652
видеотерминал 12825
видеоусилитель 12823
видимое излучение лазера 7154
вид привода манипулятора 7795
визуальная индикация 12842

диапазон показаний 6462

диапазон пропускания фильтра 5490

диапазон рабочих температур 9007

диапазон регулирования 3320

диапазон скоростей 11752

диапазон статической настройки 11867

диапазон точных значений тока измерительного прибора 211

диапазон усиления 5815

диапазон частот 5700

диапазон чувствительности 10618

диапазон электронной настройки 4940

диафанометр 4085

диафрагма 4086

дигитайзер 4244

дигитальное моделирование 4229

дизъюнктивная нормальная форма 4373

дизъюнкция 4372

динамика линейной следящей системы 7375

динамика манипулятора 7796

динамика разветвленных систем регулирования 4605

динамика реактора 10667

динамика сопряжённых паровых систем 4604

динамика функционирования автомата 4606

динамика экспериментального обтекания 5290

динамика элементов привода 4503

динамическая выдача 4575

динамическая операция 4587

динамическая оптимизация 4589

динамическая ошибка 4577

динамическая память 4608

динамическая погрешность 4577

динамическая подпрограмма 4609

динамическая рабочая характеристика 4588

динамическая связь 4583

динамическая симуляция 4602

динамическая система 4610

динамическая система управления 4568

динамическая структура программы 4593

динамическая схема 4567

динамическая точность 4559

динамическая точность воспроизведения 4579

динамическая характеристика 4566

динамическая характеристика генератора 4581

динамическая характеристика манипулятора 4585

динамическая чувствительность 4600

динамические данные 4570

динамические характеристики автоматических измерительных приборов 4599

динамический 4558

динамический анализ 4561

динамический волномер 4613

динамический демпфер 4569

динамический диапазон 4595

динамический контроль 4611

динамический контроль проблемы 4591

динамический массспектрометр 4586

динамический одномодовый лазер 4603

динамический расчёт 4565

динамический режим 4596

динамический режим работы 4563

динамический элемент 4576

динамическое запаздывание 4582

динамическое звено 4612

динамическое краевое условие 4564

динамическое определение 4571

динамическое отклонение механических узлов 4572

динамическое отклонение синхронных движений 4573

динамическое программирование 4592

динамическое программное управление 4601

динамическое равновесие 4562

динамическое распределение 4560

динамическое смещение 4597

динамическое сопротивление 4598

динамическое состояние 4607

динамическое торможение электропривода 4756

динамическое трение 4580

динамическое устройство 4574

динамометр с воздушным тормозом 567

динамо-регулятор 4614

диодная схема 4255

диодное детектирование 4258

диодный генератор функций 4259

диодный ограничитель тока 4257

диодный счётчик 4256

диодный умножитель 4260

дисковая операционная система: ДОС 4376

дисковое кодирующее устройство 4374

дискретизация 10490

дискретизированный сигнал 10501

дискретная величина 4359

дискретная дистанционная передача сигнала 4361

дискретная логика 4352

дискретная разомкнутая система с переменными параметрами 8968

дискретная система 4364

дискретная структура 4242

дискретная схема 4346

дискретное воздействие 4351

дискретное измерение длин 4202

дискретное представление 4360

дискретное программирование 4357

дискретное распределение 4349

дискретное состояние траектории 4355

дискретное управление 4178

дискретно-непрерывная система 4348

дискретные импульсы 4358

дискретный автомат 4345

дискретный импульс 4238

дискретный компонент 4347

дискретный принцип максимума 4353

дискретный процесс 4356

дискретный случайный многошаговый процесс решения 4362

дискретный стохастический многошаговый процесс решения 4362

дискретный фильтр 4350

запоминающее устройство
кратковременного действия
11503
запоминающее устройство на
линиях задержки 3927
запоминающее устройство на
электромагнитных линиях
4811
запоминающее устройство
робота 11096
запоминающее устройство с
произвольным доступом
10579
запорное усилие 2487
запорный вентиль 8766
запорный клапан 8766
запрашивать 10969,
запрещать 6608
запрещающий вход 6610
запрещенная логика 6327
запрещенная операция 6325
запрещенное состояние 1855
запрещенное состояние
релейной цепи 10881
запрещённый уровень 5631
запрограммированный обмен
данными 10296
запрограммированный рабочий
режим 10302
запрос 6693
запрос передачи 10973
запрос программы 10319
запускать 350
запускать повторно 10978
запускающий плавкий элемент
6633
запуск регулирующих схем
11824
заранее заданная величина 9997
заранее накопленная информация 10066
зарядно-контрольное устройство 2313
затемняющий импульс 1825
затрат
не требующий больших ~
времени 12450
затухание вследствие рассогласования и симметрии 5161
затухание колебаний в регулируемом объекте 9740
затухание контура 8584
затухание передачи 12546
затухание при распространении
10341
затухание при случайно распределенных изгибах 10581
затухание сигнала 3674

затухание фильтра 5483
затухающая синусоида 3662
затухающее действие 3664
затухающие импульсы 3783
затухающие колебания 3385
захватное устройство 5989
захватывание манипулируемых
объектов 5991
захватывание частоты 5087
захватывать 7480
зашифровать 5030
защита в системе дистан-
ционного управления 10375
защита данных 3738
защита кода 2528
защита нулевой последова-
тельности 12976
защита от замыкания на землю
4616
защита от импульсного
напряжения 6400
защита от короткого замыкания
11491
защита от максимальной
мощности 9323
защита от неправильного
включения 10380
защита от несанкциониро-
ванного доступа к памяти
8029
защита от обрыва фаз 8970
защита от перегрузки 9321
защита от перенапряжения 9336
защита от превышения частоты
9306
защита от сверхтока 9300
защита от тока 3592
защита от шума 8638
защита при однофазных замыканиях на землю 11614
защита радиосвязи 10567
защита с выдержкой времени
12433
защита с заземляющей шиной
5672
защита электросети
переменного тока 10376
защита электросети
постоянного тока 10377
защитная функция 10382
защитное действие 10379
защитное приспособление10381
защитный газовый контактор
10383
защищенная от отказов система
5362
защищенный от отказов 5361
звено 1833

звено без компенсации 4996
звено времени 12424
звено неравнозначности 900
звено опережения 7209
звено пневматической
регулировки 9814
звенья передаточного механизма схвата 5998
звуковая информация 11694
звуковая связь 226
звуковой сигнал 1112
звуковые колебания 11690
зинеровский пробой 12951
знак 2278
знак кода 2279
знакопеременные ряды 714
знание 7064
значащая цифра 12742
значение входного сигнала
12746
значение параметра 9404
значительная по объёму
информация 3035
значительный отказ 7769
зона 12994
зона линейности 12997
зона насыщения 12998
зона неоднозначности 12996
зона нечувствительности 3765
зона памяти 8052
зона помех 6861
зона пропорциональности10349
зона пропорциональности
регулирования 10349
зона регулирования 12995
зона управления 3234
зонная теория 1637
зубчатая передача для
непрерывного регулирова-
ния скорости 6533

игнитронная регулировка 6324
идеализированная система 6288
идеализированная теория 6290
идеальное значение величины
6291
идеальный импульс 6289
идентификатор 6307
идентификатор программы
10274
идентификационный блок 6296
идентификационный код 6298
идентификационный контроль
6297
идентификация 5204
идентификация линейных
непрерывных систем 6310
идентификация модели 8241

импульсное умножающее
устройство 6398
импульсное управление
двигателем 8359
импульсное управление с
помощью лазера 7172
импульсное управление
электродвигателем 8359
импульсное ускорение 6351
импульсно-кодовая модуляция
оптического сигнала 9113
импульсный аттенюатор 10412
импульсный
выравниватель 10436
импульсный выходной
усилитель 6399
импульсный генератор с
обратной связью 10811
импульсный инжекционный
лазер 10425
импульсный код 6377
импульсный контур 6376
импульсный лазерный локатор
10431
импульсный метод телеметрии
10459
импульсный модулятор 6397
импульсный накопитель 6374
импульсный повторитель 10446
импульсный повторитель со
смещением частоты
сигнала 5712
импульсный регулятор 9522
импульсный режим 10432
импульсный режим генератора
5915
импульсный сигнал 6392
импульсный спектрограф 10451
импульсный усилитель 10409
импульсный широкополосный
усилитель 12915
импульсный элемент 6381
импульс отклонения 9327
импульс ошибочного сигнала
5196
импульс переходного процесса
12526
импульс полузаписи 6021
импульс предварительной
установки 1705
импульс прерывания 1926
импульс сброса 10990
импульс сложения 408
импульс с плоской вершиной
5549
импульс сформированный
линией задержки 3929
импульс считывания 10674

импульс управления мани-
пулятором 7792
импульс частичного ввода 9447
импульс частичной выборки
9442
импульсы от обратного хода
развёртки 5622
импульсы с перекрытием
9313
инвариантная по времени
система 12428
инвариантная система
регулирования 6968
инвариантность 6964
инвариантность кибернети-
ческой системы 6965
инвариантный 6967
инверсия 6981
инверсная модель манипулятора 6974
инверсная структура 6978
инверсная схема 3488
инверсный режим работы 6985
инверсный элемент сложения
429
инвертировать 6984
инвертируемый автомат 6987
инвертирующая схема 6988
инвертор питаемый от сети
3956
индекс колебательности 6452
индекс разветвления 1915
индивидуальная контрольно-
измерительная аппаратура
11606
индивидуальный привод
захватного устройства 3789
индикатор 6460
индикатор баланса 1615
индикатор вакуума 12733
индикатор загрязнённости
воздуха 577
индикатор излучения 10553
индикатор колебаний 3655
индикатор контроля данных
3696
индикатор максимального
давления 9483
индикатор маршрута 11129
индикаторная диаграмма 6468
индикаторная лампа 6461
индикаторная работа 6457
индикатор неподтверждения
216
индикаторный контроллер 6459
индикаторный контур 6458
индикаторный обратный
импульс 6469

индикаторный угол 6456
индикаторный щит 888
индикатор отклонения 4044
индикатор ошибки 5186
индикатор плотности 3952
индикатор порогового значения
11251
индикатор последовательности
фаз 9616
индикатор приоритета 10105
индикатор расхода 5595
индикатор соотношения 10650
индикатор угла 867
индикатор ускорения 130
индикация 6466
индикация неисправностей
12571
индикация разъединения 4330
индикация ситуации 11638
индуктивная импликация 6492
индуктивная связь 6483
индуктивная система
синхронной связи 12115
индуктивноемкостная линия
задержки 6479
индуктивность 6478
индуктивный датчик 6493
индуктивный делитель
напряжения 6480
индуктивный тензометр 6481
индукционная уравновешенная
схема 6482
индукционный ваттметр 6491
индукционный двигатель 6488
индукционный нагревательный
прибор 6486
индукционный нагрев токами
высокой частоты 6151
индукционный ограничитель
расхода жидкости 6485
индукционный расходомер
6484
индукционный регулятор
напряжения 6490
индукционный тахогенератор
6489
инертизация 10948
инерциальное наведение 6529
инерциальное управление 6529
инерционная синхронизация
5623
инерционность 6527
инерция 6527
инженерное приближение 5072
инженерно-технический метод
работы 5082
инженерный пульт 5074
инициализация 6621

интерпретирующая программа 6926
интерпретирующее устройство 6924
интерпретирующий код 6925
интерфейс 6846
интерфейс ввода-вывода 6673
интерфейс ввода изображения 6335
интерфейс компьютера 2918
интерфейс линии 7406
интерфейс локального управления 7472
интерфейс монитора 8331
интерфейсная логика 6852
интерфейсный модуль 6853
интерфейс процессора 10204
интерфейс с нестандартными устройствами пользователя 3611
интерфейс с системой передачи 12545
интерференционная картина в дальней зоне 5386
интерференционное реле 6874
интерференционный импульс 6872
интерференционный компаратор 6862
интерференционный микроскоп 6869
интерференционный рефрактометр 6873
интерферометрическая система 6880
интерферометрический контроль 6876
интерферометрический модулятор 6878
интерферометрический сенсор 6879
интерферометрическое измерение 6877
интерферометрия 6881
информационная база для управления 7784
информационная и управляющая система 6541
информационная селекция 6568
информационная система 6570
информационная система для обучения 4631
информационная система обеспечения эксплуатации 8987
информационная система предприятия 5084

информационная система управления 6581
информационная структура 6542
информационная техника 6551
информационная технология 6573
информационная цепь 6546
информационная электроника робота 11089
информационно-управляющая система 6557
информационный канал 6544
информационный обмен 6552
информационный поток в автоматизированной системе 1144
информационный сигнал 6569
информационный терминал 6574
информация 6539
информация, необходимая для распознавания 10722
информация, ориентированная на принятие решения 3811
информация, сгенерированная компьютером 2904
информация для коммутации 12071
информация о позиции 9904
информация о температуре окружающей среды 723
информация от камеры 2065
информация управления 7783
инфразвуковая частота 6603
инфракрасная излучательная способность 6584
инфракрасная поисковая система 6596
инфракрасная система измерения дальности 10614
инфракрасная система с импульсной модуляцией 6593
инфракрасная система формирования изображения 6334
инфракрасная спектроскопия с использованием преобразования Фурье 6586
инфракрасная установка 6585
инфракрасное излучение лазера 6591
инфракрасное сканирующее устройство 6595
инфракрасный анализатор газов 6582
инфракрасный анализатор жидкости 7434

инфракрасный газоанализатор 6587
инфракрасный сенсор 6598
инфракрасный чувствительный элемент 6597
инъектирующий контакт 6635
ИС 6782
искажать 12885
искажающая сила 4464
искажающий импульс 4465
искажающий усилитель 9302
искажение 4412
искажение графика 5963
искажение импульса 10427
искажение квантования 10503
искажение отклонения 3867
искажение от обратной связи 4416
искажение при передаче 12553
искажение формы сигнала 760
искажений
без ~ 5677
искаженная информация 5831
искаженное заначение 4462
искаженный нулевой выходной сигнал 4463
искаженный сигнал 5832
искатель зоны 12999
исключающее условие 5229
исключение 5257
искомая величина 10975
искровой генератор для индукционного нагрева 11707
искрогасящее сопротивление 10532
искусственный диэлектрик 1001
искусственный интеллект 1002
искусственный перенос 1000
искусственный язык 1003
исполнительная программа 5269
исполнительная система 5270
исполнительное звено 5268
исполнительное устройство 352
исполнительное устройство автоматического контроля 1202
исполнительные электрические органы 4720
исполнительный 5262
исполнительный механизм 360
исполнительный момент 514
исполнительный орган 5497
исполнительный элемент 5268
исполнительный элемент системы регулирования 5497

калибровка регулирующего
стержня 3326
калибровочное преобразование
5873
калибровочный импульс 2048
калориметрический анализатор
следов газа 2062
камера обработки воздуха 590
камертонная стабилизация 5652
канал 2257
канал аналоговых сигналов 791
канал воздушного охлаждения
582
канал обратной связи 5425
каналоведущая структура 2269
каналообразование 2271
канал передачи 12543
канал передачи данных 3694
канал передачи двоичной
информации 1193
канал прерываний 6937
канал прямой связи 5659
канал распределения команд
6735
канал релейной защиты 10897
канал связи 2658
канал связи с шумом 8646
канал с модулированной
несущей частотой 8282
канал сопряжёния 6847
канал телеуправления 10921
канальная лента
n- ~ 8544
канальное устройство ввода-
вывода 2270
канальный байт 2260
каноническая форма 2073
каноническое преобразование
2074
каноническое распределение
2071
каноническое уравнение 2072
капельный анализ 11755
капиллярная система 2111
капиллярное давление 2110
капиллярный вискозиметр 2112
капиллярный расходомер 2109
капиллярный электрометр 2108
капиталовложение на проведение рационализации 10656
касательная 12212
каскад 11790
каскадная операция 2156
каскадная промывная установка
2160
каскадная релейная система
10878
каскадная система 2161

каскадная система охлаждения
2158
каскадная система управления
2150
каскадное включение лазеров
7117
каскадное регулирование 2145,
5628
каскадное регулирование
скорости 2961
каскадное реле 2159
каскадное соединение 2151
каскадное соединение двух
контуров 12211
каскадное управление 2149
каскадный возбудитель 2155
каскадный процесс 2157
каскадный усилитель 2147
каскадный электрооптический
модулятор 2154
каскад преобразователя 6986
каскад стробирования 5869
каскад тактового генератора
2463
каскад усиления 735
кассификатор изменяемых
изображений по времени
2435
каталитическая активность
2164
каталитический процесс 2165
категория данных 3693
катодная защита 2182
катодная обратная связь 2169
катодная поляризация 2181
катодная реакция 2183
катодное распыление 2168
катодный детектор 2167
катодный замедлитель 2180
катодный луч 2172
катодный осциллограф 2175
катодный осциллоскоп 2176
катодный повторитель 2171
катушка индукции 6494
катушка с плавнорегулируемой
индуктивностью 12772
качаться 9498
качесественный анализ 10483
качественные методы 10484
качество измерительного
прибора 8064
качество регулирования 3225
качество управления 3312
качество упреждения 10489
Кб 7047
квадратичная интегральная
оценка 6777
квадратичная матрица 11760

квадратичный 10476
квадратичный критерий 10477
квадратурная модуляция 10480
квадратурный генератор 10481
квадратурный генератор
колебаний 10481
квазиавтомат 10519
квазигармоническая система
10523
квазигармоническое колебание
10522
квазиконтурное управление
8485
квазикритическое
демпфирование 10521
квазилинеаризация 10524
квазилинейная система 10525
квазимашинная структура
10526
квазимонохроматический лазер
8545
квазипараллельное выполнение
программы 10527
квазипериодическое поведение
678
квази-связанный 10517
квази-связанный режим 10518
квазистатический 10529
квазистационарное поведение
10531
квазистационарный 10530
квазиуравновешенный мост
10520
квазиустойчивый 10528
квалификационное испытание
10482
квантизатор 11156
квантификация 10490
квантование 1627
квантование непрерывной
системы 4363
квантование по уровню 780
квантованный сигнал 10501
квантовать 10499
квантовая система 10515
квантовая теория 10516
квантовая электроника 10509
квантовая эффективность 10508
квантовое преобразование
частоты 10510
квантовое условие 10506
квантовый детектор 10507
квантовый оптический
генератор 10514
квантовый предел 10512
квантовый уровень 10511
квантовый усилитель 10505
квантовый шум 10513

нуклеарный измерительный
    прибор 8818
нулевая вероятность 12982
нулевая дисперсия 12965
нулевая матрица 8831
нулевая мощность 12981
нулевая нагрузка 12960
нулевая погрешность 12967
нулевая последовательность
    12985
нулевая размерность 12963
нулевая скорость 12992
нулевая схема 8826
нулевая точка 12977
нулевая форма 12969
нулевая частота 12970
нулевая энергия 5060
нулевое биение 12956
нулевое значение 12990
нулевое изменение 12991
нулевое направление 12964
нулевое напряжение 12993
нулевое положение 5745, 12979
нулевое смещение 12958
нулевое состояние 12988
нулевое устройство 8829
нулевой байт 12959
нулевой дрейф 12966
нулевой метод 8825, 12972
нулевой потенциал 12980
нулевой прибор 8830
нулевой сигнал выхода 12974
нулевой уровень излучения
    12983
нулевой ход 8832
нуль 12952
нуль шкалы 11197
Ньютоновская гидромеханика
    8608

обегающая система развёртки
    9847
обзор захватных устройств
    6000
обзорная инфракрасная система
    6602
обзорная лазерная установка
    7185
обзорная схема 7205
обзор с постоянной скоростью
    3058
обзор функциональных
    возможностей 5778
область 10606, 12994
область аппаратного
    обеспечения 6059
область большого сигнала 7095
область ввода 6646

область визуального обзора
    электронного дисплея
    4888
область возмущений 10615
область детектирования 4011
область допустимых
    отклонений 10814
область допустимых ошибок
    975
область заданных значений
    10619
область затухания 1104
область интерференции 6861
область квадратичных
    отклонений 10478
область неустойчивости 6702
область определения 4474
область отклонения 4042
область памяти 8052
область параметра 9403
область перекрытия 9315
область покоя 8755
область помех 4450
область применения 927
область применения компью-
    терных систем 2897
область применения сенсоров
    924
область применения
    эффекторов 4647
область регулирования 3187
область управления 3187
область ускорения 134
область фактических
    параметров 340
область чувствительности
    10618
область сходимости 3379
обмен данными 3710
обмотка возбуждения 5251
обмотка смещения 1738
обмотка усиления 745
обнаружение дефектов 3844
обнаружение излучения 4009
обнаружение неисправностей
    5367
обнаружение неисправностей
    программы контроля 5190
обнаружение нуля 8827
обнаружение ошибок
    программы 10266
обнаружение паразитных
    колебаний 6230
обнаружение повреждений
    5367
обнаружение характерного
    инфракрасного
    излучения 2288

обнаруженная неисправность
    12489
обновление памяти 11953
обобщённая модель 5889
обобщённая передаточная
    функция 5891
обобщённая частотная
    характеристика 5888
обобщённые координаты 5887
обобщённый автомат 5886
обобщённый активный элемент
    5885
обобщённый параметр 5890
оборудование 5125
оборудование, приводимое в
    действие сигналом
    прерывания 6942
оборудование для испытания
    методом световой щели
    7250
оборудование для регулирова-
    ния температуры 12274
оборудование для увлажнения
    6221
оборудование для управления
    процессом горения 2628
оборудование для экспери-
    ментов 5289
оборудование лабораторного
    масштаба 1729
оборудование лазерной
    системы связи 7123
оборудование с числовым
    программным управлением
    5136
оборудование телеуправления
    10924
оборудованный средствами
    компьютерной техники
    2920
обрабатываемая деталь,
    преобразованная в
    цифровую форму 4199
обрабатываемая информация
    10179
обрабатывать 6025
обработка 7657, 10174
обработка видеосигнала
    12827
обработка в реальном времени
    10698
обработка геометрических
    данных 5927
обработка геометрической
    информации 10184
обработка графической
    информации 5954
обработка данных 3732

самоорганизирующийся
   автомат 11308
самописец глубинного
   давления 3960
самописец перепада давления
   4136
самописец сходимости 3382
самопишущий прибор 1361
самоприспосабливающая
   система 5931
самоприспосабливающийся
   датчик 367
самоприспосабливающийся
   захватный орган 11272
саморегулирование 11313
самосинхронизация 11286
самосинхронизирующаяся
   система 11287
самосопряжённая система
   11274
самосохраняющееся колебание
   11322
самостабилизация 11318
самотестируемая электроника
   11323
самоуправление 11297
самоустанавливающаяся
   модель 11276
самоустановка 11277
самоустановка на нуль 11305
самоустановление 11309
самофокусирование 11295
сбой 5364
сбор данных 3682
сбор данных об окружающей
   среде 5094
сбор данных о нескольких
   параметрах 7962
сбор данных о процессе
   манипулирования 254
сбор данных отображения
   6330
сбор данных отображения в
   реальном масштабе
   времени 10694
сбор измеренных значений
   7939
сборка на поточной линии
   7397
сборка с применением оптических средств управления
   9102
сборочная головка с управлением от микрокомпьютера
   8103
сборочная единица 1037
сборочная машина модульного
   типа 8258

сборочная система 1032
сборочная система с использованием роботов 11074
сборочная техника 8363
сборочное захватное устройство с сенсорным управлением 11365
сборочное испытание 1036
сборочное оборудование
   1013
сборочное устройство 1004
сборочный автомат 1007
сборочный объект 1023
сборочный процесс 1028
сборочный участок 1008
сбор сенсорных данных 11372
сварка автоматом 1312
сварка в камерах с контролируемой атмосферой 3260
сварка лазерным лучом 7115
сведение 10780
свертка распределения
   вероятностей 3411
сверхкритическая температура
   12024
сверхкритическое давление
   12023
сверхпроводимость 12022
сверхразвёртка 9326
сверхустойчивость 12623
сверхчистая поверхность 8100
сверхчувствительная система
   связи 12031
световая сигнализация 7085
световая измерительная
   техника 9083
световодная передаточная
   функция 9086
световодная система 9085
световодный интерфейс
   компьютера 5460
световое перо 7256
световой импульс 7252
световой карандаш 7256
световой сигнал 7261
светофильтр 9087
светочувствительный упра-
   вляющий элемент 7246
свободная динамическая
   система 5676
свободная составляющая 5674
свободное колебание 1449
свободное программируемое
   манипуляционное
   устройство 5560
свободнопрограммируемая
   автоматическая
   сборка 5680

свободнопрограммируемый
   малогабаритный манипу-
   лятор 5681
свободный параметр 969
свободный потенциал 5580
свободный цикл 1827
свойства передачи 12548
свойства шума 8637
свойство аддитивности 432
связанная передача сигналов
   1041
связанная сеть 6841
связанная система управления
   3418
связанное регулирование 3955
связанные контуры 3502
связанные регуляторы 3464
связанный 3016
связанный автомат 3017
связанный граф 3018
связанный синтез 2836
связанный с эксплуатацией
   дефект 9037
связи двух переменных 3028
связка 3027
связной 3016
связываемость 3029
связывание 1043
связывать 7423
связь и управление в живом
   организме и в машине 2656
связь между процессами 6928
связь между станками 7653
связь по постоянному току 4282
связь с использованием цветового кодирования 2595
связь с управляемым процессом
   10150
связью
   с активной ~ 301
   с механической ~ 7998
сглаживание 11661
сглаживать 11659
сглаживающий дроссель 11665
сглаживающий контур 11662
сглаживающий фильтр 11664
сгруппированная схема 5827
сдаточное испытание 5500
сдвиг 11483
сдвигать 11481
сдвигающий регистр с
   обратной связью 5437
сделать критическим 7776
севомеханизм прерывистого
   действия 4336
сегментация 11254
сегментирование 11254
сегментировать 11253